Dendrimers and Other Dendritic Polymers

Wiley Series in Polymer Science

Series Editor:
Dr John Scheirs
ExcelPlas
PO Box 2080
Edithvale
VIC 3196
AUSTRALIA
scheirs.john@pacific.net.au

Modern Fluoropolymers
High Performance Polymers for Diverse Applications

Polymer Recycling
Science, Technology and Applications

Metallocene-Based Polyolifins
Preparation, Properties and Technology

Polymer–Clay Nanocomposites

Forthcoming titles:

Modern Styrenic Polymers

Modern Polyesters

Dendrimers and Other Dendritic Polymers

Edited by

JEAN M. J. FRÉCHET
University of California, Department of Chemistry *and* Materials Sciences
Division, Lawrence Berkeley National Laboratory, Berkeley, CA, USA

and

DONALD A. TOMALIA
Dendritic Sciences, Inc., / Dendritic Nanotechnologies Limited,
Central Michigan University, Mt. Pleasant, MI, USA

WILEY SERIES IN POLYMER SCIENCE

John Wiley & Sons, Ltd

Chemistry Library

Other Wiley Editorial Offices

John Wiley & Sons, Inc., 605 Third Avenue,
New York, NY 10158-0012, USA

WILEY-VCH Verlag GmbH, Pappelallee 3,
D-69469 Weinheim, Germany

John Wiley & Sons Australia Ltd., 33 Park Road, Milton
Queensland 4064, Australia

John Wiley & Sons (Asia) Pte Ltd, 2 Clementi Loop #02-01,
Jin Xing Distripark, Singapore 129809

John Wiley & Sons (Canada) Ltd, 22 Worcester Road,
Rexdale, Ontario M9W 1L1, Canada

Library of Congress Cataloging-in-Publication Data

Dendrimers and other dendritic polymers / edited by Jean M. J. Fréchet and Donald A. Tomalia.
   p. cm. — (Wiley series in polymer science)
Includes bibliographical references and index.
ISBN 0-471-63850-1 (alk. paper)
1. Dendrimers.   I. Fréchet, Jean M. J.   II. Tomalia, Donald A.   III. Series.
TP1180.D45 D46 2001
668.9—dc21   2001045497

British Library Cataloguing in Publication data

A catalogue record for this book is available from the British Library

Cover art by Dr Stefan Hecht, University of California, Berkeley.

ISBN 0-471-63850-1

Typeset in 10/12pt Times from the author's disks by Vision Typesetting, Manchester
Printed and bound in Great Britain by Biddles Ltd, Guildford and King's Lynn
This book is printed on acid-free paper responsibly manufactured from sustainable forestry, in
which at least two trees are planted for each one used for paper production. acↃ✱

Contents

III PROPERTIES AND APPLICATIONS OF DENDRITIC POLYMERS

15 Dendritic and Hyperbranched Glycoconjugates as Biomedical Anti-Adhesion Agents **361**
R. Roy

1 Introduction 361
2 Adhesion Mechanisms Involved in Influenza Virus and Related Microbial Infections 363
3 Suitably Functionalized Carbohydrate Precursors 365
4 Calix[4]arene and β-cyclodextrin as Sialoside Scaffolds ... 366
5 Glycodendrimers Based on Poly(amidoamines) (PAMAM) Dendrimers 368
6 Poly(ethyleneimine) Scaffold Toward Hyperbranched Sialosides 370
7 Chitosan as Polysaccharide Scaffold Toward Hyperbranched Sialosides 372
8 Conclusion 382
9 References 382

16 Some Unique Features of Dendrimers Based upon Self-Assembly and Host-Guest Properties **387**
J.-W. Weener, M. W. P. L. Baars and E. W. Meijer

1 Introduction 387
2 Self-assembly of Dendrimers 388
   2.1 Dendrimers on Surfaces: Conformational Behaviour . 388
   2.2 Functional Thin Films using Dendrimers 392
   2.3 Amphiphilic Dendrimers 396
   2.4 Liquid Crystalline Dendrimers 401
3 Host–Guest Chemistry of Dendritic Macromolecules 403
   3.1 Do Cavities Exist in Dendrimers? 403
   3.2 Topological Encapsulation of Guest Molecules 406
   3.3 Recognition based on Hydrophobic Interactions 407
   3.4 Recognition based on Hydrogen Bonding Interactions 409
   3.5 Recognition based on Electrostatic Interactions 410
   3.6 Recognition based on Metal–Ligand Interactions ... 413
4 Conclusion and Prospects 416
5 References 417

17 Dendritic Polymers: Optical and Photochemical Properties ... **425**
D.-L. Jiang and T. Aida

1 Introduction 425

Contributors

Takuzo Aida
Department of Chemistry and
Biotechnology
Graduate School of Engineering
University of Tokyo
7-3-1 Hongo, Bunkyo-ku
Tokyo 113-8656
Japan

Eric J. Amis
Group Leader
Polymer Blends & Processing
Polymers Division,
224/B210
National Institute of Standards and
Technology
Gaithersburg, MD 20899-8542
USA

Maurice W.P.L. Baars
Laboratory of Macromolecular and
Organic Chemistry
Eindhoven University of Technology
Department of Chemical Engineering
PO Box 513
5600 MB Eindhoven
The Netherlands

James R. Baker, Jr.
University of Michigan
Center for Biologic Nanotechnology
Dept of Internal Medicine
Div of Allergy
9240 MSRB III
Ann Arbor, MI 48109
USA

Barry J. Bauer
National Institute of Standards and
Technology,
Polymers Division,
Bldg 224, Room B210
Gaithersburg, MD 20899-8542
USA

Teresa Beck
Monsanto Company
700 Chesterfield Village Parkway
North AA4C
St Louis, MO 63198
USA

Anna U. Bielinska
University of Michigan
Center for Biologic Nanotechnology
Dept of Internal Medicine
Div of Allergy
9240 MSRB III
Ann Arbor, MI 48109
USA

Kelly Botwin
Monsanto Company
700 Chesterfield Village Parkway
North AA4C
St Louis, MO 63198
USA

Wei Chen
Chemistry Department
Columbia University
3000 Broadway, MC 3119
New York, NY 10027
USA

Christopher G. Clark, Jr
Washington University
Department of Chemistry
Campus Box 1134
One Brookings Drive
St Louis, MO 63130-4899
USA

Mark Davey
Cadence Design Systems, Inc.
2655 Seely Avenue
San Jose, CA 95134
USA

Ellen M.M. de Brabander-van den
Berg
DSM Research
PO Box 18
6160 MD Geleen
The Netherlands

Brian W. Donovan
University of Michigan
Center for Biologic Nanotechnology
Dept of Internal Medicine
Div of Allergy
9240 MSRB III
Ann Arbor, MI 48109
USA

Tiffany D. Duffin
Monsanto Company
700 Chesterfield Village Parkway
North AA4C
St Louis, MO 63198
USA

Karel Dušek
Institute of Macromolecular Chemistry
Academy of Sciences of the Czech
Republic
Heyrovského nám. 2
CZ-162 06, Prague 6
Czech Republic

Miroslava Dušková-Smrčková
Institute of Macromolecular Chemistry
Academy of Sciences of the Czech
Republic
Heyrovského nám. 2
CZ-162 06, Prague 6
Czech Republic

Petar R. Dvornic
Michigan Molecular Institute
1910 W. St Andrews Road
Midland, MI 48640
USA

Jonathan D. Eichman
University of Michigan
Center for Biologic Nanotechnology
Dept of Internal Medicine
Div of Allergy
9240 MSRB III
Ann Arbor, MI 48109
USA

Roseita Esfand
Dendritic Nanotechnologies Limited
Central Michigan University
Park Library
Mt Pleasant, MI 48859
USA

Alan Ford
Debye Institute
Department of Metal-Mediated
Synthesis
Utrecht University
Padualaan 8
3584 CH Utrecht
The Netherlands

Jean M.J. Fréchet
University of California
Berkeley #1460
Department of Chemistry
718 Latimer Hall
Berkeley, CA 94720-1460
USA

Adam W. Freeman
Eastman Kodak Research
Laboratories
Building 82, Rm. C608
Rochester, NY 14650-2116,
USA

Mario Gauthier
Institute for Polymer Research
Department of Chemistry
University of Waterloo
Waterloo, Ontario
N2L 3G1
Canada

Mehrnaz Gharaee-Kermani
University of Michigan
Center for Biologic Nanotechnology
Dept of Internal Medicine
Div of Allergy
9240 MSRB III
Ann Arbor, MI 48109
USA

Theodore Goodson III
Department of Chemistry
Wayne State University
Detroit, Michigan 48202
USA

Craig J. Hawker
IBM Almaden Research Center
NSF Center for Polymeric Interfaces
and Macromolecular Assemblies
650 Harry Road
San Jose, CA 95120-6099
USA

Anders Hult
Dept. of Polymer Technology
Royal Institute of Technology
SE 100 44 Stockholm
Sweden

Henrik Ihre
Amersham Pharmacia Biotech.
Björkgatan 30
SE-751 84 Uppsala
Sweden

Robert Jansson
Monsanto Company
700 Chesterfield Village Parkway
North AA4C
St Louis, MO 63198
USA

Johann T.B.H. Jastrzebski
Debye Institute
Department of Metal-Mediated
Synthesis
Utrecht University
Padualaan 8
3584 CH Utrecht
The Netherlands

Dong-Lin Jiang
Department of Chemistry and
Biotechnology
Graduate School of Engineering
University of Tokyo
7-3-1 Hongo, Bunkyo-ku
Tokyo 113-8656
Japan

R. Andrew Kee
Institute for Polymer Research
Department of Chemistry
University of Waterloo
Waterloo, Ontario
N2L 3G1
Canada

Arjan W. Kleij
Debye Institute
Department of Metal-Mediated
Synthesis
Utrecht University
Padualuan 8
3584 CH Utrecht
The Netherlands

Jolanta F. Kukowska-Latallo
University of Michigan
Center for Biologic Nanotechnology
Dept of Internal Medicine
Div of Allergy
9240 MSRB III
Ann Arbor, MI 48109
USA

Gary Lange David Kunneman
Monsanto Company
700 Chesterfield Village Parkway
North AA4C
St Louis, MO 63198
USA

Stephen C. Lee
Monsanto Company
700 Chesterfield Village Parkway
North AA4C
St Louis, MO 63198
USA

Jing Li
Dow Chemical Company
1897 Building
Midland, MI 48667
USA

Manon H.A.P. Mak
DSM Research
PO Box 18
6160 MD Geleen
The Netherlands

Patrick R.L. Malenfant
General Electric Company
CRD Emerging Technologies
Polymeric Materials Laboratory
K1-4A49
1 Research Circle
Niskayuna, NY 12309
USA

Eva Malmström
Dept. of Polymer Technology
Royal Institute of Technology
S-100 44 Stockholm
Sweden

E.W. Meijer
Laboratory of Macromolecular and
Organic Chemistry
Eindhoven University of Technology
Department of Chemical Engineering
PO Box 513
5600 MB Eindhoven
The Netherlands

M. Francesca Ottaviani
Institute of Chemical Sciences
University of Urbino
61029 Urbino
Italy

Ranjani Parthasarathy
Monsanto Company
700 Chesterfield Village Parkway
North AA4C
St Louis, MO 63198
USA

Jacques Roovers
21 Wren Rd
Ottawa, ON
K1J 7H5
Canada

Edwin Rowold
Monsanto Company
700 Chesterfield Village Parkway
North AA4C
St Louis, MO 63198
USA

René Roy
Department of Chemistry
University of Ottawa
10 Marie Curie Street,
PO Box 450 Stn A
Ottawa, ON
K1N 6N5
Canada

Pratap Singh
Dade Behring Inc.
Mail Station 700
PO Box 6100
Glasgow Business Community
Newark, DE 19702
USA

Douglas R. Swanson
Dendritic Nanotechnologies Limited
Central Michigan University
Park Library
Mt. Pleasant, MI 48859
USA

Donald A. Tomalia
Dendritic Nanotechnologies Limited
Central Michigan University
Park Library
Mt. Pleasant, MI 48859
USA

Nicholas J. Turro
Chemistry Department
Columbia University
300 Broadway, MC 3119
New York, NY 10027
USA

Srinivas Uppuluri
Flint Ink Corporation Research Center
4600 Arrowhead Drive
Ann Arbor, MI 48197
USA

Marcel H.P. van Genderen
Laboratory of Macromolecular &
Organic Chemistry
Eindhoven University of Technology
Department of Chemical Engineering
PO Box 513
5600 MB Eindhoven
The Netherlands

Gerard van Koten
Debye Institute
Department of Metal-Mediated
Synthesis
Utrecht University
Padualaan 8
3584 CH Utrecht
The Netherlands

Charles F. Voliva
Monsanto Company
700 Chesterfield Village Parkway
North AA4C
St Louis, MO 63198
USA

Jan-Willem Weener
Laboratory of Macromolecular and
Organic Chemistry
Eindhoven University of Technology
PO Box 513
5600 MB Eindhoven
The Netherlands

Karen L. Wooley
Washington University
Department of Chemistry
Campus Box 1134
One Brookings Drive
St Louis, MO 63130-4899
USA

Chunxin Zhang
University of Michigan Medical School
Center for Biologic Nanotechnology
200 Zina Pitcher Place
Ann Arbor, MI 48109
USA

James Zobel
Monsanto Company
700 Chesterfield Village Parkway
North AA4C
St Louis, MO 63198
USA

Series Preface

The Wiley Series in Polymer Science aims to cover topics in polymer science where significant advances have been made over the past decade. Key features of the series will be developing areas and new frontiers in polymer science and technology. Emerging fields with strong growth potential for the twenty-first century such as nanotechnology, photopolymers, electro-optic polymers etc. will be covered. Additionally, those polymer classes in which important new members have appeared in recent years will be revisited to provide a comprehensive update.

Written by foremost experts in the field from industry and academia, these books place particular emphasis on structure–property relationships of polymers and manufacturing technologies as well as their practical and novel applications. The aim of each book in the series is to provide readers with an in-depth treatment of the state-of-the-art in that field of polymer technology. Collectively, the series will provide a definitive library of the latest advances in the major polymer families as well as significant new fields of development in polymer science.

This approach will lead to a better understanding and improve the cross fertilization of ideas between scientists and engineers of many disciplines. The series will be of interest to all polymer scientists and engineers, providing excellent up-to-date coverage of diverse topics in polymer science, and thus will serve as an invaluable ongoing reference collection for any technical library.

John Scheirs
June 1997

A Brief Historical Perspective

D. A. TOMALIA AND J. M. J. FRÉCHET

The dendritic architecture is perhaps one of the most pervasive topologies observed on our planet. Innumerable examples of these patterns [1] may be found in both abiotic systems (e.g. lightning patterns [1], snow crystals, tributary/erosion fractals), as well as in the biological world (e.g. tree branching/roots, plant/animal vasculatory systems, neurons) [2]. In biological systems, these dendritic patterns may be found at dimensional length scales measured in meters (trees), millimeters/centimeters (fungi) or microns (neurons) as illustrated in Figure 1. The reasons for such extensive mimicry of these dendritic topologies at virtually all dimensional length scales is not entirely clear. However, one might speculate that these are evolutionary architectures that have been optimized over the past several billion years to provide structures manifesting maximum interfaces for optimum energy extraction/distribution, nutrient extraction/distribution and information storage/retrieval.

The first inspiration for synthesizing molecular level tree-like structures evolved from a lifetime hobby enjoyed by one of the editors (D.A.T.) as a horticulturist/tree grower [3]. The first successful laboratory synthesis of such dendritic complexity did not occur until the late 1970s. It required a significant digression from traditional polymerization strategies with realignment to new perspectives. These perspectives utilized major new synthesis concepts that have led to nearly monodispersed synthetic macromolecules. The result was a new core–shell macromolecular architecture, now recognized as *dendrimers*.

The concept of repetitive growth with branching was first reported in 1978 by Vögtle [4] (University of Bonn, Germany) who applied it to the construction of low molecular weight amines. This was followed closely by the parallel and independent development of the divergent, macromolecular synthesis of true dendrimers in the Tomalia Group [5,6] (Dow Chemical Company). The first

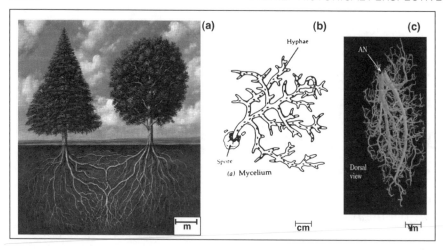

Figure 1 (a) Coniferous and deciduous trees with root systems, (b) fungal anatomy and (c) giant interneuron of a cockroach.

paper [6] describing in great detail the preparation of poly(amidoamine) dendrimers appeared in 1985, the same year a communication reported the synthesis of arborols [7] by Newkome et al. (Louisiana State University).

The divergent methodology based on acrylate monomers was discovered in 1979 and developed in the Dow laboratories during the period of 1979–85. It did not suffer from the problem of low yields, purity, or purification encountered by Vögtle in his 'cascade' synthesis, and afforded the first family of well characterized dendrimers. Poly(amidoamine) (PAMAM) dendrimers with molecular weights ranging from several hundred to over 1 million Daltons (i.e., Generations 1–13) were prepared in high yields. This original methodology was so successful that today it still constitutes the preferred commercial route to the trademarked Starburst® dendrimer family.

In contrast, the divergent iterative methodology involving acrylonitrile used by the Vögtle group [4] was plagued by low yields and product isolation difficulties and could not be used to produce molecules large enough to exhibit the unique properties that are now associated with the term 'dendrimer'. It was only a decade and a half later that two research groups Wörner/Mülhaupt [8] (Freiburg Univ.) and de Brabander-van den Berg/Meijer [9] (DSM), were able to develop a vastly enhanced modification of the Vögtle approach to prepare true poly(propyleneimine) (PPI) dendrimers. The route developed by the DSM group is particularly notable as it also constitutes a viable commercial route to this family of aliphatic amine dendrimers.

Since the 'dendrimers' discovery occurred in a Dow corporate laboratory, the period 1979–1983 was spent filing many of the original dendrimer 'composition of matter' patents [62–71]. The key Dow Starburst® dendrimer research team members associated with this initial research and development effort are shown in Figure 2. It was not until 1983, that corporate approval was given for the first public presentation of this work (by D.A.T.) at the Winter Polymer Gordon Conference in January (1983) (Santa Barbara, CA). It was after attending this Conference that de Gennes predicted the fundamental dendrimer surface congestion properties that are now referred to as the 'de Gennes [10] dense packing' phenomenon. Excitement and controversy generated at this Gordon Conference concerning this new class of monodispersed dendritic architecture led to an intense schedule of invited lectures during 1984–1985 which included: The Akron Polymer Lecture Series (April 1984), American Chemical Society Great Lakes/Central Regional Meeting (May 1984) and the 1st International Polymer Conference, Society of Polymer Science Japan, in Kyoto (August, 1984). The first use of the term 'dendrimer' to describe this new class of polymers, appeared in the form of several abstracts published during that year. The first SPSJ International Polymer Conference preprint [5] and the seminal full paper [6] that followed describe the preparation of dendrimers and their use as fundamental building blocks that may be covalently bridged to form poly(dendrimers) or so-called 'starburst polymers' as shown in Figure 3.

Figure 2 Original Dow dendrimer research team (l.-r back row: Pat Smith, Steve Martin, Mark Hall, John Ryder; front row: Jim Dewald, Don Tomalia, George Kallos, Jesse Roeck (photo taken (1982) in Dow's Functional Polymer Research Laboratory, 1710 Bldg, Midland, MI where first complete series of PAMAM dendrimers (G=1–7) were synthesized)

Polymer Journal, Vol. 17, No. 1, pp 117—132 (1985)

A New Class of Polymers: Starburst-Dendritic Macromolecules

D. A. Tomalia,* H. Baker, J. Dewald, M. Hall,
G. Kallos, S. Martin, J. Roeck,
J. Ryder, and P. Smith

*Functional Polymers/Process and *The Analytical Laboratory,
Dow Chemical U.S.A., Midland, Michigan 48640, U.S.A.*

(Received August 20, 1984)

ABSTRACT: This paper describes the first synthesis of a new class of topological macromolecules which we refer to as "starburst polymers." The fundamental building blocks to this new polymer class are referred to as "dendrimers." These dendrimers differ from classical monomers/oligomers by their extraordinary symmetry, high branching and maximized (telechelic) terminal functionality density. The dendrimers possess "reactive end groups" which allow (a) controlled moelcular weight building (monodispersity), (b) controlled branching (topology), and (c) versatility in design and modification of the terminal end groups. Dendrimer synthesis is accomplished by a variety of strategies involving "time sequenced propagation" techniques. The resulting dendrimers grow in a geometrically progressive fashion as shown: Chemically bridging these dendrimers leads to the new class of macromolecules—"starburst polymers" (*e.g.*, $(A)_n$, $(B)_n$, or $(C)_n$).

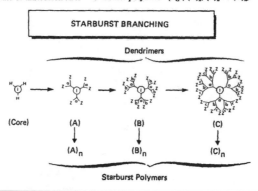

Figure 3   Abstract of first full paper (reference 6) describing dendrimers

After the appearance of the seminal 1985 paper from the Tomalia group, there was an enormous amount of intrinsic interest in dendritic polymer architecture. On the other hand, there was substantial resistance to accepting research results for publication by many of the major scientific journals, some of the reasons cited by the critics of that period included the following:

1. How can one be certain the higher molecular weight dendrimers (i.e., $> G = 2$) are as monodispersed as proposed?
2. Dendrimers are no different than 'microgels' –they are probably highly cross-linked particles akin to latexes,

3. It is difficult to believe that one can chemically advance from generation to generation (i.e., especially $G = 2$) without substantial intramolecular cyclization and crosslinking,
4. Dendrimers are not really discrete chemical structures – they are non-descript materials,
5. Dendrimers are not expected to manifest any unique properties that cannot be found in microgels or latexes,
6. Backfolding of terminal chain ends into the interior of dendrimer will prohibit any 'guest-host' properties – expectations for unimolecular micelle-like properties are absurd!
7. Since little chain entanglement would be expected from these structures, one would expect poor bulk properties compared to traditional linear random coil polymers.

In spite of this difficult acceptance, it is quite remarkable to note that by the end of 1990 about two dozen publications on dendrimers had appeared in refereed journals. By the end of 1991 the rate of publication of dendrimer papers had started to climb markedly while there still were only three papers on random hyperbranched polymers and two on dendrigraft or arborescent polymers. The courage, persistence and credibility of many key scientists listed in Table 1 during that period, set the stage for the explosive acceptance and recognition of dendritic polymers over the next decade.

Several key events also contributed to this transformation. This included an invitation by D. Seebach and H. Ringsdorf to present the 'dendritic polymer concepts' at the prestigious Bürgenstock Conference in Switzerland (May, 1987).

Table 1 Early refereed publications on dendritic molecules (1978–91)

Year	Lead authors	References
From cascade growth to dendrimers		
1978	Vögtle	4
1982	Maciejewski	11
1983	de Gennes	10
1985–1990	Tomalia/Turro/Goddard	6, 12–22
1985–1990	Newkome/Baker	7, 23–25
1990–1991	Fréchet/Hawker	26–29
1990	Miller/Neenan	30–31
Random Hyperbranched Polymers		
1988	Odian/Tomalia	32
1988/1990	Kim/Webster	33
1991	Fréchet/Hawker	34
Dendrigraft/Arborescent Polymers		
1991	Tomalia	35
1991	Gauthier/Möller	36

This lecture exposed these rather revolutionary concepts to an "elite scientific community" in Europe. Secondly, an invitation by Dr. P. Golitz (Editor, *Angew. Chem.*) to publish an important review [20] entitled '*Starburst dendrimers: molecular-level control of size, shape, surface chemistry, topology and flexibility from atoms to macroscopic matter*' provided broad exposure to the basic concepts underlying dendrimer chemistry. Finally, important contributions by key researchers significantly expanded the realm of dendrimer chemistry with the 'convergent synthesis' approach of Fréchet and Hawker [37] (Figure 4), as well as the systematic and critical photophysical characterization of Turro *et al* [38].

Influenced by Tomalia's seminal 1985 paper and stimulated by discussions with Richard Turner, then of the Eastman Kodak Company, dendrimer work at Cornell University was initiated by one of the editors (J.M.J.F.) in 1987–8. These were exciting times as the generous $2M gift by IBM Corporation to spur research in polymer chemistry had enabled the assembly of an outstanding research team within the Fréchet laboratory, leading to discoveries that included: Itsuno's polymer-supported chiral catalysts [39–40], Stover's NMR method for the characterization of crosslinked reactive polymer beads [41], Kato's self-assembly [42–43] of functional small molecules and supramolecular polymeric liquid crystals by hydrogen bonding, Cameron's photogeneration of bases for microlithography [44–45], Matuszczak's new design for chemically amplified deep-UV photoresists [46–47], and of course Hawker's convergent synthesis [26–27, 37] of dendrimers.

While repetitive syntheses of both linear and branched [4,48] small molecules and even macromolecules were not new (e.g. preparation of linear oligopeptides or branched polylysine [48]), Tomalia's dendrimers were clearly novel and had something special to offer: features and properties that develop as a function of size. We now know that the 'dendritic state', and the properties derived from it, are only accessed with certain symmetrical geometries once a critical size has been reached and the molecule adopts a globular shape encapsulating its core or focal point. Fréchet's initial 'learning' efforts were directed toward divergent syntheses of aromatic polyamide dendrimers and poly(propyleneimine) (PPI) cascade molecules [4]. These were soon abandoned due to severe problems of purification and the prevalent occurrence of stunted growth or structural defects. It was clear that only a few structures, such as Tomalia's poly(amidoamine) (PAMAM) dendrimers would lend themselves to controlled divergent growth.

The 'convergent' methodology for dendrimers synthesis was developed in the period 1988–9 soon after two very gifted postdoctoral fellows, Craig Hawker and Athena Philippides, joined Jean Fréchet at Cornell. The convergent growth approach, first demonstrated with polyether dendrimers, is probably best described as an 'organic chemist' approach to globular macromolecules as it affords outstanding control over growth, structure, and functionality. Instead of expanding a core molecule 'outwards' in divergent fashion through an ever increasing number of peripheral coupling steps, the convergent growth starts at

Figure 4 Members of the 1988–89 Cornell University team at a recent reunion: from right to left; back row, Dr Craig Hawker (IBM Almaden Research Laboratory), Prof. Takashi Kato (University of Tokyo); front row: Prof. Jean Fréchet (University of California, Berkeley), Prof. Karen Wooley (Washington University)

what will become the periphery of the molecule proceeding 'inwards' to afford building blocks (dendrons) that are subsequently coupled to a branching monomer through reaction of a single reactive group located at their focal point'. This allows for a drastic reduction in the amount of reagents used and enables intermediate purification at each step of growth, leading to single molecular entities. More importantly, the convergent growth allows unparalleled control over functionality at specified locations of the growing macromolecule and it provides access to numerous novel architectures through the attachment of dendrons to other molecules. This has led to innovative dendrimers consisting of different blocks, dendrimers with chemically varied layers or encapsulated functional entities, dendrimers with differentiated 'surface' functionalities, as well as to hybrid linear dendritic macromolecules and 'dendronized' macromolecules.

The initial presentation [37] of convergently grown dendrimers was made in 1989 at the IUPAC Symposium on Macromolecules in Seoul, Korea. Here again, following initial patent filings, publication of the work was delayed very significantly, by the thoroughly negative reception of the work by one referee, said to be 'an expert in the field' who thought it 'improbable that such precise molecules could actually have been prepared by the process described'. Soon after initial publication [26–27] of the work by Hawker and Fréchet that finally took place in 1990, the convergent synthesis of an aromatic polyester was reported by Neenan and Miller [30–31], while Hawker, working with a bright young graduate student, Karen Wooley, demonstrated the unique versatility of the convergent method with the preparation of dendrimers having differentiated functionalities [28,29]. Within a few months, the Cornell 'dendrimer' group now including Hawker, Wooley, Uhrich, Gitsov, Boegeman and Lee made use of the convergent synthesis to prepare and polymerize the first dendritic macro-monomers [49,50], develop the first true hybrid macromolecules [51,52] consisting of a linear polymer block with either one or two dendrimers chain ends, develop the first double stage convergent synthesis [53] and a variety of novel polymer architectures based on dendritic building blocks [54,55]. Closely related work also produced the first solid-phase synthesis [56] of a dendritic molecule, as well as the first hyperbranched polyester [34] obtained by one-step polycondensation.

Today the convergent approach to dendrimer synthesis has taken its place alongside Tomalia's divergent approach as one of the two seminal routes to this important new family of macromolecules and, within the past decade alone, hundreds of publications making use of the convergent synthesis, frequently with building blocks now known as 'Fréchet-type dendrons', have appeared. Figure 5 illustrates dendrimer growth by both the divergent and the convergent methodologies.

Among numerous events that contributed to the further acceptance of the

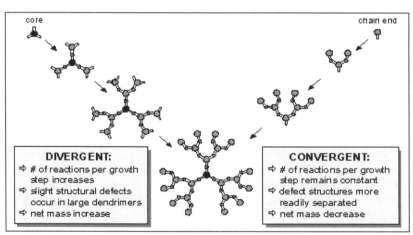

Figure 5   Schematics of dendrimer growth by the divergent and the convergent methods

dendrimers as discrete entities with remarkable structural precision was the development of mass spectrometric techniques for application in protein characterization. Mass spectrometry (MS) was shown to be useful for the precise determination of protein molecular weights (up to 10^6 Daltons) using electrospray and later MALDI-TOF techniques. By utilizing these techniques, it was possible to demonstrate unequivocally that all dendrimer constructions obeyed mathematically defined mass growth rules that could be documented routinely by MS techniques. This technological breakthrough as well as critical size exclusion chromatography [22,27,57], light scattering/viscosity [58], photophysical [38], electron microscopy [59], gel/capillary electrophoresis [60], atomic force microscopy [61], and other assorted measurements have exhaustively verified the principles of 'dendritic growth and amplification' while also illustrating some of the unusual properties that result from the dendritic state.

Beginning in the early 1990s, an overwhelming international interest in the dendritic polymer field has become apparent as manifested by research publications, reviews, monographs and patents that numbered in the 100s during the period 1990–95 then grew to thousands since 1995. While a decade ago, lectures on dendrimers were still rather scarce, the last five or six years have witnessed two major 'dendrimer' symposia at meetings of the American Chemical Society (ACS) in Chicago (1995) and Las Vegas (1998) that gathered very large international audiences. In 1999 the 'First International Dendrimer Conference' was held in Frankfurt (Germany) under the auspices of DECHEMA. The year 2001 saw another international symposium including more than 150 invited lectures and communications devoted to dendritic polymers at the San Diego ACS meeting while the 'Second International Dendrimer Conference' will take place in Tokyo.

In bringing this historical perspective to closure, it is important to share an extraordinary moment that Donald Tomalia experienced at the First Society of Polymer Science Japan International Polymer Conference in Kyoto (August 1984). Professor Paul Flory, who was not only the most prominent polymer scientist in attendance, but also presented the key plenary lecture for the conference. All invited speakers, were lodged at the Kyoto Grand Hotel. As such, many of us had the extraordinary opportunity to walk and talk with this celebrity on our many trips to the Kyoto Kaikan (lecture hall). On the other hand, I was one of the many eager, young scientists who had just presented some very intriguing, but nevertheless, 'non-traditional' dendrimer data to an audience of largely traditional polymer scientists. Needless to say, during these group walks there was considerable discussion. Many questions were raised during these discussions. For example, 'Is a dendrimer really a polymer?' 'How could we possibly force monomers to bond according to mathematically defined rules?' Because of their dimensions, 'Are dendrimers hazardous?' 'Do we really need a polymer such as a dendrimer?' 'Do dendrimers really exist?' Although I knew Flory attended the dendrimer lecture and he listened to these questions with interest, his comments were very sparse during these discussions. This troubled me, until on one very special occasion as we were making the walk alone, he shared with me two memorable perspectives that have remained with me until this day. First, he consoled me by advising me not to be troubled by many of these questions. As he stated it, historically, few revolutionary findings in science are ever accepted without a predictable period of rejection. With a grin, he said dendrimers certainly qualify on that issue. Secondly, and perhaps more profound were his perspectives on polymeric architecture. He stated it simply – 'Architecture is a consequence of special atom relationships and just as observed for small molecules, different properties should be expected for new polymeric architectures.' As such, dendrimers and other highly branched topologies should be expected to exhibit new and perhaps unexpected physical/chemical properties. He then challenged me with the following comment – 'If you have indeed synthesized these new dendritic architectures and you believe in them – then your job and your destiny will be to demonstrate these new properties, understand them and then attempt to predict the relationship between these parameters. Unfortunately, Prof. Flory passed away unexpectedly in the autumn of 1986 and the opportunity for further discussions was lost.

Some 17 years later, many of these predictions are turning into experimental reality as many of these questions are being answered in each new publication or patent that appears on dendritic architecture. Presently, dendritic polymers are recognized as the fourth major class of polymeric architecture consisting of three subsets that are based on degree of structural control, namely: (a) random hyperbranched polymers, (b) dendrigraft polymers and (c) dendrimers (Figure 6).

Hopefully, the present collection of insights on dendritic polymers will serve to assist and enlighten those who are in quest of such new architecturally driven properties and behavior.

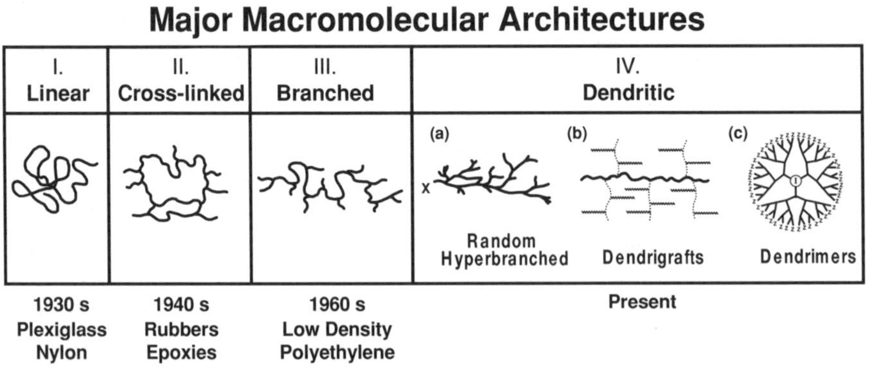

Major Macromolecular Architectures

Figure 6 Representation of the four major classes of macromolecular architectures

REFERENCES

1. Thompson, D. *On Growth and Form*, Cambridge University Press, London, 1987.
2. Mizrahi, A., Ben-Ner, E., Katz, M. J., Kedem, K., Glusman, J. G. and Libersat, F. *The Journal of Comparative Neurology*, **422**, 415 (2000).
3. Tomalia, D. A., *Sci. Am.*, **272**, 62 (1995).
4. Buhleier, E., Wehner, W. and Vögtle, F. *Synthesis*, 155 (1978).
5. Tomalia, D. A., Dewald, J. R., Hall, M. J., Martin, S. J. and Smith, P. B. *Preprints of the 1st SPSJ International Polymer Conference*, Soc. of Polym. Sci., Japan, Kyoto, 1984, p. 65.
6. Tomalia, D. A., Baker, H., Dewald, J., Hall, M., Kallos, G., Martin, S., Roeck, J., Ryder, J. and Smith, P. *Polym. J.*, Tokyo, **17**, 117 (1985).
7. Newkome, G. R., Yao, Z. -Q., Baker, G. R. and Gupta, V. K. *J. Org. Chem.*, **50**, 2003 (1985).
8. De Brabander-van den Berg, E. M. M. and Meijer, E. W., *Angew. Chem.*, **105**, 1370 (1993).
9. Wörner, C. and Mühlhaupt, R., *Angew. Chem.*, **105**, 1367 (1993).
10. de Gennes, P. G. and Hervet, H. J., *J. Physique-Lett.*, Paris, **44**, 351 (1983).
11. Maciejewski, M., *Macromol. Sci. Chem.*, **A17**, 689 (1982).
12. Tomalia, D. A., Baker, H., Dewald, J., Hall, M., Kallos, G., Martin, S., Roeck, J., Ryder, J. and Smith, P., *Macromolecules*, **19**, 2466 (1986).
13. Tomalia, D. A., Hall, M. and Hedstrand, D. M., *J. Am. Chem. Soc.*, **109**, 1601 (1987).
14. Tomalia, D. A., Berry, V., Hall, M. and Hedstrand, D. M., *Macromolecules*, **20**, 1164 (1987).
15. Padias, A. B., Hall Jr., H. K., Tomalia, D. A. and McConnell, J. R., *J. Org. Chem.*, **52**, 5305 (1987).
16. Smith, P. B., Martin, S. J., Hall, M. J. and Tomalia, D. A. 'A characterization of the structure and synthetic reactions of polyamidoamine 'STARBURST®' polymers, Mitchell, J., Jr(ed.), in *Appl. Polym. Analysis. Characterization*, Hanser Publishers, Munich, Vienna, New York, 357 (1987).
17. Friberg, S. E., Podzimek, M., Tomalia, D. A. and Hedstrand, D. M., *Mol. Cryst. Liq. Cryst.*, **164**, 157 (1988).
18. Naylor, A. M., Goddard III, W. A., Kiefer, G. E. and Tomalia, D. A. *J. Am. Chem. Soc.*, **111**, 2339 (1989).
19. Moreno-Bondi, M. C., Orellana, G., Turro, N. J. and Tomalia, D. A. *Macromolecules*, **23**, 910 (1990).
20. Tomalia, D. A., Naylor, N. M. and Goddard III, W. A. *Angew. Chem. Int. Ed. Engl.* **29**, 138 (1990).
21. Caminati, G., Turro, N. J. and Tomalia, D. A. *J. Am. Chem. Soc.*, **112**, 8515 (1990).
22. Roberts, J. C., Adams, Y. E., Tomalia, D. A., Mercer-Smith, J. A. and Lavallee, D. K. *Bioconjugate Chem.*, **1**, 305 (1990).
23. Newkome, G. R., Yao, Z. -Q., Baker, G. R., Gupta, V. K., Russo, P. S. and Saunders, M. J. *J. Am. Chem. Soc.*, **108**, 849 (1986).
24. Newkome, G. R., Baker, G. R., Saunders, M. J., Russo, P. S., Gupta, V. K., Yao, Z. -Q., Miller, J. E. and Bouillion, K. *J. Chem. Soc., Chem. Commun.*, 752 (1986).
25. Newkome, G. R., Baker, G. R., Arai, S., Saunders, M. J., Russo, P. S., Theriot, K. J., Moorefield, C. N., Rogers, L. E., Miller, J. E., Lieux, T. R., Murray, M. E., Phillips, B. and Pascall, L. *J. Am. Chem. Soc.*, **112**, 8458 (1990).
26. Hawker, C. J. and Fréchet, J. M. J. *J. Chem Soc. Chem. Commun.*, 1010 (1990).
27. Hawker, C. J. and Fréchet, J. M. J. *J. Am. Chem. Soc.*, **112**, 7638 (1990).
28. Hawker, C. J. and Fréchet, J. M. J. *Macromolecules*, **23**, 4726 (1990).

29. Wooley, K. L., Hawker, C. J. and Fréchet, J. M. J. *J. Chem. Soc. Perkin I*, 1059 (1991).
30. Miller, T. M. and Neenan, T. X. *Chem. Mat.*, **2**, 346–9 (1990).
31. Kwock, E. W., Neenan, T. X. and Miller, T. M., *Chem. Mat.*, **3**, 775 (1991).
32. Gunatillake, P. A., Odian, G. and Tomalia, D. A., *Macromolecules*, 21, 1556 (1988).
33. Kim, Y. H. and Webster, O. W., *Polymer Preprints*, **29**, 310 (1988); *J. Am. Chem. Soc.*, **112**, 4592 (1990).
34. Hawker, C. J., Lee, R. and Fréchet, J. M. J., *J. Am. Chem. Soc.* 1991, **113**, 4583. See also US Patent 5,514,764 issued 1996.
35. Tomalia, D. A., Hedstrand, D. M. and Ferritto, M. S. *Macromolecules*, **24**, 1435 (1991).
36. Gauthier, M. and Möller, M., *Macromolecules*, **24**, 4548 (1991).
37. Fréchet, J. M. J., Jiang, Y., Hawker, C. J. and Philippides, A. E. *Proc. IUPAC Int. Symp., Macromol.*, Seoul, 1989, pp. 19–20. See also US Patent 5,041,516 issued 1991.
38. Turro, Barton, J. K. and Tomalia, D. A., *Acc. Chem. Res.* , **24**, 332 (1991).
39. Itsuno, S. and Fréchet, J. M. J., *J. Org. Chem.*, **52**, 4140 (1987).
40. Itsuno, S., Sakurai, Y., Ito, K., Maruyama, T., Nakahama, S. and Fréchet, J. M. J., *J. Org. Chem.*, **55**, 304 (1990).
41. Stöver, H. D. H. and Fréchet, J. M. J. *Macromolecules*, **22**, 1574 (1989).
42. Kato, T. and Fréchet, J. M. J. *Macromolecules*, **22**, 3818 (1989).
43. Kato, T. and Fréchet, J. M. J. *J. Am. Chem. Soc.* 111, 8533 (1989).
44. Cameron, J. F. and Fréchet, J. M. J. *J. Org. Chem.*, **55**, 5919 (1990).
45. Cameron, J. F. and Fréchet, J. M. J. *J. Am. Chem. Soc.* 113, 4303 (1991).
46. Fréchet, J. M. J., Matuszczak, S., Stöver, H. D. H., Reck, B. and Willson, C. G., *The Electrophilic Aromatic Substitution Approach.*, ACS Symposium Series, **#412** Polymers in Microlithography, p. 74 (1989).
47. Fréchet, J. M. J., Matuszczak, S., Reck, B., Stover, H. D. H. and Willson, C. G *Macromolecules*, **24**, 1746 (1991), **24**, 1741 (1991)
48. Denkewalter, R. G., Kolc, J., Lukasavage, W. J. US Patent, 4,289,872 (1981).
49. Hawker, C. J., Wooley, K. L., Lee, R. and Fréchet, J. M. J., *Polym. Mat. Sci. Eng.*, **64**, 73 (1991).
50. Hawker, C. J. and Fréchet, J. M. J., *Polymer*, **33**, 1507 (1992).
51. Gitsov, I., Wooley, K. L., Hawker, C. J. and Fréchet, J. M. J. *Polym. Prep.*, **32**(3) 631 (1991).
52. Gitsov, I., Wooley, K. L., Hawker, C. J. and Fréchet, J. M. J. *Angew, Chem. Int. Ed. Engl.* **31**, 1200 (1992).
53. Wooley, K. L., Hawker, C. J. and Fréchet, J. M. J. *J. Am. Chem. Soc.*, **113**, 4252 (1991).
54. Hawker, C. J., Wooley, K. L. and Fréchet, J. M. J. *Polym. Prep.* **32**(3) 623 (1991).
55. Hawker, C. J. and Fréchet, J. M. J. *J. Am. Chem. Soc.*, **114**, 8405 (1992).
56. Uhrich, K. E., Boegeman, S., Fréchet, J. M. J. and Turner, S. R. *Polymer Bulletin*, 25, 551 (1991).
57. Dubin, P. L., Edwards, S. L., Kaplan, J. I., Mehta, M. S. and Tomalia, D. A. and Xia, *J. Anal. Chem.* , **64**, 2344 (1992).
58. Mourey, T. H., Turner, S. R., Rubinstein, M., Fréchet, J. M. J., Hawker, C. J. and Wooley, K. L. *Macromolecules*, **25**, 2401 (1992).
59. Jackson, C. L., Chanzy, H. D., Booy, F. B., Drake, B. J., Tomalia, D. A., Bauer, B. J. and Amis, E. J. *Macromolecules*, 31, 6259 (1998).
60. Brothers II, H. M., Piehler, L. T. and Tomalia, D. A. *J. Chromatogr. A*, **814**, 233 (1998).
61. Li, J., Piehler, L. T., Qin, D., Baker Jr., J. R., Tomalia, D. A. and Meier, D. J. *Langmuir*, **16**, 5613 (2000).
62. Tomalia, D. A. and Dewald, J. R., U. S. Patent, 4,507, 466 (1985).
63. Tomalia, D. A. and Dewald, J. R., U. S. Patent, 4,558, 120 (1985).
64. Tomalia, D. A. and Dewald, J. R., U. S. Patent, 4,568, 737 (1986).

65. Tomalia, D. A. and Dewald, J. R., U. S. Patent, 4,587, 329 (1986).
66. Tomalia, D. A. and Dewald, J. R., U. S. Patent, 4,631, 337 (1986).
67. Tomalia, D. A. and Dewald, J. R., U. S. Patent, 4,694, 064 (1986).
68. Tomalia, D. A. and Dewald, J. R., U. S. Patent, 4,713, 975 (1987).
69. Tomalia, D. A. and Dewald, J. R., U. S. Patent, 4,737, 550 (1987).
70. Tomalia, D. A. and Dewald, J. R., U. S. Patent, 4,871, 779 (1989).
71. Tomalia, D. A. and Dewald, J. R., U. S. Patent, 4,857, 599 (1989).

PART I

Introduction and Progress in the Control of Macromolecular Architecture

1

Introduction to the Dendritic State

D. A. TOMALIA
Dendritic Nanotechnologies Limited, Central Michigan University,
Mt. Pleasant, MI, USA

J. M. J. FRÉCHET
University of California-Berkeley, CA, USA

1 NATURAL AND SYNTHETIC EVOLUTION OF MOLECULAR COMPLEXITY

Nature has been evolving [1–4] and enhancing the complexity of matter in our universe over the past 10–13 billion years [5]. Mankind, on the other hand, began the science of creating and enhancing molecular complexity by 'synthesis' approximately 200 years ago [6, 7]. Important milestones in the evolution of complexity by covalent synthesis include the following: 'atom synthesis' (Lavoisier, 1789), the 'molecular hypothesis' (Dalton, 1808), 'organic chemistry' (Wöhler, 1828) and the 'macromolecular hypothesis' (Staudinger, 1926). Even in the context of these major developments it is obvious that our present ability to create and control molecular complexity is in its infancy compared to that demonstrated by Nature [8].

At least three major strategies are presently available for covalent synthesis of organic and related complexity beyond the atomic level (Figure 1.1) namely: (A) *traditional organic chemistry*, (B) *traditional polymer chemistry* and more recently (C) *dendritic macromolecular chemistry*.

Broadly speaking, traditional organic chemistry leads to higher complexity by involving the formation of relatively few covalent bonds between small heterogeneous aggregates of atoms (reagents) to give well-defined small molecules. On the other hand, polymerization strategies such as (B) and (C) involve the formation of large multiples of covalent bonds between homogeneous monomers to produce large molecules or infinite networks with a broad range of structure

Dendrimers and Other Dendritic Polymers. Edited by Jean M. J. Fréchet and Donald A. Tomalia
© 2001 John Wiley & Sons Ltd

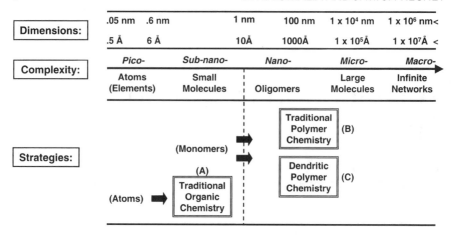

Figure 1.1   Molecular complexity as a function of covalent synthesis strategies and molecular dimensions

control [9, 10].

Within the context of comparing these three covalent synthesis strategies (i.e. (A), (B) and (C)) we wish to introduce the reader to strategy (C) and the new structural class that constitutes the 'dendritic state'.

1.1   TRADITIONAL ORGANIC CHEMISTRY

Organic synthesis, which is traditionally recognized to have originated with Wöhler in 1828, has led to the synthesis of literally millions of small molecules. Organic synthesis involves the use of various hybridization states of carbon and specific heteroatoms to produce key hydrocarbon building blocks (modules) and functional groups (connectors). These two construction parameters have been used to assemble literally millions of more complex structures by either (a) *divergent* or (b) *convergent* strategies involving a limited number of stepwise, covalent bond-forming events, followed by product isolation at each stage. Relatively small (i.e. < 1 nm) molecules are produced, which allow the precise control of shape, mass, flexibility and functional group placement. The divergent and convergent strategies are recognized as the essence of traditional organic synthesis [6]. An example of the divergent strategy (a) may be found in the Merrifield synthesis [11–14], which involves chronological introduction of precise amino acid sequences to produce a structure controlled, linear architecture.

Many examples of the 'convergent strategy' (b) can be found in contemporary approaches to natural products synthesis. Usually the routes to target molecules (e.g. I) are derived by retro-synthesis from the final product [6]. This involves transformation of the target molecule to lower molecular weight precursors (e.g. A–F).

A → B → C → D → → → I

(a) Schematic representation of the divergent strategy

A → B → E → F → I
C → D → G → H

(b) Schematic representation of the convergent strategy

Mathematically, at least one covalent bond, or in some cases several bonds, may be formed per reaction step (N_i). Assuming high-yield reaction steps and appropriate isolation stages, one can expect to obtain precise monodisperse products. In either case, the total number of covalent bonds formed can be expressed as follows:

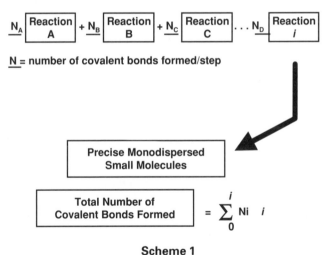

$$N_A \boxed{\begin{array}{c}\text{Reaction}\\ A\end{array}} + N_B \boxed{\begin{array}{c}\text{Reaction}\\ B\end{array}} + N_C \boxed{\begin{array}{c}\text{Reaction}\\ C\end{array}} \ldots N_D \boxed{\begin{array}{c}\text{Reaction}\\ i\end{array}}$$

$\underline{N}$ = number of covalent bonds formed/step

| Precise Monodispersed Small Molecules |

| Total Number of Covalent Bonds Formed | $= \sum\limits_{0}^{i} Ni \; i$

Scheme 1

Based on the various hybridization states of carbon, (Figure 1.2) at least four major carboskeletal architectures are known [6, 15]. They are recognized as (I) *linear*, (II) *bridged* (2D/3D), (III) *branched* and (IV) *dendritic*. In adherence with 'skeletal isomerism' principles demonstrated by Berzelius (1832) these major architectural classes determine very important differentiated physicochemical properties that define major areas within traditional organic chemistry (e.g. linear versus branched hydrocarbons). It is interesting to note that analogous

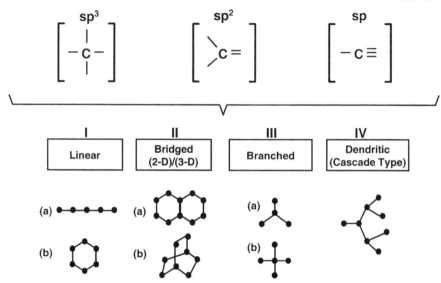

Figure 1.2 Four major small molecular architectures derived from the hybridization states of carbon

macromolecular architectural classes have also been identified and will be discussed later.

1.2   TRADITIONAL POLYMER CHEMISTRY

Over the past 70 years, a second covalent synthesis strategy has evolved based on the catenation of reactive small molecular modules or monomers. Broadly speaking, these catenations involve the use of reactive (AB-type monomers) that may be engaged to produce single large molecules with polydispersed masses. Such multiple bond formation may be driven by (a) *chain growth*, (b) *ring opening*, (c) *step-growth condensation* or (d) *enzyme catalyzed processes*. Staudinger first introduced this paradigm in the 1920s [16–19] by demonstrating that reactive monomers could be used to produce a statistical distribution of one-dimensional (linear) molecules with very high molecular weights (i.e. $> 10^6$ Daltons). As many as 10 000 or more covalent bonds may be formed in a single chain reaction of monomers. Although macro/mega-molecules with nanoscale dimensions may be attained, structure control of critical macromolecular design parameters, such as size, molecular shape, spatial positioning of atoms, or covalent connectivity – other than those affording linear or crosslinked topologies – is difficult. However, substantial progress has been made using 'living polymerization' techniques affording better control over molecular weight and some structural elements as described in Chapter 2.

n[AB] (monomers) ⟶ ᴧᴧ[AB]ₙᴧᴧ

Traditional polymerizations usually involve AB-type monomers based on substituted ethylenes, strained small ring compounds using chain reactions that may be initiated by free radical, anionic or cationic initiators [20]. Alternatively, AB-type monomers may be used in polycondensation reactions.

Multiple covalent bonds are formed in each macromolecule and, in general, statistical, polydispersed structures are obtained. In the case of controlled vinyl polymerizations, the average length of the macromolecule is determined by monomer to initiator ratios. If one views these polymerizations as extraordinarily long sequences of individual reaction steps, the average number of covalent bonds formed/chain may be visualized as shown in Scheme 2:

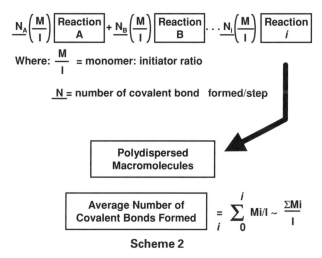

$$\underline{N_A}\left(\frac{M}{I}\right)\boxed{\begin{array}{c}\text{Reaction}\\ \text{A}\end{array}} + N_B\left(\frac{M}{I}\right)\boxed{\begin{array}{c}\text{Reaction}\\ \text{B}\end{array}} \cdots N_i\left(\frac{M}{I}\right)\boxed{\begin{array}{c}\text{Reaction}\\ i\end{array}}$$

Where: $\dfrac{M}{I}$ = monomer: initiator ratio

$\underline{N}$ = number of covalent bond formed/step

Polydispersed Macromolecules

$$\boxed{\begin{array}{c}\text{Average Number of}\\ \text{Covalent Bonds Formed}\end{array}} = \sum_{i\;0}^{i} M_i/I \sim \frac{\Sigma M_i}{I}$$

Scheme 2

Traditional polymerization strategies generally produce linear architectures, however, branched topologies may be formed either by chain transfer processes, or intentionally introduced by grafting techniques. In any case, the linear and branched architectural classes have traditionally defined the broad area of *thermoplastics*. Of equal importance is the major architectural class that is formed by the introduction of covalent bridging bonds between linear or branched polymeric topologies. These crosslinked (bridged) topologies were studied by Flory in the early 1940s and constitute the second major area of traditional polymer chemistry – namely, *thermosets*. The two broad areas of polymer science – thermoplastics and thermoset – accounting for billions of dollars of commerce support a vast array of very familiar macromolecular compositions and applications as shown in Figure 1.3.

Thermoplastics

Class I (Linear)

Polymer Types	Discovery	Production	Main Applications
Poly(methyl methacrylate)	1880	1928	Plastics (Plexiglassfi)
Poly(vinyl acetate)	1912	1930	Adhesive, poly(vinyl alcohol)
Poly(styrene)	1839	1930	Thermoplastics, foams
Poly(vinyl chloride)	1838	1931	Thermoplastics (synthetic fiber)
Poly(ethylene oxide)	–	1931	Thickeners, sizes
Poly(vinyl ethers)	1928	1936	Adhesives, plasticizers
Poly(hexamethylene adipamide)	1934	1938	Fibers, thermoplastics
Poly(vinylidene chloride)	1838	1939	Thermoplastics (packing films)
Poly(N-vinyl pyrrolidone)	–	1939	Blood plasma expander, binders
Poly(ethylene), low density	1933	1939	Thermoplastics
Poly(ε-caprolactam)	1938	1939	Fibers, thermoplastics
Polyurethanes	1937	1940	Fibers, plastics, elastomers, foams
Poly(acrylonitrile)	1940	1941	Fibers
Poly(tetrafluoroethylene)	1939	1950	Plastics, fibers
Poly(ethylene terephthalate)	1941	1953	Fibers, bottles
Bisphenol A polycarbonate	1898	1953	Thermoplastics
Poly(ethylene), high density	1953	1955	Thermoplastics, foams
Poly(propylene), isotactic	1954	1957	Thermoplastics, fibers
Poly(formaldehyde)	1839	1959	Thermoplastics
Aromatic polyamides	–	1961	High modulus fibers
Styrene-butadiene-styrene block copolymers	–	1965	Thermoplastic elastomers

Class III (Branched)

Poly(olefins), long chain branching	1980s	1990s	Elastomers, plastomers

Thermosets

Class II (Cross-Linked)

Polymer Types	Discovery	Production	Main Applications
Phenolic resins	1907	1910	Thermosets (electrical insulators)
Methyl rubbers	1912	1915	Elastomers
Alkyl resins	1847	1926	Thermosets (coatings)
Amino resins	1915	1928	Thermosets
Poly(butadiene)	1911	1929	Elastomers (number Bunas)
Poly(chloroprene)	1925	1932	Elastomers
Unsaturated polyesters	1930	1936	Thermosets
Poly(isobutylene)	–	1937	Elastomers
Styrene-butadiene rubbers	1926	1937	Elastomers (letter Bunas)
Silicone	1901	1942	Fluids, resins, elastomers
Epoxy resins	1938	1946	Adhesives
Poly(butadiene), cis, 1,4	–	1956	Elastomers

Figure 1.3 Historical discovered and production dates of commercial thermoplastic and thermoset polymers organized according to their architectural class

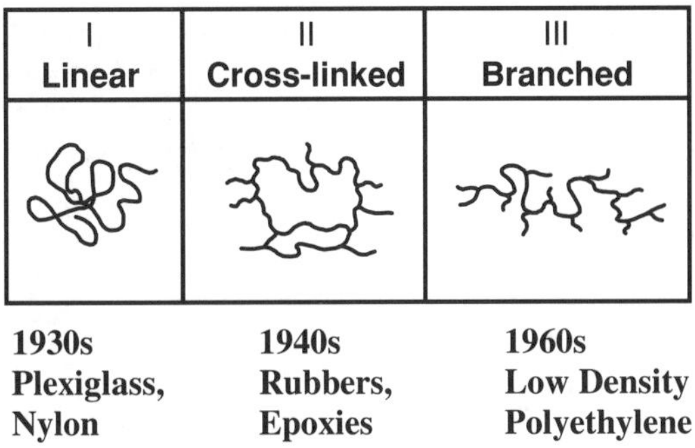

I Linear	II Cross-linked	III Branched
1930s Plexiglass, Nylon	1940s Rubbers, Epoxies	1960s Low Density Polyethylene

Figure 1.4 Traditional macromolecular architectures organized chronologically according to their commercial introduction

Therefore, approximately 50 years after the introduction of the 'macromolecular hypothesis' by Staudinger, the entire field of polymer science could simply be described as consisting of only the two major architectural classes: (i) *linear topologies* as found in thermoplastics and (ii) *crosslinked architectures* as found in thermosets. The major focus of polymer science during the time frame spanning the period of the 1920s to the 1970s was on the unique architecturally driven properties manifested by either linear or cross-linked topologies. Based on the unique properties exhibited by these topologies, many natural polymers that were critical to success in World War II were replaced with synthetic polymers for which the combination of availability and properties were of utmost strategic importance [9]. In the period covering the 1960s and 1970s, pioneering investigation into long chain branching (LCB) in polyolefins and other related branched systems began to emerge [21, 22]. More recently, intense commercial interest has been focused on new polyolefin architectures based on 'random long branched' and 'dendritic topologies' [23, 24]. These architectures are reportedly produced by 'metallocene' and 'Brookhart-type' catalysts. As a result, by the end of the 1970s, the major architectural polymer classes and commercial commodities associated with these topologies were as described chronologically in Figure 1.4.

2 THE DENDRITIC STATE

As described earlier in this book, the dendritic architecture is perhaps one of the most pervasive topologies observed at the macro and micro-dimensional length scales (i.e. μm-m). At the nanoscale (molecular) level there are relatively few natural examples of this architecture. Most notable are probably the glycogen and amylopectin hyperbranched structures that Nature uses for energy storage.

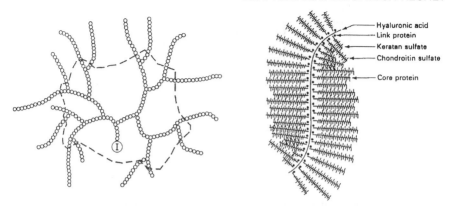

Figure 1.5 Topologies for (a) amylopectin and (b) proteoglycans

Presumably, the many chain ends that decorate these macromolecules facilitate enzymatic access to glucose for high demand bioenergy events [25]. Another nanoscale example of dendritic architecture in biological systems is found in proteoglycans. These macromolecules appear to provide energy-absorbing, cushioning properties and determine the viscoelastic properties of connective tissue (Figure 1.5).

In the past two decades, new strategies have been developed that allow the synthesis of a wide variety of such dendritic structures.

2.1 DENDRITIC BRANCHING CONCEPTS – HISTORICAL OVERVIEW

The origins of the present three-dimensional, dendritic branching concepts can be traced back to the initial introduction of infinite network theory by Flory [26–29] and Stockmayer [30–32]. In 1943, Flory introduced the term *network cell*, which he defined as the most fundamental unit in a molecular network structure [27]. To paraphrase the original definition, *it is the recurring branch juncture in a network system as well as the excluded volume associated with this branch juncture.* Graessley [33, 34] took the notion one step further by describing ensembles of these network cells as micronetworks. Extending the concept of Flory's statistical treatment of Gaussian-coil networks, analogous species that are part of an open, branched/dendritic organization are known as *branch cells* and *dendritic assemblies*. A comparison of these entities is illustrated in Figure 1.6.

Statistical modeling by Gordon *et al.* [35, 36], Dusek [37], Burchard [38] and others reduced such branched species to graph theory designed to mimic the morphological branching of trees. These dendritic models were combined with

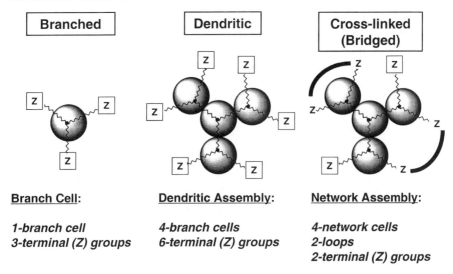

Figure 1.6   Examples of (a) branch cell, (b) dendritic assembly of branch cells and (c) crosslinked nano-networks derived from branch cells

'cascade theory' [39, 40] mathematics to give a reasonable statistical treatment for network-forming events at that time.

The growth of branched and dendritic macromolecules in the sol phase of a traditional crosslinking process may be thought of as geometric aggregations of various branch cells or dendritic/network assemblies as described above. Beginning as molecular species, they advance through the dimensional hierarchy shown in Figure 1.1, to oligomeric, macromolecular, megamolecular and ultimately to infinite network macroscale systems. Traditional network-forming systems (e.g. epoxy resins, urethanes, polyesters) progress through this growth process in a statistical, random fashion. The resulting infinite networks may be visualized as a collection of unequally segmented, Gaussian chains between f-functional branch junctures, crosslinks (loops) and dangling terminal groups.

2.2   A COMPARISON OF TRADITIONAL ORGANIC CHEMISTRY AND POLYMER SCIENCE WITH DENDRITIC MACROMOLECULAR CHEMISTRY

It is appropriate to compare the well-known concepts of covalent bond formation in traditional organic chemistry with those that apply to classical polymer chemistry and to dendritic macromolecular chemistry. This allows one to fully appreciate the differences between the three areas in the context of structure control, in concert with issues related to terminal group and mass amplification.

Covalent synthesis in traditional polymer science has evolved around the use of reactive modules (AB-type monomer) or ABR-type branch reagents that may be engaged in multiple covalent bond formation to produce large one-dimensional molecules of various lengths. Such multiple bond formation may be driven either by chain reactions, ring opening reactions or polycondensation schemes. These propagation schemes and products are recognized as Class I: *linear* or Class III: *branched* architectures. Alternatively, using combinations and permutations of divalent A–B type monomers and/or A-B_n, A_n-B polyvalent, branch cell-type monomers produces Class II, *crosslinked* (*bridged*) architectures.

A comparison of the covalent connectivity associated with each of these architecture classes (Figure 1.7) reveals that the number of covalent bonds formed per step for linear and branched topology is a multiple (n = degree of polymerization) related to the monomer/initiator ratios. In contrast, ideal dendritic (Class IV) propagation involves the formation of an exponential number of covalent bonds per reaction step (also termed G = generation), as well as amplification of both mass (i.e. number of branch cells/G) and terminal groups, (Z) per generation (G).

Mathematically, the number of covalent bonds formed per generation (reaction step) in an ideal dendron or dendrimer synthesis varies according to a power

Architectural Polymer Class	Polymer Type	Repeat Units	Covalent Connectivity
(I.) *LINEAR*	Thermoplastic	Divalent Monomers A-B	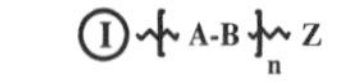
(III.) *BRANCHED*	Thermoplastic	Divalent Branch Cell Monomers 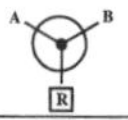	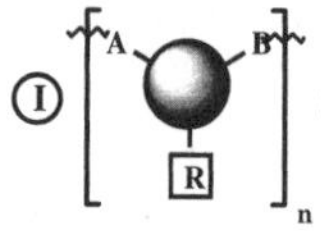
(IV.) *DENDRITIC*	Thermoplastic	Polyvalent Branch Cell Monomers	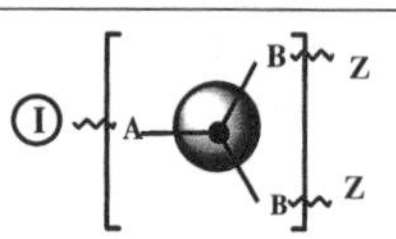 $\left(\frac{N_b^{G-1}}{N_b-1}\right)\left(\frac{N_b^{G}}{N_b-1}\right)$
(II.) *CROSS-LINKED (BRIDGED)*	Thermoset	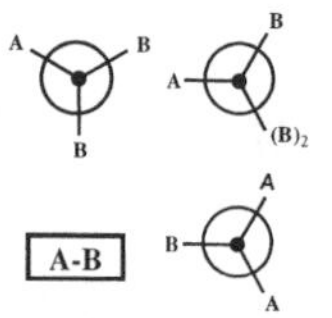	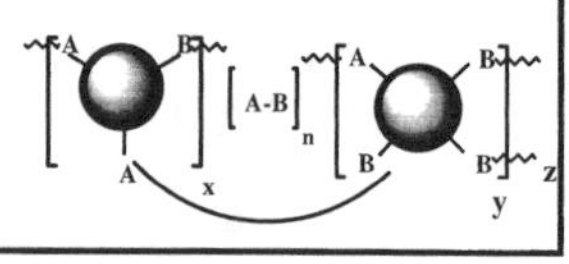

Figure 1.7 Examples of architectural polymer classses (I–IV) polymer type, repeat units and covalent connectivity associated with architectural classes

function of the reaction steps, as illustrated below. It is clear that covalent bond amplification occurs in all dendritic synthesis strategies. In addition to new architectural consequences, this feature clearly differentiates dendritic growth processes from covalent bond synthesis found in both traditional organic and polymer chemistry [8].

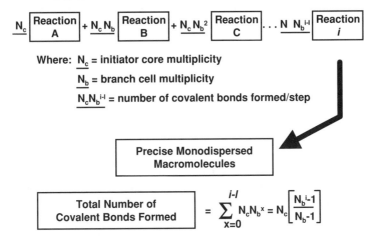

Where: N_c = initiator core multiplicity

N_b = branch cell multiplicity

$N_c N_b^{i-1}$ = number of covalent bonds formed/step

Precise Monodispersed Macromolecules

$$\text{Total Number of Covalent Bonds Formed} = \sum_{x=0}^{i-1} N_c N_b^x = N_c \left[\frac{N_b^i - 1}{N_b - 1} \right]$$

It should be quite apparent that, although all major architectural polymer classes are derived from common or related repeat units, the covalent connectivity is truly discrete and different. Furthermore, mathematical analysis of the respective propagation strategies clearly illustrates the dramatic differences in structure development as a function of covalent bond formation. It should be noted that linear, branched and dendritic topologies differ substantially both in their covalent connectivity, as well as their terminal group to initiator site ratios. In spite of these differences, these open, unlooped macromolecular assemblies clearly manifest thermoplastic polymer type behavior compared to the looped, bridged connectivity associated with crosslinked, thermoset systems. In fact, it is now apparent that these three 'open assembly-topologies (i.e. (I) linear, (III) branched, (IV) dendritic) represent a graduated continuum of architectural intermediacy between thermoplastic and thermoset behavior.

In summary, traditional organic chemistry offers exquisite control over critical molecular design parameters up to, but not including, nanoscale structural dimensions. Classical polymer science offers facile access to statistical distributions of nanoscale structures, with some control over topology, composition, flexibility or rigidity, and, in the case of living polymerization, increasingly better but still imperfect control over product size and mass distribution or polydispersity. In contrast, as will be seen below, dendritic macromolecular chemistry provides all the elements needed for unparalleled control over topology, composition, size, mass, shape and functional group placement. These are features that truly distinguish many successful, nanostructures found in nature [41].

The quest for nanostructures and devices based on the biomimetic premise of architectural and functional precision is intense and remains an ultimate challenge. One must ask – what new options or unique properties does the 'dendritic state' offer to meet the needs of nanoscale science and technology? The rest of this chapter will attempt to overview key features of the 'dendritic state' that address these and other issues.

2.3 DENDRITIC POLYMERS – A FOURTH MAJOR NEW ARCHITECTURAL CLASS

The dendritic topology has bow been recognized as a fourth major class of macromolecular architecture [42–45]. The signature for such a distinction is the unique repertoire of new properties manifested by this class of polymers [46–51]. Numerous synthetic strategies have been reported for the preparation of these materials, which have led to a broad range of dendritic structures. Presently, this architectural class consists of three dendritic subclasses, namely: (IVa) *random hyperbranched polymers*, (IVb) *dendrigraft polymers* and (IVc) *dendrimers* (Figure 1.8). The order of this subset, from a to c, reflects the relative degree of structural control present in each of these dendritic architectures.

All dendritic polymers are open covalent assemblies of branch cells. They may be organized as very symmetrical, monodispersed arrays, as is the case for dendrimers, or as irregular polydispersed assemblies that typically define random hyperbranched polymers. As such, the respective subclasses and the level of structure control are defined by the propagation methodology used to produce these assemblies, as well as by the branch cell (BC) construction parameters. The BC parameters are determined by the composition of the BC monomers, as well as the nature of the 'excluded volume' defined by the BC. The excluded volume of the BC is determined by the length of the arms, the symmetry, rigidity/flexibility, as well as the branching and rotation angles involved within each of the branch cell domains. As shown in Figure 1.8 these dendritic arrays of branch cells usually manifest covalent connectivity relative to some molecular reference marker (I) or core. As such, these branch cell arrays may be very non-ideal and polydispersed (e.g. $M_w/M_n/$ 2–10), as observed for random hyperbranched polymers (IVa), or very ideally organized into highly controlled core–shell type structures as noted for dendrons/dendrimers (IVc): M_w/M_n 1.01–1.0001 and less. Dendrigraft (arborescent) polymers reside between these two extremes of structure control, frequently manifesting rather narrow polydispersities of $M_w/M_n/$ 1.1–1.5 depending on their mode of preparation.

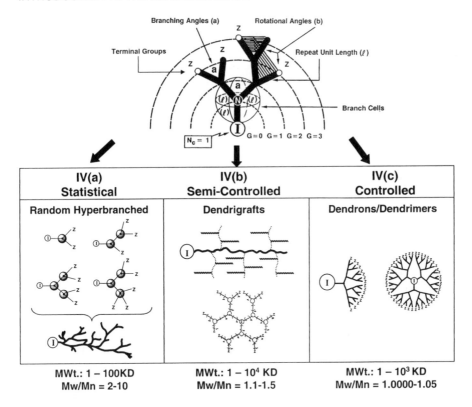

Figure 1.8 Branch cell structural parameters (a) branching angles, (b) rotation angles, (l) repeat units lengths, (Z) terminal groups and dendritic subclasses derived from branches (IVa) random hyperbranched, (IVb) dendrigrafts and (IVc) dendrons/dendrimers

2.4 RANDOM HYPERBRANCHED POLYMERS

Flory was the first to hypothesize concepts [28, 52], which are now recognized to apply to statistical, or 'random hyperbranched' polymers. However, the first purposeful experimental confirmation of dendritic topologies did not produce random hyperbranched polymers but rather the more precise, structure controlled, dendrimer architecture. This work was initiated nearly a decade before the first examples of 'random hyperbranched' polymers were confirmed independently in publications by Odian/Tomalia [53] and Webster/Kim [54, 55] in 1988. At that time, Webster/Kim coined the popular term 'hyperbranched polymers' that has been widely used to describe this type of dendritic macromolecules.

Hyperbranched polymers are typically prepared by polymerization of AB_x

monomers. When x is 2 or more, polymerization of such monomers gives highly branched polymers as shown in Figure 1.9, as long as A reacts only with B from another molecule. Reactions between A and B from the same molecule result in termination of polymerization by cyclization. This approach produces hyperbranched polymers with a degree of polymerization n, possessing one unreacted A functional group and $(x -1)n + 1$ unreacted B terminal groups. In similar fashion, copolymerization of A_2 and B_3 or other such polyvalent monomers can give hyperbranched polymers [56, 57], if the polymerization is maintained below the gel point by manipulating monomer stoichiometry or limiting polymer conversion.

Random hyperbranched polymers are generally produced by the one-pot polymerization of AB_x monomers or macromonomers involving polycondensation, ring opening or polyaddition reactions hence the products usually consist of broad statistical molecular weight distributions.

Over the past decade, literally dozens of new AB_2-type monomers have been reported leading to an enormously diverse array of hyperbranched structures. Some general types include poly(phenylenes) obtained by Suzuki-coupling [54, 55], poly(phenylacetylenes prepared by Heck-reaction [58], polycarbosilanes, polycarbosiloxanes [59], and polysiloxysilanes by hydrosilylation [60], poly(ether ketones) by nucleophilic aromatic substitution [61] and polyesters [62] or polyethers by polycondensations [63] or by ring opening [64].

New advances beyond the traditional AB_2 Flory-type branch cell monomers have been reported by Fréchet *et al.* [65, 66]. They have introduced the concept

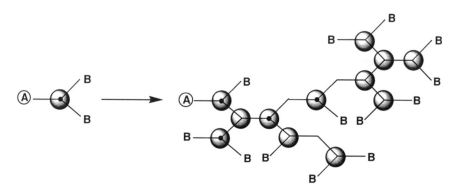

> Number of unreacted A groups = one
> Number of unreacted B groups = (functionality -1) DP + 1
> DP = degree of polymerization

Figure 1.9 Polymerization of an AB_2 monomer into a random hyperbranched polymer

of latent AB_2 monomers, referred to as 'self-condensing vinyl polymerizations' (SCVP). These monomers, which possess both initiation and propagation properties, may follow two modes of polymerization: namely, polymerization of the double bond (i.e. chain growth) and condensation of the initiating group with the double bond (i.e. step-growth). Recent progress involving the derivative process of *self-condensing ring-opening polymerizations* (SCROP) has been reviewed by Sunder *et al.* [25]. In addition, the use of enhanced processing techniques such as pseudo chain growth by slow monomer addition) [67] allow somewhat better control of hyperbranched structures [25].

2.5 *DENDRIGRAFT (ARBORESCENT) POLYMERS*

Dendrigraft polymers are the most recently discovered and currently the least well understood subset of dendritic polymers. The first examples were reported independently by Tomalia *et al.* [68] and Gauthier, *et al.* [69] in 1991. A comparison of dendrimer and dendrigraft architectures is made in Figure 1.10. Whereas traditional monomers are generally used for constructing dendrimers, *reactive oligomers* or *polymers* are used in 'protect/deprotect' or activation schemes to produce dendrigrafts.

Both hydrophilic (i.e. *dendrigraft* – poly(oxazolines)/poly(ethyleneimines)), as well as hydrophobic (i.e. *dendrigraft* – poly(styrenes) were reported in this early work. These first methodologies involved the iterative grafting of 'oligomeric reagents' derived from 'living polymerization processes' in various iterative 'graft on graft' strategies. By analogy to dendrimers, each iterative grafting step is referred to as a generation. An important feature of this approach is that *branch densities*, as well as the size of the *grafted branches* can be varied independently for each generation. Furthermore, by initiating these iterative grafting steps from a 'point-like core' versus a 'linear core' it is possible to produce spheroidal and cylindrical dendrigrafts, respectively. Depending on the graft densities and molecular weights of the grafted branches, ultra-high molecular weight dendrigrafts (e.g. $M_W > 10$ M) can be obtained at very low generation levels (e.g. $G = 3$). The dramatic molecular weight enhancements that are possible using dendrigraft techniques are compared with other propagation methodologies in Figure 1.11.

Further elaboration of these dendrigraft principles allowed the synthesis of a variety of core–shell type dendrigrafts, wherein elemental composition as well as the hydrophobic/hydrophilic character in the core can be controlled independently.

In general, the above methodologies have involved 'convergent-type' grafting principles wherein preformed, reactive oligomers are grafted onto successive branched precursors to produce semi-controlled structures. Compared to dendrimers, dendrigraft structures are less controlled since grafting may occur along the entire length of each generational branch and the exact branching

Dendrimer Architecture

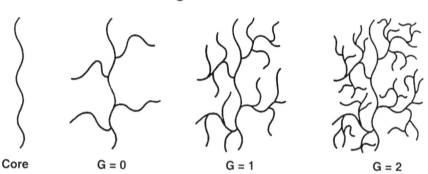

Core   G = 0   G = 1   G = 2

Dendrigraft Architecture

Core   G = 0   G = 1   G = 2

Figure 1.10 Comparison of dendrimer and dendrigraft architecture Generation: 0–2

densities are somewhat arbitrary and difficult to control.

More recently, both Gnanou [70, 71] and Hendricks [72, 73] have developed approaches to dendrigrafts that mimic dendrimer topologies by confining the graft sites to the branch termini for each generation. These methods involve so-called 'graft from' techniques and allow better control of branching topologies and densities as a function of generation. Topologies produced by these methods are reminiscent of the dendrimer architecture (Figure 1.10). Since the branch cell arms are derived from oligomeric segments, they are referred to as 'polymeric dendrimers' [21]. These more flexible and extended structures exhibit unique and different properties compared to the more compact traditional dendrimers. Fréchet and coworkers [74] have used the techniques of living polymerization and a staged polymerization process in which latent polymerization sites are incorporated within growing chains, then used to produce dendrigrafts of mixed composition and narrow polydispersity.

Another exciting development has been the emerging role that dendritic architecture is playing in the production of commodity polymers. A recent report

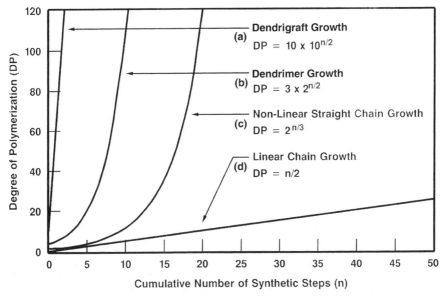

Figure 1.11 Comparison of degree of polymerization as a function of topology and growth process (a) dendrigraft, (b) dendrimer, (c) non-linear straight chain and (d) linear

by Guan *et al.* [24] has shown that ethylene monomer polymerizes to *dendrigraft*-poly(ethylene) at low pressures in contrast to high pressure conditions, which produce only branched topologies. This occurs when using late transition metal or Brookhart catalysts (Figure 1.12). Furthermore, these authors also state that small amounts of *dendrigraft poly(ethylene)* architecture may be expected from analogous early transition metal-metallocene catalysts.

2.6 DENDRONS AND DENDRIMERS

Dendrons and dendrimers are the most intensely investigated subset of dendritic polymers. In the past decade over 2000 literature references have appeared on this unique class of structure controlled polymers. The term 'dendrimer' was coined by Tomalia, *et al.* over 15 years ago in the first reports on poly(amidoamine) (PAMAM) dendrimers [75, 76]. It is derived from the Greek words *dendri-*(branch tree-like) and *meros* – part of). Poly(amidoamine) dendrimers constitute the first dendrimer family to be commercialized and undoubtedly represent the most extensively characterized and best understood series at this time. In view of the extensive literature information in this area, much of the remaining overview will focus on PAMAM dendrimers and will

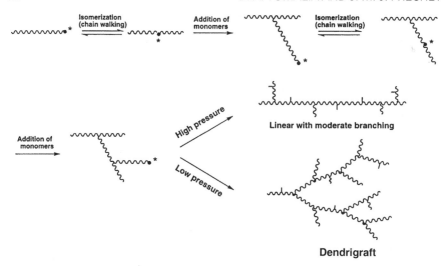

Figure 1.12 Proposed mechanism for conversion of ethylene monomer to dendrigraft polyethylene with Brookhart catalyst at low pressure

largely limit the scope of discussion to critical features offered by these fascinating structures.

2.6.1 Synthesis – Divergent and Convergent Methods

In contrast to traditional polymers, dendrimers are unique core–shell structures possessing three basic architectural components namely, (I) *a core*, (II) *an interior of shells* (*generation*) consisting of repetitive branch cell units and (III) *terminal functional groups* (i.e. the outer shell or periphery) as illustrated in Figures 1.13 and 1.14.

In general, dendrimer syntheses involves hierarchical assembly strategies that require the following construction components:

Many methods for assembling these components have been reported, however, they can be broadly categorized as either 'divergent' or 'convergent' strategies. Within each of these major approaches there may be variations in methodology for 'branch cell construction' (i.e. *in situ* versus *preformed*) or dendron construction (i.e., divergent versus convergent), as outlined in Figure 1.13.

Historically, early developments in the field were based on divergent methods.

Vögtle *et al.* (University of Bonn) first reported the synthesis of several low molecular weight (< 900 Daltons $G = 0-2$) cascade structures [77] using the divergent, *in situ* branch cell method. This method involves the use of traditional monomers to construct branch cells 'in situ' according to dendritic branching rule. This synthesis was based on a combination of acrylonitrile and reduction chemistry. As Vögtle reported later, higher generation cascade structures, and indeed dendrimers, could not be obtained by this process due to synthetic and analytical difficulties [78]. Nearly simultaneously, a completely characterized series of high molecular weight (i.e. > 58 000 Daltons $G = 0-7$), poly(amido-amine) (PAMAM) dendrimers was synthesized by Tomalia *et al.* [75, 76]. Success with his approach, was based on more facile combination of acrylate Michael addition and amidation chemistry. Since the discovery (1979) [79] occurred in the Dow Chemical research laboratories, publication was delayed until 1984–85 awaiting corporate approval and appropriate patent filing. Historically, this methodology provided the first commercial route to dendrimers, as well as the first opportunity to observe unique dendrimer property development that occurs *only* at higher generations (i.e. $G = 4$ or higher) and therefore could not be observed with smaller cascade molecules. Many of these observations were described in a seminal full length publication that appeared in 1985 [76] and later reviewed extensively in 1990 [48, 80].

The first published use of 'preformed branch cell' methodology was reported in a communication by Newkome *et al.* [81]. This approach involved the geometric coupling of 'preformed branch cell' reagents around a core to produce low molecular weight (i.e. < 2000 Daltons, $G = 3$) *arborol structures*. This approach has been used to synthesize many other dendrimer families including *dendri-poly(ethers)* [82], *dendri-poly(thioethers)* [8] and others [83]. Each of these methods involved the systematic divergent growth of 'branch cells' that defined shells within the 'dendrons' being initiated from the core. The multiplicity and directionality of the initiator sites (N_c) on the core, determine the number of dendrons and the ultimate shape, respectively, of the dendrimer. In essence, dendrimers propagated by this method constitute groups of molecular trees (i.e. two or more) that are propagated outwardly from their roots (cores). This occurs in stages (generations), wherein the functional leaves of these trees become reactive precursor templates or scaffolding upon which to assemble the next generation of branches. This methodology can be used to produce multiples of trees – dendrimers- or single trees – dendrons – as shown in Figure 1.13.

Using a totally novel approach, Hawker and Fréchet [84–87] followed by Miller and Neenan [88] reported the convergent construction of such molecular trees by first starting with the leaves or surface branch cell reagents. By amplifying with these reagents in stages (generations) one produces a dendron possessing a single reactive group at the root or focal point of the structure. If desired, subsequent coupling of these reactive dendrons through their focal point to a common 'anchoring core' yields the corresponding dendrimers. Because of the

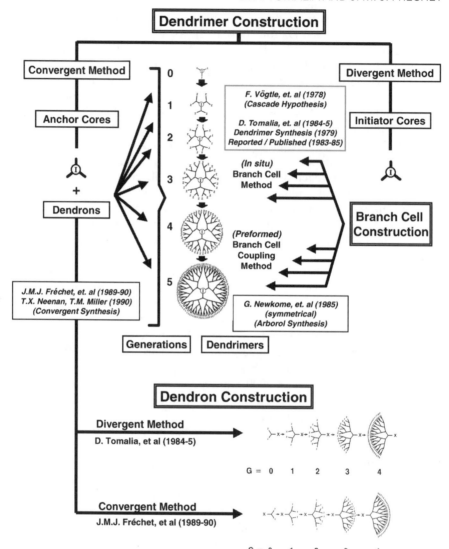

Figure 1.13 Overview of synthetic strategies for (a) branch cell construction, (b) dendron construction and (c) dendrimer construction annotated with discovery scientists

availability of orthogonal functional groups at the focal point and periphery of the dendrons, the convergent synthesis is particularly useful for the preparation of more complex macromolecular architectures [89] such as linear dendritic hybrids, block dendrimers or dendronized polymers (see Chapter 7). Another significant difference between the divergent and convergent approaches is that where the former sees an exponential increase in the number of coupling steps required for generation growth, the latter only involves a constant (typically two to three) number of reactions at each stage of the synthesis. Today, several hundred reports utilizing the original polyether Fréchet-type dendrons [85] make this the most studied, best understood, and most structurally precise family of convergent dendrimers.

Overall, each of these dendrimer construction strategies offers advantage and disadvantages. Some of these issues, together with experimental laboratory procedures, are viewed in more detail in several of the following chapters.

2.6.2   Dendrimer Features

Dendrimers may be thought of as unique nanoscale devices. Each architectural component manifests a specific function, while at the same time defining properties for these nanostructures as they are grown generation by generation. For example, the *core* may be thought of as the molecular information center from which *size, shape, directionality and multiplicity* are expressed via the covalent connectivity to the outer shells. Within the *interior*, one finds the *branch cell amplification region*, which defines the type and amount of interior void space that may be enclosed by the terminal groups as the dendrimer is grown. Branch cell multiplicity (N_b) determines the density and degree of amplification as an exponential function of generation (G). The interior composition and amount of solvent filled void space determines the extent and nature of guest–host (endoreceptor) properties that are possible with a particular dendrimer family and generation. Finally, the surface consists of reactive or passive terminal groups that may perform several functions. With appropriate function, they serve as a *template polymerization region* as each generation is amplified and covalently attached to the precursor generation. Secondly, the surface groups may function as passive or reactive gates controlling control entry or departure of guest molecules from the dendrimer interior. These three architectural components essentially determine the physicochemical properties, as well as the overall sizes, shapes and flexibility of dendrimers. It is important to note that dendrimer diameters increase linearly as a function of shells or generations added, whereas, the terminal functional groups increase exponentially as a function of generation. This dilemma enhances 'tethered congestion' of the anchored dendrons, as a function of generation, due to the steric crowding of the end groups. As a consequence, lower generations are generally open, floppy structures, whereas higher generations become robust, less deformable spheroids, ellipsoids or cylinders depending on the shape and directionality of the core.

PAMAM dendrimers are synthesized by the divergent approach. This methodology involves the *in situ* branch cell construction in stepwise, iterative stages (i.e. generation = 1, 2, 3 . . .) around a desired core to produce mathemat-

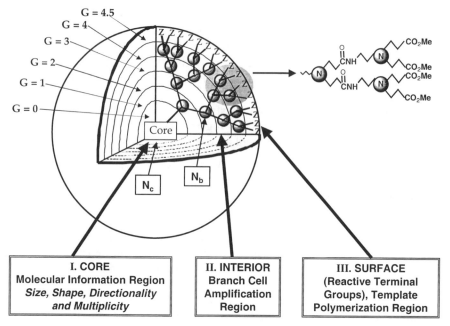

I. CORE	II. INTERIOR	III. SURFACE
Molecular Information Region *Size, Shape, Directionality and Multiplicity*	Branch Cell Amplification Region	(Reactive Terminal Groups), Template Polymerization Region

Figure 1.14 Three-dimensional projection of dendrimer core–shell architecture for G = 4.5 poly(amidomine) (PAMAM) dendrimer with principal architectural components (I) core, (II) interior and (III) surface

ically defined *core-shell* structures. Typically, ethylenediamine ($N_c = 4$) or ammonia ($N_c = 3$) are used as cores and allowed to undergo reiterative two-step reaction sequences involving: (a) exhaustive alkylation of primary amines by Michael addition with methyl acrylate, and (b) amidation of amplified ester groups with a large excess of ethylenediamine to produce primary amine terminal groups as illustrated in Scheme 3.

This first reaction sequence on the exposed dendron (Figure 1.14) creates $G = 0$ (i.e. the core branch cell), wherein the number of arms (i.e. dendrons) anchored to the core is determined by N_c. Iteration of the alkylation/amidation sequence produces an amplification of terminal groups from 1 to 2 with the *in situ* creation of a branch cell at the anchoring site of the dendron that constitutes $G = 1$. Repeating these iterative sequences (Scheme 3), produces additional shells (generations) of branch cells that amplify mass and terminal groups according to the mathematical expressions described in the box opposite.

It is apparent that both the core multiplicity (N_c) and branch cell multiplicity (N_b) determine the precise number of terminal groups (Z) and mass amplification as a function of generation (G). One may view those generation sequences as quantized polymerization events. The assembly of reactive monomers [48, 78], branch cells [48, 83, 89] or dendrons [85, 90] around atomic or molecular cores

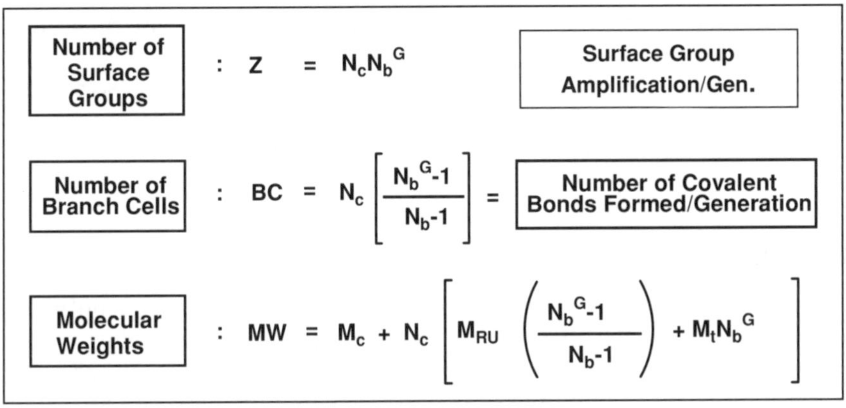

$$\text{Number of Surface Groups} \quad : \quad Z = N_c N_b{}^G \qquad \boxed{\begin{array}{c} \text{Surface Group} \\ \text{Amplification/Gen.} \end{array}}$$

$$\text{Number of Branch Cells} \quad : \quad BC = N_c \left[\frac{N_b{}^G - 1}{N_b - 1} \right] = \boxed{\begin{array}{c} \text{Number of Covalent} \\ \text{Bonds Formed/Generation} \end{array}}$$

$$\text{Molecular Weights} \quad : \quad MW = M_c + N_c \left[M_{RU} \left(\frac{N_b{}^G - 1}{N_b - 1} \right) + M_t N_b{}^G \right]$$

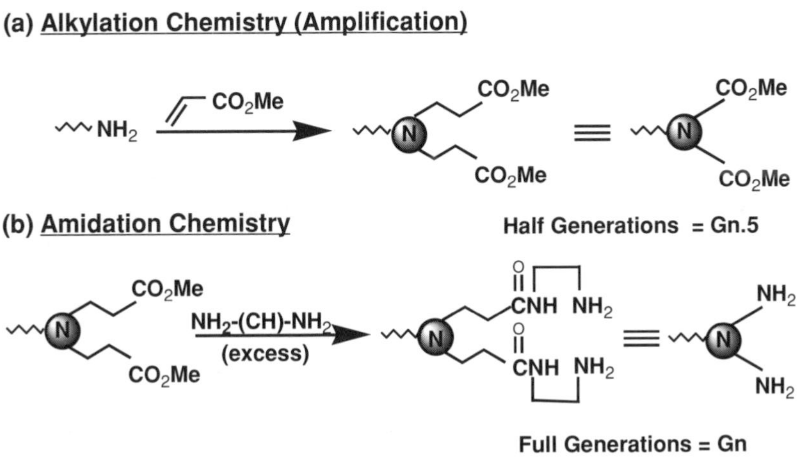

(a) Alkylation Chemistry (Amplification)

Half Generations = Gn.5

(b) Amidation Chemistry

Full Generations = Gn

Scheme 3

to produce dendrimers according to divergent/convergent dendritic branching principles, has been well demonstrated. Such systematic filling of space around cores with branch cells, as a function of generational growth stages (branch cell shells), to give discrete, quantized bundles of mass has been shown to be mathematically predictable [91]. Predicted molecular weights have been confirmed by mass spectroscopy [92–94] and other analytical methods [85, 95]. Predicted numbers of branch cells, terminal groups (Z) and molecular weights as a function of generation for ammonia core $(N_c = 3)$ PAMAM dendrimers are described in Figures 1.15. It should be noted that the molecular weights approximately double as one progresses from one generation to the next. The surface groups (Z) and branch cells (BC) amplify mathematically according to a power function, thus producing discrete, monodispersed structures with precise molecular weights as described in Figure 1.15. These predicted values can be verified by mass spectroscopy for the earlier generations however, with divergent dendrimers, minor mass defects are often observed for higher generations as congestion-induced *de Gennes dense packing* begins to take affect.

2.6.3 Dendrimer Shape Changes

As illustrated in Figure 1.15, dendrimers undergo 'congestion induced' molecular shape changes from flat, floppy conformations to robust spheroids as first predicted by Goddard *et al.* [96]. Shape change transitions were subsequently confirmed by extensive photophysical measurements, pioneered by Turro *et al.* [97–100] and solvatochromic measurements by Hawker *et al.* [101]. Depending upon the accumulative core and branch cell multiplicities of the dendrimer family under consideration, these transitions were found to occur between G = 3 and G = 5. Ammonia core, PAMAM dendrimers ($N_c = 3$, $N_b = 2$) exhibited a molecular morphogenesis break at G = 4.5 whereas, the ethylenediamine (EDA) PAMAM dendrimer family ($N_c = 4$ $N_b = 2$) manifested a shape change break around G = 3–4 [96] and the Fréchet-type convergent dendrons ($N_b = 2$) around G = 4 [101]. It is readily apparent that increasing the core multiplicity to $N_c = 4$ accelerates congestion and forces a shape change at least one

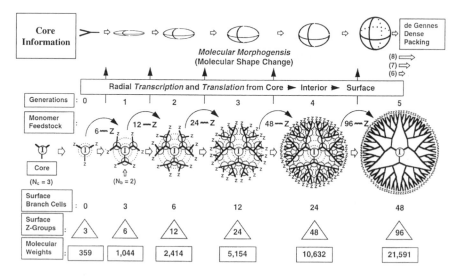

Figure 1.15 Comparison of molecular shape change, two-dimensional branch cell amplification surface branch cells, surface groups (*Z*) and molecular weights as function of generation: G= 0–6

generation earlier. Beyond these generational transitions, one can visualize these dendrimeric shapes as nearly spheroidal or slightly ellipsoidal *core–shell type architecture*.

2.6.4 De Gennes Dense Packing

As a consequence of the excluded volume associated with the core, interior and surface branch cells, steric congestion is expected to occur due to tethered connectivity to the core. Furthermore, the number of dendrimer surface groups, Z, amplifies with each subsequent generation (G). This occurs according to geometric *branching laws*, which are related to core multiplicity (N_c) and branch cell multiplicity (N_b). These values are defined by the following equation:

$$Z = N_c N_b^G$$

Since the radii of the dendrimers increase in a linear manner as a function of G, whereas the surface cells amplify according to $N_c N_b^G$, it is implicit from this equation that generational reiteration of branch cells ultimately will lead to a so-called *dense-packed state*.

As early as 1983, de Gennes and Hervet [102] proposed a simple equation derived from fundamental principles, to predict the dense-packed generation, for PAMAM dendrimers. It was predicted that at this generation ideal branching can no longer occur since available surface space becomes too limited for the mathematically predicted number of surface cells to occupy. This produces a 'closed geometric structure'. The surface is 'crowded' with exterior groups, which although potentially chemically reactive, are sterically prohibited from participating in ideal dendrimer growth.

This 'critical packing state' does not preclude further dendrimer growth beyond this point in the genealogical history of the dendrimer preparation. On the contrary, although continuation of dendrimer step-growth beyond the dense-packed state cannot yield structurally ideal, next generation dendrimer, it can nevertheless occur, as indicated by further increases in the molecular weight of the resulting products. Predictions by de Gennes [102] suggested that the PAMAM dendrimer series should reach a critical packing state at generations 9–10. Experimentally, we observed a moderate molecular weight deviation from predicted ideal values beginning at generations 4–7 (Figure 1.17). This digression became very significant at generations 7–8 as dendrimer growth was continued to generations 12 [103]. The products thus obtained are of 'imperfect' structure because of the inability of all surface groups to undergo further reaction. Presumably a fraction of these surface groups remain trapped under the surface of the newly formed dendrimer shell, yielding a unique architecture possessing two types of terminal groups. This new surface group population will consist of both those that are accessible to subsequent reiteration reagents and those that will be

sterically screened. The total number of these groups will not, however, correspond to the predictions of the mathematical branching law, but will fall between
that value, which was mathematically predicted for the next generations (i.e. G +
1), and that expected for the precursor generation (G). Thus, a mass defective
dendrimer 'generation' is formed.

Dendrimer surface congestion can be appraised mathematically as a function of generation, from the following simple relationship:

$$A_z = \frac{A_D}{N_z} \alpha \frac{r^2}{N_c N_b^G}$$

where A_z is the surface area per terminal group Z, A_D the dendrimer surface area
and N_z the number of surface groups Z per generation. This relationship predicts
that at higher generations G, the surface area per Z group becomes increasingly
smaller and experimentally approaches the cross-sectional area or van der Waals
dimension of the surface groups Z. The generation G thus reached is referred to
as the '*de Gennes*' *dense-packed generation*. Ideal dendritic growth without
branch defects is possible only for those generations preceding this dense-packed
state. This critical dendrimer property gives rise to self-limiting dendrimer
dimensions, which are a function of the branch cell segment length (*I*), the core
multiplicity N_c the branch cell juncture multiplicity N_b, and the steric dimensions of the terminal group Z. Whereas, the dendrimer radius *r* in the above
expression is dependent on the branch cell segment lengths *l*, large *l* values delay
this congestion. On the other hand, larger N_c, N_b values and larger Z dimensions
dramatically hasten it.

Additional physical evidence supporting the anticipated development of congestion as a function of generation is shown in the composite comparison in
Figure 1.16. Plots of intrinsic viscosity [η] [104], density z, surface area per Z
group (A_z), and refractive index *n* as a function of generation clearly show
maxima or minima at generations = 3–5, paralleling computer-assisted molecu-
lar-simulation predictions [96] as well as extensive photochemical probe experiments reported by Turro *et al.* [97–100].

The intrinsic viscosities [η] is expected to increase in a very classical fashion as
a function of molar mass (generation), but should decline beyond a certain
generation because of a change from an extended to a globular shape [48]. In
effect, once this critical generation is reached, the dendrimer begins to act more
like an Einstein spheroid. The intrinsic viscosity is a physical property that is
expressed in dL/g – the ratio of a volume to a mass. As the generation number
increases and transition to a spherical shape takes place, the volume of the
spherical dendrimer roughly increases in cubic fashion while its mass increases
exponentially, hence the value of [η] must decrease once a certain generation is
reached. This prediction has now been confirmed experimentally [104].

The dendrimer density z (atomic mass units per unit volume) clearly minimizes

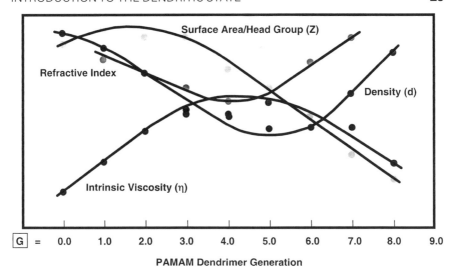

Figure 1.16 Comparison of surface area/head group (Z), refractive index, density (d) and intrinsic viscosity (η) as a function of generation: G = 1–9

between generations 4 and 5, then begins to increase as a function of generation due to the increasingly larger, exponential accumulation of surface groups. Since refractive indices are directly related to density parameters, their values minimize and parallel the above density relationship.

Clearly, this de Gennes dense-packed congestion would be expected to contribute to (a) sterically inhibited reaction rates, and (b) sterically induced stoichiometry [48]. Each of these effects was observed experimentally at higher generations. The latter would be expected to induce dendrimer mass defects at higher generations which we have used as a diagnostic signature for appraising the 'de Gennes dense packing' effect.

Theoretical dendrimer mass values were compared to experimental values by performing electrospray and MALDI-TOF mass spectral analysis on the respective PAMAM families (i.e. $N_c = 3$ and 4) [103]. Note there is essentially complete shell filling for the first five generations of the (NH_3) core ($N_c = 3\,N_b = 2$) poly(amidoamine)(PAMAM) series (Figure 1.17a). A gradual digression from theoretical masses occurs for G = 5–8, followed by a substantial break (i.e., $\Delta = 23\%$) between G = 8 and 9. *This discontinuity in shell saturation is interpreted as a signature for de Gennes dense packing.* It should be noted that shell saturation values continue to decline monotonically beyond this breakpoint to a value of 35.7% of theoretical at G = 12. A similar trend is noted for the EDA core, PAMAM series ($N_c = 4\,N_b = 2$) however, the shell saturation inflection point occurs at least one generation earlier (i.e. G = 4–7, see Figure 1.17b). This suggests that the onset of de Gennes dense packing may be occurring between G = 7 and 8.

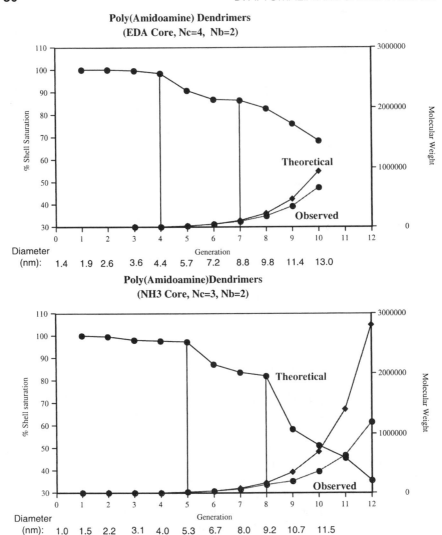

Figure 1.17 (a) Comparison of theoretical/observed molecular weights and % shell filling for EDA core poly(amidaomine) (PAMAM) dendrimers as a function of generation: G = 1–10. (b) comparison of theoretical/observed molecular weights and % shell filling for NH_3 core poly(amidoamine) (PAMAM) dendrimers as a function of generation: G = 1–12.

In the case of convergent dendrimers evidence of de Gennes dense packing rests on the inability to grow the dendrons beyond a certain size. For example the polyether Fréchet-type dendrons can be grown in high yield to the sixth generation but the yield of coupling drops dramatically for the subsequent G7 generation as growth by coupling of large dendrons becomes sterically inhibited [85].

3 NEW PROPERTIES DRIVEN BY THE DENDRITIC STATE

Throughout much of the early growth and evolution of polymer science, the quest for new properties was focused primarily on the two traditional architectures that defined thermoplastics (linear) and thermosets (crosslinked). Within each of these areas, there was intense activity to evaluate and optimize certain critical parameters. These parameters included various macromolecular chemical compositions, copolymer compositions, molecular weight effects, molecular weight distributions and crosslink densities, just to mention a few. Relatively little attention was given to the influence of architecture until the 1970s and 1980s. During that time, the first stirring of interest began concerning the influence of long chain branching on polymer properties [21]. Significant activity ensued thereafter, as it became apparent that single site metallocene/Brookhart catalysts were producing unique poly(olefin) families with completely new, commercially valuable properties [23, 24]. It is now recognized that both branched and dendritic architecture, in addition to molecular weight control, are key parameters influencing these new properties. These successful commercial developments, together with the rapid evolution of many new synthetic strategies to branched and dendritic architectures, have intensified the interest that macromolecular architecture may offer for the discovery of new properties.

3.1 COMPARISON OF TRADITIONAL AND DENDRITIC POLYMER PROPERTIES

The affect of architecture on small molecular properties has been recognized since the historical Berzelius (1832) discovery that defined the following premise: 'substances of identical compositions but different architectures – "skeletal isomers" – will differ in one or more properties' [15]. These effects are very apparent when comparing the fuel combustion benefits of certain isomeric octanes or the dramatic property differences observed in the three architectural isomers of carbon namely: graphite, diamond and buckminsterfullerene.

Similar patterns of property differentiation are clearly recognized at the macromolecular level. For example, dramatic changes in physical and chemical properties are observed by simply converting a linear topology of common composition to a cross-linked architecture. In traditional macromolecular science, these issues were considered apparent and obvious. However, as novel architectures emerged, new architecture–property relationships have not been so clearly articulated and exploited. Prompted by the synthetic accessibility of many new polymeric architectures based on common compositional monomers (i.e. branch cell monomers), this perspective was more clearly defined as early as 1994 in experiments by Fréchet and co-workers aimed at determining the influence of shape on the reactivity and physical properties of a series of comparable macromolecules including a dendrimer, a random hyperbranched polymer, and a linear aromatic polyesters all obtained from analogous building blocks [105]. This work clearly demonstrated the very significant shape-related changes in chemical reactivity as well as solubility that exist for polymers that have the same average molecular weight and composition but differ in their architecture and polydispersity. Following this report Tomalia introduced in 1996 the concept of 'macromolecular (architectural) isomerism'. Simply stated – 'macromolecular substances derived in the same proportions from the same monomer compositions, but in different architectural (configurations) will be expected to manifest different chemo/physical properties' [44, 106]. This hypothesis proposed a unique strategy for obtaining new polymeric properties by simply converting cost-effective traditional monomers into new macromolecular topologies (architectures). In 1997 Hawker et al. [107] provided the ultimate validation of this concept by preparing exact, size monodisperse, linear and dendritic polyethers analogs with the same composition. Their study revealed significant physical property differences between the two 'architectural isomers' confirming the earlier work of Fréchet and coworkers [105]. Most notable were substantially smaller hydrodynamic volumes (i.e. 40% smaller), as well as amorphous character (i.e. significantly more solvent soluble) for the dendritic isomers compared to the linear analog.

Parallel studies on PAMAM dendrimers, the Fréchet type polyether dendrons, and other dendrimer families have generated an extensive list of unique properties driven by the 'dendritic state.' Figure 1.18 compares several significant physical property differences between the linear and dendritic topologies related to conformations, crystallinity, solubilities, intrinsic viscosities, entanglement, diffusion/mobility and electronic conductivity.

In contrast to linear polymers, that obey the Mark–Houwink–Sakurada equation, the intrinsic viscosities of dendrimers do not increase continuously with molecular weight, but reach a maximum at a certain dendrimer generation. These maxima were predicted by Tomalia et al. for poly(amidoamine) dendrimers [48] and first measured for poly(arylethers) [104], and later for poly(propyleneimine) dendrimers [108], thus indicating they were not composition dependent. As indicated above, this is presumably due to the fact that once the dendrimer molecule becomes spherical, its volume grows by first approximation as n^3, whereas, mass grows as 2^n (where n = generation number). Since the intrinsic viscosity $[\eta]$ is expressed in volume per mass the quotient of the foregoing volumes and mass functions is indeed expected to display a maximum. A study of the melt viscosity of convergently grown Fréchet-type polyether dendrimers [109] also demonstrated their unique behavior, quite unlike that of comparable linear polymers. It is clear that the lack of entanglement of globular

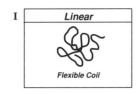

I | Linear

Flexible Coil

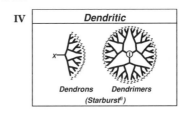

IV | Dendritic

Dendrons   Dendrimers
(Starburst[fi])

Linear

1. Random coil configurations

2. Semicrystalline/crystalline materials
   -higher glass temperatures

3. Lower solubility
   -decreases with Mwt.

4. Intrinsic viscosity follows logarithmic
   -increase with Mwt.

5. Entanglement directed rheological properties
   -shear sensitivity

6. Mobility by reptation
   -segmental and molecular mobility

7. Anisotropic electronic conductivity

Dendritic

1. Predictable shape changes as a function of Mwt. and core ①
   -robust spheroids; breathing deformability

2. Non-crystalline, amorphous materials
   -lower glass temperatures

3. Increased solubility
   -increases with Mwt.

4. Exhibits viscosity maximum and minimum plateau with Mwt.
   -low viscosities

5. Newtonian-type rheology
   -no shear sensitivity, considerably lower viscosity

6. Mobility involving whole dendrimer as the kinetic flow unit
   -virtually no reptation

7. Isotropic electronic conductivity

Figure 1.18 Comparison of properties for (I) linear and (IV) dendritic architecture

dendrimers – another attribute of the dendritic state – is largely responsible for their most unusual melt viscosity behavior [109–110].

Fréchet [49, 89] was the first to compare viscosity parameters for (A) linear topologies, as well as (B) random hyperbranched polymers and (C) dendrimers. More recently, we reported such parameters for (D) dendrigraft polymers [111] as shown in Figure 1.19. It is clear that all three dendritic topologies behave differently than the linear. There is, however, a continuum of behavior wherein random hyperbranched polymers behave most nearly like the linear systems. Dendrigrafts exhibit intermediary behavior, whereas dendrimers show a completely different relationship as a function of molecular weight.

Important physical property subtleties were noted within the dendrimer subset. For example, dendrimers possessing asymmetrical branch cells (i.e. Denkewalter type) exhibit a constant density versus generation relationship (Figure 1.20). This is in sharp contrast to symmetrical branch cell dendrimers (Tomalia-type PAMAM) that exhibit a minimum in density between $G = 4$ and $G = 7$ (NH_3 core) [48, 96]. This is a transition pattern that is consistent with the observed development of 'container properties' described in Figure 1.21.

Unique features offered by the 'dendritic state', that have no equivalency in the linear topologies, are found almost exclusively in the dendron/dendrimer subset or to a slightly lesser degree in the dendrigrafts. They include:

1. Nearly complete monodispersity.
2. The ability to control unimolecular container/scaffolding properties.

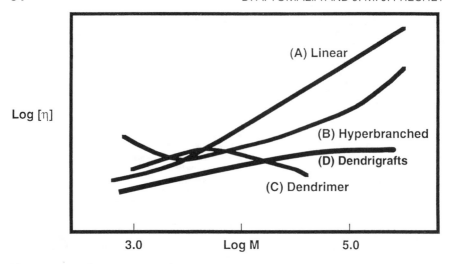

Figure 1.19 Comparison of intrinsic viscosities (log (η)) versus molecular weight (log M) for (A) linear, (B) random hyperbranched, (C) dendrimers and (D) dendrigraft topologies. Data for A, B, C adapted from Fréchet *et al.*, Ref. 49.

3. Exponential amplification of terminal functional groups.
4. Persistent nanoscale dimensions/shape as a function of molecular weight (generation).

These features are captured to some degree with dendrigraft polymers, but are either absent or present to a vanishing small extent for random hyperbranched polymers.

3.2 OVERVIEW OF UNIQUE DENDRIMER PROPERTIES – MONODISPERSITY

The monodispersed nature of dendrimers has been verified extensively by mass spectroscopy, size exclusion chromatography, gel electrophoresis and electron microscopy (TEM). As is always the case, the level of monodispersity is determined by the skill of the synthetic chemist, as well as the isolation/purification methods utilized.

In general, convergent methods produce the most nearly isomolecular dendrimers. This is because the convergent growth process allows purification at each step of the synthesis and therefore no cumulative effects of failed couplings are found [85, 89]. Appropriately purified convergent dendrimers are probably the most precise synthetic macromolecules that exist today.

As discussed earlier, mass spectroscopy has shown that PAMAM dendrimers (Figure 1.17) produced by the 'divergent method' are very monodisperse and

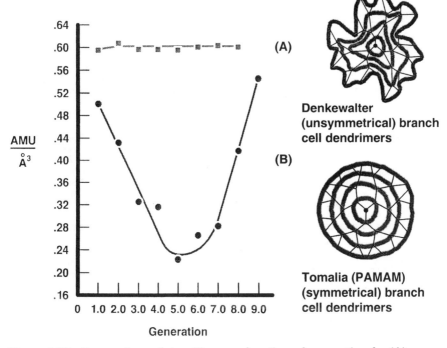

Figure 1.20 Comparison of densities as a function of generation for (A) asymmetrical branch cell in Denkewalter-type dendrimers, (B) symmetrical branch cell in Tomalia-type dendrimers ([densities calculated from experimental hydrodynamic diameters and theoretical, D.A. Tomalia, M. Hall, D.M. Hedstrand, *J. Am. Chem. Soc.*, **109**, 1601 (1987))

have masses consistent with predicted values for the earlier generations (i.e. G = 0–5). Even at higher generations, as one enters the de Gennes dense packed region, the molecular weight distributions remain very narrow (i.e. 1.05) and consistent in spite of the fact that experimental masses deviate substantially from predicted theoretical values. Presumably, de Gennes dense packing produces a very regular and dependable effect that is manifested in the narrow molecular weight distribution.

3.3 UNIMOLECULAR CONTAINER/SCAFFOLDING PROPERTIES

Unimolecular container/scaffolding behavior appears to be a periodic property that is specific to each dendrimer family or series. These properties will be determined by the size, shape, and multiplicity of the construction components that are used for the core, interior and surface of the dendrimer. Higher multiplicity components and those that contribute to 'tethered congestion' will hasten the

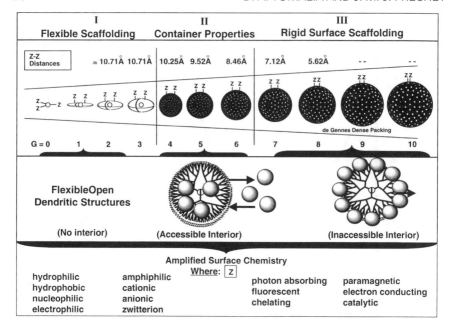

Figure 1.21 Periodic properties for poly(amidoamine) (PAMAM) dendrimers as a function of generation G = 0–10 (I) flexible scaffolding (G = 0–3) (II) container properties (G = 4–6) and (III) rigid surface scaffolding (G = 7–10) various chemo/physical dendrimer surfaces amplified according to: $Z = N_c N_b^G$ where: N_c = core multiplicity, N_b = branch cell multiplicity, G = generation

development of 'container properties' or rigid surface scaffolding as a function of generation. Within the PAMAM dendrimer family, these periodic properties are generally manifested in three phases as shown in Figure 1.21.

The earlier generations (i.e. G = 0–3) exhibit no well defined interior characteristics, whereas interior development related to geometric closure is observed for the intermediate generations (i.e. G = 4–6/7). Accessibility and departure from the interior is determined by the 'size and gating properties' of the surface groups. At higher generations (i.e. G = > 7) where de Gennes dense packing is severe, rigid scaffolding properties are observed, allowing relatively little access to the interior except for very small guest molecules. The site-isolation and encapsulation properties of dendrimers have been reviewed recently by Hecht and Fréchet [47].

3.4 AMPLIFICATION OF TERMINAL SURFACE GROUPS

Dendrimers within a generational series, can be expected to present their terminal groups in at least three different modes, namely: flexible, semi-flexible or rigid functionalized scaffolding. Based on mathematically defined dendritic branching rules (i.e. $Z = N_c N_b^G$ the various surface presentations are expected to become more congested and rigid as a function of generation level. It is implicit that this surface amplification can be designed to control gating properties associated with unimolecular container development. Furthermore, dendrimers may be viewed as versatile, nanosized objects that can be surface functionalized with a vast array of features (Figure 1.21). The ability to control and engineer these parameters provides an endless list of possibilities for utilizing dendrimers as modules for the design of nanodevices [91, 112]. Recent publications have begun to focus on this area [47, 113–117].

3.5   PERSISTENT NANOSCALE DIMENSIONS AND SHAPES

In view of the extraordinary structure control and nanoscale dimensions observed for dendrimers, it is not surprising to find extensive interest in their use as globular protein mimics. Based on their systematic, dimensional length scaling properties (Figure 1.22) and electrophoretic/hydrodynamic behavior [95], they are sometimes referred to as *artificial proteins*. These fundamental properties have in fact led to their commercial use as globular protein replacements for gene therapy [118] and immunodiagnostics [119–121]. Substantial effort has been focused recently on the use of dendrimers for 'site isolation' mimicry of proteins [47], enzyme-like catalysis [122], as well as other biomimetic applications [90, 123], drug delivery [130], surface engineering [131], and light harvesting [132]. Additional properties and applications for these dendritic polymers are reviewed throughout subsequent chapters of this book.

4   INTERMEDIATE ARCHITECTURES BETWEEN THERMOPLASTICS AND THERMOSETS

In the early days of polymer science, two major domains were defined, which were associated with certain distinguishing properties and architecture. One domain included linear, random coil thermoplastics such as poly(styrenes) or poly(acrylates). They were characterized as one-dimensional chains possessing two terminal groups per molecule, specific molecular weight distributions, reasonable solvent solubility, melt flow characteristics, chain entanglements consisting of inter- and intra-molecular knots and loops, mobility via snakelike reptation, and they exhibited expanded, large molecular volumes when immersed in 'good solvents'. The second domain included crosslinked thermosets such as vulcanized rubber, epoxies, and melamine resins that were all recognized as insoluble macromolecules. They exhibited rubber-like elasticity, and no melt flow features, yet they were semipermeable and susceptible to diffusion and pronounced swelling in certain solvents.

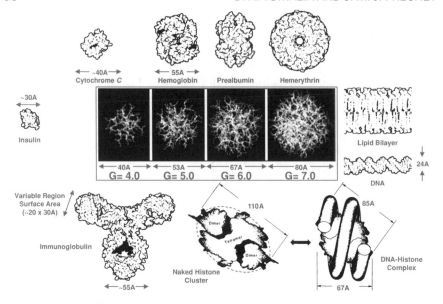

Figure 1.22 A comparison of dimensional length scales (Å) for poly(amido-amine) (PAMAM) dendrimers N_c = 3, N_b = 2 (NH_3 core) and various biological entities (e.g. proteins, DNA and lipid bilayers)

It is now recognized that a continuum of architecture and properties, which begins with the classical branched polymers, resides between these two classes. Typical branched structures such as starch or high pressures polyethylene are characterized by more than two terminal groups per molecule, possessing substantially smaller hydrodynamic volumes and different intrinsic viscosities compared to linear polymers, yet they often exhibit unexpected segmental expansion near the 'theta state'.

Completing this continuum, we may now focus on the intermediary role that (Class IV) dendritic polymers play both in architecture and properties as penultimate thermosplastic precursors to (Class II), crosslinked thermoset systems. Within the realm of traditional architectures, branched (Class III) and random hyperbranched structures (Class IVa) may be viewed as penultimate statistical precursors residing between thermoplastic structures and thermoset architectures as illustrated in Figure 1.23 [124]. The dendritic state may be visualized as advancement from a lower order (i.e. Class I–III) to a somewhat higher level of structural complexity [80]. Recent developments now demonstrate that certain dendritic subsets are manifestations of higher level structural control. In contrast to random hyperbranched polymers, the dendrimer subset, and to a lesser extent, the dendrigraft subset, represent a unique combination of high complexity with extraordinary structure control. As such, covalent bridging or crosslinking of these preformed modules would be expected to give rise to a completely new

Class V of more ordered complexity. Examples of this new architecture have been synthesized and we have coined these new topologies *megamers*.

5 MEGAMERS – A NEW CLASS OF MACROMOLECULAR ARCHITECTURE?

In the first full paper published on dendritic polymers [76], dendrimers were defined as 'reactive, structure-controlled macromolecular building blocks'. It was proposed that they could be used as repeat units for the construction of a new class of topological macromolecules referred to as 'starburst polymers'. Although there is intense activity in the field of dendrimer science, there are relatively few references focused on this specific concept [79, 91, 103, 125, 126]. Meanwhile, the term 'starburst' has been claimed as a registered trademark of the Dow Chemical Company. In view of these events, the generic term, 'megamer' has been proposed to describe those new architectures that are derived from the combination of two or more dendrimer molecules (see Figure 1.23) [103, 125].

Examples of both statistical, as well as structure controlled megamer assemblies have been reported and reviewed recently [103, 125]. Covalent oligomeric assemblies of dendrimers (i.e. dimers, trimers, etc.) are well-documented

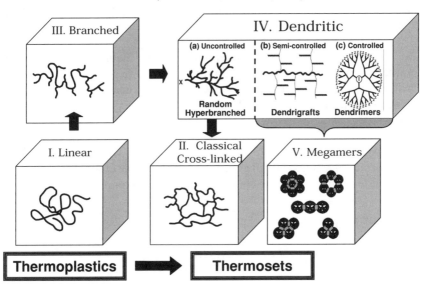

Figure 1.23 Intermediary of (III) branched and (IV) dendritic architecture in the conversion of (I) linear thermoplastics to (II) crosslinked thermoset polymers. Intermediary of (IVb) dendrigrafts and (IVc) dendrimers in the formation of megamers

examples of low molecular weight megamers. Statistical megamer assemblies have been reported as both *supramacromolecular* [114] and *supermacromolecular* (covalent) topologies. Many reports on the supramacromolecular self-assembly of these structures leading to dendrimeric clusters and monolayers are prime examples of supramacromolecular megamers. Simple, low DP covalent dendrimeric oligomers such as '[dendrimer]$_n$' where $n = 2$–10, and dendrimeric gels for which $n > 10$ represent a continuum of statistical covalent megamers that are possible.

More recently, mathematically defined, structure controlled, covalent megamers have been reported. They are a major subclass of megamers also referred to as *core–shell tecto*(*dendrimers*) [126–128]. Synthetic methodologies to these new architectures have been reported to produce precise megameric structures that adhere to mathematically defined bonding rules [91, 129]. It appears that structure controlled complexity beyond dendrimers is now possible. The demonstrated structure control within the dendrimer modules, and now the ability to mathematically predict and synthesize precise assemblies of these modules, provide a broad concept for the systematic construction of nanostructures with dimensions that could span the entire nanoscale region (Figure 1.24).

In summary, the ability to synthesize mathematically predicted megamer structures based on structure controlled dendrimer modules allows new synthesis options that are not presently available with traditional polymer science. In fact, the demonstrated structure control that is possible with the 'dendritic state' raises two important questions:

1. Do reactive dendrimer modules represent a key enabling technology for synthesizing controlled complexity in the nanoscale region?
2. Will these emerging megameric structures of poly(dendrimers) represent a new class of macromolecular architecture with unique properties and characteristics?

It is hoped the following chapters will inspire and provide some answers to these important questions, as well as further define the role of the 'dendritic state' in the quest for higher, structure controlled, molecular complexity.

6 REFERENCES

1. Eigen, M. *Naturwissenschaften* **10**, 465 (1971).
2. Eigen, M., Gardiner, W., Schuster, P., Winkler-Oswatitsch, R. *Evolution Now*, Freeman: New York, 1982.
3. Kuhn, H. and Waser, J. *Angew. Chem. Int. Ed.*, **20**, 500 (1981).
4. Prigogine, I. *Physics Today*, **25** (12), 38, (1972).
5. Mason, S. F. *Chemical Evolution*, Clarendon Press, Oxford, (1991).
6. Corey, E. J. Cheng, X.-M. *The Logic of Chemical Synthesis*, John Wiley & Sons: New York, (1989).
7. Pullman, B. *The Atom in the History of Human Thought*, Oxford University Press,

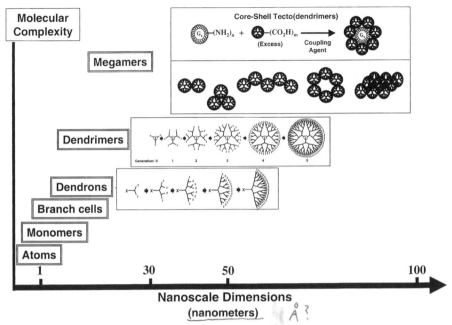

Figure 1.24 Nanoscale molecular dimensions: from atoms to megamers

New York, (1998).

8. Lothian-Tomalia, M. K., Hedstrand, D. M., Tomalia, D. A. *Tetrahedron* **53**, 15,495–15513 (1997).

9. Morawetz, H. *Polymers. The Origin and Growth of a Science*, John Wiley & Sons: New York, (1985).

10. Elias, H.-G. *An Introduction to Polymer Science*, VCH, Weinheim, (1997).

11. Merrifield, R. B. and Barany, G. *The Peptides*, Vol. 2, Academic Press, New York, (1980).

12. Merrifield, R. B. *Angew. Chem. Int. Ed. Engl.*, **24**, 799, (1985).

13. Bodanszky, M. *Principles of Peptide Synthesis*, Springer-Verlag, Berlin-Heidelberg, (1984).

14. Bodanzky, M. and Bodanzky, A. Springer-Verlag, Berlin-Heidelberg, New York, NY, (1984).

15. Berzelius, J. *J. Fortsch. Phys. Wissensch*, **11**, 44 (1832).

16. Staudinger, H. *Schweiz. Chem. Z.*, **105**, 28–33, 60 (1919).

17. Staudinger, H. *Ber.*, **53**, 1073 (1920).

18. Staudinger, H. *From Organic Chemistry to Macromolecules, A Scientific Autobiography*, John Wiley & Sons, New York, (1961).

19. James, L. K. *Hermann Staudinger 1881–1965*, James, L. K. (ed.), History of Modern Chem. Sci. Series, Am. Chem. Soc., Washington, DC, p. 359 (1994).

20. Elias, H.-G. *Mega Molecules*, Springer-Verlag, Berlin, (1987).

21. Roovers, J. *Advances in Polymer Science, Branched Polymers I*, Vol. 142, Springer-Verlag, Berlin, (1999).

22. Roovers, J. *Advances in Polymer Science, Branched Polymers II*, Vol. 143, Springer-Verlag, Berlin, (2000).

23. *Metallocene-Based Polyolefins*, Vols 1 and 2, John Wiley & Sons, Brisbane, (2000),
24. Guan, Z., Cotts, P. M., McCord, E. F. and McLain, S. J. *Science*, **283**, 2059 (1999).
25. Sunder, A., Heinemann, J. and Frey, H. *Chem. Eur. J.*, **6**, 2499 (2000).
26. Flory, P. J. *J. Am. Chem. Soc.*, **63**, 3083, 3091, 3096 (1941).
27. Flory, P. J. and Rehner, J. *J. Chem. Phys.*, **11**, 512 (1943).
28. Flory, P. J. *J. Am. Chem. Soc.*, **74**, 2718 (1952).
29. Flory, P. J. *Ann. N.Y. Acad. Sci.*, **57**, 327 (1953).
30. Stockmayer, W. H. *J. Chem. Phys.*, **11**, 45 (1944).
31. Stockmayer, W. H. *J. Chem. Phys.*, **12**, 125 (1944).
32. Zimm, B. and Stockmayer, W. H. *J. Chem. Phys.*, **17**, 1301 (1949).
33. Graessley, W. W. *Macromolecules*, **8**, 185 (1975).
34. Graessley, W. W. *Macromolecules*, **8**, 865 (1975).
35. Gordon, M. and Malcolm, G. N. *Proc. R. Soc. (London)*, **A295**, 29 (1966).
36. Gordon, M. and Dobson, G. R. *J. Chem. Phys.*, **43**, 35 (1975).
37. Dusek, K. *Makromol. Chem, Suppl.*, **2**, 35 (1979).
38. Burchard, W. *Adv. Polym. Sci.*, **48**, 1 (1988).
39. Good, I. J. *Proc. Cambridge Phil. Soc.*, **45**, 360 (1948).
40. Good, I. J. *Proc. R. Soc. (London)*, **A263**, 54 (1963).
41. Goodsell, D. S. *American Scientist*, **88**, 230 (2000).
42. Tomalia, D. A. *Macromol. Symp.*, **101**, 243 (1996).
43. Tomalia, D. A., Brothers II, H. M., Piehler, L. T., Hsu, Y. *Polym. Mater. Sci. & Eng*, **73**, 75 (1995).
44. Dvornic, P. R. and Tomalia, D. A. *Science Spectra*, **5**, 36 (1996).
45. Naj, A. K. *Persistent Inventor Markets a Molecule*. New York, p. B1, (1996).
46. Tomalia, D. A. and Esfand, R. *Chem. Ind.*, **11**, 416 (1997).
47. Hecht, S. and Fréchet, J. M. J. *Angew. Chem. Int. Ed.*, **40**, 74 (2001).
48. Tomalia, D. A., Naylor, A. M. and Goddard III, W. A. *Angew. Chem. Int. Ed. Engl.*, **29**, 138 (1990).
49. Fréchet, J. M. J., Hawker, C. J., Gitsov, I. and Leon, J. W. *J.M.S. – Pure Appl. Chem.*, **A33** (10), 1399 (1999).
50. Vögtle, F. and Fischer, M. *Angew. Chem. Int. Ed.*, **38**, 884 (1999).
51. Voit, B. I. *Acta Polymer.*, **46**, 87 (1995).
52. Flory, P. J. *Principles of Polymer Chemistry* Cornell University Press, Ithaca, NY, (1953).
53. Gunatillake, P. A., Odian, G. and Tomalia, D. A. *Macromolecules*, **21**, 1556 (1988).
54. Kim, Y. H. and Webster, O. W. *Polym. Prepr.*, **29**, 310 (1988).
55. Kim, Y. H. and Webster, O. W. *J. Am. Chem. Soc.*, **112**, 4592 (1990).
56. Emrick, T. and Chang, H. T. Fréchet, J. M. J. *Macromolecules*, **32**, 6380 (1999).
57. Emrick, T., Chang, H. T. and Frechet, J. M. J. *J. Poly Sci. A*, **38**, 4850 (2000).
58. Bharati, P. and Moore, J. S. *J. Am. Chem. Soc.*, **119**, 3391 (1997).
59. Muzafarov, A. M., Rebrov, E. A., Gorbatsevich, O. B., Golly, M., Gankema, H. and Moller, M. *Macromol. Symp.*, **102**, 35 (1996).
60. Miravet, J. F. and Fréchet, J. M. J. *Macromolecules*, **31**, 3461 (1998).
61. Chu, F. and Hawker, C. J. *Polym. Bull.*, **30**, 265 (1993).
62. Hawker, C. J., Lee, R. and Fréchet, J. M. J. *J. Am. Chem. Soc.*, **113**, 4583 (1991).
63. Uhrich, K. E., Hawker, C. J., Fréchet, J. M. J. and Turner, S. R. *Macromolecules*, **25**, 4583 (1992).
64. Liu, M., Vladimirov, N. and Fréchet, J. M. J. *Macromolecules*, **32**, 6881 (00).
65. Fréchet, J. M. J., Henni, M., Gitsov, I., Aoshima, S., Leduc, M. R. and Grubbs, R. B. *Science*, **269**, 1080 (1995).
66. Hawker, C. J., Fréchet, J. M. J., Grubbs, R.B. and Dao, J. *J. Am. Chem. Soc.*, **117**, 10763 (1995).

67. Gong, C., Miravet, J. and Fréchet, J. M. J. *J. Polym. Sci. A.*, **37**, 3193 (2000).
68. Tomalia, D. A., Hedstrand, D. M. and Ferrito, M. S. *Macromolecules*, **24**, 1435 (1991).
69. Gauthier, M. and Moller, M. *Macromolecules*, **24**, 4548 (1991).
70. Six, J.-L. and Gnanou, Y. *Macromol. Symp.*, **95**, 137 (1995).
71. Taton, D., Cloutet, E. and Gnanou, Y. *Macrmol. Chem. Phys.*, **199**, 2501 (1998).
72. Trollsas, M. and Hendrick, J. L. *J. Am. Chem. Soc.*, **120**, 4644 (1998).
73. Trollsas, M. and Hendrick, J. L. *Macromolecules*, 31, 4390 (1998).
74. Grubbs, R. B., Hawker, C. J., Dao, J. and Fréchet, J. M. J. *Angew. Chem. Int. Ed. Engl.*, **36**, 270 (1997).
75. Tomalia, D. A., Dewald, J. R., Hall, M. J., Martin, S. J. and Smith, P. B. Kyoto, Japan, August 1984, p. 65.
76. Tomalia, D. A., Baker, H., Dewald, J., Hall, M., Kallos, G., Martin, S., Roeck, J., Ryder, J. and Smith, P. *Polym. J.*, **17**, 117 (1985).
77. Buhleier, E., Wehner, W. and Vögtle, F., *Synthesis*, 155 (1978).
78. Moors, R. and Vögtle, F. *Chem. Ber.*, **126**, 2133 (1993).
79. Tomalia, D. A. *Sci. Am.*, **272**, 62 (1995).
80. Tomalia, D. A., Hedstrand, D. M. and Wilson, L. R. *Dendritic Polymers* (Ed.) Kroschwitz, J. (2nd edn), John Wiley & Sons, New York, 1990, pp. 46–92 and 251.
81. Newkome, G. R., Yao, Z.-Q., Baker, G. R. and Gupta, V. K. *J. Org. Chem.*, **50**, 2003 (1985).
82. Padias, A. B., Hall Jr., H. K. and Tomalia, D. A. *J. Org. Chem.*, **52**, 5305 (1987).
83. Newkome, G. R., Moorfield, C. N. and Vögtle, F. *Dendritic Molecules*, VCH, Weinheim, (1996).
84. Fréchet, J. M. J., Jiang, Y., Hawker, C. J. and Philippides, A. E. *Proc. IUPAC Int. Symp. Macromol. Seoul*, 19 (1989).
85. Hawker, C. J. and Fréchet, J. M. J. *J. Am. Chem. Soc.*, **112**, 7638 (1990).
86. Hawker, C. J. and Fréchet, J. M. J. *J. Chem. Soc. Chem. Commun.*, 1010 (1990).
87. Hawker, C. J. and Fréchet, J. M. J. *Macromolecules*, **23**, 4726 (1990).
88. Miller, T. M. and Neenan, T. X. *Chem. Mat.*, **2**, 346 (1990).
89. Fréchet, J. M. J. *Science*, **263**, 1710 (1994).
90. Zeng, F. and Zimmerman, S. C. *Chem. Rev.*, **97**, 1681 (1997).
91. Tomalia, D. A. *Adv. Mater.*, **6**, 529 (1994).
92. Kallos, G. J., Tomalia, D. A., Hedstrand, D. M., Lewis, S. and Zhou, J. *Rapid Commun. Mass Spectrom.*, **5**, 383 (1991).
93. Dvornic, P. R. and Tomalia, D. A. *Macromol. Symp.*, **98**, 403 (1995).
94. Hummelen, J. C., van Dongen, J. L. J. and Meijer, E. W. *Chem. Eur. J.*, **3**, 1489 (1997).
95. Brothers II, H. M., Piehler, L. T. and Tomalia, D. A. *J. Chromatogr.*, **A814**, 233 (1998).
96. Naylor, A. M., Goddard III, W. A., Keifer, G. E. and Tomalia, D. A. *J. Am. Chem. Soc.*, **111**, 2339 (1989).
97. Turro, N. J., Barton, J. K. and Tomalia, D. A. *Acc. Chem. Res.*, **24** (11), 332 (1991).
98. Gopidas, K. R., Leheny, A. R., Caminati, G., Turro, N. J. and Tomalia, D. A. *J. Am. Chem. Soc.*, 113, 7335 (1991).
99. Ottaviani, M. F., Turro, N. J., Jockusch, S. and Tomalia, D. A. *J. Phys. Chem.*, **100**, 13675 (1996).
100. Jockusch, J., Ramirez, J., Sanghvi, K., Nociti, R., Turro, N. J. and Tomalia, D. A. *Macromolecules*, **32**, 4419 (1999).
101. Hawker, C. J., Wooley, K. L. and Fréchet, J. M. J. *J. Am. Chem. Soc.*, **115**, 4375 (1993).
102. de Gennes, P. G. and Hervet, H. J. *J. Physique-Lett.* (*Paris*), **44**, 351 (1983).
103. Tomalia, D. A., Esfand, R., Piehler, L. T., Swanson, D. R. and Uppuluri, S. *High*

Performance Polymers, in press (2001).
104. Mourey, T. H., Turner, S. R., Rubinstein, M., Fréchet, J. M. J., Hawker, C. J. and Wooley, K. L. *Macromolecules*, **25**, 2401 (1992).
105. Wooley, K. L., Fréchet, J. M. J. and Hawker, C. J. *Polymer*, **35**, 4489 (1994).
106. Tomalia, D. A., Dvornic, P. R., Uppuluri, S., Swanson, D. R. and Balogh, L. *Polym. Mater. Sci. & Eng.*, **77**, 95–96 (1997).
107. Hawker, C. J., Malmstrom, E. E., Frank, C. W. and Kampf, J. P. J. *J. Am. Chem. Soc.*, **119**, 9903 (1997).
108. de Brabander-van den Berg, E. M. M. and Meijer, E. W. *Angew. Chem. Int. Ed. Engl.*, **32**, 1308 (1993).
109. Hawker, C. J., Farrington, P. J., Mackay, M. E., Wooley, K. L. and Fréchet, J. M. J. *J. Am. Chem. Soc.*, 117, 4409 (1995).
110. Farrington, P. J., Hawker, C. J., Fréchet, J. M. J. and Mackay, M. E. *Macromolecules*, **31**, 5043 (1998).
111. Qin, D. Yin, R., Li, J., Piehler, L., Tomalia, D. A., Durst, H. D. and Hagnauer, G. *Polym. Prepr.* (*ACS Div. Polym. Chem.*), **40**, 171 (1999).
112. de A.A. Soler-Illia, G. J., Rozes, L., Boggiano, M. K., Sanchez, C., Turrin, C.-O., Caminade, A.-M. and Majoral, J.-P. *Angew. Chem. Int. Ed.*, **39**, 4250 (2000).
113. Tomalia, D. A. and Durst, H. D. *Topics in Current Chemistry*, Vol. 165; Weber, E. W. (ed), Springer-Verlag: Berlin-Heidelberg, 1993, p. 193.
114. Tomalia, D. A. and Majoros, I. *Dendrimeric Supramolecular and Supramacromolecular Assemblies*, Vol. 0, Chapter 9, Ciferri, A., Marcel Dekker, New York, 1999, pp. 359–434.
115. Balogh, L. ,Tomalia, D. A. and Hagnauer, G. L. *Chemical Innovation*, **30**, 19 (2000).
116. Crooks, R. M., Lemon III, Sun, L., Yeung, L. K. and Zhao, M. *Dendrimer-Encapsulated Metals and Semiconductors: Synthesis, Characterization, and Applications*, Springer-Verlag:,Berlin-Heidelberg, (2001).
117. Freeman, A. W., Koene, S. C., Malenfant, P. R. L., Thompson, M. E. and Fréchet, J. M. J. *J. Am. Chem. Soc.*, **122**, 1285 (2000).
118. Kukowska-Latallo, J. F., Bielinska, A. U., Johnson, J., Spindler, R., Tomalia, D. A. and Baker Jr, J. R. *Proc. Natl. Acad. Sci. USA*, **93**, 4897 (1996).
119. Singh, P. *Bioconjugate Chem.*, **9**, 54 (1998).
120. Singh, P., Moll III, F., Lin, S. H., Ferzli, C., Yu, K. S., Koski, K. and Saul, R. G. *Clin. Chem.*, **40** (9), 1845 (1994).
121. Singh, P., Moll III, F., Lin, S. H. and Ferzli, C. *Clin. Chem.*, **42** (9), 1567 (1996).
122. Piotti, M. E., Rivera, F., Bond, R., Hawker, C. J. and Fréchet, J. M. J. *J. Am. Chem. Soc.*, **121**, 9471 (1999).
123. Bieniarz, C. *Dendrimers: Applications to Pharmaceutical and Medicinal Chemistry*, Vol. 18, Marcel Dekker, p. 55, (1998).
124. Dusek, K. *TRIP*, **5** (8), 268 (1997).
125. Tomalia, D. A., Uppuluri, S., Swanson, D. R. and Li, J. *Pure Appl. Chem.*, in press (2001).
126. Uppuluri, S., Piehler, L. T., Li, J., Swanson, D. R., Hagnauer, G. L. and Tomalia, D. A. *Adv. Mater.*, **12** (11), 796 (2000).
127. Uppuluri, S., Swanson, D. R., Brothers II, H. M., Piehler, L. T., Li, J., Meier, D. J., Hagnauer, G. L. and Tomalia, D. A. *Polym. Mater. Sci. & Eng.* (*ACS*), **80**, 55 (1999).
128. Li, J., Swanson, D. R., Qin, D., Brothers II, H. M., Piehler, L. T., Tomalia, D. A. and Meier, D. J. *Langmuir*, **15**, 7347 (1999).
129. Mansfield, M. L., Rakesh, L. and Tomalia, D. A. *J. Chem. Phys.*, **105**, 3245 (1996).
130. Liu, M. and Fréchet, J. M. J. *Pharmaceut. Sci. Technol. Today*, **2**, 393 (1999).
131. Tully, D. C. and Fréchet, J. M. J. *Chem. Commun.* 1229 (2001).
132. Adronov, A. and Fréchet, J. M. J. *Chem. Commun.* 1701 (2000).

2

Structural Control of Linear Macromolecules

C. J. HAWKER

IBM Almaden Research Center, NSF Center for Polymeric Interfaces and Macromolecular Assemblies, San Jose CA, USA

1 INTRODUCTION

The development of linear polymers and their impact on all aspects of modern life is without question one of the major achievements of the twentieth century. As this field has matured it is becoming increasing apparent that further developments will likely arise, not from the synthesis of totally new linear polymers, but from more accurately controlling the preparation of linear polymers from currently available monomers. Just as nature routinely uses linear macromolecules whose molecular weight, monomer sequence, etc. are precisely controlled, similar concepts are being developed for synthetic linear macromolecules in an effort to induce a myriad of physical properties similar to natural systems. The possibilities for such an approach are substantial since it should be remembered that nature employs a very limited range of monomer units compared to the vast selection of monomers that are synthetically available. The focus of this chapter will be to detail the advances that have been made in recent years for controlling the structure of linear macromolecules. It should be noted that many of these techniques are also used in the construction of complex macromolecular architectures such as hybrid dendritic-linear block copolymers, dendrigraft macromolecules, etc. Such three-dimensional architectures will be the focus of following chapters.

2 LIVING POLYMERIZATIONS

The present interest in linear polymers of defined structure can be traced back to

---

Dendrimers and Other Dendritic Polymers. Edited by Jean M. J. Fréchet and Donald A. Tomalia
© 2001 John Wiley & Sons Ltd

the pioneering work of Szwarc [1] who developed the concept of *living* polymerizations and employed it successfully in the anionic polymerization of vinyl monomers such as styrene. While the term, 'living polymerization' has now been applied to a wide range of polymerizations, for which it is not strictly true in most cases, the underlying principles and criteria remain the same. For a true living system, all growing polymer chains are initiated at the same time and grow at the same rate with no termination of the growing chain end. As shown in Scheme 1, a consequence of this is the degree of polymerization of the macromolecule being directly proportional to the relative concentration of monomer to initiator. The absence of termination reactions also allows the reactive chain end to be functionalized in a variety of different ways to give chain-end, or telechelic macromolecules.

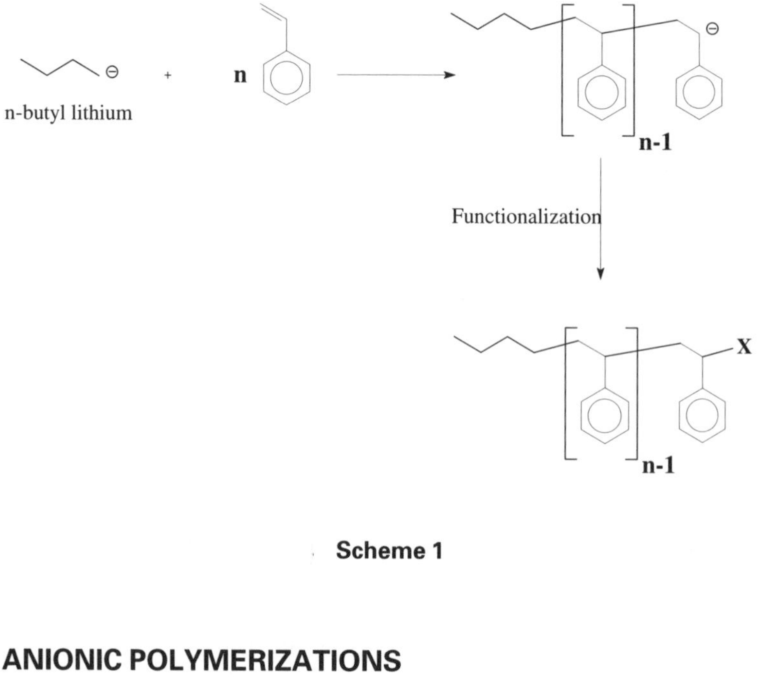

n-butyl lithium

Functionalization

Scheme 1

3 ANIONIC POLYMERIZATIONS

As the first, and perhaps the most well-studied form of living polymerizations, anionic procedures have attracted considerable attention both academically and industrially [2, 3]. While numerous examples of living procedures have been developed, they can be classified into two main families; carbanionic polymerization of vinyl monomers and anionic ring opening polymerizations.

For vinyl monomers two methods can be used to initiate polymerization, both involve alkali metal derivatives, or more rarely alkaline–earth metal derivatives, and differ only by the mechanism of formation of the primary carbanionic

non-living behavior and in some cases no polymerization at all. To overcome these difficulties a considerable amount of effort has been devoted to the development of modified reaction conditions for the successful polymerization of these polar monomers. In the case of methyl methacrylate, living anionic polymerization is best accomplished at low temperatures ($T < -75\,^{\circ}C$) and in polar solvents with large counterions such as cesium instead of lithium [14]. In the case of acrylates and acrylonitrile [15] the situation is even more difficult and truly living polymerizations are a challenge though Teyssie has reported the dramatic effect of added lithium chloride on the polymerization of t-butyl acrylate [16].

4   BLOCK COPOLYMERS

One of the principal features of living anionic procedures is that the carbanion chain end is very stable in the absence of terminating species. This permits the

$$I^{\ominus} + nM_1 \longrightarrow I\text{-}(M_1)_n^{\ominus} + mM_2 \longrightarrow I\text{-}(M_1)_n\text{-}(M_2)_m^{\ominus}$$

Block Copolymers

Scheme 4

formation of block copolymers by the initial polymerization of one monomer to give a stable, well-defined initial block which, due to the living chain end can be further extended by addition of a second monomer to give the desired diblock copolymer (Scheme 4). In fact one of the major uses of living anionic polymerizations is in the synthesis of block copolymers. Once again many of the features and concerns related to the preparation of linear homopolymers by anionic techniques are equally applicable to block copolymers. Molecular weights of each block can be accurately controlled by the relative ratios of monomers to initiator, polydispersities of each block can be very low and the chain ends are directly defined by the structure of the initiator and/or functionalization reactions.

The additional complexity present in block copolymer synthesis is the order of monomer polymerization and/or the requirement in some cases to modify the reactivity of the propagating center during the transition from one block to the next block. This is due to the requirement that the nucleophilicity of the initiating block be equal or greater than the resulting propagating chain end of the second block. Therefore the synthesis of block copolymers by sequential polymerization generally follows the order: dienes/styrenics *before* vinylpyridines *before* meth(acrylates) *before* oxiranes/siloxanes. As a consequence, styrene–MMA block copolymers should be prepared by initial polymerization of styrene followed by MMA, while PEO–MMA block copolymers should be prepared by initial polymerization of MMA followed by ethylene oxide. Attempts to circumvent this requirement by 'increasing' the nucleophilicity of the initiating chain end with silyl derivatives has been reported and shows significant promise [17].

5 ANIONIC RING OPENING

The general aspects of anionic polymerizations can also be equally applied to anions other than carbanions, and monomers other than vinyl monomers. A significant amount of work has appeared in the area of living ring-opening polymerization of cyclic monomers. For these examples the polymerization is influenced by the ring size, the atoms constituting the ring, the initiator and the reaction conditions with the main cyclic monomers being epoxides, episulfides, lactones, lactams, anhydrides and carbonates [18]. In these cases the propagating anion is oxygen, nitrogen or sulfur and a much greater range of functional groups can be incorporated into the monomer since the reactivity of the heteroatom propagating centers is less than for carbanions.

6 CATIONIC POLYMERIZATION

The underlying requirements for the polymerization of vinyl and cyclic monomers by cationic procedures are very similar to anionic systems. Both approaches require highly purified monomers and inert, controlled atmosphere reaction conditions. In each case, well-defined and low polydispersity linear polymers are obtained. The obvious difference between the two strategies is the nature of the propagating center. While both anionic and cationic centers are highly reactive, they undergo different reactions and are generated using different methodologies. Therefore the range of monomers and functionalization reactions are different for cationic procedures when compared to anionic procedures.

While numerous vinyl monomers can be polymerized/oligomerized in an uncontrolled fashion by carbocationic methods, those that give well-defined linear polymers of high molecular weight is limited. Carbocationic procedures have found wide application primarily in the polymerization of isobutylene, α-methylstyrene, vinyl ethers, and N-vinylcarbazole [19]. Initiation of these systems by protic acids has been shown to be problematic in most cases [20], with the majority of efforts being devoted to the use of Friedel–Crafts halides, such as $AlCl_3$, $TiCl_4$, BCl_3, etc [21]. The latter systems have proved to be very successful in a number of industrial processes, such as the preparation of butyl rubber [22]. However neither of these methods are particularly suited for the preparation of well-defined polymeric materials such as telechelics. To overcome these difficulties a bimolecular initiating system, consisting of a alkyl halide and a Friedel–Crafts halide has been introduced (Scheme 5) [23]. The ability to have a well defined initiating system, controlled propagation with reduced termination and transfer gives rise to a living polymerization, whereas the tertiary chloro

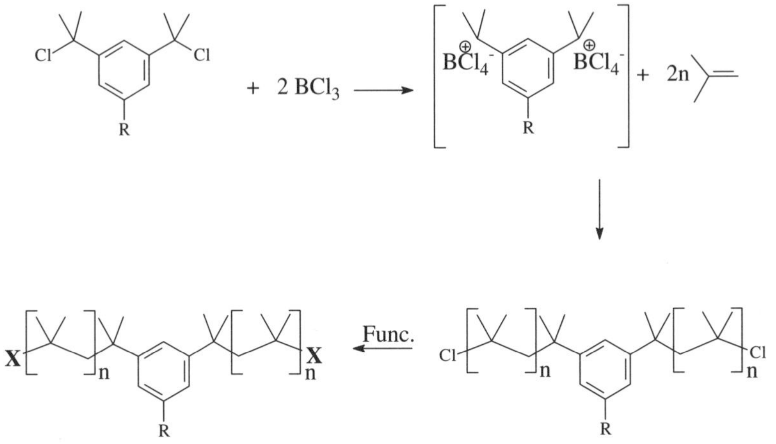

Scheme 5

end-groups can be easily converted into other functional groups such as vinyl and primary alcohol groups.

7   CATIONIC RING OPENING

The range of cyclic monomers that can undergo cationic ring opening polymerizations is large; however, the tendency of cationic systems for unwanted side reactions again limits the ability to control these polymerizations. Perhaps the best studied of these systems is the polymerization of tetrahydrofuran for which many mono, di, and multifunctional initiators have been developed. One of the most interesting is the use of trifluoromethanesulphonic anhydride for the preparation of difunctional poly(tetrahydrofuran) derivatives. Use of this reagent transforms the THF monomer into a difunctional initiator which loses its identity as an initiator fragment and becomes a repeat unit in the linear polymer chain (Scheme 6) [24].

In a similar fashion, the cationic polymerization of 2-oxazolines has been extensively studied and was found to provide the first verified entry to *linear-poly(alkyleneimine)* architectures. These acylated polymers were first recognized as precursors to linear poly(ethyleneimines) in the early 1960s [25]. Hydrolysis experiments demonstrated that deacylation of these products to linear PEI was possible. The original polymerization mechanism proposed by Tomalia *et al.*

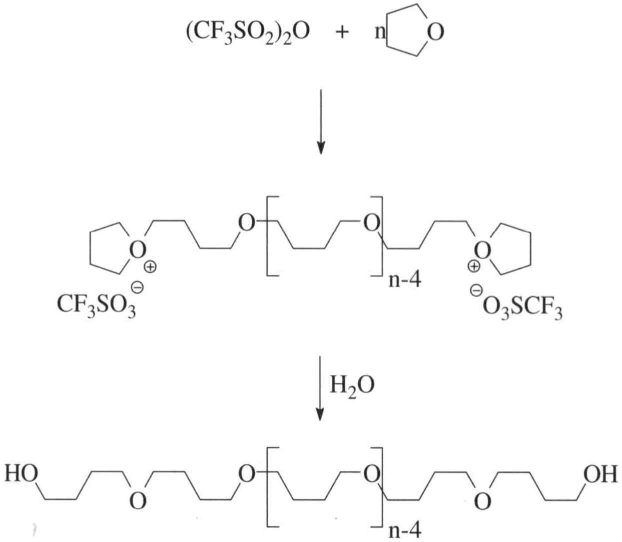

$$(CF_3SO_2)_2O \quad + \quad n \, \boxed{} O$$

$$CF_3SO_3^{\ominus} \qquad\qquad\qquad\qquad \overset{\ominus}{O_3SCF_3}$$

$\downarrow H_2O$

Scheme 6

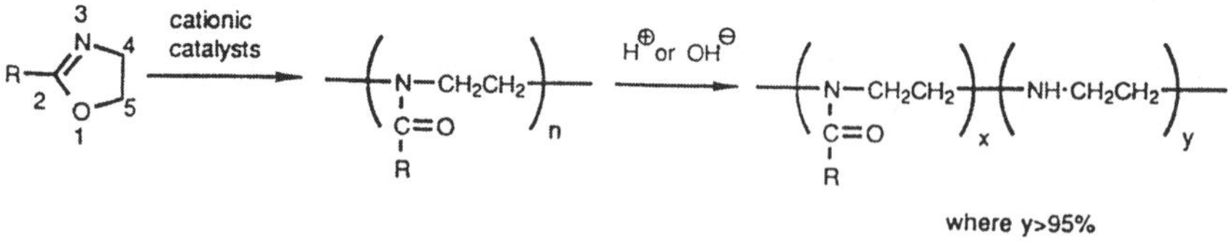

where y>95%

Scheme 7

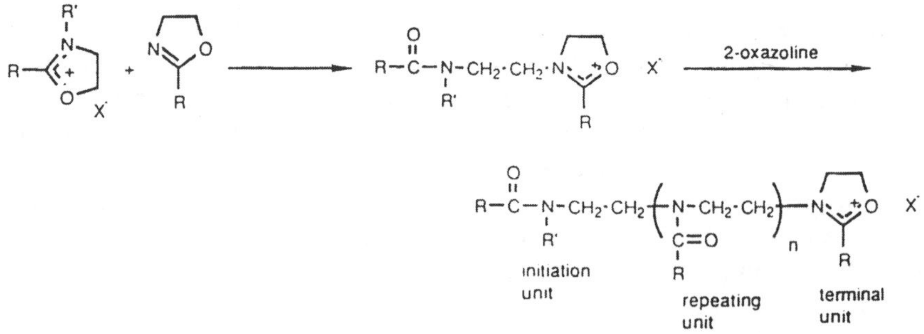

Scheme 8

[25] has since been corroborated [26–28]. By analogy to aziridine polymeriz-ation, the 2-oxazoline ring becomes activated by the cationic reagent to yield an oxazolinium ion. Subsequent reaction of this cation with unactivated oxazoline

gives rise to ring opening and propagation in a linear fashion; the low nucleophilicity of the amide nitrogen precludes branching. The final polymer structure consists of three moieties: the inhibition unit, the repeating unit and the terminal unit. These polymers are generally assumed to possess living oxazolinium end groups, especially if the cationic counterion is iodide ion or a nonnucleophilic ion such as tosylate. One of the first routes to dendrigraft (Comb-burst®) type architecture involved the grafting of oxazolinium terminal groups onto linear-poly(ethyleneimine) (PEI) cores, followed by hydrolysis to give comb-branches (PEIs). Subsequent grafting of oxazolinium terminated oligomers and hydrolyses provided a reiteration sequence by which to building very high molecular weight dendrigrafts (see Chapter 9 for further details).

8   LIVING FREE RADICAL POLYMERIZATIONS

As evidenced in the preceding discussion, the versatility of living anionic or cationic polymerizations for the preparation of well-defined linear polymers is limited by incompatibility of the growing ionic chain ends with many different functional groups. Synthetically demanding experimental requirements such as rigorous exclusion of water/oxygen and the use of ultrapure reagents and solvents further reduce the general applicability of these techniques, while a significant amount of expertise is also demanded. An alternative to these living procedures for the preparation of well-defined linear polymers has therefore been a synthetic goal of long standing.

In many respects, free radical procedures are the opposite of living ionic polymerizations since they are synthetically robust, compatible with a wide range of functional groups, but offer little or no control over macromolecular structure [29]. Despite this drawback, free radical procedures are the main route to vinyl polymers and are of substantial economic importance. One strategy for achieving the above stated goal is to develop a 'living free radical' procedure, which combines the desirable attributes of both traditional free radical systems and living polymerizations. The main difficulty with such a strategy is to mediate the reactivity of the free radical propagating center and this challenge has precluded its development until recently.

Early attempts to realize a 'living' free radical procedure involved the concept of reversible termination of growing polymer chains by iniferters [30, 31]; however, this strategy suffered from high polydispersities and poor control over molecular weight and chain ends. Moad and Rizzardo adopted a subtly different approach and introduced the reversible end-capping of the propagating chain ends with stable nitroxide free radicals, such as 2,2,6,6-tetramethylpiperidinyl-1-oxy (TEMPO) [32]. The use of TEMPO in 'living' free radical polymerization was subsequently refined and significantly extended by Georges *et al.* [33] who demonstrated that polystyrene with low polydispersity could be prepared using bulk polymerization conditions. These two seminal reports resulted in the true development of 'living' free radical polymerization procedures as an exciting research field, which has attracted substantial interest, both industrially and academically.

The field of living free radical polymerization [34] has expanded rapidly in recent years with the development of two major areas of research; nitroxide mediated processes [35], as well as atom transfer radical procedures (ATRP) [36]. The basic theme underlying the success of both approaches is the same, the reversible termination of the growing polymeric radical by a mediating species, either a nitroxide or in the case of ATRP a metal complex, to give a dormant, or inactive species in which the mediating species is covalently bound to the polymer chain end. The key to the success of both strategies is that this termination reaction is reversible and in the case of the nitroxide process, at elevated temperatures, typically 80 °C or greater, the C–ON bond of the alkoxyamine, **4**, is homolytically unstable and undergoes fragmentation to regenerate the stable free nitroxide free radical, **5**, and the polymeric radical, **6**. The polymeric radical can then undergo chain extension with monomer to yield a similar polymeric radical, **7**, in which the degree of polymerization has increased. Recombination of **7** with the nitroxide then gives dormant species **8**, which essentially has the same structure as **4** and the cycle of homolysis–monomer addition–recombination can

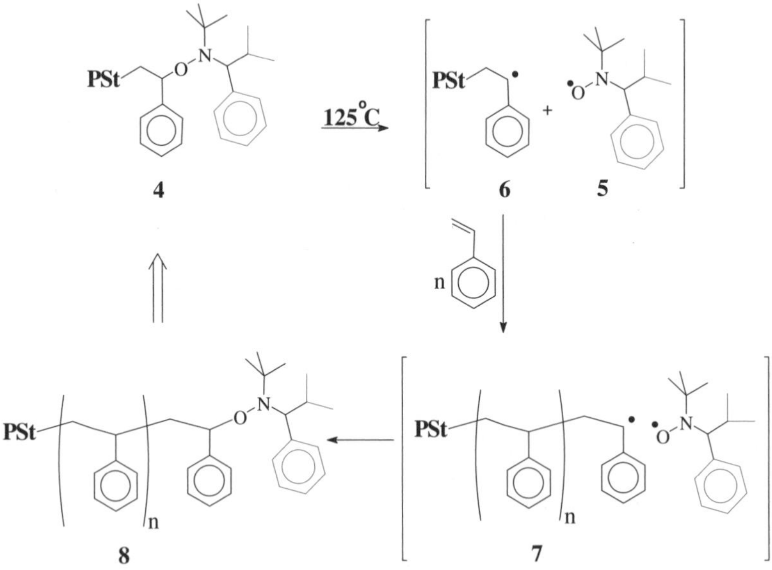

Scheme 9

be repeated (Scheme 9). Unlike the early iniferter work, the mediating nitroxide free radical does not initiate the polymerization of vinyl monomers, therefore no additional propagating centers are created.

A similar process is present in the case of atom transfer radical polymerization (ATRP) systems, except that a metal complex replaces the mediating nitroxide radical, and the dormant group is typically a halide such as Cl-, or Br-. The role of the metal complex is to transfer the halide from the dormant chain end, **9**, to give the propagating radical, **11**, and after monomer insertion to react with the carbon center free radical, **12**, to give a new dormant chain end, **13**, and the original metal complex (Scheme 10). While the presence of the metal complex does add complexity to the ATRP process the ability to tailor the reactivity of the metal center by adjusting the ligand structure can lead to significant versatility in reaction temperature and monomer selectivity.

An extremely favorable consequence of both strategies is the presence of significant amounts of covalent, or inactive, chain ends. This substantially lowers the overall concentration of reactive chain ends which results in a decrease in the occurrence of unwanted side reactions such as termination, disproportionation, or combination. This enables the polymer chain to grow in a controlled fashion, exhibiting many of the attributes typically associated with a living polymerization. However, it should be pointed out that the occurrence of these side reactions is not eliminated and in the strictest sense, the polymerizations are not truly living.

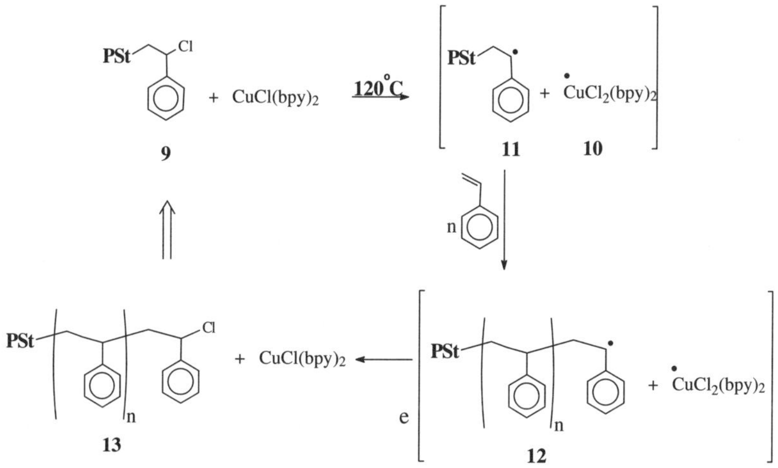

Scheme 10

9 MOLECULAR WEIGHT CONTROL

Initially a bimolecular initiating system consisting of a traditional radical initiator, such as benzoyl peroxide, in combination with TEMPO was introduced by Georges [33]. However, the difficulty in controlling the macromolecular structure, i.e. chain ends, molecular weight, etc. prompted the development of alternatives initiating strategies. Inspired by the concept of well-defined initiating species in anionic polymerizations, a range of unimolecular initiators for living free radical systems were developed [35–37]. The unimolecular initiators are all covalent adducts based on either the propagating chain end or a related structure which will rapidly generate an initiating radical under the polymerization conditions. For example, alkoxyamines readily decompose at elevated temperatures to give an initiating benzylic radical and a nitroxide radical in the desired 1:1 stoichiometry and have found wide use in the nitroxide arena. Similarly, α-haloesters, sulphonyl halides, benzylic halides, etc. are standard unimolecular initiators for the preparation of well-defined polymers by ATRP (Scheme 11).

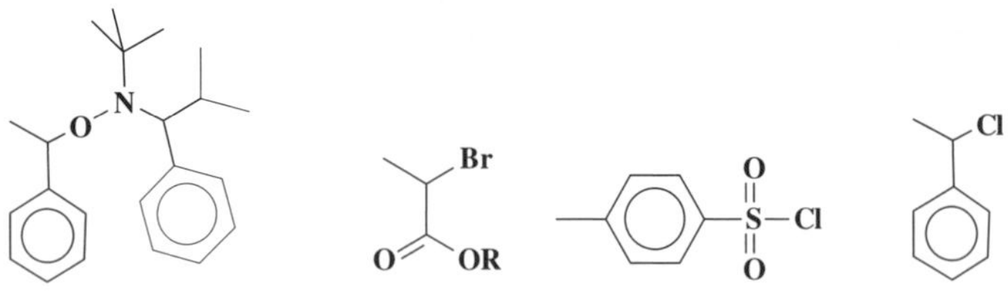

Scheme 11

The advantages of using simple organic molecules such as **14–17** as initiators for living free radical systems are many. Firstly, they permit accurate control over the molecular weight and polydispersity of linear macromolecules in a similar way to living anionic procedures [38]. As before, molecular weight, or degree of polymerization is controlled by the ratio of monomer to initiator and the functionality at the chain ends of the macromolecule is dictated by the initial structure of the initiator (Figure 2.1) [39].

The living nature of these polymerizations has also been rigorously demonstrated by a variety of techniques and shown to hold true even at high conversion. For example, the evolution of molecular weight for the polymerization of styrene by the alkoxyamine, **14**, is linear up to 90% conversion (Figure 2.2).

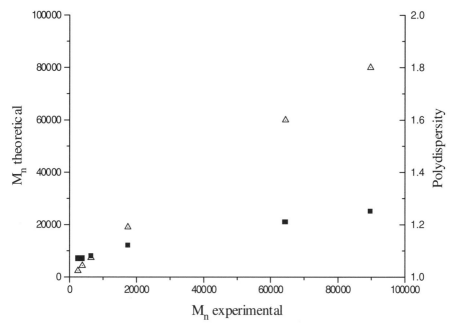

Figure 2.1 Evolution of experimental molecular weight, M_n, and polydispersity with theoretical molecular weight for the polymerization of styrene and **14** at 123°C for 18 h with no degassing or purification

10 FUNCTIONAL GROUP CONTROL

A second, and potentially more useful feature is the stability of these unimolecular initiators to a wide variety of reaction and polymerization conditions which is in sharp contrast to traditional initiators for anionic procedures, such as n-butyl lithium. This allows the initiators to be fully characterized, purified and handled by normal techniques, thus simplifying the polymerization process. It also permits a variety of chemical transformations to be performed on the initiator prior to polymerization, which greatly facilitates the preparation of chain end functionalized macromolecules. For example, the chloromethyl functionalized alkoxyamine, **18**, can be readily converted in high yield to the corresponding aminomethyl derivative, **19**, followed by polymerization to give well-defined linear polymers, **20**, with a single primary amine at the chain end (Scheme 12).

Another prime advantage of living free radical procedures is the compatibility of both nitroxide-mediated and ATRP procedures with functionalized monomers. An excellent example of this is the preparation of poly(2-hydroxyethyl methacrylate) with controlled molecular weight and low polydispersity by the ATRP of HEMA (Scheme 13) [40]. In contrast to normal monomers the

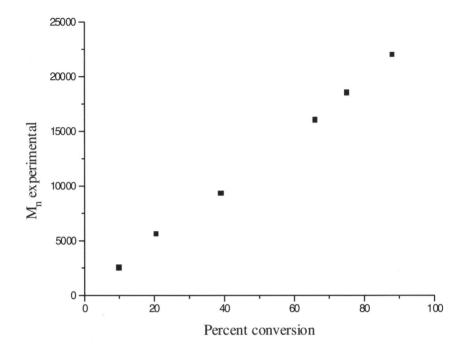

Figure 2.2 Evolution of molecular weight, M_n, with percent conversion for the polymerization of styrene (250 equivs) in the presence of **14** (1.0 equivs) at 123°C for 8 h

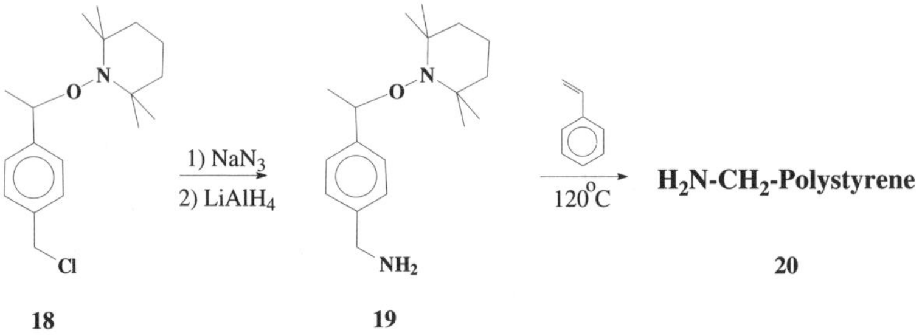

Scheme 12

polymerization of these functional, polar monomers typically require adjustment of the reaction conditions to achieve optimal control over the synthesis. In the above case, the polymerization temperature was lowered to 50°C, a mixed solvent system consisting of 7:3 methyl ethyl ketone/1-propanol was used with a slightly less reactive initiator, ethyl 2-bromo-2-methyl propionate and CuCl and bpy as the metal co-mediator. Such an adjustment in reaction conditions for

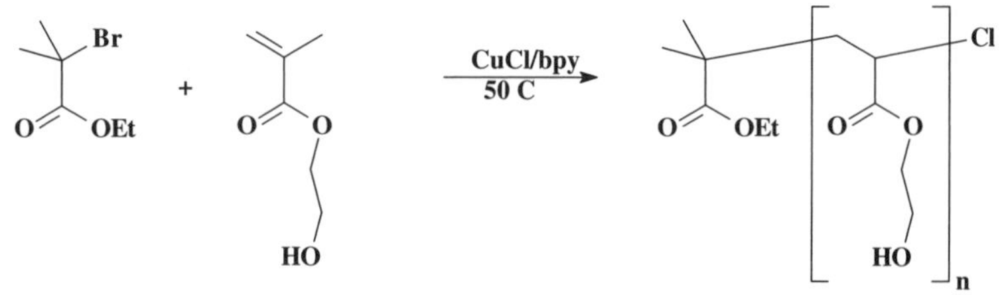

Scheme 13

functionalized monomers has been found for both ATRP and nitroxide mediated processes.

11 BLOCK COPOLYMERS

As with other living techniques, one of their fundamental advantages over other synthetic approaches is the ability to prepare block copolymers, and in this regard living free radical procedures again offer advantages compared to other living techniques. These arise from the ability to polymerize a wide variety of monomers while at the same time tolerating functional groups and polar/aqueous reaction conditions. This permits an extremely wide variety of block copolymers to be easily prepared under commercially viable reaction conditions such as emulsion, or bulk procedures. An added advantage of both ATRP and nitroxide procedures is that the second block need not be grown directly after the first block as in other living systems. Owing to the presence of a stable, dormant chain end, the first linear block can be isolated, stored and characterized before growth of the second block, which greatly facilitates the formation of well-defined block copolymers.

In contemplating the synthesis of block copolymers by living free radical methods a number of issues have to be addressed. Firstly, the homopolymerization of the monomers must be a living process under the specific polymerization method. For example, poly(acrylamide) block copolymers are best prepared by nitroxide mediated processes since the homopolymerization of acrylamides by ATRP is troublesome. In contrast, the preparation of poly(methyacrylate) block copolymers is most facile by ATRP procedures since the homopolymerization of methacrylates by nitroixde mediated processes has not been shown to be a living process. An excellent example of the latter is the preparation of ABA, PMMA-*b*-PBA-*b*-PMMA triblock copolymers from the difunctional initiator, **21** (Scheme 14). Once again subtle features such as reaction conditions, nature of the end groups, i.e. Br- or Cl-, and the nature of the metal center are critical for the successful block copolymer formation [41].

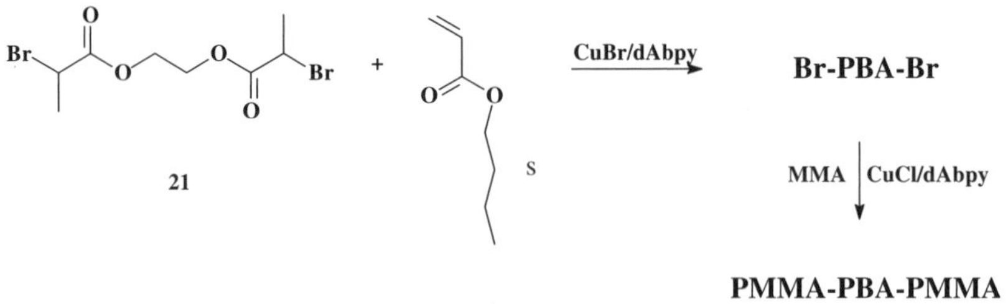

Scheme 14

12 RANDOM COPOLYMERS

While there have been a number of studies on the preparation of linear random copolymers by living anionic or cationic procedures the structural versatility available is limited due to the incompatibility of various monomer pairs with the polymerization conditions or by the grossly different reactivity ratios under living ionic conditions [42]. The carbanionic copolymerization of polar monomers, such as methacrylates with diene or styrene type monomers is difficult and not very successful, though copolymerization of various methacrylates can be achieved [43]. In contrast, it is well known that the preparation of random copolymers from an extremely wide variety of vinyl monomers is a facile process using traditional free radical procedures. The unique ability to prepare well-defined linear random copolymers by living free radical procedures has therefore attracted much interested and proven to be highly successful using either nitroxide mediated or ATRP conditions [44]. Interestingly, it has been shown by a number of authors that the reactivity ratios and triad/tetrad ratios are essentially the same for the living process as has been observed for the traditional systems [45].

A prime example of these features can be found in the synthesis of styrene/(meth)acrylate random copolymers. By controlling the initiator/total monomer ratio, the molecular weight can be accurately controlled for both styrene/methyl methacrylate and styrene/butyl acrylate random copolymers. As can be seen in Figure 2.3 the polydispersity for both systems is essentially 1.10–1.25 over comonomer ratios ranging from 1/9 to 9/1.

13 RING OPENING METATHESIS POLYMERIZATION

Since the discovery by Ziegler and Natta that transition metal complexes, in the presence of aluminum alkyl compounds, can efficiently catalyze the polymerization of ethylene and propylene, significant efforts have been devoted to the development of new catalytic systems for polymerization of olefins. One of the

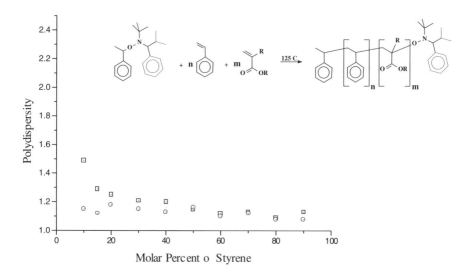

Molar Percent o Styrene

Figure 2.3 Relationship between polydispersity of the resulting random copolymers and mole percent of styrene in the feed mixture for the copolymerization of (i) styrene and n-butyl acrylate (○), and (ii) styrene and methyl methacrylate (□) mediated by **14**

ultimate aims is the discovery of catalysts which allow control over the structure, molecular weight and physical/chemical properties of the polymeric products. This desire to develop living systems has resulted in the evolution of olefin metathesis chemistry, a process wherein the carbon–carbon double bond of an olefin is broken and reformed in the presence of an organometallic catalyst. When the starting material for this process is a cyclic olefin, a ring opening reaction takes place and if the cyclic olefin is sufficiently strained, the ring opening releases energy. This release of energy can be the driving force for the polymerization of the cyclic olefin and has led to the development of living ring opening metathesis polymerizations (ROMP) (Scheme 15).

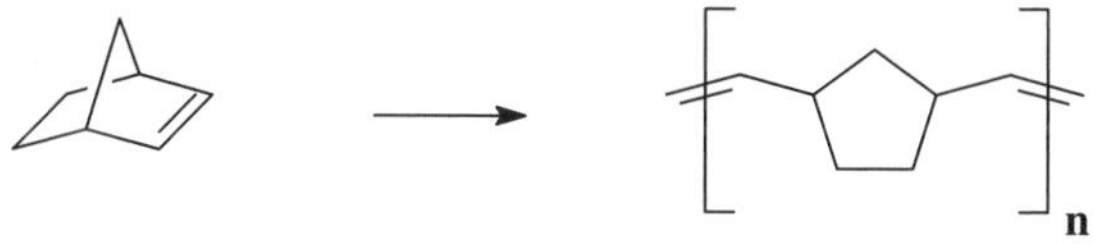

Scheme 15

While the majority of initial work in living ROMP procedures was performed with tungsten and molybdenum catalysts, the high reactivity of these systems and their low tolerance for functional groups hampered their wide adoption. In 1992, Grubbs *et al.* [46] reported the synthesis of the ruthenium–vinylalkylidene complex **22**. This new generation of metathesis catalysts, in addition to the high activity in ROMP and ring closing metathesis, presents a high stability towards air, water and acids. Catalyst **23** can be easily prepared by a simple phosphine exchange from **22** and shows an even higher activity for ROMP, being able to polymerize relatively low strain cyclic olefins while still retaining its stability in protic media [47]. Arguably the most versatile and exciting catalyst systems for ROMP that are presently available are those based on ruthenium.

14   SINGLE SITE CATALYSIS

While ROMP procedures are applicable to only a select group of monomers, the development of well-defined single '*site catalysts*' for the polymerization of a variety of different monomers have become an area of significant research

22   **23**

interest with potentially greater future commercial impact. The ability to tailor the electronic and steric nature of a single metal center allows the preparation of catalysts with tolerance for functional groups while providing control over the polymeric structure and stereochemistry. Two excellent examples of the potential for this field can be found in the recent work of Brookhart [48] and Coates [49].

In a series of publications by Johnson and Brookhart [48], cationic α-diimine complexes of nickel and palladium, **24**, were prepared and investigated as olefin polymerization catalysts (Scheme 16). Significantly, these catalysts were not only able to polymerize ethylene and α-olefins to high molecular weight but dramatic differences in the microstructure and properties of the resulting polymers were observed when compared to either Ziegler-Natta or metallocene technology. One of the unique features of this polymerization was the observation that changes in the polymerization conditions or catalyst structure resulted in macromolecular architectures varying from linear to highly branched (dendritic). The potential for using the same reactor system and catalyst system for the production of poly(olefins) with properties ranging from elastomeric to rigid thermoplastic is particularly intriguing and of considerable commercial interest.

The general formula L_nMR of a single site catalyst, where L is a ligand set, M the active metal center and R a group that may initiate polymerization can also

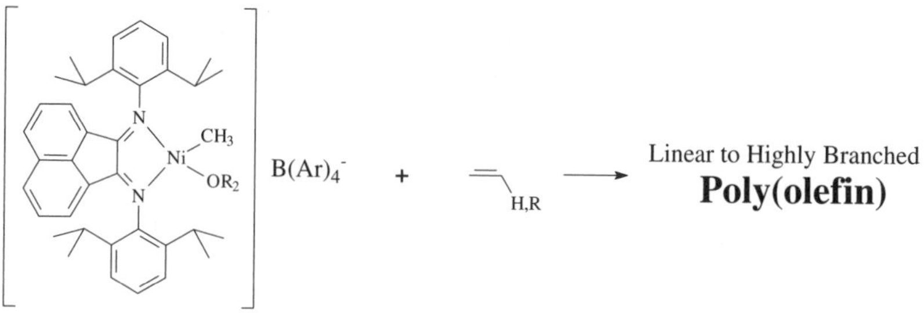

24

Scheme 16

be recognized in the work of Coates [49] for the synthesis of heterotactic poly(lactic acid) from *rac*-lactide. The zinc alkoxide complex, **25**, was shown to act as a single site catalyst for the living polymerization of *rac*-lactide with the initiating chain end being derived from the isopropoxide group (Scheme 17). The ability to obtain racemic enchainments (alternating lactide units of opposite stereochemistry) of 90–94% is a surprising result and demonstrates that by careful control of the steric environment of the metal center, high stereoselectivity of monomer addition can be obtained.

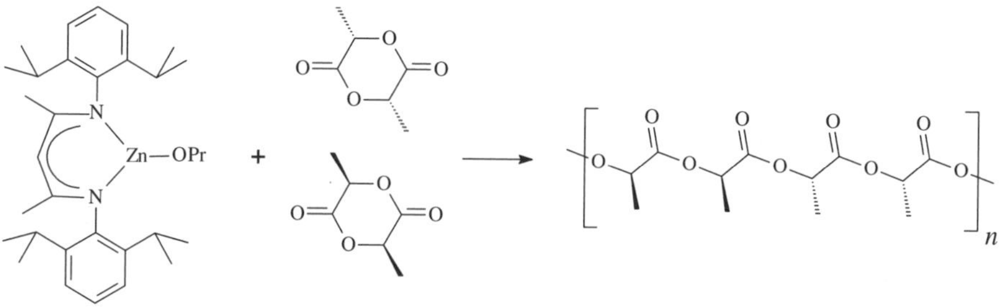

25

Scheme 17

15   METALLOCENE CATALYSTS

The advent of homogeneous olefin polymerization based on metallocenes has also ushered in a new era in olefin polymerization. In fact, the recent commercialization of numerous poly(olefins) prepared using metallocene strategies is testament to the importance of this field from both an academic and an industrial viewpoint. One of the main intellectual driving forces is the realization that as a specific example of single site catalysts, these systems are also amenable to structural and electronic manipulation. Therefore, subtle changes in the catalyst structure can lead to dramatic changes in the architecture of the poly(olefin) produced, which in turn can lead to new and/or improved physical properties. A prime example of this philosophy is the synthesis of thermoplastic elastomeric poly(propylene) by the oscillating stereocontrol induced by a metallocene catalyst. By carefully designing their catalyst, Coates and Waymouth [50] were able to prepare poly(propylenes) containing both atatic and isotactic blocks. Unlike other zirconium metallocenes, **26**, is not bridged and so can oscillate between two forms, one which produces an isotactic block from propylene and one which produces an atatic block (Scheme 18). The steric control inherent in **26** is such that the conversion between these two active forms is slow enough such that long runs of each block are formed which then leads to blocks of sufficient length to give thermoplastic and elastomeric properties.

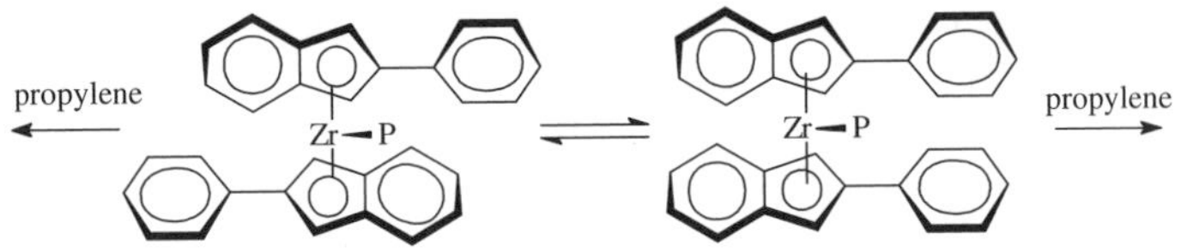

Scheme 18

16 CONCLUSION

The underlying theme throughout this discussion on the synthesis of linear polymers is control. This includes control over macromolecular architecture, molecular weight, stereochemistry, etc. It is anticipated that this quest to prepare well-defined linear macromolecules with control over various facets of the polymeric structure will continue and be refined even further as our understanding of the mechanism of the polymerizations develops and the role that the initiating or catalytic systems play in controlling structures. It is also apparent that these developments will be greatly assisted by multidisciplinary studies at the interface of polymer chemistry with organic chemistry, inorganic chemistry, physical chemistry, etc.

17 REFERENCES

1. Szwarc, M. *Nature*, **178**, 1168 (1956) Szwarc, M., Levy, M. and Milkovich, R. *J. Am. Chem. Soc.*, **78**, 2656 (1956).
2. Fontanille, M. *Comprehensive Polymer Science*, Vol. 3, G. Allen (ed.), Pergamon, 1989, p. 365.

3. Muller, A. H. E. *Comprehensive Polymer Science*, Vol. 3, G. Allen (ed.), Pergamon, 1989, p. 387.
4. Overberger, C. G. and Yamamoto, N. *J. Polym. Sci., Part A*-1, **4**, 3101 (1966).
5. Szwarc, M. *Adv. Polym. Sci.*, **49**, 1 (1983).
6. Fontanille, M., Helary, G. and Szwarc, M. *Macromolecules*, **21**, 1532 (1988).
7. Nakayama, H., Yamasawa, Y., Higashimura, T. and Okamura, S. *Kobunshi Kagaku*, **24**, 296 (1967).
8. Bywater, S. and Worsfold, D. J. *Can. J. Chem.*, **45**, 1821 (1967).
9. Broske, A. D., Huang, T. L., Hoover, J. M., Allen, R. D. and McGrath, J. E. *Polym. Prep.*, **25**, 85 (1984).
10. Quirk, R. P. and Chen, W. C. *Makromol. Chem.*, **183**, 2071 (1982).
11. Morton, M., Fetters, L. J., Inomata, J., Rubio, D. C. and Young, R. N. *Rubber Chem. Technol.*, **49**, 303 (1976).
12. Asami, R., Takaki, M., and Hanahata, H. *Macromolecules*, **16**, 628 (1983).
13. Milkovich, R. *Polym. Prep.*, **21**, 40 (1980).
14. Muller, A. H. E. *Recent Advances in Anionic Polymerization*, Hogen-Esch, T. E. and Smid, J. (eds.), Elsevier, New York, p. 205.
15. Berger, W. and Alder, H. J. *Makromol. Chem. Macromol. Symp.*, **3**, 301 (1986).
16. Wang, J. S., Jérôme, R., Warin, R. and Teyssié, Ph. *Macromolecules*, **27**, 1691–1696 (1994).
17. Zundel, T., Baran, J., Mazurek, M., Wang, J. S., Jerome, R. and Teyssie, P. *Macromolecules*, **31**, 2724 (1998).
18. Tsuruta, T. and Kawakami, Y. *Comprehensive Polymer Science*, Vol. 3, G. Allen (ed.), Pergamon, 1989, p. 457.
19. Kennedy, J. P. *Cationic Polymerization, A Critical Inventory*, Wiley-Interscience, New York, (1975).
20. Miyamoto, M., Sawamoto, H. and Higashimura, T. *Macromolecules*, **17**, 265 (1984).
21. Gandini, A. and Cheradame, H. *Adv. Polym. Sci.*, **34**, 35 (1980).
22. Nuyken, O. and Pask, S. D. *Comprehensive Polymer Science*, Vol. 3, G. Allen (ed.), Pergamon,, 1989, p. 619.
23. Kennedy, J. P., Guhaniyogi, S. and Percec, V. *Polym. Bull.*, **20**, 2089 (1982).
24. Smith, S. and Hubin, A. J. *Macromol. Sci., Chem.*, **A7**, 1399 (1973).
25. Tomalia, D. A. and Sheetz, D. P., *J. Polym. Sci. Part A-1*, **4**, 2253 (1966).
26. Kagiya, T., Matusda, T. and Hirata, R., *J. Macromol. Sci. Chem.*, **6**(3), 491 (1972).
27. Arora, K. S. and Overberger, C. G., *J. Polym. Sci. Polym. Lett. Ed.*, **20**(8), 403 (1982).
28. Saegusa, T., Kobayashi, S. and Yamada, A., *Macromolecules*, **177**, 2271 (1976).
29. Moad, G. and Solomon, D. H. *The Chemistry of Free Radical Polymerization*, Pergamon Press, NY, (1995).
30. Quirk, R. P., Kinning, D. J. and Fetters, L. J. *Comprehensive Polymer Science*, Aggarwal, S. L. (ed.), Vol. 7, Pergamon Press, London, 1989, p. 1.
31. Fradet, A. *Comprehensive Polymer Science*, 2nd Supp., Pergamon Press, London, Aggarwal, S. L. and Russo, S. (eds.), 1996, p. 133.
32. Moad, G., Rizzardo, E. and Solomon, D. H. *Macromolecules*, **15**, 909 (1982); Solomon, D. H., Rizzardo, E. and Cacioli, P. US Patent 4,581,429, 27 March (1985).
33. Georges, M. K., Veregin, R. P. N., Kazmaier, P. M. and Hamer, G. K. *Macromolecules*, **26**, 2987 (1993).
34. Matyjaszewski, K. *Controlled Radical Polymerization*, ACS Symp. Ser. vol. 685 (1998), Hawker, C. J. *Acc. Chem. Res.*, **30**, 373 (1997); Colombani, D. *Prog. Polym. Sci.*, **22**, 1649 (1997).
35. Hawker, C. J., Barclay, G. G., Orellana, A., Dao, J. and Devonport, W. *Macromolecules* **29**, 5245 (1996); Hammouch, S. O. and Catala, J. M. *Macromol. Rapid*

Commun., **17**, 149 (1996); Li, I. Q., Howell, B. A., Koster, R. A. and Priddy, D. B. *Macromolecules*, **29**, 8554 (1996); Fukuda, T., Terauchi, T., Goto, A., Ohno, K., Tsujii, Y. and Yamada, B. *Macromolecules* **29**, 6393 (1996); Moad, G. and Rizzardo, E. *Macromolecules* **28**, 8722 (1995); Odell, P. G., Veregin, R. P. N., Michalak, L. M.and Georges, M. K. *Macromolecules*, **30**, 2232 (1997); Hawker, C. J. *J. Am. Chem. Soc.*, **116**, 11314 (1994); Li, D. and Brittain, W. J. *Macromolecules*, **31**, 3852 (1998); Puts, R. D. and Sogah, D. Y. *Macromolecules*, **29**, 3323 (1996).

36. Wang, J. S. and Matyjaszewski, K. *Macromolecules*, **28**, 7901 (1995); Wang, J. S. and Matyjaszewski, K. *J. Am. Chem. Soc.*, **117**, 5614 (1995); Patten, T. E., Xia, J., Abernathy, T., Matyjaszewski, K. *Science* **272**, 866 (1996); Kato, M., Kamigaito, M., Sawamoto, M. and Higashimura, T. *Macromolecules*, **28**, 1721 (1995); Percec, V. and Barboiu, B. *Macromolecules*, **28**, 7970 (1995); Granel, C., DuBois, P., Jerome, R. and Teyssie, P. *Macromolecules*, **29**, 8576 (1996); Wayland, B. B., Basickes, L., Mukerjee, S., Wei, M. and Fryd, M. *Macromolecules*, **30**, 8109 (1997); Ando, T., Kamigaito, M. and Sawamoto, M. *Macromolecules*, **31**, 6708 (1998); Collins, J. E. and Fraser, C. L. *Macromolecules*, **31**, 6715 (1998); Percec, V., Barboiu, B. and van der Sluis, M. *Macromolecules*, **31**, 4053 (1998); Moineau, G., Granel, C., Dubois, Ph., Jerome, R. and Teyssie, Ph. *Macromolecules*, **31**, 542 (1998); Xia, J., Gaynor, S. G. and Matyjaszewski, K. *Macromolecules*, **31**, 5958 (1998); Takahashi, H., Ando, T., Kamigaito, M. and Sawamoto, M. *Macromolecules*, **32**, 3820 (1999).

37. Benoit, D., Chaplinski, V., Braslau, R. and Hawker, C. J. *J. Am. Chem. Soc.*, **121**, 3904 (1999); Gaynor, S. G., Qiu, J. and Matyjaszewski, K. *Macromolecules*, **31**, 5951 (1998); Shipp, D. A., Wang, J. L. and Matyjaszewski, K. *Macromolecules*, **31**, 8005 (1998); Howell, B. A., Dineen, M. T., Kastl, P. E., Lyons, J. W., Meunier, D. M., Smith, P. B. and Priddy, D. B. *Macromolecules*, **30**, 5194 (1997).

38. Hawker, C. J. *J. Am. Chem. Soc.*, **116**, 11314 (1994).

39. Hawker, C. J. and Hedrick, J. L. *Macromolecules*, **28**, 2993 (1995).

40. Beers, K. L., Boo, S., Gaynor, S. G. and Matyjaszewski, K. *Macromolecules*, **32**, 5772 (1999).

41. Shipp, D. A., Wang, J. L. and Matyjaszewski, K. *Macromolecules*, **31**, 8005 (1998).

42. Chen, J. and Fetters, L. J. *Polym. Bull.*, **4**, 275 (1981).

43. Yuki, H., Okamoto, Y., Ohta, K. and Hatada, K. *J. Polym. Sci., Polym. Chem. Ed.*, **13**, 1161 (1975).

44. Kotani, Y., Kamigaito, M. and Sawamoto, M. *Macromolecules*, **31**, 5582 (1998); Haddleton, D. M., Crossman, M. C., Hunt, K. H., Topping, C., Waterson, C. and Suddaby, K. G. *Macromolecules*, **30**, 3992 (1997); Greszta, D. and Matyjaszewski, K. *Polym. Prepr.* (*Am. Chem. Soc., Div. Polym. Chem.*), **37**(1), 569 (1996); Listigovers, N. A., Georges, M. K., Odell, P. G. and Keoshkerian, B. *Macromolecules*, **29**, 8992 (1996).

45. Hawker, C. J., Elce, E., Dao, J., Volksen, W. Russell, T. P. and Barclay, G. G. *Macromolecules*, **29**, 4167 (1996).

46. Nguyen, S. T., Johnson, L. K. and Grubbs, R. H., *J. Am. Chem. Soc.*, **114**, 3974–75 (1992).

47. Nguyen, S. T., Grubbs, R. H. and Ziller, J. W. *J. Am. Chem. Soc.*, **115**, 9858–9 (1993).

48. Gates, D. P., Svejda, S. A., Onate, E., Killian, C. M., Johnson, L. K., White, P. S. and Brookhart, M. *Macromolecules*, **33**, 2320 (2000) and references therein.

49. Cheng, M., Attygalle, A. B., Lobkovsky, E. B. and Coates, G. W. *J. Am. Chem. Soc.*, **121**, 11583 (1999).

50. Coates, G. W. and Waymouth R. M. *Science*, **267**, 217 (1995).

3

Progress in the Branched Architectural State

J. ROOVERS

Institute for Chemical Process and Environmental Technology, National Research Council, Ottawa, Ontario, CANADA

1 INTRODUCTION

A perfectly linear polymer molecule is an idealized, rarely encountered species. Most polymeric chains have chemical and/or structural defects. When these defects are limited to a few atoms on the polymer backbone they can best be considered as copolymerization units. However, when the length of the dangling chains is of the same order of magnitude as the other chains in the polymer, we speak of long-chain branching (LCB) [1–4] see Figure 3.1. Although a relatively small number of defect (branch) points may be involved in LCB, it always has a considerable effect on the global properties of the polymer.

Different types of LCB are distinguished. Star polymers are the simplest branched polymers because they have only one branch point. Regular star polymers have a branch point with a constant number (functionality, f) of arms and every arm has the same molecular weight. They are therefore monodisperse polymers. Star polymers may also have arms with a most probable distribution [5]. Star polymers can also be polydisperse due to a variable functionality. Palm tree [6] or umbrella polymers [7] that contain a single arm with different molecular weight (MW) than the other arms are classified under the asymmetric star [8] polymers, see Figure 3.2.

Comb polymers consist of a backbone and several branches [9]. Backbone and branches are usually recognized from the synthetic procedure. The branching functionality is usually limited to 3 or 4. The backbone and/or the branches can be monodisperse. In a regular comb polymer the number of branch points per backbone is constant and the branches are equidistantly spaced along the

Dendrimers and Other Dendritic Polymers. Edited by Jean M. J. Fréchet and Donald A. Tomalia
© 2001 John Wiley & Sons Ltd

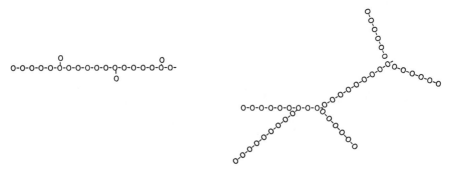

Figure 3.1 Short and long chain branching

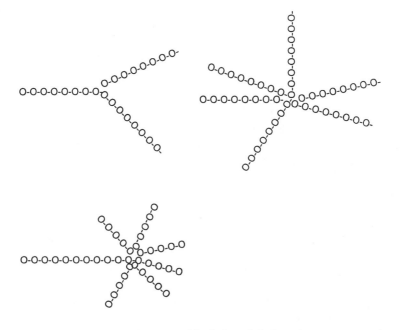

Figure 3.2 Regular star polymers with $f=3$ and $f=6$ and an asymmetric palm tree or umbrella polymer.

backbone [9]. Other comb polymers have a molecular weight polydispersity due to the polydispersity of the backbone or of the branches or due to the variable (statistical) number and/or placement of the branches. Some comb polymer architectures are shown in Figure 3.3. A comb polymer with exactly two branch points [10] is called a H-polymer [11] when there is one branch at each branch point or it is called a super-H [12], pom-pom [13] or dumb-bell [6] polymer

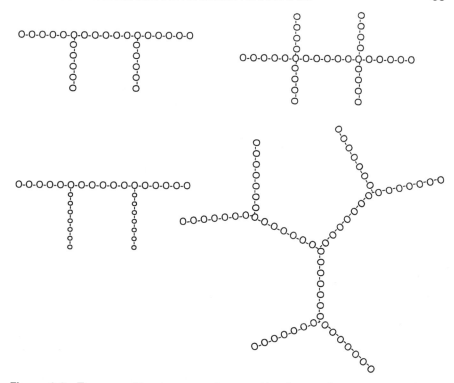

Figure 3.3 Top row: H-polymer and super H-polymer; bottom row: π-block copolymer and dendritic polymer

when there is more than one branch at each branch point. When the mass contained within the branches becomes a large fraction of the total mass of the comb polymer, the comb polymer resembles a star polymer [9].

Since the successful synthesis of dendritic macromolecules from low MW monomers [14a], research has also been performed in the area of dendritic polymers [14b, 15]. Dendritic polymers are obtained by attaching or growing several end-standing arms onto a central regular star polymer. This procedure is repeated in a generational manner, see Figure 3.3.

Finally, some polymerization processes lead to a class of randomly branched polymers. The condensation of multifunctional monomers or polymers leads to randomly branched polymers. A feature of such step-growth processes is that the largest molecules in the batch have the largest number of functional groups and therefore the highest probability of reaction and growth. As a result, the molecular weight distribution grows very rapidly with conversion and $(M_w/M_n) \propto M_w$ [10]; M_w diverges at the critical extent of reaction (α_{cr}) of the functional groups given by [1]:

$$\alpha_{cr} = 1/(f-1)(g-1) \tag{1}$$

for the case of the step-growth polymerization of an equimolar mixture of two monomers A_f and B_g or of an A_fB_g monomer or polymer. It is assumed that a functional group A reacts only with a B group without side reactions. The critical extent of reaction is macroscopically characterized by chemical gelation. Equation (1) is based on equireactivity of all functional groups throughout the polymerization process [16]. Usually, the experimental value of α_{cr} is larger than the theoretical one due to intramolecular ring formation that consumes functional groups and to other factors thus introducing nonrandomness in the process [17, 18]. The special case of A_fB ($g = 1$) cannot lead to gelation. In this case, hyperbranched polymers are formed. The architecture of hyperbranched polymers is described in greater detail in other chapters of this volume.

2 PHYSICAL PROPERTIES AFFECTED BY LONG-CHAIN BRANCHING

The fundamental effect of LCB on the polymer is a reduction of the size of the polymer relative to the size of a linear polymer with the same MW as measured, for example by its mean-square radius of gyration:

$$< R^2 > \ = (1/2N^2) \, \Sigma_i \, \Sigma_j < x_{ij}^2 > \tag{2}$$

where x_{ij} is the distance between any two segments i and j measured along the chain of N segments. The shrinkage of the radius of gyration due to LCB is quantified by the ratio

$$g = \ < R^2 >_{br} / \ < R^2 >_{lin}; \quad 0 < g < 1 \tag{3}$$

where the subscripts br and lin stand for the branched and linear polymer with the same molecular weight. Zimm and Stockmayer, in a seminal paper [10], taught us how to calculate the branching factor g for a variety of different LCB architectures. These calculations are for the Gaussian conformation or unperturbed random walk chain. For example, in the case of regular star polymers, g is given by:

$$g = (3f - 2)/f^2 \tag{4}$$

Another often used relation

$$g = [(1 + n_j/7)^{\frac{1}{2}} + 4n_j/9\pi]^{-\frac{1}{2}} \tag{5}$$

is used for monodisperse fractions of randomly branched polymers with n_j trifunctional branch points. This value of g is an average over all possible architectures. Similar calculations have been performed for other architectures (combs, dendritic polymers, etc.) [9, 19–21]. Knowledge of g alone does not yield an unambiguous value of the number of branches without independent knowledge about the branch architecture. Experimental studies on model regular star polymers have shown that equation (4) is only valid for low values of f [22]. The experimental shrinkage factor agrees with equation (4) when $f < 10$ and the arm MW is not very short. Large values of f [22] cause increased arm stretching especially near the central core where the segment density is high [23, 24] and cause an overall expansion of the polymer coil. The blob theory for branched polymers developed by Daoud and Cotton [25] gives a particularly good scaling account of this size expansion of branched polymers in good solvents. Finally, LCB polymers with their reduced coil dimensions and linear polymers have different segment density distributions in the polymer coil. Segment density distributions have been studied most extensively in the case of regular star polymers with a wide range of functionalities by means of small angle neutron scattering [26–28]. The validity of the assumptions in the Daoud–Cotton model have been nicely confirmed by these results. The tendency of branched polymers to become more spherical with increased branching makes them resemble monomolecular colloids and give them a strong tendency to order in semi-dilute solutions. The experimental work in this field has been reviewed recently [29].

The dilute solution hydrodynamic properties of polymers depend also on the type and extent of LCB. Parallel to equation (3) a viscosimetric branching factor can be defined by the ratio of intrinsic viscosities:

$$g_{[\eta]} = [\eta]_{\mathrm{br}} / [\eta]_{\mathrm{lin}} \qquad (6)$$

but $g_{[\eta]}$ decreases in a complex manner with LCB and no analytical theory is available for relating $g_{[\eta]}$ to g that is universally valid despite early attempts to find such a relationship [30, 31]. Computer simulations of the intrinsic viscosity of regular stars polymers have shown how different static and hydrodynamic factors enter into this difficult problem [32]. Nevertheless, the intrinsic viscosity of branched polymers can be measured easily and is useful to estimate the size of the polymer by means of the Einstein equivalent sphere model according to

$$R_{\mathrm{V}} = \{(10\pi/3N_{\mathrm{A}})\,[\eta]\,M\}^{1/3} \qquad (7)$$

with N_{A} as Avogadro's number. This model becomes progressively better as branching increases and the polymer coil becomes more spherical. For regular star polymers, when f is small $R_{\mathrm{V}} < R$. However, when $f \approx 18$ $R_{\mathrm{V}} = R$ [22, 33] and for stars and combs with many branches $R_{\mathrm{V}} \approx 1.29R$ [22], the theoretical relationship for the equal density hard sphere. This limit is expected to hold roughly for a variety of highly branched polymers, including spherical dendritic molecules.

Measurement of the translational diffusion coefficient, D_0, provides another measure of the hydrodynamic radius. According to the Stokes–Einstein relation

$$R_{\mathrm{h}} = kT/6\pi\eta_{\mathrm{s}}\,D_0 \qquad (8)$$

where k is the Boltzmann constant, T absolute temperature and η_s the solvent viscosity. In general it is found that R_V and R_h are nearly identical, the agreement improving as the LCB density increases [22, 34].

Other dilute solution properties depend also on LCB. For example, the second virial coefficient (A_2) is reduced due to LCB. However, near the Flory θ temperature, where $A_2 = 0$ for linear polymers, branched polymers are observed to have apparent positive values of A_2 [35]. This is now understood to be due to a more important contribution of the third virial coefficient near the θ point in branched than in linear polymers. As a consequence, the experimental θ temperature, defined as the temperature where $A_2 = 0$ is lower in branched than in linear polymers [36, 37]. Branched polymers have also been found to have a wider miscibility range than linear polymers [38]. As a consequence, high MW highly branched polymers will tend to coprecipitate with lower MW more lightly branched or linear polymers in solvent/non-solvent fractionation experiments. This makes fractionation according to the extent of branching less effective.

The MW dependence of the radius of gyration of linear polymers is given by

$$< R^2 >^{\frac{1}{2}} = a\, M^v \qquad (9)$$

where v is $\frac{1}{2}$ in a θ solvent or in the melt and 0.588 (0.6) in good solvents. Similar scaling laws also describe other static and the hydrodynamic properties. In the case of polymers with LCB the same exponents as for linear polymers are found only when the set of branched polymers is uniform. A set of branched polymers is uniform when all the polymers in the set have the same architecture and the MWs of all the subchains have a constant proportionality. For example, regular star polymers with the same functionality and increasing arm lengths form a uniform set. Also, H-polymers in which the five discernible subchains have a constant MW ratio form a uniform set and their dilute solution properties will scale with the same exponent as found for the linear polymer [11]. On the contrary, a set of comb polymers with fixed backbone and constant number of branches but with increasing branch length is not a uniform set, nor is a set of comb polymers with a fixed backbone but increasing number of branches. The physical properties of such a set of polymers will scale differently from the linear polymer. This is illustrated in Figure 3.4.

Randomly branched polymers have very wide MW distributions, especially, when prepared near the gelation point. The properties of the whole, unfractionated, randomly branched polymer are strongly affected by this wide MW distribution [18]. For example, the experimental radius of gyration, which is a z-average value, is dominated by the largest molecules in the sample, and as a result $< R^2 >_z/M_w \geq\, < R^2 >/M$ of the monodisperse linear polymer and the experimental branching factor appears to be slightly larger than unity [17, 18]. In randomly branched polymers the scaling behavior of equation (3.9) appears to hold due to a compensation of shrinkage and polydispersity effects [18]. The

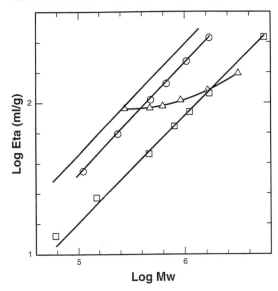

Figure 3.4 Intrinsic viscosity-molecular weight relation in toluene at 35 °C for ○: regular H-polystyrene; □: 12-arm star polystyrene; △: comb polystyrene with $f = 30$, $M_{bb} = 2.75 \times 10^5$ and increasing branch MW. Top line; linear polystyrene

same is true for the hydrodynamic radius from dynamic light scattering [18]. However, the intrinsic viscosity, when compared at constant weight average MW, is lower the wider the MW distribution of a randomly branched polymer [17].

Poly(macromonomers) with moderately long side chains attached to every few (second) atom along the backbone are very densely branched polymers. When the degree of polymerization of the backbone is low then the poly(macromonomers) tend to resemble star polymers [39, 40]. When the degree of polymerization is very high the poly(macromonomer) acquires a cylindrical conformation (bottlebrush), due to the stretching and linearization of the backbone [40].

Branched polymers are often analyzed by means of size exclusion chromatography (SEC). It should always be kept in mind that fractionation in SEC is on the basis of the hydrodynamic volume, i.e. $[\eta]_i M_i$ [41]. Analysis based on a mass detector and a calibration with linear polymers does not provide a true MW for a branched polymer. It is easily shown that, for the same reason, SEC does not provide a true MW distribution for a mixture of branched polymers. This has often led to very optimistic interpretations of chromatograms. The addition of a MW sensitive detector remedies this difficulty. However, the logarithmic nature of the hydrodynamic fractions requires an important correction to the chromatograms of very broad (randomly) branched samples [42].

It has often been proposed that the melt properties could be a convenient method to characterize branched polymers. However, the complex dynamic interactions observed and now understood in the simplest mixtures of model branched polymers [43] make this a goal unlikely to be achieved for more complex mixtures.

3 SYNTHETIC STRATEGIES FOR LONG-CHAIN BRANCHED POLYMERS

LCB polymers can be formed by chemically linking preformed polymers (*arm first* or *polymer first method*) or by growing polymer chains from a multifunctional initiatior (*core first method*). In both cases living polymerization techniques are preferred because they provide better control over MW, MW distribution and the final branching architecture. However, highly selective coupling reactions e.g. with multifunctional isocyanates, or dicyclohexylcarbodiimide (DCC) coupling, have also been successful.

In the *arm first method*, when living polymer molecules are linked to a multifunctional coupling agent, a star polymer is formed. Coupling agents with a compact dendritic architecture have been used to increase the functionality of the star polymer. The linking agent can be formed *in situ* by addition of a difunctional monomer. In the first step this difunctional monomer adds to the living polymer. In the second step, repeated intermolecular reactions lead to star-like polymers. The classic example is the coupling of living anionic polymer with divinylbenzene:

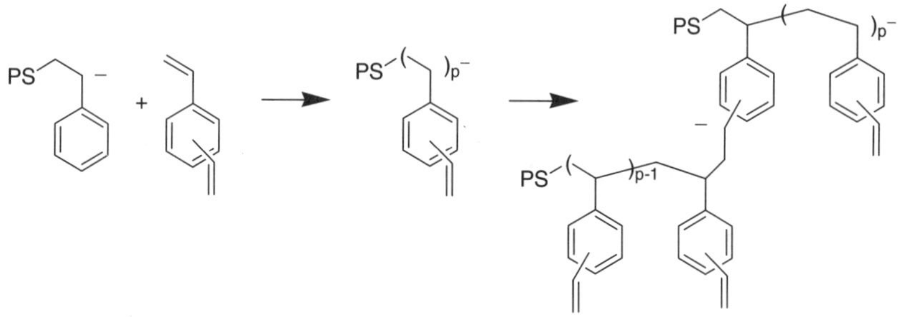

The reaction scheme is very general, but control over the extent of the intermolecular reactions and the distribution of the number of arms in the star is limited. The arm first method includes the polymerization (to form star polymers) or copolymerization (to form comb or graft copolymers) of macromonomers. The technique provides a handy simplification if the arm MW need not be very high and the MW control of the branched polymers is not very important.

If the multifunctional coupling agent is a linear polymer, a comb polymer is formed. Dendrigraft arborescent polymers are obtained when the multifunctional polymer is a star or a comb polymer. While the placement of the functional groups is often random along the chain, reaction schemes for end-standing or centrally placed functional groups have been devised so that better architectural control is available.

The *core first method* starts from multifunctional initiators and simultaneously grows all the polymer arms from the central core. The method is not useful in the preparation of model star polymers by anionic polymerization. This is due to the difficulties in preparing pure multifunctional organometallic compounds and because of their limited solubility. Nevertheless, considerable effort has been expended in the preparation of controlled divinyl- and diisopropenylbenzene living cores for anionic initiation. The core first method has recently been used successfully in both cationic and living radical polymerization reactions. Also, multiple initiation sites can be easily created along linear and branched polymers, where site isolation avoids many problems.

It should be mentioned that the 'polymer first' and 'core first' methods can be used consecutively. For example, a star polymer first produced by the divinylbenzene coupling method contains a number of active sites equal to the number of arms. In a second step, each active site can be used to initiate a new arm either of the same or from a different monomer. The resulting polymer is a double star. Mixing different polymerization mechanisms and various monomers has created entirely new combinations of branched and graft copolymers. Some of the recent progress is part of a large effort that is collectively labeled 'Macromolecular engineering' [44]. This overview must be short and can only highlight recent developments, especially from the point of view of control over the branching architecture.

3.1 BRANCHED POLYMERS VIA ANIONIC POLYMERIZATION

3.1.1 Carbanionic Star Polymers

Schematically, model regular star polymers are obtained directly from living anionic polymers where $(Si-Cl_m)_p$ is a multifunctional carbosilane coupling agent, $MeSiCl_3$, $SiCl_4$, $Cl_3SiCH_2CH_2SiCl_3$, etc. including dendritic carbosilanes

$$s\text{-BuLi} + n\,M \longrightarrow s\text{-Bu-M}_n^-\text{Li} \xrightarrow{(SiCl_m)_p} \left[(s\text{-Bu-M}_n)_m\,Si \right]_p$$

with 32, 64 and 128 silicon chlorine bonds [22, 45, 46]. These coupling agents provide better control over LCB than the commercially used divinylbenzene route [47]. The alternative method, polymerization from a multifunctional initiator, is used less in anionic polymerization because the required multifunctional organometallic compounds are usually insoluble in the polymerization media. Recent exceptions are the polymerization of ethylene oxide (EO), cyclic esters and anhydrides from multifunctional dendrimers [48]:

$$R\,(OH)_n \;+\; m\,K^+ \;\longrightarrow\; R\,(OK)_m(OH)_{n-m} \xrightarrow[\text{2. } H^+]{\text{1. } p\ EO} R[(\,OCH_2CH_2)_{p/n}\text{-}OH\,]_n$$

Ederlé and Mathis, in a very careful study, recently showed by SEC with combined mass and molecular weight (light scattering) detectors that exactly six polystyryllithium chains can be coupled with pure fullerene (C_{60}) in toluene. Furthermore, comparison of the mass and UV (set at 320 nm sensitive to the presence of C_{60}) traces proved the uniformity of the C_{60} core in the star polymer [49].

$$6\ PSLi \;+\; C_{60} \xrightarrow[\text{Toluene}]{} (PS)_6\,C_{60}^{\,6-}\,(Li^+)_6$$

Control of the stoichiometry of the reagents allows also for the preparation of fairly pure three, four and five-arm stars by this method. This conclusion was confirmed [50]. Addition of THF complicates the coupling reaction between carbanions and C_{60}. First a two electron reduction of C_{60} is observed. This is

$$2\ PSK \;+\; C_{60} \xrightarrow[\text{THF}]{} PS\text{-}PS \;+\; C_{60}^{\,2-}\,(K^+)_2$$

$$4\ PSK \;+\; C_{60}^{\,2-}\,(K^+)_2 \xrightarrow[\text{THF}]{} (PS)_4\,C_{60}^{\,6-}\,(K^+)_6$$

followed by the addition of a maximum of four polystyrene chains [51]. Further studies showed that $(PS)_6\,C_{60}^{\,6-}(Li^+)_6$ can selectively initiate the anionic polymerization of one additional polystyrene or polydiene chain, but initiates two poly(methylmethacrylate) chains [52]. By coupling these new active chain ends with α,α'-dibromoxylene the synthesis of dumb-bell (super-H) polymers with two C_{60} branch points was achieved [52]. A different route is also possible. It is known that the azide group (N_3) adds to C_{60}. A polystyrene chain capped with azide forms a mono-adduct with C_{60}. This can then be treated with excess PSLi or PILi to form the homo- or co- mikto-arm star polymer of the type $A'A_6$ and AB_6, respectively [53]. The same azide reaction was used recently to prepare star polymers with end-standing C_{60} molecules [54, 55]. In the first example, the living end-groups of a three-arm star poly(ethylene oxide) are converted to azido groups [54] and in the second case a six-arm star prepared] cationically, is chemically modified at the free ends to azide groups and then reacted with C_{60} [55].

3.1.2 Asymmetric Star Polymers by Anionic Polymerization

The controlled synthesis of star polymers with arms different in MW, chemical composition or topology requires the sequential addition of polymer chains to a central coupling agent, see Figure 3.2. These architecturally different polymers may be referred to as 'mikto-arm star polymers' and their synthesis and properties have recently been reviewed in detail [8]. The great interest in mikto-arm star copolymers of the AB_n type stems from their ability to break the composition-morphology relation that rules the self assembly of linear di- and triblock copolymers [56]. For example, if the number of arms (n) in AB_n is sufficiently large a 50/50 mikto-arm star can form a cylindrical or even a spherical A phase in a B matrix instead of the classic lamellar morphology of diblock copolymers [57]. Mikto-arm star polymers with three chemically different arms can form uniquely different morphologies not observed elsewhere [58].

There are two gereral routes to mikto-arm star polymers. The first method makes use of the stepwise addition of living polymers to multifunctional chlorosilane compounds [59–62]. The Athens group uses the sequential addition of living polymers to multifunctional chlorosilane compounds under tight stoichiometric control [63, 64].

$$PILi + excess\ (CH_3)\ Si\ Cl_3 \longrightarrow PI\ (CH_3)\ Si\ Cl_2\ +\ LiCl\ +\ (CH_3)\ Si\ Cl_3 \uparrow$$

$$PSLi\ +\ PI\ (CH_3)\ Si\ Cl_2 \xrightarrow{titration} (PI)\ (PS)\ (CH_3)\ Si\ Cl\ +\ LiCl$$

$$(PI)\ (PS)\ (CH_3)\ Si\ Cl \xrightarrow[\substack{2.\ H^+ \\ 3.\ fractionation}]{1.\ excess\ PBLi} (PI)\ (PS)\ (PB)\ (CH_3)\ Si\ +\ LiCl$$

One of the most sophisticated architectures prepared by this method has the general formula $(AB)_2(BA)_2$, where A and B are polystyrene and polyisoprene chains, respectively; two arms are linked by the styrene block, the other two arms are linked *via* the isoprene block to the central core. Moreover, the ratio of inner and outer blocks has been kept constant [65].

The second technique makes extensive use of multifunctional diphenylethylene compounds to produce e.g. A_2B_2 star copolymers with A and B polystyrene and polybutadiene chains, respectively [66].

This method seems not so versatile as the chlorosilane method, but is somewhat easier to apply and some interesting materials with archictectures like $A_2(BA)_2$ have been produced [67, 68].

Several groups have engaged in the study of linear polymer–dendrimer conjugates (i.e. architectural copolymers). These structures combine block copolymer and dendrimer branching features within one molecular architecture [69–71].

The properties of these polymers have similarities with mikto-arm star block copolymers.

3.1.3   H-, Super-H-, and π-(co) Polymers

Well-defined polymers with two branch points are called H-polymers [11]. They can be abbreviated by $A_2A'A_2$ or B_2AB_2 for the copolymer case; see Figure 3.3. They are called super-H-polymers when there are more than two branches on each branch point [12, 72], e.g. $A_3A'A_3$ [72], and B_3AB_3 [12] and B_5AB_5 [73]. The homopolymers are of particular interest for the relaxation mechanism of the bridge segment and for their diffusion and rheological properties [43, 72]. They are the smallest members of the comb polymer family and their properties are thought to be representative for this wide class of polymers. The copolymers can be considered double mikto-arm star copolymers. In the case of π-copolymers the legs are chemically different from the body of the copolymer [74]. They are the smallest possible graft copolymer and are abbreviated by (A,B)A(A,B); see Figure 3.3.

There are specific difficulties in the controlled synthesis of H-type polymers. Roovers started with the controlled synthesis of the branches, followed by insertion of a difunctional bridge [11, 73, 74].

$$2\ s\text{-BuLi} + 2\ m\ S \longrightarrow 2\ s\text{-Bu } S_m\ Li + CH_3\ Si\ Cl_3 \longrightarrow (s\text{-Bu } S_m)_2\ Si\ (CH_3)\ Cl$$

$$(1)$$

$$2\ \text{Naphthalene}^{\cdot-}\ Na^+ + n\ S \longrightarrow Na\ S_n\ Na \quad (2)$$

$$(2) + 2\ (1) \longrightarrow (s\text{-BuS}_m)_2\ (CH_3)\ Si\ S_n\ Si\ (CH_3)\ (s\text{-BuS}_m)_2$$

A homogeneous series of polymers with $n \approx m$ and increasing MW has been prepared and characterized. Fractionation is required to remove some low MW polymers. Alternatively, the bridge can be made first [12, 13, 72].

$$2 \text{ Naphthalene}^{-} \text{Na}^{+} + m \text{ S} \longrightarrow \text{Na S}_m \text{Na} + \underset{\text{excess}}{\text{SiCl}_4} \longrightarrow \text{Cl}_3 \text{ Si S}_m \text{Si Cl}_3 + \text{Si Cl}_4\uparrow$$
$$(1)$$

$$\text{s-BuLi} + n \text{ I} \longrightarrow \text{s-Bu I}_n \text{Li} + (1) \longrightarrow (\text{s-Bu I}_n)_3 \text{ Si S}_n \text{Si (s-Bu I}_n)_3$$

All the polymer fragments are monodisperse and the whole polymer has a defined architecture and an overall narrow MW distribution.

3.1.4 Branched Poly(methacrylate)s

The anionic polymerization of methylmethacrylate at room temperature (originally called group transfer polymerization) [75–77] has provided a means for preparing star poly(methylmethacrylate) via the block polymerization with ethyleneglycoldimethacrylate:

Further repeated intermolecular addition of the methacrylate ion onto the pendant methacrylate double bonds yields star-like polymers. Like the divinylbenzene process in other anionic and in carbocationic living systems 'designed' star polymers are difficult to obtain [78–80]. A recent detailed characterization of the PMMA stars has found that the number of arms typically varies between 10 and 100 and that each sample has a fairly wide distribution in the number of arms [78].

The 'core first method' has been applied to prepare four-arm star PMMA. In this case selective degradation of the core allowed unambiguous proof of the star structure. However, the MWD is a little too large to claim that only four-arm star polymers are present [81]. Comb PMMAs with randomly placed branches have been prepared by anionic copolymerization of MMA and monodisperse PMMA macromonomers [82]. A thorough dilute solution characterization revealed monodisperse samples with 2 to 13 branches. A certain polydispersity of the number of branches has to be expected. This was not detected because the branch length was very short relative to the length of the backbone [83]. Recently, PMMA stars (with 6 and 12 arms) have been prepared from dendritic polyesters with peripheral 2-bromo-2-methylpropyl groups and Ni(II) catalyst at 95 °C under ATRP conditions [84].

3.1.5 Branched Aliphatic Polyethers

Substantial recent progress has been made in the synthesis of poly(ethyleneoxide)s (PEO) with new architectures. Originally, polydivinylbenzene cores initiated with K^+ counterions have been used to form star PEO with 10 to 100 arms [85]. More recently, increased control over the arm functionality has been achieved. Starting from multifunctional initiators, three-arm, six-arm and eight-arm PEOs have been prepared from trimethylolpropane [86], hexahydroxy derivatives of hexamethylbenzene, or from a t-butylcalix[8]eneoctahydroxy derivative [87] respectively. In all cases, the initiating hydroxy function is activated by potassium and the THF solution is heterogeneous till several ethylene oxide units have added to the multifunctional initiator. In DMSO, the initial heterogeneity is avoided and narrow MW distribution polymers are obtained [87]. Star PEOs with 4, 8, 16 [48] and 32 arms [88] have also been prepared starting from hydroxy substituted carbosilane dendrimers. The dilute solution properties have been compared with those of standard linear PEOs. The experimental branching factors combined with SEC results provide a strong argument for the architecture of these monodisperse star polymers. In the arm first method, preformed PEO with a suitably activated single end-standing group is reacted with a multifunctional coupling agent. Yen and Merrill successfully prepared star poly(ethylene oxide)s with up to 32 arms using a central poly(amidoamine) dendrimer core [89]. Higher functionality star polymers suffered from incomplete coupling [89]. Both methods have advantages and disadvantages. In the case of multifunctional initiators, it is difficult to prove the monodispersity of the arms. A definitive structure proof would require a chemical degradation of the core and analysis of the detached arms. PEO stars prepared with multifunctional initiators result in arms with terminal hydroxy functional groups that are amenable to chemical modification.

Like traditional dendrimers, dendritic PEOs, consist of successive, increasingly branched shells of PEO chains. Although the chain length between the branch points is less controlled than in classical dendrimers, there is a corresponding rapid increase in the MW. Gnanou et al. started by preparing anionically a three-arm star PEO [14]. Each arm was then reacted with 2,2-dimethyl-5-ethyl-5-halomethyl-1,3-dioxane according to the following reaction:

where X is Br, I or $OSO_2C_6H_4$–CH_3. After isolation of the polymer, the acetal end groups are hydrolyzed and the new hydroxy groups are used as the initiating sites for the next generation of PEO chains. The crucial step is the introduction of the branching point. This reaction has been extensively studied [90, 91].

In this section, it should be mentioned that star poly(tetrahydrofuran) has been prepared by coupling *cationically* polymerized THF with multifunctional diethylenetriamine [92] in the presence of 2,2′6,6′tetramethylpiperidine as a proton trap. When the MW of poly(THF) is 1600 seven chains are added to the triamine, when the MW is 8000 a five-arm star has been obtained.

3.1.6 Branched Aliphatic Polyesters

Poly(ε-caprolactone) (poly(ε-CL) and poly(lactide)s can be prepared by living polymerization methods [93]. In recent years this has led to the explosive development of several types of branched polyesters with narrow MW distributions. The first star poly(ε-CL)s were made by coupling living polymer with trimesic acid chloride [94]. The polymerization is initiated with Et_2AlOCH_2X, where the XCH_2O-group becomes the terminal group on each arm of the star. The formation of the star is confirmed by SEC and, when the MW is low, by NMR. The presence of traces of dimeric and monomeric linear polymer cannot be excluded, because f varies between 2.6 and 3.3 and the MW distribution varies between 1.2 and 1.35. The 'core first method', starting from multifunctional initiators with f = 2, 4 and 6 hydroxy groups based on bis(2,2′-hydroxymethyl)propionic acid has been performed by Trollsås *et al.* [95].

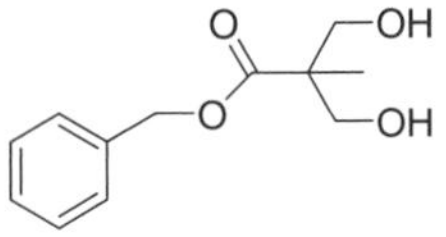

The preferred initiator is stannous octoate in catalytic amount [95, 96] and polymerizations occur in bulk at 110°C or in toluene. The synthesis was expanded to stars with 40 and 48 arms by means of a hyperbranched poly(2,2′-bis(hydroxymethyl)propionic acid) or classical dendrimer, respectively [97].

The molecular design was subsequently expanded to dendritic poly(ε-CL). As in the case of dendrimers, alternating steps are used to introduce linear polymeric segments and branching segments [98]. The end-groups of a hexa-functional poly(ε-CL) star polymer are functionalized with a benzilidene protected 2,2′-bis(hydroxymethyl)propionic acid according to the following reaction sequence:

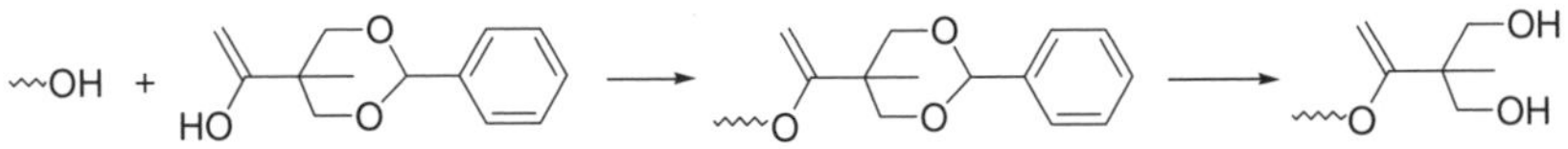

The protecting group is removed by hydrogenation. The deprotected hydroxymethyl groups are used as initiators for polymerization of new ε-CL. The process was repeated to the third generation to give a narrow molecular weight final product with 24 poly(ε-CL) arms with and 24 terminal hydroxy groups. The corresponding hyperbranched poly(ε-CL) was synthesized from an α-carboxylic-ω-dihydroxy linear polymer [99].

$$HO-\overset{O}{\overset{\|}{C}}-(CH_2)_5-O-(-CO-(CH_2)_5-O-)_n-\overset{O}{\overset{\|}{C}}-\overset{-OH}{\underset{-OH}{<}}$$

which is a AB_2 macromonomer. In this case the intermolecular esterification is performed with dicyclohexylcarbodiimide.

It should be mentioned that when a hexa(hydroxyl) initiator is used for the lipase catalyzed polymerization of ε-CL, only one hydroxy function is active [100]. This leaves five remaining OH groups for polymerization of new or another monomers. Comb poly(ε-CL)s have also been prepared [101, 102] starting from a copolymer of ε-CL and 5-ethylene ketal-ε-caprolactone as shown below:

$$R-[CO-(CH_2)_5-O-]_n--[CO-(CH_2)_2-\overset{}{\underset{O\ \ \ \ O}{\overbrace{\ \ \ \ }}}-(CH_2)_2-O-]_m$$

After deprotection the secondary hydroxy groups on the main chain can be activated for the polymerization of L-lactide or ε-CL. The resulting polymer is a comb (co)polymer [102].

3.2   BRANCHING VIA LIVING CARBOCATIONIC POLYMERIZATION

The discovery of living cationic polymerization reactions has provided many new avenues to polymers with a high degree of control over MW, MW distribution and has allowed the preparation of branched polymers with a wide variety of controlled architectures. Living cationic polymerization makes it possible to use isobutylene, alkoxystyrenes and a variety of vinylether monomers that were hitherto excluded as sources of designed polymers. In general, the living nature of the cationic process is very sensitive to the type of activator, solvent, additives and to the temperature. The degree of polymerization is often limited to about 100 so that effects of chain transfer and termination reactions stay undetectable. The different processes for producing branched polymers via cationic polymerization have been critically reviewed recently [103].

Postpolymerization of difunctional monomers to effect star branching has been successfully applied in cationic polymerization, e.g. in the case of poly-isobutylene initiated with 2-chloro-2,4,4,-trimethylpentane/TiCl$_4$. Addition of divinylbenzene leads to star polymers [104]. Vinyl ethers, when polymerized with HI/ZnI$_2$ in toluene at $-40°C$, can be copolymerized with divinylether monomers and star branched polymers result [105]. A sufficient separation of the two vinylether groups is beneficial for the star formation.

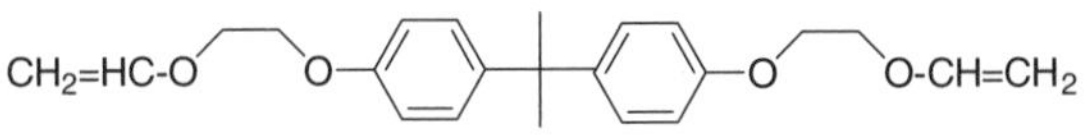

CH₂=HC-O⌒⌒O-⟨⟩—⟨⟩-O⌒⌒O-CH=CH₂

Multifunctional initiators are found to be more effective in carbocationic than in carbanionic polymerization, because of the enhanced solubility of the less polar dormant initiating complexes. For example, the formation of a six-arm star polystyrene starts from

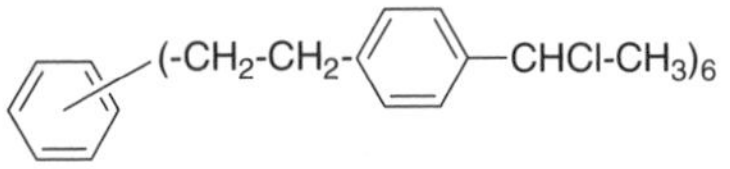

⟨⟩—(-CH₂-CH₂-⟨⟩—CHCl-CH₃)₆

in the presence of $SnCl_4$ in CH_2Cl_2 at̆ 15 − C [106]. A calixarene core with eight initiating groups has also been used for the cationic polymerization of isobutylene [107].

The completely cationic synthesis of comb or graft copolymers have yet to be realized [103]. However, numerous backbone polymers, branches and macro-monomers have been prepared separately via cationic polymerization and these have been combined with other grafting and polymerization processes to pre-pare (co)polymers that cannot easily be prepared otherwise [103].

3.3 RING-OPENING METATHESIS POLYMERIZATION (ROMP)

Bazan and Schrock were the first to use ROMP of norbornene to prepare star polymers [108]. As the coupling agent of the living polymer they used a norbor-nadiene dimer that plays the role of difunctional core-forming monomer as shown below:

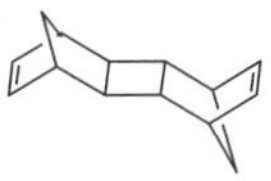

They observed a marked increase in the MW and a slight broadening of the apparent MW distribution (to 1.18–1.25) that they correctly ascribed to the random coupling of living chains. Doumis and Feast polymerized cyclopentene by ROMP and attempted to prepare three-arm stars via coupling with 1,3,5-benzenetricarboxaldehyde, but no pure star polymer was obtained [109].

ROMP is well suited for the controlled living polymerization and copolymer-ization of macromonomers because it is a system that tolerates numerous functional groups. Feast et al. prepared monodispersed double polystyrene (PS) macromonomers with a norbornene head group [110] via reaction of propyleneoxide capped polystyryllithium with the diacid chloride:

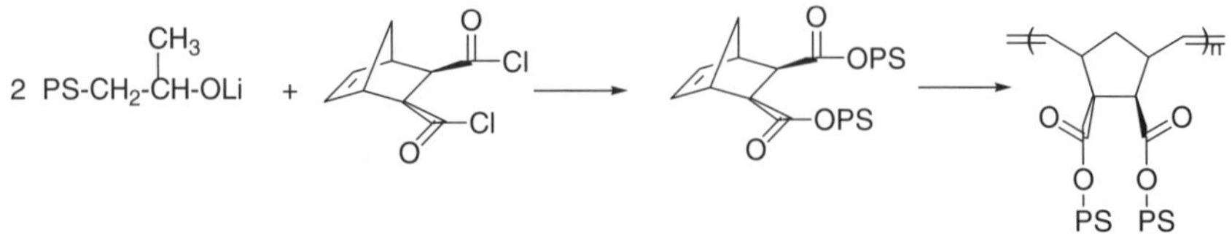

Short PS macromonmers (DP = 4 or 9) polymerized completely under ROMP conditions. Macromonomers with DP = 14 to 46 invariably led to incomplete polymerization, suggesting that the dense polymacromonomer imposes steric limitations on the polymerization. Heroguez *et al.* studied the ROMP of poly(ethylene oxide) macromonomers [111, 112]. The macromonomers are prepared from a norbornene derivative according to the following sequence:

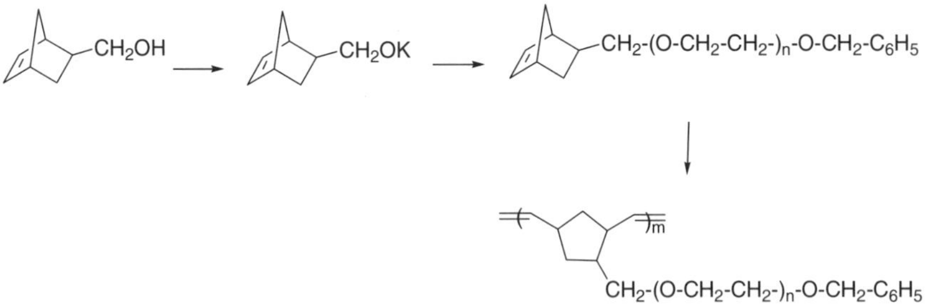

They obtained moderately monodispersed (1.2 ± 0.1) polymacromonomers with 30% initiator efficiency when short macromonomers (DP = 21 to 75) are polymerized. Higher MW macromonomers polymerized only partially. Evidence for interaction of the PEO ether groups with the catalytic center is given and assumed to be responsible for the shortcomings of the living system. Random and block copolymers of PS and PEO macromonomers, as well as of P(EO-b-S) and P(S-b-EO) macromonomers have also been made [112]. The same group successfully prepared PS macromonomers with a norbornene group in the α position [113].

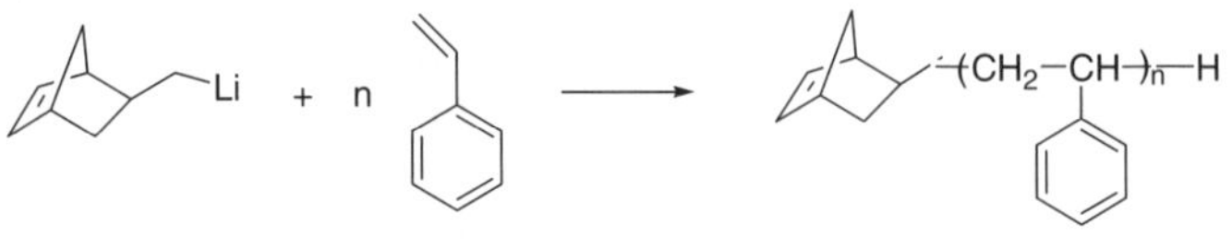

Living ROMP of a macromonomer with MW = 2600 yields star-like polymers with 10 to 100 arms on average possessing a somewhat broad MW distribution (i.e., 1.2–1.4).

3.4 LIVING RADICAL POLYMERIZATION

Hawker was the first to use a multifunctional TEMPO (2,2,6,6-tetramethyl-pyperidinyloxy-) initiator for the synthesis of a three-arm star polystyrene by the living free radical mechanism [114].

He also prepared a poly(styrene-g-styrene) polymer by this technique [114]. The lack of crosslinking in these systems is indeed proof of the control achieved with this technique. An eight-arm star polystyrene has also been prepared starting from a calixarene derivative under ATRP conditions [115]. On the other hand, Sawamoto and his coworkers used multifunctional chloroacetate initiator sites and mediation with Ru^{2+} complexes for the living free-radical polymerization of star poly(methylmethacrylate) [116, 117]. More recent work by Hedrick *et al.* [84] has demonstrated major progress in the use of dendritic initiators [98] in combination with ATRP and other methodologies to produce a variety of structure controlled, starlike poly(methylmethacrylate).

3.5 METAL-CENTERED BRANCHING

Recently, metal-centered coordination has been introduced to construct star polymers [118]. The principle is outlined for bipyridyl capped PEO when coordinated with Ru^{2+}.

The three-arm star polymers are recognized by the UV-Vis spectrum typical of the Ru^{2+} complex as well as by their SEC elution volume. Similar trifunctional complexes with Ni^{2+} and Co^{2+} are labile, subject to rapid exchange. In fact, SEC experiments show that the dissociated form is the dominant eluting species. A

3   CH$_3$-(O-CH$_2$-CH$_2$)$_n$-CH$_2$ [bipyridine structure] + RuCl$_3$ $\longrightarrow$ [Ru(bipyridine)$_3$]$^{2+}$ with CH$_2$-(CH$_2$-CH$_2$-O)-CH$_3$ 2 Cl$^-$

four-arm star has been obtained by means of a *trans*-tetrapyridyl complex of Ru^{2+} [119].

The metal-centered complexes can also be used as multifunctional initiators. For example, Fe^{2+}(4,4'dichloromethyl-2,2'-bipyridine)$_3$ or the Ru^{2+} complex have been used as initiators for the living cationic polymerization of 2-ethyl-2-oxazoline [120].

[Fe(CH$_2$Cl-bipyridine)$_3$]$^{2+}$ + 2n [ethyloxazoline] $\longrightarrow$ [Fe(bipyridine)$_3$]$^{2+}$ with CH$_2$-(CH$_2$-CH$_2$-N-)$_n$-O-CH$_3$ bearing Et C=O substituents

Interestingly, the Fe^{2+} ion in the core can be easily removed by base, the complex dissociates and the individual polymer dimers can be analyzed. Block copolymers of 2-ethyl-2-oxazoline with other substituted oxazolines have also been made [121]. Ru^{2+}(4,4'dichloromethyl-2,2'bipyridine)$_3$ has also been used as the multifunctional initiator for the ATRP of styrene at 110°C [122]. It is interesting to note that the Cu$^+$ ions necessary for the polymerization reaction are solubilized via complexation with other bipyridine species.

4   CONCLUSION

The synthesis of well-defined LCB polymers have progressed considerably beyond the original star polymers prepared by anionic polymerization between 1970 and 1980. Characterization of these new polymers has often been limited to NMR and SEC analysis. The physical properties of these polymers in dilute solution and in the bulk merit attention, especially in the case of completely new architectures such as the dendritic polymers. Many other branched polymers have been prepared, e.g. rigid polymers like nylon [123], polyimide [124] poly(aspartite) [125] and branched poly(thiophene) [126]. There seems to be ample room for further development via the use of dendrimers and hyperbranched polymers as multifunctional initiators. Finally, the rigorous synthesis of comb (and graft co-) polymers with full control over the backbone, branch MWs and branch sites along the backbone remains as an interesting synthetic challenge.

5 REFERENCES

1. Flory, P. J. *Principles of Polymer Science*, Cornell University Press, 1953.
2. Roovers, J., in *Encyclopedia of Polymer Science and Engineering*, Vol. 2, 2nd edn, J. Wiley & Sons Inc., NY, 1985, p. 478.
3. Roovers, J. in *Polymeric Materials Encyclopedia*, Vol. 1, CRC Press, Boca Raton, 1996, p. 850.
4. Small, P. A. *Adv. Polym. Sci.*, **18**, 1 (1975).
5. Schaefgen, J. R. and Flory, P. J. *J. Am. Chem. Soc.*, **70**, 2709 (1948).
6. Bayer, U. and Stadler R. *Macromol. Chem. Phys.*, **195**, 2709 (1994).
7. Wang, F., Roovers, J. and Toporowski, P. M. *Macromol. Symp.* 95, 255 (1995).
8. Hadjichristidis, N., Pispas, S., Pitsikalis, M., Iatrou, H. and Vlahos, C. *Adv. Polym. Sci.*, **142**, 71 (1999).
9. Orofino, T. A. *Polymer*, **3**, 295, 305 (1961).
10. Zimm, B. H. and Stockmayer, W. H. *J. Chem. Phys.*, **17**, 1301 (1949).
11. Roovers, J. and Toporowski, P. *Macromolecules*, **14**, 1174 (1981).
12. Iatrou, H., Avgeropoulos, A. and Hadjichristidis, N. *Macromolecules*, **27**, 6232 (1994).
13. Hakiki, A., Young, R.N. and McLeish, T.C.B. *Macromolecules*, **29**, 3639 (1996).
14. (a) Tomalia, D. A., Baker, H., Dewald, J., Hall, M., Kallos, G., Martin, S., Roeck, J., Ryder, J., Smith, P. *Polym. J.* (*Tokyo*), **17**, 117 (1985); (b) Six, J. L. and Gnanou, Y. *Macromol. Symp.*, **95**, 137 (1995).
15. Roovers, J. and Comanita, B. *Adv. Polym. Sci.*, **142**, 179 (1999).
16. Stockmayer, W. H. *J. Chem. Phys.*, 11, 45 (1943); **12**, 125 (1944).
17. Weissmüller, M. and Burchard, W. *Macromol. Chem. Phys.*, **200**, 541 (1999).
18. Burchard W. *Adv. Polym. Sci.*, **143**, 111 (1999).
19. Weissmüller, M. and Burchard W. *Acta Polym.*, **48**, 571 (1997).
20. Kurata, M. and Fukatsu, M. *J. Chem. Phys.*, **41**, 2934 (1964).
21. Burchard, W. *Adv. Poly. Sci.*, **48**, 1 (1983).
22. Roovers, J., Zhou, L.-L., Toporowski, P.M., van der Zwan, M., Iatrou, H. and Hadjichristidis, N. *Macromolecules*, **26**, 4324 (1993).
23. Grayce, C. J. and Schweizer, K. *Macromolecules*, **28**, 7461 (1995).
24. Hutchinger, L. R. and Richards, R. W. *Macromolecules*, **32**, 880 (1999).
25. Daoud, M. and Cotton, J. P. *J. Phys.* (*Paris*), **43**, 531 (1988).
26. Willner, L., Jucknischke, O., Richter, D., Roovers, J., Zhou, L.-L., Toporowski, P. M., Fetters, L. J., Huang, J. S., Lin, M. Y. and Hadjichristidis, N. *Macromolecules*, **27**, 3821 (1994).
27. Dozier, W. D., Huang, J. S. and Fetters, L. J. *Macromolecules*, **24**, 2810 (1991).
28. Huber, K., Burchard W., Bantle, S. and Fetters, L. J. *Polymer*, **28**, 1997 (1987).
29. Grest, G. S., Fetters, L. J., Huang, J. S. and Richter, D. *Adv. Chem. Phys.*, **94**, 65 (1996).
30. Zimm, B. H. and Kilb, R. W. *J. Polym. Sci.*, **37**, 19 (1959).
31. Stockmayer, W. H. and Fixman, M. *Ann. N.Y. Acad. Sci.*, **57**, 334 (1953).

32. Freire, J. J. *Adv. Polym. Sci.*, **143**, 35 (1999).
33. Bauer, B. J., Fetters, L. J., Graessley, W. W., Hadjichristidis, N. and Quack, G. F. *Macromolecules*, **22**, 2237 (1989).
34. Roovers, J. and Martin, J. E., *J. Polym. Sci.: Part B: Polym. Phys.*, **27**, 2513 (1989).
35. Okumoto, M., Terao,K., Nakamura, Y., Norisuye, T. and Teramaoto, A. *Macromolecules*, **30**, 7493 (1997).
36. Candau, F., Rempp, P. and Benoit, H. *Macromolecules*, **5**, 627 (1972).
37. Roovers, J. and Bywater, S., *Macromolecules*, **7**, 443 (1974).
38. Sato, S., Okada, M. and Nose, T. *Polym. Bull.*, **13**, 277 (1985).
39. Roovers, J., Toporowski, P. M. and Martin, J. E. *Macromolecules*, **22**, 1897 (1989).
40. Wintermantel, M., Gerle, M., Fischer, K., Schmidt, M., Watada, I., Urakawa, H., Kajiwara, K. and Tsukahara, Y. *Macromolecules*, **29**, 978 (1996).
41. Rudin, A. *The Elements of Polymer Science and Engineering*, Academic Press, NY, 1982, pp. 107–117.
42. Weissmüller, M. and Burchard, W. *Polymer Intern.*, **44**, 380 (1997).
43. McLeish, T. C. B. and Milner, S. T. *Adv. Polym. Sci.*, **143**, 195 (1999).
44. Mishra, M. K. (ed.) *Macromolecular Design: Concept and Practice*, Polymer Frontiers International Inc., Hopewell Jct, NY, 1994.
45. Hadjichristidis, N., Guyot, A. and Fetters, L. J. *Macromolecules*, **11**, 668 (1978).
46. Hadjichristidis, N. and Fetters, L. J. *Macromolecules*, **13**, 191 (1980).
47. Worsfold, D. J., Zilliox, J.-G. and Rempp, P. *Can. J. Chem.*, **47**, 3379 (1969).
48. Comanita, B., Noren, B. and Roovers, J. *Macromolecules*, **32**, 1069 (1999); Atthoff, B., Trollsås, M., Claesson, H. and Hedrick, J. L. *Macromol. Chem. Phys.*, **200**, 1333 (1999). Aoi, K., Hatanaka, T., Tsutsumiuchi, K., Okada, M. and Imae T. *Macromol. Rapid Commun.*, **20**, 378 (1999).
49. Ederlé, Y. and Mathis, C. *Macromolecules*, **30**, 2546 (1997).
50. Meleneskaya, E. Yu., Vinogradova, L. V., Litvinova, L. S., Kever, E. E., Shibaev, L. A., Antonova, T. A., Bykova, E. N., Klenin, S. I. and Zgonnik, V. N. *Polym. Sci., Ser. A*, **40**, 115 (1998).
51. Ederlé, Y. and Mathis, C. *Macromolecules*, **30**, 4262 (1997).
52. Ederlé, Y. and Mathis, C. *Macromolecules*, **32**, 554 (1999).
53. Ederlé, Y. and Mathis, C. *Macromol. Rapid Commun.*, **19**, 543 (1998).
54. Taton, D., Angot, S., Gnanou, Y., Wolert, E., Setz, S. and Duran, R. *Macromolecules*, **31**, 6030 (1998).
55. Cloutet, E., Fillaut, J.-L., Astruc, D. and Gnanou, Y. *Macromolecules*, **32**, 1043 (1999).
56. Milner, S. T. *Macromolecules*, **27**, 2333 (1994).
57. Tselikas, Y., Iatrou, H., Hadjichristidis, N., Liang, K. S., Mohanty, K. and Lohse, D. *J. J. Chem. Phys.*, **105**, 2456 (1996).
58. Sioula, S., Hadjichristidis, N. and Thomas, E. L. *Macromolecules*, **31**, 5272 (1998).
59. Roovers, J. and Bywater, S. *Macromolecules*, **5**, 384 (1972).
60. Pennisi, R. W. and Fetters, L. J. *Macromolecules*, **21**, 1094 (1988).
61. Lee, H. C., Chang, T., Haville, S. and Mays, J. W. *Macromolecules*, **31**, 690 (1998).
62. Lee, H. C., Lee, W., Chang, T., Yoon, J. S., Frater, D. J. and Mays J. W. *Macromolecules*, **31**, 4114 (1998).
63. Iatrou, H. and Hadjichristidis, N. *Macromolecules*, **25**, 4649 (1992).
64. Iatrou, H. and Hadjichristidis, N. *Macromolecules*, **26**, 2479 (1993).
65. Tselikas, Y., Hadjichristidis, N., Lescanec, R. L., Honeker, C. C., Wohlgemuth, M. and Thomas, E. L. *Macromolecules*, **29**, 3390 (1996).
66. Quirk, R. P. and Ignatz-Hoover, F. in Hogen-Esch, T. E. and Smid, J. (eds.), *Recent Advances in Anionic Polymerization*, Elsevier, NY, 1987, p. 393.

67. Quirk, R. P., Lee, B. and Schock, L. E. *Macromol. Chem. Symp.*, **52**, 201 (1992).
68. Quirk, R. P., Yoo, T. and Lee, B. *J. Macromol. Sci., Pure and Appl. Chem.*, **A31**, 911 (1994).
69. Gitsov, I., Wooley, K. L. and Fréchet, J. M. J. *Angew. Chem., Int. Ed. Engl.*, **31**, 1200 (1992).
70. van Hest, J. C. M., Delnoye, D. A. P., Baars, M. W. P. L., Elissen-Román, C., van Genderen, M. H. P. and Meijer, E. W. *Chem. Eur. J.*, **2**, 1616 (1996).
71. Román, C., Fischer, H. R. and Meijer, E. W. *Macromolecules*, **32**, 5525 (1999).
72. Archer, L. A. and Varshney, S. K. *Macromolecules*, **31**, 6348 (1998).
73. Velis, G. and Hadjichristidis, N. *Macromolecules*, **32**, 534 (1999).
74. Pispas, S., Hadjichristidis, N. and Mays, J. W. *Macromolecules*, **29**, 7378 (1996).
75. Quirk, R. P. and Bidinger, G. B. *Polymer Bull.*, **22**, **63**, 63 (1989).
76. Quirk, R. P., Ren, J. and Bidinger, G. *Makromol. Chem., Macromol. Symp.*, **67**, 351 (1993).
77. Martin, D.T. and Bywater, S. *Macromol. Chem.*, **193**, 1011 (1992).
78. Lang, P., Burchard, W., Wolfe, M. S., Spinelli, H. J. and Page, L. *Macromolecules*, **24**, 1306 (1991).
79. Haddleton, D. and Crossman, M. C. *Macromol. Chem. Phys.*, **198**, 871 (1997).
80. Estrafiadis, V., Tselikas, G., Hadjichristidis, N., Li., J., Yunan, W. and Mays, J. W. *Polym. Intern.*, **33**, 171 (1994).
81. Zhu, Z., Rider, C. Y., Gilmartin, M. E. and Wnek, G. E. *Macromolecules*, 25, 7330 (1992).
82. DeSimone, J. M., Hellstern, A. M., Siochi, E. J., Smith, S. D., Ward, T. C., Gallagher, P. M., Krukonis, V. J. and McGrath, J. E. *Makromol. Chem., Macromol. Symp.*, **32**, 21 (1990).
83. Siochi, E. J., DeSimone, J. M., Hellstern, A. M., McGrath, J. E. and Ward, T. C. *Macromolecules*, **23**, 4696 (1990).
84. Heise, A., Hedrick, J. L., Trollsås, M., Miller, R. D. and Frank, C. W. *Macromolecules*, **32**, 231 (1999).
85. Lutz, P. and Rempp, P. *Makromol. Chem.*, **189**, 1051 (1988).
86. Gnanou, Y., Lutz, P. and Rempp, P. *Makromol. Chem.*, **189**, 2885 (1988).
87. Taton, D. E. and Gnanou, Y. *Intern. Symp. Ionic Polym.* (*Paris*), 348 (1997).
88. Comanita, B. and Roovers, J. Unpublished results.
89. Yen, D. R. and Merrill, E. W. *Polym. Prepr.* **38** (1), 531 (1997).
90. Bera, T. K., Taton, D. and Gnanou, Y. *Polym. Mater. Sci. Eng.*, **77**, 126 (1997).
91. Labeau, M. P., Cramail, H. and Deffieux, A. *Polym. Intern.*, **41**, 453 (1996).
92. Van Caeter, P. and Goethals, E. L. *Macromol. Rapid Commun.*, **18**, 393 (1997).
93. Penczek, S., Duda, A., Kowalski, A. and Libiszowski, J. *Polym. Mater. Sci. Eng.*, **80**, 95 (1999).
94. Tian, D., Dubois, Ph., Jérôme, R. and Teyssié, Ph. *Macromolecules*, **27**, 4134 (1994).
95. Trollsås, M., Hedrick, J. L., Mecerreyes, D., Dubois, Ph., Jérôme, R., Ihre, H. and Hult, A. *Macromolecules*, **30**, 8508 (1997).
96. Kowalski, A., Duda, A. and Penczek, S. *Macromol. Rapid Commun.*, **19**, 567 (1998).
97. Trollsås, M., Hawker, C. J., Remenar, J. F., Hedrick, J. L., Johannson, M., Ihre, H. and Hult, A. *Polymer*, **36**, 2793 (1998).
98. Trollsås, M. and Hedrick, J. L. *J. Am. Chem. Soc.*, **120**, 4644 (1998).
99. Trollsås, M. Althoff, B., Claesson, H. and Hedrick, J. L. *Macromolecules*, **31**, 3439 (1998).
100. Córdova, A., Hult, A., Hult, K. Ihre, H., Iversen, T. and Malmström, E. *J. Am. Chem. Soc.*, **120**, 13521 (1998).
101. Tian, D., Dubois, Ph., Grandfils, C. and Jérôme, R. *Macromolecules*, **30**, 406 (1997).

102. Trollsås, M., Hedrick, J. L., Mecerreyes, D., Dubois, Ph., Jérôme, R., Ihre, H. and Hult, A. *Macromolecules*, **31**, 2756 (1998).
103. Charleux, B. and Faust, R. *Adv. Polym. Sci.*, **142**, 1 (1999).
104. Marsalko, T. M., Majoros, I. and Kennedy, J. P. *Polymer Bull.*, **31**, 665 (1993).
105. Kanaoke, S., Sawamoto, M. and Higashimura, T. *Macromolecules*, **24**, 2309 (1991).
106. Cloutet, E., Fillaut, J. L. Gnanou, Y. and Astruc, D. *J. Chem. Soc., Chem. Commun.*, 2433 (1994).
107. Jacob, S., Majoros, I. and Kennedy, J. P. *Macromolecules*, **29**, 8631 (1996).
108. Bazan, G. C. and Schrock, R. R. *Macromolecules*, **23**, 817 (1991).
109. Dounis, P. and Feast, W. J. *Polymer*, **37**, 2547 (1996).
110. Feast, W. J., Gibson, V. C., Johnson, A. F., Khosravi, E. and Mohsin, M. A. *Polymer*, **35**, 3542 (1994).
111. Heroguez, V., Breunig, S., Gnanou, Y. and Fontanille, M. *Macromolecules*, **29**, 4459 (1996).
112. Heroguez, V., Gnanou, Y. and Fontanille, M. *Macromolecules*, **30**, 4791 (1997).
113. Heroguez, V., Gnanou, Y. and Fontanille, M. *Macromol. Rapid Commun.*, **17**, 137 (1996).
114. Hawker, C. J. *Angew. Chem. Int. Ed. Engl.*, **34**, 1456 (1995).
115. Angot, S., Murthy, K. S., Taton, D. and Gnanou, Y. *Macromolecules*, **31**, 7218 (1998).
116. Ueda, J., Matsuyama, M., Kamigaito, M. and Sawamoto, M. *Macromolecules*, **31**, 557 (1998).
117. Ueda, J., Kamigaito, M. and Sawamoto, M. *Macromolecules*, **31**, 6762 (1998).
118. Chujo, Y., Naka, A., Krämer, M., Sada, K. and Saegusa, T. *J.M.S.-Pure Appl. Chem.*, **A32**, 1213 (1995).
119. Naka, K., Kobayashi, A. and Chujo, Y. *Macromol. Rapid Commun.*, **18**, 1025 (1997).
120. Lamba, J. J. S. and Fraser, C. L. *J. Am. Chem. Soc.*, **119**, 1801 (1997).
121. McAlvin, J. E. and Fraser, C. L. *Macromolecules*, **32**, 1341 (1999).
122. Collins, J. E. and Fraser, C. L. *Macromolecules*, **31**, 6715 (1998).
123. Warakomski, J. M. *Chem. Mater.*, **4**, 100 (1992).
124. Nguyen, B. N., Eby, R. K. and Meador, M. A. *Polym. Prep.*, **40 (2)**, 626 (1999).
125. Inoue, K., Sakai, H., Ochi, S., Itaya, T. and Tanigaki, T. *J. Am. Chem. Soc.*, **116**, 10783 (1994).
126. Wang, F., Wilson, M. S., Rauh, R. D., Schottland, P. and Reynolds, J. R. *Macromolecules*, **32**, 4272 (1999).

4

Developments in the Accelerated Convergent Synthesis of Dendrimers

A. W. FREEMAN AND J. M. J. FRÉCHET
Department of Chemistry, University of California, Berkeley, CA, USA

Since its introduction a decade ago, the convergent approach to dendrimer synthesis has been successfully exploited for the preparation of a great variety of dendrimers possessing a breadth of cores, interior monomers and peripheral moieties. Over time, a number of clever innovations, adaptations, and 'improvements' have appeared in an attempt to increase the throughput and efficacy of the convergent method, including the development of larger building blocks (e.g. 'hypermonomers' and 'hypercores'), double-stage convergent syntheses, double-exponential growth strategies, and orthogonal monomer systems. These developments, all related to the acceleration of convergent dendrimer syntheses, are described and exemplified in this brief chapter.

1 INTRODUCTION

Synthetic polymer chemists now have a great variety of macromolecular architectures from which to choose when engineering molecules to address particular materials requirements. The different types of polymers can be viewed as part of a continuum of structures with perfectly linear polymers on the left, and perfectly branched dendrimers on the right (Figure 4.1). Progression from left to right within this continuum reveals several salient features. For example while bulk properties are usually most important when dealing with linear polymers, the properties and behavior of individual molecules becomes the prime consideration when dealing with dendrimers. In part, this is due to the fact that unlike

Dendrimers and Other Dendritic Polymers. Edited by Jean M. J. Fréchet and Donald A. Tomalia
© 2001 John Wiley & Sons Ltd

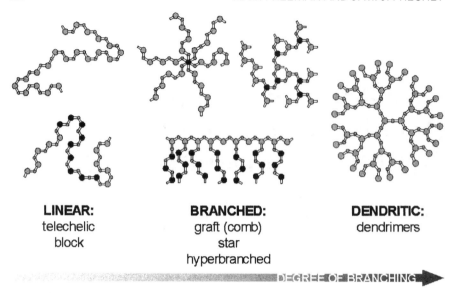

LINEAR: **BRANCHED:**   **DENDRITIC:**
telechelic   graft (comb)   dendrimers
block   star
                     hyperbranched

DEGREE OF BRANCHING ▶

Figure 4.1

high molecular weight linear polymers, large dendrimers generally do not interpenetrate [1] and can thus be assumed to act as discrete entities [2]. In addition
to increasing chemical reactivity, movement to the right within the structural
continuum is also accompanied by an abrupt discontinuity in the amount of time
and effort required for production of the desired macromolecules. Linear and
comparatively simple branched polymers such as stars, grafts, and hyperbranched polymers of moderate molecular weight (< 30 000), are usually prepared in
processes involving from one to four synthetic steps for the most complicated
structures, whereas dendrimers of comparable size may require as many as
10–15 steps with intermediate purification of intermediates.

   In spite of the synthetic challenges associated with large dendrimers, their
precisely controlled structures and unique properties such as reduced solution
viscosities [1, 2], internalization (sequestering) [3], as well as site isolation [4, 5]
and antenna effects [6–8], have fueled vigorous research programs by several
investigators over the past 16 years [9]. For the most part, dendrimers are
prepared according to either the divergent or convergent synthetic strategies,
both of which rely on a repetitive and alternating series of coupling and activation reactions to further dendrimer growth. In analogy to Vögtle's cascade
method of synthesis [10] the divergent approach [11] of Tomalia and Newkome
generates dendrimers by successive addition of layers of monomers, first to a
central core molecule, then to the growing dendrimer proceeding radially in
outwards fashion (Figure 4.2). Conversely, the convergent approach [12], which
constructs these macromolecules from the chain ends and proceeds toward the

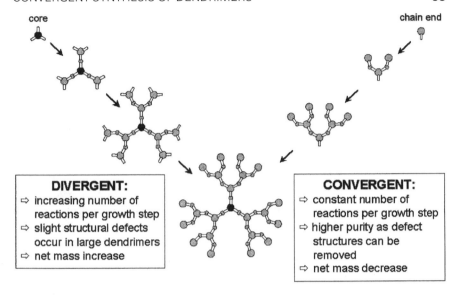

Figure 4.2

center, begins by adding the 'peripheral' moieties to the monomer to generate a small dendritic fragment or dendron. Repetitive activation of the intermediate dendron followed by coupling to additional monomer affords larger intermediates, which, if desired, can be attached to a central core to yield the final dendrimer. Several researchers have attempted to circumvent the usual iterative, two-step protocol employed for dendrimer synthesis, and have devised clever, accelerated synthetic strategies to produce large dendrimers faster, and in higher yields than ever before. Interestingly, nearly all such synthetic developments/ improvements have pertained to the convergent approach, and the following sections summarize some of the most significant or versatile phenotypes. Although the divergent approach is not discussed in further detail here, several key references are provided [11, 13].

2   CONVERGENT SYNTHESIS

The convergent growth approach to dendrimer synthesis, introduced by Hawker and Fréchet in 1989–90 [12], has received considerable attention. Just like the divergent growth strategy, it requires an iterative sequence of activation and coupling steps for the growth of successive generations. However, unlike its divergent counterpart, the number of such reactions remains constant with generation (in most cases two to three reactions), as opposed to increasing exponentially. This greatly facilitates control over the purity of the final product, reducing drastically the amount of reagent required for complete coupling, while also providing access to differentiated reactivity at the focal point and the chain ends of the growing dendrimers, two features that remain the primary attractions of the strategy.

Convergent growth starts with what will become the eventual periphery of the macromolecule and proceeds inward toward the focal point or core [12]. Initially, the monomer (typically AB_2) is end-capped with the peripheral 'groups' to yield the first-generation dendron. The unique focal point of the intermediate dendron is then activated and coupled to additional monomer to afford the [G-2] dendron. Coupling and activation steps are then repeated to yield higher generation monodendrons. If desired, these fragments can be coupled to multifunctional cores to yield multi-dendron *dendrimers.*

The first convergently prepared dendrimers were the poly(benzyl ether) dendrimers, based on the monomer 3,5-dihydroxy benzyl alcohol **1**, described by Hawker and Fréchet (Scheme 1) [12]. Synthesis begins by alkylating the phenolic moieties of **1** under mild and selective conditions to yield the [G-1] alcohol **2**. Conversion of the benzylic alcohol focal point to the corresponding bromide activates the dendron for coupling with additional monomer. Repetition of these steps then affords monodendrons of increasing size. Tridendron dendrimers, such as **7** are obtained by coupling the appropriate dendritic bromides, to tris-phenolic core **6** under the same coupling conditions used to prepare the dendrons themselves. The simplicity, mild reaction conditions, and chemical robustness of the polyether backbone makes this family of dendrimers one of the most popular and most often reproduced. A tremendous variety of core and peripheral moieties have been reported for applications ranging from redox-active dendrimers [14], to those that self-assemble [15] and polymerize [16]. In addition, the scope of the convergent growth approach has been broadened to include other polyethers [17], polyesters [18], polyamides [19], polyphenylenes [20], polyphenylacetylenes [21], polysiloxanes [22], as well as unusual block and surface-block copolymers [23] and dendritic-linear hybrids [24].

While the convergent strategy, as described above, has been exceptionally successful, a considerable amount of effort has been devoted to improving its speed and synthetic efficiency. The remainder of this chapter reviews the major methodological developments and improvements in the accelerated, *covalent,* convergent synthesis of dendrimers. Dendrimers constructed via self-assembly and noncovalent interactions, as well as inorganic dendrimers [25], are beyond the scope of this brief chapter.

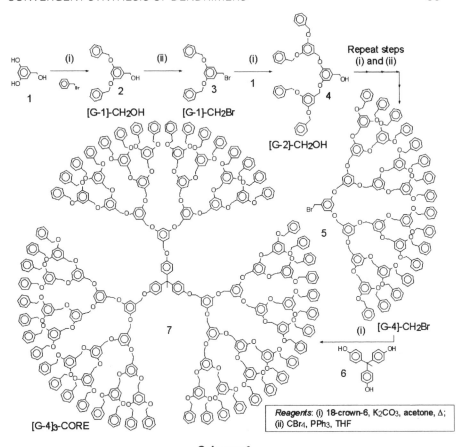

Scheme 1

3 DOUBLE-STAGE CONVERGENT GROWTH STRATEGIES: HYPERMONOMERS AND HYPERCORES

One of the first efforts to accelerate the synthesis of high molecular weight dendrimers was the development of the double-stage convergent approach first reported by Wooley *et al.* (Scheme 2) [17a]. The thrust of this approach was to decrease the number of linear synthetic steps and the amount of time required to prepare large dendrimers, while concomitantly preserving or increasing the overall yields. Therefore, the final dendrimer was disconnected into large synthons that were prepared separately and simultaneously, dramatically reducing the total number of linear synthetic steps. The target macromolecule was envisioned to arise from the coupling of large [G-4] dendrons with densely functionalized, large, flexible 'hypercores'. The polyphenolic cores, based on the 4,4'-bis(4'-hydroxyphenyl)pentanol repeat units, were designed to be spacious and flexible to better accommodate the steric bulk of the surrounding monoden-drons. Cores possessing 6, 12, and 24 phenolic moieties, such as **8** and **10**, were coupled to dendritic bromides **5** in good yields using the same efficient etherifica-tion conditions used in the preparation of the dendrons themselves [12]. In this

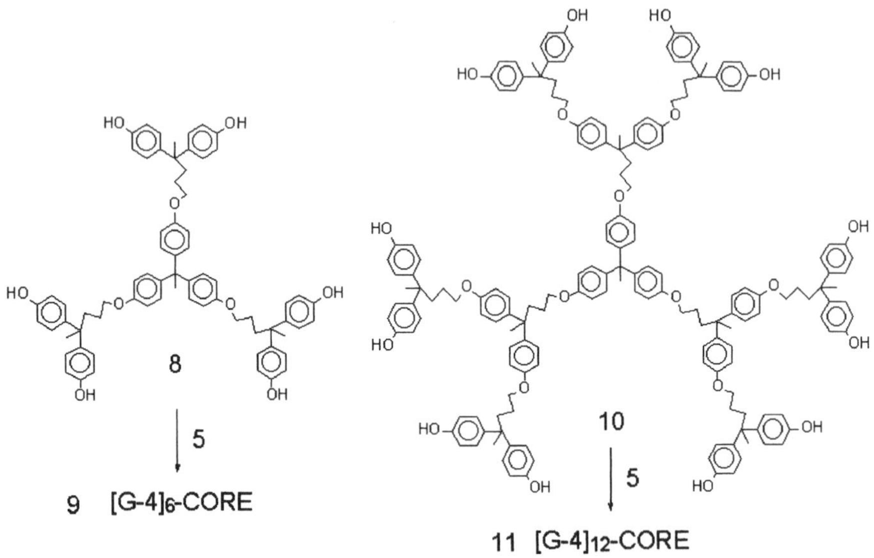

Scheme 2

way, very high molecular weight, monodisperse dendrimers having molecular weights of approximately 21 000, 42 000 and 84 000, respectively, were prepared in short order.

A variation of this accelerated approach was applied, not to the development of 'hypercores,' but to the preparation of larger monomeric building blocks, AB_4 synthons termed branched-monomers or 'hypermonomers' (Scheme 3) [23c]. These larger building blocks preserve the features of the convergent growth approach (i.e. unique focal point) while maintaining the branching pattern and flexibility of the final dendrimer. Hypermonomer **12**, based on 3,5-dihyd-roxybenzyl alcohol, possessing four terminal carboxylic acid moieties, was esteri-fied with four equivalents of [G-1]–CH_2Br **3** under mildly basic conditions to afford [G-3] alcohol **13** in a single step. Activation of the [G-3] intermediate by conversion to the corresponding bromide, followed by coupling with additional hypermonomer **12** afforded [G-5] alcohol dendron **15**, possessing alternating layers of ether and ester linkages, cleanly and in good yield.

For the preparation of large poly(benzyl ether) dendrimers, disconnection

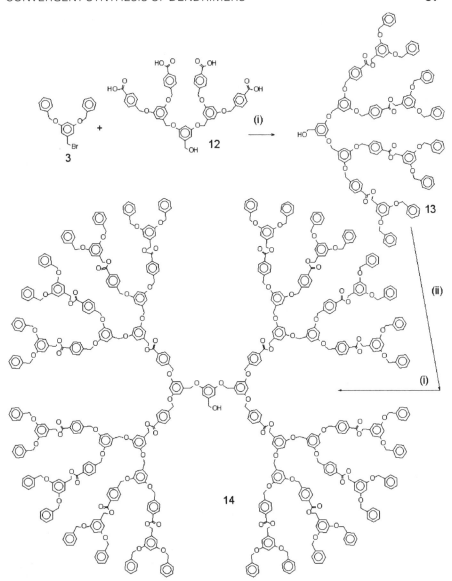

Reagents: (i) 1. *N*-TMS-TFAC, THF 2. K₂CO₃, 18-crown-6; (ii) CBr₄, PPh₃, THF

Scheme 3

eventually leads to AB_4 hypermonomer **18**. This elusive compound was prepared with some difficulty by Sanford et al. as described in Scheme 4 [26]. Key to this approach was the identification of a phenolic protecting group that was easy to install and remove, and that could survive the relatively demanding alkylation conditions required for producing compounds such as **17**. A number of common protecting groups were tried including a variety of silyl groups and phenacyl esters, and eventually the tosyl group was found to be satisfactory. Unfortunately, difficulties in achieving the high yield deprotection of **17**, as well as poor solubility of hypermonomer **18** make this an unattractive synthetic route. Once prepared, **18** could be coupled with four equivalents of [G-2]–CH_2Br to afford [G-4]–CH_2OH on a single step and in good yields.

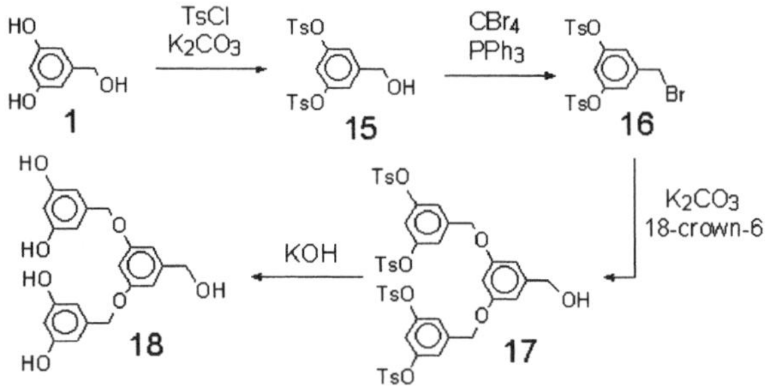

Scheme 4

An improved approach to similar AB_4 synthons was later reported by L'abbé and coworkers (Scheme 5) [27]. Their synthetic approach, which started with methyl-3,5-dihydroxybenzoate, circumvented the difficulty of finding suitable base-stable protecting groups by avoiding basic conditions altogether. Coupling was achieved using the relatively mild Mitsunobu etherification conditions to prepare the tetrasilylated hypermonomer **22**. Similarly, larger silylated dendrons such as AB_8 hypermonomer **23** were prepared. An elegant feature of this approach is the versatility of the silyl-protected phenols that could be deprotected and alkylated in situ upon treatment with alkylating agents in the presence of KF. Thus coupling of [G-2]–CH_2Br with **22** and **23** quickly and efficiently afford poly(benzyl ether) dendrons [G-4]–CH_2OH **25** and [G-5]-CH_2OH **26**, respectively, in good yields. The ease of large-scale preparation of **22** and **23**, and its facile application to the widely used poly(benzyl ether) architectures has made these some of the better and most useful hypermonomers designed to date [28].

Another example of hypermonomer-based dendrimer synthesis was reported by Gilat et al. for the rapid preparation of a new family of laser dye-labeled dendrimers that were critical to our study of energy transfer in light harvesting dendrimers (Scheme 6) [29]. In order to avoid undesirable 'through-bond' energy transfer in the poly(benzyl ether dendrimers, the 'reversed' monomer unit,

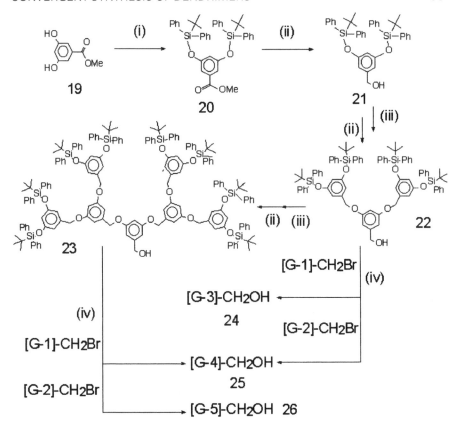

Reagents: (i) TBDPSCl, imid., DMF; (ii) LAH; (iii) **19**, PPh₃, DEAD; (iv) KF, 18-cr-6, acetone, Δ

Scheme 5

3,5-bis(hydroxymethyl)phenol, developed by Tyler and Hanson [30] was used leading to hypermonomer **27**, a hexadecanesulfonate protected tetra(benzylic bromide), designed for the attachment of monodendrons possessing a phenol focal point moiety. The use of this hypermonomer as a building block led to the rapid preparation [29] of the [G-2], [G-3] and [G-4] dye-labeled monodendrons (**30**, **31** and **32**, respectively) under very mild conditions, as well as the useful model compounds **28** and **29**. Although hypermonomer **27** proved highly successful in expediting the synthesis of the larger dendrons, difficulty in complete removal of the sulfonate protecting group under the required strongly basic conditions, in the presence of increasing number of peripheral coumarin dyes, proved problematic and led to decreased yields. The hypermonomer was later improved by changing the focal point functionality from a sulfonate-protected phenol to a silyl-protected benzylic alcohol, which could be selectively

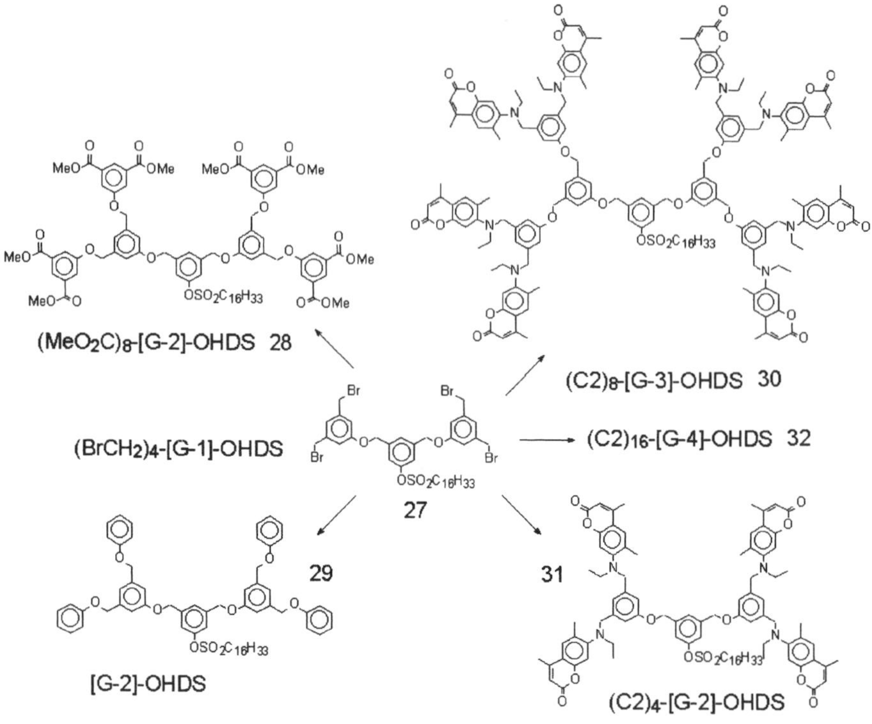

(MeO₂C)₈-[G-2]-OHDS **28**

(BrCH₂)₄-[G-1]-OHDS

27

(C2)₈-[G-3]-OHDS **30**

(C2)₁₆-[G-4]-OHDS **32**

29

[G-2]-OHDS

31

(C2)₄-[G-2]-OHDS

Scheme 6

deprotected with fluoride ion in high yields [29c]. In addition to its great value in accelerating the synthesis through a modular approach, the use of a hyper-monomer minimizes the number of synthetic steps involving the costly coumarin laser dyes.

Although the previous examples detail the preparation of aromatic polyether and polyester dendrimers, the double-stage convergent approach is not limited to this class of functional polymers. For example, Ihre and co-workers recently reported a very successful exploitation of the hypermonomer strategy [31] for the rapid synthesis of aliphatic polyester dendrimers based on the same bis-(2-hydroxymethyl)propanoic acid utilized earlier in a conventional convergent synthesis [18d]. The building blocks used in this approach benefit from their low cost and comparatively high stability resulting from the presence of a neopentyl-like branching arrangement at the vicinity of the ester moiety. However, the relatively small, compact size of the monomer mandates the preparation of higher generation dendrimers to access capabilities such as site isolation [5]. The method used for quickly accessing high generation dendrimers while reducing the number of linear steps, hinged upon the development of the orthogonally protected [G-2] dendron **36** (Scheme 7). This single hypermonomer could then be used to prepare tetraol **37** by removal of the four 'peripheral' acetonide groups under very mildly acidic conditions, and the complementary carboxylic acid **38** by hydrogenolysis of the benzylic ester focal point. Subsequent mutual coupling of the two [G-2] moieties **37** and **38** cleanly afforded the [G-4] polyester **40**. Rapid access to this family of polyesters is particularly desirable since they are being evaluated for use as initiator cores in the preparation of star poly(ε-caprolactone) [32] and as components of drug delivery agents [13g, 33].

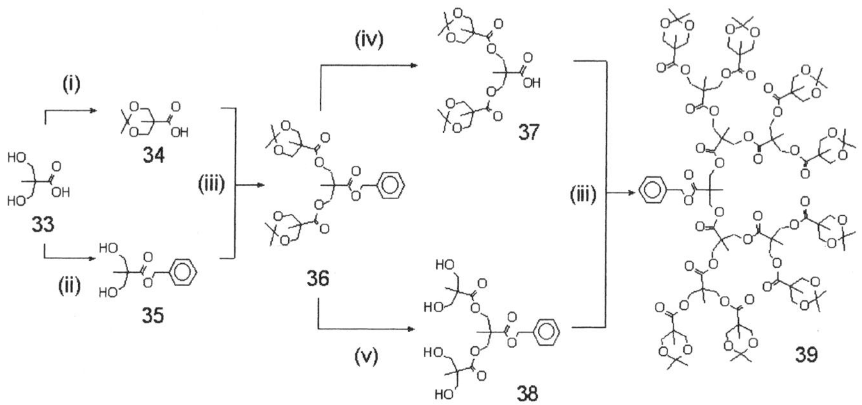

Reagents: (i) 2,2-DMP, TsOH, acetone; (ii) KOH, BnBr; (iii) DCC, DPTS, CH₂Cl₂; (iv) Pd/C, H₂, EtOAc; (v) Dowex, H⁺, MeOH

Scheme 7

4 THE DOUBLE EXPONENTIAL GROWTH STRATEGY

Moore and coworkers have reported a very elegant strategy for the extremely rapid preparation of high molecular weight dendrimers in only three to four steps. Their method, the so-called double exponential growth approach [21c], significantly extends the double-stage convergent synthesis, as exemplified by the preparation of a family of phenylacetylene dendrimers (Scheme 8). The AB$_2$ monomer **41** was first orthogonally deprotected at the 'peripheral' and focal point functionalities, followed by coupling of the resulting fragments to afford the [G-2] dendron **44**. The surface and focal point moieties of [G-2] intermediate **44** were then subjected to orthogonal deprotection conditions to generate two [G-2] fragments; one with four reactive surface groups and a protected focal point, and the other with four protected surface groups and a single reactive focal point. Combination of these fragments via Sonogashira coupling gave [G-4] dendron **47** with 16 protected surface groups and a single, protected focal point. Reiteration of the selective orthogonal deprotections/coupling sequence next

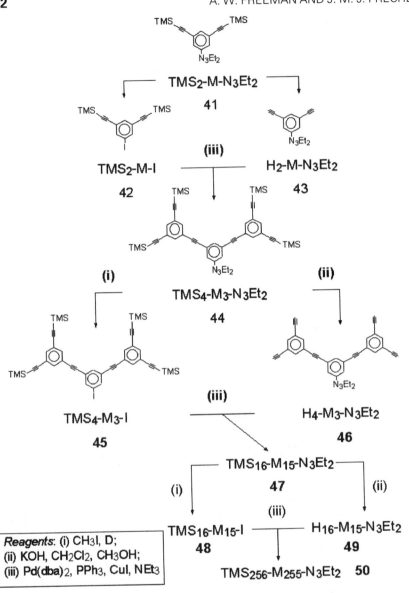

Scheme 8

afforded [G-8] dendron **50** that possesses 256 chain ends and a single focal point. In principle the sequence could be repeated several times to give dendrons of ever increasing size and molecular weight, until the limitation of steric constraints are reached. While this is perhaps the fastest known method to generate exceedingly larger dendrimers, care must be exercised to ensure good mutual solubility of the products and the intermediates to drive the coupling reactions to the highest possible conversions.

While the double-stage convergent and double exponential growth approaches have been very successful in accelerating dendrimer synthesis, they both require the use of protecting groups. One way to further increase synthetic efficiency is to switch to an orthogonal coupling strategy that obviates the need for (de)protection/activation groups. Thus each reaction becomes a coupling reaction advancing dendrimer growth by one generation. Such a strategy can markedly decrease the amount of time and number of purification steps required and consequently boost overall yields of the higher generation materials. A number of orthogonal syntheses have been reported and a selection of the various approaches used is given below.

5 ORTHOGONAL COUPLING STRATEGIES

Spindler and Fréchet [34] were first to describe in 1993 the use of orthogonal monomers for the accelerated preparation of dendrimers. Their report described the rapid synthesis of dendrimers possessing alternating ether and carbamate linkages derived from coupling of monomers **1** and **51** in a one-pot synthesis (Scheme 9). Hence, [G-1]–CH$_2$OH, derived from **1**, was coupled to bisisocyanate **51** to afford the second-generation chloride **52**. Upon completion formation of the biscarbamate, the appropriate amount of **1** and the other components for Williamson ether synthesis (18-crown-6, KI, and Cs$_2$CO$_3$) were added. The reaction was allowed to proceed until the [G-3] monodendron had formed.

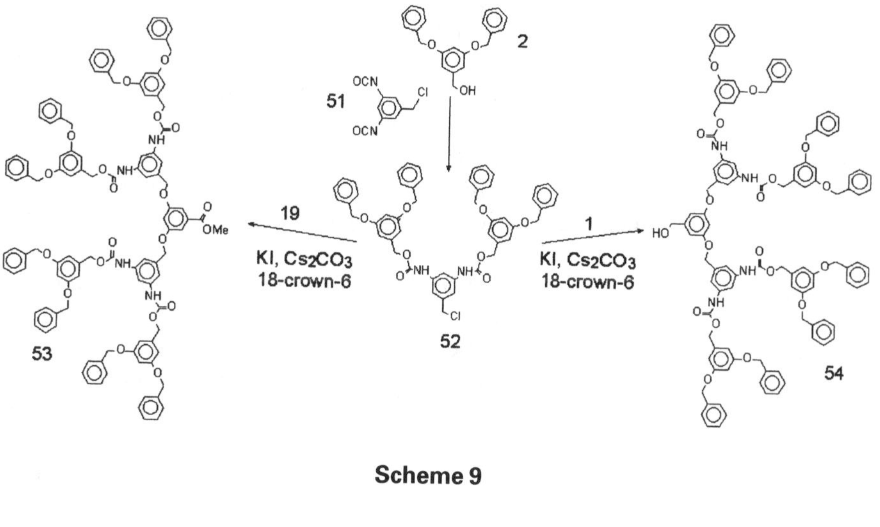

Scheme 9

Unfortunately, the presence of the benzylic alcohol moiety at the focal point of the dendrimer, along with the catalyst (used in the urethane formation) led to the formation of undesired side products, presumably due to carbamate interchange. These side reactions were avoided by switching to monomer **19**, methyl-3,5-dihydroxybenzoate. While the carbamate linkages of dendrons **53** and **54** were too unstable under the alkylation conditions required to afford larger dendrons, the merits of the concept was adequately demonstrated for this accelerated synthesis of [G-3] dendrons.

The next report of a dendrimer synthesis based upon an orthogonal set of monomers came from Zimmerman and Zeng [35], who prepared dendrons up to [G-6] based on an alternating sequence of Mitsunobu esterifications and Sonogashira reactions using monomers **56** and **58** (Scheme 10). The synthesis began by coupling alcohol **55** with **56** under Mitsunobu conditions to yield [G-1] iodide **57**. Sonogashira coupling of **57** with diyne **58** gave [G-2] alcohol **59** in good yield. Repetition of the esterification and cross couplings afforded higher generation dendrons at the rate of one generation per reaction. To further accelerate the dendrimer synthesis the authors have also described the orthogonal coupling of [G-2] 'branched monomers' **62** and **63**.

At about the same time, Yu and co-workers [36] described the synthesis of cross-conjugated poly(phenylenevinylene) dendrimers based on orthogonal monomers **65** and **67** (Scheme 11). Their approach employed an alternating series of Horner–Wadsworth–Emmons and Heck coupling reactions. Two monomers were prepared: 3,5-divinylbenzaldehyde **67** and a bisphosphite-containing aryl bromide **65**. Unlike the dendrons prepared by Spindler *et al.* [34], or by Zimmerman and Zeng [35], these dendrimers are structurally homogeneous. Monodendrons up to the fourth generation were prepared as described in Scheme 11. Unfortunately, these interesting materials could not be prepared in large quantities due to the low yields encountered in the coupling reactions (9–45%), and the low solubility of the high generation compounds. As is the case for all dendrimer syntheses, the choice of reactions and conditions that afford high coupling yields is necessary if an approach is to find widespread acceptance.

The 'activated' monomer approach used by Freeman and Fréchet [37] for the rapid synthesis of convergent poly(benzyl ester) dendrimers also involves an orthogonal coupling strategy. In this method a set of chemically orthogonal monomers, one of which is an 'activated' analog of the other, is used to rapidly synthesize large, perfect, dendrons and dendrimers with one additional generation being added in *each* reaction. The chemically orthogonal AB_2 monomers **72** and **73** are both derived from commercially available diethyl 5-(hydroxymethyl)-isophthalic acid (Scheme 12). Because these monomers are already 'activated' in terms of their ability to react with one another, once incorporated into the monodendron, growth can continue via an iterative and alternating sequence of carboxylate salt alkylations and DCC/DPTS mediated esterifications [38], respectively, without the need for further activation. The coupling

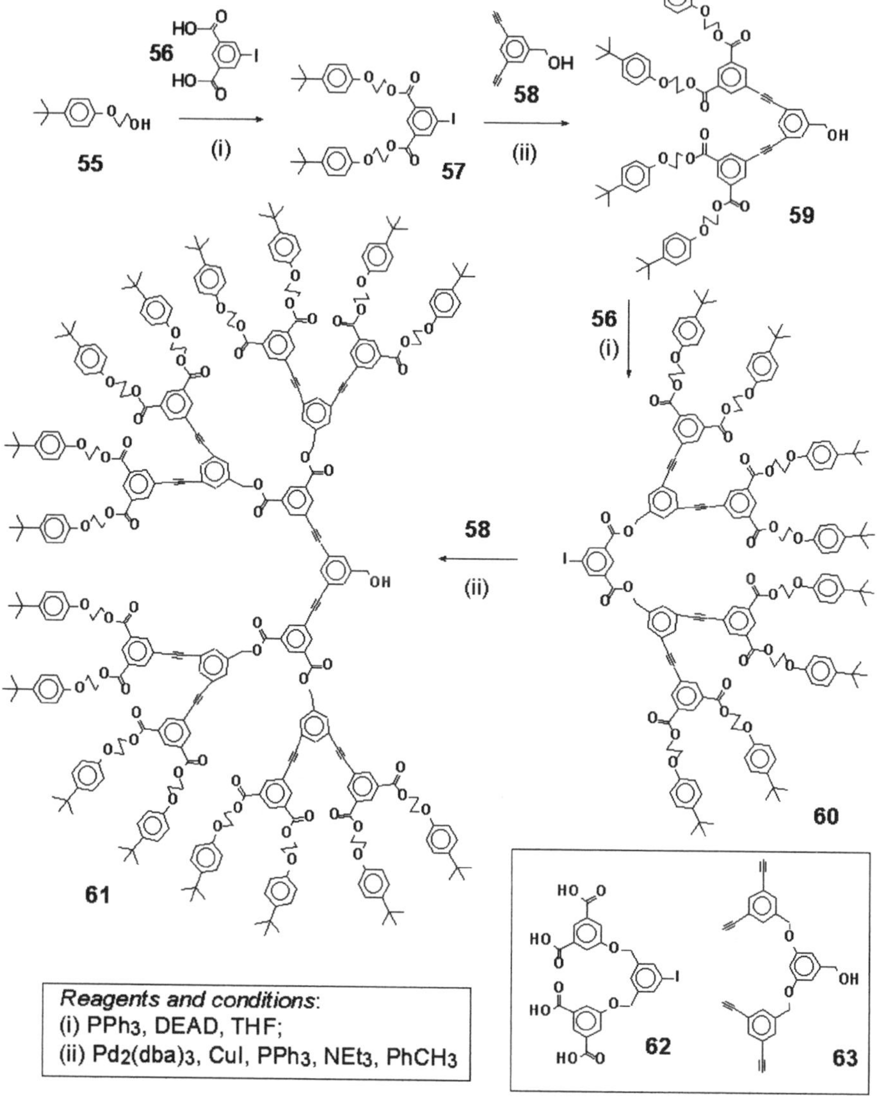

Reagents and conditions:
(i) PPh3, DEAD, THF;
(ii) Pd2(dba)3, CuI, PPh3, NEt3, PhCH3

Scheme 10

steps used in this synthesis were selected for their ability to afford high yields. For example, the overall yield of each step of the reaction sequence, from the preparation of the generation one dendron to that of the generation four dendron, **74**, averaged 90% after chromatographic purification confirming the intrinsic value of the approach.

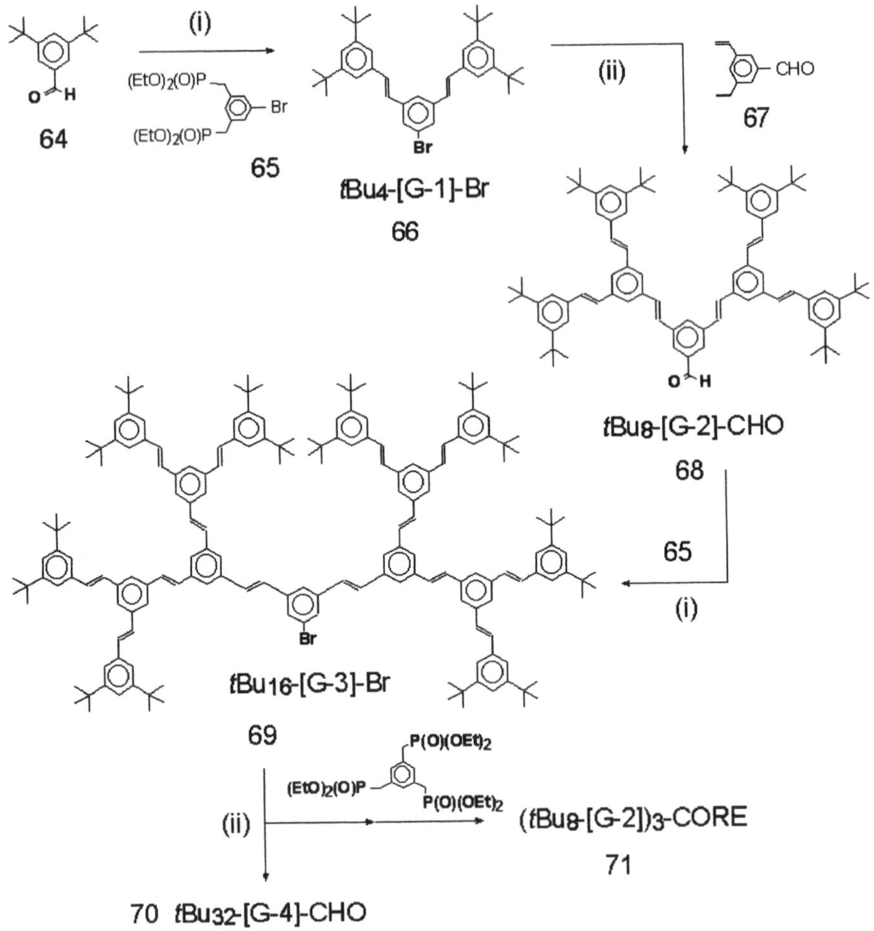

Reagents and conditions: (i) NaH, HMPA; (ii) Pd(Ac)₂, Bu₄NBr, K₂CO₃

Scheme 11

6   CONCLUSION

The convergent approach to dendrimer synthesis has evolved considerably over the past dozen years. With the benefit of hindsight, it is clear that the key criteria for a successful convergent synthesis include:

1. The use of high-yielding reactions for all steps of the synthesis.
2. Structural features that enable easy separation at various stages of growth.

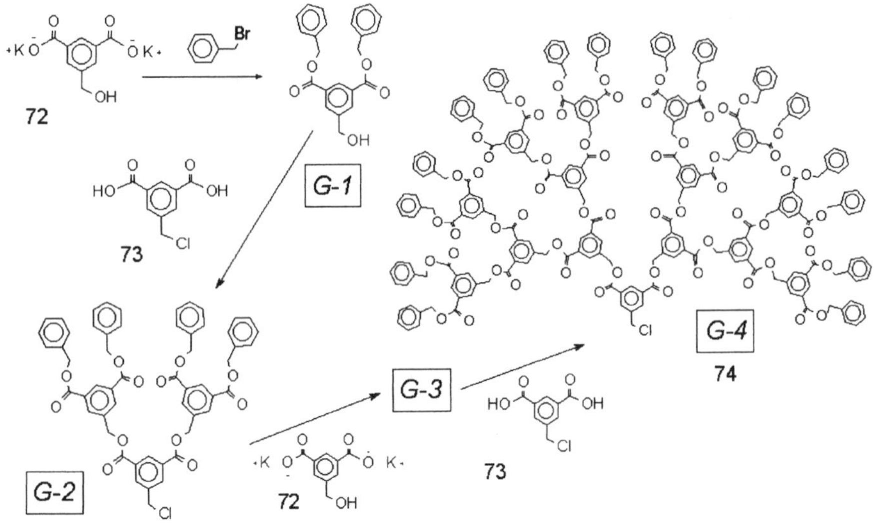

Scheme 12

3. The ability to incorporate varied functional groups at the focal point and at the chain ends of the dendrons.

While a vast number of convergent dendrimers have been reported, only a few have been used widely and some others possess features that make them particularly interesting or versatile. For example the benzyl ether 'Fréchet-type' dendrons [6] have been used most widely to date as a result of their ease of preparation and functional versatility; the phenylacetylene 'Moore-type' dendrons provide a conjugated backbone and can reach very large sizes while remaining relatively rigid; and the aliphatic polyesters derived from bis-(2-hydroxymethyl)propanoic acid [24] show excellent promise in biological applications. It is clear that significant development in dendrimer synthesis will focus on materials that can be obtained relatively easily, preferably through accelerated modular syntheses, and on functional dendrimers that deliver unique properties related to their shape, size and architecture. The potential of functional dendrimers for applications in areas such as nanotechnology [39] or medicine [40] is enormous as few other nanosized building blocks can match the intrinsic versatility of dendrimers. In all of these emerging areas, methodological developments, such as those described herein, are expected to continue not only to push the frontiers of the field, but also to increase the applicability of these materials.

The authors thank the National Science Foundation (DMR) for the continued support of research in the field of dendrimer synthesis. Additional support by the ARO (MURI program) is also acknowledged with thanks.

7 REFERENCES

1. Mourey, T. H., Turner, S. R., Rubinstein, M., Fréchet, J. M. J., Hawker, C. J. and Wooley, K. L. *J. Am. Chem. Soc.*, **25**, 2401 (1992).
2. Tomalia, D. A., Naylor, A. M. and Goddard III, W. A. *Angew. Chem. Int. Ed. Engl.*, **29**, 138 (1990).
3. (a) Jansen, J. F. G. A., Brabander-van den Berg, E. M. M. and Meijer, E. W. *Science*, **266**, 1226 (1994); (b) Cooper, A. I., Londono, J. D., Wignall, G., McClain, J. B., Samulski, E. T., Lin, J. S., Dobrynin, A., Rubinstein, M., Burke, A. L. C., Fréchet, J. M. J. and DeSimone, J. M. *Nature*, **389**, 368 (1997); (c) Balogh, L. and Tomalia, D. A. *J. Am. Chem. Soc.*, **120**, 7355 (1998); (d) Zhao, M., Sun, L., Crooks, R. M. *J. Am. Chem. Soc.*, **120**, 4877 (1998); (e) Tominaga, M., Hosogi, J., Konishi, K. and Aida, T. *Chem. Commun.*, **719** (2000).
4. Hecht, S. and Fréchet, J. M. J. *Angew. Chem. Int. Ed.*, **40**, 74 (2001).
5. Hecht, S., Vladimirov, N. and Fréchet, J. M. J. *J. Am. Chem. Soc.*, **123**, 18 (2001).
6. Adronov, A. and Fréchet, J. M. J. *Chem. Commun.*, 1701 (2000).
7. Jiang, D. L. and Aida, T. *Nature* **388**, 454 (1997).
8. (a) Kawa, M. and Fréchet, J. M. J. *Chem. Mater.*, **10**, 286 (1998); (b) Hecht, S., Ihre, H. and Fréchet, J. M. J. *J. Am. Chem. Soc.*, **121**, 9239 (1999).
9. For general reviews of dendrimers see: reference 2 and (a) Newkome, G. R., Moorefield, C. N. and Vögtle, G. *Dendritic Molecules: Concepts, Syntheses, Perspectives*. VCH, Weinheim, (1996); (b) Fréchet, J. M. J., Hawker, C. J. and Wooley, K. L. *J. Macromol. Sci. Pure Appl. Chem.*, **A31**, 1627 (1994); (c) Fréchet, J. M. J. *Science*, **263**, 1710 (1994); (d) Tomalia, D. A. *Adv. Mater.*, **6**, 529 (1994). (e) Newkome, G. R., Moorefield, C. N., Vögtle, F. *Dendritic Macromolecules*, VCH, New York, 1996; (f) Fréchet, J. M. J., Hawker, C. J. in *Comprehensive Polymer Science*, 2nd Supp., Aggarwal, S. L., Rosso, S. (eds.), Pergamon Press. London, 1996, p. 71; (g) Matthews, O. A., Shipway, A. N., Stoddart, J. F. *Prog. Polym. Sci.*, **23**, 1 (1998); (h) Fischer, M. and Vögtle, F. *Angew. Chem. Int. Ed. Engl.*, **38**, 884 (1999).
10. Buhleier, E., Wehner, W. and Vögtle, F. *Synthesis* **155** (1978).
11. (a) Tomalia, D. A., Dewald, J. R., Hall, M. J., Martin, S. J. and Smith, P. B., *Preprints of the 1st SPSJ Int'l Polymer Conf.*, Soc. of Polym. Sci. Japan, Kyoto, 1984, p. 65, later published as Tomalia, D. A., Baker, H., Dewald, J., Hall, M., Kallos, G., Martin, S., Roeck, J., Ryder, J. and Smith, D. *Polym. J.*, **17**, 117 (1985). (b) Newkome, G. R., Zao, Y., Baker, G. R. and Gupta, V. K. *J. Org. Chem.*, **50**, 2003 (1985).
12. (a) Fréchet, J. M. J., Jiang, Y., Hawker, C. J. and Philippides, A. E. *Preprints of the IUPAC International Symposium of Functional Polymers; The Polymer Society of Korea*, **19** (1989); later published as Hawker, C. J. and Fréchet, J. M. J. *J. Am. Chem. Soc.*, **112**, 7638 (1990); (b) Hawker, C. J., Fréchet, J. M. J. *J. Chem. Soc., Chem. Commun.*, **1010** (1990).
13. (a) Newkome, G. R. and Lin, X. *Macromolecules*, **24**, 1443 (1991); (b) Newkome, G. R., Moorefield, C. N. and Baker, G. R. *Aldrichimica, Acta*, **25**, 31 (1992); (c) Newkome, G. R., Nayak, A., Behera, R. K., Moorefield, C. N. and Baker, G. R. *J. Org. Chem.*, **57**, 358 (1992); (d) Tomalia, D. A. and Durst, H. D. *Top. Curr. Chem.*, **165**, 193 (1993); (e) Tomalia, D. A. *Aldrichimica Acta*, **26**, 91 (1993); (f) Tomalia, D. A., *Sci. Am.*, **272**, 62 (1995); (g) Ihre, H., Padilla de Jesùs, O. L. and Fréchet, J. M. J. *J. Am. Chem. Soc.*, **123**, 5908 (2001).
14. (a) Jin, R., Aida, T. and Inoue, S. *J. Chem. Soc., Chem. Commun.*, **1260** (1993). (b) Sadamoto, R., Tomioka, N. and Aida, T. *J. Am. Chem. Soc.* **118**, 3978 (1996); (c) Jiang, D.-L. and Aida, T. *J. Am. Chem. Soc.*, **120**, 10895 (1998); (d) Pollak, K. W., Leon, J. W., Fréchet, J. M. J., Maskus, M. and Abruna, H. D. *Chem. Mater.*, **10**, 30 (1998).

15. (10)(a) Zimmerman, S. C., Zeng, F. W., Reichert, D. E. C. and Kolotuchin, S. V. *Science*, **271**, 5252 (1996); (b) Wang, Y., Zeng, F. W. and Zimmerman, S. C. *Tetrahedron lett*. **38**, 5459 (1997); (c) Suarez, M. Lehn, J.-M. Zimmerman, S. C.,Skoulios, A. and Heinrich, B. *J. Am. Chem. Soc*., **120**, 9526 (1998); (d) Freeman, A. W., Vreekamp, R. H. and Fréchet, J. M. J. *Polym. Mat. Sci. Eng*., **77**, 138 (1997); (e) Percec, V., Johansson, G., Ungar, G. and Zhou, J. *J. Am. Chem. Soc*., **118**, 9855 (1996); (f) Balagurusamy, V. S. K., Ungar, G., Percec, V. and Johansson, G. *J. Am. Chem. Soc*., **119**, 1539 (1997); (g) Percec, V., Cho, W.-D., Mosier, P. E., Ungar, G. and Yeardley, D. J. P. *J. Am. Chem. Soc*., **120**, 11061 (1998); (h) Enomota, M. and Aida, T. *J. Am. Chem. Soc*., **121**, 874 (1999).

16. (a) Hawker, C. J. and Fréchet, J. M. J. *Polymer*, **33**, 1507 (1992); (b) Karakaya, B., Claussen, W., Gessler, K., Saenger, W. and Schluter, A.-D. *J. Am. Chem. Soc*., **119**, 3296 (1997); (c) Neubert, I. and Schluter, A.-D. *Macromolecules*, **31**, 9372 (1998); (d) Bo, Z. and Schluter, A.-D. *Macromol. Rapid Commun*., **20**, 21 (1999); (d) Bo, Z. Zhang, C., Severin, N., Rabe, J. and Schluter, A.-D. *Macromolecules*, **33**, 2688 (2000); (e) Schluter, A.-D. and Rabe, J. *Angew. Chem. Int. Ed. Engl*., **39**, 864 (2000); (f) Stewart, G. M. and Fox. M. A. *Chem. Mater*., **10**, 860 (1998); (g) Percec, V., Heck, J., Tomazos, D., Falkenberg, F., Blackwell, H. and Ungar, G. *J. Chem. Soc*., *Perkin Trans. I*, 2799 (1993); (h) Percec, V., Tomazos, D., Heck, J., Blackwell, H. and Ungar, G. *J. Chem. Soc*., *Perkin Trans. I*, **31** (1994).

17. (a) Wooley, K. L., Hawker, C. J. and Fréchet, J. M. J. *J. Am. Chem. Soc*., **113**, 4252 (1991); (b) Jayaraman, M. and Fréchet, J. M. J. *J. Am. Chem. Soc*., **120**, 12996 (1998).

18. (a) Hawker, C. J. and Fréchet, J. M. J. *J. Chem. Soc*., *Perkin Trans. I*, 2459 (1992); (b) Hawker, C. J. and Fréchet, J. M. J. *J. Am. Chem. Soc*., **114**, 8405 (1992); (c) Miller, T. X., Kwock, K. E. and Neenan, T. X. *Macromolecules*, **25**, 3143 (1992); (d) Ihre, H., Hult, A. and Soderlind, E. *J. Am. Chem. Soc*., **118**, 6388 (1996).

19. (a) Miller, T. M. and Neenan, T. X. *Chem. Mater*., **2**, 346 (1990); (b) Ulrich, K. E. and Fréchet, J. M. J. *J. Chem. Soc*., *Perkin Trans. I*, 1623 (1992); (c) Bayliff, P. M., Feast, W. J. and Parker, D. *Polym. Bull*. **29**, 265 (1992).

20. (a) Miller, T. M., Neenan, T. X., Zayas, R. and Bair, H. E. *J. Am. Chem. Soc*., **114**, 1018 (1992); (b) Sakimoto, Y., Suzuki, T., Miura, A., Fujikawa, H., Tokito, S. and Taga, Y. *J. Am. Chem. Soc*., **122**, 1832 (2000).

21. (a) Xu, Z. and Moore, J. S. *Angew. Chem. Int. Ed. Engl*., **32**, 1354 (1993); (b) Xu, Z. F., Kahr, M., Walker, K. L. and Moore, J. S. *J. Am. Chem. Soc*., **116**, 4537 (1994); (c) Kawaguchi, T., Walker, K. L., Wilkins, C. L. and Moore, J. S. *J. Am. Chem. Soc*., **117**, 2159 (1995).

22. Morikawa, A., Kakimoto, M. and Imai, Y. *Macromolecules*, **25**, 3247 (1992).

23. (a) Wooley, K. L., Hawker, C. J. and Fréchet, J. M. J. *J. Chem. Soc*., *Perkin Trans. I*, 1059 (1999); (b) Wooley, K. L., Hawker, C. J. and Fréchet, J. M. J. *J. Am. Chem. Soc*., **115**, 11496 (1993); (c) Wooley, K. L., Hawker, C. J. and Fréchet, J. M. J. *Angew. Chem. Int. Ed. Engl*., **33**, 82 (1994); (d) Grayson, S. M. and Fréchet, J. M. J. *J. Am. Chem. Soc*., **122**, 10335 (2000); (e) Grayson, S. M., Jayaraman, M. and Fréchet, J. M. J. *J. Chem. Soc. Chem. Commun*., 1329 (1992).

24. (a) Gitsov, I., Wooley, K. L. and Fréchet, J. M. J. *Angew. Chem. Int. Ed. Engl*., **31**, 1200 (1992); (b) Gitsov, I. and Fréchet, J. M. J. *Macromolecules* **26**, 6536 (1993); (c) Gitsov, I., Wooley, K. L., Hawker, C. J., Ivanova, P. T. and Fréchet, J. M. J. *Macromolecules*, **26**, 5621 (1993); (d) Fréchet, J. M. J. and Gitsov, I. *Macromol. Symp*., **98**, 441 (1995); (e) Leduc, M. R., Hawker, C. J., Dao, J. and Fréchet, J. M. J. *J. Am. Chem. Soc*., **118**, 11111 (1996).

25. Majoral, J. P. and Caminade, A. M. *Chem Rev*., **99**, 845 (1999).

26. (a) Sanford, E. M., Fréchet, J. M. J., Wooley, K. L. and Hawker, C. J. *Polym. Prepr*. **34**,

654 (1993); (b) Sanford, E. M., Spindler, R. and Fréchet, J. M. J. Unpublished results.
27. (a) L'abbé, G., Forier, B. and Dehaen, W. *Chem. Commun.*, 2143 (1996); (b) Forier, B. and Dehaen, W. *Tetrahedron*, **55**, 9829 (1999).
28. Junge, D. M. and McGrath, D. V. *Tetrahedron Lett.*, **39**, 1701 (1999).
29. (a) Gilat, S. L., Adronov, A. and Fréchet, J. M. J. *J. Org. Chem.*, **64**, 7474 (1999); (b) Gilat, S. L., Adronov, A. and Fréchet, J. M. J. *Angew. Chem. Int. Ed. Engl.*, **38**, 1422 (1999); (c) Adronov, A., Gilat, S. L., Fréchet, J. M. J., Ohta, K., Neuwahl, F. V. R. and Fleming, G. *J. Am. Chem. Soc.*, **122**, 1175 (2000).
30. (a) Tyler, T. L. and Hanson, J. E. *Chem. Mater.*, **11**, 3453 (1999); (b) Höger, S. *Synthesis*, 20 (1997).
31. Ihre, H., Hult, A., Fréchet, J. M. J. and Gitsov, I. *Macromolecules*, **31**, 4061 (1998).
32. (a) Trollsas, M., Hedrick, J. L., Mecerreys, D., Dubois, P., Jerome, R., Ihre, H. and Hult, A. *Macromolecules*, **30**, 8508 (1997); (b) Trollsas, M., Hedrick, J. L., Mecerreys, D., Dubois, P., Jerome, R., Ihre, H. and Hult, A. *Macromolecules*, **31**, 2756 (1998); (c) Trollsas, M. and Hedrick, J. L. *J. Am. Chem. Soc.*, **120**, 4644 (1998); (d) Trollsas, M., Claesson, H., Atthoff, B. and Hedrick, J. L. *Angew. Chem. Int. Ed. Engl.*, **37**, 3132 (1998); (e) Cordova, A., Hult, A., Hult, K., Ihre, H., Iversen, T. and Malmstrom, E. *J. Am. Chem. Soc.*, **120**, 13521 (1998).
33. Fréchet, J. M. J. and Ihre, H. US patent application.
34. Spindler, R. and Fréchet, J. M. J. *J. Chem. Soc., Chem. Commun.*, 913 (1993).
35. Zeng, F. and Zimmerman, S. C. *J. Am. Chem. Soc.*, **118**, 5326 (1996).
36. Deb, S. K., Maddux, T. M., Yu, L. *J. Am. Chem. Soc.*, **119**, 9079 (1997).
37. Freeman, A. W., Fréchet, J. M. J. *Org. Lett.*, **1**, 685 (1999).
38. Moore, J. S. and Stupp, S. I. *Macromolecules*, **23**, 65 (1990).
39. (a) Tully, D. C. and Fréchet, J. M. J. *Chem Commun.* 1229 (2001); (b) Dykes, G. M. *J. Chem. Technol. Biotechnol.* **76**, 903 (2001).
40. (a) Liu, M. and Fréchet, J. M. J. *Pharmaceut. Sci. Technol. Today* **2**, 393 (1999); (b) Malik, N., Wiwattanapatapee, R., Klopsch, R., Lorenz, K., Frey, H., Weener, J. W., Meijer, E. W., Paulus, W. and Duncan, R. *J. Control. Rel.* **65**, 133 (2000).

5

Formation, Structure and Properties of the Crosslinked State Relative to Precursor Architecture

K. DUŠEK AND M. DUŠKOVÁ-SMRČKOVÁ
Institute of Macromolecular Chemistry Academy of Sciences of the Czech Republic, Prague 6, Czech Republic

LIST OF SYMBOLS AND ABBREVIATIONS

f	functionality of a precursor
g	functionality of a precursor
n_A, n_B, n_C	mole fractions of components A, B, C
r_H	initial molar ratio of hydroxy to isocyanate groups
w_g	gel fraction
w_s	sol fraction
w_{BC}	weight fraction of backbone chains
w_{DC}	weight fraction of dangling chains
x_j	mole fraction of units with j reacted functional groups
z_X	auxiliary (dummy) variable of probability generation function identifying bonds extending to group X
z_{YX}	auxiliary (dummy) variable of probability generation function identifying bonds extending from group Y to group X
z	vector of auxiliary (dummy) variables of probability function (pgf) identifying bonds
$F_{0n}(Z, z)$	probability generating function of vectors of auxiliary (dummy) variables Z and z
G_e	equilibrium shear modulus
M_c	molecular weight of chains between crosslinks
M_n	number-average molecular weight

M_w	weight-average molecular weight
M_0	molecular weight of monomer unit
R	gas constant
T	temperature in K
T_g	glass transition temperature
Z_{Xu}	auxiliary (dummy) variable in probability generating identifying unreacted groups X
$\mathbf{Z}$	vector of auxiliary (dummy) variables in probability generating function identifying unreacted groups
$\alpha, \alpha_A, \alpha_B$	degree of conversion of functional groups (A or B)
$\beta, \beta^*, \beta^{**}$	branching index
v_e	concentration of EANC's per unit volume
$(\phi_{Af})_n$	number-average functionality for unreacted A groups
$(\phi_{Af})_w$	weight-average functionality for unreacted A groups
$(\phi_{Af})_2$	second-moment average of the functionality distribution for unreacted A groups
pgf	probability generating function
BC	backbone chain (an EANC without dangling chains)
C	crosslinker
DC	dangling chain
EAC	elastically active crosslink
EANC	elastically active network chain
HP	hyperbranched polymer
PAMAM	poly(aminoamine) dendrimer
PAMAMOS	poly(aminoamine) organosilicon dendrimer
POSS	polyhedral oligomeric silsesquioxanes
TBP	theory of branching processes

1 INTRODUCTION

Covalent polymer networks or (Class II) crosslinked macromolecular architecture polymers rank among the largest molecules known. Their molecular weight is given by the macroscopic size of the object; for instance, a car tire made of vulcanized rubber or a crosslinked layer of protective coating can be considered one crosslinked molecule. Such networks are usually called macronetworks. On the other hand, micronetworks have dimensions of several nanometers to several micrometers (e.g. siloxane cages or microgels).

Classifying polymers in their crosslinked state according to end-use properties, polymer networks include: vulcanized rubbers, crosslinked thermosetting materials, protective coatings, adhesives, polymeric sorbents, microelectronics materials, soft gels, etc. Polymer networks in contrast to uncrosslinked polymers, are distinguished by their dimensional stability, increased thermal and chemical resistance and ability to store information concerning their shape (shape memory) and formation history when the gel point is surpassed.

Polymer networks are formed from functional precursors by covalent bond formation [1]. As a result, molecular weights and polydispersity increase and the system passes through a critical point, the gel point. At this point, an 'infinite' structure (molecule) is formed for the first time. Beyond the gel point, the fraction of the 'infinite' structure (the gel) increases at the expense of finite (soluble) molecules (the sol). The sol molecules become gradually bound to the gel and eventually all precursor molecules can become a part of the gel – the network. This is not always the case for different reasons; sometimes sol is still present after all functional groups have reacted. In passing from the gel point to the final network not only the gel fraction increases, but also the network becomes 'denser' containing increasing amounts of crosslinks and strands between them called elastically active network chains.

The word 'infinite' used in connection with the structure existing beyond the gel point deserves clarification. For macronetworks, the term 'infinite' structure means a structure large enough that their structure development and their properties are virtually indistinguishable from those of any larger system. Specific properties of the system exhibit different sensitivity to the system size. For instance, the gel point, characterized by time or conversion of functional groups, at which an infinite structure (sequence of units connected-by covalent bonds) appears for the first time, is less sensitive than weight- and higher averages of molecular weight distribution. Also, the molecular weight averages higher than the number average and steady shear viscosity diverge. In a finite system, infinite distance means an average distance between any pair of points on the surface of a macroscopic sample. 'Macroscopic' is related to dimensions of the order of micrometer and higher. In a finite system, 'gel' is identified with a largest molecule in the system. If the system is sufficiently large, the size of this molecule at one point starts growing much faster than the second largest molecule and this change tends to have a character of a discontinuous transition. Also, in the gel circuits (large cycles) are formed and the crosslinking density can be expressed by the *cycle rank* related to the number of circuits (cycles). For a crosslinked particle of the order of magnitude of 10^6 monomer units in size, the difference in the gel point conversion with respect to that obtained for sizes $\gg 10^6$ is virtually within experimental error (below 1%, simulation results [2, 3]). However, the molecular weight distributions in the experimentally measurable range are quite different and fluctuate from system to system. Figure 5.1 shows the simulated dependence of the gel fraction on conversion for various system sizes [2, 3]. The gel point is identified with the break observed on the dependence of the weight fraction of the largest molecule on conversion.

For small microgels, like crosslinked particles of a microemulsion (diameter below 10^2 nm), the definition of the gel point is no longer clear. For instance, a

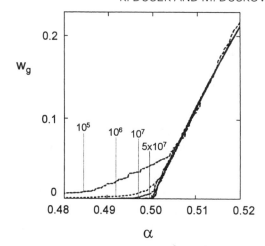

Figure 5.1 Simulated dependence of the weight fraction of the largest molecule, w_g, on conversion of functional groups, α. Varying systems size counted in the numbers of monomer molecules used in simulation is indicated: 10^5, 10^6, 10^7, 5×10^7. Modified Figure 5b of ref. [3]; reprinted with permission

siloxane cage structure (POSS) cannot be called a microgel because its molecular weight is of the order of 10^3.

Polymer networks are built up from functional precursors of various architectures carrying various numbers of functional groups [4, 5]. The functional groups of the precursor can be of the same or different type; groups of the same type can have the same or different reactivities. Usually, reactions of type $A + B \rightarrow A - B$ are used for the crosslinking of precursors into a network, but the $A + A \rightarrow A - A$ reactions are also important (e.g. vinyl type polymerizations, polyetherification, etc.). Usually, multifunctional precursors are crosslinked with low-functionality ($f = 2, 3$) molecules but reactions between two precursors of similar or different architectures are also possible. The variation of precursor architectures and the number of ways to combine them with crosslinkers makes it possible to use a modular approach to network build-up and to vary processing and materials properties over a wide range. The original precursor structures and their combinations can be traced in the network structure. They represent distinct *substructures* in the network and are more or less organized. In the modular approach, the preformed precursors of desired structure – modules – are linked in a macronetwork by crosslinking. Of not negligible significance is the build-up of substructures (chemical clusters) in polymer networks performed *in situ*.

Introduction of new precursor architectures brings about new challenges in description and modeling of network formation because the apparent reactivities of functional groups become dependent on the size and shape of the precursors or substructures.

Special features of crosslinking related to precursor architectures will be demonstrated on a few examples of precursors – telechelics and hyperbranched polymers.

Throughout the chapter, the importance of network formation theories in understanding and predicting structural development is stressed. Therefore, a short exposé on network formation theories is given in this chapter. Although the use of theoretical modeling of network build-up and comparison with experiments play a central role in this chapter, most mathematical relations and their derivation are avoided and only basic postulates of the theories are stated. The reader can always find references to literature sources where such mathematical relations are derived.

2   NETWORK FORMATION

2.1   GENERAL FEATURES OF NETWORK FORMATION

To form a polymer network, at least one of the starting components must have functionality, f, (equal to the number of functional groups per molecule) larger than two ($f > 2$). This is a necessary but often not a sufficient condition. The precursors of networks differ in two ways: (1) they are of low or of high functionality, (2) they bear functional groups that are engaged in bond formation either by stepwise or chain mechanisms .

The functionality of precursors varying between $f = 2$ and $f = 6$ is considered to be low (Figure 5.2). Polyurethane networks prepared from bifunctional telechelics and trifunctional triisocyanates, diepoxide ($f = 2$)-diamine ($f = 4$) systems, diepoxide ($f = 4$)-cyclic anhydride ($f = 2$) systems, phenol ($f = 3$)-formaldehyde ($f = 4$) resins, or melamine ($f = 6$)-formaldehyde ($f = 2$) resins are in this category.

High functionality precursors of $f > 10$–20 are primary chains if they participate in crosslinking by vulcanization: each monomer unit of primary chain is a potential site for crosslinking. Also, dendrimers and random hyperbranched polymers of higher molecular weight, rank among high-functionality precursors.

In stepwise reactions, all functional groups take part in bond formation. Their reactivity can be considered independent of the size and shape of the molecules or substructures they are bound to (Flory principle). If such a dependence exists, it is mainly due to steric hindrance. In chain reactions only activated sites participate in bond formation; if propagation is fast relative to initiation, transfer and termination, long multifunctional chains are already formed at the beginning of the reaction and they remain dissolved in the monomer. Free-radical copolymerization of mono- and polyunsaturated monomers can serve as an example. The primary chains can carry a number of pendant C=C double bonds

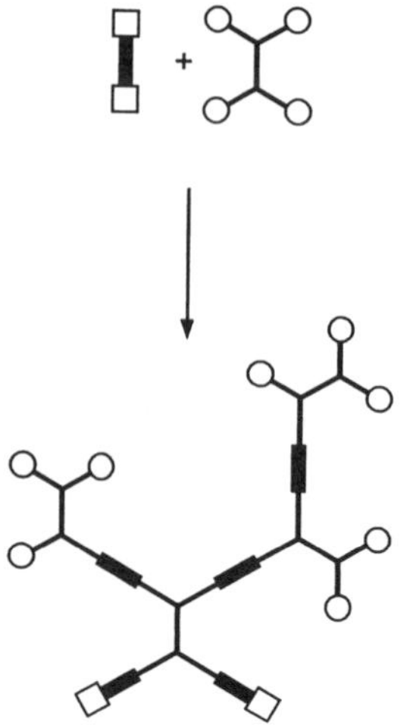

Figure 5.2 Formation of branched molecules from tetrafunctional and bifunctional monomers

which are potential sites for inter- and intramolecular crosslinking. It is a special feature of the free-radical crosslinking copolymerization that the intramolecular reaction is strong especially at the beginning of polymerization. Initially, compact and internally crosslinked molecules – micronetworks (microgels) are formed. Experimentally observed macrogelation is a result of intermolecular chemical linking of these compact molecules. The peripheral pendant double bonds of the micronetworks and still unreacted monomers are involved in this reaction [6, 7].

Despite the differences in starting components and the reaction mechanism, network formation has certain common features characteristic of the structure development:

- Increase in molecular weights.
- Increase in polydispersity.
- Existence of a gel point characterized by divergence of the weight-average and higher-average molecular weight.
- Transformation of finite molecules (sol) into the network structure (gel).
- Decrease in molecular weights and polydispersity of molecules in the sol.
- Build-up of the gel structure.
  – formation of closed circuits – increase in the cycle rank from a zero value at the gel point
  – increase in the degree of crosslinking (concentration of elastically active network chains (EANC))
  – decrease in the fraction of units in *dangling chains*; the dangling chains are

composed of units only single-connected to network structure

Any system in which network structure is developing can be characterized by the states of its building units. A building unit is usually represented by a monomer unit but it can be smaller or larger. The building units in a network exist in various reaction states given by the types and numbers of bonds by which the unit is bonded to neighboring units. An example of all possible reaction states of an BA_2 monomer in the corresponding hyperbranched polymer is shown in Figure 5.3. Beyond the gel point, each bond can represent a connection either to finite substructure or to the gel (infinite) structure (looking out of the given unit). This classification is very helpful for characterization of the gel structure [8]. To determine whether a bond issuing from a unit has a finite continuation, one has to look in the direction *out of the unit* through the bond. The bond has a finite continuation if the substructure in the direction out of the unit is finite; it has an infinite continuation if the unit is connected through the bond to the gel structure. For a trifunctional monomer, the various structural elements are assigned to the specific states of the units in Table 5.1.

The structure elements in a system undergoing crosslinking are illustrated by Figure 5.4.

As the reaction proceeds beyond the gel point, the molecular weight of EANCs decreases and the fraction of material in the EANCs increases. The fraction of material in dangling chains passes through a maximum but their molecular weight decreases. Figure 5.5 characterizes the behavior of simple polyurethane systems.

Table 5.1 Reaction states of units of a trifunctional monomer characterized by the number of reacted groups and their assignment to various structure elements in a system undergoing crosslinking

Unreacted groups	Groups engaged in bonds	With infinite continuation	With finite continuation	Unit is a part of
3	0	0	0	Sol
2	1	0	1	Sol
2	1	1	0	Gel, dangling chain
1	2	0	2	Sol
1	2	1	1	Gel, dangling chain
1	2	2	0	Gel, ENAC
0	3	0	3	Sol
0	3	1	2	Gel, dangling chain
0	3	2	1	Gel, ENAC
0	3	3	0	Gel, elastically active branch point

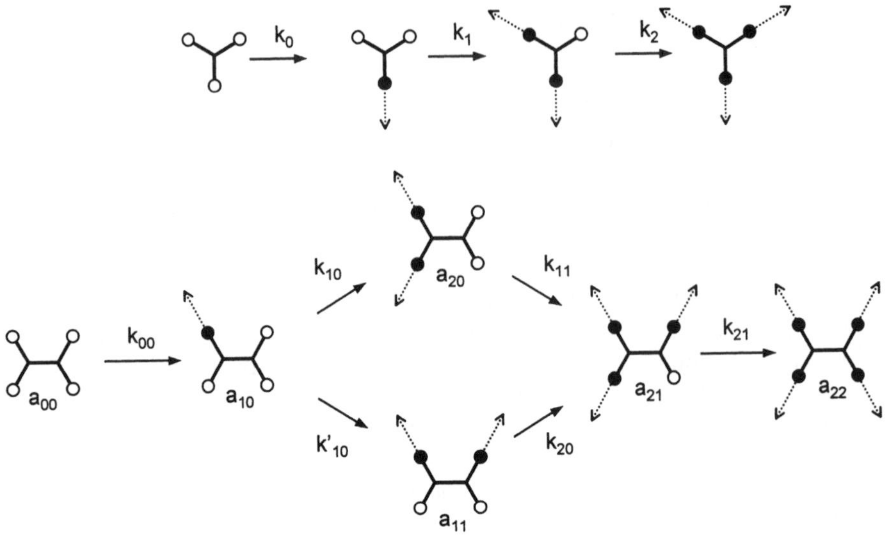

Figure 5.3 Trifunctional and tetrafunctional monomer units in different reaction states and their transformation

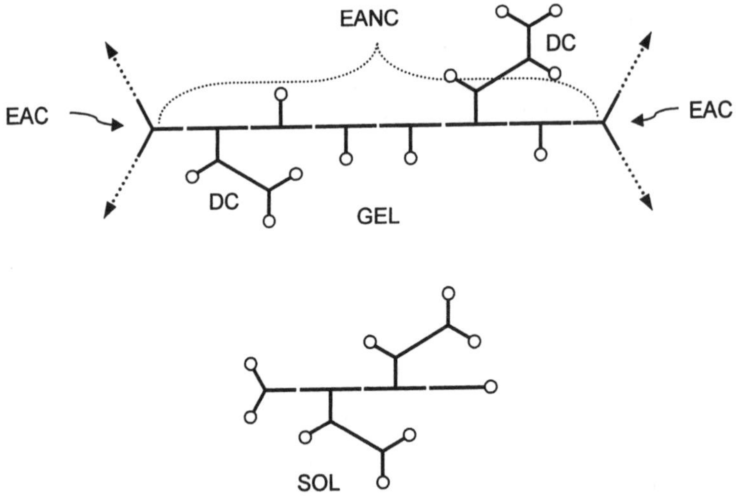

Figure 5.4 Schematic representation of sol and a part of the gel: DC dangling chains, EANC elastically active network chains, EAC elastically active crosslinks

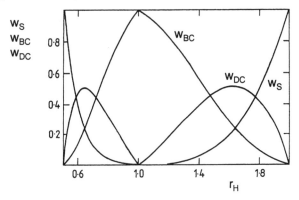

Figure 5.5 Calculated dependence of weight fractions of various substructures in the crosslinking system of $H_3 + I_2$ type on the initial molar ratio of H-groups to I-groups, r_H (a polyurethane system): DC – dangling chains, BC – backbone chains, S – sol (backbone chains are elastically active network chains without dangling chains)

2.2 PRECURSORS OF VARIOUS ARCHITECTURES

In earlier times, polymer networks were generally prepared by reactions between small monomers and by vulcanization of primary chains. Successively, the notion of a *precursor* developed for a preformed polymer molecule carrying functional groups, or a distribution of them. Synthesis of precursors and their use in network build-up have been motivated by the following needs:

1. Adjustment of viscosity and viscosity build-up before gelation.
2. Control of the critical gel conversion and time.
3. Lowering of shrinkage by preforming some bonds in the liquid state.
4. Incorporation into the network structure of specific groupings that affect network properties.
5. Incorporation into the network of specific substructures determining the network functions.

   The most frequent precursors are shown in Figure 5.6 and are listed below:

- *Telechelic polymers* usually bear monofunctional groups at each of their extremities. However, sometimes each end-group is bifunctional, such as in α, ω-bis-unsaturated telechelics, or trifunctional as in α, ω-bis(trialkoxysilyl) telechelics; wherein they participate in crosslinking by the sol–gel reactions (hydrolysis and condensation of alkoxysilane groups).
- *Macromonomers* carrying mono-, bi- or trifunctional end groups are sometimes important constituents of a network. Through them, terminating, chain-extending, or crosslinking units are introduced to modify the network

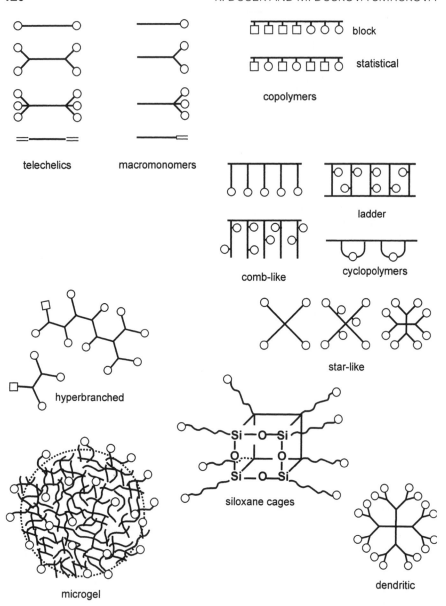

Figure 5.6 Precursor of polymer networks of various architecture

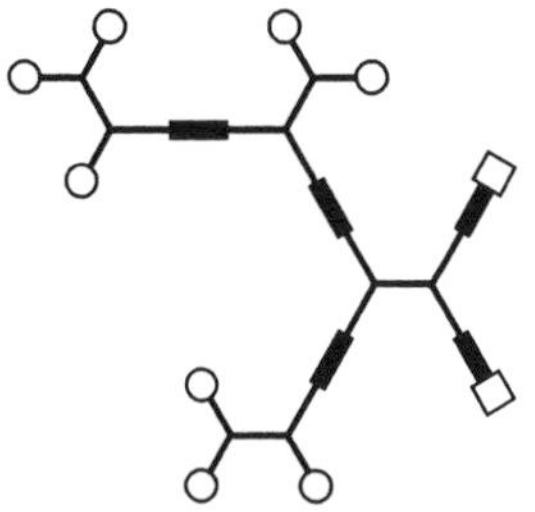

Figure 5.6   (*cont.*)   randomly branched

formation and properties. For instance, chains terminated with a trialkoxysilyl group at one extremity can gel and a network may be formed.

- *Functional stars* are small molecules compared with dendrimers. Some stars have simple and some dendritic structures. Their functional groups are placed at arms extremities or distributed along the arms.
- *Functional copolymers*: are quite frequently used as precursors. One of the monomers carries a functional group active in crosslinking. The glass transition temperature, T_g and some other properties are adjusted by the other comonomer. Modern polymerization methods make it possible to control the sequence distribution; the monomer units carrying functional groups may be arranged in one or several blocks, distributed statistically or in an alternating fashion. The sequential arrangement of monomer units carrying functional groups has an effect on crosslinking kinetics (neighbor group effect).
- *Functional cyclopolymers*: some bis-unsaturated monomers carrying an active group cyclopolymerize yielding polymers containing cycles. These cycles contain a functional grouping active in crosslinking reaction [10]. Polymers of acrylic anhydride can serve as an example: the anhydride groups can react with hydroxy or amine groups. However, copolymers of maleic anhydride are more important precursors carrying a cyclic anhydride function.
- *Functional comb polymers* can be prepared by several techniques, e.g. by polymerization or copolymerization of functional macromonomers. The functional groups can be placed on side-chain extremities or distributed along the side chains. Also the backbone chain can carry functional groups. Functional combs are used as modifiers of surface activity.
- *Functional ladder and cage precursors* are formed by hydrolysis and condensation of trifunctional silanes, like trichloro-or trialkoxysilanes carrying one functional group, $(RO)_3$ Si–X–A, where R is usually an alkyl group and X is a bridge connecting Si with the functional group A by a Si–C bond. They are also called polyhedral oligomeric silsesquioxanes (POSS). A number of cage

structures carrying functional groups have been prepared recently either as simple cages or mixtures with their oligomers [11, 12].

• *Reactive microgels and micronetworks*: microgels are prepared in several ways: by suspension, emulsion, or microemulsion polymerization. Their sizes vary between tens of micrometers and about 10 nanometers. Surfactant or surfactant-free polymerizations of oil-in-water or water-in-oil types are employed. Free-radical crosslinking polymerization in solution can also be used for microgel preparation. Microgel particles contain functional groups not only on their surface but also in their interior. The latter are generally less reactive because of limited accessibility, which may cause some problems during and after crosslinking. Particularly remarkable are the functional nano- to micrometer particles obtained through a sol–gel precipitation technique where the particle surface was stabilized against agglomeration, so that the particles are redispersible from its powder form [13]. Hybrid microgels, with structures that vary between those of a microgel and a multiarm star, have been obtained by selective chemical or radiation-induced crosslinking of micelle assemblies or microseparated solids of diblock copolymers [14]. These core–shell structures range from 1 μm down to tens of nanometers and exhibit interesting ordering and phase separation phenomena in solution. No studies of system with functionalized arms are available until recent reports on core–shell tecto-dendrimers (see Chapter 1).

• *Dendrimers with terminal functional groups* represent mode compact precursors that are spherical and almost monodisperse, with reactive groups placed on their periphery. Their synthesis, structure and properties have been reviewed in monographs and review articles often together with hyperbranched polymers (cf., e.g. [15–20]), as well as in this book. Application of dendrimers as precursors for conventional materials is limited at this time by their relatively high cost.

• *Random hyperbranched polymers* prepared from monomers of type BA_f ($f \geq 2$), $A + B \rightarrow A - B$, are much less expensive than dendrimers. However, they exhibit molecular weight and shape (symmetry) distributions (distribution of topological isomers of the same molecular weight differing in shape which ranges from dendritic to linear structures). Also cyclization may take place [15, 21]. Because of the distribution in molecular weights and shapes, the segment density distribution of hyperbranched polymers in solution as revealed by scattering behavior [22] is closer to that of randomly branched systems rather than dendrimers. Modification of the end-groups of hyperbranched polymers by reactions of soft or hard molecules allows one to obtain core–shell structures with hard core and soft shell and vice versa. Multi-shell organic or organometallic structures may also be prepared [23].

• *Randomly branched precursors*: any crosslinking system before the gel point can potentially be used as a polymer network precursor provided that
  – the reactive groups are stable under storage conditions

– for the A + B type reaction, only one type of groups is present in the precursor

If the crosslinking reaction is interrupted before the gel point, the molecular weight and functionality distributions of such functional precursors are wider but not basically different from that of polymers of BA_f monomer. It was stressed recently that they resemble hyperbranched polymers [24] in a certain respects. The pre-gel polymers are generally not stable because the crosslinking reaction can occur during storage. Stable precursors, e.g. for $RA_f + R'B_g$, can be obtained in two ways:

- by stopping the reaction before the gel point either by lowering the temperature, or by blocking one type of reactive group (e.g. isocyanate groups in the hydroxy-isocyanate system),
- by using excess of one-type groups, such that gelation does not occur. Reaction products of diepoxides with excess of diamines carrying amine functional groups may serve as an example. The critical molar ratio of amine:epoxide groups below which the system does not gel even at full conversion of the minority (epoxide) groups was predicted theoretically and verified experimentally [25–29]. The critical molar ratio in polyurethane systems has also been determined [30].

These subgel prepolymers have been manufactured by industry for some time and used in two-component adhesives or coating materials. For one-component materials, storage below the actual T_g, whenever applicable, is the most efficient method for blocking the reactivities. This kind of blocking is used in powder coatings.

2.3 EFFECT OF PRECURSOR STRUCTURE ON NETWORK BUILD-UP

Many precursors are prepared as individual chemical compounds with fixed molecular weight and functionality. Many others exhibit a distribution of molecular weights and, if branched, distribution of topological isomers of the same molecular weight (i.e. so-called shape distribution). Also, distribution in the number of functional groups per molecule (functionality distribution) is an important characteristic. For crosslinking, information concerning these distributions is very important, since the sensitivity of various properties of the crosslinked systems relative to particular distribution is different. For instance, the gel point conversion is determined by the second moment of the functionality distribution, the weight average molecular weight of the branched polymer or polymer in the sol is determined both by weight average molecular weight of the precursor and its second moment functionality distribution. There is no straightforward dependence of the average number of EANC's on the distribution, since the first as well the second moment of the functionality distribution plays a role. Also the sensitivity of various structural parameters to uncertainties in the distributions are different; for instance, the concentration of EANC's is strongly influenced by small changes in the functionality distribution with low functionality systems here gelation occurs at a relatively high conversion.

Thus, the functionality averages and functionality distribution in precursors play the most important role in network build-up. The functionality distributions should be characterized and controlled as much as possible. There is no general way of controlling the distributions, it largely depends on the particular chemistry of the precursor preparation involved. Thus, α, ω-bifunctional telechelics exhibit only a negligible deviation from $f = 2$, whereas for the trifunctional systems the deviation from $f = 3$ becomes more important. For functional siloxane cages with a few exceptions, the functionality distribution determined by formation of higher oligomers is always a problem and is difficult to control. In hyperbranched polymers, the functionality distribution is the determining factor in network formation. Several strategies exist to narrow the distribution, such as addition of a core, or controlled addition of the monomer (see section 3.2).

In a wider sense, functionality distribution also means combination of reactive groups of one kind (for instance, hydroxy, or carboxyl, or isocyanate groups) of higher and lower reactivity in one precursor molecule. By this combination, the network build-up can be effectively controlled. For instance, one less reactive A group out of three in a trifunctional monomer in a $RA_3 + R'B_2$ system promotes chain extension in contrast to branching and shifts the gel point to higher values [30].

In general, wider polydispersity in functionality or group reactivity makes the gel time shorter and critical conversion lower. This conclusion is based on the results of theoretical and experimental studies concerning the effect of distributions in functionality and in group reactivity on crosslinking of functional stars [31–33].

2.4 FORMATION OF SUBSTRUCTURES IN SITU

Covalently bonded substructures having compositions distinguishable from their surroundings are formed in multicomponent systems; they are called *chemical clusters*. The adjective *chemical* defines covalency of bonds between units in the cluster. To be a part of a cluster, the units must have a common property. For example, *hard clusters* are composed of units yielding T_g domains. Hard chemical clusters are formed in three-component polyurethane systems composed of a macromolecular diol (soft component), a low-molecular-weight triol (hard component) and diisocyanate (hard component). Hard clusters consist of two hard

components – a triol and diisocyanate. A mixture of long chains (soft component) and short chains (hard component) crosslinked with a trifunctional crosslinker (hard component) is another example [34–36] (Figure 5.7).

The degree of polymerization of hard clusters increases with evolution of the
system as a whole. The hard clusters already exist in pregel molecules. Before the
macroscopic gel point of the system is reached they remain usually small. Later
on, the hard clusters grow faster and eventually a gel point (percolation threshold) of the hard structure is reached. Below this point, clusters are embedded in
the soft matrix; beyond the percolation threshold, the hard and soft structures
interpenetrate (Figure 5.7). *Below the percolation threshold, hard clusters are
essentially dendritic; when the percolation threshold is surpassed, circuits (cycles)
develop within the hard structure.*

The degree of polymerization, polydispersity and percolation threshold of the
hard clusters can be controlled by

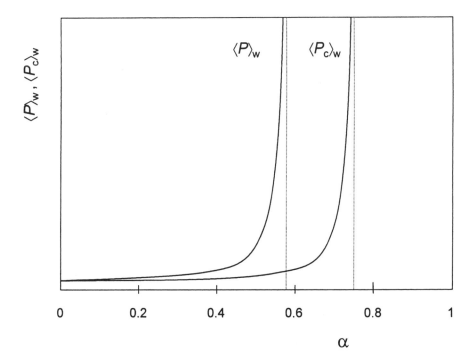

Figure 5.7 Calculated dependence of the weight-average degree of polymeriz-
ation of molecules, $\langle P \rangle_w$, and hard clusters, $\langle P_c \rangle_w$, on conversion in a stoichiometric $A_4(h) + B_2(h) + B_2(s)$ system (h – hard, s – soft). The system corresponds to a
mixture of short and hard chains crosslinked with a tetrafunctional crosslinking
agent

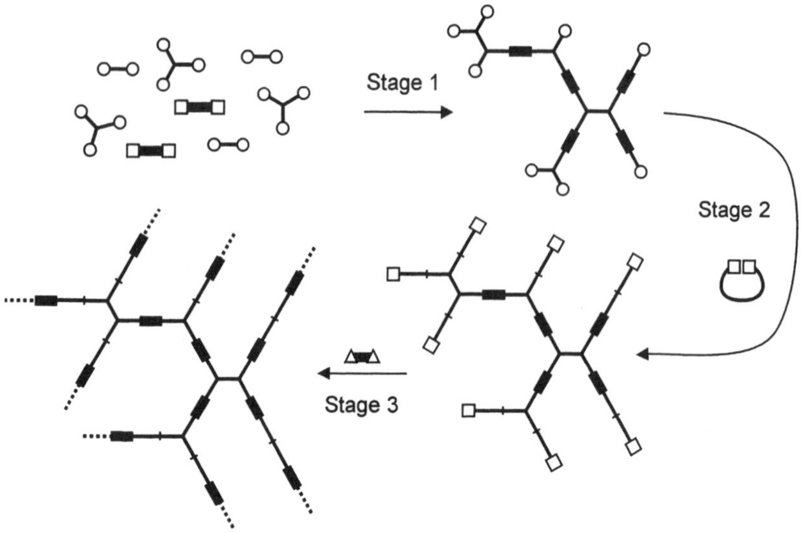

Figure 5.8 Scheme of a multistage process: first stage: a distribution of branched polymer is formed with end groups ○; second stage: the end groups are modified by addition of the ring compound – ○ → □; third stage: the distribution with □ endgroups is crosslinked with a bifuctional compound bearing △ groups

- Fractions and functionalities of hard components in initial system.
- Stoichiometry or off-stoichiometry.
- Relative reactivities of similar groups in hard and soft units (e.g. in polyurethanes: differences in reactivities of OH groups of macromolecular diol and triol).
- Changes in reactivities of functional groups of hard components (substitution effect).

In a three-component polyurethane system with OH groups, wherein crosslinker is a triol, one finds them substantially more reactive than those of the macromolecular diol. The hard clusters grow substantially larger, at relatively low conversions. Their size remains almost constant because all triol units have been used up in the reaction. In the opposite case, the lower reactivity of OH groups found in macromolecular diol compared with that of triol, the clusters remains small throughout the reaction and grows larger only at its end.

Some other ways of controlling the formation of chemical clusters can be developed and the existing clusters can be chemically modified *in situ*. The concept of hard clusters makes it possible to explain dependences of various properties, such as T_g or ultimate mechanical properties on composition and the extent of reaction [36, 37].

2.5 MODELING OF NETWORK FORMATION

Throughout this chapter, predictions are made concerning the effect of external variables on the network structure evolution and these predictions are compared with experimental results. The predictions are based on branching theories. The purpose of this section is to outline the basis of present network formation theories and the underlying assumptions.

The theoretical methods describing the network build-up can be grouped into three categories:

1. Statistical methods based on generation of branched and crosslinked structures from units in different reaction states.
2. Kinetic methods describing the evolution of distributions of molecules by systems of kinetic differential equations (obeying either the classic mass action law of chemical kinetics or the generalized Smoluchowski coagulation process).
3. Simulation in finite three-dimensional space using various assumptions and techniques.

Computer simulation in space (method 3) can in principle take into account most interactions (i.e. chemical reactivities, physical and chemical interactions in space, mobilities of structures and substructures) but, at present, quantitative knowledge of these interactions and tools to implement them into efficient algorithms remains limited. Also certain limits are imposed on the system size by the available operation time. In particular, the properties of the critical region are quite sensitive to the system size. At present, the major problem is the incorporation of proper dynamics of the structures.

The first two approaches work essentially with 'infinite' systems. The statistical theories are fully mean-field, the classic kinetic theory as well. However, in the kinetic simulations, many non-mean field effects can be taken into account making the rate constant for a pair of groups dependent on the size or shape or other structural characteristics of the reacting molecules. In the critical region, this dependence also results in a non-mean-field scaling of properties of the branching system against the distance from the critical point. Moreover, the kinetic approach correctly takes into account correlations originating from the network formation history, such the reaction sequences; initiation–propagation–transfer–termination, or staging in multistage preparation of a network.

The simplest are statistical theories, where the input information is reduced to the distribution of *units in different reaction states*. The *reaction state* of a unit is defined by the number and type of bonds issuing from the unit. In a reacting system, the distribution fraction of units in different reaction states is a function of the reaction time (conversion) (cf. e.g. [7, 8, 29, 30] and can be obtained either experimentally (e.g. by NMR) or calculated by solution of a few simple kinetic differential equations. An example of reaction state distribution of an AB_2 unit is shown in Figure 5.3. To proceed to larger structures, distribution of the branched molecules and the gel, as well as the reacted functional groups (representing in fact half bonds) are assembled randomly into bonds. The rules of combination of various reacted groups directed by chemistry must be respected.

Despite its simplicity, the statistical method has been quite successful in predicting the effect of various chemical variables on network formation (cf. e.g. [29, 30, 34–37]). Since the internal structure of the gel can be characterized to a certain degree by the statistical method (e.g. average size of dangling chains and weight fraction of material in them), these methods offer a basis for correlations between structure and viscoelastic properties.

Within the group of *kinetic methods* it is the distribution of molecules distinguished by the numbers and types of monomer units and unreacted functional groups, which develops in time, so that the effect of long-range correlations is respected. In principle, the molecules can also be distinguished by their shapes; namely, topological isomers. The time evolution of the distribution of concentration gradients of distinguishable molecules is described by an infinite set of kinetic differential equations [38–53]. These chemical kinetic equations are governed by mass action law, where the reaction rate is proportional to the product of concentrations of the two reacting molecules, and the rate constant is directly proportional to the product of the number of reacting groups. Possible differences in reactivities of functional groups in a monomer unit can be taken into account.

Alternatively, one can make the reactivity of groups dependent on the size and shape of the reacting molecule. In such a way, for instance, the effect of steric hindrances, cyclization, and diffusivities of the molecules can be modeled using generalized Smoluchowski coagulation differential equations.

In the case of classic chemical kinetics equations, one can get in a few cases analytical solution for the set of differential equations in the form of explicit expressions for the number or weight fractions of *i*-mers (cf. also treatment of distribution of an ideal hyperbranched polymer). Alternatively, the distribution is stored in the form of generating functions from which the moments of the distribution can be extracted. In the latter case, when the rate constant is not directly proportional to number of unreacted functional groups, or the mass action law are not obeyed, Monte-Carlo simulation techniques can be used (cf. e.g. [2, 3, 47–52]). This technique was also used for simulation of distribution of hyperbranched polymers [21, 51, 52].

Combined kinetic and statistical theories are particularly useful when the long-range, reaction-mechanism-determined effects are effective in building up primary linear and branched structures, but crosslinking occurs randomly. The primary structures are generated kinetically and the formed structures are linked together (crosslinked) using the statistical theory [44]. This is the case, for instance, in free-radical polymerization or copolymerization of bis-unsaturated monomers when the two double bonds are separated by a long bridge and cyclization is weak [17]. Polyetherification accompanying curing of diepoxide–diamine systems can serve as another example [46].

The branching theories can be applied to the characterization of molecular weight and functionality distributions of various precursors, i.e. to systems in a state below the gel point. Hyperbranched polymers may serve as an example. The branching theories make possible to keep track of reacted as well as unreacted functional groups and are thus well adapted for treatment of a multistage process [54, 55] (cf. scheme in Figure 5.8). In a three-stage process, a branched polymer bearing certain functional groups is formed; these groups are transformed in the second stage in other functional groups, and this precursor of the second stage is then crosslinked by reaction with a crosslinker into the final network. Theoretically, the distributions obtained in the first stage are used as input information for the second stage reaction, and the modified distribution as input information for the crosslinking stage 3.

The difference between the statistical and kinetic methods is visualized in Figure 5.9.

Lattice percolation models were the first spatial simulation models applied to the network build-up. Classic lattice or off-lattice percolation modeling is based on random introduction of bonds between components placed randomly on the lattice or in space [56–58]. They suffer from the rigidity of the system and disregard of conformational changes accompanying the structure growth. These assumptions implicitly mean that the bond formation is much faster than conformation changes. Such assumption is somewhat closer to reality for fast bond-

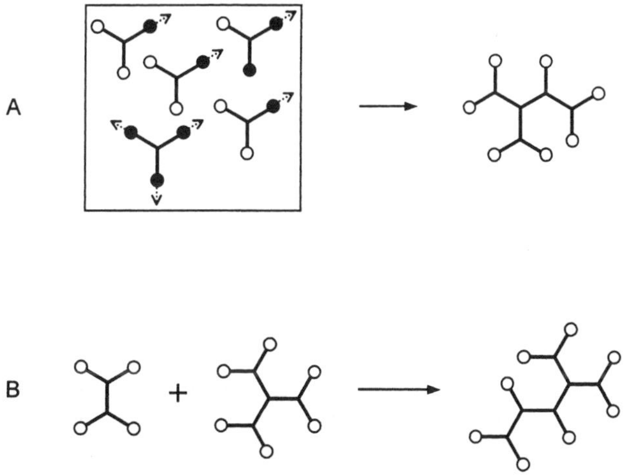

Figure 5.9 Difference in concepts of statistical (A) and kinetic (B) network formation theories

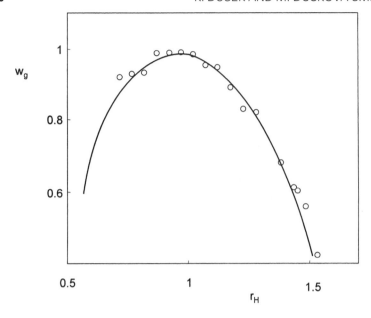

Figure 5.10 Dependence of the gel fraction, w_g, on molar ratio of [OH]/[NCO] groups, r_H, for poly(oxypropylene)triol (Niax LG 56)4,4'-diisocyanatodiphenyl-methane system. The dependence has been reconstructed from data of ref. [78]

forming reactions, such as free-radical chain copolymerizations [57]. Recently a simulation method has been reported [59] in which mobility and structural relaxation can be taken into account.

3 SPECIFIC EXAMPLES

3.1 TELECHELIC POLYMERS: CONTROL OF PROPERTIES THROUGH DANGLING CHAINS

Telechelic polymers rank among the oldest designed precursors. The position of reactive groups at the ends of a sequence of repeating units makes it possible to incorporate various chemical structures into the network (polyether, polyester, polyamide, aliphatic, cycloaliphatic or aromatic hydrocarbon, etc.). The cross-linking density can be controlled by the length of precursor chain and functionality of the crosslinker, by molar ratio of functional groups, or by addition of a monofunctional component. Formation of elastically inactive loops is usually weak. Typical polyurethane systems composed of a macromolecular triol and a diisocyanate are statistically simple and when different theories listed above are applied the results are very similar and in some cases identical, except for their scaling properties in the critical region near the gel point.

Bond formation leading to crosslinking is based on the –NCO + HO– → –NHCOO– reaction which belongs to the category of the A + B → AB type. Therefore, a maximum of crosslinking density and minimum of sol fraction is predicted for the stoichiometric system for which the ratio of concentrations of hydroxy to isocyanate groups, r_H, is equal to unity:

$$r_H = \frac{[OH]_0}{[NCO]_0}$$

Here, $[OH]_0$ and $[NCO]_0$ are initial concentrations of OH and NCO groups, respectively. Figures 5.10 and 5.11 show the calculated dependences of the gel fraction, w_g, and concentration of EANCs, v_e on r_H [30, 60–62] and their comparison with experimental results.

One can observe positive deviations in the region of $r_H < 1$ (excess of isocyanate groups) which are due to side reactions (allophanate, urea and biuret groups). In the region of $r_H > 1$ the agreement of w_g values is good. In the case of v_e, the predicted curves depend not only on the results of the branching theory but also

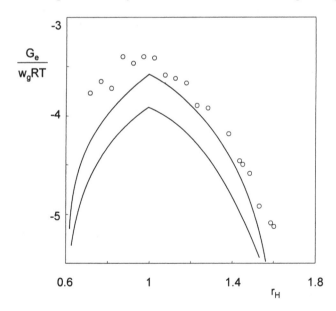

Figure 5.11 Dependence of the reduced equilibrium shear modulus, $G_e/w_g/RT$ on the molar ratio of [OH]/[NCO] groups, r_H, for poly(oxypropylene)triol (Niax LG 56)-4,4'-diisocyanatodiphenylmethane system; (----) limits of the Flory–Erman junction fluctuation rubber elasticity theory. The dependence has been reconstructed from data of ref. [78]

on the chosen rubber elasticity theory. In Figure 5.11 the dependences were calculated for the Flory–Erman junction fluctuation theory; the limits for the phantom and affine states of the network are shown. Details of interpretation can be found in refs [30, 60, 61]. Here, it can be stated that the predicted trends are obeyed. Again, in the region where isocyanate groups are in excess, the experimental values of v_e are higher than the predicted ones. This is caused by additional crosslinking due to allophanate and biuret formation.

With increasing off-stoichiometry, the fraction of the material in dangling chains increases which has an effect on viscoelastic properties. It has been found that the presence of dangling chains affects the motions of substructures in the network; it is manifested by changes in viscoelastic properties in the main transition region [60–63]. The length of dangling chains and the fraction of materials in them can be controlled by varying the ratio higher-molecular-weight triol/ diol/monofunctional alcohol. An example of the dangling chains is shown in Figure 5.12.

Increasing length and number of dangling chains make the glass transition region wider and the widening can be correlated with the fraction of material in the dangling chains and the size of dangling chains.

In ternary systems, amorphous hard clusters can be formed. As explained above, at certain fraction of hard units and a certain conversion of functional groups, percolation threshold of the hard structure is reached. It has been found experimentally by analyzing the ultimate behavior of three- and four-component

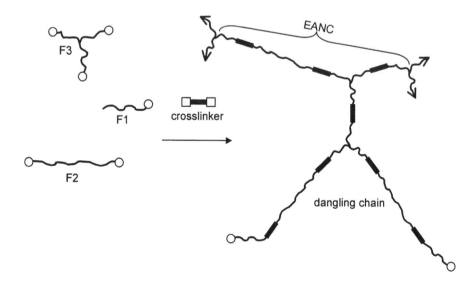

Figure 5.12 Structure of a polyurethane network with dangling chains prepared from F1 + F2 + F3 components

polyurethane systems [37] that the tensile strength, σ_b, is markedly higher and the elongation at break, ε_b, substantially smaller for the systems where the percolation threshold has been exceeded. Near the percolation threshold, σ_b and ε_b change steeply, but the true strength given by the product $\sigma_b\varepsilon_b$ decreases. Therefore, for obtaining a crosslinked material with good toughness, it is preferable to keep the system slightly below the percolation threshold of the hard clusters.

3.2 DENDRIMERS, HYPERBRANCHED POLYMERS AND DERIVED NETWORKS

Dendrimers represent a model for compact multifunctional precursor of polymer networks. Polymer networks prepared by crosslinking of dendrimers were suggested several years ago [64]. Since then, some experimental work has been performed, but there are still many points in structural interpretation of network formation and network properties that are not well understood.

There are two aspects of linking the dendritic or hyperbranched polymers into a network:

1. Assemblage of the dendritic molecules by chemical crosslinking into an array, which is more or less organized. The specific functions of dendrimers can be possibly amplified when they are assembled into superstructures [65].
2. Use of dendritic molecules (mainly hyperbranched polymers) as precursors of a crosslinked (engineering) material in order to modify its processing and materials properties.

In the first case, the details of network build-up and modification of network structure described above are not very important. The main aim of crosslinking is to keep the dendritic structures together permanently. Formation of three-dimensional nanostructures by metal-mediated self-assemblage can serve as an example: Exo-tridentate tripyridyl compounds self-assemble upon treatment with (en)Pd(NO$_3$)$_2$ [66].

The most extensive work in this direction concerns crosslinked dendrimers obtained by modification of functional groups of a dendrimer using silane chemistry. PAMAM dendrimers were functionalized with organosilicon acrylate and alkyl(halo)silanes, so that their surface activity could be significantly changed. When the surface was (partly) modified by alkoxysilyl groups, the exposure to moisture of the modified dendrimers resulted in crosslinking and low T_g elastomeric networks were obtained [23, 67, 68]. The glass transition temperature can be changed by variation of the thickness and composition of the siloxane interlayer in the cured sample. Thus addition of tetraalkoxysilane as comonomer in the sol–gel reaction increases the T_g, while addition of a dialkoxy(dimethyl)silane makes the polymer more flexible. The type of material from which networks can be prepared was called PAMAMOS (poly(amidoamine)-organosilicon) dendrimers.

Similar organic–inorganic systems, which were ultimately crosslinked by sol–gel chemistry, were prepared with cores composed of high- T_g dendrimers [69, 70]. Tough materials with high heat resistance were obtained. Also, core-shell structures were prepared via silylation, or hyrosilylation. The resulting structures were further crosslinked to give supramolecular assemblies [71].

Dendrimers with a carbon–silicon core were prepared by a divergent hydrosilylation – vinylation method followed by methoxylation [72]. Hydrolysis and condensation resulted in gels only in the case of generation two dendrimers, while the sol–gel reaction products of generation one were still soluble, apparently due to strong cyclization characteristics of the sol–gel reactions.

Crosslinking of amine- or hydroxy-terminated PAMAM dendrimers using cyclic anhydride – amine or cyclic anhydride – hydroxy addition reactions was employed for preparation of crosslinked thin films of very low permeability [73]. Polyanhydrides, such as maleic anhydride-methyl vinyl ether copolymers, were used as crosslinking components. In the case of amine-terminated PAMAM, crosslinking and chemical stability were further increased by imidization of the maleamic acid groups; retro-Michael eliminations were followed by Michael additions to further crosslink the film.

The classic application of dendrimers terminated with functional groups as precursors of polymer networks to obtain products such as coatings, composites, or RIM materials, with advanced technological and materials properties are relatively scarce. The higher cost of dendrimers is undoubtedly one of the main reasons.

Zukas et al. (74) used the amine-terminated fourth generation dendrimer as the polyamine component in a diepoxide–diamine formulations. It was found that a significant fraction of terminal amine groups did not take part in the reaction with epoxide groups. Also, polyetherification involving epoxide groups runs parallel with epoxide-amine addition. This study exposes the existing and expected problems when using polyfunctional spherical molecules as polymer network precursors. One can benefit from the good rheological behavior of nearly spherical precursor and their uniformity. Their spatial ordering in liquid media is an interesting phenomenon that can positively influence network properties. The relatively high functionality of these precursors (of the order of 10^1–10^2) can, however, cause problems manifested by early gelation and incomplete reaction of functional groups even after long reaction times. The unreacted functional groups are sites of potential chemical reactions with substances occurring in the environment and, on the other hand, interaction with radiation can make the service life shorter. Partial chemical transformation of these groups into inactive moieties in the crosslinking reaction is one of the ways out by which the functionality is decreased. At the same time, other properties of the precursor can be modified, such as T_g, polarity, or surface activity. Chemical transformation of reactive groups by chain extension results in formation of harder or softer shells or in a change of the nature of end-groups (cf. reaction of OH group with cyclic anhydride or lactone, respectively). The same modification is used with hyperbranched polymers (see below).

Hyperbranched polymers have been considered somewhat less spherical but less expensive materials that may perform some of the functions normally associated with dendrimers. However, their polydispersity was often under-estimated and methods for their control overestimated. Sometimes, hyperbran-ched polymers are designated with a 'generation number' suggesting that they resemble dendrimers. This is an illusion because of their polydispersity in mol-ecular weights and isomeric shapes. In this section, some points related to functionality and molecular weight distributions, as well as their implications for crosslinking are discussed. The use of hyperbranched polymers as precursors of polymer networks has been described in several papers, but mainly their effect on final properties was determined without deeper structural understanding. Sev-eral reviews devoted to hyperbranched polymers and their functions were pub-lished recently (e.g. [15–19]). Among non-aromatic hyperbranched polymers, poly[2,2-bis(hydroxymethyl)propanoic acid] is offered as a large-quantity inex-pensive commercial material. Recently, another competing material – a poly(esteramide) resin with terminal hydroxy groups – has become available [75].

For applications of hyperbranched polymers as precursors the polymer net-works, the following structural features are important:

1. Chemical structure: groups and bonds introduced into the networks structure which determine the network physical properties, such as thermal, mechan-ical, or aging properties.
2. Average values of molecular weight and functionality of the hyperbranched polymer.
3. Molecular weight and functionality distributions.

The contribution of hyperbranched architecture to the chemical structure can be readily changed by modification of the end-groups or by changing the structure of the core. By increasing the size of the hyperbranched precursor, the domain structure of the network is enhanced; however, reaction-induced phase separation is undoubtedly involved. The phenomenon of phase separation is utilized in formation of microseparated phases which are known to be effective in toughening of thermosetting materials like epoxy resins [76, 77]. High function-ality of the precursor causes early gelation – lower gel-point conversions and shorter gelation times.

In order to prolong the pot life of the system, a reduction in polydispersity is quite important. Shape polydispersity of the polymer, which is determined by the distribution of topological isomers, is expressed through a quantity Fréchet and coworkers originally termed 'degree of branching' [78], which reaches unity for perfectly spherical dendritic structures. The molecules are then composed only of fully reacted units and terminal units with only one reacted group. The degree of branching or branching index, β, is a function of the unit fractions with j reacted functional groups, x_j. It has been defined in several ways. For a polymer from a BA_2 monomer, it was expressed in the form [78, 79]:

$$\beta = \frac{x_3}{x_3 + x_2} \tag{1}$$

or alternatively [51]

$$\beta^* = \frac{2x_3}{2x_3 + x_2} \tag{2}$$

The latter definition can be extended to hyperbranched polymers with a higher number of A groups in the monomer, $f > 2$. Frey and Hölter [80] used the definition

$$\beta^{**} = \frac{\sum_{j=2}^{m} (j-1)x_{J+1}}{\frac{f-1}{f}\sum_{j=2}^{f} jx_j} \tag{3}$$

The branching index depends on the following:

- Conversions of A and B groups, α_A and α_b; for acyclic molecules $\alpha_A = \alpha_B/f$.
- Difference in reactivities of A groups and on the dependence of group reactivities on the reaction state of the unit (substitution effect) [78, 79].
- Addition of a core molecule, or any other comonomers (chain extending AB, terminating A or B).
- In the case of network formation controlled by (irreversible) kinetics: programmed polymerization regime (starved feed conditions, etc.).

If rings are formed, the branching coefficient must be redefined.

In the ideal case, all reactive groups have the same reactivity irrespective of the shape and size of the hyperbranched molecules and no rings are formed. Then, the distribution of units in different reaction states is expressed by the following probability generating function (pgf), $F_{0n}(Z, z)$:

$$F_{0n}(Z, z) = [(1 - \alpha_A)Z_{Au} + \alpha_A z_B]^2[(1 - \alpha_B)Z_{Bu} + \alpha_B z_A] \tag{4}$$

This is a procedure which is common in the theory of branching processes. The auxiliary variables of the pgf Z_{Au} and Z_{Bu} identify unreacted functional groups, and z_B and z_A bonds extending from reacted group A to reacted group B and from reacted group B to reacted group A, respectively; the variables z_B and z_A also identify reacted A and B groups. This probability generating function is also useful as a starting point for the description of crosslinking of hyperbranched molecules.

Thus, for this random case one can obtain by expansion of $F_{0n}(Z, z)$ the values of x_j needed for calculation of β, β^*, or β.

Thus, the expression given in Table 5.2 can be substituted into the definition equations for β and a dependence of the branching coefficient on conversion (i.e. on molecular weight averages) can be calculated. For $\alpha_B \to 1$, $\alpha_A \to 1/2$, $\beta = 1/3$ and $\beta^* = 1/2$.

The pgf is a source for calculation of various properties of the hyperbranched precursor under the simplifying conditions defined above. The algorithms have been described elsewhere (cf. e.g. [29–30]), here only the resulting equations are given.

The number- and weight-average molecular weights are given by the following equations:

$$M_n = M_0 \frac{1}{1 - \alpha_B} \tag{5}$$

Thus, the number-average molecular weight is independent of f and diverges for $\alpha_B = 1$, irrespective of the number of A groups in the monomer, f.

$$M_w = M_0 \frac{1 - \alpha_B^2/f}{(1 - \alpha_B)^2} \tag{6}$$

where M_0 is the molecular weight of the monomer unit. The polydispersity is characterized by

$$\frac{M_w}{M_n} = \frac{1 - \alpha_B^2/f}{1 - \alpha_B} \tag{7}$$

Thus, the molecular weight averages diverge at full conversion ($\alpha_B \to 1$), and the polydispersity does as well.

For the application of hyperbranched polymers as precursors in network formation, the functionality averages are important. The number-average functionality is given by the equation

$$(\phi_{Af})_n = \sum_{\varphi,x} \phi n_{\varphi,x} = \frac{f(1 - \alpha_A)}{1 - \alpha_B} \tag{8}$$

Table 5.2 Fractions of units differing in the number of reacted functional groups, x_j, for BA$_2$ units

x_0	x_1	x_2	x_3
$(1 - \alpha_A)^2(1 - \alpha_B)$	$(1 - \alpha_A)^2\alpha_B +$ $2(1 - \alpha_A)\alpha_A(1 - \alpha_B)$	$2\alpha_A(1 - \alpha_A)\alpha_B +$ $\alpha_A^2(1 - \alpha_B)$	$\alpha_A^2\alpha_B$

Higher-order averages are also important for network formation: the number average value of functional groups per weight-average degree of polymerization is defined by

$$(\phi_{Af})_w = \sum_{\varphi,x} \phi w_{\varphi,x} = \frac{f(1 - \alpha_A)(f - \alpha_B)}{(1 - \alpha_B)^2} \tag{9}$$

and the second-moment average of the functionality distribution $(\Phi_{af})_2$

$$(\phi_{Af})_2 = \frac{\Sigma_{\varphi,x}\phi^2 n_{\varphi,x}}{\Sigma_{\varphi,x}\phi n_{\varphi,x}} = 1 + \frac{(f - 1)(1 - \alpha_A)}{(1 - \alpha_B^2)} \tag{10}$$

determine the onset of gelation.

In reality, the polydispersity of the hyperbranched polymer even in the absence of core is lower than that predicted for the ideal case. Cyclization and steric hindrance during polymerization can be the reasons. Polydispersity can also be lowered intentionally, for instance, by introduction of core molecules or by programmed addition of the monomers.

Cyclization is inherent to the formation of hyperbranched polymers because any acyclic molecule contains one B group and several A groups. Whether or not cyclization is important depends on the probability that a B group may closely approach any of the A groups in the molecule. The ring formation probability depends on the geometry and conformational properties of the bond sequences of connecting the A groups with the B group and on the number of available A groups. Once a cycle is formed, further cyclization reactions are impossible and the cyclic molecule can participate only in intermolecular reactions (cf. reaction scheme in Figure 5.12).

It has been found, that in the case of poly[2,2-bis(hydroxymethyl)propanoic acid] cyclization is quite important and, at high conversions, almost all large molecules contain a ring structure and have no unreacted B group [21]. The number of molecules with B groups is a function of the core fraction and synthesis history. It follows from simulation as well as from experiments that the addition of a core and gradual addition of the monomer lowers the cyclization. In some systems, cyclization is important [21, 81], in other it is negligible [82]. Also, more groups may be sterically hindered in large molecules than in the smaller ones, which is another reason for lowering the polydispersity.

Addition of a core compound even in a batch wise polymerization makes the distribution narrower. This problem can simply be handled by TBP. If one starts with a core and the monomer is gradually added, the distribution can be narrowed even more [51, 83–87]. This follows from kinetic simulations. However, the monomer addition method has certain limits. It is effective for low molecular weights, so that B groups are almost absent in all initial larger molecules. Unless the monomer addition is infinitesimally slow, which is impractical, condensates by reactions between the monomers are formed containing B groups. Later, they self-condense thus making the distribution wider. Bimodal distributions can develop due to parallel reactions leading to cluster–monomer and monomer–monomer connectivity. By simulation and experiments with PDMPA [21], it was found that addition of a core and gradual addition of the monomer lowered cyclization.

To adjust processing properties during crosslinking and physical properties of the resulting networks, the functional groups of hyperbranched polymers are modified either with monofunctional agents (the functionality is lowered), or with bifunctional compounds by which the functional groups of the same type (but different reactivity) are recovered or the type of functional groups is changed. In this way, hyperbranched core is provided with a shell. Lactones such as poly(ε-caprolactone) are often used to modify hyperbranched polymers with hydroxy functional groups [88–90]. In these modifications, differences in hydroxyl groups in dendritic and hyperbranched molecules have been found: some groups in less symmetric hyperbranched polymer molecules are apparently more shielded [90]. End capping groups in PDMPA molecules with aliphatic or aromatic groups, changes thermomechanical properties of the hyperbranched polyesters [91]; in dependence on the nature of the substituent, T_g increases or decreases.

Crosslinking. Manifestation of high functionality in the high-molecular weight precursors and polydispersity in functionality distribution result in early gelation. For instance, when a hyperbranched polymer with functional groups of equal and independent reactivities is crosslinked with a g-functional crosslinking agent C_2 with groups C also of equal and independent reactivities, the gel point conversions are given by the following:

$$(\alpha_A)_{\text{crit}}(\alpha_C)_{\text{crit}} = \frac{1}{((\phi_{Af})_2 - 1)(g_C - 1)} \tag{11}$$

One can see that for a polymer obtained from BA$_2$ at $\alpha_B = 0.95$, $(\Phi_{Af})_2 = 191$ while $(\Phi_{Af})_n = 19$. If this polymer is crosslinked with a bifunctional crosslinking agent C_2 under stoichiometric conditions, the gel point conversion is about 0.07. However, the gel point conversion is expected to be somewhat higher not only because of the lower polydispersity of the hyperbranched polymer, but also because some cyclization can occur or multiple crosslinks can be formed during crosslinking.

For a theoretical description of crosslinking and network structure, network formation theories can be applied. The results of simulation of the functionality and molecular weight distribution obtained by TBP, or by off-space or in-space simulations are taken as input information. Formulation of the basic pgf characteristic of TBP for crosslinking of a distribution of a hyperbranched polymer is shown as an illustration. The simplest case of a BA$_f$ monomer corresponding to equation (4) is considered:

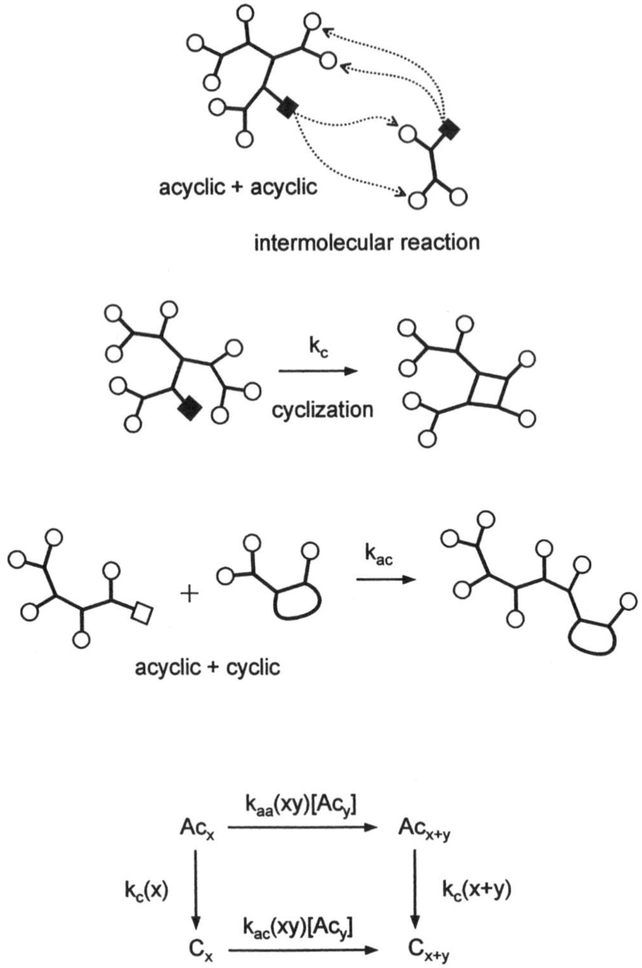

acyclic + acyclic

intermolecular reaction

k_c

cyclization

k_{ac}

acyclic + cyclic

$$\text{Ac}_x \xrightarrow{k_{aa}(xy)[\text{Ac}_y]} \text{Ac}_{x+y}$$
$$k_c(x) \downarrow \qquad\qquad\qquad \downarrow k_c(x+y)$$
$$\text{C}_x \xrightarrow{k_{ac}(xy)[\text{Ac}_y]} \text{C}_{x+y}$$

Figure 5.13   Basic reactions operative in formation of a hyperbranched polymer by irreversible reactions and a corresponding reaction scheme (cf. ref. [21])

$$F_{0n}(Z, z) = n_{BAf}[(1 - \alpha_A)Z_{Au} + \alpha_A z_{AB}]^f[(1 - \alpha_B)Z_{Bu} + \alpha_B z_{BA}]$$
$$+ n_{C2}[1 - \alpha_C + \alpha_C(\phi_A z_{CA} + \phi_B z_{CB})]^2 \tag{12}$$

where transformations are to be made:

$$Z_{Au} \rightarrow 1 - \alpha'_A + \alpha'_A z_{AC} \tag{13}$$

$$Z_{Bu} \rightarrow 1 - \alpha'_B z_{BC} \tag{14}$$

In the structure generation process, six types of bonds formed are distinguished –

AB, BA, AC, CA, BC, CB (the sequence of letters expresses the direction of the bonds looking out of the given units: $A \rightarrow B$, $B \rightarrow A$, etc.); α'_A and α'_B are conversions of the end-groups of the hyperbranched polymer and α_C is the conversion of C groups of the crosslinker. The pgf contains information on both the hyperbranched polymer and crosslinker C_2; its components are weighted by the mole fractions n_{BAf} and n_{C2}; the factors ψ_A and ψ_B express relative participation of A and B groups in the reaction with C groups, respectively.

Several applications of hyperbranched polymers as precursors for synthesis of crosslinked materials have been reported [91–97] but systematic studies of crosslinking kinetics, gelation, network formation and network properties are still missing. These studies include application of hyperbranched aliphatic polyesters as hydroxy group containing precursors in alkyd resins by which the hardness of alkyd films was improved [94]. Several studies involved the modification of hyperbranched polyesters to introduce polymerizable unsaturated C=C double bonds (maleate or acrylic groups). A crosslinked network was formed by free-radical homopolymerization or copolymerization.

The main advantage for using hyperbranched structures may rest in their rheology (film formation from high-solids systems) and in some cases added value in properties (hardness), but deeper knowledge should be accumulated before a more comprehensive evaluation can be made.

4 CONCLUSION

Advances in methods of synthetic polymer chemistry have made it possible to vary and tune the architecture of precursors of polymer networks. Many of them are available at the moment; however, a full understanding of the influence of architecture on network formation and network properties is incomplete. Relatively clear is the connection between the compactness and functionality of precursor, viscosity build-up, and onset of gelation. It is clear that precursors leading to gelation are essentially dendritic in nature and include dendrimers, hyperbranched polymers or pregel polycondensates; however, cyclized structures such as microgels are also interesting. The fact that the structural features influence the ability of a functional group to react and form a bond, makes the crosslinking kinetics and structure development more complicated. The neighboring group effect, steric obstruction and complex molecular weight/functionality distributions can serve as examples of the many complicating factors. If these factors are not characterized and taken into account, interpretation of the results may be obscured and the ability to predict the values of a particular architecture can be impeded.

ACKNOWLEDGEMENT

The authors wish to thank the Grant Agency of the Academy of Sciences of the Czech Republic (grant No. A4050808), Grant Agency of the Czech Republic (grant No. 203/99/D062) and EC INCO-Copernicus programme (grant No. IC-15-CT98-0822) for financial support.

5 REFERENCES

1. Dušek, K. in Meijen, H. E. H. (ed.), *Network Formation*, in *Processing of Polymers, Material Science and Technology*, Vol. 18, Verlag Chemie, Weinheim, 1997, pp. 401–428.
2. Šomvársky, J. and Dušek, K., *Polym. Bull.*, **33**, 369 (1994).
3. Šomvársky, J. and Dušek, K., *Polym. Bull.*, **33**, 377 (1994).
4. Dušek, K., *Trends Polym. Sci.*, **5**, 268 (1997).
5. Dušek, K., *Polym. Mater. Sci. Eng.*, **74**, 170 (1996).
6. Dušek, K., *Network formation by chain crosslinking (co)polymerization*, in Haward, R. N. (ed.), *Development in Polymerization*, Vol. 3, Applied Science Publ., Barking, 1982, pp. 143–206.
7. Dušek, K., Vol. 3, Network formation involving polyfunctional polymer chains, in Stepto, R. F. T. (ed.), *Polymer Networks: Principles of their Formation, Structure and Properties*, Thomson Science, London, 1998, Chap. 3, pp. 64–92.
8. Dušek, K., *Macromolecules*, **17**, 716 (1984).
9. Krakovský, I., Havránek A., Ilavský, M. and Dušek, K., *Colloid Polym. Sci.*, **266**, 324 (1988).
10. Corfield, G. C. and Butler, G. B., Cyclopolymerisation and cyclocopolymerisation, in Haward, R. N. (ed.), *Development in Polymerization*. Vol. 3, Applied Science Publ., Barking, 1982, pp. 1–55.
11. Baney, R. H., Itoh, M., Sakikabara, A. and Suzuki, T., *Chem. Rev.*, **95**, 1409 (1995).
12. Provatas, A. and Matisons, J. G., *Trends Polym. Sci.*, **5**, 327 (1997).
13. Schmidt, H., *5th Int. Conf. Frontiers of Polymers and Advanced Materials*, 1999, 72 (available from Inst. Natural Fibers, Poznan, Poland).
14. Ishizu, K., Star polymers by immobilizing functional block copolymers, in (Mishra, N. K. and Kobayashi, S. (eds), *Stars and Hyperbranched Polymers*, Marcel Dekker, New York 1999.
15. Fréchet, J. M. J., Hawker, C. J., Synthesis and properties of dendrimers and hyperbranched polymers, in Aggarwal, S. L. and Russo, S. (eds), *Comprehensive Polymer Science*, Second Supplement, Elsevier Science, Oxford 1996, pp. 71–133.
16. Hult, A., Johansson, M. and Malmström, E., *Adv. Polym. Sci.*, **143**, 1 (1999).
17. Voit, B., *Acta Polym.*, **46**, 87 (1996).
18. *Stars and Hyperbranched Polymers*, Mishra, N.K. and Kobayashi, S., eds, M. Dekker, New York 1999.
19. Newkome, G. R., Moorefield, C. N. and Vögtle, P., *Dendritic Molecules*, VCH, Weinheim, 1996.
20. Kim, Y. H., *J. Polym. Sci., Polym. Chem., Ed.*, **36**, 1685 (1998).
21. Dušek, K., Šomvársky, J., Smrčková, M., Simonsick, W. J. Jr and Wilczek, L., *Polym. Bull.*, **42**, 489 (1999); Gong, C., Miravet, J. and Fréchet, J. M. J. *J. Polym. Sci.* A37, 3193 (1999).

22. Bauer, B. J., Topp, A., Prosa, T. J., Amis, E. J., Yin, R., Qin, D. and Tomalia, D. A., *Polym. Mater. Sci. Eng.*, **77**, 87 (1997).
23. De Leuze-Jallouli, A. M. D., Swanson, P. R., Dvornic, S. V., Parz, M. J. and Owen, M. J., *Polym. Prep.* (*Am. Chem. Soc., Div. Polym. Chem.*, **39**(1), 475 (1998).
24. Emrick, T., Chang, H.-T. and Fréchet, J. M. J., *Macromolecules*, **32**, 6380 (1999); Emrick, T., Chang, H.-T. and Fréchet, J. M. J., *J. Poly Sci.* **A38**, 4850 (2000).
25. Luňák, S. and Dušek, K., *J. Polym. Sci., Polym. Symposia*, **53**, 45 (1975).
26. Dušek, K., Ilavský, M. and Lunňák, S., *J. Polym. Sci., Polym. Symposia*, **53**, 29 (1975).
27. Dušek, K. and Ilavský, M., *J. Polym. Sci., Polym. Phys. Ed.*, **21**, 1323 (1983).
28. Ilavský, M., Bogdanova, L. and Dušek, K., *J. Polym. Sci., Polym. Phys. Ed.*, **22**, 265 (1984).
29. Dušek, K., *Adv. Polym. Sci.*, **78**, 1 (1986).
30. Dušek, K., Networks from telechelic polymers: theory and application to polyurethanes, in Goethals, E. J. (ed.), *Telechelic Polymers: Synthesis and Applications*, CRC Press, Boca Raton, FL, (1989), pp. 289–360.
31. Huybrechts, J. and Dušek, K., *Surf. Coat. Int.*, **81**, 117 (1998).
32. Huybrechts, J. and Dušek, K, *Surf. Coat. Int.*, **82**, 172 (1998).
33. Huybrechts, J. and Dušek, K., *Surf. Coat. Int.*, **82**, 234 (1998).
34. Dušek, K., and Šomvársky, J., *Faraday Discuss. Chem. Soc.*, **101**, 147 (1995).
35. Dušek, K., and Šomvársky, J., *Macromol. Symp.*, **106**, 119 (1996).
36. Nabeth B., Pascault J.P. and Dušek, K., *J. Polym. Sci., Polym. Phys. Ed.*, **34**, 1031 (1996).
37. Dušek, K., Pascault, J.-P., Špírková, M. and Šomvársky, J., *Polyurethane networks and chemical clusters*, in Kresta, J., and Eldred, W. (eds), 60 Years of Polyurethanes, Technomic, Lancaster, 1998, pp. 143–160.
38. Kuchanov, S. I. and Povolotskaya, E. S., *Vysokomol. Soedin.*, **A24**, 2179 (1982).
39. Galina, H. and Szustalewicz, A., *Macromolecules*, **23**, 3833 (1990).
40. Galina, H. and Lechowicz, J.B., *Adv. Polym. Sci.*, **137**, 135 (1998).
41. Dušek, K., *Rec. Trav. Chim. Pays-Bas.*, **110**, 507 (1991).
42. Tobita, H. *Macromolecules*, **26**, 5427 (1993).
43. Tobita, H. *J. Polym. Sci., Polym. Phys.*, **36**, 24233 (1998).
44. Dušek, K., and Šomvársky, J., *Polym. J. Int.*, **17**, 185 (1985).
45. Dušek, K., and Šomvársky, J., *Polym. Int.*, **44**, 225 (1997).
46. Dušek, K., and Šomvársky, J., Ilavskỳ, M. and Matějka, L. *Comput. Polym. Sci.*, **1**, 90 (1991).
47. Mikeš, J. and Dušek, K. *Macromolecules*, **15**, 93 (1982).
48. Šomvársky, J., Dušek, K., Smrčková, M. *Comput. Theor. Polym. Sci.*, **8**, 201 (1998).
49. Tobita, H., *Macromolecules*, **28**, 5119 (1995).
50. Tobita, H., *Polymer*, **36**, 2585 (1995).
51. Hanselmann, R., Hölter, D., Frey, H., *Macromolecules*, **31**, 3790, (1998).
52. Lee, Y. U., Jang, S. S., Yang, J. S. and Jo, W. H., *Polym. Mater. Sci. Eng.*, **80**, 163 (1999).
53. Yan, D. and Zhou, Z., *Macromolecules*, **32**, 819 (1999).
54. Dušek, K., Scholtens, B. J. R. and Tiemersma-Thoone, G. P. J. M., *Polym. Bull.*, **17**, 239 (1987).
55. Tiemersma-Thoone, G. P. J. M., Scholtens, B. J. R., Dušek, K. and Gordon, M., *J. Polym. Sci., Part B: Polym. Phys.*, **29**, 463 (1991).
56. Stauffer, D., Coniglio, A. and Adam, M., *Adv. Polym. Sci.*, **44**, 103 (1981).
57. Boots, H. M. J. and Kloosterboer, J. G., *Br. Polym. J.*, **17**, 219 (1985).
58. Leung, Y. K. and Eichinger, B. E., *J. Chem. Phys.*, **80**, 3877, 3885 (1984).
59. Pakula, T., *19th Discuss. Conf.: Rheology of Polymers*, Prague 1999, (available from the Institute of Macromolecular Chemistry, Prague).

60. Ilavský, M. and Dušek, K., *Polymer*, **24**, 981 (1983).
61. Ilavský, M. and Dušek, K., *Macromolecules*, **19**, 2139 (1986).
62. Dušek, K. and Ilavský, M., *Progr. Coll. Polym. Sci.*, **75**, 11 (1987).
63. Fedderly, J. J., Lee, G. F., Lee, J. D., Hartmann, B., Dušek, K., Šomvársky, J. and Smrčková, M., *Macromol. Symp.*, accepted.
64. Dušek, K., *J. Macromol. Sci.-Chem.*, **A28**, 843 (1991).
65. Aida, T., Jiang, D.-L., Choi, M.-S. and Enomoto, M., *Polym. Mater. Sci. Eng.*, **80**, 257 (1999).
66. Fujita, M., *Polym. Mater. Sci. Eng.*, **80**, 27 (1999).
67. Dvornic, P. R., de Leuze-Jallouli, A. M., Perz, S. V. and Owen, M. J., *5th Int. Conf. Frontiers of Polymers and Advanced Materials*, 1999, p. 68 (available from Inst. Natural Fibers, Poznan, Poland).
68. Dvornic, P. R., de Leuze-Jallouli, A. M., Owen, M. J., Dalman, D. A., Parkham, P., Pickelman, D. and Perz, S. V., *Polym. Mater. Sci. Eng.*, **81**, 187 (1999).
69. Hedrick, J. L., Hawker, C. J., Miller, R. D., Twieg, R., Srinivasan, S. A. and Trollsas, M., *Macromolecules*, **30**, 7607 (1997).
70. Hedrick, J. L., Hawker, C. J., Twieg, R., Srinivasan, S. A., Harbison, M., Trollsas, M., Miller, R. D., Kim, S. M. and Yoon, D. Y., *Polym. Mater. Sci. Eng.* **77**, 202 (1997).
71. Niu, Y., Chai, M., Rinaldi, P. L., Tessier, C. A. and Youngs, W. J., *Polym. Prepr. Am. Chem. Soc., Div. Polym. Chem.*, **39**, 629 (1998).
72. Boury, B., Corriu, R. J. P. and Nunez, R., *Chem. Mater.*, **10**, 1795 (1998).
73. Zhao, M., Liu, Y., Crooks, R. M. and Bergbreiter, D. E., *J. Am. Chem. Soc.*, **121**, 923 (1999).
74. Zukas, W. X., Wilson, P. M. and Gassner, J. J., *Polym. Mater. Sci. Eng.*, **77**, 232 (1997).
75. Van Benthem, R. A. T. M., Muscat, D. and Stanssens, D. A. W., *Polym. Mater. Sci. Eng.*, **80**, 72 (1999).
76. Boogh, L., Pettersson, B. and Manson, J-A. E., *Polymer*, **40**, 2249 (1999).
77. Wu, H., Xu, J. and Heiden, P., *J. Appl. Polym. Sci.*, **72**, 151 (1999).
78. Hawker, C. J.; Lee, R.; Fréchet, J. M. J. *J. Am. Chem. Soc.* **113**, 4583 (1991).
79. Dušek, K., Šomvársky, J., Smrčková, M., Wilczek, L. and Simonsick, W. J., Jr., *Polym. Mater. Sci. Eng.*, **80**, 102 (1999).
80. Frey, H. and Hölter, D., *Polym. Mater. Sci. Eng.*, **80**, 266 (1999).
81. Gooden, J. K., Gross, M. L., Mueller, A., Stefanescu, A. D. and Wooley, K. L., *J. Am. Chem. Soc.*, **120**, 10180 (1998).
82. Chu, F., Hawker, C. J., Pomery, P. J. and Hill, D. J. T., *J. Polym. Sci., Part A: Polym. Chem.*, **35**, 1627 (1997).
83. Radke, W., Litvinnenko, G. and Muller, A.H.E., *Macromolecules*, **31**, 239 (1998).
84. Beginn, U., Drohman, C. and Moeller, M., *Macromolecules*, 30, 4112 (1997).
85. Hanselmann, R., Hölter, D. and Frey, H., *Polym. Mater. Sci. Eng.*, **77**, 165 (1997).
86. Hölter, D. and Frey, H., *Acta Polym.*, **48**, 298 (1997).
87. Burgath, A., Mock, A., Hanselmann, R. and Frey, H., *Polym. Mater. Sci. Eng.*, **80**, 126 (1999).
88. Trollsas, M., Hedrick, J., Mecerreyes, D., Dubois, P., Jerome, R., Ihre, H. and Hult, A., *Macromolecules*, **30**, 8508 (1997).
89. Trollsas, M., Hedrick, J., Mecerreyes, D., Jerome, R. and Dubois, P., *J. Polym. Sci., Part A: Polym. Chem..*, **36**, 3187 (1998).
90. Trollsas, M., Hawker, C. J., Remenar, J. F., Hedrick, J. L., Johansson, M., Dubois, P., and Hult, A., *J. Polym. Sci., Part A: Polym. Chem. Ed.*, **36**, 2793 (1998).
92. Malmström, E., Hult, A., Gedde, U. W. and Boyd, R. H., *Polymer*, **38**, 4873 (1997).
93. Johansson, M., Malmström, E. and Hult, A., *J. Polym. Sci., Part A: Polym. Chem.*, **31**, 619 (1993).

94. Johansson, M. and Hult, A., *Coat. Technol.*, **67**, 37 (1995).
95. Pettersson, B. and Sorensen, K., *Proc.* 21st *Waterborn, High Solids & Powder Coat. Technol.*, New Orleans 1994, p. 753.
96. Johansson, M. and Hult, A., Proc. **16**th *Waterborne, High-Solids & Radcure Technol. Conf.*, Frankfurt 1994, p. 1.
97. Johansson, M., Rospo, G. and Hult, A., *Polym. Mater. Sci. Eng.*, **77**, 124 (1997).
98. Emrick, T., Chang, H. T., Fréchet, J. M. J., Woods, J. and Baccei, L. *Polymer Bulletin*, **46**, 1 (2000).

6

Regioselectively-Crosslinked Nanostructures

C. G. CLARK, JR. AND K. L. WOOLEY
Washington University, Department of Chemistry, One Brookings Drive,
St Louis, MO, USA

1 INTRODUCTION

Regioselectivity and *regiospecificity* are useful terms for describing the site at which a chemical reaction takes place, and they are common to the vocabulary of organic chemists. *Regioselectivity* is observed in reactions that yield a major product when two or more isomers are possible, differing in the orientational or directional preference for the chemical reaction, whereas the term *regiospecificity* is reserved for those reactions that give exclusively a single product of two or more possible isomers [1]. Regioselectivity and regiospecificity apply to many types of chemical reactions including additions, eliminations, ring openings, and cycloaddition reactions. As an example (Figure 6.1), the alkylation of de-protonated 1,3-cyclohexanedione occurs regiospecifically with methyl iodide to give exclusively the C-alkylated product, but reaction with *n*-butyl iodide occurs with O-alkylation regioselectivity [2]. In cases involving ambient nucleophiles, such as the enolate chemistry of Figure 6.1, the factors that govern regio-control are principally the polarizability of the nucleophile, the nature of the counterion, the extent of solvation, and in extreme cases, steric effects. A classic example of regioselectivity in polymer chemistry is the anionic polymerization of isoprene [3], which has six potential regiochemistries (arising from 1,4-, 4,1-, 1,2-, 2,1-, 3,4-, and 4,3-additions) and a number of different stereochemical possibilities as well. Polymerization with lithium as the counterion, in a polar solvent, tetrahydrofuran, yields polyisoprene with a majority of repeat units bearing side chain vinyl groups (kinetic product) [4], however, similar polymerization of isoprene in a non-polar solvent, hexane, produces polyisoprene of mainly *cis*-1,4-micro-

Dendrimers and Other Dendritic Polymers. Edited by Jean M. J. Fréchet and Donald A. Tomalia
© 2001 John Wiley & Sons Ltd

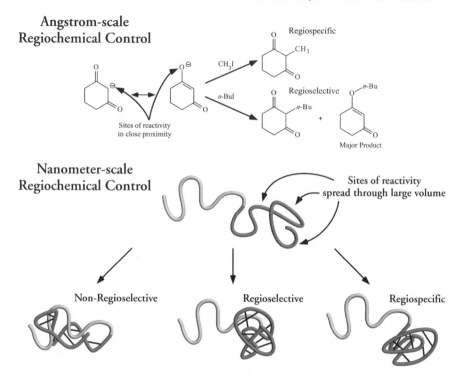

Figure 6.1 A review of regiochemical control is presented for the typical 'angstrom-scale reactivity', with an example taken from enolate chemistry (top). This shows that for *regiospecificity*, a single regiosomer is strictly obtained (steroisomers are also possible); whereas, the term *regioselective* can be applied for reactions that give a majority of one product. The extension of these terms to nanoscale reactions requires consideration of only the basic elements of regiochemical control – the sites at which reaction takes place (bottom). Nanometer-scale reactions that occur throughout the entire molecule exhibit no region selectivity; whereas isolation of the reactive sites of the macromolecule can produce regioselective and regiospecific products.

structure [5] (thermodynamic product). The defining regiochemical terms have thus far been limited to consideration of reaction sites of close proximity, and we now propose the extension of the terminology regioselectivity and regiospecificity to the nanoscale regime.

In broad definition, regio-control over reactivity (regiochemical control) involves the demonstration of locale preferences for a chemical reaction. Generally, regio-control is thought of in terms of through bond separation of reactive sites, as in the resonance forms for the cyclohexanedionyl anion of Figure 6.1.

Regio-control is also found to globally involve the entire molecule or portions of the molecule, as in the nucleophilic attack upon isoprene being at the 1-, 2-, 3-, or 4-positions. For polymeric systems, where there are a large number of sites at which chemical reactions can occur, the assignment of regio-control can become quite complicated, since there are several levels at which regio-control can be considered. The local regio-control involves the exact location of the chemical reaction along the polymer backbone and is at higher resolution than the global regio-control, resulting from identification of the overall regions at which reaction has taken place. In this initial analysis, the more simplified cases will be discussed, where consideration of regiochemistry is based upon the nanoscale region in which chemical reactivity is observed. For example, Figure 6.1 includes three possibilities for intramolecular chain–chain coupling reactions (i.e. cross-linking) of a linear diblock copolymer containing reactive sites along the backbone of one of the blocks: non-regioselectivity, regioselectivity and regiospecificity. In the non-regioselective case, crosslinks are found throughout the macromolecule from intramolecular crosslinking of the reactive segment when the entire diblock copolymer is well solvated. If instead, spatial differentiation of the reactive and non-reactive portions of the polymer chain occurs, then the intramolecular crosslinking can be found in a more selective region of the macromolecule. In the extreme case, there would be no mixing of the non-crosslinked and crosslinked regions, and this would be defined as regiospecificity. Therefore, organizationally driven control over the location of chemically reactive functional groups is proposed as a means to effect regional control over chemical reactivity, to allow for regioselectively- and regiospecifically crosslinked nanostructures. Since it is difficult to imagine that the rigidly defined regiospecificity requirements of an exclusive product could be met for nanoscale assemblies, especially given the potential for mixing at the interfacial boundary between regions of reactivity and non-reactivity, the term 'regioselective' will be used for the remainder of the chapter.

Globally- and locally-regiospecific intramolecular and intermolecular crosslinks are abundant in nature: crosslinks buried within proteins stabilize the tertiary and quaternary structural assemblies; crosslinking of bacterial cell walls maintains mechanical integrity and allows them to withstand osmotic pressure changes; and there are numerous other instances. Throughout this chapter, several examples of synthetic, nanoscale, organized assemblies that contain domain- or location-selective crosslinking will be highlighted. In most cases, the exact local nature of the crosslinks is not known. It is expected, however, that these kinds of covalently-stabilized crosslinked assemblies [6] are less advanced than biological systems in their local regiochemical specificity. The excitement toward the preparation and study of well-defined nanoscale materials is based across multiple disciplines, in which the development of novel synthetic methods is generating unique nanostructured materials. The application of characterization tools to interesting new systems often requires advances in analytical techniques, and the potential that these materials possess toward application in nanotechnology and biotechnology is driving (bio)engineering studies.

2 REGIOSELECTIVE BULK CROSSLINKING

As is illustrated in Figure 6.2, crosslinked materials have traditionally included homogeneously crosslinked matrices of macroscopic dimensions. Crosslinked materials are prepared by two principle methods: preparation from the polymerization of monomer(s) with average functionality greater than two can involve a one-step process; or a multi-step approach can be used, in which the preparation of a prepolymer containing functionalities that provide for crosslinking is followed by a subsequent curing step. Because the covalently bonded crosslinks extend throughout the sample volume, these materials are essentially macroscopic molecular networks, and as such, they are insoluble and intractable. The properties associated with crosslinking reinforcement (elastic recovery, dimensional stability, and rigidity) often lead to excellent physical and mechanical properties for application as elastomers, engineering materials, organic-based ceramics, and other applications. Methods involving the prepolymer chain-chain coupling approach for crosslinking have been developed in recent years to enable control over the regions in which the chemical crosslinking takes place within a sample matrix. This control has lead to interesting nanostructured thermoset materials, and to the generation of unique nanoscale objects, as discussed below.

Phase separation of block copolymers is well known to result in homogeneously dispersed, highly ordered domains, in which the morphologies undergo phase transitions from domains of spheres of one component within the other, to cylinders, to a bicontinuous phase, to perforated layers, to lamellae, depending upon the relative amounts of each of the components in the copolymer composition. Thus, the phase separation of block copolymers offers a versatile and powerful means by which to segregate reactive block segments from nonreactive segments in order to accomplish regio-selective chemical crosslinking reactions, isolated within nanodomains of specified shape and volume.

Crosslinking of the minor component that exists as spheres within phase separated diblock copolymer bulk samples, followed by dissolution of the matrix has yielded core-crosslinked nanospheres having covalently attached linear polymer surface chains (Figure 6.2a). As shown in Figure 6.3a, the phase separation of poly(styrene-*b*-4-vinylpyridine) containing *c.* 10 mol% 4-vinylpyridine yielded spheres of 4-vinylpyridine within a polystyrene matrix. Reaction with the vapor of 1,4-dibromobutane gave quaternization of the 4-vinylpyridine groups and effected crosslinking within the poly(4-vinylpyridine) core domains [7]. This resulted in crosslinked nanospheres in which the quaternized and hydrophilic

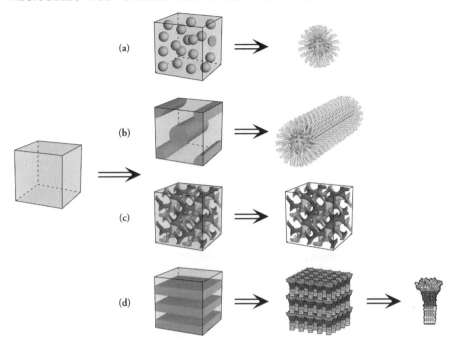

Figure 6.2 Illustrations of the morphologies generated (for one half of the phase diagram) for bulk block copolymer phase separation are shown. Beginning with a matrix of pure component, crosslinking yields a solid mass of insoluble networked material of macroscopic dimensions. Progressive incorporation of a second component, in the form of a two-component block copolymer with increasing volume percentages of one component of the block copolymer gives: (a) spheres of the minor component within a matrix of the major component; (b) cylinders of the minor component surrounded by the major component; (c) a double gyroid morphology; perforated layers (not shown); (d) lamellae at or near equal volumes of the respective blocks. The other half of the phase diagram has the same morphologies, with the major and minor components reversed. Either block may be crosslinked to stabilize the phase-separated morphology. To the right of each morphology is the macromolecular object that results from the establishment of connectivity throughout the minor component domains, followed by solubilization (a, b, d) or removal of the material between the minor domains (c). In the case of the double gyroid (c), the connectivity is two fold, first by physical mixing of adjacent chains in the major component as in (a) and (b), and second the double gyroid structure consists of two interpenetrating networks which cannot be removed from one another, resulting in a nanoporous, macroscopic solid rather than a macromolecular object. In the case of the lamellae (d), removal of one component produces two-dimensional objects, but crosslinking within confined domains results in smaller, narrowly disperse 'mushroom-shaped' objects

core domains were surrounded by polystyrene (Figure 6.3a), where each core chain was covalently attached to one polystyrene chain, for a high density of polystyrene surface coverage. The two- and three-dimensional macrocrystalline packing of these core-crosslinked spheres has been examined [8]; monolayers of the spherical particles were found to pack in a hexagonal arrangement upon a carbon substrate by transmission electron microscopy, whereas small-angle X-ray scattering indicated a face-centered cubic lattice within the bulk samples. Other immiscible diblock copolymer compositions have been employed in this synthetic method [9], and in each case, the nanosphere dimensions are of narrow size dispersity with diameters ranging from *c*. 30 to 100 nm, depending upon the diblock composition, particularly the relative volume of the minor component that forms the spheres (core domain) within the matrix.

In a similar fashion, nanofibers have been produced, by the regioselective crosslinking of the cylindrical domains formed via phase separation of diblock copolymer bulk samples, followed by their dissolution (Figure 6.2b) [10]. For poly(styrene-*b*-cinnamoylethyl methacrylate), PS-*b*-PCEMA, at 13 mol% (21 vol%) PCEMA and overall degrees of polymerization of 1400, the PCEMA minor component forms cylinders within a PS matrix. Irradiation of disks (∼ 1 mm thick) with UV light passed through a 310 nm cut-off filter for 30 min per each side resulted in crosslinking via photo-induced [2 + 2]-cycloaddition of

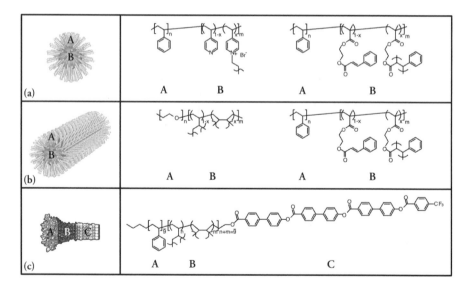

Figure 6.3 The chemical compositions for macromolecular objects that result from crosslinking within nanodomains of bulk phase separated block copolymers include: (a) core-crosslinked spheres; (b) core-crosslinked rods; (c) 'mushroom'-shaped objects.

~ 20% of the CEMA units within the cylindrical domains, and afforded a 50% yield of tetrahydrofuran-dispersable nanofibrous product (Figure 6.3b). Transmission electron microscopy (TEM) imaging, following selective staining of the PCEMA component with OsO_4, revealed nanofibers of several micrometers in length with a preferentially stained PCEMA core (26 nm diameter) surrounded by a PS layer (33 nm thickness). By light scattering studies, the persistence length was measured to be 400 nm and the diameter was 86 nm. As with many biological and synthetic rod-like macromolecules and nanostructures, the nanofibers exhibited shear-induced lyotropic liquid crystalline ordering with a banded texture when visualized by polarized optical microscopy in bromoform.

Intricately shaped nanostructured materials can also be prepared by this general self-assembly and covalent stabilization approach. Interconnecting gyroid-like void channels have been constructed within a ceramic matrix by an approach that involves the self-assembly of A_1–B–A_2 triblock copolymers of polyisoprene (A_1 = 24 kg/mol, A_2 = 26 kg/mol) and poly(pentamethyldisilyl-styrene) (PPMDSS, B = 100 kg/mol), followed by an oxidation process that gave concomitant and selective removal of the PI regions and transformation of the silicon-containing regions to silicon oxycarbide ceramic [11]. Alternatively, the gyroid nanochannels can be transformed into the ceramic material, following the formation of inverse double gyroid morphologies (PI, A_1 = 44 kg/mol, A_2 = 112 kg/mol; PPDMSS, B = 168 kg/mol) to generate the nano-relief structure (Figure 6.2c). This general approach offers the ability to conveniently prepare either the nanoporous or nanorelief ceramic materials, which have exhibited interesting properties, including high-temperature stability, low dielectric constant, low density and uniformity of channel pore dimensions that suggest potential applications in membrane separation, catalyst supports, interconnects or as photonic band gap materials.

Crosslinked two-dimensional sheets have been produced by the regio-crosslinking of selective layers within bulk samples of lamellar phase separation [12, 13]. In fact, more recently, the lamellar morphology was further dissected into nanoscopic volumes of segregated reactive domains within the layer, by assembly of triblock rod-coil molecules into mushroom-shaped nanostructures (Figure 6.2d) [14]. When the triblock rod-coil molecule was composed of butyl-(styrene)$_9$-b-(butadiene)$_9$-b-(biphenyl ester unit)$_3$-CF_3, the oligostyrene segment sterically limited the aggregation number, the oligobutadiene offered reactive carbon–carbon double bonds for regioselective crosslinking reactions, the rigid biphenyl ester component assembled by π-stacking interactions, and the trifluoromethyl chain end offered additional driving force for assembly (Figure 6.3c). The resulting 'mushroom-shaped' assemblies organized (Figure 6.2d) by a head-to-tail fashion and with well-defined pre-aggregate dimensions to provide for orientational and directional preference for the chemical reactions, while also isolating the crosslinking chemistry to selective domain volumes. Annealing samples of the triblock rod-coil lamellar samples at 250 °C afforded crosslinking through c. 70% of the butadiene double bonds to yield macromolecular species of highly anisometric shape (2 × 8 nm) composed of 23 ± 2 rod-coil molecules assembled in a parallel manner [15]. As monitored by gel permeation chromatography (GPC) over a 20 h period of time, the amount of triblock oligomer decreased as the amount of the nanoscale molecular object, of narrow size distribution increased and with only a small fraction of intermediately sized products and no larger aggregates or long-range coupled products (i.e. insoluble crosslinked gel) being observed. Characterization of the discrete nanostructured macromolecules by small-angle X-ray scattering (SAXS) and transmission electron microscopy (TEM) indicated that the pre-assembled form of the precursor triblock rod-coil was maintained without disruption by the intermolecular coupling reactions. These structures are the most unique nanoscopic objects that have been synthesized to date. Their unique qualities are associated with the avoidance of a centrosymmetric structure, which is inherent for other covalently-stabilized self-assemblies. Therefore, these materials most closely emulate the capabilities for the production of complex macromolecules of asymmetrical structure found in nature, and they bear a striking resemblance to mushroom-shaped protein aggregates, for example α-hemolysin or ATP synthase. Each of these biological nanostructures relies upon the narrow 'shaft' of the 'mushroom' to insert into membranes. For α-hemolysin, a pore through the center mediates transport of molecules and ions across cell membranes, while ATP synthase [16] serves as a rotating machine for the synthesis of adenosine triphosphate (ATP), the energy storage mechanism for all levels of life forms. These similarities suggest that further advances in the chemical composition of the synthetic macromolecular 'mushrooms' may advance their function to the point that they could substitute for biological systems in cellular transport mechanisms, biological energy conversion, or other important processes.

3 REGIOSELECTIVE CROSSLINKING WITHIN SUPRAMOLECULAR ASSEMBLIES

The examples discussed above illustrate the importance of block copolymer chain segment incompatibilities for the phase separation of bulk materials, combined with the ability to perform chemistry within specific nanoscale domains to impose permanence upon those self-assembled nanostructured morphologies. Each is limited, however, to crosslinking of internal domains within the solid-state assemblies in order to create discrete nanoscale objects. To advance the level of control over regioselective crosslinking and offer methodologies that allow for the production of additional unique nanostructured materials, the pre-assembled structures can be produced in solution (Figure 6.4), as isolated 'islands' with reactivity allowed either internally or on the external

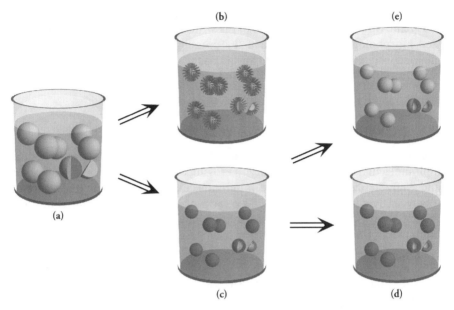

Figure 6.4 The preparation of nanostructured materials in solution evolves from (a) the classic examples of suspension, dispersion, or emulsion polymerization, to the methods that include the covalent crosslinking of select domains within supramolecular polymer assemblies; (b) core crosslinking of polymer micelles; (c) shell crosslinking of polymer micelles (SCKs); (d) nanocages from core-eroded SCKs; (e) 'shaved' hollow nanospheres from outer shell/core-eroded vesicles.

surface, without the threat of inter-object crosslinking into a macroscale networked matrix (gel).

Suspension, dispersion or emulsion polymerization processes are traditional methods for the preparation of polymeric particulate materials (Figure 6.4a) [17]. While each of these methods serve important purposes for the preparation of polymer microparticles, scaling down to 10–100 nm on a routine basis and with complicated nanostructural partitioning of compositions is not feasible by these means. Block copolymers that are composed of block segments of differing solubility parameters assemble when placed into a solvent system that is selective for a portion of the chain length. A range of morphologies is possible, including micelles, inverse micelles, cylinders, and vesicles through variation in the block copolymer composition and the solvent system used for organization of the assemblies [18]. Crosslinking of the polymer assemblies reinforces the structures and produces mechanically robust materials from fragile supramolecular assemblies. In each case, functional groups along the backbone of the block copolymer and located within a specific domain of the nanoscale assembly undergo reactions in order to bind the entity into a single nanoscopic macromolecule, thus removing the dynamic nature of the structure while maintaining the organized morphology. The location of the crosslinking functionalities can be regioselectively controlled to yield several nanoscale materials of very different compositions, structures, properties, and functions.

3.1   CORE-CROSSLINKED NANOSTRUCTURES

Polymer micelles exist as an equilibrium between monomolecular micelles and multimolecular micelles; the ability to crosslink both structures and then isolate and characterize the crosslinked monomolecular micelles ('tadpole' structure) was recently demonstrated [19]. This form of single chain intramicellar crosslinking is consistent with the schematic drawings of Figure 6.1, where it is unknown to what extent mixing of the two phases occurs, and thus assignment of regioselectivity vs. regiospecificity has not been made. These 'tadpole' nanostructures are expected to resemble the dendritic globular-linear chain hybrid structures that have been generated by covalent monocoupling of dendritic macromolecules to linear polymer chains [20]. Although only 5–25% 'tadpole' structures were obtained within the samples of larger core-crosslinked, multimolecular micelles, and their isolation required a tedious fractional gel permeation chromatography procedure, the 'tadpole' macromolecules may prove interesting as asymmetrically ordered, core–shell structures of angstrom dimensions.

Crosslinking of the core domains within multimolecular polymer micelles has been accomplished using several different chemistries (Figure 6.4b and 6.5a). Nucleation of the butadiene chain segments of poly(styrene-b-butadiene) in a solvent selective for polystyrene, followed by crosslinking of the polybutadiene core by reaction with free radicals, generated *in situ* from irradiation in the presence of dibenzoyl peroxide, resulted in stabilized micelles that, unlike the self-assembled precursors, exhibited no critical micelle concentration [21]. Other examples of the preparation of core-crosslinked nanospheres in solution via intramicellar reactions [22, 23], have also been reported and, with one exception [24]; these structures were prepared in mixed organic solvents. The initially prepared core-crosslinked nanoparticles consist of crosslinked and insoluble core domains encapsulated within linear polymer chains. When those surface linear polymer chains were composed of polyisoprene, treatment of the core-crosslinked polymer assemblies under ozonolysis conditions cleaved the surface chains to yield 'shaved' crosslinked nanospheres bearing reactive surface functionalities [25]. The porosity of the crosslinked core domains could also be modified, through the use of a homopolymer porogen [26]. When oligomeric 2-octanoylethyl methacrylate (OEMA) was present during the assembly of poly[(2-cinnamoylethyl methacrylate-*co*-2-octanoylethyl methacrylate)-*b*-

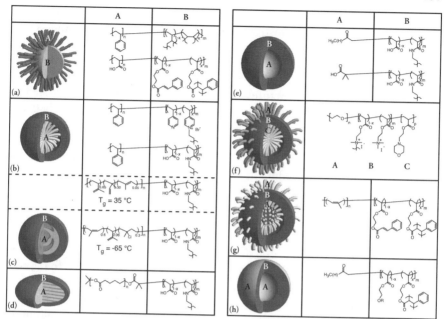

Figure 6.5 Illustrations of nanoscale spherical assemblies resulting from block copolymer phase separation in solution are shown, along with the chemical compositions that have been employed to generate each of the nanostructures: (a) core crosslinked polymer micelles; (b) shell crosslinked polymer micelles (SCKs) with glassy cores; (c) SCKs with fluid cores; (d) SCKs with crystalline cores; (e) nanocages, produced from removal of the core of SCKs; (f) SCKs with the crosslinked shell shielded from solution by an additional layer of surface-attached linear polymer chains; (g) crosslinked vesicles; (h) 'shaved' hollow nanospheres produced from cleavage of the internally and externally attached linear polymer chains from the structure of (g)

(acrylic acid)], it partitioned into the core domain and did not participate in the crosslinking of the core upon irradiation. Therefore, following the crosslinking of the cinnamoylethyl methacrylate chain segments, the OEMA oligomers were extracted to yield porous, crosslinked core domains, with presumably lowered core density. These porous nanospheres exhibited a greater ability to contain guest molecules, as they sorbed higher quantities of perylene from aqueous solutions than did their non-porous counterparts.

Regioselective crosslinking of the core domain of cylindrically shaped, worm-like micelles composed of poly[(butadiene)$_{45}$-b-(ethylene oxide)$_{55}$] and assembled in aqueous solution at $< 5\%$ block copolymer concentrations, was performed using radical coupling of the double bonds throughout the poly(butadiene) phase [27] (Figure 6.3b). This resulted in a 13% reduction in the core diameter, from 14.2 to 12.4 nm, as measured by small-angle neutron scattering, and transformed the wormlike micelles from fluid-like, deformable, 'living' assemblies to 'permanent' nanostructures, the effects of which were manifested in the viscoelastic characteristics of the nanomaterials. Due to the poly(ethylene oxide) surface chains, these core-crosslinked wormlike micelles of several micrometers in length were dispersible in aqueous solutions. The large aspect ratio of these cylindrical nanostructures suggests that they may be considered as nanoscale linear polymer chains, that bridge the gap in dimensionality between angstrom scale linear polymers consisting of molecular-level straight chain bonding sequences and fabricated micro- and macroscale objects. Moreover, this solution-based approach to their synthesis offers an alternative method to the bulk phase-separation procedure described above (Figure 6.2b).

3.2 SHELL-CROSSLINKED NANOSTRUCTURES

There are several advantages associated with the covalent crosslinks being regioselectively placed in the shell layer [28] (Figure 6.4c), as opposed to the core or interface regions. By 'tying together' the peripheral portion of the nanostructure, the particle surface consists of a reinforcing network that provides robust character and stability under the influence of changing environmental conditions. This network also serves as a membrane layer, the permeability of which can be tailored to control the transport of guests to and from the particle core. The core is essentially a nanoscale domain of internally surface-grafted polymer chains. Furthermore, the lack of crosslinks in the core region maintains chain mobility and access to the core volume. Perhaps the most significant aspect of the shell crosslinking, however, is the opportunities that are then available by subsequent modification of the initial SCK nanostructures, for example the conversion to solvent-filled nanocages [29] (Figure 6.4d).

The first family of shell crosslinked knedel-like [30] nanoparticles (SCKs) were assembled from amphiphilic diblock copolymers based upon polystyrene (PS) as the hydrophobic, inert segment from which the core nucleated, and poly(4-vinyl pyridine) (P4VP) that was partially quaternized with p-chloromethylstyrene (ClMeS) (Figure 6.5b) [31]. The overall degrees of polymerization were typically 100–130 and the quaternization extents ranged from 10 to 50%. The quaternization procedure introduced water-soluble salts along the P4VP(ClMeS) backbone, and also introduced reactive styrenyl side chain moieties that provided for crosslinking of the P4VP(ClMeS) chain segments. Therefore, assembly of the PS-b-P4VP(ClMeS) diblocks in a mixture of 30% tetrahydrofuran (THF)/H_2O at a concentration of 10^{-5} M yielded spherical polymer micelles (critical micelle concentration (cmc) of c. 10^{-7} M), which were then stabilized through crosslinking reactions between the styrenyl groups within the peripheral shell by reaction with a water-soluble radical initiator, 4,4'-azobis(4-cyanovaleric acid), under irradiation at 254 nm for 24 h to yield the shell crosslinked nanospheres (no observable cmc). The diameters of the polymer micelles increased with increasing relative volume of the hydrophobic core block segment, and this was used as a method to alter the SCK particle size [32]. Styrene-to-vinyl pyridine molar compositions of 1:2, 1:1.2, and 1.9:1 gave SCKs of 9 ± 3, 15 ± 2 and 27 ± 5 nm solid-state diameters, respectively by atomic force microscopy (AFM) height analysis, and solution-state hydrodynamic diameters of 14 ± 1, 21 ± 1, 33 ± 2 nm for greater than 95 vol% of the samples by dynamic light scattering (DLS) measurement. Analytical ultracentrifugation (AU) experiments measured the SCK molecular weights (M_w) as 244 ± 36, 1046 ± 78 and 6336 ± 75 kg/mol and aggregation numbers (the number of linear polymer chains that originally assembled to yield the polymer micelles and then were 'trapped' through shell crosslinking) of 12 ± 2, 71 ± 5 and 439 ± 5, respectively [33]. However, AU measurements for the polymer micelles were complicated by nonideal solution properties, attributable to an inherently less stable structure of the micelle which was subject to deformation and distortion under ultracentrifugation forces. In addition, the SCKs retained their height and appeared to be of uniform spherical shape, in contrast, the micelle precursors exhibited deformation upon adsorption onto the mica substrate used as a support for AFM imaging. The presence of the positively charged pyridinium salts on the surface of the SCKs was confirmed by measurements of electrophoretic mobility and zeta potential via electrophoretic light scattering, which indicated a partial burial of charge beneath the particle surface. The accessibility of the positively charged groups for interactions with negatively charged small dye molecules [34] and the negatively charged phosphodiester backbone of DNA [35] was demonstrated.

The importance of the shell composition and crosslink density motivated alternative methods for the production of SCKs, utilizing different diblock compositions and also different crosslinking chemistries. The assembly of poly(styrene-*b*-acrylic acid) into micelles within an aqueous solution allowed for the shell crosslinking to be performed through amidation chemistry (Figure 6.5b), providing crosslinks through a condensation mechanism and incorporating modifiable shell and surface properties through selection of the crosslinking agent. Following micelle formation and activation of the carboxylic acid side chain groups by reaction with a water-soluble carbodiimide, 1-(3-dimethylamino)propyl)-3-ethylcarbodiimide methiodide, a diamino crosslinker was then added to complete the intramicellar crosslinking of the shell layer. A number of diamino (and polyamino) crosslinkers have been studied and found to be effective in the formation of SCKs [36, 37]. For instance, functionalities that promote specific binding to cell surfaces are being explored as a method by which to tailor the SCK targeting.

Loading of guests within the SCKs (for potential delivery) is modeled after lipoproteins, which are composites of cholesterol, cholesteryl ester, phospholipids, and protein forming biological structures of core-shell morphology and overall diameters of 10–100 nm, whose role is to transport insoluble esterified cholesteryl ester. Some of the most important aspects of any transport device are the loading capacity and the mechanism of uptake and release, both of which rely upon the location of the guests within the host system. The locations of guests within the SCK nanostructure were accurately determined by rotational-echo double-resonance (REDOR) [38] NMR experiments. It was found that amphiphilic small molecules partition to the amphiphilic region of the SCK, near the core–shell interface [39] whereas hydrophobic guests reside in the hydrophobic PS core domain [40]. Moreover, it was found that at high amidation extents (an indirect measure of the crosslink density), the guests were prohibited from passing across the shell membrane layer, and were excluded from entering the SCKs.

The assembly and crosslinking of polymer micelles in an aqueous solution generally produces SCKs with a hydrophobic core and hydrophilic shell. Interestingly, SCKs possessing tunable core hydrophilicity [41] were prepared from poly(N-(morpholino)ethyl methacrylate) as the hydrophobic nucleating-core segment, as it exhibits hydrophobic character at 60 °C in a solution of 0.1 M sodium sulfate at pH 10, and becomes hydrophilic when the temperature is brought to 25 °C. Therefore, placement of N-(morpholino)ethyl methacrylate (MEMA) segments within the core domain of the SCK nanospheres yielded stable, nanostructured particles with temperature-variable core hydrophilicity. In the construction of these SCK nanospheres, a diblock copolymer of MEMA and 2-(dimethylamino)ethyl methacrylate, partially quaternized with methyl iodide (DMAEMA) (number average molecular weight 36 000 g/mol, 65 mol% MEMA) was allowed to assemble into polymer micelles at 60 °C, and the shell was then cross-linked via further quaternization of the shell DMAEMA segments with 1,2-bis-(2-iodoethoxy)ethane. The resulting SCK nanosphere possessed a hydrophobic core, until the temperature was decreased to 25 °C. Upon core hydration, the intensity-average SCK diameter measured by dynamic light scattering increased from 28 nm to 30 nm. These structures are highly unique, as the corresponding entirely hydrophilic polymer micelles cannot exist without the shell cross-links. Moreover, it was suggested that the temperature-dependent core hydration may be useful as a mechanism for controlled release of guest species. SCKs composed of zwitterionic core-shell compositions have also been prepared [42].

The ability of the SCK shell to affect the overall properties of the materials is quite obvious, given the fact that the shell encounters the initial contact with the environment; however, the effects of core modifications upon the SCK properties and behavior are less obvious. SCKs containing glassy polystyrene cores (Figure 6.5b and discussed above) exhibit a rigid spherical shape, without undergoing deformation upon adsorption onto a solid substrate at room temperature. SCKs prepared from poly(isoprene-b-acrylic acid), PI-b-PAA, can have varying degrees of glassy or fluid character, depending upon the PI microstructure (Figure

6.5b and 6.5c) [43, 44]. Polymer micelles prepared from (PI-*b*-PAA) with predominantly *cis*-4,1-isoprene microstructure and block lengths of 130 isoprene and 170 acrylic acid repeat units, were difficult to handle due to the fluid-like character of the polyisoprene core domain ($T_g = -65\,°C$). In contrast, crosslinking of the PAA shell of the PI-*b*-PAA micelles to form the corresponding SCK (Figure 6.5c) provided stabilization of the nanostructure through the covalent crosslinking and produced robust nanostructured materials. Essentially, once the shell was crosslinked, it served as a nanoscale containment device, in this case as a fluid-filled membrane. The nanoparticle, therefore, exhibited shape adaptability, and this characteristic is currently being exploited in our laboratories.

The incorporation of poly(ε-caprolactone), PCL, into the core domain of the SCKs possessing the poly(acrylic acid-*co*-acrylamide) crosslinked shell yielded nanoscale isolated crystalline phases, contained within the crosslinked membrane layer (Figure 6.5d) [45]. This shell layer effectively limited the PCL crystallization process so that the T_m of the PCL decreased with decreasing core volume. These SCKs exhibited unusual behavior, related to their flattened, lamellar shape from the crystalline PCL core domain ($T_m = 45\text{--}60\,°C$), which gave them a high aspect ratio in solution and upon adsorption onto a substrate. Characteristic features included diameter to height ratios from AFM measurements of *c.* 10: 1, which agreed with the solution-state hydrodynamic diameters from DLS measurements that were *c.* ten fold greater than the nanoparticle heights measured by AFM. The fundamental properties associated with these unique nanoscale crystallites, consisting of a crystalline polymer confined to the SCK core of nanoscale volume, are under current investigation.

The hydrophobic block segments are required for the initial self-assembly to form the polymer micelles; however, once the crosslinks are established throughout the shell layer the core is no longer required. Degradation and extraction of the core material has been accomplished for the PI core as well as the PCL core, by ozonolysis [46] and hydrolysis chemistries [45b] respectively (Figure 6.5e). This resulted in the production of nanocage structures, composed of the poly(acrylic acid-*co*-acrylamide) hydrogel material, which was originally present as the shell layer of the SCK. Since SCK structures rely upon a combination of hydrophobic interactions within the core domain (which initially provided for the micellar assembly) and reinforcing covalent shell crosslinks for maintenance of their structural features, removal of the core material resulted in a significant expansion of the hydrogel shell material to give an overall diameter for the nanocage that was much greater than that occupied by the SCK. This expansion was believed to occur due to a filling of the nanocage with water, and expansion of the structure to the extent that the crosslinked network allowed. The porosities and nanocage shell thicknesses are presently unknown. Ozonolysis of the PI core leaves ketones (and aldehydes) on the inner nanocage surface, whereas the remainder of the nanocage is composed of carboxylic acid and amide functional groups. This offers the ability to chemically derivatize specific regions of the nanocage, to provide for further regiochemical control in the preparation of advanced materials. Hydrophobic nanocages have been prepared in a similar manner, by using triblock copolymer micelles [47].

A significant advance in the practical aspects of SCK synthesis has involved the use of triblock copolymers, in order to perform the stabilizing crosslinking chemistry at the core–shell interface under conditions of high nanostructure concentration, with reduced potential for intermicellar coupling reactions (Figure 6.5f) [48]. In a similar manner, crosslinking of the central (bi-layer) region of polymer vesicles has led to the preparation of 'hollow' nanoscale crosslinked shells having polymer chains bound to the internal and external surfaces (Figure 6.5g). When the vesicles were composed of poly(isoprene-b-cinnamoylethyl methacrylate) diblocks, crosslinking was accomplished by photochemical [2+2]-cyclization reactions between the cinnamoylethyl methacrylate groups, with poly(isoprene) extending, internally and externally, from this crosslinked membrane. Ozonolytic cleavage of the polyisoprene removed the non-crosslinked polymer chain segments, while leaving behind unique aldehyde and ketone interior and exterior surface chemistries (Figures 6.4e and 6.5h) [49]. Although the use of vesicles for interior shell crosslinking followed by cleavage of internal and external surface chains is a promising route for the construction of hollow nanocages, the present chemistry, involving the photochemical crosslinking of CEMA groups in the presence of PI chains has potential for spurious crosslinking throughout the structure, and thus loss of regioselectivity. The authors do state that intervesicular crosslinking between the PI chains occurred with storage time [50].

4 REGIOSELECTIVE COUPLING/CROSSLINKING WITHIN MACROMOLECULES

4.1 INTRAMOLECULAR CROSS-LINKING OF DENDRIMER SURFACES

Perhaps the smallest example of a surface-attached, excavated nanostructure is that produced from the chain-end connection of dendritic macromolecules, followed by core unit removal [51]. The pre-attachment of dendritic fragments ('wedges') to a central core unit that acts as a template for establishment of the intramolecular chain end connectivity allows for the chain end coupling chemistry to be performed at high dilution to avoid intermolecular coupling reactions that could produce insoluble and uncontrolled networks. This assures that the monodispersity will be maintained without the possibility of intermolecular exchange of components that can result in the crosslinking of supramolecular assemblies. Crosslinking of the chain end functionalities of dendritic macromolecules closely resembles the polymerization of small surfactant molecules [52–54], in which there is only one polymerizable group per chain end of the dendrimer or per surfactant molecule in the assembly. However, because a covalent assembly has been established through the initial attachment to the core unit, the dendrimer case does not experience the difficulties associated with changes that can occur during polymerization of the surfactant functional groups that are part of a dynamic self-assembled system [55]. The dendritic macromolecule of Figure 6.6 was used to demonstrate the chain end coupling, accomplished through ring-closing metathesis reactions between the 1,3-homoallyl substituted phenyl rings. Model studies suggested a low probability of intrachain-end cyclization, providing for optimization of connections between dendrimer chain ends. It was found that c. 90% of the homoallyl chain ends underwent the ring-closing metathesis reaction. The central core was then removed by hydrolytic cleavage of the ester linkages. Although 10^6 isomers are possible from ring-closing metathesis couplings between the twelve chain ends to give six intra-dendrimer macrocycles, a much smaller number of those couplings will result in a structure that has the three dendritic fragments connected so that the overall structure is maintained following core removal. Figure 6.6 illustrates several modes of coupling that allow for maintenance of the overall molecular weight following hydrolytic core removal. Although the chemistry occurs regioselectively at the dendrimer chain ends, even with these rather small molecules (in comparison to the nanoscale assemblies described throughout most of this chapter), the location of the chain ends with respect to the dendritic molecule, and thus, the actual regiospecificity of the products is not yet known.

4.2 CORE-SHELL TECTO(DENDRIMERS)

Intermolecular dendrimer-dendrimer chain end coupling reactions under conditions that limit the assembly process to the attachment of smaller dendrimer units upon a larger central, templating dendrimer core has led to the construction of core-shell nanostructures, called core-shell tecto(dendrimers) (Figure 6.7). Core-shell tecto(dendrimers) are actually a subclass of relatively well defined poly(dendrimers) referred to as 'megamers' [56] (see Chapter 1). Nanostructures of varying dimensions can be synthesized from combinations of poly(amidoamine) (PAMAM) dendrimers of various generation levels. Two different approaches to these structures have been reported. The first strategy, referred to as the 'direct covalent bond formation method', produces partially filled shell structures and involves the reaction of a nueclophilic dendrimer core reagent with an excess of electrophilic dendrimer shell reagent [57]. Generally amine terminated dendrimer core reagents are combined with an excess of ester terminated dendrimer shell reagents to produce amide linkages at the dendrimer-dendrimer interface. The resulting core-shell structures are relatively

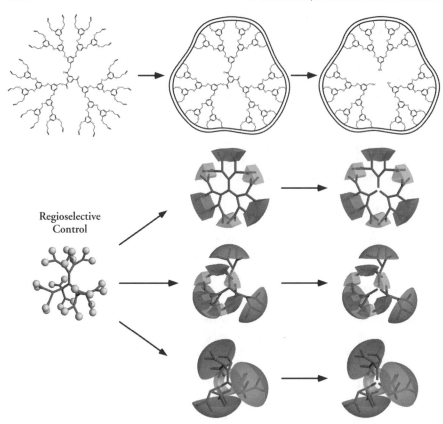

Figure 6.6 Peripheral crosslinking of a poly(benzyl ether) dendrimer through ring-closing olefin metathesis reactions, followed by hydrolysis of the core results in an intact final structure (having no net loss of individual dendritic fragments). A variety of crosslinking motifs may be seen: crosslinking all along the periphery with adjacent interdendron crosslinking (blue shading) (top); crosslinking of a majority of the chain ends on one face of the dendrimer, leaving the rest of the chain ends for intra-dendritic fragment crosslinking (magenta shading) (middle); inward-folding and crosslinking of a chain end from each dendritic fragment, with the remainder of the chain ends participating in intra-dendritic fragment crosslinking (bottom). Combinations of these motifs are also possible, as long as the final structure remains intact after core hydrolysis. Fortunately, olefin metathesis is reversible, so that a thermodynamic minimum may eventually be reached, possibly leading to only one final structure, depending on the global and local minima on the energy surface

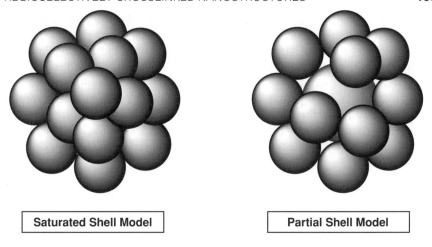

| **Saturated Shell Model** | **Partial Shell Model** |

Figure 6.7   Illustration of a core–shell tecto(dendrimer): partial shell and saturated shell models.

well defined as determined by MALDI-TOF mass spectrometry and gel electrophoresis (PAGE), however, only partially filled shells are obtained. Usually shell-filling levels of 40–60% are observed. These values are determined by comparison to theoretical values that may be predicted mathematically [58] as a function of the generational levels of the core and shell reagents, respectively.

More recently, core-shell tecto(dendrimers) possessing very high shell filling values have been reported [59]. The chemistry used in this approach involved the combination of an amine terminated core dendrimer with an excess of carboxylic acid terminated shell reagents dendrimer. These two charge differentiated species were then allowed to equilibrate and self-assemble into the electrostatically driven core-shell tecto(dendrimer) architecture. This equilibration is then followed by covalent fixing of these charge neutralized contact sites with a carbodiimide reagent. These well defined megameric structures have been shown to have shell filling values of 75–87% as determined by mass spectrometry and gel electrophoresis.

A 'scaled-up' version of this central template-concentric sphere surface assembly approach has been demonstrated for the growth of multi-layer core-shell nano- and microparticles, based upon the repeated layer-by-layer deposition of linear polymers and silica nanoparticles onto a colloidal particle template (Figure 6.8) [60]. In this case, the regioselective chemistry occurs via electrostatic interactions, as opposed to the covalent bond formation of most of the examples in this chapter. The central colloidal 'seed particle' dictates the final particle dimension, in magnitude and dispersity, and the number of adsorbed layers controls the shell thickness. When the colloidal template was present in solution at a few weight percent, electrostatically driven deposition of alternating layers of positively charged poly(diallyldimethylammonium chloride) and negatively charged silica nanoparticles was limited to individual template surface coverage, without extensive aggregation. Moreover, subsequent removal of the central colloidal core allowed for the fabrication of hollow capsules of submicrometer to micrometer diameters and 10–200 nanometer thicknesses (the shell thickness depends upon the number of adsorbed layers), composed of the surface adsorbed species.

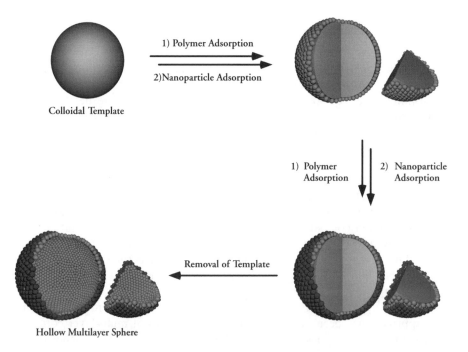

Figure 6.8 Illustration of colloid-templated nanoparticle assemblies. The process involves the layer-by-layer adsorption of charged polymers and oppositely charged nanoparticles onto the surfaces of the colloidal template. The colloidal core particles may then be removed to generate a hollow sphere of nanoparticles, held together by electrostatic interactions with the linear polymer 'glue'

5 CONCLUSION AND OUTLOOK

The challenges associated with the preparation of well-defined nanoscopic objects by a facile route are being addressed with rapidly increasing attention and progress. From the initial stages, it was recognized that the construction of such large objects (on the molecular scale) could not involve the direct, singular use of small molecule precursors, but rather that such syntheses would be most easily facilitated by the selective ordering and cementing of macromolecular building blocks. Therefore, the general theme that has been adopted in the current synthetic methods is to rely upon the self-assembly processes of multi-block copolymers, along with covalent crosslinking to provide reinforcement and produce individual, crosslinked, nanoparticulate entities.

The examples contained throughout this chapter illustrate the importance of incompatibilities between segments of block copolymers, which allows for their phase separation into reactive and nonreactive domains, in the bulk, in solution, or within a macromolecule, as a means to limit crosslinking reactions to selective regions of these polymer assemblies. This approach has provided a general and versatile method to impose permanence upon self-assembled nanostructured morphologies and to create complex materials with structural features of nanoscale dimensions. These materials possess the features introduced via the self-organization of polymer components; however, they are reinforced through intra-assembly crosslinks. Covalent reinforcement provides stability to the assembly and offers new sites of connectivity, which can be utilized in several ways.

1. Identification of the regions that segregate during the phase separation process in the bulk or in solution can be made by covalently connecting those regions containing reactive functional groups. This can be used, for example, to kinetically trap intermediate stages in the phase transition processes, to gain a better understanding of the migration events that lead to phase separation.
2. Advanced characterization of the structure, properties and function of the self-assembled precursor can be extrapolated from studies on the more robust crosslinked material, especially in changing or challenging environments, in which the assemblies would not remain intact. The introduction of crosslinks has aided in the maintenance of native conformations as a powerful technique during studies to determine the order and structure of biological assemblies [61, 62]. Moreover, the robust characteristics that the crosslinks provide, combined with the ability to define their regioselectivity, are expected to expand the realm of possible applications for nanoscale materials.
3. The new connective sites provide the opportunity to destroy or disconnect other regions of the nanostructure without destruction of the entire nanoscale entity. This was demonstrated by the excavation of the core of shell crosslinked polymer micelles, by the removal of the colloid from colloidally templated

nanoparticle assemblies, and in a molecular-level system, in which the single core unit of chain-end linked dendritic molecules was removed. In each case, the connectivity established through the crosslinked assembly remained intact.

6 REFERENCES AND NOTES

1. Hassner, A. *J. Org. Chem.*, **33**, 2684 (1968).
2. Stetter, H. and Dierichs, W. *Ber.*, **85**, 61 (1952).
3. Morton, M. 'Current status of anionic polymerization', in J. E. McGrath (ed.), *Anionic Polymerization: Kinetics, Mechanisms and Synthesis, ACS Symposium Series* **166**, American Chemical Society, Washington DC, 1981, pp. 27–31.
4. Pham, Q.-T. *Polymer Lett.* **8**, 723–729 (1970).
5. Sato, H. and Tanaka, Y. *J. Polym. Sci., Part A: Polym. Chem.* **17**, 3551–3558 (1979).
6. Clark, C. G. Jr. and Wooley, K. L. *Curr. Opin. Colloid & Interface Sci.* **4**, 122–129 (1999).
7. Ishizu, K. and Fukutomi, T. *J. Polym. Sci., Part C: Polym. Lett.* **26**, 281–286 (1988).
8. Ishizu, K., Sugita, M., Kotsubo, H. and Saito, R. *J Colloid Interface Sci.* **169**, 456–461 (1995).
9. Ishizu, K. and Saito, R. *Polym.-Plast. Technol. Eng.*, **31**, 607–633 (1992).
10. Liu, G., Ding, J., Qiao, L., Guo, A., Dymov, B. P., Gleeson, J. T., Hashimoto, T. and Saijo, K. *Chem. Eur. J.*, **5**(9), 2740–2749 (1999).
11. Chan, V. Z.-H., Hoffman, J., Lee, V. Y., Iatrou, H., Avgeropoulos, A., Hadjichristidis, N., Miller, R. D., Thomas, E. L. *Science*, **286**, 1716–1719 (1999).
12. Stupp, S. I., Son, S., Lin, H. C. and Li, S. *Science* **259**, 59–63 (1993).
13. Stupp, S. I., Son, S., Li, L. S., Lin, H. C. and Keser, M. *J. Am. Chem. Soc.*, **117**, 5212–5227 (1995).
14. Stupp, S. I., BeBonheur, V., Walker, K., Li, L. S., Huggins, K. E., Keser, M. and Amstutz, A. *Science*, **276**, 384–389 (1997).
15. Zubarev, E. R., Pralle, M. U., Li, L. and Stupp, S. I. *Science*, **283**, 523–526 (1999).
16. Stock, D., Leslie, A. G. W. and Walker, J. E. *Science*, **286**, 1700–1705 (1999).
17. Odian, G. *Principles of Polymerization*, 2nd edn, John Wiley & Sons, New York, 1981, pp. 87–288 and 319–337.
18. Zhang, L. and Eisenberg, A. *Science*, **268**, 1728–1731 (1995).
19. Tao, J. and Liu, G. *Macromolecules* **30**, 2408–2411 (1997).
20. Leduc, M. R., Hawker, C. J., Dao, J. and Fréchet, J. M. J. *J. Am. Chem. Soc.*, **118**, 11111–11118 (1996).
21. Procházka, K., Baloch, M. K. and Tuzar, Z. *Makromol. Chem.*, **180**, 2521–2523 (1979).
22. Wilson, D. J. and Riess, G. *Eur. Polym. J.*, **24**, 617–621 (1988).
23. Saito, R., Ishizu, K. and Fukutomi, T. *Polymer*, **33**, 1712–1716 (1992).
24. Henselwood, F. and Liu, G. *Macromolecules*, **30**, 488–493 (1997).
25. Tao, J., Liu, G., Ding, J. and Yang, M. *Macromolecules* **30**, 4084–4089 (1997).
26. Henselwood, F. and Liu, G. *Macromolecules*, **31**, 4213–4217 (1998).
27. Won, Y.-Y., Davis, H. T. and Bates, F. S. *Science*, **283**, 960–963 (1999).
28. Thurmond, K. B. II, Huang, H., Clark, C. G. Jr., Kowalewski, T. and Wooley, K. L. *Colloids and Surfaces, B: Biointerfaces*, **16**, 45–54 (1999).
29. Bergbreiter, D. E. *Angew. Chem. Int. Ed.*, **38**(19), 2870–2872 (1999).
30. Knedel is a Polish term for 'dumpling', and the shell crosslinked nanoparticles

resemble macroscopic knedels in the core-shell morphology, with the shell providing the reinforcement of the particle and containment of the core material.

31. Thurmond, K. B. II, Kowalewski, T. and Wooley, K. L. *J. Am. Chem. Soc.*, **118**, 7239–7240 (1996).
32. Thurmond, K. B. II, Kowalewski, T. and Wooley, K. L. *J. Am. Chem. Soc.*, **119**, 6656–6665 (1997).
33. Remsen, E. E., Thurmond, K. B. II and Wooley, K. L. *Macromolecules*, **32**, 3685–3689 (1999).
34. Thurmond, K. B. II and Wooley, K. L. *ACS Symposium Series on Materials for Controlled Release Applications*, Ch. 13, American Chemical Society, Washington, DC, 1998.
35. Thurmond, K. B. II, Remsen, E. E., Kowalewski, T. and Wooley, K. L. *Nuc. Acids. Res.* **27**, 2966–2971 (1999).
36. Huang, H., Kowalewski, T., Remsen, E. E., Gertzmann, R. and Wooley, K. L. *J. Am. Chem. Soc.*, **119**, 11653–11659 (1997).
37. Huang, H., Remsen, E. E. and Wooley, K. L. *Chem. Commun.*, 1415–1416 (1998).
38. Gullion, T. and Schaefer, J. *Adv. Magn. Reson.*, **13**, 57 (1989).
39. Baugher, A. H., Goetz, J. M., McDowell, L. M., Huang, H., Wooley, K. L. and Schaefer, J. *Biophys. J.*, **75**, 2574–2576 (1998).
40. (a) Kau, H.-M., O'Connor, R. D., Mehta, A. K., Huang, H., Poliks, B., Wooley, K. L., Schaefer, J. *Macromolecules*, **34**, 544–546 (2001); (b) Huang, H., Wooley, K. L., Schaefer, J. *Macromolecules*, **34**, 547–551 (2001).
41. Bütün, V., Billingham, N. C. and Armes, S. P. *J. Am. Chem. Soc.*, **120**, 12135–12136 (1998).
42. Bütün, V., Lowe, A. B., Billingham, N. C. and Armes, S. P. *J. Am. Chem. Soc.*, **121**, 4288–4289 (1999).
43. Huang, H. and Wooley, K. L. *ACS Polym. Prepr.*, **39**(1), 239 (1998).
44. Wooley, K. L., Huang, H. and Kowalewski, T. *ACS Polym. Matl. Sci. Eng.*, **80**, 13 (1999).
45. (a) Zhang, Q. and Wooley, K. L. *ACS Polym. Prepr.*, **40**(2), 986 (1999); (b) Zhang, Q., Remsen, E. E. and Wooley, K. L. *J. Am. Chem. Soc.*, **122**, 3642–3651 (2000).
46. Huang, H., Remsen, E. E., Kowalewski, T. and Wooley, K. L. *J. Am. Chem. Soc.*, **121**, 3805–3806 (1999).
47. Stewart, S. and Liu, G. *Chem. Mater.*, **11**, 1048–1054 (1999).
48. Bütün, V., Wang, X.-S., de Paz Banez, M. V., Robinson, K. L., Billingham, N. C., Armes, S. P. and Tuzar, Z. *Macromolecules*, **33**, 1–3 (2000).
49. Ding, J. and Liu, G. *Chem. Mater.*, **10**, 537–542 (1998).
50. Ding, J. and Liu, G. *J. Phys. Chem. B*, **102**, 6107 (1998).
51. Wendland, M. S. and Zimmerman, S. C. *J. Am. Chem. Soc.*, **121**, 1389–1390 (1999).
52. Fendler, J. H. *Surfactants in Solution* [Proc. Int. Symp.], K. L. Mittal and B. Lindman, (eds), Plenum Press, New York, (Pub. 1984), **3**, 1947–1989 (1982).
53. Paleos, C. M. and Malliaris, A. *J. Macromol. Sci. Rev. Macromol. Chem. Phys.*, **C28**(3–4), 403 (1988).
54. Ringsdorf, H., Schlarb, B. and Venzmer, J. *Angew Chem. Int. Ed. Engl.*, **27**(1), 113–158 (1988).
55. Hamid, S. and Sherrington, D. *J. Chem. Soc., Chem. Commun.*, **12**, 936–937 (1986).
56. Tomalia, D. A. and Majoros, I. Chapter 9, Dendrimeric Supramolecular and supra-macromolecular Assemblies in *Supramolecular Polymers* (A. Ciferri, ed.), Marcel Deker, New York, 359–434 (2000).
57. Li, J., Swanson, D. R., Qin, D., Brothers, H. M., Piehler, L. T., Tomalia, D. and Meier, D. J. *Langmuir* **15** 7347–7350 (1999).

58. Mansfield, M. L., Rakesh, L., Tomalia, D. A. *J. Chem. Phys.*, **105**, 3245–3249 (1996).
59. Uppuluri, S., Piehler, L. T., Li, J., Swanson, D. R., Hagnauer, G. L. and Tomalia, D. A. *Adv. Mater.*, **12**(11), 796–800 (2000).
60. Caruso, F. *Chem. Eur. J.*, **6**(3), 413–419 (2000).
61. Muhlberg, A. B., Warnock, D. E. and Schmid, S. L. *EMBO J.*, **16**(22), 6676–6683 (1997).
62. Leroux, M. R., Melki, R., Gordon, B., Batelier, G. and Candido, E. P. M. *J. Biol. Chem.* **272**(39), 24646–24656 (1997).

7

Hybridization of Architectural States: Dendritic-linear Copolymer Hybrids

P. R. L. MALENFANT AND J. M. J. FRÉCHET
Department of Chemistry, University of California, Berkeley, CA, USA

1 INTRODUCTION

As their name suggests, dendritic-linear copolymers are hybrid structures that combine two very different types of macromolecular architectures. Given the relationship between molecular architecture and properties, combining one or more perfectly branched globular component with one or more linear chain in a single macromolecule can have a profound effect on the ultimate properties of the hybrid material that results. These properties are affected not only by the relative proportion of the architectural components but also by their own intrinsic properties, their placement within the macromolecule and the presence of functional groups at specified locations of the hybrid entity. It is only within the past decade that a significant number of reports of such true hybrid copolymers have appeared [1], as their widespread study was made possible [2] by the introduction of the highly versatile convergent method of synthesis of dendrimers in 1989–90 [3].

For simplicity, this unique class of 'hybrid' macromolecules may be divided into four major families shown schematically in Figure 7.1.

- AB diblock copolymers.
- ABA triblock copolymers, sometimes referred to as 'dumbbell' shaped copolymers when B is a linear macromolecule.
- Side-chain functionalized or 'dendronized' [4] copolymers.
- Linear-dendritic star copolymers [5], most frequently obtained via processes in which dendrimers function as multifunctional initiator cores for the poly-

Dendrimers and Other Dendritic Polymers. Edited by Jean M. J. Fréchet and Donald A. Tomalia
© 2001 John Wiley & Sons Ltd

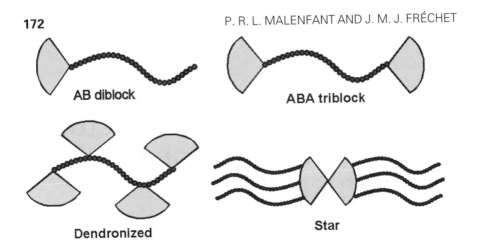

Figure 7.1   Types of hybrid dendritic–linear macromolecules

merization of linear polymers. These star copolymers are not within the scope of this chapter and will therefore not be discussed.

Regardless of the desired architecture, there are three main synthetic approaches to the preparation of diblock, dumbbell or dendronized hybrid copolymers. These involve grafting, polymerization or stepwise dendritic growth processes as follows:

- Grafting: hybrids are obtained by coupling one or more reactive groups of a pre-formed linear polymer chain to the reactive focal moiety of a convergently grown dendritic block or dendron. An analogous grafting process, leading to star-like architectures, involves the grafting of linear chains to the multiple chain ends of a dendrimer.
- Polymerization: hybrids are obtained by the polymerization of a linear chain initiated from the focal point of a convergent dendrimer or the chain ends of a dendrimer, or by polymerization or copolymerization of a dendronized monomer.
- Dendritic growth: hybrids are prepared using one or more functional groups of a preformed linear chain to effect the stepwise growth of one or more dendritic blocks via divergent synthesis.

Although, at first glance, these approaches may seem interchangeable, the structural precision of the final molecules varies greatly with the synthetic route selected. In addition, the compatibility of the blocks as well as the specific reaction conditions used during the preparation of the hybrid may restrict the choice of synthetic approach. For instance, the multistep synthetic design used for divergent [6] dendritic growth from a linear polymer must be tailored to accommodate the linear fragment of the block copolymer, while also allowing intermediate purifications during growth. In contrast, the attachment of a linear chain to a single focal point of a convergent [3] dendrimer involves a single step, thereby facilitating synthetic design, though compatibility of the two blocks and accessibility of reactive sites may be a significant problem. In the end, the specific features of the target material and its inherent properties, as well as those of its precursors, will determine the feasibility of the ultimate synthetic protocol.

This chapter will not attempt to present encyclopedic coverage of the field, but it will only provide representative examples of the processes used to access each of the main types of hybrid block architectures, then it will take a somewhat closer look at specific families of amphiphilic or electroactive hybrid dendritic-linear macromolecules with broad potential for practical applications.

2 DIBLOCK HYBRIDS PREPARED BY POLYMERIZATION FROM A DENDRITIC INITIATOR

While the three synthetic approaches shown in Figure 7.2 have been used to prepare diblock hybrids, this section will only focus on their preparation via polymerization of a linear chain from a dendron used as the initiator (route B, Figure 7.2). In particular, this approach has exploited the functional versatility [1a, 7] of the poly(benzyl ether) dendrons [3] frequently referred to as 'Fréchet-type' dendrons, that are prepared by convergent growth. The single reactive group located at the focal point of a convergent dendron is ideally suited not only for the grafting of a linear chain, but also for growth of a linear chain in cases where it can act as a polymerization initiator.

Taking advantage of this feature, Gitsov *et al.* used the benzylic alcohol focal

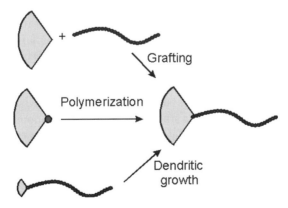

Figure 7.2 Preparation of diblock hybrids

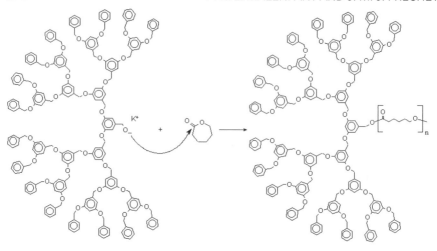

Figure 7.3 Growth of poly(caprolactone) linear block from a dendritic macro-initiator

functionality of a Fréchet-type dendron to polymerize ε-caprolactone (Figure 7.3) [8]. It is well known that the anionic polymerization of ε-caprolactone is frequently plagued by side reactions such as back-biting that severely limit the molecular weight of the resulting polymers. Two valuable observations were made while studying this dendritic macroinitiator system. First, the successful polymerization of ε-caprolactone initiated with a generation four (G-4) dendron showed that the alcohol focal point of the dendrimer is very accessible and can be used as an efficient site for initiation as evidenced by the high yield of copolymer obtained in a very short time. Second, the use of a G-4 dendritic initiator also suppressed the backbiting reactions typically seen in the anionic polymerization of ε-caprolactone initiated with *t*-butoxide or a lower generation G-1 initiator, thus leading to high molecular weight materials with minimal oligomer formation as confirmed by the very low polydispersity index (PDI) of the final hybrid.

Using a similar approach, Leduc *et al.* have prepared diblock hybrids in which a polystyrene block is grown from the focal point of a poly(benzyl ether) dendron [9]. The polymerization process made use of the 'living' free radical polymerization methods employing either a stable nitroxide [10] compound as initiator or a metal catalyzed process such as 'atom transfer radical polymerization' (ATRP) [11], with the convergent dendron as the macroinitiator. Once again the versatility of the convergent approach is illustrated by the ease of preparation of the macroinitiators through the simple functionalization of the focal point of suitable dendrons. In the case of the nitroxide-mediated polymerization, we made use of the seminal work of Hawker [10b] with unimolecular initiators, preparing a dendritic nitroxide based macroinitiator [9a]. Based on the pioneering work of

Matyjaszewski *et al.* on the development of ATRP [11], the benzylic halide focal functionality of Fréchet-type dendrons was later used to initiate a Cu-catalyzed polymerization via ATRP (Figure 7.4) [9b]. With both types of macroinitiators, the process was well controlled when the polymerization was used to produce linear blocks of polystyrene with number average molecular weight (Mn) of up to 3×10^4 as determined by size exclusion chromatography (SEC). For higher molecular weight linear blocks, control of the process appeared to deteriorate slightly as more significant deviations from theoretical molecular weights were observed.

While the nitroxide macroinitiator afforded best results, it should be noted that the benzylic halide focal point of the dendrons used to initiate the metal-catalyzed ATRP process was not optimized since a primary rather than a secondary benzylic halide was used to initiate polymerization. Overall, copolymer hybrids with relatively narrow molecular weight distributions (PDI = 1.1–1.5) could be obtained in high yields. It is interesting to note that no polystyrene homopolymer that might have resulted from a thermally initiated styrene autopolymerization [12] was observed despite the high temperatures used in these living radical polymerizations. All of the polymers prepared were found to have a single glass transition (T_g) as assessed by differential scanning calorimetry (DSC), which suggests good miscibility between the dendritic and linear blocks.

The properties of the hybrid diblock structures can be altered drastically by simply taking advantage of the high terminal functionality of the dendritic block. For example unusual diblock structures useful for the modification of surfaces have been prepared by ATRP of polystyrene (PS) initiated from the benzylic halide focal point of Fréchet-type dendrons with terminal isophthalate ester groups [9b]. Well-defined copolymers with narrow molecular weight distributions were obtained and excellent agreement was observed between calculated

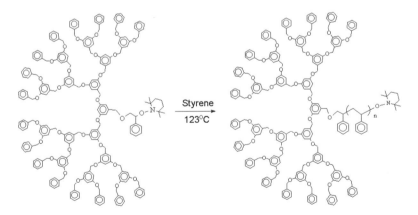

Figure 7.4 Living radical polymerization from the focal point of a dendron

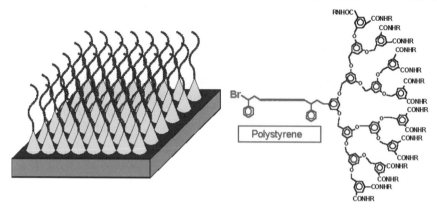

Figure 7.5 Hybrid copolymers with functionalized dendrons and their assembly on a surface

and experimental molecular weights as assessed by SEC up to Mn values of *ca.* 4×10^4. Subsequent modification of the dendritic isophthalate ester surface provided high yields of materials with a wide range of polarities. The use of such diblock hybrids as brush-type surface modifiers can be advantageous as the high local concentration of potentially 'interacting' groups on the dendritic surface can provide strong, multi-prong, anchoring anchor to the surface, while the lengthy, brush-like, linear block will afford novel surface properties such as compatibility, adhesion or repellent-action, etc. In such applications, the functional periphery of the dendritic block provides a means of easily tailoring the hybrid material for use with a variety of substrate surfaces. For example, the dendritic amide functionalized chain ends [9b] of a hybrid are capable of strong interactions with a complementary surface such as nylon through the formation of multiple hydrogen bonds (Figure 7.5).

Since this work was performed significant advances have been made in the area of living radical polymerization with the introduction of novel, better controlled, initiators as well as reaction conditions that enable the use of lower polymerizations temperatures with a broader choice of monomers. It is clear that these advances could easily be applied to the preparation of a broader array of well-defined hybrid dendritic-linear structures.

3   TRIBLOCK ABA COPOLYMER HYBRIDS

ABA triblock hybrids with Fréchet-type dendrons as terminal A blocks and polystyrene as the central B block have been prepared by anionic bidirectional growth of polystyrene followed by attachment of reactive dendrons at the chain

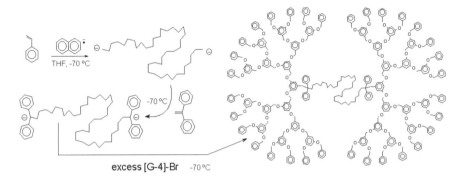

Figure 7.6 Preparation of triblock hybrid from two-ended anionic polystyrene

ends of the linear polymer (Figure 7.6) [13]. In order to achieve bidirectional growth of a reactive polystyrene telechelic structure, the anionic polymerization of styrene was initiated with potassium naphthalenide. The well-controlled living anionic polymerization process was then followed by termination with 1,1-diphenylethylene to afford a polymer with two anionic chain ends that is incapable of further propagation with styrene, but can participate in nucleophilic coupling with the reactive focal point (benzylic halide, aldehyde, or ester) of convergent dendrons. As was the case for the analogous hybrid diblocks, these materials exhibit a single T_g indicating that, unlike physical mixtures of their individual components, the linear and dendritic blocks are miscible on the molecular level. Studies of solutions of the ABA hybrid triblocks by SEC and viscometry show that they are not entangled and that as the PS block increases in length, the molecules undergo a transition from an extended globule to a statistical coil. Therefore, when the PS block is short, the dumbbell-shaped structure behaves just like a classical dendrimer, whereas the hybrid assumes a statistical coil confirmation that is reminiscent of linear PS homopolymers when the PS block is sufficiently long.

While polymerization routes are convenient for the controlled preparation of diblock hybrids, the same cannot be said for ABA triblocks. Nevertheless it is possible to use a living free radical approach to prepare such triblock hybrids. For example, this might involve the use of a unimolecular alkoxyamine initiator with a dendron on either side of the central cleavable C-O–N bond (Figure 7.7) [14]. While this approach has significant limitations imparted by the thermal lability of the alkoxyamine itself, it provides some insight into the reaction dynamics of this living polymerization. Polymerization of styrene initiated by this bis-dendritic alkoxyamine affords a product mixture consisting mainly of the desired dumbbell-shaped ABA triblock together with some undesired AB diblock impurity. Although initial SEC studies had suggested a monomodal molecular weight distribution, detailed proton nuclear magnetic resonance

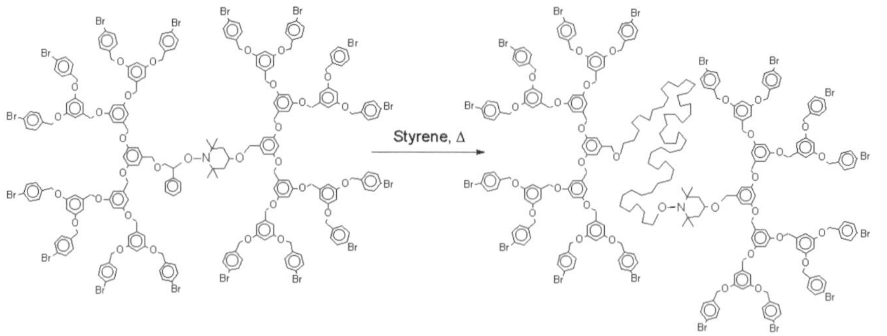

Figure 7.7 Growth of ABA triblock hybrid by living free radical polymerization

(^{1}H NMR) studies confirmed the presence of the AB diblocks in the product. This determination was facilitated by the fact that the dendritic nitroxide could be differentiated from the nonnitroxide-bearing dendron by ^{1}H NMR spectroscopy. Careful analytical studies confirmed that the pure ABA copolymers could be separated by column chromatography and that the undesired diblock impurity resulted mainly from the loss of the dendritic nitroxide during the course of the reaction. Obviously, this approach to ABA triblocks has rather limited practical value since the thermal stability of the final product is quite low.

4 SIDE-CHAIN FUNCTIONALIZED OR 'DENDRONIZED' COPOLYMER HYBRIDS

Another type of architecture featuring a linear main chain surrounded by dendritic side-chains has emerged over the last decade [4]. The highly descriptive term 'dendronized', coined by Schlüter [4] aptly describes this novel type of macromolecular architecture. Though three separate routes can be used to prepare such dendronized hybrids (Figure 7.8), the most successful approach to date has generally involved the polymerization of dendronized monomers.

The concept of making brush-type polymers in which a linear polymer is funtionalized with dendritic side-chains was first suggested by Tomalia in a 1987 patent, though actual experimental work on his approach was only reported recently recently [15]. Hawker and Fréchet were first to document the preparation of a vinyl copolymer containing a few pendant Fréchet-type dendrons (Figure 7.9).

These were obtained by copolymerization of styrenic macromonomers containing dendritic pendant groups with styrene [16a]; with a later extension to methacrylate-type copolymers [16b]. As these studies were carried out with

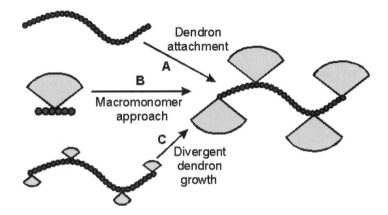

Figure 7.8   Preparation of 'dendronized' macromolecules

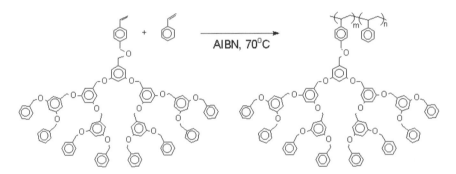

Figure 7.9   Copolymerization of dendronized styrenic monomer and styrene

macromonomers containing rather large G-3 or G-4 Fréchet-type pendant dendrons, homolpolymerization of these monomers was not successful.

Other efforts based on the macromonomer approach to homopolymers having dendritic side chains, include the work of Draheim and Ritter on acrylate and methacrylate derived structures having dendritic chiral side chains based on L-aspartic esters [17a], and of Xi and coworkers with poly(methacrylate) structures containing very small benzyl ether dendritic side-chains [17b]. Unfortunately, both of these approaches met with limited success due to a significant drop in degree of polymerization (DP) when the size of the dendron used as pendant group in the macromonomers increased from G-1 to G-2.

Much better results were obtained when dendronized structures were prepared via ring opening metathesis polymerization (ROMP) of norbornene monomers functionalized with Fréchet-type dendrons [18]. Using a rhodium catalyzed polymerization, Oikawa and coworkers successfully prepared side-chain functionalized hybrids [19] based on Moore-type phenylacetylene dendrons [20] and a linear polyacetylene backbone.

Attempts to prepare polymers that are shape persistent in solution and in the solid state have led to dendronized structures in which the steric bulk provided by the dendrons is maximized in order to permanently influence the shape of the linear polymeric backbone. In seminal work, Percec and coworkers have explored the polymerization of methacrylate and stryrenic macromonomers [21] (Figure 7.10) having pendant benzyl ether dendrons that contain alkyl chains at their periphery. Percec elegantly overcame the problem of steric hindrance that usually limits the degree of polymerization (DP) achievable with bulky macro-monomers, by discovering that, at a critical concentration, the self-assembly [22] of dendritic macromonomers leads to the formation of spherical 'nano-reactors' in which a dramatic, yet well-behaved, self-accelerated free radical polymerization takes place [23a]. As a result, bulky dendritic macromonomers could easily be polymerized without the use of a spacer that would adversely influence the driving force for shape persistence. Furthermore, the DP could be controlled leading to a variety of nanoscale structures with unique shapes For example, at relatively low DP, a spherical polymer is obtained in which the linear fragment is 'bundled' as a random-coil at the center of the sphere. At higher DP, the dendronized copolymer minimizes its free energy by adopting a cylindrical shape (Figure 7.10) in which the linear polymer is now extended or helical. Simultaneously, the dendrons, adopt a less conical and flatter tapered geometry, thus behaving as 'quasi-equivalent' building blocks [23b].

The supramolecular transformation from sphere to cylinder is supported by X-ray data indicating that the spherical polymers adopt a cubic phase, whereas the cylindrical polymers adopt a hexagonal phase [23b]. Further studies involving a library of dendritic macromonomers led to the conclusion that the effect of DP on polymer shape is a general phenomenon [24]. More recently, scanning

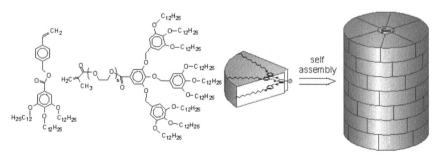

Figure 7.10   Tapered macromonomers and their cylindrical polymerized assemblies

force microscopy (SFM) revealed that branching density on the dendritic segment, as well as intramolecular and surface interactions, leads to unusual conformations and molecular ordering on substrate surfaces [25].

In a related study, Jahromi *et al.* have recently shown that self-assembly can also be achieved with macromonomers having Fréchet-type pendant dendrons, but no LC-phase inducing substituents (such as the peripheral alkyl chains of Percec's dendrons) [26]. Careful analysis of the intrinsic viscosity of side chain copolymer hybrids based on benzyl ether dendrons attached to a polyurethane backbone suggests that a conformational change from spherical to cylindrical occurs at a certain DP as the generation number changes from G-3 and G-4. In contrast to Percec's system, it appears that in the absence of LC inducing substituents, larger dendrons are required to induce a conformational change. In recent work, Tomalia has also reported the formation of rods in hybrids based on a 2-ethyl-2-oxazoline backbone and G-4 PAMAM dendrons grown divergently from the linear polymer [27].

In one of several important studies on dendronized polymers [4c, 4d]. Schlüter and coworkers explored the stiffening of polystyrene chains through the incorporation of Fréchet-type dendrons as side chains [28, 29]. While the G-1 and G-2 dendrons were not sufficiently bulky to effectively stiffen the polystyrene chain, the G-3 dendron provides enough steric bulk to force the hybrid polymer into adopting a cylindrical shape in solution [28b]. In a complementary study, Neubert and Schlüter demonstrated that adding charges to the dendritic wedges leads to an expansion of the chains of the hybrid copolymer in aqueous solution [29].

In their continuing search for nanoscale objects that are shape persistent in solution, Schlüter and co-workers used a different and very imaginative approach to prepare dendronized macromolecules consisting of poly([1.1.1]propellanes) and poly(*p*-phenylene) (PPP) functionalized with pendant Fréchet-type dendrons [30]. The dendronized poly(*p*-phenylene)s were obtained using two different synthetic approaches: (1) attachment of the dendrons to an existing PPP backbone, and (2) polymerization of macromonomers via a Suzuki cross-coupling polycondensation reaction. The second reaction is preferred as it leads to more regular structures with dendron incorporation at each repeat unit of the macromolecule (Figure 7.11) [31]. Using this approach, copolymers with G-3 pendant dendrons and as many as 110 repeat units were obtained attesting to the excellent accessibility of the focal point of the dendritic macromonomers as well as the quality of the Suzuki cross-coupling reaction conditions used by Schlüter [32].

As was the case for Percec's hybrid structures [25] (see above), studies of the G3-dendronized PPP by scanning force microscopy revealed a high degree of dimensional ordering in which the three-fold symmetry of the substrate surface is effectively recognized by the nanocylinders over several layers of polymer [32]. More recently, Schlüter extended his study to the preparation of a PPP hybrid

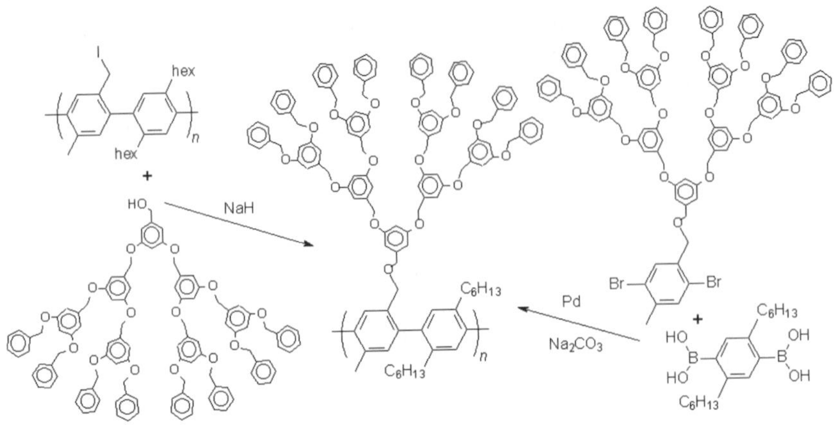

Figure 7.11   Two routes for the preparation of dendronized poly(p-phenylene)

dendronized with even larger G-4 dendrons. In this case, the steric requirements of the bulky G-4 Fréchet-type dendrons prevented the formation of all but oligomeric materials [33]. An excellent review of dendronized copolymer hybrids has appeared [4b].

5   AMPHIPHILIC HYBRIDS

The incorporation of dendritic moieties into amphiphilic structures is attractive as the dendritic component provides a unique opportunity to vary, in a controlled fashion, 'head group' properties such as size and polarity, by simply changing the generation or the nature of its surface groups. Similarly, the length of the linear fragment can easily be modified to adjust properties.

Although well outside the scope of this brief review, it should be noted that dendritic unimolecular micelles based on a classical dendrimer architecture have been prepared by several research groups including ours [1d, 34]. The solubilizing and container properties of such dendritic 'micelle-like' molecules form the basis of their application in novel drug delivery systems [35]. While also not true, dendritic-linear hybrids, given their lack of a macromolecular linear component, the small, barbel-like, amphiphilic dendrimers of Newkome *et al.* consisting of two arborol end groups linked by a variety of very short spacers display gel-forming properties [36]. Such properties, resulting from medium-dependent supramolecular assembly, are not unique as they are also found in numerous small molecule and macromolecular gelators [37].

In order to explore the properties that may be obtained by hybridizing the linear and dendritic architectural states, both diblock and triblock copolymer amphiphiles have been investigated. We were the first to prepare and explore true AB diblock and ABA triblock copolymer amphiphiles. A particularly successful design involved dendritic polyether A blocks and rather lengthy macromolecular linear B blocks based on poly(ethylene oxide) (PEO) or poly(ethylene glycol) (PEG) [38]. The synthesis of these materials by a fragment (block) coupling process is quite remarkable as essentially quantitative coupling of the macromolecular fragments is observed in the Williamson coupling of hydroxy-terminated PEO or PEG with the benzylic halide focal point of Fréchet-type dendrons (Figure 7.12). Surprisingly, the rate constant for the coupling reaction actually increased as the size of the individual macromolecular blocks increased [39]. Such anomalous behavior probably reflects the increased reactivity of the alkoxide anions as a result of solvation processes involving the polyether components, and the relative affinity of the two polyether components for each other and the various ions present in the reaction mixture. Clearly, the environment of the dendron must be energetically favorable for the anionic end of the PEO block and its counterion, thus bringing the reactive ends of the two blocks into close proximity. Solid-state studies revealed that the ABA triblocks could crystallize into axialites, spherulites or novel dendritic structures as a function of copolymer composition and casting solvent [40]. Thermal analysis of the hybrid materials reveals that all have a single glass transition, which indicates that the A and B blocks effectively plasticize each other. As will be seen in Figure 7.12, all of the hybrid copolymers behave as nonionic surfactants and their water solubility depends on the PEO or PEG to dendron ratio.

Not unexpectedly in view of their composition that includes two globular 'A' components, the polystyrene equivalent molecular weights of the ABA structures, as assessed by SEC, were lower than expected [38–40]. Although this is

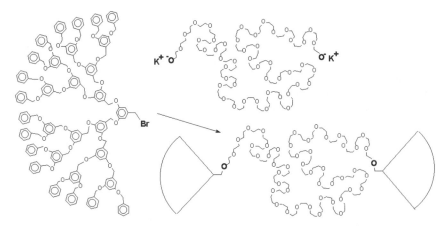

Figure 7.12 Preparation of an amphiphilic ABA triblock hybrid by Williamson ether synthesis

consistent with the lower hydrodynamic radius of dendrimers vs. linear polymers, it could also be the result of monomolecular micelle formation. ^{1}H NMR studies reveal that these unique triblock structures can form micelles in solvents where both blocks are poorly soluble, or when only one block is selectively soluble [39]. Investigation of the behavior of ABA triblocks in which the dendron generation is varied from G1 through G4 and four different PEG lengths are used (1K, 2K, 5K and 11K) [40] in methanol/water solutions using SEC with coupled viscometric detection (VISC) showed that both mono- and multimolecular micelles can form as one varies the concentration and dendron generation. In contrast, only monomolecular micelles are observed in THF. A more elaborate study [41] of similar AB and ABA hybrid macromolecules revealed that the critical micelle concentration (cmc) of such molecules in water depends on both the dendron generation and the length of the linear aliphatic polyether block. Architecture also has a clear effect on the cmc value. For instance, changing the architecture from diblock to triblock for structures with similar A and B blocks leads to a tenfold decrease in cmc. The ABA triblock molecules have an interesting ability to change drastically the properties of surfaces on which they are spread. For example, they are capable of increasing significantly the hydrophilicity of a cellulose surface, thereby demonstrating potential for surface protection applications [41].

A related star-like stimulus-responsive system consisting of a central four-arm PEG terminated by Fréchet-type dendrons, was shown to change both its shape and 'external' functionality in response to changes in its environment [42]. In CHCl$_3$, a good solvent for both blocks, the star polymer adopts an extended conformation, while in THF, a good solvent for the dendrons but not for the PEG chains, a more compact structure is formed as assessed by SEC/VISC analysis. This type of medium-dependent change in shape, volume and 'external' functionality is useful for the design of sensors and macromolecular drug-delivery systems [43].

Chapman and co-workers have used a totally different approach to prepared dendritic-linear diblock amphiphilic structures. This involved divergent dendritic growth of a poly(lysine) dendritic block [44] from an amine terminated PEO. The resulting 'hydraamphiphile' (Figure 7.13A) had cmc values that were low, but still one to two orders of magnitude higher than those of the more contrasted amphiphilic hybrids consisting of poly(benzyl ether) dendrons and PEO or PEG blocks described above [41]. Aqueous solutions of the G4 hydraamphiphile were able to solubilize the water-insoluble dye Orange-OT.

Meijer and co-workers also used a divergent dendrimer synthesis to prepare AB diblock structures (Figure 7.13B) in which the polystyrene linear block is used to initiate growth of the poly(propylene imine) dendritic block [45]. An amino end group had to be introduced in the polystyrene as a core for subsequent growth of the dendritic fragment via an iterative protocol of sequential

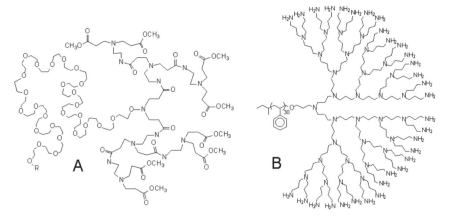

Figure 7.13 (A) Hydraamphiphile and (B) polystyrene-poly(propylene imine) diblock structures

cyanomethylation and hydrogenation reactions. Although this approach appears straightforward, the poor accessibility of the reactive end of the polystyrene block, and the lack of compatibility of the dissimilar blocks made this approach experimentally challenging. We experienced similar difficulties in our own research while attempting to couple hydroxy terminated polystyrene with the benzylic halide focal point of polyether dendrons [46]. Despite these difficulties, Meijer *et al.* have reported well-defined hybrid structures with dendritic blocks as large as G5. These unique copolymers were found to mimic the properties of both small molecule surfactants as well as linear block copolymer amphiphiles. For instance, in contrast to more conventional linear block copolymer amphiphiles, which typically form spherical micelles, these macromolecular amphiphiles were found to form highly stable aggregates [45] at very low concentrations (cmc as low as 5×10^{-7} M) even at elevated temperatures. Forming stable aggregates at low concentration is a very important characteristic, since in a controlled-release 'delivery' application, the micellar structure is subjected to an infinitely dilute environment and must retain its aggregated integrity in order to avoid premature release of the payload [43].

Using what is perhaps the most efficient laboratory-scale synthesis to-date, Fréchet *et al.* have prepared a family of ABA hybrids with aliphatic ester dendron A blocks and PEG B block. These non-toxic hybrids are being tested as drug delivery vectors [47].

Meijer and co-workers have also explored hybrids based on polystyrene and poly(propylene imine) dendrons with carboxylic acid functionalities on the surface of the dendrons [48]. In these materials the polarity of the head group can easily be modified in a controlled manner leading to both pH and generation-dependent aggregation behavior [48]. For this type of hybrids, it is the surface carboxylic acid moieties of the dendritic blocks that contributes most to incompatibility between blocks and microdomain formation with morphological features that vary with dendron generation [49].

Similarly, Hammond and co-workers have prepared hybrids by divergent growth [6a] of a PAMAM dendron from an amino-terminated linear PEO backbone [50]. Purification of the hybrid copolymer was facilitated by its lack of solubility in diethyl ether enabling its isolation by precipitation. In agreement with earlier findings by Fréchet and co-workers [51], the glass transition temperatures of the hybrids were affected by the nature of the chain ends and generally increased with the dendron generation. For example, amine-terminated dendritic hybrids were found to have T_g values c. 30 °C higher than those of their ester-terminated analogs. Consistent with the earlier observations of Gitsov et al. [39] (see above), phase segregation occurs when the PEO content reaches a threshold value and the crystalline PEO segment exhibits a melting point. Intrinsic viscosity studies in water indicates that diblocks with short PEO segments behave similarly to linear PEO homopolymers, whereas diblocks with longer PEO segments form unimolecular micelles as demonstrated earlier [38–41]. In a separate study, Hammond also observed that modifying the surface functionality of the dendrons with stearate groups, led to materials with good film forming properties [52].

In our studies on PEO-poly(benzyl ether) dendritic hybrids, we had determined that the PEO to dendron ratio, and hence the PEO block length, had a significant effect on the area of macromolecules at the air–water interface, thus suggesting that both blocks are present at the interface [41]. Hammond comes to the same conclusion when examining surface pressures for stearate functionalized hybrids at the air–water interface [52]. In a related study, Hawker and co-workers examined poly(benzyl ether) dendrimers with oligo(ethylene glycol) chains of exact lengths tethered to their focal point [53]. This rigorous study shows conclusively that the molecular area occupied by the molecules at the air–water interface increases linearly with increasing oligomer length.

Okada and co-workers have also prepared hybrids based on poly(2-methyl-2-oxazoline) and a PAMAM dendrimer by living ring-opening polymerization of 2-methyl-2-oxazoline followed by end-capping with ethylene diamine and subsequent iterative dendritic growth [6a] of the PAMAM dendritic block [54]. In this system, the linear segment is the more hydrophilic component, whereas the dendron is only somewhat more hydrophobic. As a result of the low hydrophilic–hydrophobic contrast between the two blocks, the hybrids have very high cmc values of 0.49 wt% for the G5.5 hybrid and 2.2 wt% for the even less hydrophobic G3.5 analog having ester surface groups.

Schlüter and co-workers have reported dendronized polystyrene with dendrons having amine, ammonium and hydroxyl surface functionalities [29, 55]. Ammonium functionalized materials are soluble in water, methanol and

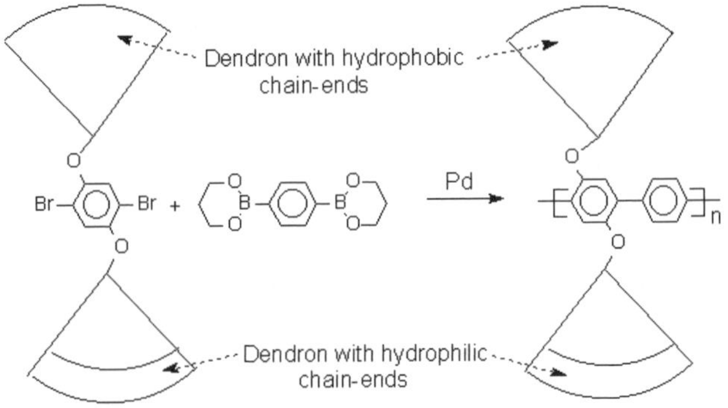

Figure 7.14 Dendronized amphiphilic triblock structure

dimethylformamide, whereas the hydroxyl terminated materials are insoluble in water. Both the hydroxy- and amine-terminated dendronized amphiphiles exhibit a strong tendency to hydrogen bond, and, once dried, the materials cannot be redissolved. Even more unusual amphiphilic hybrid cylinders that can segregate lengthwise into two different halves have also been reported by Schlüter and co-workers [55, 4b]. The architecture of these copolymers is quite different from that of most other polymeric amphiphiles that are typically based on a block-type architecture. In this case, the PPP-based material is prepared via the Suzuki cross-coupling polymerization of a bi-polar macromonomer (Figure 7.14) not unlike the surface-block dendrimers of Hawker *et al.* [56]. Standard Fréchet-type dendrons were used as the source of hydrophobicity while analogous dendrons with short oxyethylene chain ends provided the required hydrophilicity. Langmuir monolayers of these amphiphilic macromolecules at the air–water interface are well behaved and measurements of area per molecule clearly suggest that the hybrids are close-packed with their long axes in the plane of the monolayer. This constitutes the first example in which dendritic substituents segregate lengthwise into separate domains [55].

6 ELECTROACTIVE HYBRID COPOLYMERS

The field of conjugated oligomers and polymers has experienced significant activity driven by the progress made toward their application in practical devices such as organic light-emitting diodes (OLED), organic field effect transistors (OFET), and photovoltaic (PV) cells [57]. Numerous research groups have studied the properties of short conjugated oligomers as model compounds for their polydisperse polymeric analogs [58]. The preparation of well-defined, lengthy conjugated oligomers that should behave as molecular wires has been the major objective in this field [59]. As the unique properties of conjugated materials not only depend on their intrinsic effective conjugation but also on interactions between individual chains, dendrimer chemistry provides many opportunities to alter or manipulate these particular interactions in a controlled fashion. Dendritic substitution can lead to efficient shielding of the conducting main chain, and therefore the dendritic component of hybrid structures can act as an insulating layer. Furthermore, the dendrimer backbone dramatically improves the processability of the obtained hybrid material facilitating device manufacturing processes. In addition, the prospects of using dendrimers to induce a higher order, self-assembled, architecture incorporating opto-electronic moieties are exciting and could lead to new materials with superior properties. In this section we will focus briefly on some recent efforts that involve dendrimers in the preparation of new, nanometer scale, optoelectronic materials having partially encapsulated linear cores [60].

The encapsulation of conjugated linear cores can be achieved either via terminal dendritic substitution leading to triblock dumbbell-like structures or via 'dendronization' giving rise to cylindrical architecture [4]. The first synthesis of a triblock hybrid dendrimer based on poly(benzyl ether) dendrons and an oligothiophene core with 11 2,5-linked thiophene repeat units (G3-11T-G3) has recently been reported by our own research group [61]. The triblock hybrid structure was obtained in high yield via a procedure that combined halogenation and Stille cross-coupling reactions. An analogous series of dumbbell-shaped materials with a central oligothiophene block, including a G-3 heptadecamer dumbbell (G3-17T-G3, Figure 7.15A) was also prepared together with a phenyl end-capped G-3 pentamer di-block (G3-5T-Ph) [62]. A significant enhancement in solubility is observed upon functionalization of the oligothiophene moieties with the dendritic blocks. For example, while the benzyl ester telechelic heptadecathiophene is only very slightly soluble in hot CS_2, the G-3 functionalized analog is easily dissolved in common organic solvents such as THF, $CHCl_3$, and CH_2Cl_2. In addition, incorporation of the dendritic blocks causes significant alteration of the thermal characteristics of the oligothiophenes due to phase segregation of the dissimilar blocks. By examining the various redox states of G3-11T-G3 and G3-17T-G3, the di-cationic state was found to consist of two individual polarons, as opposed to a single bipolaron. Furthermore, oxidation of G3-5T-Ph resulted in the formation of a dendritic supramolecular assembly arising from π-dimerization of two radical cations.

Roncali and co-workers have also prepared dumbbell structures based on modified poly(benzyl ether) dendrimers (G-1, G-2 and G-3) with oligothienylvinylene cores (Figure 7.15B) [63]. Their synthetic approach first involves the functionalization of the dendron focal point with a phosphonate functionality that is later coupled to the electrophore via a double Wittig–Horner olefination of the dialdehyde core. Consistent with our own observations, UV-vis spectrometry does not suggest any significant change in the planarity of the electrophore that might have been caused by the steric bulk of the higher generation

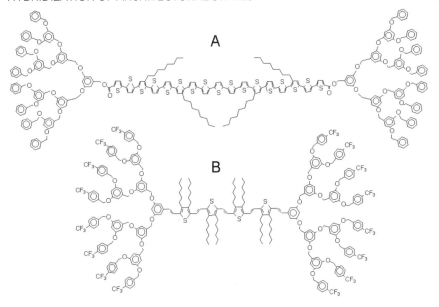

Figure 7.15   Encapsulated conjugated oligomers

dendrons. Cyclic voltammetry studies indicate that the dendron size has no effect on the kinetics of electron transfer. This is in contrast to earlier findings of a kinetic limitation to electron transfer in tetrathiophene cored dendrimers [62] and a variety of related dendritically encapsulated redox-active cores [60]. Roncali suggests that, as a result of its increased length, the oligothienylvinylene electrophore is no longer adequately shielded by the dendritic blocks.

Miller and co-workers have recently prepared 'nanometer-scaled molecular dumbbells' based on poly(benzyl ether) dendrons (G1–G4) and oligoimide spacers [64]. Their synthetic approach involved the coupling of amine-terminated oligoimides to dendrons with a carboxylic acid focal point. The resulting hybrid materials were found to be quite soluble thus allowing their analysis by cyclic voltammetry in DMF. Consistent with Roncali's observation, the kinetics of reduction of the oligoimide core was not found to be limited by the presence of the dendritic wedges.

Stoddart and co-workers have made use of Fréchet-type dendrons as dendritic stoppers for self-assembled [n]rotaxanes [65]. The solubility enhancement that results from incorporating dendritic wedges at the termini facilitated the purification of these materials by column chromatography despite the polycationic nature of their bipyridinium backbone. Again, the dendritic wedges did not alter the electrochemical characteristics of the viologen subunits. However, the enhanced solubility resulting from the presence of the dendritic components enabled a study of the influence of solvent on the shuttling dynamics along the backbone.

A conceptually different approach utilizes dendritic substitution of the side chain to wrap a conjugated backbone in an isolating shell, hence affording an insulated molecular wire. In an attempt to access isolated poly(triacetylene) (PTA) molecular wires with both enhanced processability and a shielded conjugated backbone, Diederich and co-workers prepared phenylacetylene endcapped PTA oligomers several nanometers in length that were dendronized with Fréchet-type dendrons (G-1, G-2 and G-3) with *tert*-butyl chain ends (Figure 7.16A) [66]. Dendritic trans-enediyne monomers were found to be unusually stable and could be stored in air at room temperature for extended periods of time. Oxidative Hay coupling was used to form extended structures that were purified by SEC. Diederich observed that the π-conjugation of the tubular backbone is not altered by the presence of dendritic side-chains of various generations. While these hybrid materials could not be oxidized electrochemically due to the similar oxidation potential of the linear chain and the dendrons, they could effectively be reduced in several irreversible steps. The increase in irreversibility with generation was attributed to steric hindrance.

In a separate study, we have reported the use of aliphatic ether dendrimers as a solubilizing platform for the preparation of soluble oligothiophenes with minimal main-chain substitution (Figure 7.16B) [67]. Aliphatic ether dendritic blocks

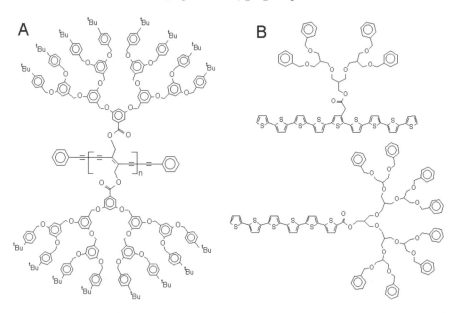

Figure 7.16 (A) Dendron-capped poly(triacetylene) and, (B) soluble oligothiophenes obtained by dendrimers assisted synthesis

were used since the activated aromatic nature of the poly(benzyl ether) dendrons precludes their use in the N-bromosuccinimide-mediated bromination reaction that is necessary to activate the oligothiophenes prior to coupling. Attachment of the dendrons at either the α or β position provided oligothiophene moieties with extended conjugation lengths when compared to analogous materials functionalized with alkyl chains along the conjugated backbone. This dendron supported synthesis provides a unique avenue toward the preparation of normally insoluble materials having minimal substitution, thus affording access to novel electrophores.

From a more applied perspective, OLED fabrication suffers from the dilemma that high luminescence output is critically dependent on chromophore content. Simply increasing the amount of emissive material is not a viable option since chromophore aggregation can lead to excimer formation, thus resulting in less efficient and bathochromically shifted photo- and electroluminescence. Several groups have tried to overcome this obstacle by encapsulating the dye within a dendritic shell [60]. For instance, Bao et al. have prepared a poly(phenylenevinylene) (PPV) having Percec type poly(benzyl ether) dendrons end-functionalized with LC inducing moieties (Figure 7.17A) [68]. In this example, a macromonomer approach, similar to that developed by Schlüter et al. [4], was used to prepare dendronized PPV via Heck coupling. As expected, the dendritic end blocks of the dendronized material prevent π-stacking as evidenced by the similarity of the solution and solid-state UV-vis spectrum. In addition, the LC character of the dendrons was exploited and films of the annealed dendronized materials were found be nematic, regardless of the nature of the casting solvent. Such liquid crystallinity could, in principle, be used to align the liquid crystalline phase, thereby enhancing photoconductivity [69] or affording polarized light emission.

A similar shielding effect has been reported by Aida and co-workers, who investigated poly(phenyleneethynylene)s dendronized with Fréchet-type dendrons obtained via a macromonomer route (Figure 7.17B) [70]. When a high generation was used to dendronize the backbone, the luminescence efficiencies did not drop with increasing concentration, suggesting the absence of aggregation. In addition, the antenna properties [71, 72] of the dendron subunits could be used to sensitize the core emission leading to efficient UV-photon harvesting [70].

Similar results have been presented by Miller and co-workers, who capped the ends of oligo(dihexylfluorene)s and poly(dihexylfluorene)s with Fréchet-type dendrons (Figure 7.17C) [73]. Annealing experiments coupled with emission studies revealed that G-3 and G-4 dendrons were effective at preventing excimer formation, even when the poly(fluorene) spacer was 50–80 repeat units long.

Ultimately, the encapsulated chromophore must be interfaced with hole- and electron-transporting elements to afford an actual OLED device. In this context, a dendrimer is ideally suited since it not only provides steric protection (site

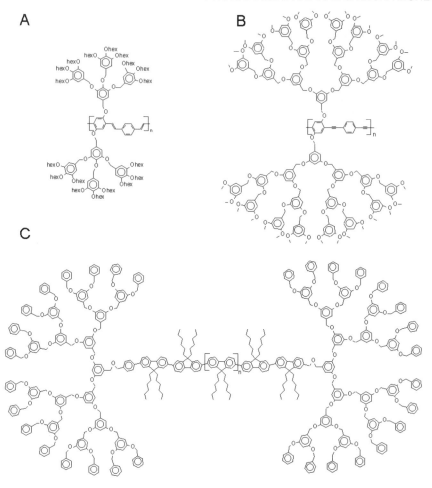

Figure 7.17 Luminescent hybrid materials with encapsulated chromophores

isolation) [60], but, in addition, the large number of peripheral groups can be used to attach efficient charge carrier moieties. Upon exciton formation these are able to funnel their energy to the core lumophore [72]. This concept has been exploited by Moore and co-workers [74] to realize single-layer OLED devices that have the potential of cost-effective device fabrication. Our own research group [75] has pushed the concept further by demonstrating that color tunability can be achieved by varying the core chromophore [75]. Site isolation of chromophores by dendrimers encapsulation enables their coexistence in fully functional form as each chromophore is capable of emitting on its own. Clearly, this constitutes a key step towards single-layer multichromophoric white LEDs, while also demonstrating the pivotal and optimized role of dendrimer chemistry in this rather applied area of research.

7 PERSPECTIVE

As this brief overview demonstrates, novel copolymers obtained by hybridization of the linear and globular architectural states are readily prepared through a variety of synthetic approaches. In general the dendritic components of the hybrid copolymers are well defined, with unique molecular and structural characteristics. In contrast, all the linear components prepared polymerization are less precisely defined and are polydisperse. Only the very short linear components, themselves prepared by stepwise synthesis just like the dendrons, are monodisperse and can be used to prepare well-defined, monodisperse hybrids. While architectural and structural precision may be of great importance for the determination of ultimate properties, some degree of structural variation is quite acceptable for practical applications in many areas including, for example, surface modification, sensing, or encapsulated delivery.

It is clear that the combination of different architectures and the precise localization of functionalities within a single macromolecule provide unique opportunities for the control of molecular shape as well as molecular, optical, and electronic properties. A significant hurdle that still remains today is the relatively demanding multistep process used to prepare dendrons and hybrids. This, in turn, translates into limited availability; but, as high added-value applications emerge, it is clear that current, as well as yet-to-be-developed, syntheses will be used to prepare specialty materials that benefit from the unique properties derived from the combination of dendritic and linear architectures.

Financial support of this research by the National Science Foundation (DMR-9816166) and AFOSR-MURI is acknowledged with thanks.

8 REFERENCES

1. (a) Fréchet, J. M. J. *Science* **263**, 1710–15 (1994); (b) Fréchet, J. M. J. and Gitsov, I. *Macromol. Symp.* **98**, 441–465 (1995); (c) Fréchet, J. M. J., Hawker, C. J. in *Comprehensive Polymer Science, Supplement Vol. 2*, Elsevier Science, Oxford, 1996, pp. 71–132; (d) Newkome, G. R., Moorefield, C. N. and Vögtle, G. *Dendritic Molecules: Concepts, Syntheses, Perspectives.* VCH, Weinheim, 1996; (e) Emrick, T. and Fréchet, J. M. J. *Curr. Opin. Colloid Interface Sci.*, **4**, 15–23 (1999).
2. (a) Gitsov, I., Wooley, K. L. and Fréchet, J. M. J. *ACS Polym. Prep.*, **32**(3) 631–632 (1991); (b) Wooley, K. L. and Fréchet, J. M. J. *ACS Polym. Mat. Sci. Eng.* **67**, 90–1 (1992).
3. (a) Fréchet, J. M. J., Jiang, Y., Hawker, C. J. and Philippides, A. E. *Proc. IUPAC Int.*

Symp., Macromol., Seoul, 19–20 (1989); (b) Hawker, C. J. and Fréchet, J. M. J. *J. Am. Chem. Soc.*, **112**, 7638–7647 (1990); (c) Hawker, C. and Fréchet, J. M. J. *J. Chem Soc. Chem. Commun.*, 1010–1012 (1990).

4. (a) Schlüter, A. D. *Top. Curr. Chem.* **197**, 165–191 (1998); (b) Schlüter, A. D. and Rabe, J. P., *Angew. Chem. Int. Ed.* **39**, 864–883 (2000).

5. Roovers, J. and Comanita, B. *Adv. Polym. Sci.*, **142**, 179–228 (1999).

6. (a) Tomalia, D. A., Baker, G. R., Dewald, J. R., Hall, M. Kallos, G., Martin, S., Roeck, J., Ryder, J. and Smith, P. *Polymer J.*, **17**, 117–141 (1985); (b) Newkome, G. R., Yao, Z., Baker, G. R., Gupta, V. K. *J. Org. Chem.*, **50**, 2003–2004 (1985); (c) de Brabander-van den Berg, M. M. E. and Meijer, E. W. *Angew. Chem. Int. Ed.*, **32**, 1308–1311 (1993); (d) Majoral J. P. and Caminade, A. M. *Chem. Rev.*, **99**, 845–880 (1999).

7. Hawker, C. J. and Fréchet, J. M. J. *Macromolecules*, **23**, 4726 (1990); Lochmann, L., Wooley, K. L., Ivanova, P. T. and Fréchet, J. M. J. *J. Am. Chem. Soc.*, **115**, 7043 (1993).

8. Gitsov, I., Ivanova, P. T. and Fréchet, J. M. J. *Macromol. Rapid. Commun.*, **15**, 387 (1994).

9. (a) Leduc, M. R., Hawker, C. J., Dao, J. and Fréchet, J. M. J. *J. Am. Chem. Soc.*, **118**, 11111 (1996); (b) Leduc, M. R., Hayes, W. and Fréchet, J. M. J. *J. Polym. Sci. A*, **36**, 1 (1998).

10. (a) Georges, M. K., Veregin, R. P. M., Kazmaier, P. M. and Hamer, G. K. *Macromolecules*, **26**, 2987 (1993); (b) Hawker, C. J. *J. Am. Chem. Soc.*, **116**, 11185 (1994); (c) Hawker, C. J. *Acc. Chem. Res.*, **30**, 373 (1997).

11. Patten, T. E. and Matyjaszewski, K. *Adv. Mater.*, **10**, 901 (1998).

12. Mayo, F. R. *J. Am. Chem. Soc.*, **90**, 1289 (1968).

13. Gitsov, I. and Fréchet, J. M. J. *Macromolecules*, **27**, 7309–7315 (1994).

14. Emrick, T., Hayes, W. and Fréchet, J. M. J. *J. Polym. Sci. A*, **37**, 3748 (1999).

15. Tomalia, D. A., Kerchoff, P. M. US Patent 4,694,064, 1987.

16. (a) Hawker, C. J. and Fréchet, J. M. J. *Polymer* **33**, 1507–1511 (1992); (b) Fréchet, J. M. J. and Gitsov, I. *Macromol. Symp.* **98**, 441 (1995).

17. (a) Draheim, G. and Ritter, H. *Macromol. Chem. Phys.* **196**, 2211–2222 (1995); (b) Chen, Y.-M., Chen, C.-F., Liu, W.-H., Li, Y.-F. and Xi, F. *Macromol. Rapid Commun.*, **17**, 401–407 (1996).

18. (a) Percec, V., Schlueter, D., Ronda, J. C., Johansson, G., Ungar, G. and Zhou, J. P. *Macromolecules*, **29**, 1464–1472 (1996); (b) Percec, V. and Schlueter, D. *Macromolecules*, **30**, 5783–5790 (1997); (c) Stewart, G. M. and Fox, M. A. *Chem. Mater.*, **10**, 860–863 (1998).

19. Moore, J. S. and Xu, Z., *Macromolecules*, **24**, 5893 (1991); Xu, Z., Kahr, M., Walker, K. L., Wilkins, C. L. and Moore, J. S. *J. Am. Chem. Soc.*, **116**, 4537 (1994): Kawaguchi, T., Walker, K. L., Wilkins, C. L. and Moore, J. S. *J. Am. Chem. Soc.*, **117**, 2159 (1995).

20. Kaneko, T., Horie, T., Asano, M., Aoki, T. and Oikawa, E. *Macromolecules*, **30**, 3118–3121 (1997).

21. (a) Percec, V., Heck, J., Tomazos, D., Galkenberg, F., Blackwell, H. and Ungar, G. *J. Chem. Soc. Perkin Trans.*, **1**, 2799–2811 (1993); (b) Percec, V., Ahn, C.-H., Cho, W.-D., Johansson, G. and Schlueter, D. *Macromol. Symp.*, **118**, 33–43 (1997).

22. Hudson, S. D., Jung, H. T., Percec, V., Cho, W.-D., Johansson, G., Ungar, G. and Balagurusamy, V. S. K. *Science*, **278**, 449–452 (1997); Percec, V., Johansson, G., Ungar, G. and Zhou, J. *J. Am. Chem. Soc.* **118**, 9855–9866 (1996).

23. (a) Percec, V., Ahn, C.-H. and Barboiu, B. *J. Am. Chem. Soc.*, **119**, 12978–12979 (1997); (b) Percec, V., Ahn, C.-H., Ungar, G., Yeardley, D. J. P., Möller, M. and Sheiko, S. S. *Nature*, **391**, 161–164 (1998).

24. Percec. V., Ahn, C.-H., Cho, W.-D., Jamieson, A. M., Kim, J., Leman, T., Schmidt, M., Gerle, M., Möller, M., Prokhorova, S. A., Sheiko, S. S., Cheng, S. Z. D., Zhang, A., Ungar, G. and Yeardley, D. J. P. *J. Am. Chem. Soc.*, **120**, 8619–8631 (1998).

25. Prokhorova, S. A., Sheiko, S. S., Ahn, C.-H., Percec, V. and Möller, M. *Macromolecules*, **32**, 2653–2660 (1999).
26. Jahromi, S., Coussens, B., Meijerink, N. and Braam, A. W. M. *J. Am. Chem. Soc.*, **120**, 9753–9762 (1998).
27. Yin, R., Zhu, Y. and Tomalia, D. A. *J. Am. Chem. Soc.*, **120**, 2678–2679 (1998).
28. (a) Neubert, I., Amoulong-Kirstein, E. and Schlüter, A.-D. *Macromol. Rapid Commun.*, **17**, 517–527 (1996); (b) Stocker, W., Schürmann, B. L., Rabe, J. P., Förster, S., Lindner, P., Neubert, I. and Schlüter, A.-D. *Adv. Mater.*, **10**, 793–797 (1998).
29. (a) Neubert, I. and Schlüter, A.-D., *Macromolecules*, **31**, 9372–9378 (1998); (b) Förster, S., Neubert, I., Schlüter, A.-D. and Lindner, P. *Macromolecules*, **32**, 4043 (1999).
30. (a) Freudenberger, R., Claussen, W., Schlüter, A.-D. and Wallmeier, H. *Polymer*, **35**, 4496–4501 (1994); (b) Claussen, W., Schulte, N. and Schlüter, A.-D. *Macromol. Rapid Commun.* **16**, 89–94 (1995).
31. Karakaya, B., Claussen, W., Gessler, K., Saenger, W. and Schlüter, A.-D. *J. Am. Chem. Soc.*, **119**, 3296–3301 (1997).
32. Stocker, W., Karakaya, B., Schürmann, B. L., Rabe, J. P. and Schlüter, A.-D. *J. Am. Chem. Soc.*, **120**, 7691–7695 (1998).
33. Bo, Z., Schlüter, A.-D. *Macromol. Rapid Commun.*, **20**, 21–25 (1999).
34. Kono, K., Liu, M. and Fréchet, J. M. J., *Bioconjugate Chem.*, **10**, 1115–1121 (1999); Liu, M. and Fréchet, J. M. J. *Polym. Bull.*, **43**, 379–86 (1999); Liu, M., Petro, M., Fréchet, J. M. J., Haque, S. A. and Wang, H. C. *Polym. Bull.*, **43**, 51–58 (1999).
35. Liu, M., Kono, K. and Fréchet, J. M. J., *J. Controlled Release*, **65**, 121–131 (2000); Hawker, C. J., Wooley, K. L. and Fréchet, J. M. J. *J. Chem. Soc. Perkin I*, 1287–1297 (1993).
36. (a) Newkome, G. R., Lin, X., Yaxiong, C. and Escamilla, G. H. *J. Org. Chem.*, **58**, 3123–3129 (1993); (b) Newkome, G. R., Baker, G. R., Saunders, M. J., Russo, P. S., Gupta, V. K., Yao, Z.-Q., Miller, J. E. and Bouillon, K. *J. Chem. Soc., Chem. Commun.* 752–753 (1986); (c) Newkome, G. R., Baker, G. R., Arai, S., Saunders, M. J., Russo, P. S., Theriot, K. J., Moorefield, C. N., Rogers, L. E., Miller, J. E., Lieux, T. R., Murray, M. E., Phillips, B. and Pascal, L. *J. Am. Chem. Soc.*, **112**, 8458–8465 (1990); (d) Newkome, G. R., Moorefield, C. N., Baker, G. R., Behera, R. K., Escamillia, G. H. and Saunders, M. J. *Angew. Chem. Int. Ed. Engl.*, **31**, 917–919 (1992).
37. Abdallah, D., Weiss, R. *Adv. Mat.*, **12**, 1237–1247 (2000).
38. Gitsov, I., Wooley, K. L. and Fréchet, J. M. J. *Angew. Chem. Int. Ed. Engl.*, **31**, 1200–1202 (1992).
39. Gitsov, I., Wooley, K. L., Hawker, C. J., Ivanova, P. T. and Fréchet, J. M. J. *Macromolecules*, **26**, 5621–5627 (1993).
40. Gitsov, I. and Fréchet, J. M. J. *Macromolecules*, **26**, 6536–6546 (1993).
41. Fréchet, J. M. J., Gitsov, I., Monteil, T., Rochat, S., Sassi, J.-F., Vergelati, C. and Yu, D. *Chem. Mater.*, **11**, 1267–1274 (1999).
42. Gitsov, I. and Fréchet, J. M. J. *J. Am. Chem. Soc.*, **118**, 3685–3786 (1996).
43. Liu, M. and Fréchet, J. M. J., *Pharmaceut. Sci. Technol. Today*, **2**, 393–401 (1999); Liu, M., Kono, K. and Fréchet, J. M. J., *J. Polym. Sci.*, **A37**, 3492–3503 (1999).
44. Chapman, T. M., Hillyer, G. L., Mahan, E. J. and Shaffer, K. A. *J. Am. Chem. Soc.*, **116**, 11195–11196 (1994).
45. (a) van Hest, J. C. M., Delnoye, D. A. P., Baars, M. W. P. L., van Genderen, M. H. P. and Meijer, E. W. *Science*, **268**, 1592–1595 (1995); (b) van Hest, J. C. M., Delnoye, D. A. P., Baars, M. W. P. L., Elissen-Roman, C., van Genderen, M. H. P. and Meijer, E. W. *Chem. Eur. J.*, **2**, 1616–1626 (1996).
46. Schipor, I. and Fréchet, J.M.J. *ACS Polym. Prep.*, **35**, 480–481 (1994).
47. Ihre, H., Padilla de Jesùs, O. L., Fréchet, J. M. J. *J. Am. Chem. Soc.*, **123**, 5908 (2001).
48. van Hest, J. C. M., Baars, M. W. P. L., Elissen-Roman, C., van Genderen, M. H. P. and

Meijer, E. W. *Macromolecules*, **28**, 6689–6691 (1995).
49. Roman, C., Fischer, H. R. and Meijer, E. W. *Macromolecules*, **32**, 5525–5531 (1999).
50. Iyer, J., Fleming, K. and Hammond, P. T. *Macromolecules*, **31**, 8757–8765 (1998).
51. Wooley, K. L., Hawker, C. J., Pochan, J. M. and Fréchet, J. M. J. *Macromolecules*, **26**, 1514–1519 (1993).
52. Iyer, J. and Hammond, P. T. *Langmuir*, **15**, 1299–1306 (1999).
53. Kampf, J. P., Frank, C. W., Malmström, E. E. and Hawker, C. J. *Langmuir*, **15**, 227–233 (1999).
54. Aoi, K., Motoda, A. and Okada, M. *Macromol. Rapid Commun.*, **18**, 645–952 (1997).
55. Bo, Z., Rabe, J. P. and Schlüter, A. D. *Angew. Chem. Int. Ed.*, **38**, 2370–2372 (1999).
56. (a) Hawker, C. J. and Fréchet, J. M. J. *J. Am. Chem. Soc.*, **114**, 8405–8413 (1992); (b) Fréchet, J. M. J., Hawker, C. J. and Wooley, K. L. *J.M.S. – Pure Appl. Chem.*, **A31**, 1627–1645 (1994); (c) Hawker, C., Wooley, K. and Fréchet, J. M. J. *Macromol. Symp.*, **77**, 11–20 (1994).
57. (a) Skotheim, T. A., Elsenbaumer, R. L. and Reynolds, J. R. (eds) in *Handbook of Conducting Polymers* (2nd edn), Marcel Dekker. New York, 1998; (b) Feast, W. J., Tsibouklis, J., Pouwer, K. L., Groenendaal, L. and Meijer, E. W. *Polymer*, **37**, 5017–5047 (1996); (c) Kraft, A., Grimsdale, A. C. and Holmes, A. B. *Angew. Chem., Int. Ed. Engl.*, **37**, 402–428 (1998).
58. (a) Wegner, G. and Müllen, K. (eds), in *Electronic Materials: The Oligomer Approach*; VCR, Weinheim, 1997; (b) Roncali, J. *Chem. Rev.* **97**, 173–205 (1997).
59. (a) Tour, J. M. *Chem. Rev.*, **96**, 537–553 (1996); (b) Martin, R. E. and Diederich, F. *Angew. Chem., Int. Ed. Engl.*, **38**, 1350–1377 (1999); (c) Tour, J. M. *Acc. Chem. Res.* **33**, 791 (2000).
60. Hecht S. and Fréchet, J. M. J. *Angew. Chem. Int. Ed.*, **40**, 74–91 (2001).
61. Malenfant, P. R. L., Groenendaal, L. and Fréchet, J. M. J. *J. Am. Chem. Soc.*, **120**, 10990–10991 (1998).
62. Apperloo, J. J., Jansen, R. A. J., Malenfant, P. R. L., Groenendaal, L. and Fréchet, J. M. J. *J. Am. Chem. Soc.*, **122**, 7042–7051 (2000).
63. Jestin, I., Levillain, E. and Roncali, J. *Chem. Commun.*, 2655–2556 (1998).
64. Miller, L. L., Zinger, B. and Schlechte, J. S. *Chem. Mater.*, **11**, 2313–2315 (1999).
65. Amabilino, D. B., Ashton, P. R., Balzani, V., Brown, C. L., Credi, A., Fréchet, J. M. J., Leon, J. W., Raymo, F. M., Spencer, N., Stoddart, J. F. and Venturi, M. *J. Am. Chem. Soc.*, **118**, 12012–12020 (1996).
66. Schenning, A. P. H. J., Martin, R. E., Ito, M., Diederich, F., Boudon, C., Gisselbrecht, J.-P. and Gross, M. *Chem. Commun.*, 1013–1014 (1998).
67. Malenfant, P. R. L., Jayaraman, M. and Fréchet, J. M. J. *Chem. Mater.* **11**, 3420–3422 (1999).
68. Bao, Z., Amundson, K. R. and Lovinger, A. J. *Macromolecules*, **31**, 8647–8649 (1998).
69. Percec, V., Cho, W.-D., Singer, K. D. and Zhang, J. *Polym. Mat. Sci. Eng.*, **80**, 262–263 (1999).
70. Sato, T., Jiang, D.-L. and Aida, T. *J. Am. Chem. Soc.*, **121**, 10658–10659 (1999).
71. Kawa, M. and Fréchet, J. M. J. *Chem. Mater.*, **10**, 286–296 (1998).
72. Adronov, A. and Fréchet, J. M. J. *Chem Commun.*, 1701–1710 2000).
73. Klaerner, G., Miller, R. D. and Hawker, C. J. *Polym. Mat. Sci. Eng.*, **79**, 1006–1007 (1998).
74. Wang, P.-W., Liu, Y.-J., Devadoss, C., Bharathi, P. and Moore, J. S. *Adv. Mater.*, **8**, 237–241 (1996).
75. Freeman, A. W., Koene, S. C., Malenfant, P. R. L., Thompson, M. E., Fréchet, J. M. J. *J. Am. Chem. Soc.*, 2000, in press.

8

Statistically Branched Dendritic Polymers

E. MALMSTRÖM AND A. HULT

Dept. of Polymer Technology, Royal Institute of Technology, Stockholm, Sweden

1 INTRODUCTION

In 1952, Paul Flory described theoretical implications involved in the production of hyperbranched polymers obtained by condensation of AB_x-monomers in a statistical growth process. [1] Flory pointed out that such a multifunctional monomer would have one A group and two or more B groups, wherein, the A and B groups would be reactive with each other. Owing to the high branching and expected absence of chain entanglement, it was predicted that these new architectures would exhibit poor mechanical properties. The synthesis of hyperbranched polymers remained an unsolved synthetic challenge for nearly 35 years, until the concept was revisited by Odian/Tomalia et al. [2] in 1988 and reawakened by Kim/Webster who coined the term 'hyperbranched polymers' [3].

Since that time, synthetic chemists have explored numerous routes to these statistically hyperbranched macromolecular structures. They are recognized to constitute the least controlled subset of structures in the major class of dendritic polymer architecture. In theory, all polymer-forming reactions can be utilized for the synthesis of hyperbranched polymers; however, in practice some reactions are more suitable than others.

Dendrimers, the most precise subset of 'structure-controlled' dendritic polymers preceeded by nearly half a decade, the more recent attention focused on hyperbranched polymers. It is notable that literature reports describing dendrimers far exceed the number of investigations published on random hyperbranched polymers.

The scope of this chapter is to examine some of the most extensively utilized

Dendrimers and Other Dendritic Polymers. Edited by Jean M. J. Fréchet and Donald A. Tomalia
© 2001 John Wiley & Sons Ltd

synthetic routes to hyperbranched polymers. Special attention will be given to new synthetic strategies. This chapter is divided into three main sections; namely: (1) random hyperbranched polymers, (2) hyperbranched polymers by self-condensing vinyl polymerization; (3) hyperbranched polymers by proton transfer polymerization.

2   RANDOM HYPERBRANCHED POLYMERS

The step-growth polymerization of AB_x-monomers is by far the most intensively studied synthetic pathway to hyperbranched polymers. A number of AB_2-monomers, suitable for step-growth polymerizations, are commercially available. This has, of course, initiated substantial activity in hyperbranched condensation polymers and a wide variety of examples have been reported in the literature [4].

A typical condensation procedure involves a one-step reaction where the monomer and suitable catalyst/initiator are mixed and heated to the required reaction temperature. To accomplish a satisfactory conversion, the low molar mass condensation products formed throughout the reaction have to be removed. This is most often accomplished by using a flow of inert gas and/or by reducing the pressure in the reaction vessel. The resulting polymer is generally used without any purification or, in some cases, after precipitation of the dissolved reaction product from a nonsolvent.

When polymerizing highly functional monomers, one must always consider the occurrence of unwanted side reactions leading to the onset of gelation. For example, in the reaction of an AB_x-system the preferred reaction should be A reacting with B. Unwanted side reactions have to be suppressed. Even very low levels of A–A or B–B reactions can be detrimental and would inevitably lead to gelation. The one-pot polymerization of AB_2-monomers offers no control over molecular weight, and consequently, gives rise to highly polydispersed polymers [1]. The co-polymerization of AB_2-monomers with B_y-molecules introduces a tool to control not only the molecular weight but also to reduce the molecular weight distribution.

In a classical step-growth polymerization of AB-monomers, backbiting usually occurs, resulting in the formation of intramolecular cyclics. This, of course, terminates linear molecular growth since the functional terminal groups are lost. When polymerizing AB_2-monomers, there is the possibility of losing the unique focal point functionality at the root of the molecular tree due to intramolecular cyclics. This leads to the loss of the reactive A group at the focal point, however, this cyclized molecule still possesses reactive B groups which can still react to further enhance molecular weight. The maximum molecular weight and rate of polycondensation, however, are reduced by the occurrence of intramolecular cyclization reactions.

One way to reduce the intramolecular cycle formation, is to add AB_2-monomer successively throughout the reaction in a so-called 'concurrent slow-addition'. Several authors have shown that slow addition of monomer leads to a reduction in side reactions and increased molecular weight [5], while others have studied the occurrence of cyclization in hyperbranched systems [6].

Assuming that all B groups have the same reactivity, the chemical reaction giving rise to a branched molecule is identical to the reaction resulting in a linear polymer. Statistically this will eventually result in a hyperbranched polymer. However, dependent on the chemical structure of the monomer, steric effects might favor the growth of linear polymers. Computer simulations of of AB_x-monomer condensation and AB_x-monomers co-condensed with B_y-functional cores have been published. Only a few papers deal with the experimentally studied structure build-up in hyperbranched polymers [7].

A shortcoming with condensation polymers, is their sensitivity towards hydrolysis, which might restrict the use of such polymers in certain applications. For that reason, some hyperbranched polymers are synthesized via substitution or ring opening reactions that provide more hydrolytically stable polymers.

It is well known that dendritic polymers possess substantially different properties compared to their linear analogues [8]. The exploration of dendritic polymers has been intensified by the desire to identify synthetic polymers with fundamentally new properties.

Dendritic polymers (i.e. dendrimers, dendrigrafts and hyperbranched polymers), are substantially different than traditional random coil polymers in that they have a highly branched structures and a multitude of end groups. Owing to their highly branched architecture, dendritic polymers are generally recognized to be amorphous. The high branching also leads to minimal chain entanglements. The lack of entanglements generally produce lower melt and solution viscosities compared to linear polymers. The nonlinear architecture reduces crystallinity and renders hyperbranched systems more soluble than linear polymers.

Dendrimers/dendrons are synthesized almost exclusively via elaborate synthetic procedures that are usually more costly than hyperbranched processes. Even though they display many unique properties, their higher costs do not always justify use in some applications. Consequently, hyperbranched polymers may serve as a more cost-effective alternative when optimum properties are not required.

3 CONDENSATION STRATEGIES TO HYPERBRANCHED POLYMERS – COMMERCIAL PRODUCTS

A key monomer, namely, 2,2-bis(methylol)propionic acid (bis-MPA) has been used extensively for the preparation of hyperbranched aliphatic polyesters. Hult *et al.* described the condensation of bis-MPA and a tetra-functional polyol (di-trimethylolpropane) resulting in hydroxy-functional hyperbranched polyesters [9]. The degree of branching is usually around 0.45 [10]; whereas, the molecular weight and number of terminal hydroxyl groups can be varied by altering the stoichiometric ratio between the polyol core and bis-MPA.

Similar materials, hyperbranched polyesters based on bis-MPA and a polyol are now commercially available [11] from Perstorp AB under the trade name Boltorn® [12], Figure 8.1. The average number of hydroxyl groups per molecule can be tailored between 8 and 64 and molecular weight can be varied between *c.* 2000 and 11 000. The co-polymerization of bis-MPA and a polyol core keeps the molecular weight distribution fairly low, typically below 2.

Commercially available hyperbranched polymer, a poly(ester-amide) is currently being marketed by DSM under the product name Hybrane™ [13] (Figure 8.2). It is also a hydroxyl-functionalized product, but contains both amide and ester linkages. The synthesis is accomplished in two steps: cyclic anhydrides are reacted with diisopropanolamine to give an amide-intermediate, possessing two hydroxyl groups and one carboxylic acid. The subsequent polymerization takes places via an oxazolinium intermediate which results in the formation of a

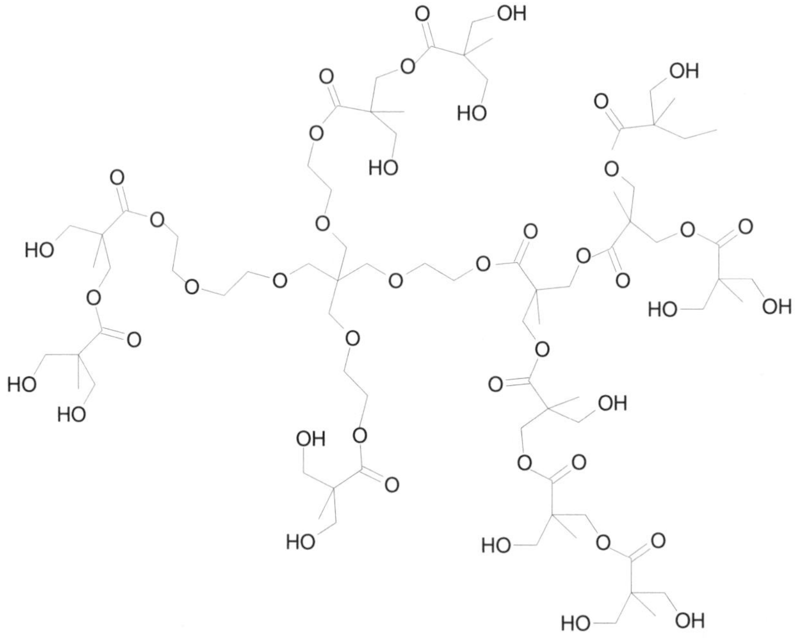

Figure 8.1 A schematic representation of the hydroxy-functional hyperbranched polyester Boltorn® H20, based on ethoxylated pentaerythritol and 2,2-bis(methylol)propionic acid [10]

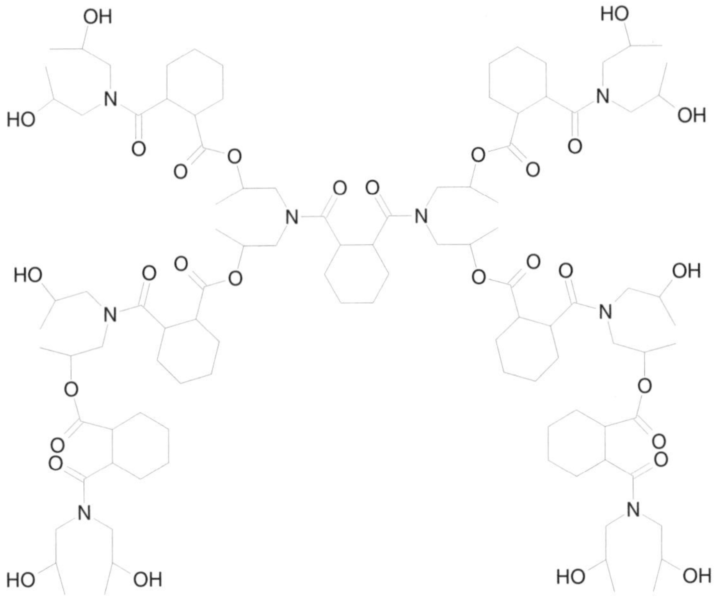

Figure 8.2 A schematic representation of hydroxy-functional hyperbranched poly(ester-amide) Hybrane™ based on diisopropanolamine and an anhydride [11]

hydroxyl-functionalized hyperbranched polymer. The properties of Hybrane™ can be altered by the choice of anhydride compounds used in the process.

4 RING-OPENING STRATEGIES TO HYPERBRANCHED POLYMERS

The use of ring-opening polymerization for the synthesis of hyperbranched polymers has been somewhat limited. Conceptually, ring-opening polymerizations hold an advantage over ordinary step-growth reactions in that that no lower molecular weight compounds have to be removed, thus facilitating the formation of higher molecular weight products.

Although the first examples of hyperbranched polymers proposed by Flory involved condensation-type polymerization strategies, the first well-characterized hyperbranched example involved the ring-opening polymerization of 2-carboxylic-2-oxazoline derivatives. As early as 1988, Odian and Tomalia [2] reported the ring-opening polymerization of these derivatives to form random hyperbranched products possessing approximately one branch juncture per five monomer units (i.e. degree of branching $\cong$ 20%). The degree of branching was confirmed by ^{1}H-NMR spectroscopy which was in agreement with hydrolysis experiments that produced 'branch cell products' that were analyzed by HPLC. The polymerization involved the ring opening of an AB-type monomer to produce an AB_2-type oligomer. This occurs *in situ* when the nitrogen of an unprotonated oxazoline ring competes with carboxylate anion as a nucleophilic agent in the ring opening of a protonated oxazoline ring as described in Figure 8.3. Subsequent propagation of the AB_2 oligomer produces AB_x oligomers ultimately leading to the random hyperbranched products.

Suzuki *et al.* [14] reported the Pd-catalyzed ring-opening polymerization of a cyclic carbamate in the presence of an initiator, which also acts as a core molecule, to afford a hyperbranched polyamine. The polymerization was proposed to be an *in situ* multibranching process, wherein the number of propagating chain ends increase with the progress of the polymerization.

Ring-opening polymerization of hydroxyl-functionalized cyclic ethers could by analogy with hydroxyl-functionalized lactones, give rise to hyperbranched polyethers. An example of such a polymerization is glycidol, an oxirane-ring substituted with a hydroxymethyl group. Extensive studies by Vandenberg *et al.* [15] on the anionic and cationic polymerizations of glycidol concluded that branched polymers were formed. More recently, studies by Frey *et al.* [16] have confirmed that hyperbranched systems are formed via the anionic ring-opening polymerization of glycidol.

More recently, Kim attempted unsuccessfully to anionically polymerize 2-hydroxymethyloxetane [17]. The failure of such a reaction is most likely due to the fact that oxetanes are not known to ring-open under basic conditions. The

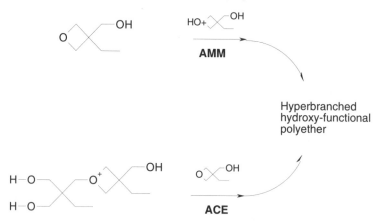

Figure 8.3   Two mechanisms have been proposed for the cationic ring-opening polymerization of oxetanes, the ACE route and the AMM [16]

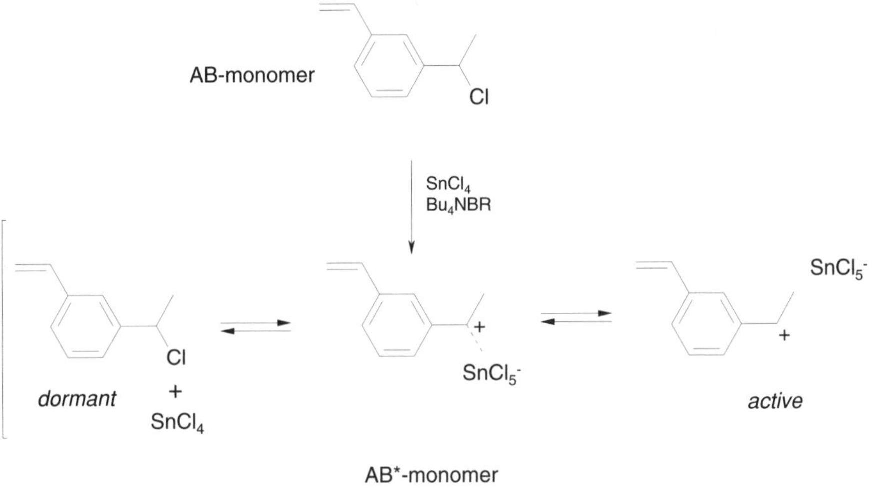

Figure 8.4 Self-condensing vinyl polymerization of 3-(1-chloroethyl)-ethenyl-benzene requires an external activator, in this case the activator is SnCl$_4$ [19]

successful cationic ring-opening polymerization of 3-ethyl-3-(hydroxymethyl)oxetane was reported by Penczek *et al.* [18] and Hult *et al.* [19] to give hydroxyl-functionalized hyperbranched polyethers. This cationic polymerization can proceed according to two different mechanisms: namely, an activated chain end (ACE) or activated monomer mechanism (AMM) as suggested by Penczek *et al.* [18] (Figure 8.3). Based on our own findings, we believe that the ring-opening polymerization of a hydroxyl-functionalized oxetane occurs by a combination of both mechanisms.

Fréchet *et al.* [20] reported on the ring-opening polymerization of an AB-monomer, 4-(2-hydroxyethyl)-ε-caprolactone. ε-Caprolactone is easily polymerized using ring-opening polymerization under facile conditions and the primary hydroxyl group can be used to initiate the polymerization. The polymers were reported to have molecular weights in the range 65 000–85 000 (PDI *c.* 3.2) as determined by SEC.

5 SELF-CONDENSING, VINYL POLYMERIZATION STRATEGIES

The first strategies to random hyperbranched polymers involved exclusively step-growth polymerizations. This limited the potential applications for these architectures to areas where only condensation-type polymers are acceptable. Fréchet *et al.* [21] presented the first example of a hyperbranched vinyl polymerization in 1995,] initiating the birth of a 'second generation' of hyperbranched

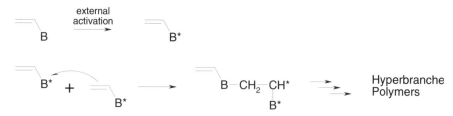

Figure 8.5 Schematic representation of the self-condensing vinyl polymerization of an AB*-monomer to give a hyperbranched vinyl polymer [19]

polymers. 3-(1-chloroethyl)-ethenylbenzene was cationically polymerized in the presence of SnCl$_4$. The polymerization was termed '*self-condensing vinyl polymerization*' (SCVP) since the polymerization was found to proceed by repeated step-wise couplings of otherwise chain-growing species. 3-(1-chloroethyl)-ethenylbenzene is an AB-monomer wherein the A-group is a readily polymerized vinyl group and the B group (i.e. a benzyl halide), functions as the latent initiator moiety. External activation of the B group was affected by the addition of SnCl$_4$ (Figure 8.5).

AB* represents the activated monomer. The polymerization is initiated by addition of B* to an A group, which produces a dimer possessing one double bond and two active sites, B* (Figure 8.6). Given the chemical structure of the monomer it can be assumed that the reactivities of A* and B* are similar and that is why both the initiating B* group and the newly created propagating cation can react with the vinyl group of another molecule (monomer or polymer) in the same way. These events eventually lead to a hyperbranched polymer.

The polymerization of AB*-functional vinyl monomers is fundamentally different from the step-growth polymerization of AB$_2$-monomers. Condensation of AB$_2$-monomers results immediately in the formation of hyperbranched polymers since the reactivity of the end-groups are the same, regardless of what type of repeat unit (linear or dendritic) that is formed.

In the case of AB*-monomers it is not obvious how the chain growth takes place. Depending on the chemical structure of the monomer there will be a competition between conventional, linear, chain-growth polymerization, via the double bond, and the branching reaction, i.e. where the group capable of initiation, B*, reacts with a vinyl group. If the reactivities of the two different propagating species are exactly the same one envision that a randomly branched system will be the result. However, all monomers utilized in SCVP so far possess unequal reactivity of the propagating sites. Fréchet *et al.* undertook a systematic investigation on how the branching could be maximized by altering the reaction conditions when polymerizing 4-chloromethylstyrene using metal catalyzed 'living' free radical polymerization [22].

The introduction of SCVP initiated extensive research focused on the use of vinyl monomers for the synthesis of hyperbranched polymers. More recently,

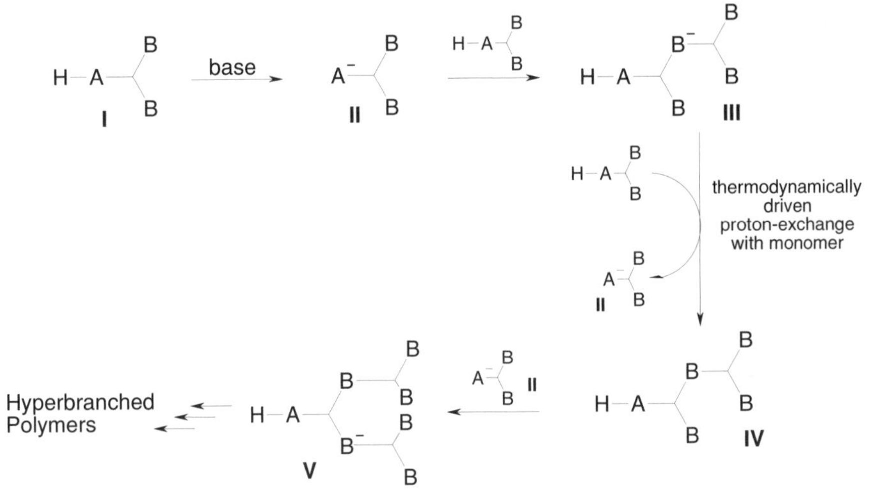

Figure 8.6 Schematic representation of the newly developed proton-transfer polymerization as a route to hyperbranched polymers [27]

there has been a keen interest in 'living' free radical procedures that allow accurate control over molecular weight, molecular weight distribution and chain ends. The SCVP concept was further developed by Hawker et al. [23] by using TEMPO-initiated 'living' free radical polymerization methodology to produce hyperbranched polystyrenes. The extensive development of metal-catalyzed 'living' free radical polymerizations has produced new possibilities for using radical polymerization as a means to obtain 'structure controlled' macromolecular architectures. Matyjaszewski et al. [24] developed atom-transfer radical polymerization (ATRP) techniques to obtain hyperbranched polystyrenes; whereas Müller et al. [25] reported the first use of group-transfer polymerization to obtain hyperbranched methacrylates.

Since these reports, a number of new approaches based on vinyl monomers and various initiating systems have been explored to yield hyperbranched polymers; such as, poly(4-acetylstyrene) [26], poly(vinyl ether) [27] and polyacrylates [28]. In view of the fact that free radical polymerizations are most widely used in industrial polymerization processes the development of these procedures for vinyl monomers has opened a very important area for hyperbranched polymers.

6 PROTON-TRANSFER POLYMERIZATION

Recently, Fréchet and Chang [29] reported proton-transfer polymerization (PTP) (Figure 8.7), as a versatile route to hyperbranched polymers. Conceptually, PTP is an acid-base controlled reaction where the nucleophilicity and basicity of the monomers/intermediates play important roles. The monomer should be an AB_2-type, containing an acidic proton, H-AB_2, **I**, (Figure 8.7). A base serves as an initiator and abstracts the labile proton from the monomer, forming a reactive nucleophilic species, $[-]AB_2$, **II**. This species **II**, adds rapidly to the B-group on a monomer leaving an anionic site in the dimer, **III**. Intermediate **III**, is less nucleophilic than **II** and undergoes a rapid, thermodynamically driven, proton exchange with monomer, instead of a nucleophilic addition. This produces a new nucleophile, **II**, and an inactive dimer **IV**. Ultimately reactive B-groups are amplified as the molecule **V** grows; thus, ensuring formation of a hyperbranched polymeric product.

The usefulness of the PTP concept was further demonstrated by Fréchet *et al.* [30] in a study where hyperbranched aliphatic polyethers were synthesized from a diepoxide and a tri-functional alcohol, utilizing an A_2–B_3 monomer concept.

7 CONCLUSION

Despite the simplicity of chemistries presently employed for the commercially available hyperbranched polymers, these materials are still relatively high priced compared to traditional commodity polymers. This is undoubtedly related to the early stage, low volume demands for these products. Hyperbranched polymers will evolve as substitutes for traditional polymers as their unique properties are used to greatly enhance products on a cost-performance basis. In summary, the enhanced used of hyperbranched polymers in engineered products will depend upon many of the following prerequisites:

- Availability and price of suitable monomers.
- Ease and versatility of polymerization (to produce desired yields and molecular weights).
- The degree of control over molecular weight and polydispersity.
- The nature of the end groups.
- The unique architecturally driven properties of the final products.

Unique architecturally driven properties that may be expected from hyperbranched polymers will be largely derived from their (a) amplified number of terminal functional groups, (b) new rheological properties based on less chain entanglement and (c) new architectural arrangements that may modulate crystallinity, flow characteristics and glass transition properties in designed systems.

The development of novel hyperbranched synthesis strategies now offers new options for assembling 'old' commodity monomers into unique, unprecedented polymeric architectures. There are high expectations that these new architectures will produce substantially different properties that will be utilized in a variety of 'niche' or even commodity markets. It is appropriate to be optimistic about these possibilities based on recent reports that 'metallocene' or 'Brookhart-type' catalysts may in fact be used to produce so-called hyperbranched or dendrigraft-type polyolefins [31]. One must ask – are many of the new properties already noted in this area due to dendritic architecture in combination with the well recognized monodispersity?

8 REFERENCES

1. Flory, P. J. *J. Am. Chem. Soc.*, **74**, 2718 (1952).
2. Gunatillake, P. A., Odian, G., Tomalia, D. A. *Macromolecules*, **21**, 1556 (1988).
3. (a) Kim, Y. H. and Webster, O. W. *ACS Polym. Prepr.* **29**(2), 310 (1988); (b) Kim, Y. H. and Webster, O. W. *J. Am. Chem. Soc.*, **112**, 4592 (1990).
4. For reviews see: (a) Hult, A., Johansson, M. and Malmström, E. *Adv. Pol. Sci.*, **46**, 1 (1999); (b) Kim, Y. H. *J. Polym. Sci., Polym. Chem.*, **36**, 1685 (1998); (c) Voit, B. I. *Acta Polym.*, **46**, 87 (1995); (d) Fréchet, J. M. J., Hawker, C. J. *React. Polym.*, **26**, 127 (1995).
5. (a) Hanselmann, R., Hölter, G. and Frey, H. *Macromolecules*, **31**, 3790 (1998); (b) Gong, G., Miravet, J. and Fréchet, J. M. J. *J. Pol. Sci., Pol. Chem.*, **37**, 3193 (1999).
6. (a) Dušek, K., Šomavársky, J., Smrčková, Simonsick, W. J. and Wilczek, L. *Polym. Bull.*, **42**, 489 (1999); (b) Kambouris, P., Hawker, C. J. *J. Chem. Soc., Perkin Trans.*, **1**, 2717 (1993).
7. (a) Schmaljohann, D., Komber, H. and Voit, B. I. *Acta Polym.* **50**, 196 (1999); (b) Magnusson, H., Malmström, E. and Hult, A. *Macromolcules*, 2000, in press.
8. Fréchet, J. M. J. *Science*, **263**, 1710 (1994); (b) Fréchet, J. M. J., Hawker, C. J., Gitsov, I. and Leon, J. W. *J. Macrmol. Sci., Pure-Appl. Chem.*, **A33**(10) 1399 (1996).
9. Johansson, M., Malmström, E. and Hult, A. *J. Pol. Sci.,Part A: Polym. Chem.*, **31**, 619 (1993).
10. The degree of branching was initially reported to be close to 0.8 [1], but was recently reevaluated after it was shown that the hydroxy-functional hyperbranched polyesters undergo facile acetal formation. The acetal formation was catalyzed by residual trace amounts of acid remaining in the sample. After reevaluation in DMSO the degree of branching was close to 0.45 which is in accordance with most other hyperbranched polymers. (1. Malmström, E., Johansson, M. and Hult, A. *Macromolecules*, **28**, 1698 (1995); 2. Malmström, E., Trollsås, M., Hawker, C.J., Johansson, M. and Hult, A. *Polym. Mat. Sci. Eng.*, **77**, 151 (1997).
11. Perstorp AB, www.perstorp.com.
12. PCT WO 93/17060 1993.
13. http://www.hybrane.com.
14. (a) Suzuki, M., Ii, A. and Saegusa, T. *Macromolecules*, **25**, 7071 (1992); (b) Suzuki, M., Yoshida, S., Shiragar, K. and Saegusa, T. *Macromolecules*, **31**, 1716 (1998).
15. (a) Vandenberg, E. J. *J. Pol. Sci., Part A: Polym. Chem.*, **23**, 915 (1985); (b) Vandeberg, E. *J. Pol. Sci., Part A: Polym. Chem.*, **27**, 3113 (1989).
16. Sunder, A., Hanselmann, R., Frey, H. and Mülhaupt, R. *Macromolecules*, **32**, 4240 (1999).
17. Kim, Y. H. *J. Polym. Sci., Polym. Chem.*, **36**, 1685 (1998).
18. Bednarek, M., Biedron, T., Helinski,, J., Kaluzynski, K., Kubisa, P. and Penczek, S. *Macromol Rapid. Commun.*, **20**, 369 (1999).
19. Magnusson, H., Malmström, E. and Hult, A. *Macromol. Rapid. Commun.*, **20**, 453

(1999).
20. Liu, M., Vladimirov, N. and Fréchet, J. M. J. *Macromolecules*, **32**, 6881 (1999).
21. Fréchet, J. M. J., Henmi, M., Gitsov, I., Aoshima, S., Leduc, M. R. and Grubbs, R. B. *Science*, **269**, 1080 (1995).
22. Weimer, M. W., Fréchet, J. M. J. and Gitsov, I. *J. Pol. Sci., Part A: Polym. Chem.*, **36**, 955 (1998).
23. Hawker, C. J., Fréchet, J. M. J., Grubbs, R. B. and Dao, J. *J. Am. Chem. Soc.*, **117**, 10763 (1995).
24. Gaynor, S. G., Edelman, K. and Matyjaszewski, K. *Macromolecules*, **29**, 1079 (1996).
25. Simon, P. F. W., Radke, W. and Müller, A. H. E. *Macromol. Rapid Commun.*, **18**, 865 (1997).
26. Lu, P., Paulasaari, J. K. and Weber, W. P. *Macromolecules*, **29**, 8583 (1996).
27. Zhang, H. and Ruckenstein, E. *Polym. Bull.*, **39**, 399 (1997).
28. Matyjaszewski, K., Gaynor, S. G., Kulfan, A. and Podwika, M. *Macromolecules*, **30**, 5192 (1997).
29. Chang, H. T. and Fréchet, J. M. J. *J. Am. Chem. Soc.*, **121**, 2313 (1999).
30. Emrick, T., Chang, H. T. and Fréchet, J. M. J. *Macromolecules*, **32**, 6380 (1999).
31. Guan, Z., Cott, P. M., McCord, E. F. and McLain, S. J. *Science*, **283**, 2059 (1999).

9

Semi-Controlled Dendritic Structure Synthesis

R. A. KEE AND M. GAUTHIER
Institute for Polymer Research, Department of Chemistry, University of
Waterloo, Waterloo, Ontario, Canada

D. A. TOMALIA
The University of Michigan Medical School, Center for Biologic
Nanotechnology, Ann Arbor, MI, USA

1 DENDRITIC POLYMERS: HYPERBRANCHED, DENDRIGRAFTS AND DENDRIMERS

Dendritic polymers, the fourth major architectural class of macromolecules, can be divided into three subclasses. These subclasses may be visualized according to the degree of structural perfection attained, namely: (1) hyperbranched polymers (statistical structures, Chapter 7), (2) dendrigraft polymers (semi-controlled structures, reviewed in this chapter) and (3) dendrimers (controlled structures, Chapter 1).

The statistical subset, hyperbranched polymers are derived from one-pot condensation reactions of AB_n monomers. These propagations generally produce moderate to high molecular weight structures. Control over molecular weight and the branching process by this approach is limited, since molecular growth relies on random condensation reactions. Molecular weight distributions of hyperbranched polymers tend to approach those proposed by Flory, with a polydispersity index, $M_w/M_n \approx 2$.

The precise structure subset, dendrimers are prepared using iterative protection–condensation–deprotection reaction cycles. These reiterative cycles incorporate AB_n monomers (i.e. branch cell units) into structural domains referred to as dendrons. Assembly of these dendrons can proceed in a divergent [1] (core

Dendrimers and Other Dendritic Polymers. Edited by Jean M. J. Fréchet and Donald A. Tomalia
© 2001 John Wiley & Sons Ltd

first) or a convergent [2] (core last) manner. Dendrimers are precisely defined macromolecules with a low polydispersity ($M_w/M_n < 1.01$), exhibiting an exact mathematically predictable molecular weight. Since small molecules are used as building blocks for the construction of these macromolecules, many synthetic cycles (generations) must be completed to obtain polymers with a high molecular weight. Between these two subclasses resides a subclass of semi-controlled structures, referred to as dendrigraft (arborescent) polymers.

Dendrigraft (arborescent) polymers, obtained from ionic polymerization and grafting schemes, combine features common to dendrimers as well as to random hyperbranched polymers. Dendrigrafts are most commonly synthesized from polymeric chains that are assembled according to a dendrimer-like generational scheme, consisting of functionalization and grafting cycles (Figure 9.1). Grafting polymer chains onto a linear (core) polymer randomly functionalized with reactive sites yields a comb-branched (or generation G0) architecture. Repetition of the functionalization and grafting reactions subsequently leads to higher generation dendrigraft macromolecules (i.e. G1, G2, etc.). Since these materials

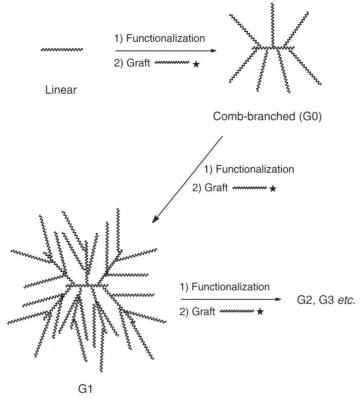

Figure 9.1 Generic synthetic route to dendrigraft polymers

are prepared by linking macromolecular building blocks, very high molecular weights are attained in few cycles. Various names have been used to describe these dendritic graft polymers. The term 'dendrigraft polymers' was coined by Tomalia *et al.* in deference to the analogous dendrimer polymers [3]. Gauthier and Möller used the designation 'arborescent graft polymers' [4], whereas others have used the term 'polymeric dendrimers' [5] to describe the tree-like architecture of these branched polymers. Originally, the term Comb-burst polymers [6] was used by Tomalia *et al.* by analogy to Starburst® polymers to identify these materials. However, in view of this present use as trademark designations, the term 'dendrigraft' polymers is preferred.

The architecture of dendrigrafts resembles that of dendrimers. Each system consists of a core, dendritically branched units forming the interior of the molecule, and multiple chain termini. A unique feature found in dendrimers, is that the branching sites leading to the next generation occur strictly at the periphery (chain ends) of the molecule. In dendrigrafts, grafting sites are usually distributed randomly along the dendritic chains of the dendrigraft interior. This variation in the grafting sites leads to a 'diffuse layer growth' mechanism [4, 6, 7]. Nevertheless, the method still provides extensive control over the size, shape, flexibility and chemical functionality (critical molecular design parameters) of the molecules [1].

2 SYNTHETIC ROUTES TO DENDRIGRAFT POLYMERS

Preparation of dendrigraft polymers has been achieved by three methods, namely: (a) 'grafting onto', (b) 'grafting from', and (c) 'grafting through'.

The 'grafting onto' method is most commonly used for the preparation of dendrigraft polymers with tailored structure and topology. It relies on the introduction of grafting sites on a polymeric substrate, followed by coupling with reactive 'living' polymer chains. In this approach, characterization of the branches is easily achieved by removing a sample of the side chains from the reactor prior to the grafting reaction. By measuring the molecular weight of the substrate, the side chains and the graft polymer, the number of grafts introduced and the average spacing between grafts can be calculated.

In a 'grafting from' scheme, a polymeric substrate is first functionalized to bear a number of accessible reactive groups. Activation of these groups to provide initiating sites, followed by addition of a monomer, results in the growth of side chains from the backbone polymer. Unfortunately, the exact number of grafts and the molecular weight of the side chains usually cannot be determined independently in this method. Solubility problems are also often encountered for polymeric substrates bearing multiple charges. This can lead to heterogeneous reaction conditions and usually a broader molecular weight distribution.

'Grafting through' refers to a method used to produce comb-branched copolymers by the copolymerization of vinyl monomers with macromonomers (telechelic oligomers bearing a terminal vinyl group). For the synthesis of dendritic graft polymers, the 'grafting through' technique is based on the reaction of living polymeric chains with a capping agent that also contains a polymerizable moiety (e.g. vinyl benzyl chloride). This leads to spontaneous coupling reactions analogous those encountered in random hyperbranched polymer syntheses.

3   SYSTEMS WITH RANDOMLY DISTRIBUTED BRANCHING POINTS

Prime examples of systems that have randomly distributed grafting sites, yet reasonable structure control, include dendrigraft (Comb-burst$^{®}$ or arborescent) type polymers. The general synthetic route to dendrigraft polymers is described in Figure 9.1. Reactive sites are randomly introduced along a linear functionalized polymer core. Living polymeric chains are then grafted to the reactive sites, to give a comb-branched (G0) graft polymer. The introduction of functional groups onto the 'teeth' of the comb-branched structure, followed by grafting, yields the G1 structure. Repeating the cycle of functional group introduction and grafting leads to G2 and higher generations.

A synthetic method based on grafting reactions must meet a number of requirements to yield well-defined dendrigraft polymers. The ionic (cationic or anionic) propagating centers derived from the monomer must possess sufficient reactivity, yet good living characteristics. It must be possible to incorporate reactive functional (grafting) sites along the polymer chains grafted during the previous reaction cycle, without inducing crosslink events. The grafting reaction should ideally proceed without side reactions and in high yield. If living polymerization techniques are employed, good control over the molecular weight and the molecular weight distribution is possible, thus resulting in well-defined structures.

3.1   DENDRIGRAFT (COMB-BURST®) POLYMERS

The synthesis of *dendrigraft-copolymers* consisting of [core]poly(ethylenimine)-*graft*-poly(2-ethyl-2-oxazoline) copolymers and *dendrigraft* poly(ethylenimine) homopolymers was reported by Tomalia *et al.* in 1991 [6]. The synthetic scheme involves successive grafting reactions of living poly(2-ethyl-2-oxazoline) (PEOX) oligomers onto *linear*-poly(ethylenimine) substrates (Scheme 1). The poly(oxazoline) side chains used for the dendritic grafting are prepared by cationic polymerization, yielding a narrow molecular weight chain possessing a reactive oxazolin-

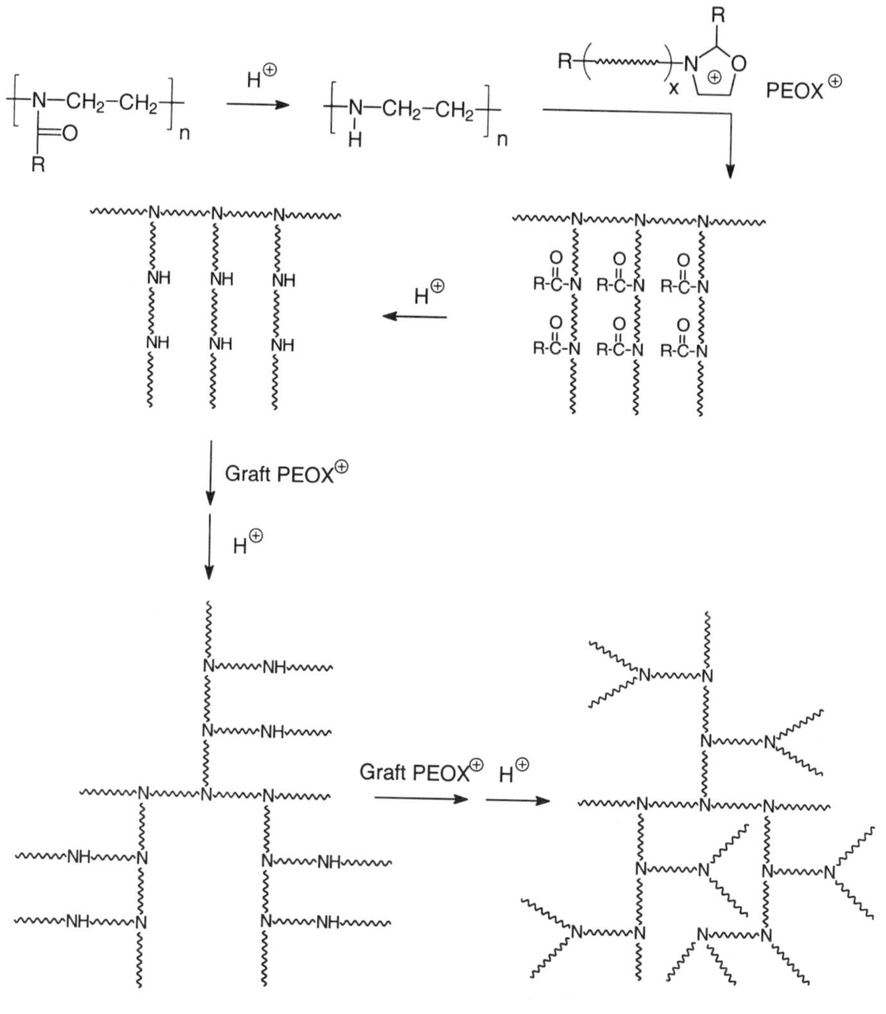

Scheme 1

ium cation terminus. A comb-branched or G0 poly(ethylenimine)-poly(oxazo-line) graft copolymer is first prepared by grafting these reactive poly(oxazoline) oligomers onto a preformed *linear*-poly(ethylenimine) core. Subsequent de-protection (i.e. deacylation) of the poly(oxazoline) side chains under acidic conditions yields the G0 poly(ethylenimine). Secondary amine functionalities generated along the side chains of the comb polymer then serve as coupling sites for subsequent grafting of poly(oxazoline) chains leading to generation G1.

Synthesis of higher generation dendrigraft polymers is achieved by repetition of the deprotection and grafting cycles.

These reaction cycles lead to a geometric increase in the molecular weight of the polymers for successive generations. The number of repeat units assembled (or overall degree of polymerization, N_{RU}) can be predicted as a function of the number of grafting sites on the core polymer (core multiplicity N_c), the branch cell multiplicity N_b and the generation G using equation (1). Defining M_c, M_{RU} and M_t as the molecular weight of the core, the repeat units and the terminal units, respectively, the theoretical attainable molecular weight (MW) can be calculated from equation (2). It should be noted that these equations assume that every repeat unit of the core and side chains can react to produce a branching point, and therefore N_c and N_b also correspond to the degree of polymerization (DP) of the core and the branches, respectively.

$$N_{RU} = N_c \left[\frac{N_b^{G+1} - 1}{N_b - 1} \right] \qquad (1)$$

$$MW = M_c + N_B \left[M_{RU} \left(\frac{N_b^{G+1} - 1}{N_b - 1} \right) + M_t N_b^{G+1} \right] \qquad (2)$$

It is clear from equations (1) and (2) that if high molecular weights are desired with a minimum number of grafting reactions, N_c and N_b must be large. When polymeric building blocks are used, N_c and N_b are much greater (5–100) than for dendrimers, wherein values of 2–4 are normally encountered. In other words the number of repeat units N_{RU} assembled as a function of generation G is dependent upon (1) the degree of polymerization of the grafted side chains, and (2) the number of chains grafted to the substrate. Since living polymerization techniques are used to generate the poly(oxazoline) side chains, the degree of polymerization can be controlled by varying the ratio of monomer to initiator. The number of chains grafted on the core polymer may be controlled by varying the degree of polymerization of the substrate [8].

Characterization data for *dendrigraft*-poly(ethylenimine) samples have been reported for up to the third generation (G3) (Table 9.1). In this example, short poly(oxazoline) chains (with a degree of polymerization $N_b = 10$) were grafted onto a linear ($N_c = 20$) poly(ethylenimine) core to produce a G0 graft polymer with a high branching density. The length of the poly(oxazoline) chains was increased ($N_b = 100$) for subsequent generations. Weight-average molecular weights (from light scattering measurements) of $M_w = 10^3$–10^7 are thus obtained, while a relatively narrow apparent [*linear*-poly(ethylene oxide) equivalent] molecular weight distribution ($M_w/M_n = 1.1$–1.5) is maintained. The M_w and branching functionality (*f*) of the polymers are seen to increase essentially geometrically for successive generations, as predicted by equation 2. The efficiency of the grafting reaction is reported to range between 65 and 80%, and to be sensitive to a number of factors. The grafting efficiency decreases as the

Table 9.1   Characterization data for polyethylenimine comb-burst polymers (adapted from ref. 8)

Sample	M_w^{LSa}	M_w/M_n^b	f^c
Core	1000	1.05	—
G0	2500	1.22	5
G1	138 000	1.34	26
G2	1 080 000	1.47	176
G3	10 400 000	1.20	745

[a] Weight-average molecular weight from light scattering measurements.
[b] Apparent polydispersity index for the graft polymers, from size exclusion chromatography (SEC) analysis.
[c] Number of branches added, based on M_w^{LS} and N_b.

poly(oxazoline) chain length is increased, or as the size of the *linear*-poly(ethylenimine) core increases. This effect is attributed to steric congestion resulting from a high segmental density within the core, particularly for higher generations. As the structure fills in, the number of readily accessible sites decreases at the expense of the grafting efficiency. Other parameters like the reaction time, additives such as diisopropylethylamine (a proton trap), and the ratio of poly(oxazoline) chains to grafting sites also influence the grafting efficiency [8].

A range of molecular topologies is accessible using these dendrigraft branching strategies. It was mentioned that the number of coupling sites available on the core polymer chain (N_c) and the grafted branches (N_b) can be varied independently. It should be noted that the molecular topology also varies with the ratio N_c / N_b. For example, the synthesis of rod-shaped *dendrigraft*-poly(ethylenimine) can be achieved by successive grafting reactions of poly(oxazoline) side chains with $N_b = 5$ onto a *linear*-poly(ethylenimine) core with $N_c = 200$ [9]. When $N_c \gg N_b$ a rod-like topology is expected. Conversely, if $N_c \ll N_b$, a spherical topology should be obtained.

Dendrigraft polymer molecules are expected to become denser and more rigid as the branching functionality increases. Evidence in support of a globular morphology is found in the solution properties of these polymers: The intrinsic viscosity of *dendrigraft*- poly(ethylenimine) increases non-linearly for successive generations up to G2, then decreases slightly for the G3 polymer, reminiscent of other dendritic polymers [10].

3.2 DENDRIGRAFT (ARBORESCENT) POLY(STYRENES)

Gauthier and Möller [4] described in 1991 the use of anionic polymerization and grafting techniques to prepare poly(styrenes) with a dendritic structure. Styrene is well suited to be incorporated into a synthetic scheme aimed at producing multiply grafted polymers: The anionic polymerization of styrene yields reactive macroanions with exceptional living characteristics, and a wide range of reactive, electrophilic functional groups can be introduced on the substrates by electrophilic substitution. The synthetic steps leading to the preparation of *dendrigraft*-poly(styrenes) [4] are summarized in Scheme 2. Partial chloromethylation of a *linear*-poly(styrene), under conditions selected to minimize the occurrence of cross-linking reactions, serves to introduce coupling sites on the core polymer. The substrate is then reacted with poly(styryllithium), after capping the chains with a 1,1-diphenylethylene (DPE) unit. Subsequent chloromethylation and grafting leads to higher generation *dendrigraft*-(arborescent) poly(styrenes). The use of diphenylethylene as a reactivity modifier for the polystyryl anions constitutes a key step for avoiding side reactions in the grafting process: The direct reaction of uncapped polystyryl anions with chloromethylated polystyrene only proceeds with a yield around 50%, due to competing metal–halogen exchange reactions (Figure 9.2a). In contrast, the grafting efficiency is increased to 96% after capping with diphenylethylene (Figure 9.2b). The deep red coloration of the capped macroanions facilitates monitoring of the stoichiometry of the coupling reaction: A solution of the chloromethylated substrate can be added slowly to the macroanion solution until the coloration fades.

For a series of reactions where the molecular weight of the branches (M_b) and the number of grafting sites per backbone chain (grafting site density, f) remain constant for each generation, the molecular weight of a generation G polymer can be predicted using equation (3).

$$M = M_b + M_b f + M_b f^2 + M_b f^3 + \dots = \sum_{x=0}^{G+1} M_b f^x \qquad (3)$$

A geometric growth in branching functionality and overall molecular weight is

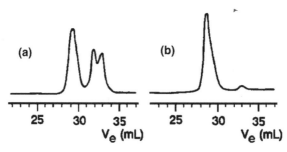

Figure 9.2 Size exclusion chromatography traces for raw grafting products formed in the coupling reaction of chloromethylated polystyrene with polystyryllithium (a) without 1,1-diphenylethylene capping, and (b) with capped polystyryl anions (adapted from ref. [4])

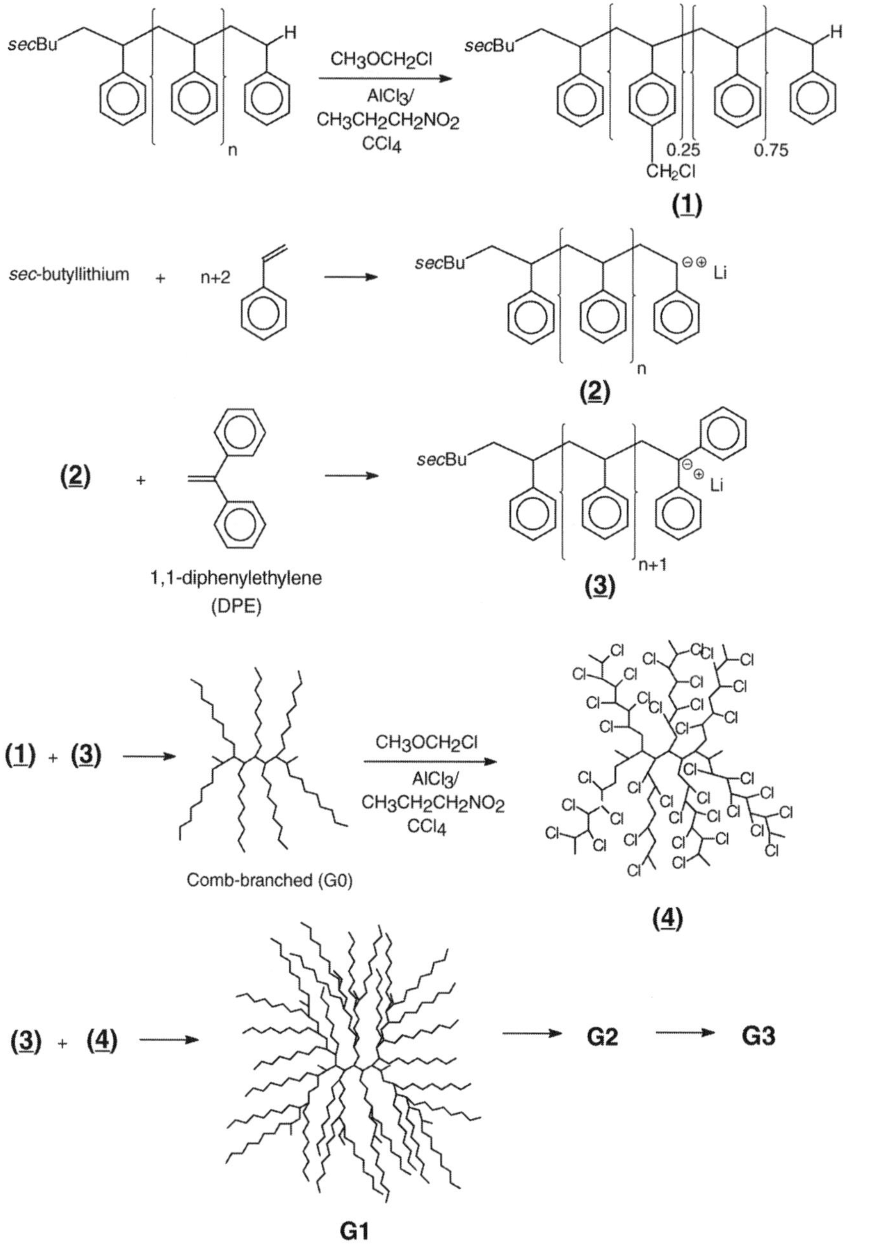
secBu
H
CH3OCH2Cl
AlCl3/
CH3CH2CH2NO2
CCl4
n
0.25
0.75
CH2Cl
(1)
sec-butyllithium
n+2
(2)
Li
1,1-diphenylethylene
(DPE)
n+1
(3)
Comb-branched (G0)
(4)
G1
G2
G3

Scheme 2

Table 9.2 Characterization results for two series of arborescent polystyrenes with different side chain molecular weights (adapted from ref. 4)

			S05 Series			
G	$M_w^{br}/10^{3a}$	M_w^{br}/M_n^{br}	M_w^{LS}	M_w^b	M_w/M_n^b	$f_w^c{}_B$
0	4.3	1.03	6.7×10^4	4.0×10^4	1.07	14
1	4.6	1.03	8.7×10^5	1.3×10^5	1.07	170
2	4.2	1.04	1.3×10^7	3.0×10^5	1.20	2900
3	4.4	1.05	9×10^7	4.5×10^5	1.15	17 500
4	4.9	1.08	2×10^8	—	—	22 000

			S30 Series			
G	$M_w^{br}/10^4$	M_w^{br}/M_n^{br}	M_w^{LS}	M_w	M_w/M_n	f_w
0	2.8	1.15	5.1×10^5	2.1×10^5	1.12	18
1	2.7	1.09	9.0×10^6	5.9×10^5	1.22	310
2	2.7	1.09	1×10^8	—	—	3400
3	2.8	1.09	5×10^8	—	—	14 300

[a] The superscript *br* refers to the branches (side chains).
[b] Apparent values for the graft polymers, from SEC analysis.
[c] Number of branches added in the last grafting reaction, based on the M_w^{LS} increase and M_w^{br}

thus expected, if all the coupling sites on the grafting substrate are consumed in the reaction.

Since anionic polymerization techniques are used, reactive oligomers with a controllable molecular weight and a narrow molecular weight distribution can be generated. The number of functional groups on the grafting substrate can also be likewise varied, which provides control over the branching density for each generation. Consequently, the side chain molecular weight and branching density can be varied independently for each generation.

Characterization data for two series of arborescent poly(styrenes) prepared from either $M_w \approx 5000$ (S05) or $M_w \approx 30\,000$ (S30) poly(styrene) side chains [11] are compared in Table 9.2. The molecular weight (M_w^{LS}) and branching functionality (f_w) increase in an approximately geometric fashion for each generation up to G2, as predicted by equation (9.3). The smaller increases observed for the G3 and G4 polymers can be explained by invoking steric effects, which undoubtedly limit the accessibility of grafting sites on the substrate. A broad range of molecular weights ($M_w^{LS} \approx 6 \times 10^4$–$5 \times 10^8$) and branching functionalities ($f_w = 14$–$22\,000$) can be achieved, while maintaining a low apparent (linear polystyrene equivalent) polydispersity ($M_w/M_n \leq 1.22$). Comparison of the apparent molecular weights M_w, from size exclusion chromatography (SEC) analysis, to the absolute molecular weights determined from light scattering measurements (M_w^{LS}) shows that SEC analysis strongly underestimates the molecular weight, due to the very compact structure of the molecules [11].

The influence of branching functionality and side chain molecular weight on the physical properties of *dendrigraft*-poly(styrenes) has been investigated in a number of studies [11–14]. The properties displayed by the polymers more closely mimic those of rigid spheres as the branching functionality of the molecules is increased, or as the size of the side chains is decreased. For example, *arborescent*-poly(styrenes) with short ($M_w \approx 5000$) side chains show almost no swelling in a good solvent (toluene) relative to a poor solvent (cyclohexane). In contrast, molecules with larger ($M_w \approx 30\,000$) side chains swell considerably in a good solvent.

3.3   DENDRIGRAFT (ARBORESCENT)-POLY(BUTADIENES)

The synthesis and characterization of a series of dendrigraft polymers based on polybutadiene segments was reported by Hempenius *et al.* [15]. The synthesis begins with a *linear*-poly(butadiene) (PB) core obtained by the *sec*-butyllithium-initiated anionic polymerization of 1,3-butadiene in *n*-hexane, to give a micro-structure containing approximately 6% 1,2-units (Scheme 3). The pendant vinyl moities are converted into electrophilic grafting sites by hydrosilylation with

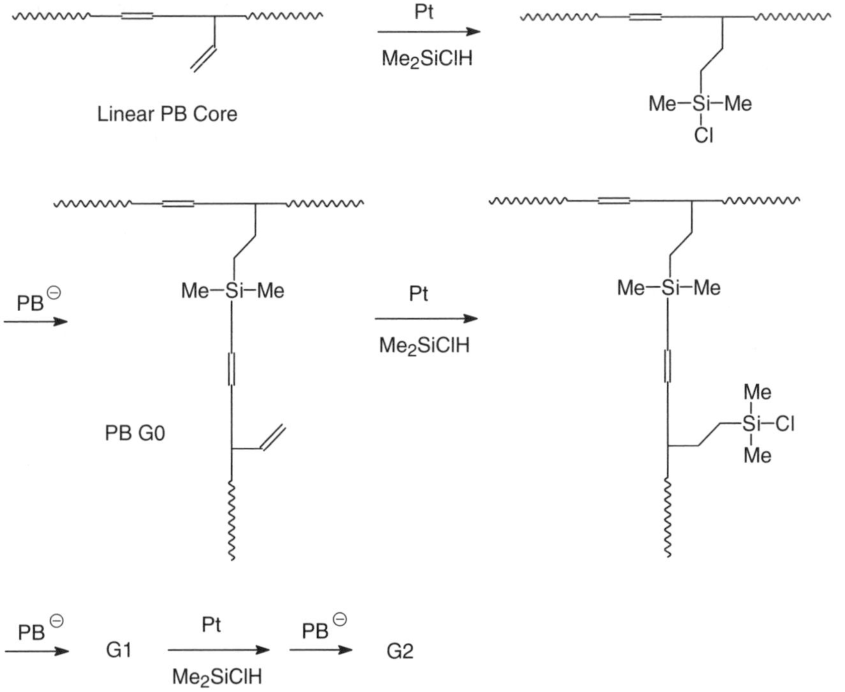

Scheme 3

chlorodimethylsilane in the presence of a platinum catalyst. The hydrosilylated polybutadiene substrate is then reacted with an excess of polybutadienyllithium, to yield a comb-branched or G0 polymer. Repetition of the hydrosilylation and grafting reactions leads to the G1 and G2 polymers. Because polybutadienyl macroanions are essentially colorless, it is difficult to monitor the reaction stoichiometry during the grafting process. Complete reaction of the chlorosilyl coupling sites is insured by using an excess of polybutadienyllithium in the reaction.

A series of samples was synthesized using a poly(butadiene) core and side chain molecular weight $M_w \approx 10\,000$, held constant for each generation. Polymerization in *n*-hexane yields side chains with a 6% 1,2-units content, corresponding to a constant branching density of approximately 10 grafts per chain. The characteristics of the series of *arborescent*-poly(butadienes) obtained are summarized in Table 9.3. With a constant poly(butadiene) branch molecular weight ($M_w \approx 10\,000$) and number of branching sites per chain, a geometric increase in the molecular weight and branching functionality of the graft polymers is observed, as predicted by equation (9.3).

The synthetic approach used for *dendrigraft*-poly(butadienes) has the potential to provide control over the composition and architecture of the molecules. The branch molecular weight is easily varied with the amount of initiator used in the polymerization reaction. Solvent polarity control in the polymerization allows variation of the proportion of 1,2-units in the side chains, and hence the branching density.

The solution properties of dendrigraft polybutadienes are, as in the previous cases discussed, consistent with a hard sphere morphology. The intrinsic viscosity of *arborescent*-poly(butadienes) levels off for the G1 and G2 polymers. Additionally, the ratio of the radius of gyration in solution (R_g) to the hydrodynamic radius (R_h) of the molecules decreases from $R_g/R_h = 1.4$ to 0.8 from G1 to G2. For linear polymer chains with a coiled conformation in solution, a ratio $R_g/R_h = 1.48-1.50$ is expected. For rigid spheres, in comparison, a limiting value $R_g/R_h = 0.775$ is predicted.

Table 9.3 Structural characteristics of arborescent graft polybutadienes (adapted from ref. 15)

G	M_w^{br}	Branches / chain[a]	M_w^{LS}	M_w/M_n	f_w
Core	9600	9.7	9600	1.1	—
0	10800	10.9	190000	1.2	10
1	11000	11.1	4500000	1.3	105
2	10500	10.7	71000000	1.3	1160

[a] Calculated from M_w^{br} and 6 mol% 1,2-polybutadiene units content.

3.4 DENDRIGRAFT (ARBORESCENT)-POLY(STYRENE)-GRAFT-POLY(ISOPRENE) COPOLYMERS

The manner in which the architecture of dendrigraft homopolymers can be systematically varied is very interesting for establishing structure–property relationships. When considering potential applications for dendrigraft polymers, however, materials with a wider range of physical and chemical properties would be more interesting. This can be achieved by incorporating other monomers in the grafting process. The basic synthetic techniques developed for *arborescent*-poly(styrenes) were thus extended to the synthesis of graft copolymers using both *grafting onto* and *grafting from* schemes. For example, a variation in the basic method used for incorporating a *grafting onto* scheme has been applied to the synthesis of graft copolymers with poly(isoprene) or poly(2-vinylpyridine) side chains. Another example based on a *grafting from* scheme is the synthesis of amphiphilic dendrigraft copolymers consisting of a poly(styrene) core with end-linked poly(ethylene oxide) segments. Since the molecular weight of dendrigraft polymers usually increases geometrically for each generation, the overall composition of dendrigraft copolymers should be dominated by the side chains grafted in the last generation. Consequently, the physical properties of the copolymers should be mainly determined by the characteristics (composition, molecular weight and number) of these terminal side chains.

The synthesis of arborescent copolymers by grafting polyisoprene side chains onto an *dendrigraft*-poly(styrene) substrate will be considered first. The techniques used for *arborescent*-poly(styrenes) (Scheme 2) can be employed with only minor modifications to prepare elastomeric isoprene copolymers using a 'grafting onto' strategy [16]. For this purpose polyisoprenyllithium chains are capped with 1,1-diphenylethylene, and grafted onto partially chloromethylated linear, G0, and G1 *dendrigraft*-poly(styrene) substrates. The method is therefore completely analogous to that used for *dendrigraft*-poly(styrenes), except that polyisoprenyl anions are substituted for the polystyryl anions in the last grafting cycle. The flexibility of the method was demonstrated by using polyisoprene side chains of different molecular weights ($M_w \approx 5000$–$100\,000$) and different microstructures (high *cis*-1,4-content or mixed microstructure) in the grafting process.

Characterization data for the grafting substrates and for some of the copolymers synthesized are provided in Tables 9.4 and 9.5, respectively. The substrates used are composed of poly(styrene) side chains with $M_w \approx 5000$, and a chloromethylation level of *c*. 25 mol%, corresponding to about 10 grafting sites per side chain or one grafting site for every four repeat units. The nomenclature used in Table 9.5 for the copolymers specifies their composition and structure. For example, G1PS–PIP5 corresponds to $M_w \approx 5000$ polyisoprene (PIP) side chains grafted onto a G1 poly(styrene) substrate. The copolymers obtained have a high content of the elastomeric component, varying from 86 mol% to over 99 mol% for longer poly(isoprene) side chains ($M_w^{br} \approx 3$–10×10^4, Table 9.5). A

Table 9.4 Characteristics of arborescent polystyrene substrates used in the preparation of isoprene graft copolymers (adapted from ref. 16)

Polymer	M_w^{br}	M_w^{br}/M_n^{br}	M_w^{LS}	f_w
Linear	—	—	4.4×10^3	—
G0 PS	4430	1.04	5.0×10^4	10
G1 PS	5570	1.03	8.0×10^5	130

marked decrease in grafting efficiency is observed for longer polyisoprene side chains and for higher generation substrates. Despite the decreased grafting efficiencies, high molecular weight copolymers ($M_w^{LS} \approx 5 \times 10^4$–$8 \times 10^6$) are still obtained while a low apparent polydispersity index ($M_w/M_n = 1.07$–1.13) is maintained.

The microstructure of the poly(isoprene) side chains can be controlled by varying solvent polarity during the polymerization. Side chains with a high (> 70%) *cis*-1,4 content are obtained by polymerization in cyclohexane; whereas, polymerization in THF yields a mixed microstructure (with a 1:1:1 ratio of 1,2-, 3,4- and 1,4-units). Variation of the dimensions of the grafted poly(isoprene) side chains relative to the poly(styrene) substrate provides control over the morphology of the molecules, as illustrated in Figure 9.3. For short poly(isoprene) side chains, a core–shell (heterogeneous) morphology is observed by atomic force microscopy in the phase contrast mode. When longer side chains are used, the molecules are similar to homopolymers with no detectable core component [16].

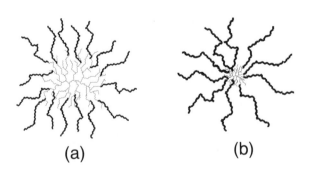

(a)   (b)

Figure 9.3 Morphology control in the preparation of arborescent copolymers: (a) core–shell morphology from short side chains, and (b) star-like morphology from long side chains

Table 9.5 Characterization data for arborescent polystyrene-*graft*-polyisoprene copolymers (adapted from ref. 16)

Sample	$M_w^{br}/10^3$	M_w^{br}/M_n^{br}	Grafting efficiency	M_w^{LS}	M_w/M_n	f_w	Composition/ mol% PIP
PS–PIP5	4.5	1.07	0.80	5.0×10^4	1.11	11	93
PS–PIP30	35	1.03	0.66	4.2×10^5	1.10	12	> 99
PS–PIP100	118	1.08	0.43	—	1.13	—	> 99
G0PS–PIP5	4.7	1.06	0.65	5.8×10^5	1.07	112	89
G0PS–PIP30	32	1.06	0.63	2.8×10^6	1.13	85	96
G0PS–PIP100	112	1.03	0.35	4.0×10^6	1.12	35	> 99
G1PS–PIP5	5.1	1.09	0.71	5.6×10^6	1.11	950	86
G1PS–PIP30	35	1.05	0.18	8.2×10^6	1.11	225	96
G1PS–PIP100	118	1.07	0.06	—	1.07	—	> 99

3.5 DENDRIGRAFT (ARBORESCENT)-POLY(STYRENE)-GRAFT-POLY(2-VINYLPYRIDINE) AND POLY(STYRENE)-GRAFT-POLY(TERT-BUTYL METHACRYLATE) COPOLYMERS

Most recently, the arborescent copolymer synthesis has been extended to the preparation of highly branched polyelectrolyte precursors incorporating ionizable poly(2-vinyl pyridine) (P2VP) or poly(*tert*-butyl methacrylate) (P*t*BMA) side chains [17]. The method used for the preparation of the vinylpyridine copolymers is analogous to the *arborescent*-poly(styrene) synthesis of Scheme 2: the side chains are synthesized by the 1,1-diphenyl-3-methylpentyllithium-initiated anionic polymerization of 2-vinylpyridine in THF, followed by direct coupling (without capping) of the macroanions with chloromethylated linear, G0, G1, or G2 polystyrene substrates. Characterization data for the *arborescent*-poly(styrene)-*graft*-polyvinylpyridine copolymers synthesized are reported in Table 9.6. Copolymers with either short ($M_w \approx 5000$, P2VP5) or long ($M_w \approx 30\,000$, P2VP30) side chains have been prepared. As in the case of isoprene copolymers, the grafting efficiency decreases for higher generation (more highly branched) poly(styrene) substrates and for longer poly(vinylpyridine) side chains. The copolymers have a low apparent polydispersity ($M_w/M_n \approx 1.07-1.13$) and very high molecular weight ($M_w^{LS} \approx 8.2 \times 10^4 - 2.7 \times 10^7$). The overall composition is dominated by the poly(vinylpyridine) component, and ranges from 81 to 87 mol% and 92–97 mol% for short ($M_w \approx 5000$) and long ($M_w \approx 30\,000$) side chains, respectively.

The preparation of copolymers incorporating poly(*tert*-butyl methacrylate) side chains can also be achieved by modification of the grafting scheme [17]. Poly(*tert*-butyl methacrylate) anions are considerably less nucleophilic than the poly(vinylpyridine) anions, and direct coupling with chloromethylated poly(styrene) substrates only proceeds in low yields. The efficiency of the grafting reaction is considerably improved by quantitative conversion of the chloromethyl sites to more reactive bromomethyl groups. Graft copolymers are obtained by coupling poly(*tert*-butyl methacrylate)lithium side chains (in THF at $0\,°C$) with linear and G0 poly(styrene) substrates bearing bromomethyl groups.

Characterization data for *dendrigraft*-poly(styrene)-*graft*-poly(*tert*-butyl methacrylate) copolymers synthesized with either short ($M_w \approx 5000$, P*t*BMA5) or long ($M_w \approx 30\,000$, P*t*BMA30) branches are provided in Table 9.7. The low apparent polydispersities ($M_w/M_n = 1.13-1.20$) obtained for the graft polymers are indicative of a narrow molecular size distribution.

The poly(vinylpyridine) and poly(*tert*-butyl methacrylate) copolymers can easily be converted to either cationic or anionic polyelectrolytes by protonation of the pyridine rings or by base hydrolysis of the *tert*-butyl ester units, respectively. The highly branched structure of the molecules, in combination with the polyelectrolyte effect, should confer useful properties to these materials in solution for applications such as pH-sensitive reversible gels.

Table 9.6 Characterization data for arborescent polystyrene-*graft*-poly(2-vinylpyridine) copolymers (adapted from ref. 17)

Sample	M_w^{br} /10³	M_w^{br}/M_n^{br}	Grafting efficiency	M_w^{LS}	M_w/M_n	f_w	P2VP /mol%
PS–P2VP5	5.2	1.08	0.92	8.2×10^4	1.11	14	86
G0PS–P2VP5	5.8	1.08	0.89	7.2×10^5	1.07	112	87
G1PS–P2VP5	5.0	1.06	0.81	5.7×10^6	1.12	991	83
G2PS–P2VP5	5.2	1.12	0.76	2.5×10^7	1.11	3880	81
PS–P2VP30	27.2	1.07	0.80	4.2×10^5	1.10	14	95
G0PS–P2VP30	28.6	1.08	0.65	3.2×10^6	1.13	111	97
G1PS–P2VP30	34.4	1.09	0.26	2.7×10^7	1.11	758	92

Table 9.7 Characterization data for arborescent polystyrene-*graft*-poly(*tert*-butyl methacrylate) copolymers (Adapted from ref. 17)

Sample	$M_w^{br}/10^3$	M_w^{br}/M_n^{br}	Grafting efficiency	$M_w/10^{3a}$	M_w/M_n^a
PS–P*t*BMA5	5.6	1.10	0.79	3.4×10^4	1.16
G0PS–P*t*BMA5	5.9	1.07	0.67	1.2×10^5	1.20
G0PS–P*t*BMA30	28.5	1.15	0.32	4.7×10^5	1.13

[a] Apparent values for the graft polymers, from SEC analysis.

3.6 DENDRIGRAFT (ARBORESCENT) POLY(STYRENE)-GRAFT-POLY(ETHYLENE OXIDE) COPOLYMERS

A variation of the method developed for *dendrigraft*-poly(styrenes) was suggested for the synthesis of amphiphilic copolymers incorporating a *dendrigraft*-poly(styrene) core with end-linked poly(ethylene oxide) (PEO) segments [18]. For this purpose, *dendrigraft*-poly(styrene) (G0-G3) substrates are first synthesized as described in Scheme 2. The substrates are then partially chloromethylated and coupled with polystyrene side chains prepared using a bifunctional alkyllithium initiator (6-lithiohexyl acetaldehyde acetal, LHAA) as shown in Scheme 4. Cleavage of the acetal functionalities under mildly acidic conditions, yields a core carrying hydroxyl groups at the chain ends (i.e. close to the surface) of the molecules. The end-linked poly(ethylene oxide) segments are introduced by titration of the hydroxyl groups with a strong base (potassium naphthalide), and addition of purified ethylene oxide. A shell of hydrophilic, poly(ethylene oxide) is thus 'grown' by a chain extension reaction from the outer chains of the core polymer.

The synthesis of amphiphilic copolymers, based on G1 and G4 poly(styrene) cores with $M_w \approx 5000$ side chains [18], and of copolymers based on G1 poly(styrene) cores with $M_w \approx 30\,000$ side chains [19] has been demonstrated. Since living polymerization techniques are used, the structure and composition of the molecules are easily controlled. For example, the size and flexibility of the poly(styrene) core can be changed by selecting different core generations, and by varying the chloromethylation level and/or the side chain molecular weight in the grafting reactions. The amount of ethylene oxide incorporated in the chain extension reaction determines the thickness of the hydrophilic shell. Evidence for a core–shell morphology is found in the solubility behavior of the molecules: copolymers with sufficiently long poly(ethylene oxide) side chains are freely soluble in polar solvents such as water and methanol [18].

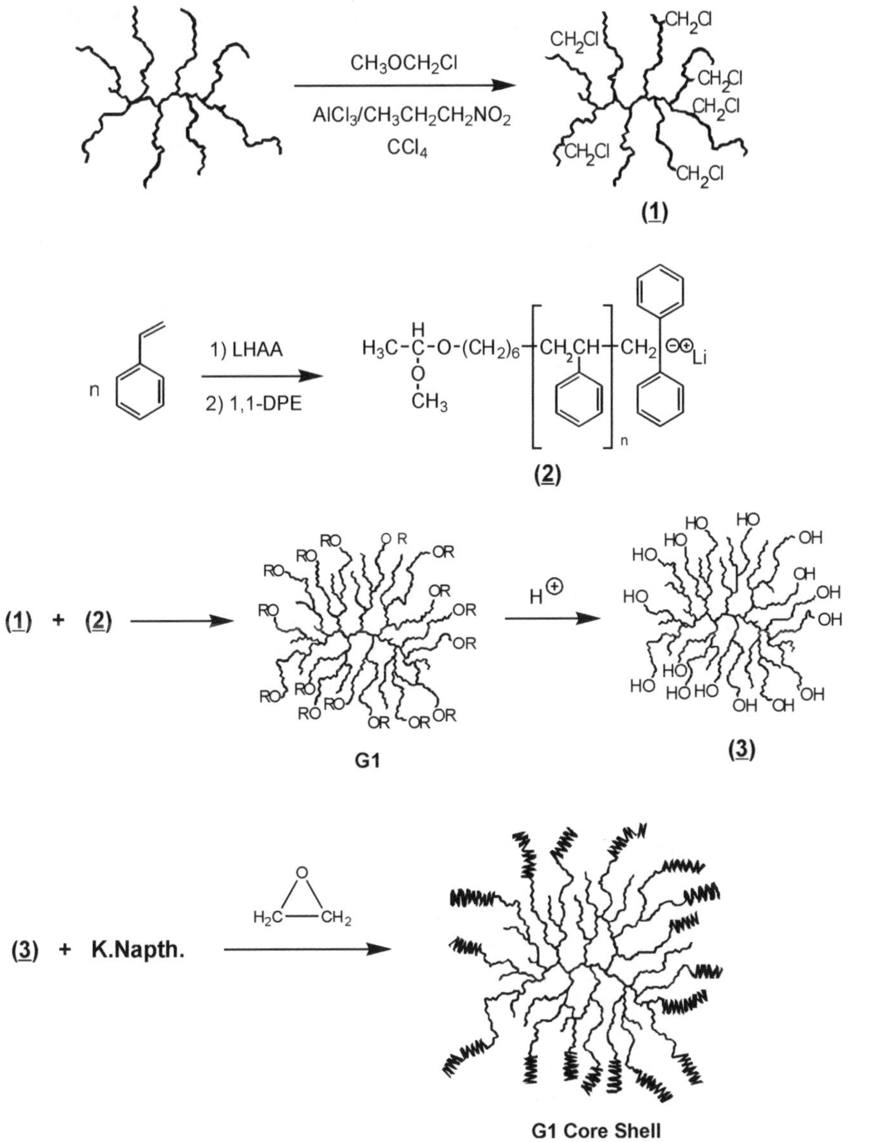
CH3OCH2Cl
AlCl3/CH3CH2CH2NO2
CCl4
CH2Cl
(1)
n
1) LHAA
2) 1,1-DPE
H3C-C-O-(CH2)6-CH2CH-CH2 Li
H
O
CH3
(2)
(1) + (2)
RO
OR
H+
HO
OH
G1
(3)
(3) + K.Napth.
O
H2C CH2
G1 Core Shell

Scheme 4

4   SYSTEMS WITH BRANCHING POINTS AT THE CHAIN ENDS

The examples discussed so far are polymer architectures based on grafting sites randomly distributed on the side chains of the substrate. One consequence of this approach is that the position of the branching points in the molecules will vary, resulting in a 'diffuse layer growth' mechanism, or partial interpenetration of the layers added in each grafting reaction [4, 6, 7]. Branched polymer architectures, displaying a lower degree of interpenetration between successive layers, could be obtained if the grafting sites on the substrate are located strictly at the end of the side chains (in analogy to the dendrimer syntheses), or at least preferentially located toward the outer end of the side chains. Synthetic schemes leading to both types of architectures will now be considered.

4.1   *DENDRIGRAFT-POLY(ETHYLENE OXIDE) WITH DENDRIMER-LIKE TOPOLGIES BY TERMINAL GRAFTING*

The assembly of dendrigraft polymer structures using 'grafting onto' methodologies has received far more attention than procedures based on 'grafting from' schemes. Even in the case of the arborescent ethylene oxide copolymers discussed in section 3.6, the poly(styrene) cores were synthesized by grafting onto chloromethylated substrates, and only *linear*-poly(ethylene oxide) chains were grown from the substrate in the last step. A completely different approach, in which new chains of increasing branching functionality are successively generated from the previous generation, was proposed by the group of Gnanou. This strategy was used to produce poly(ethylene oxide)-based dendrigraft polymers with a *dendrimer-like* structure [20]. For example, a triarm star-branched polymer with living poly(ethylene oxide) arms is first prepared from a trifunctional alcohol initiator, by titration with diphenylmethylpotassium and addition of ethylene oxide (EO), as shown in Scheme 5. The polymer is then amplified to a hexafunctional substrate by treatment with trifluoroethanesulfonyl chloride and a cyclic tin-based ketal derivative, bearing two protected hydroxyl groups in a six-membered ring. After hydrolysis under mildly acidic conditions, the hexafunctional triarm polymer is again used to initiate the polymerization of ethylene oxide. Repetition of the amplification and ethylene oxide polymerization reactions, yields *dendrigraft*-poly(ethylene oxide) structures containing up to 12 poly(ethylene oxide) chains (generation = 2).

Characterization results have been reported for *dendrigraft*-poly(ethylene oxide)s of generation G1, prepared by adding different amounts of ethylene oxide to the initiator core in the side chain growth reaction (Table 9.8). Because of the 'grafting from' method used, the molecular weight of the poly(ethylene oxide) branches cannot be accurately determined, and hence it is impossible to confirm

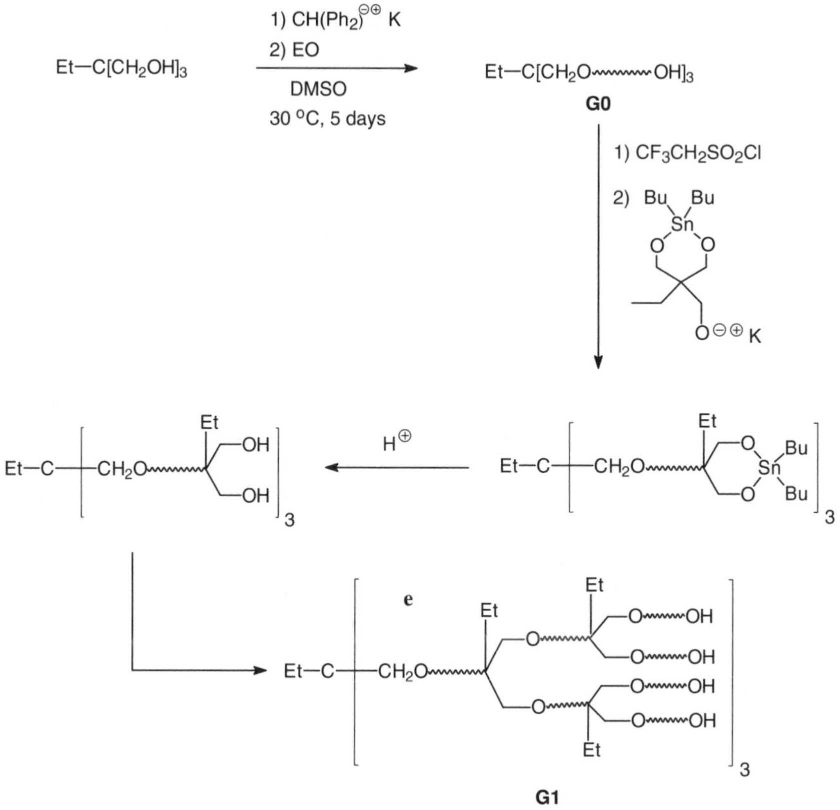

Scheme 5

Table 9.8 Structural characteristics of G1 dendritic poly(ethylene oxides) obtained by terminal grafting (adapted from ref. 20)

M_w^{LSa}	M_w^b (SEC)	M_w/M_n^b	$V_{H,dend}/V_{H,lin}$
9400	4200	1.17	0.51
14 000	6500	1.14	0.49
21 000	10 000	1.10	0.47
28 000	11 300	1.30	0.48

[a] Absolute M_w for the graft polymers from light scattering.
[b] Apparent values for the graft polymers, from SEC analysis.

that the branches grown during each cycle are of uniform length. Nonetheless, the apparent molecular weight distribution [based on linear poly(ethylene oxide) calibration standards] is surprisingly narrow for the generations characterized ($M_w/M_n = 1.1$–1.3). Comparison of the molecular weight data obtained from light scattering and SEC indicates that the SEC values are significantly under-estimated, as observed for other dendrigraft polymers. Comparison of the hydrodynamic volume ratio for the branched polymers and *linear*-poly(ethylene oxide) samples of comparable molecular weight gives a ratio close to the $V_{H,dend}/V_{H,lin} \approx 0.5$ value predicted by a modified Zimm–Stockmayer equation.

4.2 DENDRIGRAFT-*POLY(STYRENE)*-GRAFT-*POLY(ETHYLENE OXIDE)* COPOLYMERS BY TERMINAL GRAFTING

Using a variation of the method described in Scheme 5, dendrigraft block copolymers consisting of poly(styrene) and poly(ethylene oxide) segments can also be prepared [21]. In this case a hexafunctional compound, hexa[4-(1-chloroethyl)phenethyl]benzene is activated with $SnCl_4$ to initiate the living cationic polymerization of styrene. Termination of the six polystyrene chains using the cyclic tin-based ketal derivative of Scheme 5, followed by hydrolysis under mildly acidic conditions, yields a six-arm star-branched polystyrene with 12 reactive hydroxyl groups at the chain ends. Anionic polymerization of ethylene oxide after titration of the hydroxyl groups with diphenylmethylpotassium yields a G1 *dendrigraft*-poly(styrene)-poly(ethylene oxide) copolymer with 12 poly(ethylene oxide) outer arms. More recently, this terminal grafting strategy was also extended to a variety of 'dendrimer-like' dendrigraft topologies reported by Hedrick *et al.* [22–25].

4.3 DENDRITIC POLY(STYRENES) BY GRAFTING ONTO POLY(CHLOROETHYL VINYL ETHER)

The synthesis of dendrigraft polymers based on poly(chloroethyl vinyl ether) (PCEVE) and polystyrene macromolecular building blocks was recently reported by Deffieux and Schappacher [26]. The synthetic path used for the preparation of these compounds is shown in Scheme 6. Living cationic polymerization is first used to prepare a *linear*-poly(vinyl ether) backbone. The anionic polymerization of styrene is then initiated with a lithioacetal compound, to generate polystyryllithium with a protected hydroxyl group at the chain end (RO–PSLi). Coupling of the side chains with the poly(vinyl ether) backbone generates a comb-branched structure. Subsequent treatment of the acetal chain termini with trimethylsilyl iodide (TMSI) yields α-iodoether groups, that can be activated with zinc chloride to initiate the cationic polymerization of a new aliquot of chloroethyl vinyl ether. The resulting comb-branched (G0) structure with poly(styrene)-*block*-poly(vinyl ether) side chains can be subjected to further grafting with acetal-terminated polystyryllithium, and so on, to prepare the higher generation polymers. The architecture obtained from the grafting process described in Scheme 6 has branching points located only in the terminal (outer)

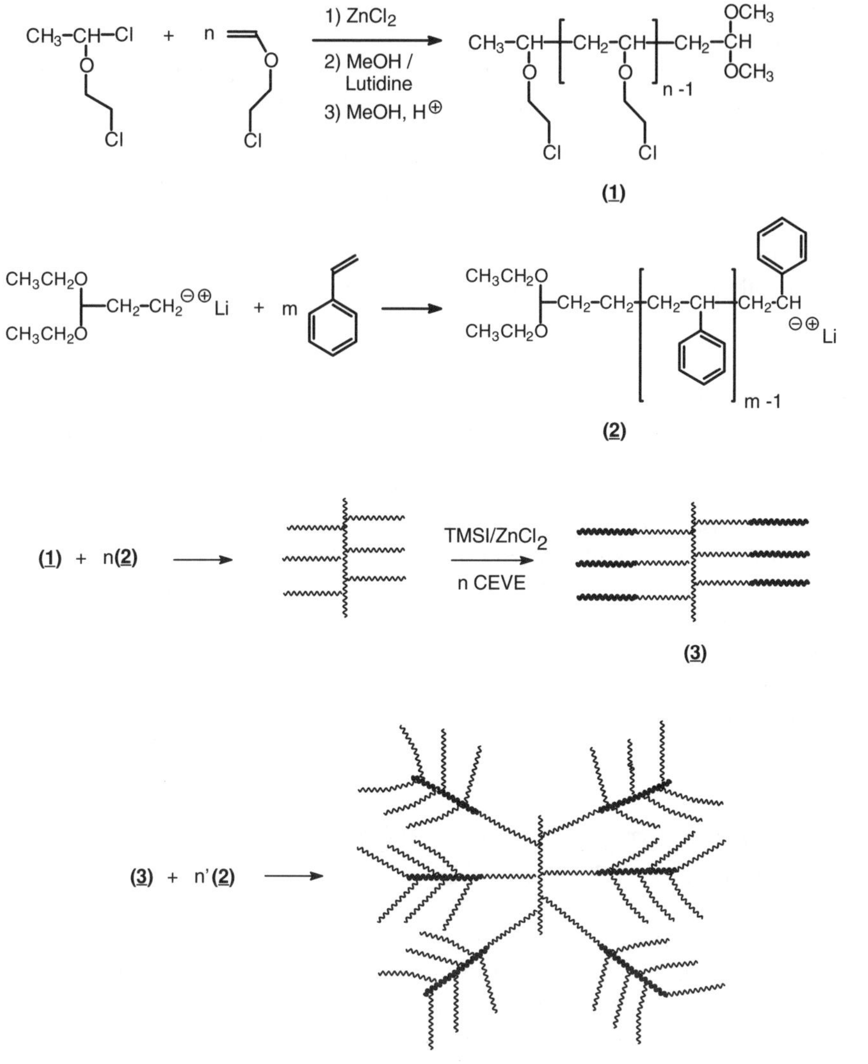

Scheme 6

portion of the side chains on the substrate, rather than randomly distributed along the backbone polymer as described earlier for dendrigraft (arborescent) polymers. While the layer structure obtained by this method is not expected to be as well defined as in the case for terminal grafting (section 4.2), the increase in molecular weight for each generation is larger, since many chains can be grafted in a single reaction.

Variations in the graft polymer architectures are possible by controlling the

Table 9.9 Characteristics of $PCEVE_1$-*graft*-(PS_1-*block*-$PCEVE_2$-*graft*-PS_2) copolymers (adapted from ref. 26)

$PCEVE_1$ $DP_n{}^a$	PS_1 $DP_n{}^a$	$PCEVE_2$ $DP_n{}^a$	PS_2 $DP_n{}^a$	$M_n{}^{th}$ /10^{3b}	M_n /10^{3a}	$M_w{}^{LS}/$ 10^3	$M_w/M_n{}^a$	R_g (nm)
26	51	82	52	11 606	460	9500	1.06	25
26	146	50	52	7300	—	5700	1.33	30
60	30	55	37	13 250	500	10 350	1.05	28

^{a} From SEC analysis using linear PS calibration.
^{b} Theoretical molecular weight based on the PS and PCEVE block lengths.

length of the poly(vinyl ether) and poly(styrene) blocks used in the synthesis. The length of the poly(styrene) block influences the overall size of the molecules, whereas the branching multiplicity of the side chains is primarily determined by the length of the poly(vinyl ether) blocks.

The synthetic route described was used for the synthesis of three first-generation dendrigraft copolymers with molecular weights reaching 10^7 and narrow apparent molecular weight distributions ($M_w/M_n = 1.05–1.33$, Table 9.9). The branched polymers have a radius of gyration (R_g) in solution, 5–10 times smaller than *linear*-poly(styrene) samples of comparable molecular weight. Absolute molecular weight of the products determined by light scattering ($M_w{}^{LS}$) is in good agreement with the theoretical M_n calculated for complete consumption of the chloroethyl vinyl ether units in the grafting reaction.

5 CONVERGENT (SELF-BRANCHING) ANIONIC POLYMERIZATION METHOD

The schemes described so far are all based on divergent (core first) syntheses, using either 'grafting onto' or 'grafting from' methodologies. A different approach, analogous to the convergent random hyperbranched polymer syntheses, uses self-branching condensation reactions of macroanions in a 'grafting through' scheme. The method has been suggested by Knauss *et al.* for the synthesis of dendritic poly(styrenes) [27, 28] and poly(isoprenes) [29] of high molecular weight. An interesting feature of self-branching anionic polymerization is that it is a one-pot reaction providing structures somewhat comparable to arborescent (dendrigraft) polymers. The graft polymers obtained also have a narrow molecular weight distribution in some cases.

A procedure used in the self-branching polymerization reaction for the preparation of dendritic polystyrenes is outlined in Scheme 7. Oligomeric polystyryllithium chains are reacted with a coupling agent such as 4-(chlorodimethyl-silyl)styrene (CDMSS), which contains both a polymerizable double bond as well

Table 9.10 Structural characteristics of dendritic polystyrenes obtained by the self-convergent grafting method (adapted from ref. 29)

Initial chain $M_n{}^a$	$M_w{}^{LS\ a}/10^3$	$M_w/M_n{}^a$	Average generation
620	117	1.24	5.8
950	106	1.22	5.4
4779	131	1.21	4.1
9114	222	2.15	3.3
17 570	225	1.16	3.3
35 490	282	1.25	2.6
57 660	433	1.18	2.6

^{a} Determined by SEC analysis using a multi-angle laser light scattering detector.

as a reactive moiety that preferentially reacts with the oligomer. Slow addition of the coupling agent to a solution of living poly(styrene) oligomers (unimer I) leads to the formation of some CDMSS-capped chains (unimer Ia). These chains subsequently react with the poly(styrene) oligomers to give living dimers (II, living dendrons composed of two poly(styrene) oligomers). Continued addition of the coupling agent results in capping of a portion of the living dimers with CDMSS (dimer IIa). Capped (unsaturated) and uncapped (living) species may subsequently react with each other and with CDMSS to give higher generation products of increasing branching functionatities.

At each step of the reaction, every molecule carries either a single anionic reactive site or a vinyl bond at the focal point. After a few coupling cycles (generations), accessibility to reactive sites is expected to decrease, ultimately limiting the growth of the molecules. This problem can be minimized by adding styrene monomer along with the coupling agent, to increase spacing between coupling points within the structure. This modified approach enables the formation of dendritic molecules up to the sixth (average) generation, using a CDMSS to styrene ratio of 1 : 10 in the coupling reaction and primary chains of different molecular weights (Table 9.10) [30]. The average generation attained in the coupling reactions is clearly influenced by the length of the primary chains used. The apparent molecular weight distribution obtained in these reactions remains relatively narrow ($M_w/M_n = 1.14$–1.66), presumably because of the role of steric crowding in limiting the growth of the molecules. The narrow molecular weight distributions obtained were rationalized in terms of a kinetic model, relating steric crowding effects to the individual rate constants of the coupling reactions involved [28].

The variation in intrinsic viscosity with molecular weight was examined for a series of *dendrigraft*-poly(styrenes) obtained by varying the ratio of styrene monomer to coupling agent in the grafting reaction [28]. The polymers obey the Mark–Houwink–Sakurada equation, and are characterized by lower intrinsic viscosity values relative to linear poly(styrenes) of comparable molecular weight, in agreement with a compact structure.

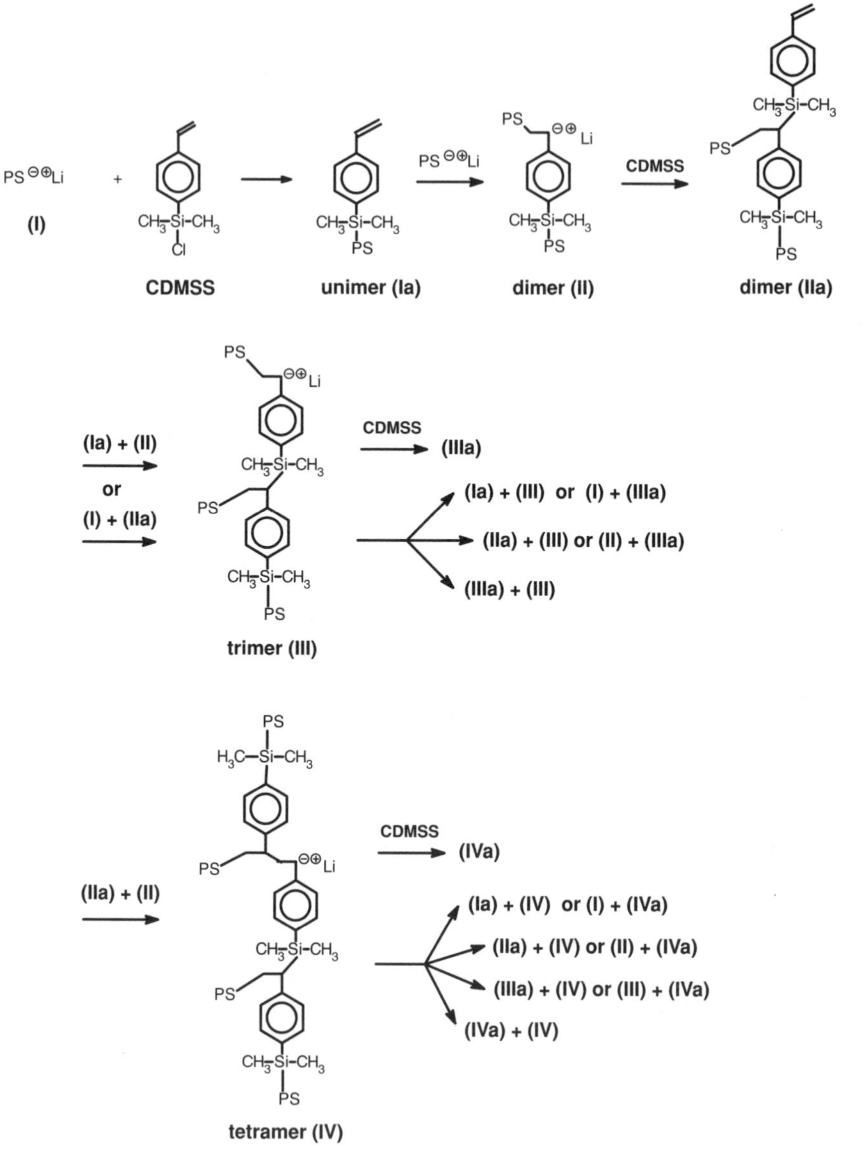

Scheme 7

The approach outlined in Scheme 7 was directly applied to the synthesis of dendritic poly(isoprenes), by simple substitution of isoprene for the styrene monomer [29]. Two coupling agents were examined in this case, namely CDMSS and 2-chlorodimethylsilyl-1,3-butadiene. Other interesting architectures have been prepared based on these reactive dendron structures. For example, the focal anion present on the molecules (after full consumption of the coupling agent) can be used to initiate the polymerization of monomers such as styrene to prepare more complex structures such as linear-dendritic hybrids, or poly(styrene) stars with dendritic poly(styrene) end blocks.

6 CONCLUSION

It should be apparent that dendrigraft (arborescent) methodologies described in this chapter offer new options for macromolecular structure control that surpass those available for producing random hyperbranched polymers. On the other hand, one cannot attain the precise control that is possible from dendrimer synthesis. Many of the core-shell type topologies reminiscent of dendrimers can be synthesized with these methods, producing in some cases 'container-like' properties with relatively monodispersed character (i.e. $M_w/M_n = 1.1\text{--}1.5$). A major advantage offered by these strategies is the ability to attain very high molecular weight products in relatively few iterations (i.e. $G = 1\text{--}3$). The dramatic amplifications/generation and the ability to use relatively inexpensive monomers for 'reactive oligomer' grafting, offers the potential to manufacture and develop dendrigraft polymer products, possessing dendrimer-like properties, at a much lower cost. Commercial development in this area will undoubtedly be receiving substantial attention in the future.

7 REFERENCES

1. Tomalia, D. A., Naylor, A. M. and Goddard III, W. A. *Angew. Chem. Int. Ed. Engl.*, **29**, 138 (1990).
2. Hawker, C. J. and Fréchet, J. M. J. *J. Am. Chem. Soc.*, **112**, 7638 (1990).
3. Tomalia, D. A., Dvornic, P. R., Uppuluri, S., Swanson, D. R. and Balogh, L. *Polym. Mater. Sci. Eng.*, **77**, 95 (1997).
4. Gauthier, M. and Möller, M. *Macromolecules*, **24**, 4548 (1991).
5. Roovers, J. and Comanita, B. *Adv. Polym. Sci.*, **142**, 179 (1999).
6. Tomalia, D. A., Hedstrand, D. M. and Ferritto, M. S. *Macromolecules*, **24**, 1435 (1991).
7. Gauthier, M. in Puskas, J. E. (ed.), *Ionic Polymerizations and Related Processes*, NATO ASI Ser. E-359, Kluwer Academic, Dordrecht (1999), p. 239.
8. Yin, R., Swanson, D. R. and Tomalia, D. A. *Polym. Mater. Sci. Eng.*, **73**, 277 (1995).
9. Tomalia, D. A., Swanson, D. R. and Hedstrand, D. M. *Polym. Prepr.*, **33**(1), 180 (1992).
10. Yin, R., Qin, D., Tomalia, D. A., Kukowska-Latallo, J. and Baker Jr., J. R. *Polym. Mater. Sci. Eng.*, **77**, 206 (1997).
11. Gauthier, M., Li, W. and Tichagwa, L. *Polymer*, **38**, 6363 (1997).
12. Gauthier, M., Möller, M. and Burchard, W. *Macromol. Symp.*, **77**, 43 (1994).
13. Gauthier, M., Möller, M. and Sheiko, S. *Macromolecules*, **30**, 2343 (1997).
14. Hempenius, M. A., Zoetelief, W. F., Gauthier, M. and Möller, M. *Macromolecules*, **31**, 2299 (1998).

15. Hempenius, M. A., Michelberger, W. and Möller, M. *Macromolecules*, **30**, 5602 (1997).
16. Kee, R. A. and Gauthier, M. *Macromolecules*, **32**, 6478 (1999).
17. Kee, R. A. and Gauthier, M. *Polym. Prepr.*, **40**(2), 165 (1999).
18. Gauthier, M., Tichagwa, L., Downey, J. S. and Gao, S. *Macromolecules*, **29**, 519 (1996).
19. Gauthier, M., Cao, L., Rafailovich, M. and Sokolov, J. *Polym. Prepr.*, **40**(2), 114 (1999).
20. Six, J.-L. and Gnanou, Y., *Macromol. Symp.*, **95**, 137 (1995).
21. Taton, D., Cloutet, E. and Gnanou, Y. *Macromol. Chem. Phys.*, **199**, 2501 (1998).
22. Trollsås, M., Atthof, B., Würsch, A. and Hedrick, J. L. *Macromolecules*, **33**, 6423 (2000).
23. Heise, A., Nguyen, C., Malek, R., Hedrick, J. L., Frank, C. W. and Miller, R. D. *Macromolecules*, **33**, 2346 (2000).
24. Trollsås, M., Kelly, M. A., Claesson, H., Siemens, R. and Hedrick, J. L., *Macromolecules*, **32**, 4917 (2000).
25. Heise, A., Hedrick, J. L., Trollsås, M., Miller, R. D. and Frank, C. W. *Macromolecules*, **32**, 231 (1999).
26. Deffieux, A. and Schappacher, M. Macromol. *Symp.*, **132**, 45 (1998).
27. Al-Muallem, H. A. and Knauss, D. M. *Polym. Prepr.*, **38**(1), 68 (1997).
28. Knauss, D. M., Al-Muallem, H. A., Huang, T. and Wu, D. T. *Macromolecules*, **33**, 3557 (2000).
29. Knauss, D. M., Al-Muallem, H. A. and Huang, T., *Polym. Mater. Sci. Eng.*, **80**, 153 (1999).
30. Al-Muallem, H. and Knauss, D. M. *Polym. Prepr.*, **39**(2), 623 (1998).

PART II
Characterization of Dendritic Polymers

10

Gel Electrophoretic Characterization of Dendritic Polymers

C. ZHANG AND D. A. TOMALIA
Center for Biologic Nanotechnology, University of Michigan, Ann Arbor, MI,
USA

1 INTRODUCTION

Gel electrophoresis is widely used in the routine analysis and separation of many well-known biopolymers such as proteins or nucleic acids. Little has been reported concerning the use of this methodology for the analysis of synthetic polymers, undoubtedly since in many cases these polymers are not soluble in aqueous solution – a medium normally used for electrophoresis. Even for those water-soluble synthetic polymers, the broad molecular weight dispersities usually associated with traditional polymers generally preclude the use of electrophoretic methods. Dendrimers, however, especially those constructed using semi-controlled or controlled structure synthesis (Chapters 8 and 9), possess narrow molecular weight distribution and those that are sufficiently water solubile, usually are ideal analytes for electrophoretic methods. More specifically, poly(amidoamine) (PAMAM) and related dendrimers have been proven amendable to gel electrophoresis, as will be discussed in this chapter.

We begin with a short introduction to provide polymer chemists who may be new to the field of electrophoresis with a brief background concerning electrophoretic separation and characterization. Those who wish to obtain an in-depth understanding of the theory or detailed practical techniques are referred to textbooks/monographs on this subject [1–6]. Next, the advantages of gel electrophoresis as an analytical tool, and the structural requirements regarding dendrimers as electrophoretic analytes are discussed. Finally, studies directed at

Dendrimers and Other Dendritic Polymers. Edited by Jean M. J. Fréchet and Donald A. Tomalia
© 2001 John Wiley & Sons Ltd

gel electrophoretic characterization of dendritic polymers and related materials are reviewed in the rest of the chapter.

2 GEL ELECTROPHORESIS: BASIC CONCEPTS

A particle with a net charge moves under the influence of an electric field toward an oppositely charged electrode. This phenomenon, defined as *electrophoresis*, is the basis for various electrophoretic methods in biochemistry research. Separation of particles via electrophoresis is achieved through the difference in their migration distances, which is dependent upon their electrophoretic mobility.

The electrophoretic driving force for a particle to move, depending on the electric field strength (E) and the net charge on the particle (z), is balanced by the frictional resistance (f) which the particle must overcome to migrate.

$$f = E \cdot z \tag{1}$$

In free solution, the frictional force obeys Stokes' law so that

$$f = 6\pi r v \eta \tag{2}$$

where r is the particle radius, v its velocity and η the viscosity of the medium. In gels, frictional resistance is a complex function of gel pore size and particle size.

Electrophoretic mobility (m) is defined as the particle's migration distance (d) in time t under the influence of unit electric field strength

$$m = \frac{d}{E \cdot t} = \frac{v}{E} \tag{3}$$

Based on equation (3), when $E \cdot t$ is kept constant, migration distances can be used directly for comparing electrophoretic mobility of different particles. Combining equations (1)–(3), one can see that the electrophoretic mobility of charged particles in free solution depends on their net charge, particle size and nature of the medium. The situation in solid support gels is much more complicated, and particle electrophoretic mobility is determined by its *charge density* (charge-to-mass ratio) rather than net charge, in addition to its size and nature of the medium.

In the following, we will take a brief look at additional factors that affect the electrophoretic mobility and thus separation of charged particles.

2.1 *INFLUENCE OF MEDIUM PH*

Proteins possess both anionic and cationic groups as part of their primary structure and form zwitterions. Since the dissociation constants (pK values) of these groups are different, the charges on proteins may vary widely depending upon pH of the medium. As a result, pH has a profound influence on protein mobility.

Nucleic acids possess sugar–phosphate backbones, whose net charges do not change over a relatively wide range of pH. Thus, charge densities are nearly constant for different nucleic acids, as their net charge is proportional to the number of residues (i.e. mass). Therefore, the pH of the medium is not as critical in nucleic acid electrophoretic characterization.

Structure controlled dendritic polymers that have been studied using gel electrophoresis generally behave as mimics of either proteins or nucleic acids, and possess similar ionic groups such as $-NH_3^{\oplus}$, $-COO^{\ominus}$, or $PO_4^{\ominus}$ functionality. Dendrimer structures may be widely modified as a function of their interior composition and as well as the nature of their surface groups. Depending on their structure, the influence of pH may vary dramatically for different dendrimers.

2.2   IONIC STRENGTH OF THE MEDIUM

In general, lower ionic strength of the separation medium leads to higher migration rates of analytes, while higher ionic strength yields lower migration rates but sharper separation. Thus, the choice of buffer ionic strength is an important parameter in determining the time and resolution of an electrophoretic analysis.

2.3   SUPPORT MEDIA

Though electrophoretic separations were historically first studied in free solutions, more recent developments have extended its application to solid supports, including polyacrylamide, agarose, and starch gels. The purpose of a solid support is to suppress convection current and diffusion so that sharp separations may be retained. In addition, support gels of controlled pore sizes can serve as size-selective molecular sieves to enhance separation – smaller molecules experience less frictional resistance and move faster, while larger molecules move slower. Therefore, separation can be achieved based on molecular size.

Polyacrylamide gel is the most commonly used type of support medium for gel electrophoresis, and polyacrylamide gel electrophoresis is simply known as PAGE. The gel is usually formed by polymerization of acrylamide and the cross-linking agent N, N'-methylene-bis-acrylamide (Bis) in the presence of ammonium persulfate (APS, initiator) and N, N, N', N'-tetramethyl-ethylenediamine (TEMED, accelerator). The total concentration of acrylamide

$$\left(T(\%) \, \frac{\text{acrylamide(g)} + \text{bis(g)}}{100 \, \text{ml (solution)}} \right)$$

and the proportion of Bis

$$\left(c(\%) = \frac{\text{Bis(g)}}{\text{acrylamide(g)} + \text{Bis(g)}} \times 100 \right)$$

are two important factors in determining such physical properties of the gel such as mechanical strength, density, elasticity and pore size. In fact, a major advantage associated with poly(acrylamide) gels is the ease with which the pore sizes and hence the degree of molecular sieving can be altered by simply changing T and c. For this reason, PAGE is effective in separation of molecules of different sizes, yet possessing similar charge densities.

Particles are subjected to molecular sieving during their passage through the gel, and the molecular sieving effect is, to a large extent, determined by pore sizes. Therefore, careful consideration must be given to the choices of an appropriate acrylamide concentration for the optimal separation of particles of certain sizes. For example, if the gel cross-link density is too high, the particles may be totally excluded from the gel or if too low, all the particles will run together and no separation may be achieved.

Gels of a uniform cross-link density are referred to as 'homogeneous gels', while those possessing cross-link density gradient are 'gradient gels'. Gradient gels are normally used to extend the particle size range that may be separated, and to sharpen the separation.

Agarose is a purified linear galactan hydrocolloid, isolated from agar or recovered directly from agar-bearing marine algae. Gels are formed through physical crosslinking of the linear polymers simply achieved by cooling down their hot solutions. Agarose gels have larger pore sizes than polyacrylamide gels, and are usually used for separation of large molecules or complexes (e.g. dendrimer/DNA complexes shown later in this chapter).

Two kinds of gel configurations are usually used, either slab-gels or cylindrical rod gels. Slab-gels have the advantage of easy manipulation and better reproducibility and they are currently the method of choice for most electrophoretic analyses. Shown in Figure 10.1 is an illustration of a vertical slab-gel setup. Normal polarity (with the anode at the bottom) is used for negatively charged analytes (e.g. half generation PAMAM dendrimers possessing anionic $-COO^{\ominus}$ surface groups), while reverse polarity is used for positively charged analytes (e.g. full generation PAMAM dendrimers possessing cationic $-NH_3^{\oplus}$ surface groups).

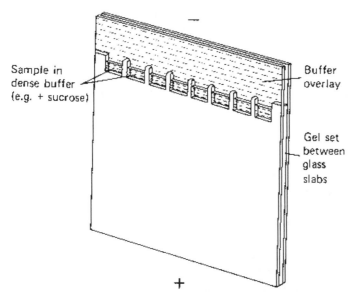

Figure 10.1 An illustration of a vertical slab-gel apparatus. Samples are applied in the wells set in the gel using a 'comb'. Side-by-side comparison of samples is possible in slab systems (ref. 7). (Reprinted with permission from reference 7, Figure 11.2, Page 295, copyright (1994) Springer-Verlag.)

2.4 GEL ELECTROPHORESIS UNDER NATIVE OR DENATURING CONDITIONS

As discussed above, electrophoretic separation is based on both the charge density and size of the analyte particles. When the influence of charge is eliminated, gel electrophoresis may be used to determine molecular size or mass. In protein chemistry, this is achieved by running PAGE in the presence of additives, such as sodium dodecylsulfate (SDS). SDS is an anionic detergent that complexes with proteins at a rough ratio of one SDS per two amino acid residues, providing protein–SDS complexes whose net charge is proportional to molecular mass. Therefore, SDS denatured proteins have nearly identical charge density which is independent of molecular mass, and for this reason, SDS–PAGE is commonly employed to measure protein molecular weight.

Nucleic acids of different molecular mass possess very similar charge density, as mentioned previously, so they are also separated as a function of particle size. A non-ionic additive such as urea is usually used to prevent multimolecular aggregation.

This introduction was intended to merely provide basic concepts and principles of electrophoretic separation. It should be noted that a wide variety of electrophoretic techniques, such as isoelectric focusing and two-dimensional gel electrophoresis, in addition to various visualization and quantification methods, are often used in more sophisticated analyses, however, these techniques will not be reviewed in this chapter.

3   WHY GEL ELECTROPHORESIS IS USEFUL IN ANALYZING DENDRIMERS

Since the first reports on dendritic architecture appeared in the mid-1980s [8–10], the number of new synthetic dendritic polymers has increased almost exponentially [11–14]. As the efforts in dendrimer synthesis intensifies, characterization of these large yet complex molecules have offered new challenges to scientists in this field. From the previous chapter, one can see that chromatographic methods such as size exclusion chromatography (SEC) and HPLC can provide basic information about the purity, homogeneity and size (molecular mass) of these dendrimers. Yet a major advance in characterizing dendrimers should probably be attributed to the recent progress in the development of two important mass spectroscopic techniques as discussed in Chapter 13, electrospray ionization (ESI) and matrix-assisted laser desorption ionization (MALDI) mass spectroscopic (MS) methods, which are capable of detecting fine structural defects at relatively high molecular weights [15, 16]. However, all these methods, although they are very useful, require expensive instrumentation, which is sometimes cost-prohibitive to many laboratories, especially for routine analysis.

Gel electrophoresis, on the other hand, offers many benefits, since it does not require sophisticated instruments and has been used as a routine lab tool by biochemists for decades. In many respects, gel electrophoresis for dendrimer chemists is comparable to thin layer chromatography (TLC) for synthetic organic chemists. Furthermore, gel electrophoresis has remarkably high resolving power, compared to that of SEC. This is primarily due to a minimum zone spreading resultant from migration in a homogeneous medium, and efficient fractionation aided by focusing the sample in a stacking gel prior to electrophoretic separation [17, 18]. Gel electrophoresis may be used qualitatively, or semi-quantitatively, if it is combined with densitometric analysis. On the other hand, preparative separations of important dendritic species are possible on a very small scale.

In the following section, some of the basic requirements for dendritic polymers as electrophoretic analytes are discussed.

3.1   NARROW DISPERSITY

Proteins and nucleic acids are structure controlled monodispersed compounds that typically exhibit narrow bands as analytes on gels. Synthetic polymers, on the other hand, generally form smears rather than bands because of their traditionally broad molecular weight distributions (see Figure 10.5). For this reason, electrophoretic analysis has been reported for relatively few synthetic polymers so far [18, 19]. Dendrimers, as illustrated in Chapters 8 and 9, are usually constructed through controlled or semi-controlled processes that lead to precise molecular architecture with very narrow molecular weight distributions. This is the fundamental reason they may be analyzed using gel electrophoresis methodology to yield useful characterization results.

3.2 SOLUBILITY

Electrophoresis is normally run in aqueous media, hence the analytes must be soluble in water. Presently only three types of water-soluble dendrimers have been successfully analyzed using gel electrophoresis techniques. The list includes Starburst® PAMAM dendrimers [21], nucleic acid dendrimers [21] and poly(lysine) dendrimers [23, 24] (see Figures 10.2, 10.4 and 10.6). However, in each case appropriate water solubilizing terminal groups are required (i.e. $-NH_2$, $-OH$ or $-CO_2H$ groups) for suitable electrophoretic analysis.

3.3 CHARGE

Particles must possess a net charge to migrate in a electrophoretic field. All the dendrimers that have been analyzed using gel electrophoresis have either cationic or anionic groups that may be ionized at appropriate pH values and form charged particles.

In summary, dendrimers are a unique class of monodispersed synthetic molecules reminiscent of proteins or nucleic acids. If they can be functionalized to be soluble in water with appropriately charged terminal groups, they are generally ideal candidates for gel electrophoretic analyses.

4 GEL ELECTROPHORESIS IN ANALYZING DENDRITIC POLYMERS AND RELATED MATERIALS

Just as gel electrophoresis methodology is commonly used to determine homogeneity, as well as molecular weights of proteins these techniques may be used similarly for dendrimer analysis. Other uses have included the assessment of DNA/dendrimer binding constants which have been studied using gel electrophoresis. Application of gel electrophoresis in dendritic science will be discussed from these three aspects.

4.1 PURITY AND HOMOGENEITY ASSESSMENT

Poly(amidoamine) (PAMAM) dendrimers have been envisioned as globular covalent assemblies reminiscent of folded, three-dimensional proteins. The mimicry is not confined to the globular topology, but also includes their β-alanine interior composition, amide linkages, cationic or anionic surface groups and monodispersity. This family is probably the most thoroughly studied class of dendrimers using gel electrophoresis [10, 11, 19]. PAMAM dendrimers are constructed via a controlled divergent approach, which includes alternating reiterations of exhaustive Michael addition followed by amidation as shown in Figure 10.2 [8, 9]. The initiator core may be either NH_3, ethylenediamine (EDA) or other variations. Full generations are produced at the amidation stage and possess terminal NH_2 groups, while half generations are formed at the Michael addition stage to yield terminal methyl ester groups. Hydrolysis of half generation PAMAMs in the presence of group I metal hydroxides affords PAMAM dendrimers with $-COOH$ terminal [9].

With each iteration of the reaction sequence (Michael addition and amidation) leading to formation of the next generation, the number of terminal groups doubles, and the mass approximately doubles. As a result, PAMAM dendrimers of different generations possess approximately identical charge density (charge-to-mass ratio) assuming that all the terminal groups carry a charge. Thus,

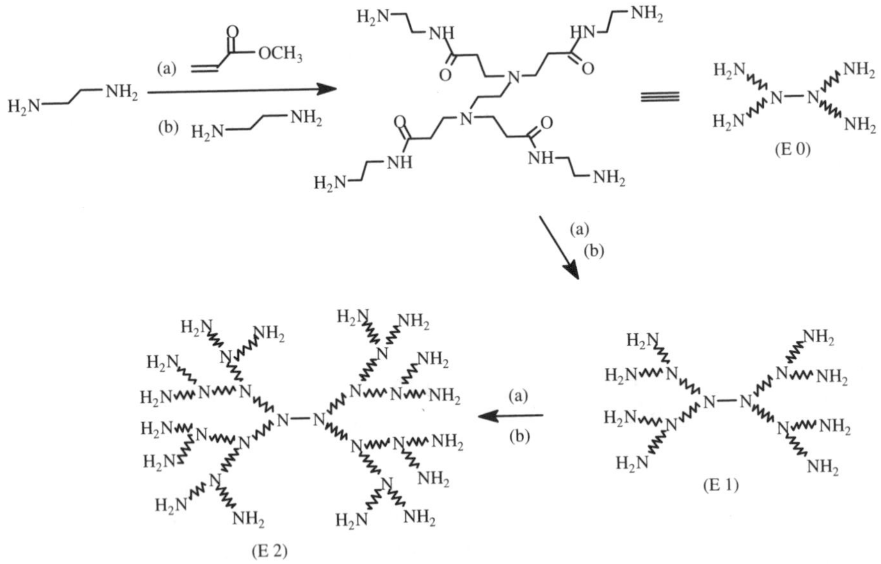

Figure 10.2 Schematic illustration of synthesis of (EDA) core PAMAM dendrimers (E 0–2). Higher generations can be obtained by successive reiterations.

PAMAM dendrimers provide a series of nearly precise macromolecules with almost equivalent charge density, yet a wide molecular weight range (a few hundred to almost a million), of which gel electrophoretic analysis can be carried out under native conditions [21].

In Figure 10.3 is shown the electrophoretogram of (EDA-core) PAMAM dendrimers from generation 2 to 10 on a 5–40% T gradient gel. In the running buffer at pH = 3, both the primary and the tertiary amines are presumably protonated, leading to positively charged particles that move to the cathode on a reversed polarity gel. The lower generations, (i.e. small sized particles) move further than their higher and larger analogs.

As shown in Figure 10.3, PAMAMs of each generation are flanked by its precursors (G − 1) and its next higher (G + 1) generations on the electrophoretogram. Referring to the synthetic scheme in Figure 2, it is apparent that dimers and higher oligomers are possible. Such is the case as observes 'higher flanking species' which have been shown by mass spectroscopy to be dimers that resulted from a double amidation of EDA between two dendrimer molecules. The similar migratory properties are consistent with the expectation that they should have nearly identical molecular weights to that of the next higher generation. The 'lower flanking species' are formed as a result of incomplete removal of EDA after the amidation step. The residual EDA can then act as a new initiator core then to form a 'so-called' (G − 1) trailing generation. In this way, the purity or homogeneity of each generation is clearly demonstrated using gel electrophoresis. In addition, semi-quantitative or quantitative assessment can be conducted to determine purity at a specific generation as long as appropriate linearity studies are performed.

For comparison, linear and dendritic poly(lysine)s were analyzed in a parallel study using PAGE [21]. The linear poly(lysine), with a reported polydispersity of

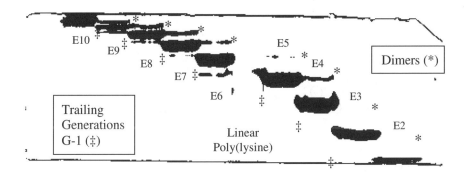

Figure 10.3 Electrophoretogram of several generations of (EDA-core) PAMAM dendrimers (E 0–10). The unlabeled smear in the middle of the gel is a conventional polydispersed linear poly(lysine) sample as a comparison

1.3 and molecular weight of 37 200, exhibits a broad smear. On the other hand, the so-called 'Denkewalter dendrimer' poly(lysine) [22] constructed according to dendritic rules as shown in Figure 10.4 was found to possess substantial molecular weight monodispersity, as manifested by the narrow bands observed in its PAGE (Figure 10.5).

In another study, nucleic acid dendrimers were synthesized *via* both convergent and divergent approaches, and their purity assessed using PAGE [23, 24]. In Figure 10.6 is shown convergent synthetic route, which involves:
(1) synthesis of thymidine (T) oligonucleotides on the surface of controlled-pore glass;
(2) introduction of branching-points by coupling of two adjacent oligomer chain ends with an adenosine (A) derivative;
(3) repetitive chain elongation and branching steps to form higher generations (G = 1–3).

The resultant nucleic acid dendrimer can be cleaved from the support and analyzed using PAGE. Shown in Figure 10.7 is the electrophoretogram of crude dendron **4** [23]. The least mobile band has been identified as the full-length dendrimer, while other 'fingerprint' bands resulted from incomplete reactions at the branching step. Preparative PAGE purification was conducted to afford pure dendrimer **4** with purity greater than 98%.

In a recent report, the divergent solid-phase synthesis of nucleic acid dendrimers was also reported and the purity assessed using PAGE [24], however, the details are not included here due to space constraints.

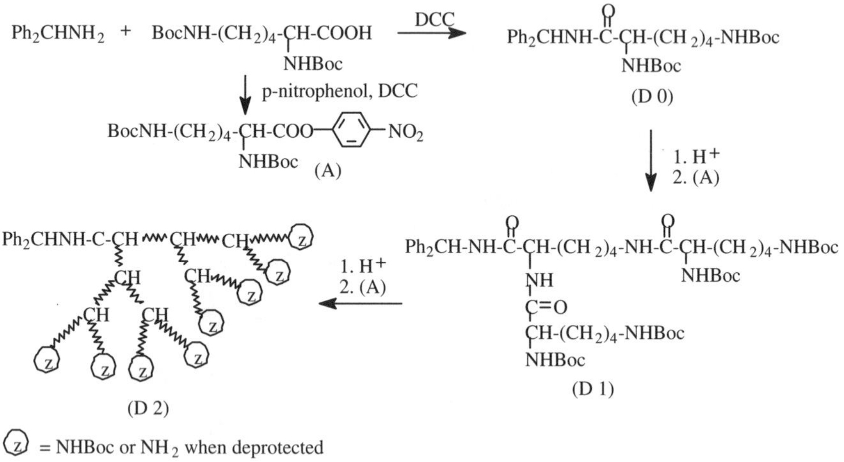

(Z) = NHBoc or NH$_2$ when deprotected

Figure 10.4 Synthetic scheme for 'Denkewalter' dendrimers of generations 0 to 2 (D 0–2).

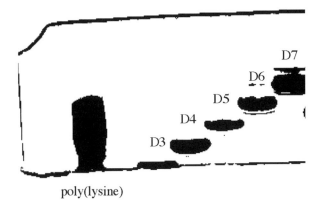

poly(lysine)

Figure 10.5 Electrophoretogram of linear poly(lysine) (left) and dendritic poly(lysine)s (D 3–7).

4.2 MOLECULAR WEIGHT ESTIMATIONS

PAMAM dendrimers have been used as nanoscale building blocks (tectons) to construct more complex dendritic topologies referred to as core–shell tecto(dendrimers)[31] (see Chapter 5). In this case, amine terminated dendrimers are used as a core, around which carboxylic acids or their esters terminated dendrimers are covalently bonded to form core-shell tecto(dendrimers) [25]. Using PAMAM dendrimers (G = 2–10) as calibration standards, molecular weight of the resultant tecto(dendrimers) were estimated using PAGE. The results are in excellent agreement with molecular weights obtained both from MALDI-TOF mass spectrometry and calculated from AFM observed molecular dimensions, as summarized in Table 10.1.

4.3 STUDY OF DNA/DENDRIMER COMPLEXES

Dendrimers with positively charged surfaces can complex with DNA and have shown great potential in gene transfection [11, 26]. Studies have used agarose gel electrophoresis to detect the binding between dendrimers and DNA, by examining the charge neutralization, which retards the migration of the DNA in electrophoresis.

In the first report, PAMAM dendrimers with primary amine surface groups were used [27]. It was found that complex formation is dependent both upon the size (generation) of the dendrimers used and the charge ratio between the (cationic and anionic species, i.e. ammonium groups on PAMAM to phosphate groups on DNA). Retardation of DNA migration was not observed with

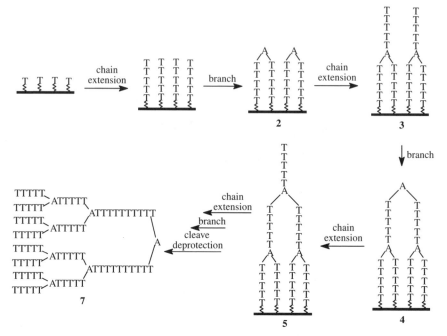

Figure 10.6 Synthetic scheme of nucleic acid dendrimers

Table 10.1 Core-shell tecto(dendrimer) molecular weight analysis results obtained from MALDI-MS, AFM and PAGE.

Core–Shell compounds	1	2	3	4
MALDI–MS (MW)	114 000	172 000	238 000	403 000
AFM (MW)	—	195 000	369 000	469 000
PAGE (MW)	120 000	168 000	250 000	670 000

PAMAM dendrimer G3 or lower generations at any charge ratio, while complete retardation with higher generations occurred only when an equivalent (charge ratio of 1) or excess dendrimer amine groups are present.

In a similar study, poly(ethylene glycol)-*block*-poly(L-lysine) dendrimers were used to complex with DNA [28]. It was found that retardation of migration increases with increasing charge ratios, and complete retardation occurs at cationic/anionic ratio of 2.

In summary, gel electrophoresis is a very convenient method to assess the purity of dendrimers, to estimate molecular weight of similar materials, or to probe the interaction between various dendrimers and important biopolymers such as DNA. It is noteworthy that capillary electrophoresis (CE), which offers analogous features to gel electrophoresis, also has shown great potential in

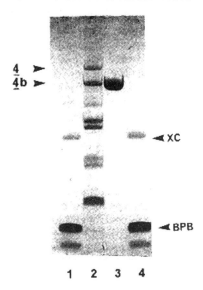

Figure 10.7   Electrophoretogram of dendritic nucleic acid **4** (lane 2) and a linear analog of identical base composition (lane 3). Lanes 1 and 4: marker dyes xylene cyanol (XC) and bromophenol blue (BPB). (Reprinted with permission from reference 23, copyright (1993) American Chemical Society.)

analyzing dendritic polymers [21, 30]. In view of the fact that dendrimers have manifested important benefits in a variety of biomedical fields, such as drug delivery, gene transfection or immunodiagnostics, one can expect that biomedical tools, such as electrophoresis will become increasingly valuable for dendrimer scientists as significant methodology for characterization and purity assessment.

5   CONCLUSION

It is widely recognized that dendrimers are the first synthetic 'structure control-led polymers' that rival the control and monodispersity observed for proteins. Their precise masses and behavior as analytes on electrophoretic gels demonstrate the contention that dendrimers (i.e. specifically poly(amidoamines)) may be viewed as 'artificial proteins'.

Gel electrophoresis, a common analytical method in biological and medical research, can be conveniently applied in the analysis of dendritic polymers, owing primarily to the unique structure of the dendrimer analytes. It has currently been used to assess purity, measure molecular weight of dendrimers, and probe the binding between dendrimers and DNA. As the number of new biomedical applications for dendrimers continues to grow, the use of electrophoretic methods for characterization of dendritic polymers and related materials are also expected to expand.

6 REFERENCES

1. Andrews, A. T. *Electrophoresis: Theory, Techniques, and Biochemical and Clinical Applications*, Oxford University Press, New York, 1986.
2. Allen, R. C. *Gel Electrophoresis of Proteins and Nucleic Acids: Selected Techniques*, W. de Gruyter, New York, 1994.
3. Mosher, R. A. *The Dynamics of Electrophoresis*, VCH, New York, 1992.
4. Dunn, M. J. *Gel Electrophoresis: Proteins*, Bios Scientific Publishers in Association with the Biochemical Society, Oxford, 1993.
5. Gersten, D. M. *Gel Electrophoresis – Proteins: Essential Techniques*, John Wiley and Sons, New York, 1996.
6. Dunn, M. J. *Gel Electrophoresis of Proteins*, Wright, Bristol, 1986.
7. Scopes, R. K. *Protein Purification: Principles and Practice*, Springer-Verlag, New York, 1994.
8. Tomalia, D. A., Baker, H., Dewald, J. R., Hall, M., Kallos, G., Martin, S., Roeck, J., Ryder, J. and Smith, P. *Polym. J. (Tokyo)*, **17**, 117 (1985).
9. Tomalia, D. A., Baker, H., Dewald, J. R., Hall, M., Kallos, G., Martin, S., Roeck, J., Ryder, J. and Smith, P. *Macromolecules*, **19**, 2466 (1986).
10. Newkome, G. R., Yao, Z.-Q., Baker, G. R. and Gupta, K. *J. Org. Chem.*, **50**, 2003 (1985).
11. Tomalia, D. A., Naylor, A. M. and Goddard, W. A. III *Angew. Chem., Int. Ed. Engl.*, **29**, 148 (1990).
12. Tomalia, D. A. and Durst, H. D. in *Topics in Current Chemistry*, Springer-Verlag, Berlin, 1993, p. 193.
13. Matthews, O. A., Shipway, A. N. and Stoddart, J. F. *Prog. Polym. Sci.*, **23**, 1 1998.
14. Bosman, A. W., Janssen, H. M. and Meijer, E. W. *Chem. Rev.*, **99**, 1665 (1999).
15. Kallos, G. J., Tomalia, D. A., Hedstrand, D. M., Lewis, S. and Zhou, J. *Rapid Commun. Mass Spectrom.*, **5**, 383 (1991).
16. Schutz, B. L., Rockwood, A. L., Smith, R. D., Tomalia, D. A. and Spindler, R. *Rapid Commun. Mass Spectrom.*, **9**, 1552 (1995).
17. Ornstein, L. *Ann. N. Y. Acad. Sci.*, **121**, 321 (1964).
18. Chen, J.-L. and Morawetz, H. *Macromolecules*, **15**, 1185 (1982).
19. Hoagland, D. A. and Muthukumar, M. *Macromolecules*, **25**, 6696 (1992).
20. Sayed-Sweet, Y., Hedstrand, D. M., Spindler, R. and Tomalia, D. A. *J. Mater. Chem.*, **9**, 1199 (1997).
21. Brothers, H. M. II, Piehler, L. T. and Tomalia, D. A. *J. Chromat. A*, **814**, 233 (1998).
22. Denkewalter, R. G., Kole, J. F. and Lukasavage, W. J. US Pat. **4**, 410, 688 (1983).
23. Hudson, R. H. E. and Damha, M. J. *J. Am. Chem. Soc.*, **115**, 2119 (1993).
24. Hudson, R. H. E., Robidoux, S. and Damha, M. J. *Tet. Let.*, **39**, 1299 (1998).
25. (a) Uppuluri, S., Swanson, D. R., Brothers, H. M., Piehler, L. T., Li, J., Meier, D. J., Hagnauer, G. L. and Tomalia, D. A. *Polym. Mater. Sci. Eng.*, **80**, 55 (1999); (b) Tomalia, D. A., Uppuluri, S., Swanson, D. R., Brothers, H. M., Piehler, L. T., Li, Meier, D. J., Hagnauer, G. L. and Balogh, L. *Proc. Boston Mater. Res. Soc. Meeting*, 1998, in press.

26. Haensler, J. and Szoka, F. C. Jr. *Bioconjugate Chem.*, **4**, 372 (1993).
27. Kukowska-Latallo, J. F., Bielinska, A. U., Johnson, J., Spindler, R., Tomalia, D. A. and Baker, J. R. Jr. *Proc. Natl. Acad. Sci., USA* **93**, 4897 (1996).
28. Choi, J. S., Lee, E. J., Choi, Y. H., Jeong, Y. J. and Park, J. S. *Bioconjugate Chem.*, **10**, 62 (1999).
29. Choi, Y. H., Liu, F., Park, J. S. and Kim, S. W. *Bioconjugate Chem.*, **9**, 708 (1998).
30. Stockigt, D., Lohmer, G. and Belder, D. *Rapid Commun. Mass Spectrom.*, 10, 521 (1996).
31. Uppuluri, S., Swanson, D. R., Piehler, L. T., Li, J., Hagnauer, G. L. and Tomalia, D. A. *Adv. Mater.*, **12**(11), 796–800 (2000).

11

Characterization of Dendritically Branched Polymers by Small Angle Neutron Scattering (SANS), Small Angle X-Ray Scattering (SAXS) and Transmission Electron Microscopy (TEM)

B. J. BAUER AND E. J. AMIS
National Institute of Standards and Technology Gaithersburg, MD, USA

1   THE UNIQUENESS OF DENDRITIC STRUCTURES

Since the introduction of synthetic techniques to produce dendritic polymers, it has become widely recognized that they produce structures unlike any previous synthetic polymers [1, 2]. The stepwise layering of shells (generations) around a core with uniform branching in each step produces molecules of elegant symmetry that have become immediately recognizable. Figure 11.1 illustrates a two-dimensional projection of a classical dendrimer that is instructive of the synthetic steps required to develop such topology and architecture.

At the center of the dendrimer is the core that represents the first synthetic step in divergent synthesis, or the last step in convergent synthesis. Each concentric shell of branched units is a generation (X), designated (GX), of identical repeat units. Hundreds of chemical repeat units have been reported [3, 4], but our characterization has been focused primarily on poly(amidoamine) (PAMAM), $-(CH_2CH_2CONHCH_2CH_2N) <$ [1], or poly(propyleneimine)

Dendrimers and Other Dendritic Polymers. Edited by Jean M. J. Fréchet and Donald A. Tomalia
© 2001 John Wiley & Sons Ltd

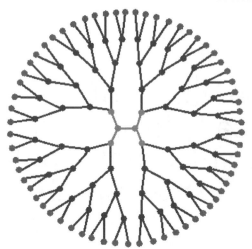

Figure 11.1   Classical two-dendrimer projection showing core, branch points and terminal groups

(PPI), $-(CH_2CH_2CH_2N) < $ [5, 6] dendrimers. The discontinuous lines at the perimeter of the circle in Figure 11.1 are the terminal units of the dendrimer which are $-NH_2$ for the full generations of both PAMAM and PPI dendrimers. We are interested in the structure of a generic dendrimer, however, whose properties result from the connectivity within a dendrimer and not primarily from the chemistry involved in the synthesis.

Although the synthetic techniques of dendrimer production have been extensively studied [3, 4], characterization of the size, shape and interactions of dendrimers have lagged behind. While the cartoon of Figure 11.1 is very instructive for learning about the covalent connectivity resulting from the synthesis, it can be very misleading if it is used to describe how a dendrimer actually distributes itself in space. If dendrimers were shaped as depicted by Figure 11.1, they would be round (circular in two dimensions, spherical in three dimensions), have a lower segment density on the inside than at the outside and have all of the terminal units at the outside. Since these factors have very important application implications, it is imperative that measurements be made to characterize the reality of dendrimers.

The doubling or amplification that is inherent in dendrimer chemistry is the dominant process that controls the dendrimer shape [1]. With each generation, the number of terminal units usually doubles. Each shell (generation) enhances at approximately a constant value, whereas the total molecular mass approximately doubles with each generation as does the number of branch points. While the dendrimer mass doubles with generation, the space to fit the units increases at a much slower rate. The contour length of any chain from the core to the terminal units is proportional to the number of chemical bonds and hence the number of generations. Equation (1a) and (1b) show the relationship between the molecular mass, M, and the molecular volume, V, for a dendrimer of generation G. The ratio of M to V gives a relative local density ρ and the inverse of this quantity gives an intrinsic viscosity $[\eta]$.

$$M \propto 2^{G} \tag{1a}$$

$$V \propto G3 \tag{1b}$$

$$\rho \propto [\eta]^{-1} \propto M/V \propto 2^{G}/G^{3} \to \infty \tag{1c}$$

The increase of the molecular mass proceeds much more rapidly than the space available and means that after a few generations, there will not be enough room to fit in all of the required units of the next generation. The result is that the chemistry at a certain generation cannot be complete and less than 100% chemical conversion will result. It is of interest to study dendrimers up to this limit to find the effect of such extreme crowding.

2   HISTORY OF DENDRIMER CHARACTERIZATION

The earliest work on dendrimer characterization was concerned with aspects of the organic chemistry – did the proposed chemical reaction take place without side reactions and what was the conversion? Since near-100% conversion and near-perfect removal of excess reactants is required for making pure dendrimer, common methods of spectroscopy and chromatography can be used to verify the structure. In a wide variety of dendrimer chemistries, nearly perfect structures have been produced, at least for earlier generations wherein the techniques are more quantitative.

The development of mass spectroscopic techniques such as matrix assisted laser desorption (MALDI) and electrospray mass spectrometry has allowed the absolute determination of dendrimer perfection [7, 8]. For divergent dendrimers such as PAMAM and PPI, single flaws in the chemical structure can be measured as a function of generation to 'genealogically' define an unreacted site of or a side reaction producing a loop at a particular generation level. Mass spectrometric results on dendrimers, not only demonstrate the extreme sensitivity of the technique, but also demonstrate the uniformity of the molecular mass. The polydispersity index of M_{w}/M_{n} for a G6 PAMAM dendrimer can be $\cong 1.0006$ which is substantially narrower than that of 'living' polymers of the same molecular mass [7].

Hydrodynamic sizes from intrinsic viscosity (IV), gel permeation chromatography (GPC), and holographic relaxation spectroscopy (HRS) map the changes in size with generation. For example, (IV) of PAMAM [9] and PPI [6] dendrimers goes through a maximum as a function of generation suggesting that the dendrimers become more densely packed. While these techniques give valuable information, they are actually communicating more about the size of the dendrimers rather than about their internal structure.

Light-scattering techniques measure the radius of gyration (R_g) of dendrimers which is an average of the spatial distribution of all of the units. Aharoni and Murthy [10] have used scattering methods to measure the R_g of dendrimers with lysine repeat units and conclude that they are spherical. These molecules differ from conventional dendrimers, however, in that the branch cell (units) are asymmetric and therefore the internal segment distributions are modified. Other laboratories have begun to publish results of small angle neutron scattering (SANS) [11] and small angle X-ray scattering (SAXS) [12, 13] of dendrimers, but only a few studies have had sufficient resolution to resolve the higher order features characteristic of uniform dendrimers [14–17].

Transmission electron microscopy (TEM) [1, 18–20] has been used to image individual dendritic molecules, usually the larger generations. More recently, atomic force microscopy (AFM) [21] has also been used to image dendritic molecules.

3   IMPORTANT TECHNOLOGICAL QUESTIONS

The unique topology and architectural components of dendrimers suggest many important applications that may be possible. We have identified at least five important factors that pose questions concerning these technological applications. Table 11.1 lists these factors along with the applicability of SANS, SAXS and TEM to address the questions.

The first question concerns the segment density distribution (SDD) of the atoms in the interior of a dendrimer. Predictions vary from the extremes of a very hollow structure with a densely packed exterior [22] to a high central segment density with a gradually tapering concentration of units [23] and a variety of intermediate structures [24–28]. If dendrimers have a unimolecular micelle-like structure, they may be able to solubilize molecules in the interior for important applications such as drug delivery [29], phase transfer [30], etc. If dendrimers have tapered distributions, they may not be significantly different from linear or slightly branched polymers for these applications.

Another important issue is the difference between various branching types such as random hyperbranched [31], dendrigrafts and dendrimers. The complexity of synthesis requirements manifested by the statistical dendritic polymers versus the more structurally controlled dendrimers could make the former orders of magnitude more expensive than hyperbranched. Are the structures as significantly unique and of sufficient value effectiveness to justify the higher costs?

The terminal groups of a dendrimer are large in number and can have functionalities capable of chemical reactions. If the terminal reactive terminal groups were near the periphery, they would be readily accessible for attachment to surfaces or to reagents. Block copolymers or networks with dendrimers as crosslink points would benefit from having them on the outside.

When dendrimers are exposed to forced intermolecular contact by increasing their concentration in solution, or by placing them on a surface, do they freely pass through one another or do they avoid interpenetration and align themselves? Dendrimers have sizes in the range 1–20 nm and would be excellent candidates for nanoscopic structures if such ordering does occur.

Finally, do dendrimers change greatly in size when placed in different solvents? For applications as size standards or molecular probes dendrimers possessing a relatively fixed size would be preferable. For applications using the release of stored guest molecules, however, it would be preferable to 'open' and 'close' dendrimers by designing appropriate 'container release strategies'.

4 MEASUREMENT METHODS USED

Dendritically branched polymers have been characterized in our laboratory primarily by SANS, SAXS and TEM. All three techniques are capable of measuring and defining the sizes of dendritic materials. SAXS and SANS probe a size scale that reveals information about the average size of the molecules, namely, the radius of gyration (R_g) [14, 15, 17, 32, 33]. More complex treatment of the scattering data can reveal information about the internal structure of the entire dendrimer or individual components of dendrimers [17, 32, 34]. Scattering data may also give information concerning the intermolecular spacing between dendrimers [32, 35, 36]. TEM can be used to image individual polymer molecules, giving direct information on size, shape, polydispersity and alignment [20].

Both SANS and SAXS probe the same size length scale characterized by the scattering vector q (with $q = (4\pi/\lambda) \sin(\theta/2)$, θ being the scattering angle and λ being the wavelength of the probing radiation). The intensity of the scattering as a function of angle (q) gives information on the arrangement and spacing of the polymer segments. The scattering from a collection of particles can be broken into contributions from a single particle, $P(q)$, as well as between particles, $S(q)$.

$$I(q)) \propto P(q)S(q) \tag{2}$$

The form factor term, $P(q)$, contains information on the distribution of segments within a single dendrimer. Models can be used to fit the scattering from various types of particles, common ones being a Zimm function which describes scattering from a collection of units with a Gaussian distribution (equation (3a)), a

Guinier function which describes the scattering from sphere-like objects (equation (3b)), and the Sphere function which gives the exact scattering from a perfect sphere, where $R = (5/3)^{\frac{1}{2}} R_g$ (equation (3c)).

$$P(q) = (1 + q^2 R_g^2/3)^{-1} \qquad (3a)$$

$$P(q) = \exp(-q^2 R_g^2/3) \qquad (3b)$$

$$P(q) = 9(\sin(qR) - qR\cos(qR))^2/(qR)^6 \qquad (3c)$$

Therefore, a fit of the scattering data not only gives information on an average size, R_g but also information on the segment density distribution within the dendrimers.

Table 11.1 compares the usefulness of the three techniques in addressing the questions posed. SANS has the ability to analyze labeled parts of the molecule by replacing hydrogens with deuterium. This allows for the location of dendrimeric components, such as the end groups. SAXS offers the advantage of working with very low background radiation, thus providing the best resolution of the higher q features which are normally weak in intensity. TEM is the best technique of the three for viewing individual molecules and offers direct visualization.

Descriptions of the experimental scattering and microscopy conditions have been published elsewhere and are referenced in each section. Throughout this report certain conventions will be used when describing uncertainties in measurements. Plots of small angle scattering data have been calculated from circular averaging of two-dimensional files. The uncertainties are calculated as the estimated standard deviation of the mean. The total combined uncertainty is not specified in each case since comparisons are made with data obtained under

Table 11.1 Major technological issues. Key to symbols, + +, best technique for measurement; +, acceptable technique for measurement; −, unsuitable for measurement

Structure Issues	Critical Questions	Influence on Applications	SANS	SAXS	TEM
Segment distribution	Uniform or diffuse?	Solubilization	+	+ +	+
Type of branching	Important differences?	Major cost differences	+	+	+
Terminal group location	Surface or interior?	Attachment networks	+ +	+	—
Dendrimer–dendrimer interactions	Interpenetration or collapse?	Ordering nanostructures	+	+ +	+
Size variation with solvent	Large changes of stable size?	Standards release	+	+	—

the same conditions. In cases where the limits are smaller than the plotted symbols, the limits are left out for clarity. Data plots produce uncertainties larger than the symbols, representative confidence limits are plotted at appropriate places. Fits of the scattering data are made by a least-squares fit of the data giving an average and a standard deviation to the fit, this is the case for fit values such as radius of gyration and exponents.

5 DENDRIMER SIZE VS GENERATION

SANS and SAXS have been used to measure the average (radius of gyration), R_g of dendrimers in dilute solution [14, 15, 17, 33]. Figure 11.2 is a plot of R_g vs theoretical M_w for both PPI and PAMAM dendrimers. The theoretical R_g of spheres with a density of 1.23 g/cc, based on the value reported for PAMAM dendrimers, is plotted for comparison. Therefore, these values provide a limit for PAMAM dendrimers that have excluded all solvent and shrunken to single molecules without any interior void space. The distance from these PAMAM reference points to the line gives the relative segment density within the dendrimers. These data points roughly parallel the line, with the points getting slightly closer to the line for the highest generations.

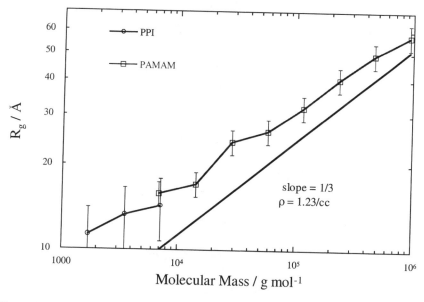

Figure 11.2 SAXS R_g of PPI and PAMAM dendrimers. Straight line is the limit of PAMAM dendrimers collapsed to bulk density. The 1/3 power law suggests uniform spherical shape

A slope of 1/3 would suggest that the dendrimers are spherical with similar internal density profiles. There seems to be a slight increase in internal segment density as the dendrimers get larger. For the highest PAMAM generations, it is known that the theoretical molecular mass is not reached due to crowding [1]. This would shift the high generation data points to the right and may indicate that the internal densities are uniform after a certain generation.

TEM has been used to measure the average diameter of PAMAM dendrimers from G5 to G10 [20]. Figure 11.3 shows micrographs of G5, G6, G8, and G10.The G5 micrograph also has a small amount of G10 dendrimers added to help focus the microscope and to identify the relative sizes. Single dendrimers can easily be distinguished within each of the populations shown. Individual polymer molecules can easily be seen, while conventional linear polymers in this molecu-

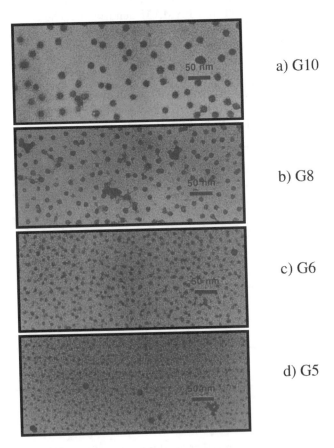

a) G10

b) G8

c) G6

d) G5

Figure 11.3 TEM of G5, G6, G8, and G10 PAMAM dendrimers. Individual molecules are seen

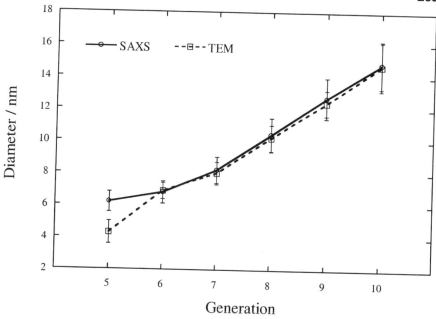

Figure 11.4 PAMAM dendrimer diameter from SAXS and TEM

lar mass range would be difficult to image. This attests to the fact that the dendrimers are indeed much more spherical in shape and possess have a very high internal segment densities compared to traditional polymers.

Size ranges are tabulated from the TEM images and are reported elsewhere [20]. Each generation has a well-defined average with a relatively narrow range of sizes. Figure 11.4 compares a plot of the average diameter from TEM [20] and from SAXS [17] assuming that the dendrimers are spheres and that therefore the diameter $D = (20/3)^{\frac{1}{2}} R_g$. It is apparent there is excellent agreement between the two types of measurements.

6 DENDRIMER INTERNAL SEGMENT DENSITY DISTRIBUTION (SDD)

While microscopy can give important information on the average size of dendrimers, scattering gives much more detailed information on their internal structure. Based on the variety of predictions concerning SDDs that may exist within a dendrimer, ranging from a hollow structure with a dense shell [22] to a dense center with a gradually tapering outward distribution [23], it is important to clarify this issue.

The angular dependence of the scattered intensity contains information on the SDD. In principle, if the SDD is known exactly, then the scattering can be calculated exactly and vice versa. The uncertainties in the scattering become translated into uncertainties in the SDD, however, and direct transformation is not practical for samples that do not scatter strongly. There are some characteristics of the scattering at intermediate or high q that can provide insights into relative structures, thus easily distinguishing the structural class.

Figure 11.5 is a plot of the SAXS from G3 through G10 PAMAM dendrimers over a very wide q range [34]. The scattering at low q is a measure of dendrimer R_g, and a gradual transition is easily seen from G3 to G10, indicating increasing R_g with generation number. The high q scattering also has a gradual transition in the limiting power law that is sensitive to the transition of segment densities at the outside of the dendrimer. The G3 dendrimer has a high q power law of $-5/3$ which is characteristic of a macromolecule in a good solvent such as linear or star polymers [32]. This indicates a relatively diffuse outer boundary. The G10 dendrimer, however, has a power law of -4 which is characteristic of an object with a very sharp outer boundary [32]. There is a gradual transition from G3 to G10, showing that dendrimers transform from star-like to sphere-

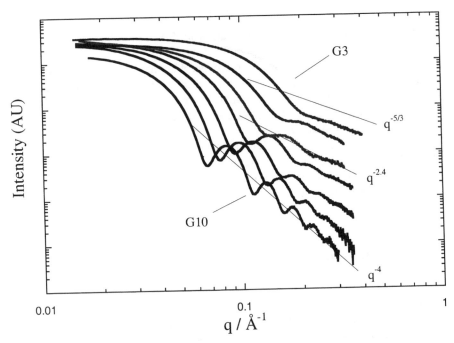

Figure 11.5 SAXS curves for PAMAM mass fraction ~1% dendrimer/methanol solutions of generation G3 (top) through G10 (bottom). Power law behavior continuously varies from $-5/3$ (-1.7) to -4

like entities [1, 34].

Another important factor is the appearance of inflections in the high q data that become multiple peaks as the generation number increases. Equation (3c), which describes scattering from a monodisperse, uniform sphere has multiple peaks at high q. These features also appear in the SAXS data, indicating that the large generation PAMAM dendrimers are becoming quite sphere-like. Even the relatively small G5 dendrimer is beginning to show this feature and the G10 dendrimer shows at least five additional peaks in the data.

The strong scattering from the G10 dendrimers and a multitude of other subtle features make it the best candidate for thorough fitting of the structure. Figure 11.6 is a plot of the G10 SAXS data in a Porod plot, Iq^4 vs q. A Porod plot is a convenient way of looking at the scattering from objects with sharp outer boundaries, since scattering with a -4 power law will be a horizontal line in this type of plot. The data points oscillate around a horizontal line showing a sphere-like structure with a sharp outer boundary. The diminished size of the oscillations with increased q is due to the polydispersity of the sizes of the objects.

The polydispersity can be fit by two limiting cases: namely, as a population of perfect spheres with a range of diameters or as a single randomly oriented

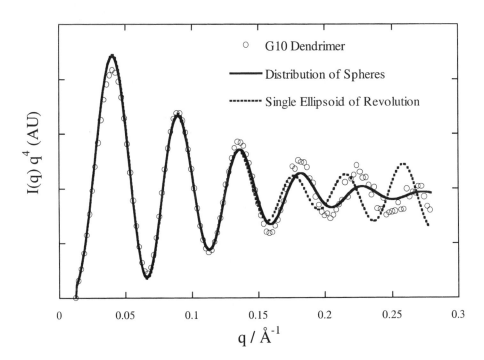

Figure 11.6 Fits of the G10 PAMAM dendrimer data to a distribution of spheres and a single ellipsoid of revolution

ellipsoid. The solid line in Figure 11.6 is the theoretical scattering from a population of spheres with a Gaussian distribution of radii with a mean of 69.9 Å and a distribution half width of 4.9 Å. The dashed line is for a randomly oriented ellipsoid with a major axis of 74.1 Å and a minor axis of 59.8 Å. It is not possible to distinguish between these two types of polydispersities, but it is likely that both types are present, resulting from a population of slightly elliptical molecules with a small size variation.

Therefore, within the PAMAM family, the shapes of dendrimers span the range from stars to spheres. Small G3 dendrimers have a diffuse, open structure while large G10 dendrimers are spheres with a uniform interior, sharp outside transition, and low polydispersity.

7 COMPARISON OF DENDRIMERS, HYPERBRANCHED, AND DENDRIGRAFT

Dendrimer synthesis involves a repetitive building of generations through alternating chemistry steps which approximately double the mass and surface functionality with every generation as discussed earlier [1–4, 18]. Random (statistical) hyperbranched polymer synthesis involves the self-condensation of multifunctional monomers, usually in a 'one-pot' single series of covalent formation events [31]. Random hyperbranched polymers and dendrimers of comparable molecular mass have the same number of branch points and terminal units, and any application requiring only these two characteristics could be satisfied by either architectural type. Since dendrimer synthesis requires many defined synthetic and process purification steps while hyperbranched synthesis may involve a 'one-pot' synthetic step with no purification, the dendrimers will necessarily be a much more expensive material to produce.

A third type of branched molecule has been described recently, known alternatively as dendrigraft [37] or arborescent [38] molecules. They are made in a stepwise synthesis as are dendrimers, however, the iterations involve conventional grafting reactions of end functionalized linear polymers onto a polymer substrate. Ungrafted linear polymer is removed by fractionation and the grafting reaction is repeated, forming the next generation of dendrigrafts. Since many linear grafts can be added to the precursor generation each time, the molecular mass increases by substantially more then the doubling that is usually observed for dendrimers. Molecular mass increases of 10 to 100 times per generation are often typical.

Figure 11.7 illustrates the three different subtypes of dendritically branched molecules that have been identified within the major architectural class of dendritic polymers. Random hyperbranched polymers, not only exhibit polydispersity in molecular mass between individual molecules, it should also be noted

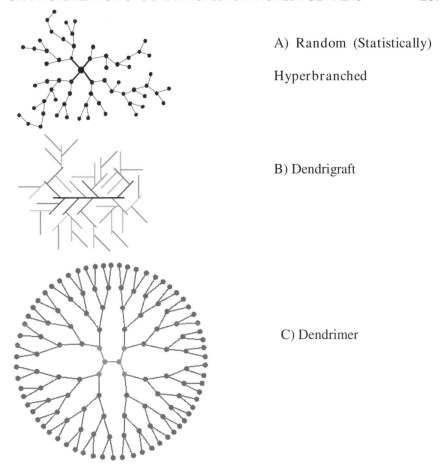

A) Random (Statistically)

Hyperbranched

B) Dendrigraft

C) Dendrimer

Figure 11.7 Cartoons representing dendritically branched polymers listed in Table 11.2

that the internal covalent connectivity between the core and the terminal units varies dramatically. A dendrigraft made by repeated grafting of 'living' linear polymers, each having a very narrow molecular mass distribution, making the overall distribution narrow. Dendrimers are the most precise, structure controlled subset of dendritic polymers as described earlier.

Table 11.2 lists some of the typical characteristics of the three types. All have some characteristics in common, such as a high degree of branching as is defined by having a low number (typically < 10) of covalent bonds between branch points and a large number of terminal groups. The most important differences between these dendritic types are in the polydispersity, the number of end groups and the number of synthetic and purification steps. The last factor is mirrored in the resultant cost that may be different by orders of magnitude. For this reason, it is important to determine property and performance differences between these three dendritic types, as well as traditional cost–performance benefits.

As was described earlier, SANS [39–41] and SAXS [17, 34] have been used successfully to define certain structural parameters of dendrimers. Different types of structures have different form factors. If the scattering fits a given model over a wide q range, then the molecular structure is well described by the distribution described by that model. Figure 11.8 is a plot of the SANS from a G4 PAMAM dendrimer and a G5 random hyperbranched polyol, both of which have the same nominal M_w, 14 000 g/mol. Both SANS curves are plotted as a Zimm plot (I^{-1}) and a Guinier plot ($\ln(I)$). It is quite clear that the dendrimer is linear in the Guinier plot and curved in the Zimm plot, with the opposite being true for the random hyperbranched polymer. A Zimm plot describes the scattering from particles such as a linear chain having a Gaussian distribution of segments, while a Guinier plot in this q range fits the scattering from a uniform spherical object. Therefore, to a first approximation, the segment density distribution of a dendrimer is relatively uniform, while that of a hyperbranched is

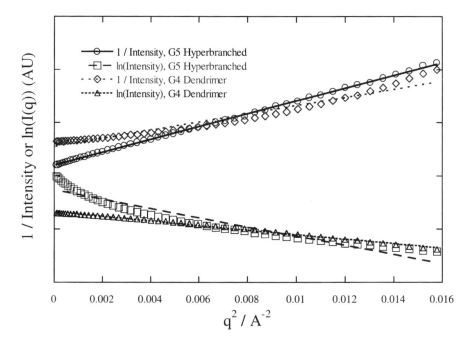

Figure 11.8 Zimm and Guinier plots of G4 PAMAM dendrimer and G5 hyperbranched polyol. The dendrimer fits Guinier (spherical) and the hyperbranched fits Zimm (Gaussian)

quite tapered, with a high density in the center and a gradual decrease in segment density as one moves outward.

The low q range shown in Figure 11.8 gives important information, but the most sensitive q range is at somewhat higher values of q. Figure 11.9 is a Kratky plot, of scattering from low-generation (G4) and high generation (G10) PAMAM dendrimers [17], along with low-generation (G1) and high-generation (G3) polystyrene dendrigrafts [42]. A Kratky plot, Iq^2 vs q, emphasizes the differences in scattering at higher q values. The uppermost curve is the theoretical scattering from a linear polymer. All of the others show a peak in the scattering which is characteristic of dense internal structure. The G1 dendrigraft has a shape characteristic of a star polymer with a broad peak and leveling off at higher q. The G4 dendrimer has a similar shape, but the peak is much narrower. This suggests a star-like structure on the outside, but more of a tendency for uniform a spherical interior than the G1 dendrigraft. As shown previously, the G10 dendrimer exhibits several higher order peaks typical of a robust sphere-like distribution. The G3 dendrigraft has a weak higher order peak, however, but the uniformity is less pronounced than the G10 dendrimer.

Figure 11.10 is a plot of the average local segment density distribution for

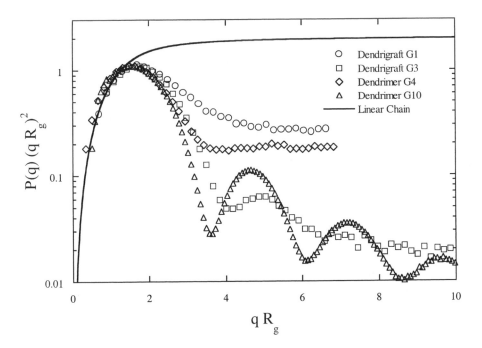

Figure 11.9 Kratky plots of G1 PS dendrigraft, G3 PS dendrigraft, G4 PAMAM dendrimer and G10 PAMAM dendrimer. Flat high q shows star-like structure, higher order features shows sphere-like structure

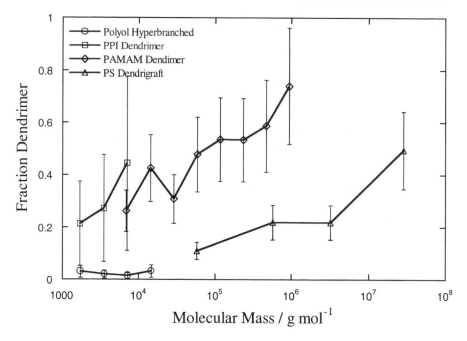

Figure 11.10 Volume fraction of interior polymer segments in solution. PPI dendrimer in methanol, PAMAM dendrimer in methanol, random hyperbranched polyol in methanol, and PS dendrigraft in cyclohexane

typical dendrimers, dendrigrafts and random hyperbranched. The average size is calculated from the R_g as measured from SAXS or SANS, assuming a spherical shape. The size of a collapsed molecule calculated from M_w and bulk density, ρ, is then used to calculate the average internal volume fraction of polymer segments, ϕ_p, as is shown in equation (4).

$$\phi_P = \frac{\dfrac{4\pi}{3}\left(\dfrac{5}{3}\left\langle R_g \right\rangle\right)^{3/2}}{M_w/N_{av}\rho} \qquad (4)$$

N_{av} is Avogadro's number. The hyperbranched molecules have the most open structure, with values of ϕ_p, in the range of (2–3)%. PPI and PAMAM dendrimers form a composite curve with ϕ_p, ranging from (20 to 75)% from the smallest to the largest dendrimers. The dendrigrafts ϕ_p, ranges from (10 to 50)% from G0 to G3.

The results are summarized in Table 11.2. A low generation dendrimer has the shape of a dense star with an interior having a more uniform density than a low generation dendrigraft. As the generation number increases, the shape becomes

Table 11.2 Comparison of different dendritically branched polymer types, see Figure 11.7

	Dendrimer	Dendrigraft	Hyperbranched
Interior branching	High	High	High
Terminal units	Small	Linear chains	Small
Molecular mass distribution	Narrow	Narrow	Broad
Shape – low generation	Dense star	Star-like	Polydisperse, Diffuse
Shape – high generation	Dense sphere	Uniform center star-like exterior	Polydisperse, Diffuse
Synthetic steps	4–20	2–5	1
Purification steps	4–20	2–5	0
Cost	Very high	Moderate	Low

more sphere-like. At the limit of a G10 PAMAM dendrimer, the molecules have a uniform interior with an abrupt transition in segment density at the outside, and low polydispersity. A G0 dendrigraft is like a conventional multiarm star, but a G3 dendrigraft has a uniform interior with a narrow, star-like zone of transition at the outside. Random hyperbranched molecules have polydispersity both in molecular mass and in the internal structure of each molecule. They have a much more open structure with a broad, tapered segment density distribution than do dendrimers of similar size.

8 LOCATION OF THE TERMINAL GROUPS

A dendrimer is illustrated in two forms, Figure 11.11a being the classical picture with all of the terminal groups (shown as circles connected with a single bond) extended in an exo-configuration to the outside of the dendrimer. Figure 11.11b shows how the terminal groups could enter the interior of the dendrimer by backfolding. SANS of unlabeled dendrimers can give detailed information on the SDD as was demonstrated in the SAXS studies [17]. However, the calculated density distribution cannot distinguish between the terminal units and all of the other units. Therefore, the distributions described in Figures 11.11a and 11.11b would be expected to give identical scattering information.

To distinguish the terminal groups from the groups of the earlier generations,

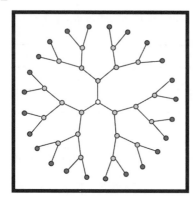
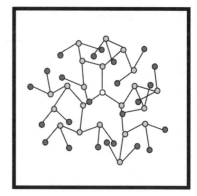

a) End groups – Outside   b) End groups Backfolded

Figure 11.11   Cartoon of possible end group locations. a) whole dendrimer end groups outside, b) whole dendrimer end groups backfolded, a) matched dendrimer end groups backfolded, a) matched dendrimer end groups backfolded. Outside end groups have a larger R_g

deuterium labeling is used. PAMAM dendrimers having a tetrafunctional core of ethylenediamine (EDA) were synthesized according to reported methods [43]. Fully hydrogenated G7 PAMAM dendrimers and partially deuterated versions were prepared. The partially deuterated dendrimer were prepared by reacting the fully hydrogeneous G6 dendrimer with partially deuterated methyl acrylate, $CD_2 = CDCO_2CH_3$ to form G6.5 dendrimer with a hydrogenated interior and deuterated terminal units. The G6.5 was converted to the partially deuterated G7(D) dendrimer by reaction with hydrogenated EDA.

Solutions for SANS experiments were prepared with methyl alcohol (CH_3OH), and its partially deuterated counterpart, CD_3OH. For the determination of the contrast match point of the unlabeled parts of the dendrimer, two stock solutions of equal G9 dendrimer concentrations were prepared in CH_3OH and CD_3OH, respectively. Various amounts of the two stock solutions were mixed to make a series of solutions with different isotopic contents. The CH_3OH / CD_3OH match point was determined as the point at which the coherent SANS scattering goes to zero. The procedure is described elsewhere [43].

After the match point was determined, solutions were made of labeled G7(D) and unlabeled G7(H) dendrimer in the 'match solvent' together with the unlabeled G7(H) in CD_3OH. Figure 11.12 shows the SANS of the three samples. The unlabeled sample in the 'match solvent' is completely flat, showing only incoherent scattering. This proves that the 'match solvent' completely masks the hydrogenous parts of the dendrimer. The unlabeled G7(H) in CD_3OH shows strong scattering characteristic of dendrimer scattering. The labeled dendrimer

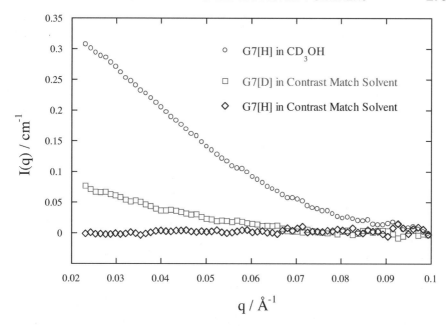

Figure 11.12 SANS of G7(H) in CD₃OH, G7(D) in contrast match methanol, and G7(H) in contrast match solvent. Labeled dendrimer in core match solvent shows scattering from labeled terminal generation

in the match solvent still shows some coherent scattering. This scattering is from the deuterated half of the last generation.

To ensure that instrumental artifacts do not perturb the results, we collected two independent data sets for G7(D) in the match solvent by measuring the solutions on two different 30 m SANS instruments at NIST [44, 45]. The absolute values of the coherent scattering intensity of the G7(D) labeled segments were reproduced to high accuracy, independent of the SANS instrument and its configuration. The values of the R_g are determined by weighted, linear least-squares fits to equation (3b) and give an R_g of non-labeled dendrimer of (34.4 $\pm$ 0.2) Å. For the deuterated segments of the partially deuterated dendrimer, the results are $R_g(G7(D)) = (38.8 \pm 1.2)$ Å and $R_g(G7(D)) = (39.8 \pm 1.2)$ Å for the data taken on NG3 and NG7, respectively (uncertainties are based on the standard deviation of the fits). We can take the average value of $R_g(G7(D))$ as (39.3 $\pm$ 1.0) Å for the deuterated segments of G6.5 in a G7 dendrimer.

If a uniform distribution of end groups were present throughout the dendrimer, we would see the same R_g for both the labeled and the whole dendrimer. This is demonstrated by the illustration in Figure 11.11, where the end groups have different locations. Instead, we see a clear difference between the R_g values for the labeled units and the unlabeled dendrimers. The distribution of dendrimer terminal groups is not uniform throughout the interior of the dendrimer, but rather *the terminal groups are localized near the periphery of the dendrimer much as predicted by de Gennes et al.* [22].

There have been solid-state NMR studies [46] on the distribution of end groups by site-specific stable-isotope-labeling, rotational-echo double-resonance (REDOR) NMR. REDOR experiments which measure dipolar couplings between C-13 atoms located near the chain ends and an F-19 label placed at the core of benzyl ether dendrimers (generations 1–5). They find that average distances are quite small, suggesting that terminal units do approach the core unit. The important difference between the NMR and the SANS studies is the moment of the average distribution. The weighting of the R_g from SANS goes as the square of the distance, R^2, so that segments with large distances from the center of mass are weighted heavily and those near the core have little weighting. The NMR technique, however, weighs the cube of the inverse distance, R^{-3} so that small distances are weighted heavily and large distances are not seen. Therefore SANS and NMR are complementary measurements with the SANS showing that most, but not all of the terminal units are predominately on the outside of the dendrimer, but some can approach the core through backfolding.

Most, but not all of the terminal units are near the outside of the dendrimer at any given time. The SAXS studies [17] of the segment density distribution have shown that there is an abrupt transition region at the outside of large PAMAM dendrimers. The combination of these two factors suggests that the terminal functionalities of dendrimers are accessible from the outside and available for chemical reactions such as attachment to surfaces, mounting of a catalyst, or for use as a crosslink junction.

9 DENDRIMER–DENDRIMER INTERACTIONS

As was described earlier, the internal SDD of dendrimers is remarkably high compared to traditional polymers and is well defined by a narrow transition zone at the outside. When linear polymers are forced together by increasing their concentration in a solvent, they freely interpenetrate each other due to their open structure. Dendrimers, on the other hand, appear to represent a different case, due to their compact size and architectural features.

At low concentrations, dendrimers may avoid interpenetrating each other by occupying the space available between other dendrimers. As the dendrimer concentration is increased, the solvent swollen dendrimers begin to encroach upon each other, first by touching, and then interacting more strongly. Two possibilities arise, first, the dendrimers could retain their dilute solution sizes and begin to interpenetrate, by mixing their outer segments. The other possibility would be for the dendrimers to collapse upon themselves, avoiding interpenetration. Figure 11.13 illustrates the three possible states with 11.13a showing a

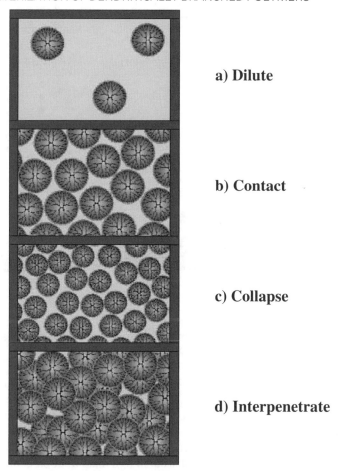

a) **Dilute**

b) **Contact**

c) **Collapse**

d) **Interpenetrate**

Figure 11.13 Cartoon showing dendrimers at different concentrations

dilute solution, 11.13b showing the overlap concentration at which point the dendrimers contact one another, 11.13c showing dendrimer collapse and noninterpenetration; whereas, 11.13d shows a concentration above the overlap concentration where the dendrimers retain their size and interpenetrate.

SANS was used to measure the scattering of G4 and G5 PPI dendrimers at concentrations up to a mass fraction of 80% [47]. Partially deuterated solvent, CD_3OH was used instead of CD_3OD to prevent exchange of deuterium with the terminal primary amine hydrogens. Scattering experiments were performed at the 30 m SANS facility at NIST [44, 45] with the spectrometer operating in a configuration which gave a wide scattering vector range, q, $0.05 \text{ Å}^{-1} \leq q \leq 0.55$ Å^{-1}. The maximum q was required to resolve the scattering at high concentrations.

Figure 11.14 is a plot of the coherent scattering from a G5 PPI dendrimer as a

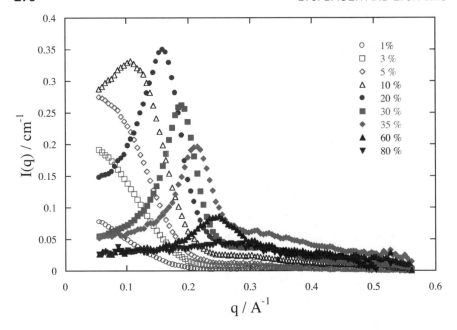

Figure 11.14 SANS of G5 PPI dendrimer at various mass fractions. Peaks move to higher q (smaller separation distance) with increased concentration

function of concentration. At concentrations below a mass fraction of 5% the scattering decreases monotonically with q. This is typical of dilute solution scattering with no interactions. A peak in the SANS appears at a mass fraction of 10% indicating that long-range fluctuations have been suppressed because the dendrimers avoid each other. The overlap concentration can be calculated from the R_g assuming a spherical shape and is equal to a mass fraction of about 25%. At the overlap concentration and higher, the peak is strong and moves to higher q as the dendrimers are forced together.

A previous SANS study of concentrated PPI dendrimers also noted that this peak appears and moves in a way consistent with increased concentration [48]. This peak behavior suggests ordering, but cannot easily distinguish between dendrimer shrinking and interpenetrating. It is interesting to note, however, that in Figure 11.14 at q values higher than the peak position, there is excess scattering over the incoherent background. As was shown earlier, higher order features in scattering from spheres can be used to fit the size of the sphere. As the sphere size decreases, this feature moves to higher q. If this feature could be resolved, the size of a dendrimer could be measured at high concentrations.

SANS often displays a large amount of incoherent scattering that is a flat background containing no structural information. The higher order peaks are in a q range that has low coherent scattering so that a large fraction of the scattering

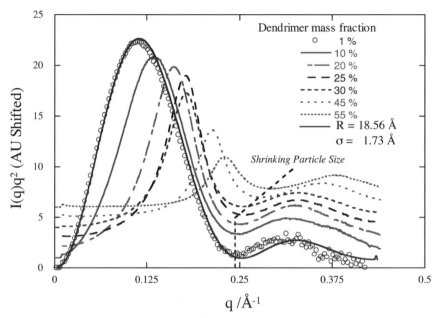

Figure 11.15 Kratky plot of SAXS of G5 PPI dendrimer at various mass fractions. Higher order feature moves to higher q, suggesting shrinkage with increased concentration

is a noisy baseline, and resolving features becomes difficult. SAXS, however, has only weak background scattering in this angular region. The experiments were repeated at the Advanced Polymers PRT beamline (X27C; SUNY Stony Brook/ NIST/GE/Allied Signal/Montell) at the National Synchrotron Light Source, Brookhaven National Laboratory.

Figure 11.15 is a Kratky plot of the scattering of a G5 PPI dendrimer at various concentrations. The mass fraction 1% scattering can be fit with the scattering from a sphere giving $R = 18.6$ Å and $R_g = 14.4$ Å which is consistent with previous results which involved fitting the low q scattering alone. It is easy to distinguish the higher order features in these plots. As the concentration increases, the minimum remains stationary until the overlap concentration of 25% is reached. Beyond this point the minimum moves to higher q, showing that the dendrimers are becoming smaller. At the highest concentration, the resolution is lost, but at moderate concentrations above the overlap concentration, the dendrimers tend to shrink in size rather than to overlap.

Figure 11.16 is a cryo-TEM image of G10 PAMAM dendrimers in water. The cryo technique involves flash freezing of the dendrimer solution as a thin film on a grid and is described elsewhere [20]. Individual dendrimers appear to be organized into an array of single dendrimer thickness. While there are complicating factors due to the sample preparation, the picture is consistent with

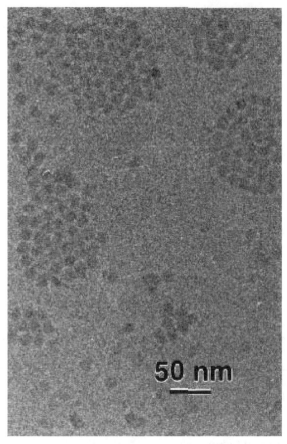

Figure 11.16   TEM of aqueous PAMAM G10. Note dendrimers are touching but not interpenetrating

den-drimer–dendrimer contact behavior which demonstrates lack of inter-penetration. AFM studies of G3 dendrigraft polystyrene have resolved individual molecules as a surface monolayer [21]. This work also is consistent with our PAMAM results and is further evidence of dendrimer or dendrigraft noninterpenetration.

Dendrimers remain discrete objects in dilute solution, avoiding interpenetration. As the concentration increases above overlap, the dendrimers preferentially shrink in size rather than interpenetrating. When dried to a solvent-free condition, the dendrimers must either deform from their spherical shape into polyhedrons, or must interpenetrate. The solvent-free condition would require deuterium labeled dendrimers, and experiments are under way to probe this last concentration regime.

10   DENDRIMER SIZE CHANGE IN DIFFERENT SOLVENTS

The effect of solvent quality on the dimensions of linear chains has been studied for many years, and chain dimensions can change a large amount, especially at high molecular mass. There are reports in the literature that the size of dendrimers is largely influenced by the solvent quality. A molecular dynamics study by Murat and Grest [26] resulted in an increasing internal segment density of dendrimers when the dendrimer–solvent interactions are less favorable, which leads to a considerable decrease of the average dimensions of the simulated structures. A holographic, relaxation spectroscopy, HRS, study by Stechemesser and Eimer [49] concluded that the radius of higher generation PAMAM dendrimers would be extremely sensitive to solvent conditions. The effect of introducing charges on a dendrimer in aqueous solution was investigated by Welch and Muthukumar [50] using Monte Carlo simulations. They found that simulating a charged PPI dendrimer can change its linear dimension by as much as a factor of 1.8.

Small-angle scattering is a direct method for determining dendrimer size, thus, SANS was used to measure dendrimer size by two methods. Dilute solutions of noninteracting dendrimers can be fit by equation (3b) to determine R_g as a function of dendrimer generation, solvent type and temperature. The effect of charging is more complex since long-range ionic interactions can impose a uniform spacing between dendrimers that affects the low-angle scattering. In this case, higher q scattering can be used to measure a size if a higher order feature can be resolved, as was the case for the high concentration dendrimer solutions discussed earlier.

The SANS experiments [51] were performed with solutions of G8 PAMAM dendrimer in D_2O, methyl-d_4, ethyl-d_6, and n-butyl-d_{10} alcohol at a temperature of $T = 20.0\,^{\circ}C$. PAMAM dendrimers do not dissolve in acetone, but they readily dissolve in methyl alcohol/acetone mixtures over a wide range of composition. Solvents of different composition, were prepared and added to a weighed amount of dried G5 or G8 dendrimer. In a separate set of experiments, the NIST NG7 30 m instrument was used to measure the effects of charging on the dendrimer size. PAMAM G8 dendrimers in D_2O were charged by addition of HCl in the presence of various amounts of NaCl to the charged dendrimers to screen the electrostatic interactions.

Guinier plots were made of the scattering from G8 PAMAM dendrimers to measure the R_g in alcohols possessing different chain lengths. Figure 11.17 is a plot of the values vs alcohol chain length. There is a small but consistent drop in R_g as the length of the hydrocarbon chain goes from 0 to 4, with the total relative change being 10%. The range of solvents goes from very good (water) to very poor (n-butanol). This is considerably lower than the range predicted from dynamic calculations. Also plotted in Figure 11.17 are results of holographic

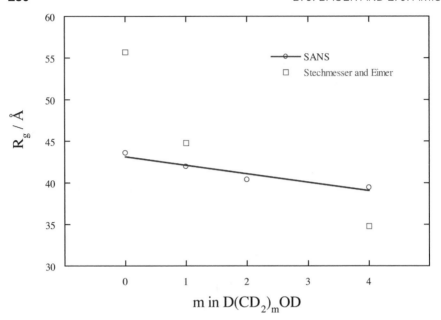

Figure 11.17 R_g of G8 PAMAM dendrimer in alcohol of various types from SANS. R_g of G8 PAMAM dendrimer calculated from HRS results in ref. 50

relaxation spectroscopy from the literature [49]. The large changes in the reported hydrodynamic size are not found in the SANS results.

Similar results were observed with mixtures of acetone-d_6 and methanol-d_4, wherein the R_g of the dendrimers is not significantly affected by the fraction of acetone in the solvent. The same result is found over the whole range of temperature studied (i.e. in the range $-10\,°C \le T \le 50\,°C$). Only data obtained in the solvent compositions, $x_s = 0.50$ show some consistent decrease of R_g, relative to the other solvent compositions. This may indicate the onset of dendrimer shrinkage in the vicinity of the solubility gap, which is around $x_s \approx 0.52$. However, the difference from the a value of R_g for $x_s = 0.50 < 5\%$ of the mean value of R_g for the other solvent compositions.

Monte Carlo calculations [50] have predicted that there should be a strong effect, with R_g varying by as much as 1.8 by charging with acid and by screening the charges by addition of salts. Ionic effects were used in an attempt to modify the size of a G8 PAMAM dendrimer. Figure 11.18 shows SANS of the dendrimer under three conditions [52]. A dendrimer solution without acid or salt addition has a pH of 10.1 and exhibits scattering typical of dilute noninteracting spheres. The R_g of the dendrimer can be calculated from the position of the higher order feature of the sphere scattering and is consistent with previous measurements. The dendrimer is then charged by addition of acid until the pH becomes 4.7. At

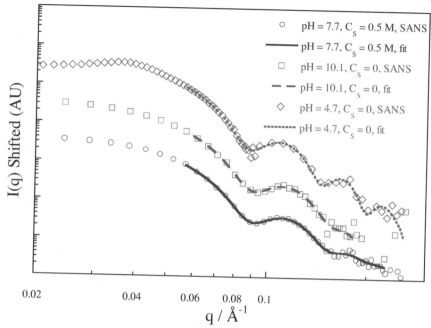

Figure 11.18 SANS from G8 PAMAM dendrimer with and without acid and salt additions. Note constant position of higher order features showing constant size

this point the low q scattering changes dramatically, developing a peak that is characteristic of dendrimer ordering by long-range ionic effects. As with the high concentration dendrimer studies, the minimum in the scattering can be used to calculate a dendrimer R_g, even when the interactions occur. The R_g has changed only very slightly, increasing by no more than 5%. Upon addition of salt to screen the long-range interactions, the peak disappears, indicating that the long-range ordering has been lost. The position of the minimum has not changed, however, indicating that charging and screening the dendrimers does not significantly modify the dendrimer dimensions.

Both large (G8) and small (G5) PAMAM dendrimers do not significantly change size when the solvent environments are changed. When the solvent is made increasingly hydrophobic by increasing the hydrocarbon chain length, only a 10% change occurs. Also the addition of a nonsolvent to a good solvent does not significantly change the dendrimer size up to the point of phase separation. Charging a dendrimer in aqueous solution does not change the size of the dendrimer by more than 5%.

11 SUMMARY

Dendritically branched molecules have shapes and interactions that are radically different from traditional linear polymers. SANS, SAXS and TEM have been used to measure the properties of dendrimers and a generic understanding of their properties has emerged. Five technologically important questions on the nature of dendrimers have been posed and generalized observations can be made.

Dendrimers are nearly spherical in shape when compared to conventional linear polymers [17, 34]. The lower generation dendrimers have internal segment density distributions similar to star molecules, but even the smallest are more compact than stars. As the generation number increases, the sphere-like characteristics increase and the largest dendrimers form a population of spheres with uniform internal density (roughly half dendrimer units–half solvent) [47], sharp interfaces on the outside and a small polydispersity in shape.

Dendrigraft molecules are star-like in the 0th generation, but have a uniform interior with an exterior transition zone for G3 [17, 34]. Random hyperbranched molecules have broad distributions both in molecular mass and in shape [17]. Scattering from hyperbranched is closer to that of linear polymers than to spheres, indicating that there is a gradually tapering distribution on units from the center to the exterior.

The terminal units of a dendrimer are extended predominately towards the exterior of the dendrimer [43]. While some terminal units may fold inwards, most of the terminal units are on or near the outside perimeter. Combined with the fact that there is a narrow interface on the outside of large dendrimers [17], most terminal groups should be accessible for chemical manipulation such as attachment to surfaces, mounting of catalysts or attachment of linear chains.

As dendrimers are forced together by increasing the concentration, they avoid interpenetrating each other and prefer to shrink in size rather than overlap [35, 47]. This should provide a strong thermodynamic force for ordering dendrimers on surfaces or in the bulk. *Dendrimers seem to be well-suited for forming well ordered nanostructures.*

The size of dendrimers in dilute solution is relatively constant. Changing the quality of the solvent or charging the dendrimers results in only a very small change in dimensions. Therefore, dendrimers seem well suited for applications as standards or as probes of other materials.

ACKNOWLEDGMENT

We would like to thank Donald A. Tomalia (University of Michigan, Center for Biologic Nanotechnology and Dendrimax, Inc., Ann Arbor, MI) for the PAMAM dendrimers, Rolf Scherrenberg (DSM) for the PPI dendrimers and

Mario Gauthier (University of Waterloo) for the PS Arborescent polymers. Special thanks to the NIST dendrimer team who provided all of the characterization work, Andreas Topp, Ty J. Prosa, Da-Wei Liu, Catheryn L. Jackson, and Giovanni Nisato. This work is supported in part by the US Army Research Office under contract number 35109-CH.

12 REFERENCES

1. Tomalia, D. A., Naylor, A. M. and Goddard, W. A., *Angewandte Chemie-International Edition in English*, **29**, 138 (1990).
2. Hawker, C. J. and Fréchet, J. M. J., *Journal of the American Chemical Society*, **112**, 7638 (1990); Hawker, C. J., Wooley, K. L. and Fréchet, J. M. J., *Journal of the Chemical Society – Perkin Transactions*, **1**, 1059 (1991).
3. Hawker, C. J. and Fréchet, J. M. J. in Ebdon, J. R. and Eastmond, G. C. (eds), *New Methods of Polymer Synthesis*, Blackie, Glasgow, UK, 1995, p. 290; Fréchet, J. M. J. *Science*, **263**, 1710 (1995).
4. Newkome, G. R., Morefield, C. N. and Vögtle, F. *Dendritic Molecules: Concepts, Synthesis, Perspective*, VCH, Weinheim, Germany, 1996.
5. Debrabandervandenberg, E. M. M. and Meijer, E. W., *Angewandte Chemie-International Edition in English*, **32**, 1308 (1993).
6. Debrabandervandenberg, E. M. M. *et al.*, *Macromolecular Symposia*, **77**, 51 (1994).
7. Dvornic, P. R. and Tomalia, D. A., *Macromolecular Symposia*, **98**, 403 (1995).
8. Hummelen, J. C., vanDongen, J. L. J. and Meijer, E. W., *Chemistry – a European Journal*, **3**, 1489 (1997).
9. Uppuluri, S., Keinath, S. E., Tomalia, D. A. and Dvornic, P. R., *Macromolecules*, **31**, 4498 (1998).
10. Aharoni, S. M. and Murthy, N. S., *Polymer Communications*, **24**, 132 (1983).
11. Potschke, D., Ballauff, M., Lindner, P., Fischer, M. and Vogtle, F., *Macromolecules*, **32**, 4079 (1999).
12. Kleppinger, R., Reynaers, H., Desmedt, K., Forier, B., Dehaen, W., Koch, M. and Verhaert, P., *Macromolecular Rapid Communications*, **19**, 111 (1998).
13. Omotowa, B. A., Keefer, K. D., Kirchmeier, R. L. and Shreeve, J. M., *Journal of the American Chemical Society*, **121**, 11130 (1999).
14. Bauer, B. J., Hammouda, B., Briber, R. M. and Tomalia, D. A., *ACS PMSE Preprints*, **69**, 341 (1993).
15. Bauer, B. J., Hammouda, B., Barnes, J. D., Briber, R. M. and Tomalia, D. A., *ACS PMSE Preprints*, **71**, 140 (1994).
16. Bauer, B. J., Topp, A., Prosa, T. J. and Amis, E. J., *ACS PMSE Preprints*, **77**, 140 (1997).
17. Prosa, T. J., Bauer, B. J., Amis, E. J., Tomalia, D. A. and Scherrenberg, R., *Journal of Polymer Science Part B – Polymer Physics*, **35**, 2913 (1997).
18. Tomalia, D. A. *et al.*, *Macromolecules*, **19**, 2466 (1986).
19. Newkome, G. R., Moorefield, C. N., Baker, G. R., Saunders, M. J. and Grossman, S. H., *Angewandte Chemie – International Edition in English*, **30**, 1178 (1991).
20. Jackson, C. L., Chanzy, H. D., Booy, F. P., Drake, B. J., Tomalia, D. A., Bauer, B. J. and Amis, E. J., *Macromolecules*, **31**, 6259 (1998).
21. (a) Sheiko, S. S., Gauthier, M. and Moller, M., *Macromolecules*, **30**, 2343 (1997); (b) Li, J., Swanson, D. R., Qin, D., Brothers, H. M., Piehler, L. T., Tomalia, D. A. and Meier, D. J. *Langmuir*, **15**, 7347 (1999); (c) Li, J., Piehler, L. T., Qin, D. Baker, Jr., J. R.,

Tomalia, D. A. and Meier, D. J. *Langmuir*, **16**, 5613 (2000).
22. deGennes, P. G. and Hervet, H., *Journal De Physique Lettres*, **44**, L351 (1983).
23. Lescanec, R. L. and Muthukumar, M., *Macromolecules*, **23**, 2280 (1990).
24. Mansfield, M. L. and Klushin, L. I., *Macromolecules*, **26**, 4262 (1993).
25. Mansfield, M. L., *Polymer*, **35**, 1827 (1994).
26. Murat, M. and Grest, G. S., *Macromolecules*, **29**, 1278 (1996).
27. Boris, D. and Rubinstein, M., *Macromolecules*, **29**, 7251 (1996).
28. Naylor, A. M., Goddard, W. A., Kiefer, G. E. and Tomalia, D. A., *Journal of the American Chemical Society*, **111**, 2339 (1989).
29. Uhrich, K., *Trends in Polymer Science*, **5**, 388 (1997).
30. Cooper, A. I. *et al.*, *Nature*, **389**, 368 (1997).
31. Kim, Y. H. and Webster, O. W., *Journal of the American Chemical Society*, **112**, 4592 (1990).
32. Higgins, J. S. and Benoît, H. C. *Polymers and Neutron Scattering* (Clarendon Press, Oxford, 1994).
33. Bauer, B. J., Briber, R. M., Hammouda, B. and A., T. D., *ACS PMSE Preprints*, **67**, 430 (1992).
34. Prosa, T. J., Bauer, B. J. and Amis, E. J., *Macromolecules*, **34**, 4897 (2001).
35. Prosa, T. J., Bauer, B. J., Topp, A., Amis, E. J. and Scherrenberg, R., *ACS PMSE Preprints*, **79**, 307 (1998).
36. Topp, A., Bauer, B. J., Amis, E. J. and Scherrenberg, R., *ACS PMSE Preprints*, **77**, 137 (1997).
37. Yin, R., Qin, D., Tomalia, D. A., KukowskaLatallo, J. and Baker, J. R., *ACS PMSE Preprints*, **77**, 206 (1997); Grubbs, R. B., Hawker, G. J., Dao, J. and Fréchet, J. M. J. *Angewandte Chemie – International Edition in English.*, **36**, 270 (1997).
38. Gauthier, M. and Moller, M., *Macromolecules*, **24**, 4548 (1991).
39. Bauer, B. J., Briber, R. M. and Han, C. C., *Macromolecules*, **22**, 940 (1989).
40. Bauer, B. J., Hanley, B. and Muroga, Y., *Polymer Communications*, **30**, 19 (1989).
41. Bauer, B. J. and Briber, R. M., *ACS Polymer Preprints*, **31**, 578 (1990).
42. Choi, S., Briber, R. M., Bauer, B. J., Topp, A., Gauthier, M. and Tichagwa, L., *Macromolecules*, **32**, 7879 (1999).
43. Topp, A., Bauer, B. J., Klimash, J. W., Spindler, R., Tomalia, D. A. and Amis, E. J., *Macromolecules*, **32**, 7226 (1999).
44. Glinka, C. J., Barker, J. G., Hammouda, B., Krueger, S., Moyer, J. J. and Orts, W. J., *Journal of Applied Crystallography*, **31**, 430 (1998).
45. Prask, H. J., Rowe, J. M., Rush, J. J. and Schroder, I. G., *Journal of Research of the National Institute of Standards and Technology*, **98**, 1 (1993).
46. Wooley, K. L., Klug, C. A., Tasaki, K. and Schaefer, J., *Journal of the American Chemical Society*, **119**, 53 (1997).
47. Topp, A., Bauer, B. J., Prosa, T. J., Scherrenberg, R. and Amis, E. J., *Macromolecules*, **32**, 8923 (1999).
48. Ramzi, A., Scherrenberg, R., Brackman, J., Joosten, J. and Mortensen, K., *Macromolecules*, **31**, 1621 (1998).
49. Stechemesser, S. and Eimer, W., *Macromolecules*, **30**, 2204 (1997).
50. Welch, P. and Muthukumar, M., *Macromolecules*, **31**, 5892 (1998).
51. Topp, A., Bauer, B. J. and Amis, E. J., *Macromolecules*, **32**, 7232 (1999).
52. Nisato, G., Ivkov, R. and Amis, E. J., *Macromolecules*, **33**, 4172 (2000).

12

Atomic Force Microscopy for the Characterization of Dendritic Polymers and Assemblies

J. LI[1] AND D. A. TOMALIA[2]
[1]Dow Chemical Company, 1897 Building, Midland, MI, USA [2]Center for Biologic Nanotechnology, Medical School, University of Michigan, Ann Arbor, MI, USA

1 INTRODUCTION

Dendritic and hyperbranched macromolecules possessing tree like architecture are recognized as promising nanoscale building blocks for both super and supramolecular constructs [1a, b]. These macromolecules are uniform in size, with a high density of functional groups located in close proximity to the surface (see Chapter 10 by Amis and Bauer) [2]. Dendrimers are produced in an iterative sequence of reaction steps, in which each additional iteration leads to a higher generation material [3]. The precisely defined size, structure, and large number of exo-presented functional groups have qualified dendrimers as ideal nano-scaffolding for the attachment of antibodies, contrast agents, radionuclides, etc. for use in biological/medical and various chemical applications [4, 5]. Recent research has shown that PAMAM dendrimers can be used as DNA-transfer vectors for gene transfection[6, 7] or drug delivery vehicles for small molecules [42].

With these emerging applications, there is a critical need for analytical techniques that will provide insights to fundamental questions concerning dendrimer characteristics and properties (e.g. their dimensions, uniformity of size, shape and degree of rigidity, etc.) [8–10]. atomic force microscopy (AFM) offers this

Dendrimers and Other Dendritic Polymers. Edited by Jean M. J. Fréchet and Donald A. Tomalia
© 2001 John Wiley & Sons Ltd

possibility, since it provides high-resolution imaging, measurement of surface topography as well as other molecular level [11] and nanoscale [12] structure properties. It has been shown to be very useful in gaining insight into the properties of nanostructures [12].

This chapter describes the use of AFM as a methodology for characterizing dendritic macromolecules and their self-assembled films. Systematic AFM studies of the PAMAM dendrimers properties (e.g. their size, shape, rigidity, packing textures and molecular weight calculations) will be fully addressed in this chapter.

2 OVERVIEW OF ATOMIC FORCE MICROSCOPY

SPM (scanning probe microscopy) is a universal term for a multitude of microscopy modes, all based on the same principle [13]. Scanning microscopy operates on principles that are substantially different than those involved with optical or electron microscopy. In traditional microscopy, lenses focus light or electrons to produce images that are substantial amplifications of the original. The resulting resolution is ultimately limited by the wavelength of the illumination source. Optical microscopes can amplify features measuring as small as approximately 250 nm; electron microscopes can image (amplify) features smaller then 1 nm, but only under vacuum and with samples that have been laboriously prepared. Scanning microscopes, on the other hand, use no lenses, need minimal sample preparation and images are produced by 'touch contact' with the sample as shown in the schematic diagram Figure 12.1.

The key to scanning microscopy is the sensitive cantilever arm. AFM senses the deflection of the cantilever beam as it interacts with the sample surface. A very sharp probe whose apex is only tens of nanometers wide is mounted on the free end of the cantilever which often has a spring constant of the order of 1 N/m. A laser beam reflects off the end of the cantilever onto a position-sensitive photodetector. A piezoelectric scanner drags the probe across the surface to be imaged. When changes in surface topography cause the probe tip to move up or down, the photodetector senses the motion, and the microscope's computer translates the deflection into three-dimensional surface information. Since the AFM is normally operated in the contact mode (the cantilever is held less than a few angstroms from the sample surface), the interactive force between the cantilever and the sample is repulsive, as shown in Figure 12.2. Here the force increases dramatically with a decrease in the tip-sample distance. In the noncontact regime, the cantilever is held on the order of tens to hundreds of angstroms from the sample surface, and the interatomic force between the cantilever and the sample is attractive. In the intermitent-contact regime, the vibrating cantilever tip is brought closer to the sample so that the tip just 'taps' the sample. This

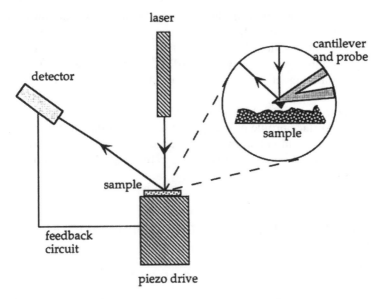

Figure 12.1   Schematic of the atomic force microscope

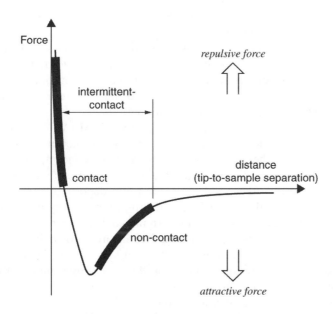

Figure 12.2   AFM operating force regions [14]

patented technique allows high-resolution topographic imaging of sample surfaces that are soft, adhesive or fragile. Changing tips and techniques can also provide different information. Whereas the light microscope can produce only visual images, a skilled researcher with an SPM can map magnetic fields, surface friction, adhesion, magnetism, electrostatism, thermal conductivity and more [15].

With the development of AFM, the capability of imaging at the molecular level allows the surface profile of each molecule to be obtained from which the molecular volume of individual molecules can be obtained. AFM has become one of the most powerful instruments for use in the characterization of surfaces and materials [16].

3 CHARACTERIZATION OF DENDRITIC MACROMOLECULES BY AFM

3.1 DENDRITIC MACROMOLECULAR FILMS

The structural state of dendritic macromolecules at air–water (Langmuir monolayers) and air–solid (adsorbed monolayers, self-assembled films and cast films) interfaces have been reviewed by Tsukruk [17]. Although this work summarizes various characterization techniques for dendritic films by AFM techniques, in this chapter, we will present recent progress on the characterization of the dendritic film surface morphologies.

Full generation poly(amidoamine) (PAMAM) dendrimers possessing amino surface groups are easily spread by spin coating on a mica surface to form films. Interdendrimer assembly events appears to influence these operations. Regardless of the generation levels examined, the film uniformity is determined primarily by the concentration of the dendrimer solution.

Figures 12.3 (a) and (b) show this effect using a G4 (PAMAM) dendrimer, in which the dendrimer is deposited on a mica surface by spin coating. The AFM image in Figure 12.3a of the film deposited from a higher concentration (0.1% w/w) solution shows randomly deposited globular structures (aggregates) of different sizes. In contrast, the AFM image of a film deposited from a lower concentration solution (0.01% w/w) Figure 12.3(b), shows a very uniform and flat surface. This indicates that these amine terminated dendrimers tend to form uniform densely packed films on a mica surface in order to maintain lower surface tension at lower concentrations. On the other hand, excess dendrimers at higher concentrations tend to aggregate to produce dendrimer films with less uniform globular structures (Figure 12.3(a)).

As the dendrimer concentration is lowered to 0.001% w/w, the cast G4 film is no longer uniform and flat. Individual dendrimer molecules are not seen, but instead the dendrimers aggregate to produce a patchy film, Figure 12.4(b).

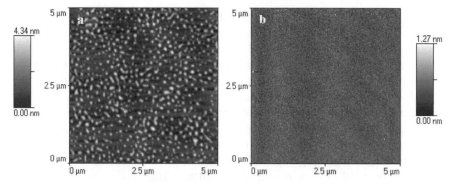

Figure 12.3 Tapping mode AFM images of G4 PAMAM dendrimers on mica surface at different concentrations. (a) 0.1% w/w; (b) 0.01% w/w (provided by Jing Li, D. A. Tomalia) [18]

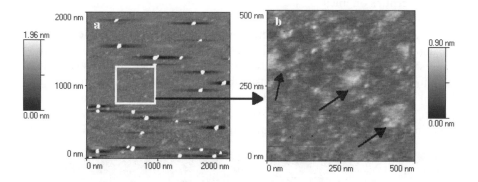

Figure 12.4 Tapping mode AFM images of G4 PAMAM dendrimers on mica surface at different magnifications. Conc.: 0.001% w/w (a) 2000 nm × 2000 nm, (b) 500nm × 500 nm (provided by Jing Li and D. A. Tomalia) [18]

Except for some regions indicated by arrows, most of the surface consists of small domains of various sizes, as seen in Figure 12.4 (b). Low generation members (i.e. G = 0–3) in the PAMAM dendrimer series appear to exist as floppy open, plate-like molecules in which the dendrimer branches can easily interpenetrate each other and establish intermolecular interactions [19].

Spin-coating films of other compositionally different dendritic macromolecules such as carbosilane dendrimers and hyperbranched poly(styrene) were studied by Sheiko *et al.* [20–22] using tapping mode AFM.

Substantial long-range order was observed for dendrigraft poly(styrene) films using two-dimensional Fourier transformation of the AFM images. The resulting image in Figure 12.5 (A and B) demonstrated ordered particles of approximately 46 nm diameter. These macromolecules appeared to manifest distinct hexagonally packed globular arrays whose size was consistent with molecular dimensions obtained from viscosity and dynamic light-scattering measurements [21]. Similarly, AFM images of carbosilane dendrimers with methyl terminal groups formed a variety of assemblies including: individual islands, monolayers, bilayers and trilayers with a globular conformation [20]. These aggregation events were visualized as follows: (1) single molecules coagulated to (2) clusters, followed by formation of (3) continuous layers on the solid substrate. Interestingly, replacement of the methyl terminal group with hydroxyl groups led to significant changes in the dendrimer aggregation behavior. Using AFM these hydroxylated carbosilane dendrimers were observed to form monolayers upon casting on mica and are highly deformed macromolecules, oblate in shape with an axial ratio close to 1:3. This reorganization of the dendritic structure is believed to be caused by the preferential adsorption of the hydroxyl terminal groups on mica [22].

An AFM study of monomolecular layers of diarylethene-containing dendrimers (i.e. G2 and G3) which were synthesized from bis-phenol derivative of the diarylethene chromophore [23] was reported by Karthaus *et al.* [24] AFM images have demonstrated that only G3 of this dendrimer family is capable of forming stable, homogeneous, transferable Langmuir films as seen in Figure 12.6(b). Generation 2 in this series does not form stable monomolecular films, as shown, which may indicate that the rigid chemical structure of these asymmetric dendrimers is critical to the preservation of their original 'half bowl' shape within

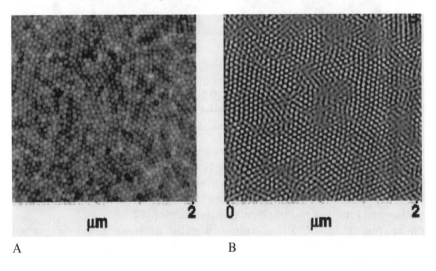

A   B

Figure 12.5 [18] Structure of the monolayer of S05-3 arborescent graft (dendrigraft) polystyrenes cast on mica as prepared (A) and Fourier filtered (B)

a   b

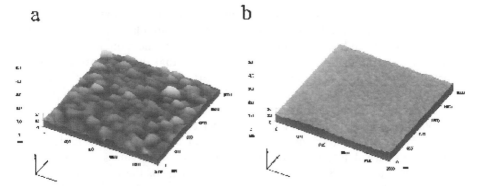

Figure 12.6 AFM images of a transferred film G2 (a) and G 3 (b) on freshly cleaved mica surface [24]

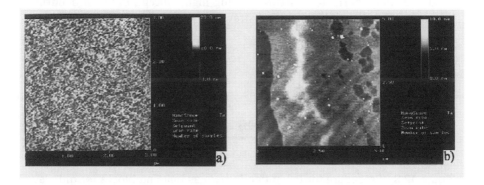

Figure 12.7 AFM images of monolayers and multilayers of dendritic self-assembled films:(a) surface morphology of G10 monolayer with light grainy structure, scan size 3 μm; (b) G6 monolayer at the edge of the film, scan size 5 μm. The edge of the film is running in the vertical direction with monolayer and bare silicon surfaces located on the right and left sides, respectively. Images were obtained with the tapping mode in air, 256 × 256 resolution, 1 Hz scanning rate. Fringes at the bottom of the images are the result of laser beam interference

monolayers at the air–water interface and after deposition on mica.

Polyamidoamine (PAMAM) dendrimers (G3.5–G10) were used by Tsukruk *et al.* [25, 26] to fabricate self-assembled monolayers by an electrostatic deposition techniques [27].

It was found that dendrimer monolayers formed on a silicon surface by all amine-terminated generations are very smooth and homogeneous, with very light, grainy surface morphologies, Figure 12.7(a). The thickness of dendrimer monolayers was measured by the AFM technique from occasional holes in the monolayer and near the edge formed by the meniscus as shown in the Figure 12.7(b). The measured thickness of the dendrimer monolayer should represent the size of the dendrimers. However, the thickness of the monolayer was much smaller than the theoretical value. This suggests that the high attractive force between 'sticky' surface groups as well as short-range van der Waals forces and long-range capillary forces are responsible for the formation of the compressed, compact monolayer dendrimer structures.

Watanabe and Regen [28] used an electrostatic layer-by-layer deposition technique to fabricate self-assembled films from alternating molecular layers of opposite charged PAMAM dendrimers and low molar mass compounds as shown in Figure 12.8. Using AFM, linear growth of the film thickness was observed which is consistent with multilayer ordering. Evensong and Badyal [29] also studied the G9.5 PAMAM dendrimers by AFM. Observations were made on a new type of cell structure, and randomly intertwined PAMAM dendrimer microfibrils obtained by casting high concentrations of dendrimer solution onto hydrophilic silicon substrates.

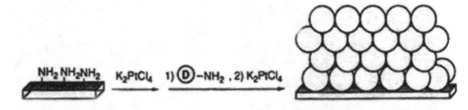

Figure 12.8 Procedure for using dendrimers as building blocks for multilayer films [28]

3.2 SHAPE CONTROL WITH QUASI-EQUIVALENT DENDRITIC SURFACES – DENDRITIC CYLINDRICAL AND SPHERICAL SHAPES

Controlling the size, shape and ordering of synthetic organic materials at the macromolecular and supramolecular levels is an important objective in chemistry. Such control may be used to improve specific advanced material properties. Initial efforts to control dendrimer shapes involved the use of appropriately shaped core templates upon which to amplify dendritic shells to produce either dendrimer spheroids or cylinders (rods). The first examples of covalent dendrimer rods were reported by Tomalia *et al.* [43] and Schluter *et al.* [44]. These examples involved the reiterative growth of dendritic shells around a preformed linear polymeric backbone or the polymerization of a dendronized monomer to produce cylinders possessing substantial aspect ratios (i.e. 15–100) as observed by TEM and AFM. These architectural copolymers consisting of linear random coil cores surrounded by dendritic shells have been referred to as 'dendronized polymers' [44].

Recent seminal work reported by Percec *et al.* [45–47] has shown that either spheroidal or cylindrically shaped 'dendronized polymers' may be obtained by the polymerization of a 'dendronized macromonomer'. The degree of polymerization of these monomers determines the ultimate shape which appears to demonstrate the 'quasi-equivalence of dendritic coats'.

The self-assembly mechanism proposed for these spherical and cylindrical polymer backbones surrounded with quasi-equivalent dendritic coats is outlined in Figure 12.9. This knowledge allows the rational design of polymers with well defined spherical and cylindrical shapes. Quasi-equivalent character of these

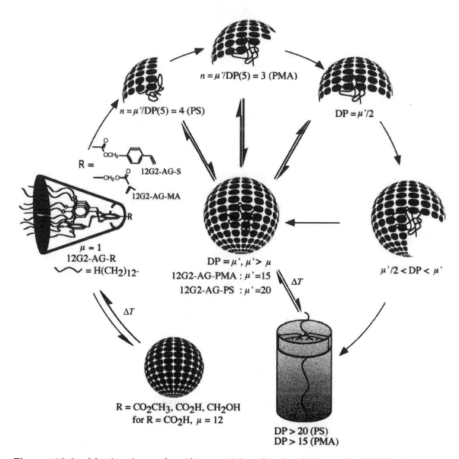

Figure 12.9 Mechanism of self-assembly of spherical and cylindrical supramolecules from polymer backbones jacketed with quasi-equivalent dendritic coats [31]

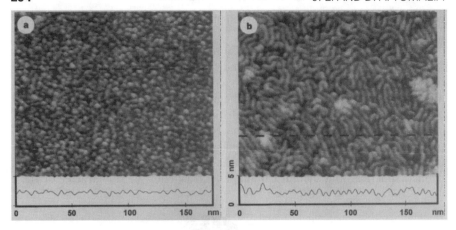

Figure 12.10 Topographic scanning force micrscopy images of monomolecular films of 12G2-AG-PS. A. Mn = 10,800, b. Mn = 186,500

dendritic structures is controlled by the degree of polymerization (DP) of the backbone. As illustrated, an increase in DP changes the backbone conformation and polymer shape, respectively, from random-coiled and spherical to extended and cylindrical. AFM is a methodology of choice.

Monolayers of both spherical and cylindrical polymers were visualized in their isotropic phase by AFM as shown in Figure 12.10. For both cylindrical and spherical 12G2-AC-PS, the monolayer thickness and the lateral diameters were about 50 ± 5 A. Topographic contrast shows strong molecular aggregation and weak interpenetration of 12G2-AC-PS consistent with high steric crowding of the dendritic coats. The geometric constraint of the flat and hard substrate, as well as the cohesive forces acting during film formation from dilute solution, appear to cause some orientational ordering of the cylindrical molecules. The average length of the straight segment is about 400 A for 12G2-AC-PS and 600 A for 12G2-AC-PMA. Hairpin folding, expected for relatively long worm-like molecules, is also visible in Figure 12.10(b).

3.3 POLY(AMIDOAMINE) (PAMAM) DENDRIMERS

3.3.1 The Packing of PAMAM Dendrimers (Generation = 9)

Polymerization in microemulsions allows the synthesis of ultrafine latex particles in the size range of 5 to 50 nm with a narrow size distribution [33]. The deposition of an ordered monolayer of such spheres is known to be increasingly difficult as the diameter of such particles decreases [34]. Vigorous Brownian motion and capillary effects create a state of disorder in the system that is difficult to control. So far, the smallest 2-d arrays of latex spheres prepared have had a periodicity of 42 nm [35]. However, successful AFM imaging of G9 PAMAM dendrimers 11.4 nm diameter is an exception which has broken the record of the smallest latex arrays.

This section will demonstrate G9 PAMAM dendrimer packing using AFM imaging techniques. G9 PAMAM dendrimers can easily form two-dimensional packing on a freshly cleaved mica surface, as seen in Figure 12.11. This top-view AFM image reveals very well-ordered particle packing except for some vacancies, dislocations and grain boundaries.

Most of the areas show very dense packing of globular G9 dendrimer, exhibiting a clear hexagonal order as indicated in the square of Figure 12.11. The driving force for this unique two-dimensional packing is considered to be chiefly a function of the specific chemical and physical properties of G9 PAMAM dendrimers. This is related to the dispersion forces, molecular interpenetration and rigidity of the dendritic structure.

It was also found that the deposition pattern changes if the amount of G9 solution placed on the mica surface is decreased from 6 μl to 3 μl as illustrated in Figure 12.12. It can be seen that the molecules assembled to make connections with each other to form an interlinked chain texture. Some hexagonal pattern

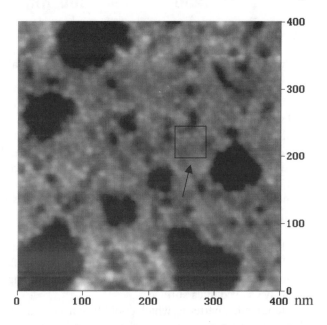

Figure 12.11   Tapping Mode AFM image of G9 PAMAM dendrimer molecules on mica surface. Sample prepared by placing 6 μl of a dilute aqueous solution, conc. $5 \times 10-3\%$ (w/w) G9 on a freshly cleaved mica surface and allowing the film to dry slowly at room temperature (provided by Jing Li and D. A. Tomalia)

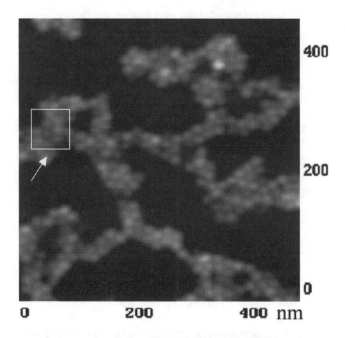

Figure 12.12 Tapping Mode AFM image of G9 dendrimers on mica surface. The sample prepared from the same solution in Figure 12.10, except that 3 μl solution was placed onto the freshly cleaved mica surface and the film was allowed to dry at room temperature (provided by Jing Li and D. A. Tomalia)

can be observed within these arrays as indicated in Figure 12.12. Very few if any isolated G9 dendrimer molecules were found even after multiple scans of several areas. These results indicate that the interfacial forces between dendrimer molecules dominated the particle pattern. The chemical and physical characteristics of G9 PAMAM dendrimer molecules, such as its spherical shape, high surface density of the functional groups with hydrogen bonding characteristics as well as the relative rigidity of the dendrimers plays a very important role in this interfacial behavior.

Figure 12.13 shows the striking perturbation of G9 PAMAM packing that is possible by applying the small force of a nitrogen gas stream flowing at an approximate 35 degree angle to the sample surface, after placing 3 μl of G9 solution on a freshly cleaved mica surface.

A variety of self-assemble patterns, which could be considered a G9 PAMAM dendritic packing library under these perturbation conditions, can be observed in Figure 12.13. One can observe including single isolated G9 dendrimer molecules, dimers, trimers and other packing patterns. These patterns may be seen more clearly in the high magnification images shown in Figure 12.14. A single G9

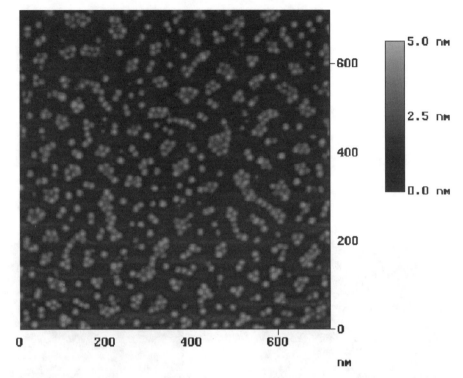

Figure 12.13 Tapping Mode AFM images of G9 PAMAM dendrimer molecules on mica surface. Sample prepared by placing 3 μl of a dilute aqueous solution, conc. $5 \times 10^{-3}\%$ (w/w) G9 dendrimer onto a freshly cleaved mica surface and allowing the film to dry with a 35 degree angle by stream of nitrogen gas at room temperature (provided by Jing Li and D. A. Tomalia)

is enclosed by the circle, the dimer by the oval and the trimer by the square.

The real surprise of the application of small flow forces is that, in addition to the dimer and trimer configurations of G9, molecules in image (a), Figure 12.14 changes in the shape of the G9 dendrimer can also observed. Some of the G9 molecules have the appearance of having been squeezed together, their shape contorted into irregular, unsymmetrical forms instead of their characteristic round shape, indicated by arrows 1 and 2 in Figure 12.14, image (b). These results suggest that dendrimers are soft, spongy and elastic, and that their shape can be changed by the application of very small flow forces.

A large variety of packing assemblies is also present, some resembling curved linear catenations as in Figure 12.14, image (c) 3, and others such as barbell type-dimers, pyramidal trimmers parallelogram-type tetromers and hexamers.

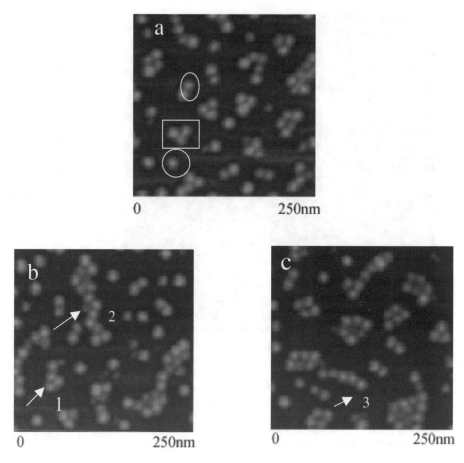

Figure 12.14   Magnification images from Figure 12.13

3.3.2   G5 to G10 PAMAM Dendrimers [18]

The spin coating technique of preparing PAMAM dendrimer samples for AFM is the best method to maintain undistorted dendrimer shape allows the visualization of isolated single dendrimer molecules. In spin coating, a dendrimer solution is rapidly spread across the sample surface and the majority of particles separate from one another.

In the AFM images of Figure 12.15, one can see many separated and randomly deposited globular particles on the mica surface. In each image, the particles appear to be substantially uniform in size, i.e. they are essentially monodisperse. This is not surprising if each bright spot represents a single dendrimer molecule.

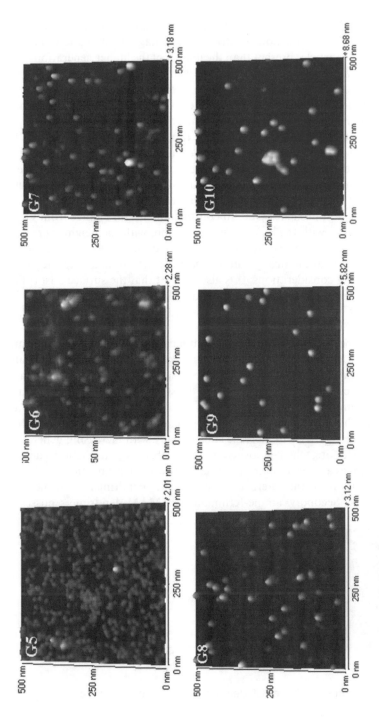

Figure 12.15 Tapping mode AFM three-dimensional images of PAMAM dendrimer molecules from G5 to G10. Dendrimer molecules (0.001% w/w) are deposited on mica surface by spin coating (provided by Jing Li and D. A. Tomalia)

There are a few large irregular clusters in the G6 image, perhaps caused by dendrimer aggregation. It is noteworthy that the images for G5 and G6 generations have more particles present than the images of the higher generations, due to their lower molecular weight (i.e. at the same weight concentration, their solutions contain more molecules than those of the higher generations).

Using computer imaging analysis (TopoMetrix SPMLab 3.06.06) [36], profile data for each dendrimer molecule was obtained as shown in Figure 12.16.

A plot of diameter and height as a function of generation (Figure 12.17) demonstrates that the measured diameters are always larger than the heights, indicating that the dendrimer molecules are no longer spherical, but instead dome-shaped when deposited on a mica surface. It also shows that the diameter from G5 to G9 increases slowly with an increase in generation, with curve flattening out as it approaches G10. In contrast, the height change from G5 to G8 increases slowly with the increase in generation, with the height increasing more rapidly after G8. These results suggest that the molecular rigidity of dendrimers increase dramatically with increasing generation, as expected from molecular packing considerations. It is also evident that the surface adsorption forces between the mica surface and individual dendrimers decreases with increasing generation, as predicted from Mansfield's model of dendrimer surface adsorption (see Figure 12.18) [37].

In fact, it has been previously observed that measured diameters of dendrimer molecules by AFM are much larger than the theoretical values, which indicates that the dendrimers 'spread out' and flatten on the surface [25, 26]. Three major factors could account for this deformation. First, the unique architecture and chemical structure of PAMAM dendrimers result in macromolecules that are not solid balls, but instead are relatively 'open' and hence soft materials. It is expected that the rigidity will increase substantially with increasing generation number [9]. Therefore, when deposited on solid substrates, they tend to deform to different degrees as a result of the interplay between their inherent rigidity and surface energetics from the interaction between the dendrimer molecules and the mica surface. Secondly, amine-terminated PAMAM dendrimers possess a

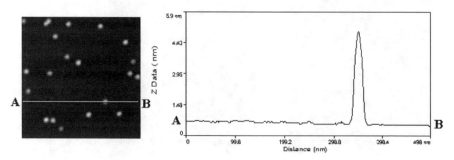

Figure 12.16   Profile section of G9 dendrimer on the mica surface

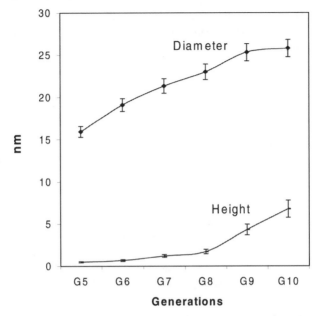

Figure 12.17 Mean dendrimer diameter and height measured by AFM from particle's profiles, as a function of generation. The error bars for AFM data represent one standard deviation

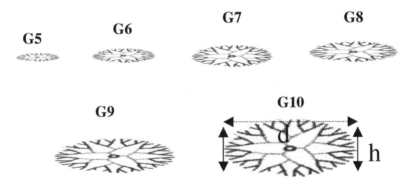

Figure 12.18 Model of PAMAM dendrimers after drying on a mica surface

positively-charged surface (in aqueous solution of pH 7.0), while mica surfaces possess the opposite polarity. Finally, we believe that the AFM tip can influence the sample surface morphology. This occurs when the tip scans a soft sample, even when using the tapping mode. Unfortunately, there is currently no established method to explore how an AFM tip acts on soft molecules to induce apparent size changes [38].

Figure 12.18 shows models representing single, isolated dendrimer molecules from G5 to G10 after drying on a mica surface. Based on the dendritic dome shape (i.e. a spherical cap), molecular volume can be calculated from the following equation [36]:

$$V = 1/6 \pi h (h^2 + 3/4 d^2) \tag{1}$$

where h is the height and d the diameter of the cap. For each generation, molecular volume can be calculated using measured diameter and height from each particle's profile. If the density of the dried dendrimer molecules is taken to be 1.0 g/cm^3, an absolute molecular weight can be estimated, with results shown for each generation in Table 12.1.

The distribution of molecular weights of each generation was determined from measurements on about 50 molecules, with results shown in Figure 12.19 (the weight fraction is the percent dendrimer in each interval of molecular weight under consideration). Based on these distributions, the polydispersity index (M_w/M_n) of G5 to G10 can be calculated, with results shown in Table 12.1 [39]. They are all less than 1.08, which means that the particle size distribution is very uniform for each generation.

Comparing the calculated and theoretical values for molecular weight, we found that there is a very good match for G5 to G8, but not for G9 and G10. The discrepancy for G9 and G10 are probably due to a breakdown of the assumption that the deposited molecular shape is that of a spherical cap (see Figure 12.18).

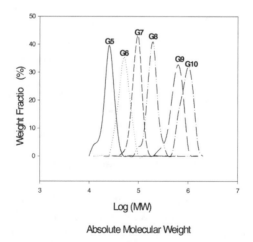

Figure 12.19 The absolute molecular weight distribution as measured by AFM

Table 12.1 Comparison of the mean molecular weight calculated by AFM and theoretical molecular weight of PAMAM dendrimers

PAMAM dendrimers	No. of molecules	MW (AFM) measured	MW theoretical	Relative error (%)	Polydispersity
G5	48	27 900	28 826	− 3.2	1.05
G6	48	56 000	58 048	− 2.7	1.06
G7	61	120 000	116 493	3.2	1.05
G8	59	242 000	233 383	3.7	1.06
G9	52	616 000	467 162	31.9	1.08
G10	47	1 198 000	934 720	28.2	1.07

3.3.3 Core-shell Tecto-dendrimers [36]

Dendrimers are a unique class of polymer that may bridge the gap between the synthetic and biological polymer fields [40]. Although, the size of most dendrimers, e.g. PAMAM dendrimers, is large enough to mimic many protein molecules, they are substantially smaller than other biological targets such as viruses. An important approach in synthesizing larger dendritic molecules without excessive synthetic steps has been recently reported [41]. These are dendritic architectures in which a PAMAM dendrimer molecule is used as a core, around which a shell of other PAMAM dendrimers are covalently attached as illustrated in Figure 12.20. Typically, the generation number of the core is larger than that of the surrounding shell dendrimers. These new architectures are referred to as core-shell tecto-(dendrimer) (shortened to tecto-(dendrimer) here). The individual tecto-(dendrimer) molecules could also be clearly observed in the AFM images as shown in Figure 12.21.

The images show many globular particles randomly deposited on the mica surfaces. In each of the images, the particles appear to be substantially uniform in size, i.e. they are essentially mono-dispersed. Also, the images apparently are of

Figure 12.20 Schematic representation of a core–shell tecto-(dendrimer) molecule in solution

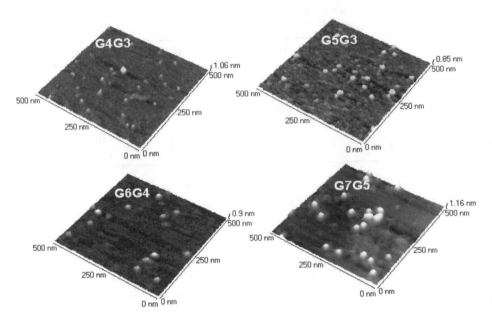

Figure 12.21 Tapping mode AFM images of tecto-(dendrimer) molecules. (Sample preparation: one drop of a 1×10^{-5} wt% solution was spread on a freshly cleaved mica surface by spin coating, and then dried at room temperature)

single molecules (a few obvious clusters or aggregates can be identified in the images). The behavior of newly prepared films was consistent, provided the casting solutions were not allowed to stand or age. Solutions of the G4G3 and G5G3 tecto-(dendrimer)s were stable for about one month, after which substantial aggregation occurred. In contrast, the solutions of the G6G4 and G7G5 tecto-(dendrimer)s remained stable for at least six months, showing that solution stability increases with increasing molecular size.

From the profile of each particle, the size and the height data were obtained using the analytical capabilities of the AFM instrument, with results shown in Table 12.2 for the average height and width of the individual molecule on the mica surface.

Based on the same equation (1), absolute molecular weight of tecto-dendrimers could be calculated. The results shown in Table 12.3 clearly demonstrate that the molecular weights calculated from the AFM images of individual molecules are very close to those obtained by the more classical MALDI-TOF MS and PAGE techniques.

There are a few requirements that must be met if this AFM technique is to be used to measure molecular volumes. First, the molecules on the surface must be 'soft' enough that the molecule flattens out completely on the surface, i.e. the

Table 12.2 The molecular size of the core–shell tecto-dendrimers

Core–shell tecto-dendrimers	Theoretical calculations	AFM measurement	
Observed shell sat. levels	Diameter[a] (nm)	Diameter (nm)	Height (nm)
G4(G3)$_4$	11.6	25.0	0.38
G5(G3)$_{8-10}$	12.6	33.0	0.53
G6(G4)$_{6-8}$	15.6	38.0	0.63
G7(G5)$_6$	19.0	43.0	1.10

[a] In the solution.

Table 12.3 Molecular weights of the tecto dendrimers

Tecto dendrimers	PAGE (MW)	MALDI-TOF MS (MW)	AFM (MW)
G4(G3)$_4$	58 048	56 496	56 000
G5(G3)$_{8-10}$	116 493	120 026	136 000
G6(G4)$_{6-8}$	233 383	227 606	214 000
G7(G5)$_5$	467 162	288 970	479 000

molecule cannot have an undercut region such that the probe cannot 'see' those portions hidden under the molecule. Also, the actual AFM image represents a convolution of the probe tip and contour of the material being examined. This is not a problem for very soft materials where the height-to-diameter ratio is quite small, but deconvolution will be required for more rigid molecules where the height is an appreciable fraction of the diameter. These experiments have demonstrated that AFM is not only useful in imaging dendritic macromolecules, but is also useful in estimating important dimensional molecular parameters, such as volume and molecular weight, which are not easily obtained by other techniques.

The authors are grateful to Drs Dale J. Meier, Lars T. Piehler, Dujie Qin and Mr. Walter Meixner for valuable discussion and acknowledge financial support from the US Army Research Laboratory and the National Cancer Institute.

4 REFERENCES

1. (a) Tomalia, D. A., Naylor, A. M. and Goddard III, W. A. *Angew. Chem. Int. Ed. Engl.* **29**, 138 (1990); (b) Fréchet, J. M. J. *Science*, **263**, 1710 (1994).
2. Topp, A., Bauer, B. J., Klimash, J. W., Spindler, R., Tomalia, D. A. and Amis, E. J. *Macromolecules*, **32**, 7226 (1999).
3. Tomalia, D. A., Baker, H., Dewald, J. R., Hall, M. J., Kallos, G., Martin, S. J., Roeck, J., Ryder, J. and Smith, P. B. *Poly. J.* (*Tokyo*), **17**, 117 (1985).
4. Zeng, F. and Zimmerman, S. C. *Chem. Rev.*, **97**, 1681 (1999).
5. Brunner, H. J. *Organomet. Chem.*, **500**, 39 (1995).
6. Kukowska-latalll, J. F., Bielinska, A.U., Johnson, J., Spindler, R., Tomalia, D. A. and

Baker, Jr, J. R. *Proc. Nalt. Acad. Sci.*, **93**, 4897 (1996).
7. Bielinska, A.U., Kukowska-latallo, J. F., Johnson, J., Tomalia, D. A. and Baker, Jr, J. R. *Nucleic Acids Research*, **24**, 2176 (1996).
8. Robert, F. *Science*, **267**, 459 (1995).
9. Uppuluri, S., Keinath, S. E., Tomalia, D. A. and Dvornic, P. R. *Macromolecules*, **31**, 4498 (1998).
10. Bosman, A. W., Janssen, H. M. and Meijer, E. W. *Chem. Rev.*, **99**, 1665 (1999).
11. Bottomley, L. *Anal. Chem.*, **70**, 425R (1998).
12. McCarty, G. S. and Weiss, P. S. *Chem. Rev.*, **99**, 1983 (1999).
13. Trausser, Y. E. and Heaton, M. G. *American Laboratory*, 1994, May.
14. *A Practical Guide to Scanning Probe Microscopy*, Park Scientific Instruments, p. 6.
15. Snyder, S. R. and White, H. S. *Anal. Chem.*, **64**, 116R (1992).
16. Kumaki. J., Nishikawa, Y. and Hashimoto, T. *J. Am. Chem. Soc.*, **118**, 3321 (1996).
17. Tsukruk, V. V. *Advanced Materials*, **10**, 253 (1999).
18. Li, J., Piehler, L. T., Qin, D., Tomalia, D. A. and Meier, D. *Langmuir*, **16**, 5613 (2000).
19. Naylor, A. J. Goddard III, W. A., Kiefer G. E. and Tomalia, D. A. *J. Am. Chem. Soc.*, **111**, 2339 (1989).
20. Sheiko, S. S., Eckert, G., Ignat'eva, G., Musafarov, A. M., Spickermann, J., Rader, H. J. and Moller, M. *Makromol, Rapid Commun.*, **17**, 283 (1996).
21. Sheiko, S. S., Gauthier, M. and Moller, M. *Macromolecules*, **30**, 2343 (1997).
22. Sheiko, S. S., Musafarov, A. M., Winkler, R. G., Getmanova, E. V., Eckert, G. and Reineker, P. *Langmuir*, **13**, 4172 (1997).
23. Hawker, C. J. and Frechet, J. M. J. *J. Am. Chem. Soc.*, **112**, 1119 (1990).
24. Karthaus, O., Ijiro, K., Shimomura, M., Helmann, J. and Irie, M. *Langmuir*, **12**, 6714 (1996).
25. Tsukruk, V. V., Rinderspacher, F. and Bliznyuk, V. N. *Langmuir*, **13**, 2171 (1997).
26. Bliznyuk, V. N., Rinderspacher, F. and Tsukruk, V. V. *Polymer*, **39**, 5249 (1998).
27. Yu, M. and Deeher, L. G. *Crystallogr. Rep.*, **39**, 628 (1994).
28. Watanabe, S. and Regen, S. L. *J. Am Chem. Soc.*, **116**, 8855 (1994).
29. Evenson, S. A. and Badyal, J. P. S. *Advanced Materials* **9**, 1097 (1997).
30. Percec, V., Ahn, C. H., Ungar, G., Yeardley, D. J. P., Moller, M. and Sheiko, S. S. *Nature*, **391**, 161 (1998).
31. Prokhorova, S. A., Sheiko, S. S., Moller, M., Ahn, C. H. and Percec, V. *Macromol. Rapid Commun.*, **19**, 359 (1998).
32. Stocker, W., Schurmann, B. L., Rabe J. P., Forster, S. Lindner, P., Neubert, I. and Schluter, A. D. *Advanced Materials*, **10**, 793 (1998).
33. Antonietti, M., Basten, R. and Lohman, S. *Macromol, Chem. Phys.*, **196**, 441 (1995).
34. Chevalier, Y., Pichot, C., Graillat, C., Joanicot, M., Wong, K., Maquet. J., Linder, P. and Cabane, B. *Colloid Polym. Sci.*, **270**, 806 (1992).
35. Micheletto, R., Fukuda, H. and Ohtsu, M. *Langmuir*, **11**, 3333 (1995).
36. Li, J., Swanson, D. R., Qin, D., Brothers, H. M., Piehler, L. T., Tomalia, D. A. and Meier, D. J. *Langmuir*, **15**, 7347 (1999).
37. Mansfield, M. L. *Polymer*, **37**, 3835 (1996).
38. Yang, J., Tamm, L. K., Somlyo, A. P. and Shao, Z. *J. of Microscopy*, **171**, 183 (1993).
39. Cowie, J. M. G. *Polymers: Chemistry & Physics of Modern Materials*, Academic & Professional, Chapman & Hall, Wester Cleddens Road, Bishopbriggs, Blackie, Glasgow, UK. 1991.
40. Roovers, J. and Comanita, B. *Advances in Polymer Science*, **142**, 180 (1999).
41. Uppuluri, S., Swanson, D. R., Brothers, H. M., Piehler, L. T., Li, J., Meier, D. J., Hagnauer, G. L. and Tomalia, D. A. *ACS Poly. Mater. Sci. Eng.*, **80**, 55 (1999).

42. Liu, M. Kono, K., Fréchet, J. M. J. *J. of Controlled Release*, **65**, 121 (2000).
43. Yin, R., Zhu, Y. and Tomalia, D. A. *J. Am. Chem. Soc.*, **120**, 2678 (1998).
44. Schlüter, A. D. and Rabe, J. P. *Angew. Chem. Int. Ed.*, **39**, 864 (2000).
45. Percec, V., Johansson, G., Ungar, G. and Zhou, J. P. *J. Am. Chem. Soc.*, **118**, 9855 (1996).
46. Hudson, S. D., Jung, H.-T., Percec, V., Cho, W.-D., Johansson, G., Ungar, G. and Balagurusamy, V. S. K. *Science*, **278**, 449 (1997).
47. Percec, V., Ahn, C.-H., Unger, G., Yeardly, D. J. P. and Moller, M. *Nature*, **391,** 161 (1998).

13

Characterization of Dendrimer Structures by Spectroscopic Techniques

N. J. TURRO[1], W. CHEN[1], M. F. OTTAVIANI[2]
[1]Columbia University, New York, USA
[2]Institute of Chemical Sciences, University of Urbino, Italy

1 INTRODUCTION

Dendrimers have precise compositional and constitutional aspects, but they can exhibit many possible conformations. Thus, they lack long-range order in the condensed phase, which makes it inappropriate to characterize the molecular-level structure of dendrimers by X-ray diffraction analysis. However, there have been many studies performed using indirect spectroscopic methods to character-ize dendrimer structures, such as studies using photophysical and photochemical probes by UV-Vis and fluorescence spectroscopy, as well as studies using spin probes by EPR spectroscopy.

Computer simulations [1, 2] of dendrimer structures revealed that the shape, morphology and surface structure of dendrimers could be controlled as a func-tion of generation and size. These simulations indicated a rather dramatic change in the dendrimer morphology and the surface characteristics as a func-tion of generation for PAMAM dendrimers (Figure 13.1). The early generations (G = 1–3) possess a highly asymmetric, open, hemispherical shape, while the later generations (G ≥ 4) possess a nearly spherical shape with dense-packed surface.

Dendrimers and Other Dendritic Polymers. Edited by Jean M. J. Fréchet and Donald A. Tomalia
© 2001 John Wiley & Sons Ltd

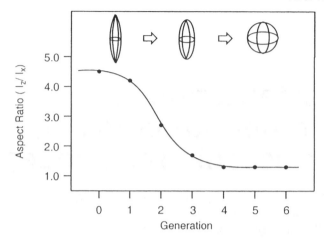

Figure 13.1   Simulated dendrimer morphology (aspect ratio, I_z/I_x) *as a function of generation for PAMAM dendrimers. Adapted from ref. 1*

2   STRUCTURAL CHARACTERIZATION BY PHOTOPHYSICAL AND PHOTOCHEMICAL PROBES

Scheme 1 summarizes four different approaches used to characterize dendrimer structures by photophysical and photochemical probes: 1. Non-covalent, inter-molecularly bound interior probes – to study the internal cavities and the encapsulation abilities of dendrimers. 2. Non-covalent, intermolecularly bound surface probes – to study surface characteristics of dendrimers. 3. Covalently linked probes on dendrimer surfaces – to study the molecular dynamics of dendrimers. 4. Covalently linked probes at the dendrimer central core – to study the site isolation of the core moiety and define the hydrodynamic volume of dendrimers by the concentric dendrimer shells. Critical literature in these four categories will be described using representative examples.

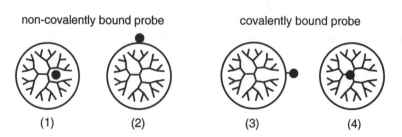

Scheme 1   Four different approaches of photophysical and photochemical probes

2.1 ENCAPSULATION OF PROBE MOLECULES

As the dendrimer generation increases, the diameters increase linearly as a function of each added shell, the number of surface groups amplifies according to a power function and as a consequence the surface area per surface group becomes increasingly smaller. de Gennes and Hervet [3] predicted that ideal dendrimer growth will reach a self-limiting generation in which the dendrimer surface becomes too dense-packed to allow further ideal growth without branch defects. Molecular simulation studies by Goddard and Tomalia [1, 2] indicated that later PAMAM dendrimer generations $(G \geq 4)$ are dense-packed spheroids with internal cavities. The existence of the internal available spaces makes it possible for symmetrically branched dendrimers to host small guest mole-cules. As shown in Scheme 2, guest molecules may bind to the dendrimer interior by either hydrophobic or hydrophilic interactions, in this case they still can diffuse out of the interior of the dendrimers at suitable conditions; in the other case, guest molecules are physically entrapped inside a so-called 'dendritic box', in which the highly dense-packed end groups of dendrimer form a 'closed shell'. Thus, guests can be released only after the 'closed shell' is removed or perturbed.

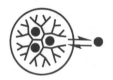

dynamic interior binding
with "open" dendrimers

physical encapsulation
with closed "Dendritic Box"

Scheme 2 Two ways to encapsulate probe molecules inside dendrimers

2.1.1 Non-covalent, Dynamic Interior Binding

In this section, references regarding the adsorption and binding of guest mol-ecules to the dendrimer interior, as well as the binding onto dendrimer surface will be reviewed. Many important studies have demonstrated the encapsulation of guest molecules by dendrimers; however, issues regarding the specific binding sites have not been well resolved yet. In most cases, the binding of guest molecules is presumably dynamic; wherein, the guest molecules can diffuse in and out of dendrimers, although there is little direct evidence.

Goddard and Tomalia [1] investigated the encapsulation of 2,4-dich-lorophenoxyacetic acid and acetylsalicylic acid into methyl ester-terminated PAMAM dendrimers by measuring the ^{13}C spin-lattice relaxation times (T_1) of the guest molecules. In the presence of dendrimer, the T_1 values of these guest molecules in CDCl$_3$ were much lower than those in pure solvent. The relaxation time decreased from G0.5 to G3.5, but remained constant for generations 3.5 to 5.5. The transition of the change in T_1 occurred at dendrimer G3.5, paralleling the morphology transition (from open hemispherical domes to closed spheroids) between G3 and G4 as predicted by computer simulation. In this study, the maximum guest: host stoichiometries was shown to be $\sim 4:1$ by weight and $\sim 3:1$ based on the concentration ratio of guest to the interior tertiary amino groups of dendrimers.

Binding of small hydrophobic probe molecules within the interior of 'unimolecular micelle' type dendrimers was first demonstrated by Newkome and coworkers [4]. This type of dendrimer (Figure 13.2) has a completely hydrophobic interior (saturated hydrocarbon branches) and polar surface groups, which mimics surfactant aggregates. The micellar properties of these dendrimers in aqueous solutions were established by fluorescence and UV-Vis spectroscopy. For example, the phase-resolved fluorescence anisotropy results of diphenylhexatriene (DPH) in the presence of **1** were similar to those previously observed with DPH in micelles, indicating that DPH molecules were predominately associated within the lipophilic interior of dendrimers. Further support of interior binding came from the observation that chlortetracycline, a probe which exhibits fluorescence only in a lipophilic environment, gave such a signal in aqueous solution of **1**. The increased solubility of naphthalene in aqueous solution of **1** evaluated by the UV-Vis spectrum of naphthalene provided additional evidence for the interior binding.

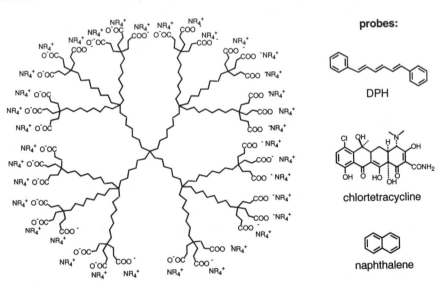

Figure 13.2 'Unimolecular micelle' type dendrimer (**1**) reported by Newkome and probes

Another 'unimolecular micelle' type carboxylic acid-terminated dendrimer (i.e. poly(aryl ether) dendrimers) (Figure 13.3) were synthesized and investigated by Fréchet and coworkers [5]. Solubilization studies of hydrophobic guest molecules in aqueous solution of **2** were performed by first sonicating the host/guest mixture at elevated temperature and then examining the UV-Vis absorbance of the guest molecule. In the presence of **2**, the saturated concentration of pyrene (9.5×10^{-5} M) increased 120-fold compared to that in pure water (8.0×10^{-7} M), revealing a solubilization property comparable to SDS micelles. It was suggested that the high solubilization of **2** might be related to the stabilizing π–π interaction of pyrene with the aryl groups of the dendrimer **2**. This concept was supported by the solubility of polycyclic aromatic guests possessing different electron densities. Compared with that in pure water, the solubility in the presence of **2** increased by 58 times for anthracene (a probe less electron deficient than pyrene), and 258 times for 2,3,6,7-tetranitrofluorenone (a probe more electron deficient than pyrene). On average, 0.45 pyrene dissolved in one single dendrimer **2** in water. The number increased to 1.9 pyrene per dendrimer by the addition of 1.5 M NaCl. This salt effect was attributed to less water within the interior of the dendrimer at higher ionic strength.

Meijer *et al.* [6] reported on inverted 'unimolecular micelle' type dendrimers (Figure 13.4) which have a hydrophilic interior and a hydrophobic shell, synthesized by modifying the end groups of hydrophilic poly(propylene imine) dendrimer with alkyl chains. It was shown that these dendrimers could host

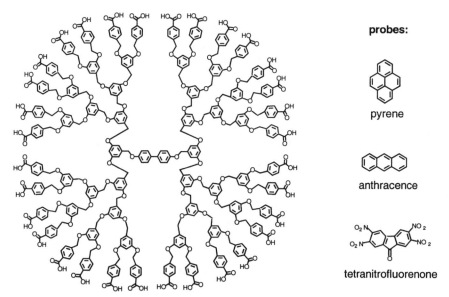

probes:

pyrene

anthracence

tetranitrofluorenone

Figure 13.3 Carboxylic acid-terminated poly(aryl ether) dendrimer (**2**) and probes

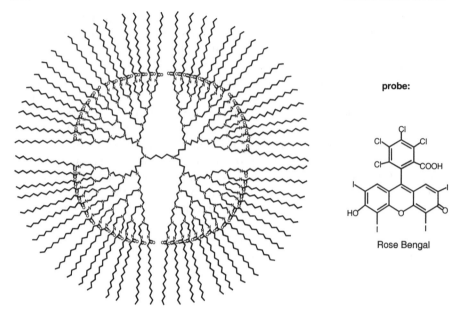

probe:

Rose Bengal

Figure 13.4 Inverted 'unimolecular micelle' type dendrimer (**3**) reported by Meijer and the probe

hydrophilic guest molecules such as Rose Bengal. As determined by UV spectroscopy, one dendrimer **3** could encapsulate up to 7 Rose Bengal molecules. Binding of hydrophilic guest molecules within the hydrophilic dendrimer interior was achieved in either ethanol or *n*-hexane solution. The encapsulated guests could be released by the addition of toluene, but not by the addition of water.

Turro and co-workers [7] studied a series of amine-terminated PAMAM dendrimers (Gn(Cm), $n = 0$–5, with G1(Cm) shown in Figure 13.5) possessing a hydrophobic aliphatic methylene core (Cm). The hydrophobic fluorescent probe, Nile Red (phenoxazon-9), fluoresces intensely in the presence of hydrophobic lipids and organic solvents, but gives weak emission in aqueous media. Judged by the fluorescence emission spectra and polarization values of the probe, it was concluded that the dendrimer in this series with methylene chain core (where (C)m = 12) dendrimers provided a hydrophobic environment to the probe, whereas such behavior was not observed for dendrimers Gn(C4) and Gn(C8) with shorter chains [8]. The accessibility of the probe to the methylene chain core of Gn(C12) was the highest for generation = 1 with core (C12) compared with higher generations. This is undoubtedly due to an idealized balance between the spacial and hydrophobic character of the core relative to the steric demands of the higher generation shells. Further, a more extended structure of Gn(C12) was achieved by changing the pH value from pH 9 to pH 7, presumably

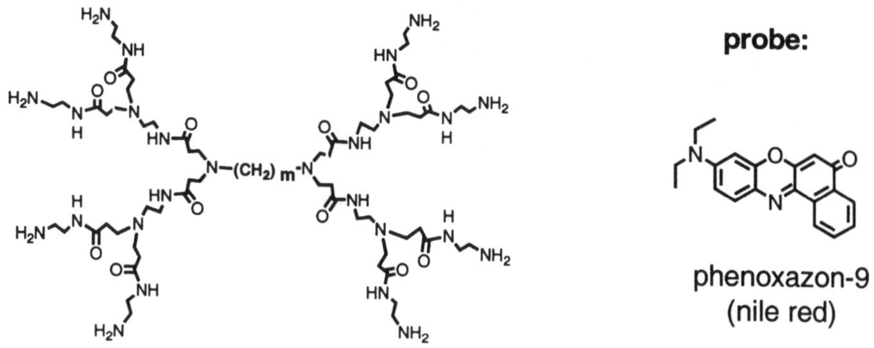

probe:

phenoxazon-9
(nile red)

Figure 13.5 PAMAM dendrimer G1(Cm) (with an aliphatic methylene chain as the central core) and the probe

caused by higher protonation of surface and internal amino groups of dendrimers.

Binding of pyrene to amine-terminated PAMAM dendrimers (G0–G2) with an ethylenediamine core was investigated by Malliaris and co-workers [9]. In the aqueous solution of these dendrimers, the solubility of pyrene increased linearly with the size of the dendrimers as determined by UV-Vis spectroscopy. In addition, the I_1/I_3 ratio of pyrene, which is the intensity ratio of the first to the third vibrational peak, in the pyrene fluorescence spectrum, was used as an index of the micropolarities at the solubilization sites of pyrene in these dendrimers. The I_1/I_3 value of pyrene decreases when the polarity of its environment decreases, from *ca.* 1.6 in water to *ca.* 0.6 in *n*-hexane. The I_1/I_3 values of pyrene in G0 and G1 solutions were close to that observed in pure water (*ca.* 1.5–1.6), whereas in G2 solution it was close to that in aqueous micelles (*ca.* 1.0–1.4). This was attributed to less water penetration into the microcavities of G2 compared to G0 and G1. It was observed that pyrene fluorescence was quenched by adsorption onto G2. This quenching was attributed to the tertiary amino groups in dendrimer interior and thus supported the proposal that pyrene was associated with the interior of these dendrimers. Finally, excimer emission of pyrene was observed in a situation where there was negligible probability of two pyrene molecules occupying a single dendrimer, possibly caused by the exciplex formation between pyrene and some part of the dendrimer interior.

Diederich and coworkers [10] synthesized so-called 'dendrophanes' (Figure 13.6) containing a paracyclophane core embedded in dendritic poly(ether-amide) shells. X-ray crystal-structure analysis indicated that these dendrimers had an open cavity binding site in the center, suitable for the binding of aromatic guests. NMR and fluorescence titration experiments revealed a site specific binding between these dendrimers and 6-(*p*-toluidino)naphthalene-2-sulfonate (TNS) with a 1:1 association. Also, the fluorescence spectral shift of TNS, which is

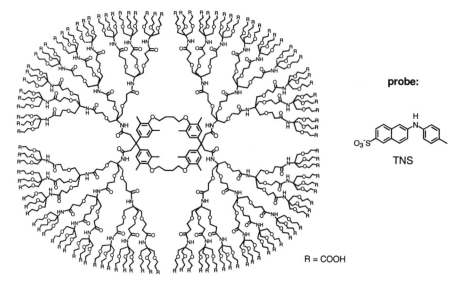

Figure 13.6   Dendrophanes synthesized by Diederich and probe TNS

sensitive to the environmental polarity, suggested that the microenvironment around the cavity binding site becomes less polar with increasing dendrimer generation shells.

2.1.2   Non-covalent Encapsulation – 'Dendritic Box'

In the preceding section, we reviewed the non-covalent dynamic encapsulation of guest probe molecules within dendrimer interiors. The second case, Scheme 2, involves the physical encapsulation of guest molecules; wherein, guest molecules are 'locked' inside dendritic containers (so-called dendritic boxes). This concept was originally proposed by Tomalia *et al.* and referred to as 'unimolecular encapsulation' [2]. More recent and well characterized examples have now been demonstrated by Meijer and co-workers [11–15].

These dendritic boxes (Figure 13.7) were synthesized by the conjugation of a chiral shell of protected amino acids onto a flexible poly(propylene imine) dendrimer with 64 amino end groups. In solution, the shell was highly hydrogen-bonded and dense-packed, displaying a solid-phase behavior, which was indicated by the low NMR relaxation time of the surface groups [11].

The encapsulation of small guest molecules was achieved by constructing the dendrimer shell in the presence of the guest molecules followed by extensive dialysis to remove free guest molecules in solution [11]. A variety of probe molecules were applied, including Rose Bengal, 7,7,8,8,-tetracyano-quino-dimethane (TCNQ) and 3-carboxy-PROXYL. UV-Vis, fluorescence and EPR

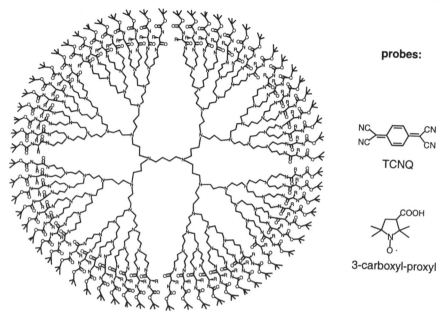

probes:

TCNQ

3-carboxyl-proxyl

Figure 13.7 Meijer's dendritic box and probes

studies of these probe molecules confirmed the entrapment of guest molecules inside the dendritic boxes. The number of entrapped guests per dendritic box could reach up to eight, depending on the architecture of the dendritic box (such as the dendrimer generation and the surface amino acid derivatives), as well as the shape of the guest.

The diffusion of entrapped guest molecules out of dendritic boxes was unmeasurably slow over a period of several months. However, Meijer and coworkers have demonstrated that the shape-selective liberation of encapsulated guests could be achieved by removing the 'closed shell' in two steps (Scheme 3) [15]. First, hydrolysis of the tBOC groups of the dense shell with formic acid gave a 'partly open dendritic box'. At this point, small guests such as p-nitrobenzoic acid and nitrophenol could diffuse out of the box by dialysis. Susequent and complete removal of the outer shell by refluxing with 12 N HCl, lead to the

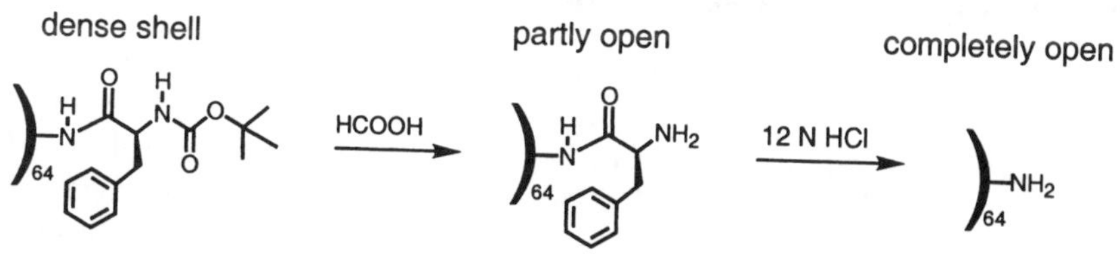

dense shell   partly open   completely open

HCOOH →   12 N HCl →

Scheme 3 Two step opening of the dense shell of dendritic boxes release of bigger guests such as Rose Bengal, Rhodamine B, and methylene violet. The starting poly(propylene imine) dendrimer was recovered in 50–70% yield.

Although each individual amino acid group in the outer shell was enantiomerically pure, the dendritic boxes displayed negligible optical activity. This was attributed to the internal compensation effects caused by the rigid, dense-packing conformation of the shell [13]. However, induced circular dichoism (CD) signals, originating from transfer of chirality from the dendritic box to an entrapped achiral guest, were observed for several entrapped achiral guests. Even more interestingly, the exciton-coupled CD spectrum was obtained in the case of 4 Rose Bengal molecules entrapped in one dendritic box. This indicated the close proximity of the entrapped Rose Bengal molecules with a certain fixed orientation. However, it was not clear why the induced chirality was not present for some of the guest molecules. Furthermore, it was noticed that the induced CD signals were dependent on the solvent used and the history of the sample [14].

2.2 SURFACE BINDING OF PROBE MOLECULES

The surface morphologies of PAMAM dendrimers have been studied extensively by Turro and co-workers [16–23]. As shown in Scheme 4, one approach was to study the adsorption of organic dye molecules and metal complexes on the dendrimer surface by UV-Vis and fluorescence spectroscopy; another approach took advantages of electron transfer processes between two adsorbed species on a single dendrimer surface or between the adsorbed species on a dendrimer surface and other species in aqueous solution.

Scheme 4 Approaches to study the surface morphologies of PAMAM dendrimers

2.2.1 Binding Properties of Probe Molecules

Pyrene was used as a fluorescent probe to sense various hydrophobic sites in the microheterogeneous architecture offered by PAMAM dendrimers, possessing an ammonia core and sodium carboxylated surface (Gn.5, $n = 0$–9) [17]. The I_3/I_1 ratio of pyrene in the presence of low generations (G0.5–G3.5) remained very similar to those in pure water. In the presence of higher generation dendrimers, however, pyrene sensed a more hydrophobic outer surface which was presumably congested (Figure 13.8). From a study of dendrimer-surfactant complexation probed by the pyrene fluorescence, lower generations behaved as ordinary electrolytes and higher generations acted as a novel type of polyelectrolytes. For the later generations, the addition of dodecyltrimethylammonium bromide (DTAB) first occurred in a noncooperative manner via electrostatic binding, but at higher concentration, cooperative binding caused aggregation of surfactants to form micellar structures on the dendrimer surface (Scheme 5). This cooperative binding was suggested to result from alkyl chain association which was induced by the closely packed charged groups on the surface of higher generation dendrimers. Overall, the break in the dendrimer behavior sensed by pyrene coincided with the predicted change in the morphology of the dendrimer structures by simulation [1, 2].

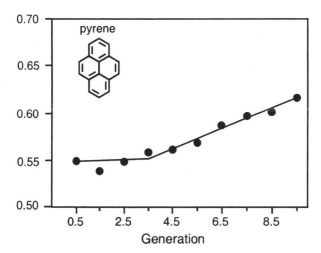

Figure 13.8 I_3/I_1 ratio of pyrene emission as a function of dendrimer generation. Modified from ref. 17

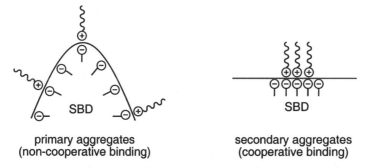

Scheme 5 Aggregates formed between DTAB and a PAMAM dendrimer

The photophysical parameters (such as luminescence lifetime τ) of ruthenium polypyridyl complexes ($Ru(bpy)_3^{2+}$ and $Ru(phen)_3^{2+}$, bpy = 2,2'-bipyridine, phen = 1,10-phenanthroline) have also been used to characterize the surface environment of these dendrimers [16, 18, 21, 23]. NMR experiments suggested that the $Ru(L)_3^{2+}$ complexes were bound to the dendrimer surface, presumably from electrostatic interactions with the negatively charged dendrimer surface. The lifetimes of $Ru(bpy)_3^{2+}$ (Figure 13.9) were indistinguishable in the presence of and absence of the earlier generation dendrimers (G0.5–G2.5); while they were strongly enhanced by the later generations (G3.5–G9.5). This indicated a qualitative difference in the binding of $Ru(L)_3^{2+}$ complexes to the earlier and later generation dendrimers. The binding constant of $Ru(phen)_3^{2+}$ to dendrimer G4.5 was 5.0×10^5 M^{-1}, determined from changes in the excited state lifetime of the probe as a function of dendrimer concentration. It decreased to 2300 and 750 M^{-1} with the addition of 0.24 and 0.48 M NaCl respectively, consistent with the electrostatic interactions involved. Similar results were found in the binding of $Ru(4,7-(SO_3C_6H_5)_2-phen)_3^{4-}$ to the positively charged PAMAM dendrimers possessing amine surface groups (Gn, $n = 1–7$) [22]. The determined binding constant of the probe to G4 was 4.7×10^6 M^{-1} in water and decreased with the addition of NaCl.

Adsorption and aggregation of organic dye molecules on anionic PAMAM dendrimers with carboxylate surface groups were studied by UV-Vis and fluorescence spectroscopy [20]. The aggregation of methylene blue (MB) depended strongly on the dendrimer generation: for later generations, the aggregation occurred more readily and the number of MB molecules involved in one aggregate was larger. It was proposed that MB molecules stacked perpendicular to the dendrimer surface. Again, a qualitative break in the binding behavior of the dye to the dendrimer surface was observed between G2.5 and G3.5 (Figure 13.10). The same trend was observed in the binding of fluorescein to the cationic

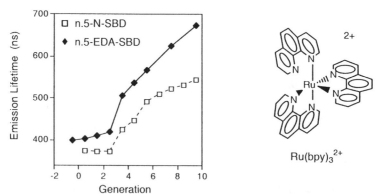

Figure 13.9 Emission lifetimes of $Ru(bpy)_3^{2+}$ (20 μM) monitored in air saturated aqueous solutions of PAMAM dendrimers (20 μM) as a function of generation. Adapted from ref. 23

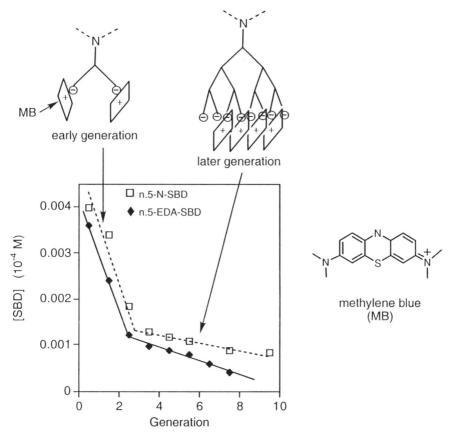

Figure 13.10 Concentration of PAMAM dendrimers in surface groups at the minimum optical absorption (666 nm) as a function of generation. Adapted from ref. 23

PAMAM dendrimers, and in the binding of probes to PAMAM dendrimers with an ethylenediamine core [23].

2.2.2 Photoinduced Electron Transfer Processes on Dendrimer Surface

Photoinduced electron transfer processes between $Ru(L)_3^{2+}$ and several quenchers (Scheme 6) have been investigated in the presence of anionic PAMAM dendrimers [16, 18, 19, 23].

In the presence of earlier generations, the luminescence quenching of aqueous solutions of $Ru(L)_3^{2+}$ by methyl viologen (Figure 13.11) was found to follow Stern–Volmer bimolecular kinetics with a quenching constant $k_q \sim 5 \times 10^9$ $M^{-1}s^{-1}$, typical for the bimolecular quenching in homogeneous solutions. This

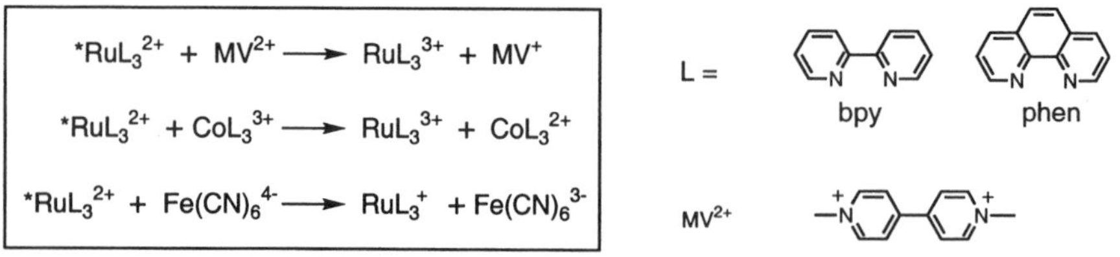

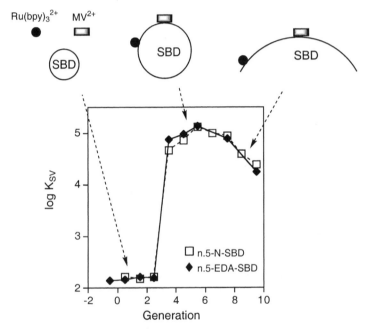

$$*RuL_3^{2+} + MV^{2+} \longrightarrow RuL_3^{3+} + MV^+$$

$$*RuL_3^{2+} + CoL_3^{3+} \longrightarrow RuL_3^{3+} + CoL_3^{2+}$$

$$*RuL_3^{2+} + Fe(CN)_6^{4-} \longrightarrow RuL_3^+ + Fe(CN)_6^{3-}$$

Scheme 6 Electron transfer quenching of the excited state of $Ru(L)_3^{2+}$ by different quenchers

Figure 13.11 Stern–Volmer constant K_{SV} for the quenching of $*Ru(bpy)_3^{2+}$ (5 μM) by MV^{2+} in air-saturated aqueous solutions of PAMAM dendrimers. Adapted from ref. 23

revealed the weak interactions between the probes and the anionic dendrimers, as well as the rapid exchange between the adsorbed probes and the probes in the aqueous phase during the lifetime of $Ru(L)_3^{2+}$. However, the quenching process in later generation dendrimer solutions (G3.5 and up) obeyed a general kinetic model employed for micellar solutions. In these cases, the quenching was found to be unimolecular (or, namely, intradendrimer) in nature, indicating that the probes did not leave the dendrimer surface during their luminescence lifetime. The quenching constant k_q (c. 10^6–10^7 s^{-1}) decreased with increasing dendrimer size, presumably because of the slower diffusion rate of the probes on the dendrimer surface. Quenching studies using other quenchers such as $[Fe(CN)_6]^{4-}$ and $Co(phen)_3^{3+}$ substantiated this conclusion.

2.3 PROBE COVALENTLY LINKED ON THE DENDRIMER SURFACE

In order to study the molecular dynamics of the outer segments of a dendrimer, one pyrene moiety was selectively and covalently attached to one dendron of poly(aryl ester) dendrimers by Adams (in total three pyrene molecules per dendrimer) [24]. The fluorescence decay of pyrene in the THF solution of the labeled dendrimers provided details of the pyrene excimer formation, such as the excimer formation rate, the excimer decomposition rate constant and the equilibrium constant of the excimer formation. These parameters were utilized to evaluate the diffusional mobility of the dendrimer branches.

In another case, a fluorescein moiety (Figure 13.12) was covalently linked to the surface of the amine-terminated PAMAM dendrimers with average labeling density less than one probe per dendrimer [25]. The diffusion coefficients and the hydrodynamic sizes of the labeled dendrimers in aqueous solutions were measured by fluorescence recovery after photobleaching, and compared with those of unlabeled dendrimers determined by dynamic light scattering. Despite the small perturbation (a small decrease in the dendrimer diameter) upon the labeling of the dendrimers, the authors suggested that these fluorescently labeled dendrimers could make suitable markers and diffusion probes in different environment.

2.4 PROBE COVALENTLY LINKED AT THE CENTER OF THE DENDRIMER

Fréchet *et al.* [26] studied the microenvironment in dendritic molecules by covalently attaching a solvatochromic probe; namely: *N*-methylamino-*p*-

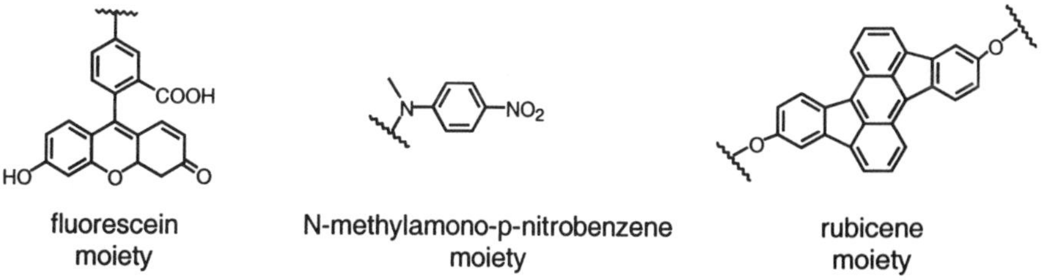

| fluorescein
moiety | N-methylamono-p-nitrobenzene
moiety | rubicene
moiety |

Figure 13.12 Functional moieties used as photophysical probes

nitrobenzene (Figure 13.12) to the center of the dendritic polyethers. They studied the solvatochromic behavior of various generations of labeled dendrimers in solvents varied as a function of polarity. For solvents of low polarity, there was a dramatic change in the absorption maximum for the probe from G3 to G4, consistent with a transition from an open, extended to a more globular dendrimer structure. The probe also sensed a high polar environment for dendrimers at generations higher than G4.

A similar approach was taken by Moore [27], utilizing poly(phenyl acetylene) dendrimers with a dimethylbenzene moiety attached at the core of the dendrimer. An anomalous shift (41 nm) in the fluorescence spectra of the probe in various nonpolar hydrocarbon solvents was observed for G5 and G6, but not for G1 to G4. This observation confirmed significant in the size and shape changes for these dendrimers between G4 and G5.

A series of poly(aryl ether) dendrimers with a Rubicene moiety (Figure 13.12) as the dendrimer core were synthesized by De Schryver and co-workers [28]. Time-resolved fluorescence anisotropy of the Rubicene core was used to determine the hydrodynamic volume of dendrimers in a variety of solvents. These dendrimers were fully extended in a good solvent – toluene, indicated by the same hydrodynamic volume between 20 and 94 °C. In a poor solvent, acetonitrile, the hydrodynamic volume of dendrimers was substantially smaller, especially for the higher generations. It was further proposed that these dendrimers could adopt a conformation where the dendrons interact with the dendrimer core. They also demonstrated the possibility of detecting a single dendrimer molecule embedded in a thin polystyrene film, utilizing poly(aryl ether) dendrimers with a fluorescent dihydropyrrolopyrroledione core.

Aida and co-workers prepared poly(aryl ether) dendrimers with a zinc tetraphenyl porphyrin as the core (Figure 13.13) [29, 30]. A clear hypochromicity was observed for the aqueous solution of $[KO_2C]_{32}L4PZn$ (**4**) upon increasing the ionic strength or lowering the pH value. This was attributed to the shrinkage of the hydrophobic dendrimer framework when the surface became less ionic. $[KO_2C]_8L2PZn$ had a very open structure, thus the porphyrin core could be accessed by quencher molecules such as methyl viologen. Interestingly, while the zinc porphyrin core within **4** was sterically shielded by the dendritic shell from access by the quenchers, the quenching behavior suggested a long-range photoinduced electron transfer process through the dendrimer framework between the porphyrin core and the electrostatically adsorbed methyl viologen on the dendrimer surface.

Furthermore, they investigated the interpenetrating interaction between methyloxy-terminated dendrimers LnPZn (i.e. **5**) and dendritic imidazoles (LnIm: i.e. **7**) [31]. Spectral changes in the Soret absorption band of LnPZn was used to calculate the binding constants. Results indicated a 1 : 1 complexation between LnPZn and LnIm. The binding constant decreased when the generation number increased, with a notable gap between L3PZn and L4PZn.

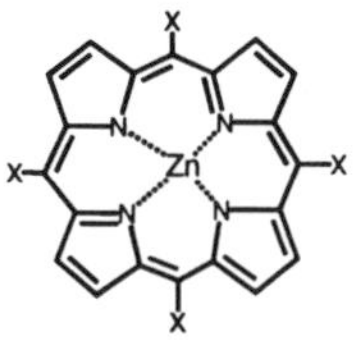

4: [KO$_2$C]$_{32}$ L$_4$PZn (R = CO$_2$K)
5: L$_4$PZn (R = OMe)
6: L$_4$PZn (R = H)

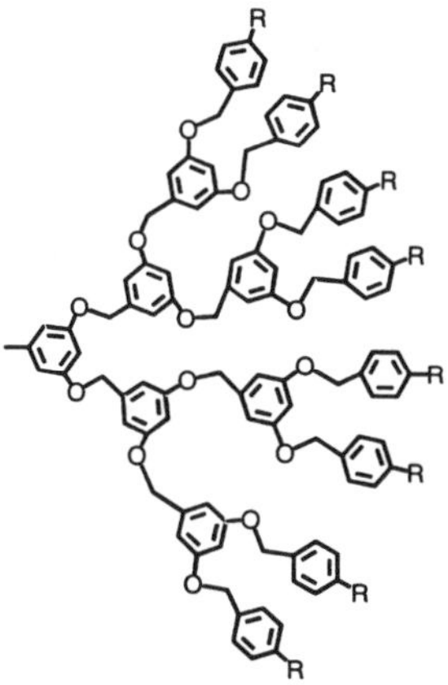

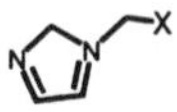

7: L$_4$Im (R = OMe)

Figure 13.13   Poly(aryl ether) dendrimers with a porphyrin core and dendritic imidazoles

Fréchet and co-workers [32] studied the ability of the dendrimer shell to provide site isolation of the core porphyrin moiety, using benzene-terminated dendrimers Zn[G-n]$_4$P (i.e. **6**). From the cyclic voltammograms in CH$_2$Cl$_2$, the interfacial electron transfer rate between the porphyrin core and the electrode surface decreased with increasing dendrimer generation. However, small molecules like benzyl viologen could still penetrate the shell of **6** to access the porphyrin core as observed from the quenching of porphyrin fluorescence. Their results also revealed that the dendritic shell did not interfere electrochemically or photochemically with the porphyrin core moiety.

3   OTHER PHOTOCHEMICAL CHARACTERIZATION

3.1   PHOTORESPONSIVE DENDRIMERS

3.1.1   Photoswitchable Dendrimers

Vögtle and co-workers first reported a photoswitchable dendrimer [33] with six peripheral azobenzene groups, which took advantage of the efficient and fully reversible photoisomerization reaction of azobenzene-type compounds (Scheme 7). In a follow-up study [34], poly(propylene imine) dendrimers bearing azobenzene moieties (p-Im-Gn, n = 1–4) on the periphery were synthesized. These dendrimers displayed similar photoisomerization properties as the azobenzene monomers. Irradiation of the all-E azobenzene dendrimers at 313 nm led to the Z-form dendrimers, while irradiation at 254 nm or heating could convert the Z-form dendrimers back to the E-form dendrimers. The observation that the

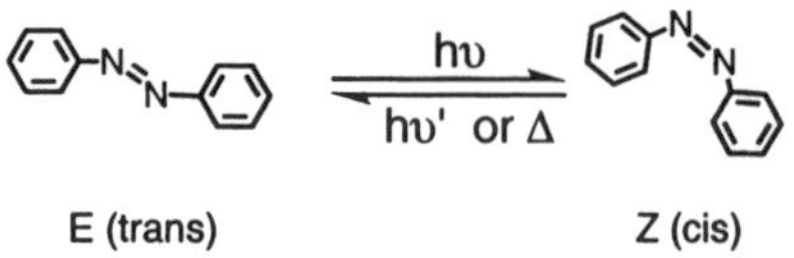

E (trans)   Z (cis)

Scheme 7   Reversible photoisomerization of azobenzene

quantum yield of the photoisomerization did not depend on the dendrimer generation, indicated the negligible steric constraint of the photoisomerization for dendrimers up to *p-Im*-G4.

They also investigated the quenching of the azobenzene dendrimer fluorescence by Eosin Y (2',4',5',7'-tetrabromofluorescein dianion) [35]. It was concluded that Eosin Y was hosted in these dendrimers and Z-form dendrimers were more efficient hosts than E-form dendrimers.

McGrath and Junge [36] reported a photoresponsive poly(aryl ether) dendrimer with azobenzene as the dendrimer core. These dendrimers exhibited reversible *trans* to *cis* photoisomerization by irradiation at 350 nm. The authors proposed the use of this type of dendrimer as novel photoswitchable transport vectors. This is based on the expected ability of dendrimers to encapsulate or eject small molecules reversibly upon light perturbation.

A similar type of azobenzene dendrimers was reported by Jiang and Aida [37]. Interestingly, the aryl groups in such a dendrimer were found to be able to simultaneously absorb five infrared photons and then transfer the energy to the azobenzene core moiety to cause *trans/cis* isomerizations.

3.1.2   Dendritic Antennae

Stewart and Fox [38] first reported a study on light harvesting antenna using chromophore-labeled dendrons (Figure 13.14). These dendrons have an electron acceptor (C_2: 3-[dimethylamino]phenoxy group) attached at the focal point and electron donors (C_1: pyreneyl or naphthyl groups) attached at the periphery. Detailed studies on quenching of the electron donor fluorescence by the acceptor suggested significant electronic coupling between appended chromophores and quencher. This electron transfer appears to occur intramolecularly across the dendrimer framework.

Moore and coworkers [39, 40] designed perylene-terminated poly(phenyl acetylene) monodendrons (Figure 13.15) which could efficiently transfer light energy absorbed by the phenylacetylene chromophores to the perylene moiety at the dendron focal point. Determined by steady-state and time-resolved fluorescence spectroscopy, the light-harvesting ability increased with generation while the energy transfer efficiency decreased with generation. Dendrimer **8** had an energy transfer quantum yield of 98% and energy transfer rate constant of 1.9×10^{11} s^{-1}. This was explained by the presence of an electronic energy

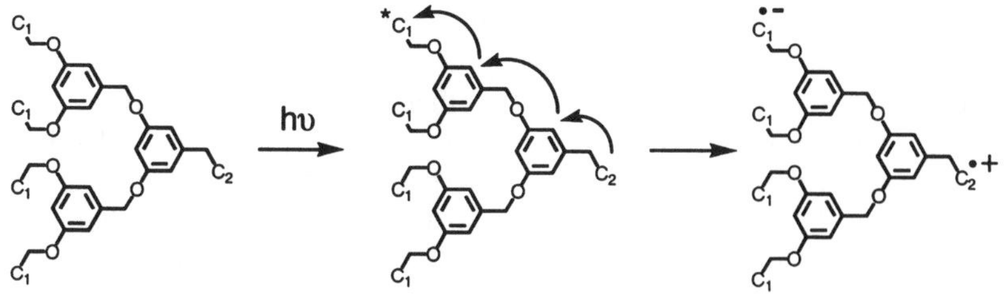

Figure 13.14 Light harvesting antenna reported by Fox and coworkers

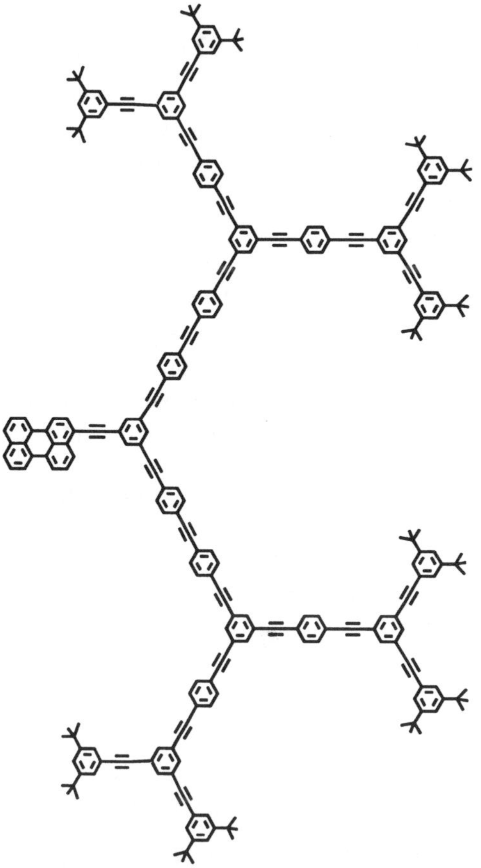

Figure 13.15 Dendritic antenna (**8**) reported by Moore

gradient in the dendrimer framework, which facilitated the directional transduction of excitation energy to the focal point.

Jiang and Aida [41] demonstrated the efficient intramolecular energy transfer from the dendron subunits (aryl groups) to the porphyrin core using dendrimer **6**. The tetrasubstituted (L5)4P with a spherical morphology exhibited a much higher energy transfer quantum yield (80.3%) compared with other partially substituted dendrimers. Kawa and Fréchet [42] reported a similar example of an 'antenna effect' observed in self-assembled lanthanide-cored dendrimer complexes. A review of dendritic antennae and light harvesting dendrimers [43] has appeared recently.

3.2 METAL NANOCOMPOSITES STABILIZED BY DENDRIMERS

UV-Vis spectroscopy has been utilized to characterize the amount of copper(II) ions bound inside PAMAM dendrimers, as well as the size of resulting Cu nanocomposites upon chemical reduction [44–46]. This work has been reviewed extensively by Balogh and Tomalia [50].

3.3 ELECTRONIC CONDUCTING DENDRIMERS

Near-infrared (NIR) absorption spectroscopy has been used to characterize the delocalized π-stacks on electronic conducting dendrimers by Miller and co-workers [47–49]. These dendrimers were prepared by peripherally modifying PAMAM dendrimers with cationically substituted naphthalene diimides, and then reduced with one electron per imide group to convert each imide into its anion radical. The π-stacking of these radical anions on these dendrimer surfaces was indicated by an absorbance band beyond 2000 nm in the NIR spectra.

This work was supported in part by the MRSEC Program of the National Science Foundation under Award Number DMR-98-09687 and by NSF grant CHE98-12676 to the authors at Columbia University.

4 REFERENCES

1. Naylor, A. M. and Goddard, W. A., III *J. Am. Chem. Soc.*, **111**, 2339 (1989).
2. Tomalia, D. A., Naylor, A. M. and Goddard, W. A., III *Angew. Chem. Int. Edit. Engl.*, **29**, 138 (1990).
3. de Gennes, P. G. and Hervet, H. *J. Phys. Lett.*, **44**, 351 (1983).
4. Newkome, G. R., Moorefield, C. N., Baker, G. R., Saunders, M. J. and Grossman, S. H. *Angew. Chem. Int. Edit. Engl.*, **30**, 1178 (1991).
5. Hawker, C. J., Wooley, K. L. and Fréchet, J. M. J. *J. Chem. Soc. Perkin Trans.*, **1**, 1287 1993.
6. Stevelmans, S., van Hest, J. C. M., Jansen, J. F. G. A., van Boxtel, D. A. F. J., de

Brabander-van den Berg, E. M. M. and Meijer, E. W. *J. Am. Chem. Soc.*, **118**, 7398 (1996).

7. Watkins, D. M., Sayed-Sweet, Y., Klimash, J. W., Turro, N. J. and Tomalia, D. A. *Langmuir*, **13**, 3136 (1997).
8. Sayed-Sweet, Y., Hedstrand, D. M., Spindler, R. and Tomalia, D. A. *J. Mater. Chem.*, **7**(7), 1999 (1997).
9. Pistolis, G., Malliaris, A., Paleos, C. M. and Tsiourvas, D. *Langmuir*, **13**, 5870 (1997).
10. Mattei, S., Seiler, P. and Diederich, F. *Helv. Chim. Acta*, **78**, 1904 (1995).
11. Jansen, J. F. G. A., de Brabander-van den Berg, E. M. M. and Meijer, E. W. *Science*, **266**, 1226 (1994).
12. Jansen, J. F. G. A. and Meijer, E. W. *J. Am. Chem. Soc.*, **117**, 4417 (1995).
13. Jansen, J. F. G. A., Peerlings, H. W. I. and de Brabander-van den, E. M. M. *Angew. Chem. Int. Edit. Engl.*, **34**, 1206 (1995).
14. Jansen, J. F. G. A., de Brabander-van den Berg, E. M. M. and Meijer, E. W. *Recl. Trav. Chim. Pays-Bas-J. Roy. Neth. Chem. Soc.*, **114**, 225 (1995).
15. Jansen, J. F. G. A., Meijer, E. W. and de Brabander-van den Berg, E. M. M. *Macromol. Symp.*, **102**, 27 (1996).
16. Morenobondi, M. C., Orellana, G., Turro, N. J. and Tomalia, D. A. *Macromolecules*, **23**, 910 (1990).
17. Caminati, G., Turro, N. J. and Tomalia, D. A. *J. Am. Chem. Soc.*, **112**, 8515 (1990).
18. Turro, N. J., Barton, J. K. and Tomalia, D. A. *Accounts Chem. Res.*, **24**, 332 (1991).
19. Gopidas, K. R., Leheny, A. R., Caminati, G., Turro, N. J. and Tomalia, D. A. *J. Am. Chem. Soc.*, **113**, 7335 (1991).
20. Jockusch, S., Turro, N. J. and Tomalia, D. A. *Macromolecules*, **28**, 7416 (1995).
21. Turro, C., Niu, S. F., Bossmann, S. H., Tomalia, D. A. and Turro, N. J. *J. Phys. Chem.*, **99**, 5512 (1995).
22. Schwarz, P. F., Turro, N. J. and Tomalia, D. A. *J. Photochem. Photobiol. A-Chem.*, **112**, 47 (1998).
23. Jockusch, S., Ramirez, J., Sanghvi, K., Nociti, R., Turro, N. J. and Tomalia, D. A. *Macromolecules*, **32**, 4419 (1999).
24. Wilken, R. and Adams, J. *Macromol. Rapid Commun.*, **18**, 659 (1997).
25. Yu, K. and Russo, P. S. *J. Polym. Sci. Pt. B-Polym. Phys.*, **34**, 1467 (1996).
26. Hawker, C. J., Wooley, K. L. and Fréchet, J. M. *J. Am. Chem. Soc.*, **115**, 4375 (1993).
27. Devadoss, C., Bharathi, P. and Moore, J. S. *Angew. Chem. Int. Edit. Engl.*, **36**, 1633 (1997).
28. De Backer, S., Prinzie, Y., Verheijen, W., Smet, M., Desmedt, K., Dehaen, W. and De Schryver, F. C. *J. Phys. Chem.*, **A**, **102**, 5451 (1998).
29. Jin, R. H., Aida, T. and Inoue, S. *J. Chem. Soc. Chem. Commun.*, 1260 (1993).
30. Sadamoto, R., Tomioka, N. and Aida, T. *J. Am. Chem. Soc.*, **118**, 3978 (1996).
31. Tomoyose, Y., Jiang, D. L., Jin, R. H., Aida, T., Yamashita, T., Horie, K., Yashima, E. and Okamoto, Y. *Macromolecules*, **29**, 5236 (1996).
32. Pollak, K. W., Leon, J. W., Fréchet, J. M. J., Maskus, M. and Abruna, H. D. *Chem. Mater.*, **10**, 30 (1998).
33. Mekelburger, H.-B., Rissanen, K. and Vögtle, F. *Chem. Ber.*, **126**, 1161 (1993).
34. Archut, A., Vögtle, F., De Cola, L., Azzellini, G. C., Balzani, V., Ramanujam, P. S. and Berg, R. H. *Chem. Eur. J.*, **4**, 699 (1998).
35. Archut, A., Azzellini, G. C., Balzani, V., De Cola, L. and Vögtle, F. *J. Am. Chem. Soc.*, **120**, 12187 (1998).
36. Junge, D. M. and McGrath, D. V. *Chem. Commun.*, 857 (1997).
37. Jiang, D. L. and Aida, T. *Nature*, **388**, 454 (1997).
38. Stewart, G. M. and Fox, M. A. *J. Am. Chem. Soc.*, **118**, 4354 (1996).

39. Devadoss, C., Bharathi, P. and Moore, J. S. *J. Am. Chem. Soc.*, **118**, 9635 (1996).
40. Shortreed, M. R., Swallen, S. F., Shi, Z. Y., Tan, W. H., Xu, Z. F., Devadoss, C., Moore, J. S. and Kopelman, R. *J. Phys. Chem. B*, **101**, 6318 (1997).
41. Jiang, D. L. and Aida, T. *J. Am. Chem. Soc.*, **120**, 10895 (1998).
42. Kawa, M. and Fréchet, J. M. *J. Chem. Mater.*, **10**, 286 (1998).
43. Adronov, A. and Fréchet, J. M. *J. Chem. Commun.*, 1701 (2000).
44. Zhao, M. Q., Sun, L. and Crooks, R. M. *J. Am. Chem. Soc.*, **120**, 4877 (1998).
45. Balogh, L. and Tomalia, D. A. *J. Am. Chem. Soc.*, **120**, 7355 (1998).
46. Esumi, K., Suzuki, A., Aihara, N., Usui, K. and Torigoe, K. *Langmuir*, **14**, 3157 (1998).
47. Miller, L. I., Hashimoto, T., Tabakovic, I., Swanson, D. R. and Tomalia, D. A. *Chem. Mater.*, **7**, 9 (1995).
48. Miller, L. L., Duan, R. G., Tully, D. C. and Tomalia, D. A. *J. Am. Chem. Soc.*, **119**, 1005 (1997).
49. Tabakovic, I., Miller, L. L., Duan, R. G., Tully, D. C. and Tomalia, D. A. *Chem. Mater.*, **9**, 736 (1997).
50. Balogh, L., Tomalia, D.A. and Hagnauer, G. L. *Chem. Innov.*, **30**(3), 19 (2000).

14

Rheology and Solution Properties of Dendrimers

P. R. DVORNIC
Michigan Molecular Institute, Midland, MI, USA

S. UPPULURI
Flint Ink Corporation, Research Center, Ann Arbor, MI, USA

1 INTRODUCTION

Rheological behavior is among the most characteristic and at the same time potentially most important technological properties of dendrimers. In many aspects it is unprecedented among macromolecular substances, as it directly reflects some of the unique fundamental features of dendrimer molecular architecture [1]. However, the understanding is still in its infancy, and it has not yet received deserving scientific attention. This is mostly because some of the required testing methods use relatively prohibitive quantities of samples that are either unavailable from typical laboratory-scale dendrimer syntheses, or rather expensive for higher generation commercially available products [2]. Nevertheless, even from a limited number of reported studies, a picture is beginning to emerge which suggests that some of the most promising applications of dendrimers may either directly result from their unique rheological behavior, or at least seriously depend on it. In materials science, these may include utilization of dendrimers for preparation of various nano-structures and/or nano-devices, high-viscosity standards, flow modifiers, lubricants and/or processing aids, nanoscopic toughening and/or reinforcing agents, specialty fluids with electrical, magnetic, or optical properties, multicomponent blend compatibilizers, sizing agents, etc. In the biomedical areas, rheological behavior of dendrimers in blood and other bodily fluids may directly or indirectly predetermine their fate as

Dendrimers and Other Dendritic Polymers. Edited by Jean M. J. Fréchet and Donald A. Tomalia
© 2001 John Wiley & Sons Ltd

potential drug carriers, release substances, gene transfer or cell-targeting agents, anti-viral decoy molecules, vaccine components, etc.

Until the present, rheology of dendrimers has developed mostly as an analytical probe of their molecular and morphological structure. Indeed, it has been shown that many rheological properties of these unique polymers clearly and directly depend on their size, shape, inherent flexibility, molecular weight, density of end-group functionality, intramolecular morphology and intermolecular interactions. Therefore, in this chapter we will also assume a structural perspective of the subject, beginning with a brief overview of selected fundamental molecular parameters that predetermine dendrimer rheological behavior. Following this, their rheological properties are discussed in three separate sections dealing with different states of interdendrimer interactions. The first of these sections treats isolated dendrimer molecules in good solvents as examined by dilute solution viscometry; the second deals with the rheology of concentrated dendrimer solutions and gradually increasing interdendrimer interactions; and the third focuses on the dendrimer bulk state where individual molecules are subjected to intermolecular forces that are strong enough to cause significant molecular deformation. With only a few notable exceptions, the relevant data are presently available for only three compositionally different dendrimer families: polyamidoamines (PAMAMs), polypropyleneimines (PPIs) and poly(benzyl ether)s (PBzEs). However, while all three of these families were examined by dilute solution viscometry and in the bulk state, only PAMAMs were studied in concentrated solutions. Throughout the chapter, rheological results are complemented with selected data from other relevant methods. However, because of serious space limitations only some of the latter could be included, and apologies are offered to all whose work had to be regretfully omitted.

2 ARCHITECTURAL FEATURES OF DENDRIMER MOLECULES THAT AFFECT THEIR RHEOLOGICAL BEHAVIOR

Dendrimers are globular, nano-scaled macromolecules that consist of two or more tree-like dendrons emanating from a single central atom or atomic group referred to as the core. The main constitutive element of their molecular architecture are branch cells, which represent the excluded volume associated with repeat units that contain at least one branch juncture [3]. A dendrimer consists of three different types of such branch cells: the core cell, the interior cells and the exterior cells. In an idealized case of complete and perfect connectivity (i.e. in ideal dendrimer), these branch cells are organized in mathematically precise, geometrically progressive arrangements that result in a series of radially concentric layers (called generations) around the core. Consequently, an ideal dendrimer has its molecular size and shape precisely defined by the generation number, chemical composition and functionality of the branch junctures.

Molecular modeling and theoretical calculations have shown that with increase in generation dendrimers evolve from open, dome-shaped molecules to relatively closed spheroids [4] (see Figure 14.1). Based on structural considerations and many experimentally observed structure-property relationships, the former may be considered as simple branched, low molecular weight precursors for truly developed dendrimers that have overgrown a certain generation stage defined by the so-called critical degree of branching [3]. This stage corresponds to the smallest molecule of the homologous series that contains all three types of fundamental branch cells of a dendrimer structure. It is specific for every compositional dendrimer family, but for many, including both PAMAMs and PPIs, it corresponds to generation 2 [3, 5].

During this hierarchical transition from simply branched molecules to fully developed dendrimers, the molecular weight and size discretely increase with generation spanning the range from only several hundred to almost 1 million Daltons, and from about 1 nm (i.e. 10 Å) to almost 7 nm (i.e. 70 Å) [3] in diameter (see Table 14.1). At the same time, the molecular shape changes to spheroidal, but does not seem to reach ideal sphericity even at very high generations (see Figure 14.1) [4, 6]. For example, molecular asphericity of tetradendron PAMAM dendrimers decreases from 0.46 at generation 1 to 0.15 at generation 6, but does not reach the value of zero, characteristic for ideal sphere, even at generation 9 [6]. Consistent with this, the ratio of the two principal

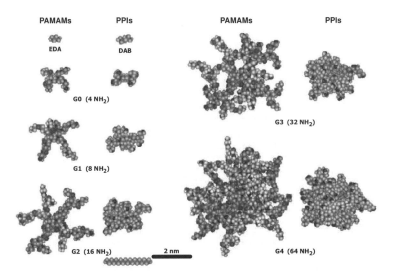

Figure 14.1 Energy minimized CPK models of generation 0 through 4 PAMAM and PPI dendrimers. EDA: ethylene diamine; DAB: diamino butane; G#: generation number; numbers of end groups given in parentheses

Table 14.1 Selected molecular properties of PAMAM, PPI and PBzE dendrimers[f]

Generation	PAMAM Number of end groups	PAMAM Molecular weight	PAMAM R_G/R_H, Å	PAMAM T_g, °C	PPI Number of end groups	PPI Molecular weight	PPI R_G/R_H, Å	PPI T_g, °C	PBzE Number of end groups	PBzE Molecular weight	PBzE R_G/R_H, Å	PBzE T_g, °C
0	4[d]	517	—/7.7[a]	−11	4[d]	317	4.4[c]/6.5[b]	−107	3[e]	576	—/7[b]	9
1	8[d]	1430	—/10.1[b]	−3	8[d]	773	6.9[c]/9.2[b]	−97	6[e]	1212	—/10[b]	25
2	16[d]	3256	—/14.4[b]	0	16[d]	1687	9.3[c]/12.1[b]	−90	12[e]	2484	—/13[b]	36
3	32[d]	6909	16.5[c]/17.5[b]	11	32[d]	3514	11.6[c]/15.4[b]	−87	24[e]	5026	—/17[b]	39
4	64[d]	14215	19.7[c]/25[b]	14	64[d]	7168	13.9[c]/19.8[b]	−84	48[e]	10126	—/21[b]	41
5	128[d]	28826	24.3[c]/27.2[a]	14					96[e]	20292	—/25[b]	42
6	256[d]	58048	30.3[c]/33.7[a]	16					192[e]	40644	—/29[b]	
7	512[d]	116493	35.8[c]/40.5[a]									
8	1024[d]	233383	—/48.5[a]									
9	2048[d]	467162	57[c]/49.2[a]									
10	4096[d]	934720	—/67.5[a]									

[a] Values from size exclusion chromatography (SEC) data obtained at 25°C in 0.1 molar citric acid in water.
[b] Values calculated from dilute solution viscometry (DSV) data: for PAMAMs obtained in the same solution as in a; for PPIs from ref. 19; for PBzEs from ref. 15.
[c] Values obtained from small angle neutron scattering (SANS) data: for PAMAMs in the same solution as in a; for PPIs in D_2O.
[d] All end groups are NH_2.
[e] All end groups are benzylether.
[f] Data for PAMAMs from refs 3, 5, 40 and from Amis, E. et al., presentation at 29th ACS Central Regional Meeting, Midland, MI, 28–30 May, 1997; for PPIs from ASTRAMOL® web page at www.dsm.nl; and for PBzEs from ref. 49.

moments of inertia of the same dendrimers undergoes a stepwise transition from 4.2 at generation 2 to 1.3 at generation 5, but does not reach unity, characteristic for an ideal sphere, even at generation 7 [4].

Thus, dendrimers exhibit a unique combination of (a) high molecular weights, typical for classical macromolecular substances, (b) molecular shapes, similar to idealized spherical particles and (c) nanoscopic sizes that are larger than those of low molecular weight compounds but smaller than those of typical macro-molecules. As such, they provide unique rheological systems that are between typical chain-type polymers and suspensions of spherical particles. Notably, such systems have not been available for rheological study before, nor are there yet analytical theories of dense fluids of spherical particles that are successful in predicting useful numerical results.

3 DILUTE SOLUTION VISCOMETRY

The first rheological study of a dendrimer family was reported in 1982–83 by Aharoni [7] and Aronini and Murthy [8]. It involved dilute solution viscometry, DSV, of poly(α,ε-L-lysine) dendrimers, which had $-\{C(O)-C(H)[(CH_2)_4-N(H)]-N(H)\}$- branch cells [9], and in N,N-dimethylformamide/LiCl showed constant intrinsic viscosity, $[\eta]$, of 2.5 mL/g over nine generations that ranged in nominal molecular weight, M, from 511 to 233 600 [7, 8]. In addition, these dendrimers also showed a linear increase of both hydrodynamic radius, R_H (calculated from viscosity data), and radius of gyration, R_G (obtained from SAXS measurements) with $M^{1/3}$, and had R_G/R_H values of about 0.88 to 1.04 [8]. Concluding that this indicated non-draining, hard-sphere-type behavior [10], and assuming uniform molecular density, the authors calculated that in the dry state they occupied about 52% of their molecular volume (i.e. three times as much as the 17% that is expected for freely draining linear macromolecules of identical molecular weight) [8].

Following this early work, a whole decade passed before new reports on dendrimer rheology started to appear in the literature. During this time, many compositionally new dendrimer families were synthesized, a number of theoretical models of intradendrimer morphology was proposed, and the first data on dendrimer physical properties were reported. Amidst the 'explosion' of interest in these new polymers, an expectation was born that because of their unique molecular architecture, dendrimers should also exhibit highly unusual physical properties, among which the rheological behavior may be a particularly extraordinary one. For example, it was predicted that in contrast to the constant intrinsic viscosity of polylysines, and monotonic, Mark–Houwink-type increase with molecular mass characteristic for more traditional macromolecular architectures such as linear, randomly branched or star-like polymers, dendrimers should exhibit a maximum in the dependence of $[\eta]$ on generation [11, 12]. This resulted from the realization that for ideal dendrimers, molecular volume scales with G^3 (where G is the generation number proportional to the molecular radius which in turn is a linear function of the number of branch cell layers around the core), while molecular mass increases with $N_B{}^G$ (where N_B is the functionality of the branch junctures). Therefore, for small values of G, the volume of dendrimer molecules should grow faster with generation than the molecular weight, until a certain generation is reached at which the rate of increase of the latter overtakes that of the former. As a consequence, generational dependence of dendrimer intrinsic viscosity, $[\eta] = $ molecular volume/molecular mass $\propto G^3/N_B{}^G$ [11], should follow a curve that falls in between those predicted for rigid rods by Zimm's equation and for flexible coils by Fixman's equation [12], and have a maximum at the generation at which G^3 becomes equal to $N_B{}^G$. To account for the obvious deviation of polylysine dendrimers from this behavior, it was suggested [13] that it resulted from the geometrical asymmetry of their branch cells, which contained two arms of unequal lengths (i.e. $-(CH_2)_4NH-$ and $-NH-$) and enabled very close segmental packing into compact interior structure of constant average molecular density. It was also pointed out that because the experimentally determined intrinsic viscosity of these dendrimers exceeded 2.12 mL/g, which is the calculated value for hard spheres resulting from $[\eta] = 5/(2\rho)$ if average molecular density ρ is assumed to be constant and equal to the reported 1.18 g/mL [8], their hydrodynamic volumes were larger than the volumes occupied by their van der Waals radii, indicating substantial solvation by the medium and not completely solid sphere character [14].

Several experimental studies confirmed the existence of maxima in $[\eta]$ vs G relationships in all three of the most investigated dendrimer families: PBzEs [15], PAMAMs [16] and PPIs [17, 18]. Figure 14.2 summarizes some of the reported data, indicating that generational locations of the respective maxima may even be compositionally specific: appearing for PBzE dendrimers at generation 3 [15], for amine terminated PAMAMs at generation 4 [16], and for amine and CN terminated PPIs at generation 5 [17–19]. It was also pointed out that if the reciprocal of intrinsic viscosity is taken as the average hydrodynamic density of solvated molecules [13, 15], dendrimers should exhibit generationally dependent minima of molecular densities at generations at which the corresponding $[\eta]$ vs G curves show their maxima [13]. In support of this view it was shown that the generational dependence of the refractive index increment, dn/dc, of PBzE dendrimers indeed passed through a minimum between generations 1 and 4 (see Figure 14.3) [15]. In addition to this, it also seems that this behavior is in general agreement with the density well model of dendrimer molecular morphology, suggested by Mansfield and Klushin [20], and to some extent by Murat and Grest [21].

Polypropyleneimines have been examined by dilute solution viscometry by several groups of investigators [17–19, 22]. For example, Scherrenberg and

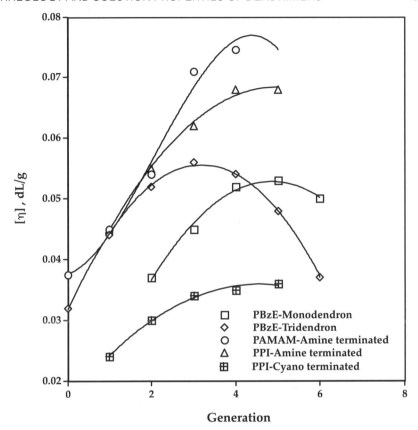

Figure 14.2 Dependence of intrinsic viscosity, $[\eta]$, on generation for PAMAM, PPI and PBzE dendrimers. Data for PBzE monodendrons are included for comparison

co-workers [18] combined DSV with small-angle neutron scattering, SANS, and molecular dynamics modeling to study both NH_2 and CN terminated derivatives in good solvents, including D_2O and methanol-d_4 for the former, and acetone-d_6 and THF-d_8 for the latter. They found linear dependencies of both hydrodynamic radii, R_H (obtained from DSV), and radii of gyration, R_G (obtained from SANS), on $M^{0.37-0.38}$, which implied that both the volume and the mass of these dendrimers increased almost equally fast with generation number. Since R_G/R_H values ranged from 0.70 to 0.79 for NH_2 terminated derivatives, and from 0.77 to 0.85 for CN terminated counterparts, the authors concluded that these dendrimers behaved as compact, space-filling structures with fractal dimensionality of three and some degree of back-folding of their chain ends into interiors. The accompanying molecular dynamics calculations also indicated substantial molecular flexibility and a relatively homogeneous intramolecular

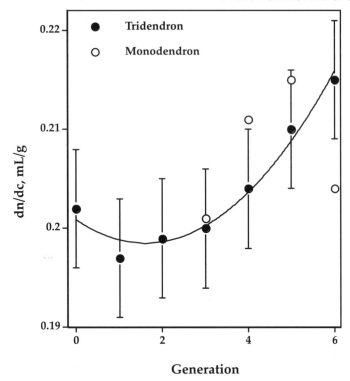

Figure 14.3 Refractive index increment, *dn/dc*, in THF of PBzE monodendrons and tridendron dendrimers as a function of generation number. (according to ref. [15])

density distribution in agreement with the Mansfield-Klushin [20] and Murat–Grest models [21].

In another recent study, Rietveld and Smit examined the first five generations of amine-terminated PPIs in methanol by combined DSV and static low-angle laser light scattering, LALLS, vapor pressure osmometry, VPO [19], and solution densitometry [22]. They found from molecular radii determined from second virial coefficients, viscosity and partial specific volumes, that in solutions about 30–40% of the dendrimer volume was occupied by solvent molecules, but that swollen dendrimers still showed average R_G/R_H ratios of about 0.81, indicating non-draining, hard-sphere-type character. Interestingly, however, the intrinsic viscosities determined in this work showed no generational maximum, but instead leveled off at about generations 4–5 in a manner similar to that predicted by Cai and Chen [23]. In agreement with this, experimental specific partial volumes decreased continuously with generations, but the $N_A R_G^3/M$ ratios calculated from experimentally determined molecular weights showed a maximum at generations 3–4, which agreed quite well with the maximum in the intrinsic viscosity data of Scherrenberg and co-workers [18] (see Figure 14.2). Comparing the generational dependencies of the molecular radii obtained from SANS, DSV and molecular modeling, and by Scherrenberg and co-workers, these authors concluded that in agreement with an earlier observation by Valachovic [24], dendrimer radii did not seem to scale with molecular weight, probably because of the lack of self-similarity between generations that is required for scaling laws to be valid [19]. Most recently, Bodnar and co-workers examined acetylated- (i.e. hydrophobic) and amine-terminated (i.e. hydrophilic) generation 4 and 5 PPIs in water, using DSV, dynamic and steady shear rheometry, solution densitometry and SANS measurements [22]. They observed considerable electrostatic interactions between the protonated amine-terminated derivatives, and from intrinsic viscosity data concluded that both acetylated and nonacetylated derivatives showed great solvent penetration (i.e. as high as about 50% of hydrodynamic volume), with the latter being somewhat more swollen than the former.

Water has been the solvent of particular interest for both PPI and PAMAM amine-terminated dendrimers for many investigators [16, 25–28]. In this medium, these highly soluble, multifunctional polyamines behave as precise, globular, nanoscopic polyelectrolytes whose high density of surface charge may be easily manipulated by varying pH and concentration [28]. Therefore, they are exceptionally convenient for testing theories and gaining a better understanding of interactions between charged particles, but also extremely important as a potential means of delivery in many proposed biomedical applications where dendrimers show promise. A DSV study of PAMAM dendrimers in water over a relatively wide concentration range from 5×10^{-3} to 13 g/dL [16] showed that generation 4 and higher dendrimers behaved in a manner similar to conventional polyelectrolytes [29, 30], while the lower homologues did not show the characteristic upswing of reduced viscosity at low concentrations [16]. In addition to this, it was also observed that in agreement with what had earlier been found for polylysine dendrimers [8], the polyelectrolyte-like behavior could be suppressed by addition of monovalent salts, such as NaCl, capable of screening electrostatic interdendrimer interactions [16]. Subsequent SANS studies showed that in dilute D_2O solutions generation 5 PAMAM behaved as noninteracting particles [28], but that local ordering through long-range repulsive electrostatic interactions occurred if amino groups were protonated by acidification with HCl. Furthermore, it was also shown that as the intensity of the charge on dendrimers increased, these electrostatic repulsions gave rise to electrostatic excluded volume, which results in the establishment of a limiting distance for the closest approach of neighboring dendrimers to each other (see Figure 14.4) [28]. Consistent with this, it was demonstrated in another study that generation 5 PPI in D_2O also underwent local ordering that was dependent on concentration [27]. In fact, interdendrimer correlations were detected by SANS already above about 5% (v/v) concentration, but diminished as the dendrimer content increased

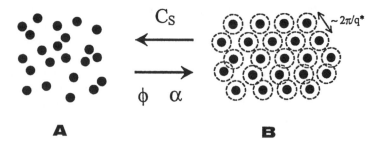

Figure 14.4 Illustration of the ordering phenomena in aqueous solutions of amine terminated PAMAM dendrimers. (A) 'Gaslike' structure forming at very low concentrations or upon addition of NaCl to screen the long-range interactions. (B) 'Organized' structure resulting from electrostatic excluded volume at higher concentrations and/or in the absence of salt. C_S: salt concentration; α: extent of dendrimer end-group ionization; ϕ: dendrimer volume fraction; $2\pi/q^*$: average interdendrimer distance where q^* is the peak in the scattering profile (according to ref. [28])

(solutions of up to 60% (v/v) were examined) similar to the behavior of many polyelectrolyte, protein and ionic micelle systems [27]. For concentrations above the overlap dendrimer concentration (which was calculated to be 36.5% (v/v) from $V_W/(4\pi/3)R_H^3 N_A$ using experimentally determined R_H values), this was explained by 'large interpenetration' which was enabled by the softness of the dendrimer molecules and the filling of the medium which caused excluded volume interactions. Assuming the Guinier–Fournet model based on the arrangement of particles in a distorted face-centered cubic lattice, the average distance between particles, D, was determined to decrease with solution concentration as shown in Figure 14.5. Nevertheless, even at the highest studied concentration, the SANS data indicated that the internal dendrimer structure was not significantly influenced by intermolecular interactions [27].

The fact that in spite of their relatively high intramolecular density which results from extensive branching within the confined molecular volume, dendrimers exhibit surprisingly high flexibility, has been pointed out by a number of different invetsigators [5, 14, 16, 22, 27, 28, 31]. For example, various calculations based on experimentally determined molecular radii and assumed spherical shape indicated that in solutions various dendrimers can fill as much as 50% of their molecular volume with solvent molecules [14, 16, 22, 31]. It was also found that in medium and concentrated (30–75 wt. %) ethylene diamine (EDA) solutions, PAMAM dendrimers exhibited considerable sensitivity of viscosity to temperature [5]. In addition to this, on dissolution in water their molecular volumes expanded by more than sevenfold relative to the bulk state depending on generation [16, 31], but also substantially shrank on the addition of NaCl to screen the long-range electrostatic interactions [16]. For example, intrinsic

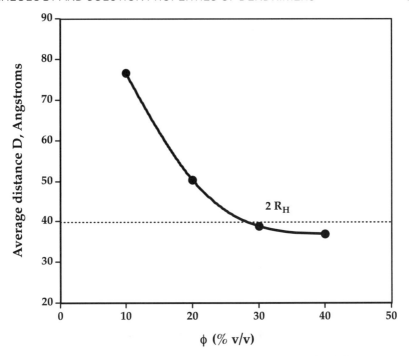

Figure 14.5 Dependence of the average distance between dendrimer centers on dendrimer solution concentration, ϕ. Interpenetration of neighboring dendrimers occurs when this distance becomes shorter than $2R_H$ (according to ref. [27])

viscosities of generation 5 and 6 PAMAMs reduced from 1.55 and 0.78 dL/g, respectively, in deionized water, to 0.22 and 0.18 dL/g, respectively, when 0.004% (w/w) of NaCl was added, and to 0.06 dL/g for both generations when the amount of salt was increased to 5% (w/w) [16]. Although the reasons for this behavior are not completely clear from these data alone, they may involve the complexation of cations by amino groups and consequent more pronounced back-folding by an increased pull of nucleophilic interior branch junctures on the charged dendrimer end-groups.

4   RHEOLOGY OF CONCENTRATED DENDRIMER SOLUTIONS

Compared to dilute solution viscometry and to some extent to bulk rheology, the flow properties of dendrimers in concentrated solutions have been the least investigated area of dendrimer rheology. In fact, with the notable exception of some recent data on generation 4 PPI in water [22] the only [32] reported detailed study of the flow of a dendrimer family in medium to highly concentrated solutions is that of PAMAM dendrimers in ethylene diamine (EDA) solvent [5]. In this work, the steady-shear flow behavior of 34 different solutions, spanning the concentration range from 30 to 75 wt%, of the first seven generations of amine terminated PAMAMs, having molecular weights from about 500 to almost 60 000, was investigated at seven different temperatures between 10 and 40 °C [5].

The main result was that regardless of dendrimer generation (i.e. molecular weight) and concentration, all of the examined solutions exhibited characteristic Newtonian flow behavior, as shown in Figure 14.6. This was in striking contrast to the typical behavior of either chain-type polymers of comparable molecular weights [33], or suspensions of spherical particles [34–37], both of which exhibit

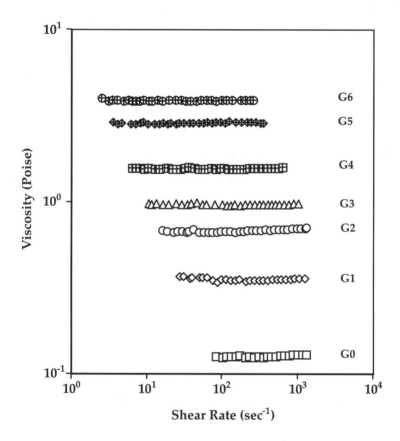

Figure 14.6 Dependence of dendrimer solution viscosity, η, on shear rate, $\dot{\gamma}$, for the 40% wt solutions of the first seven generations, G, of PAMAM dendrimers in ethylenediamine solvent at 25 °C.

characteristic non-Newtonian shear-thinning above some critical shear rate (or shear stress), molecular weight and/or concentration [38]. In chain-type polymers, including linear, randomly or regularly branched, cyclic, star-shaped and H-shaped ones [38], this behavior results from the establishment of transient quasi-networks through entanglement-type intersegmental interactions [33], while for suspensions of spherical particles, it originates from concentration-dependent particle agglomeration, which, in turn, results from the tendency to minimize surface energy. Hence, the observed behavior of PAMAM solutions indicates that under the examined conditions these dendrimers did not undergo either of these two types of interactions, even at relatively high concentrations and/or molecular weights (i.e. generations) [5]. Notably, this conclusion is in good agreement with the behavior of generation 4 PPI in water [22], and with the result of small-angle X-ray scattering (SAXS) study, which for nonprotonated generation 10 PAMAM found no interdendrimer correlation even at concentrations as high as 26 wt%. in methanol [39]. However, this is also quite different from the behavior of electrostatically charged PAMAM dendrimers in acidified water solutions [16, 28].

The dependencies of the steady-shear viscosity of PAMAM/EDA solutions on dendrimer molecular weight showed monotonically increasing viscosity with molecular weight with the slopes of $\log \eta$ vs $\log M$ curves steadily decreasing with molecular weight and the magnitude of this decrease increasing with an increase in solution concentration and a decrease in temperature (see curves A and B of Figure 14.7) [5]. In addition to this, dependencies of identical type were also found for the zero-shear viscosities from steady and dynamic oscillatory shearing of all three examined dendrimer families in the bulk, including: PPIs (see curve C of Figure 14.7) [50], PAMAMs (curve D of the same figure) [5, 40] and PBzEs (curve E of the same figure) [41]. This behavior is fundamentally different from what is generally found for typical chain-type polymers which show characteristic breaks in the slopes of the linear portions of their $\log \eta_0$ vs $\log M_w$ dependencies at the chain length at which entanglement couplings are established between chain segments, as was clearly demonstrated by direct comparison of linear and dendrimer PAMAMs within the same range of molecular weights (see F of Figure 14.7) [5]. In addition, the shape of the dendrimer viscosity vs molecular weight relationship is also distinctly different from the steep linear relationship (slope > 10) reported for suspensions of spherical microgels of comparable particle sizes (7–28 nm) [38]. For all these reasons, although more experimental data would be desired, it appears that the viscosity–molecular weight relationship is a dendrimer fingerprint property that is independent of the state (i.e. solution or bulk) and chemical composition (i.e. PAMAM, PPI or PBzE), but characteristic of their molecular architecture. It seems to reflect their rather sharp outer boundaries [39] and apparent lack of interdendrimer interactions [5].

Figure 14.8 shows the shear viscosity-concentration dependencies for EDA

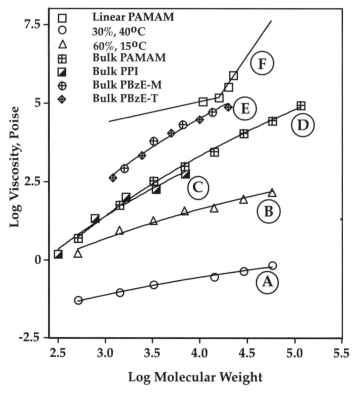

Figure 14.7 Dependence of the zero-shear viscosity, η_0, on molecular weight, *M*, for different dendrimer systems. (1) Dendrimers of different chemical composition but in the same state (i.e. PAMAM, PPI and PBzE dendrimers in bulk: D, C and E, respectively). (2) Compositionally identical dendrimers (i.e. PAMAMs) in solutions and in the bulk state (A, B and D, respectively). (3) Compositionally identical dendrimers and linear polymers of comparable molecular weights (i.e. PAMAMs in the bulk state: D and F, respectively)

solutions of generation 0–5 PAMAM dendrimers in comparison with the theoretical predictions of three frequently used models of idealized suspensions of non-draining hard spheres of uniform density, including the Krieger [42], Eilers [43] and Mooney [44] models. For this comparison, the volume fractions of dendrimer solutions were obtained from experimentally determined dendrimer bulk densities [31], and the model curves were calculated using the earlier reported value for the viscosity of EDA solvent [16]. It can be seen from Figure 14.8 that although all dendrimer solutions clearly deviate from the models, they seem to approach model behavior with an increase in dendrimer generation. This indicates a decrease in dendrimer surface permeability to solvent molecules at higher generations, which should correlate with an accompanying decrease of

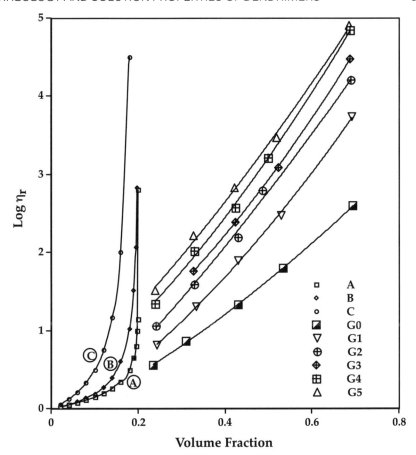

Figure 14.8 Generational dependence of relative viscosity, η_r, on solution volume fraction for the first six generations, G#, of PAMAM dendrimers in ethylenediamine (EDA) in comparison with theoretical predictions of Krieger (A), Eiler (B) and Mooney (C) 'hard sphere' models (according to ref. [5])

R_G/R_H ratio. Unfortunately, in this respect the available data are not conclusive enough (compare with Table 14.1) [45].

Within the temperature range from 10 to 40 °C, both PAMAM dendrimers in EDA [5] and PPI dendrimers in water [22] showed a linear relationship between $\ln \eta$ and $1/T$, in good agreement with the kinetic rate theory of flow [46]. The apparent activation energies of flow (E_η) were constant and independent of temperature, and it was shown for PAMAM/EDA systems that the dependence of E_η on solution concentration was linear for all generations examined [5]. This was considerably different from the typical relationships for the solutions of linear and/or randomly branched chain polymers, where a break in the slope of

E_η vs c [47] is usually associated with the establishment of entanglement couplings [48]. Thus, the observed linearity of E_η vs c dependencies of dendrimer solutions seems to be another manifestation of dendrimer reluctance to interpenetrate, either with other dendrimers or with their parts.

5   DENDRIMER BULK RHEOLOGY

Bulk rheology was studied under steady and dynamic oscillatory shear for PBzE [41, 49], PAMAM [16, 40] and PPI [50] dendrimers. Of the former, three different series of samples were examined, including six generations of monodendrons (which spanned the molecular weight, M, range from 320 to 13 464), six generations of tridendron dendrimers (M ranging from 576 to over 20 292), and three types of so-called 'hypercore' dendrimers which contained 6, 12 and 24 generation 4 dendrons (ranging in M from 20 712 to 84 219) [49]. Of the PAMAMs, eight generations of tetradendron amine-terminated dendrimers, ranging in molecular weight from 517 to 116 493 were investigated [16, 40], while of the PPIs, the first five generations of NH_2 terminated dendrimers and three high generations of their CN homologues were tested [50]. All studies were performed within the WLF temperature region [51] of up to about 100 °C above the respective dendrimer glass transition temperatures. For PBzEs this spanned the range from 60 to 120 °C, while for PAMAMs between 40 and 95 °C. All studies were also performed within the dendrimer's linear viscoelastic response range, using for PBzEs parallel plates, and for PAMAMs and PPIs the cone-and-plate geometry. In all studies, care was taken to protect the samples from thermal oxidation and, in the case of PAMAMs, also from atmospheric moisture absorption.

Consistent with the Newtonian flow of concentrated PAMAM solutions, it was found that all three types of dendrimers [40, 41, 50] under steady-shear conditions, and both PAMAMs [40] and PPIs [50] under creep [16, 50] showed typical viscous behavior at all applied stress levels and testing temperatures. For example, as illustrated in Figure 14.9 [40], all of the first seven generations of PAMAMs showed constant viscosities over the entire ranges of shear rates investigated, and in addition to this, there was no hysteresis between the forward and the reverse stress sweeps in steady shearing, indicating the absence of thixotropy.

In contrast to this, however, small-amplitude oscillatory shearing of these same dendrimers revealed rather unexpected viscoelastic responses, manifested by generationally dependent complex-viscosity thinning and finite elastic moduli at all generations [16, 40]. The former was especially prominent at lower temperatures and for higher generations, suggesting either the proximity of the respective glass temperatures or some kind of supramolecular structuralization

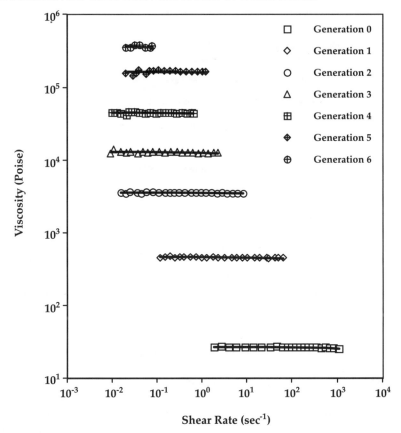

Figure 14.9 Dependence of steady shear viscosity, η, on shear rate, $\dot{\gamma}$, for the first seven generations of PAMAM dendrimers at 70°C in the bulk state

in the state of rest [40]. In addition, the Cox–Merz rule was found to hold and the results obtained from steady and oscillatory shear could be time-temperature superimposed, permitting the construction of viscosity master curves shown in Figure 14.10 [40]. A 'slight decrease' in complex viscosity at higher frequencies was also noted for higher generation PBzEs, with their terminal complex viscosities being identical to their terminal viscosities under steady shear [49]. In contrast to this, however, amine-terminated PPI dendrimers showed essentially Newtonian flow even under dynamic shear, but their CN substituted counterparts exhibited a frequency dependence of the storage and relaxation moduli consistent with Rousean behavior [50].

Bulk PAMAM dendrimers showed finite moduli of elasticity at all generations, as shown for the storage (G') and loss (G'') components in Figure 14.11 [40]. For lower generations (i.e. generations 0 through 3) these dependencies

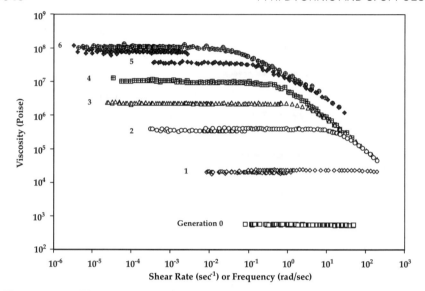

Figure 14.10 Master curves of steady shear viscosity, η (at lower shear rates) and complex viscosity, η* (at higher frequencies) for the first seven generations of PAMAM dendrimers at 40°C in the bulk state

conformed very closely at all frequencies to a single relaxation time Maxwell model, which was also representative of the steady shear data since it predicts constant steady-shear viscosity as a function of shear rate. Higher generations, however (i.e. generations 4 through 6), were much better represented by a multimode model which included a spectrum of relaxation times associated with a variety of corresponding mobility modes. The spacing of the modes was consistent with a truncated Rouse spectrum [52, 53], so that a good fit of experimental data for generation 4 required two such modes, while generations 5 and 6 needed 9 and 15 modes, respectively. Notably, generation 6 also required yet another (16th) non-Rousean low-frequency mode, to fit the observed $G' \sim \omega^{1.49}$ and $G'' \sim \omega^{0.92}$ dependencies in the terminal zone.

Comparison of the longest relaxation times calculated from the time-temperature superimposed moduli with directly measured viscosities revealed considerable internal consistency of the data over the entire range of molecular weights (i.e. generations) examined. This reconfirmed the Rouse-like behavior at higher generations, and was also in good agreement with the data for PPI and PBzE dendrimers which showed an 'almost linear relationship' between η_0 and M [49, 50]. For PBzEs, a slope of 1.10 at higher generations was obtained after the data was corrected to a constant reference fractional free volume (see Figure 14.7) [49].

Hence, it seems that bulk dendrimers change their rheological behavior from Maxwell-like at lower generations to Rousean at higher ones. For PAMAMs for

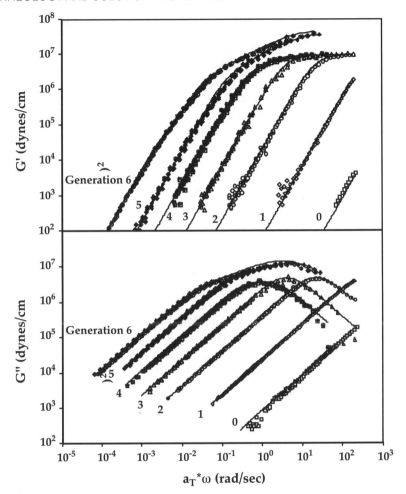

Figure 14.11 Elastic (G') and loss (G'') components of the dynamic modulus of the first seven generations of bulk PAMAM dendrimers as a function of shifted shear frequency $(a_T{}^*\omega)$ at 40°C. Solid lines indicate a fit for each curve to the generalized Maxwell model

example, this change takes place at generation 4 [40], which is about the same generation at which a major molecular conformation change seems to occur as a result of overgrowing the critical branching stage [3] at which the dendrimer outer surface closes in upon itself [5]. Therefore, in contrast to the situation found in concentrated PAMAM solutions, higher generations of these dendrimers in bulk do not act as single kinetic flow units [5]. Instead, they seem to develop more than one submolecular mode of stress relaxation through the motions of smaller parts of their structure, including individual dendrons, sub-dendrons, etc. [40].

Conversely, the appearance of an ultra-low frequency relaxation mode, found for generation 6 PAMAM [40], indicates the formation of a structure that is larger than an individual dendrimer molecule. Such supramolecular organization in the state of rest can account for the observed complex-viscosity thinning, and is consistent with relatively high dendrimer bulk densities [16, 31]. It would also suggest the establishment of interdendrimer interactions as a result of the collapse of dendrimer spheroids into flattened pancake-like entities. However, even in such conformation, the dendrimer interiors appear able to retain a surprising degree of mobility. This follows not only from the relaxation mobility modes of the Rousean model discussed above, but also from the well-known universal trend of the dendrimer glass transition temperature (T_g) vs generation [16, 54–56], and from the demonstrated ability of some dendrimer networks to accommodate low molecular weight guests inside their dendritic domains even in the solid state [57, 58]. The only reported exception to this is the dynamic experiments with bulk amine terminated PPI dendrimers by Sendijarevic and McHugh which showed $G' \sim 0$ and $G'' \sim \omega$ [50]. From a comparison with the corresponding steady shear and creep data, the authors concluded that this behavior indicated that the PPIs did not exhibit any intermolecular entanglement couplings and remained essentially Newtonian, even in the bulk state. If so, this would make these dendrimers unique among their PAMAM and PBzE counterparts, probably reflecting their smaller and more compact molecules (see Figure 14.1).

In attempts to better understand dendrimer intramolecular morphology, considerable attention was devoted to the fractional free volume near the glass temperature [40, 49, 50]. Because all of the studies were performed within the WLF temperature range, the data were analyzed using the equation

$$\log a_{\text{T}} = -C_1{}^0(T - T_0)/[C_2{}^0 + (T - T_0)],$$

where

$$C_1{}^0 = B/2.303 f_{\text{ref}}; \ C_2{}^0 = f_{\text{ref}}/\alpha_f$$

the reference free volume, f_{ref}, was taken to be f_0 the free volume at $T_0 = 40\,°C$ for PAMAMs [41] and at $T_0 = 80\,°C$ for PBzEs [49], respectively, with α_f being the thermal expansion coefficient at the same reference temperature. Values of $C_1{}^0$ and $C_2{}^0$ were obtained by a least-squares method from the slopes and intercepts of the linear fits of $(T - T_0)/\log a_{\text{T}}$ vs $(T - T_0)$ diagrams, and from the values of $C_1{}^0$, the fractional free volumes, f_0, were calculated assuming $B = 1$ [59], to give the results listed in Table 14.2. It was demonstrated for PAMAMs [40] (see curve A of Figure 14.12) that the fractional free volume initially rapidly decreased with increasing dendrimer molecular weight (i.e. generation), but then considerably

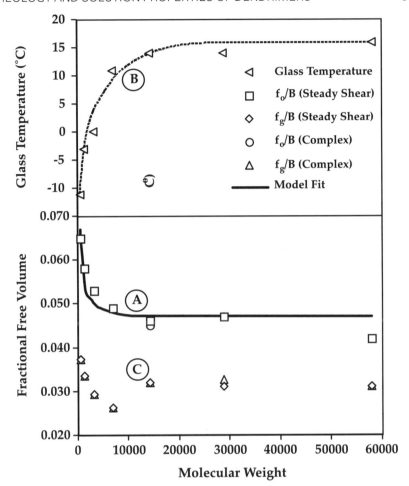

Figure 14.12 Dependencies of glass temperature (T_g) (B), fractional free volume (f_0/B) (A) and fractional free volume at glass temperature (f_g/B) (C) on dendrimer molecular weight (i.e. generation) for the first seven generations of PAMAM dendrimers (according to ref. [40])

slowed down between generations 3 and 6, in good agreement with the leveling off of these same dendrimers' glass temperatures (see curve B of the same figure), which are expected to depend on the free volume and which reach an asymptotic value of about 15 °C at generation 4 (compare with Table 14.1).

Values of the fractional free volumes at the glass temperature were calculated for PBzEs and PAMAMs from $f_g = f_{ref} + (T_g - T_{ref})/2.303C_1{}^0$ [53], and the results obtained are listed in Table 14.2 and shown (for PAMAMs) as C in Figure 14.12. It can be seen from these data that f_g appears to be independent of

Table 14.2 Values of WLF $C_1^{\,0}$ and $C_2^{\,0}$ Parameters, Fractional Free Volumes f_0 and f_g and Thermal Expansion Coefficient α_f for PAMAM, PPI and PBzE dendrimers[a]

	PAMAM[b]					PPI[c]		PBzE[d]				
Generation	$C_1^{\,0}$	$C_2^{\,0}$, K	f_0/B	f_g/B	$10^4\,\alpha_f/B$	$C_1^{\,0}$	$C_2^{\,0}$, K	$C_1^{\,0}$	$C_2^{\,0}$, K	f_0/B	f_g/B	$10^4\,\alpha_f/B$
0	6.7	120	0.065	0.037	5.4	9.19	160.2	6.33	95	0.069	0.029	7.2
1	7.5	102	0.058	0.034	5.7	12.08	165.1	6.77	89	0.064	0.032	7.2
2	8.2	90	0.053	0.029	5.9	13.72	162.5	7.16	84	0.061	0.031	7.2
3	8.9	84	0.049	0.026	5.8	14.20	177.9	7.25	83	0.060	0.031	7.1
4	9.5	87	0.046	0.032	5.2	20.39	234.8	7.32	82	0.059	0.032	7.2
5	9.3	86	0.047	0.033	6.1							
6	10.4	101	0.042	0.031	4.2							

[a] For molecular parameters of these dendrimers see Table 14.1.
[b] Data from ref. 40.
[c] Data from ref. 50.
[d] Data from ref. 49.

dendrimer composition, type and molecular weight, and equal to about 0.032 ± 0.002 [40, 49]. Notably, this is somewhat higher than the 'universal' value of 0.025 that is usually found for most linear polymers [53], and it remains an open question whether it supports the concept of the glass temperature as an iso-free volume state [49, 60]. However, it certainly seems significant that identical f_g values were obtained for two dendrimer families of very different chemical compositions and end-groups.

For linear polymers, it is well known that the fractional free volume, f, increases with $1/M$ leveling off at high molecular weights where the free volume associated with the end groups becomes an insignificant contribution to the total free volume. For dendrimers, however, this cannot be expected to hold because the number of end-groups increases in a geometrically progressive manner with generation [49, 55], so that at every dendrimer generation, the fractional free volume associated with the end groups does contribute to the total free volume of the molecule, $f(M)$, together with the free volume associated with the core cell and interior branch cells [40]. Hence, the fractional free volume for dendrimers may be represented by the following equation:

$$f(M) = (\rho/M)(V_{core} + V_{interior}N_{interior} + V_{exterior}N_{exterior})$$

where ρ is the density, V_{core}, $V_{interior}$ and $V_{exterior}$ are molar volumes of the core, interior and exterior branch cells, while $N_{interior}$ and $N_{exterior}$ are the corresponding numbers of the interior and exterior branch cells (since $N_{core} = 1$) [40]. As shown in Figure 14.13, this equation gives rise to curves A and B for bulk PPIs and PAMAMs, respectively, and straight line C for bulk PBzEs. However, over the intermediate molecular weight range from about 800 to about 15 000 (i.e. $1000/M$ between about 1.3 and 0.065), all three dendrimer families showed reasonably similar, linear $f(1/M)$ relationships (compare dashed lines in Figure 14.13), while at lower and higher values of $1/M$ (where the data are not available for PBzEs), this relationship is clearly and significantly curved. The authors of the PBzE study [49] suggested that the linearity of the dependence C in Figure 14.13 indicates that the contribution of dendrimer end-groups to the total free volume may not be as important as anticipated [55], and that the free volume theory is, therefore, flawed when applied to dendrimers. In light of the comparison of all three examined dendrimer families, however, which was not available to the authors of the PBzE study, it appears that this suggestion may not apply to very low and truly high generation dendrimers.

On the same subject, Sendijarevic and McHugh concluded that for PPIs the fractional free volume contributed by the end-groups remained constant with generations [50], indicating the possibility of backfolding [61] of the end-groups into the 'void space' in the dendrimer interior, with the relative intensity of this backfolding increasing with generation in order to keep the fractional free volume of the end groups constant. As a consequence, the 'void space' would

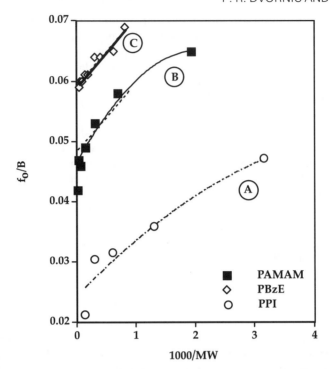

Figure 14.13 Dependencies of the fractional free volume (*f098$_0$/B*) on molecular weight for PPI (A), PAMAM (B) and PBzE (C) dendrimers. Data are from ref. [40], [50] and [49], respectively (see Table 14.2)

have to decrease, forming a more compact space-filling structure similar to the model proposed by Scherrenberg [18], or earlier by Mansfield and Klushin [20], or Murat and Grest [21].

It was also noted by Farrington and co-workers that the chemical nature of the end-groups may have a substantial effect on the viscosity of bulk dendrimers, even under iso-free volume conditions [49]. This seems consistent with the establishment of interdendrimer interactions and supramolecular organization at rest in the bulk state. However, as in most other areas of dendrimer rheology, more data are desirable before definite conclusions can be drawn.

6 CONCLUSION

Although a body of data on dendrimer rheological behavior has been accumulated, and new reports are appearing at an increasing pace, the field is still in its infancy and will remain an intriguing and important subject of research for some time to come. The picture that is emerging, however, seems to indicate that dendrimers do not behave as idealized hard spheres, but that instead they more closely resemble relatively 'soft' and flexible spheroids that are prone to penetration by solvent and other small molecules. Although under appropriate experimental conditions as much as one half of a dendrimer's molecular volume may be occupied by solvent, even highly swollen dendrimers still have an unusually high molecular density due to extensive intramolecular branching and high regularity of segmental packing. Furthermore, although decreasing with generation, this considerable solvent draining ability does not seem to disappear even at very high generations.

In many aspects different dendrimers seem to respond similarly to similar testing conditions, indicating that the dominant role in these responses is played by their molecular architecture, and not by chemical composition. For example, several studies showed that different dendrimers exhibited similar size dependencies (i.e. R_G and/or R_H) on molecular weight (i.e. generation), and that at least those that have symmetrical branch cells appear to show maxima in their intrinsic viscosities. In addition to this, all examined dendrimers indicate considerable inherent molecular flexibility as judged by their volume change on dissolution, type of solvent and temperature; show identical dependencies of viscosity, fractional free volume and glass temperature on molecular weight (i.e. generation).

Nevertheless, the existence of the famed maximum in the dendrimer $[\eta]$ vs G relationship, and whether this behavior could be a characteristic fingerprint property of this type of macromolecular architecture, was questioned recently [19, 23]. In fact, so was the linearity of the dependence of dendrimer molecular radii on $M^{1/3}$ [19], so that whether this may be the beginning of yet another controversy, remains to be seen. Perhaps the future may bring an interesting debate on these subjects, but until new data become available, one should refrain from drawing premature conclusions because the exciting architectural beauty of idealized dendrimer structures has already proven itself to easily tempt the most astonishing hypotheses that may not be readily substantiated by reality.

Inherent segmental flexibility can account for the characteristic Newtonian flow of dendrimers in solutions [5] and under steady shear in the bulk state [40, 41, 49, 50] even at unusually high molecular weights (i.e. generations) and solution concentrations. It can also explain their Rousean viscoelastic response to dynamic shearing [40, 50], and the 'universal' shape of dendrimer glass temperature vs generation curves [54]. In fact, this flexibility is probably responsible for a rather surprising degree of conformational mobility of dendrimer segments with respect to their high density of packing relative to the corresponding open chain-type macromolecules. Such conformational mobility may enable the controversial segmental bending back into the molecular interior [61], and provide for the proposed relatively constant intradendrimer density profile [20, 21]. Finally, dendrimer segmental flexibility may also be the reason for the considerable cargo space responsible for the accommodation of various guest ions or molecules (even in the solid state [57, 58]), including large macromolecules [62] in the dendrimer hosts. Its relative degree, however, would be expected to depend on dendrimer chemical composition, predetermining the intensity of secondary interactions both in inter- or intramolecular group communications.

However, perhaps the most inspiring rheological property of dendrimers is nevertheless their inability to interpenetrate (i.e. entangle) even at unusually high molecular weights and solution concentrations [5]. In this respect, they are indeed unique among high molecular weight polymers and because their entire molecules flow as a single kinetic flow unit they may rightfully be referred to as nanoscopic 'molecular ball bearings' [41].

ACKNOWLEDGEMENT

The authors gratefully acknowledge the contribution of Dr Steven E. Keinath of Michigan Molecular Institute who kindly supplied Figure 14.1 from his rich library of dendrimer models.

7 REFERENCES

1. Dvornic, P. R. and Tomalia, D. A. *Sci. Spectra*, **5**, 36 (1996).
2. Two different dendrimer families are presently commercially available: Starburst® polyamidoamine (PAMAM) dendrimers from Dendritech Inc., Midland, Michigan, and ASTRAMOL® polypropyleneimine (PPI) dendrimers from DSM, Geleen, the Netherlands.
3. Tomalia, D. A. and Dvornic, P. R. 'Dendritic polymers, divergent synthesis (starburst polyamidoamine dendrimers)', in Salamone, J. C. (ed.), *Polymeric Materials Encyclopedia*, Vol. 3, CRC Press, Boca Raton, FL, 1996, pp. 1814–1830.
4. Naylor, A. M., Goddard, W. A. III, Kiefer, G. E. and Tomalia, D. A. *J. Am. Chem. Soc.*, **111**, 2339 (1989).
5. Uppuluri, S., Keinath, S. E., Tomalia, D. A. and Dvornic, P. R. *Macromolecules*, **31**, 4498 (1998).
6. Mattice, W. L. 'Masses, sizes and shapes of macromolecules from multifunctional monomers' in Newkome, G. R., Moorefield, C. N. and Vogtle, F. (eds), *Dendritic Molecules. Concepts, Synthesis, Perspectives*, Chapter I, VCH Verlag, Weinheim, Germany, 1996.
7. Aharoni, S. M., Crosby C. R., III and Walsh, E. K. *Macromolecules*, **15**, 1093 (1982).
8. Aharoni, S. M. and Murthy, N. S. *Polym. Commun.*, **24**, 132 (1983).
9. Denkewalter, R. G., Kolc, J. F. and Lukasavage, W. J. US Patent, 4,410,688, 1982.
10. Note that for solid spheres of uniform density: $R_G/R_H = (3/5)^{1/2} = 0.775$. Values larger than this could be the result of higher segmental density near the exterior of the dendrimer. See ref. 27 for more details.
11. Meier, D. J. Unpublished data, 1984.

12. Mansfield, M. L. and Klushin, L. I., *J. Phys. Chem.*, **96**, 3994 (1992).
13. Tomalia, D. A., Hall, M. and Hedstrand, D. M., *J. Am. Chem. Soc.*, **109**, 1601 (1987).
14. Elias, H.-G. '*An Introduction to Polymer Science*', VCH Publishers, Weinheim, 1997, p. 212.
15. Mourey, T. H., Turner, S. R., Rubinstein, M., Fréchet, J. M. J., Hawker, C. J. and Wooley, K. L., *Macromolecules*, **25**, 2401 (1992); 15a. For *dn/dc* of PPI dendrimers, ref. 19 reports a mean value of 0.230 ± 0.003 for generations 1–5 and no generational dependence.
16. Uppuluri, S., Ph.D. thesis, Michigan Technological University, Houghton, MI, 1997.
17. de Brabander-van der Berg, E. M. M. and Meijer, E. W. *Angew. Chem. Int. Ed. Engl.*, **32**, 1308 (1993).
18. Scherrenberg, R., Coussens, B., van Vliet, P., Edouard, G., Brackman, J., de Brabander, E. and Mortensen, K. *Macromolecules*, **31**, 456 (1998).
19. Rietveld, I. B. and Smit, J. A. M. *Macromolecules*, **32**, 4608 (1999).
20. Mansfield, M. L. and Klushin, L. I. *Macromolecules*, **26**, 4262 (1993).
21. Murat, M. and Grest, G. S. *Macromolecules*, **29**, 1278 (1996).
22. Bodnar, I., Silva, A. S., Deitcher, R. W., Weisman, N. E., Kim, Y. H. and Wagner, N. J., *J. Polym. Sci., Part B: Polym. Phys.*, **38**, 857 (2000).
23. Cai, C. and Chen, Z. Y., *Macromolecules*, **31**, 6393 (1998).
24. Valachovic, D. E. Ph.D. Thesis, University of Southern California, Los Angeles, CA, 1997.
25. Tomalia, D. A., Baker, H., Dewald, J., Hall, M., Kallos, G., Martin, S., Roeck, J., Ryder, J. and Smith, P., *Polymer J.*, **17**, 117 (1985).
26. Dubin, P. L., Edwards, S. L., Kaplan, J. I., Mehta, K. S., Tomalia, D. A. and Xia, *J. Anal. Chem.*, **64**, 2344 (1992).
27. Ramzi, A., Scherrenberg, R., Brackman, J., Joosten, J. and Mortensen, K., *Macromolecules*, **31**, 1621 (1998).
28. Nisato, G., Ivkov, R. and Amis, E. J. *Macromolecules*, **32**, 5895 (1999).
29. Hess, W. and Klein, R., *Adv. Phys.*, **32**, 173 (1983).
30. Antonietti, M., Forster, S., Zisenis, M. and Conrad, J. *Macromolecules*, **28**, 2270 (1995).
31. Uppuluri, S., Tomalia, D. A. and Dvornic, P. R., *Polym. Mater. Sci. Eng.*, **77**, 116 (1997).
32. A paper on 'The rheological properties of poly(fluorophenylene germane) dendrimers' was published by Myasnikova and co-workers in the Russian journal *Polym. Sci., Ser. B.*, **37**, 362 (1995). The authors claimed that acetone solutions (5, 10 and 20%, specifically) of a sample that had gel permeation chromatography retention time close to that of a linear polystyrene of 1.1×10^6 molecular mass, had four decades lower viscosity than the corresponding solutions of flexible-chain linear poly(butyl methacrylate). However, in our opinion, neither the examined sample was characterized satisfactorily enough to be referred to as a dendrimer, nor the rheology was described sufficiently enough to draw any conclusions about the solution's flow behavior. Therefore, we refer to this paper here only for reasons of curiosity.
33. See for example: Tirrell, M., 'Rheology of polymeric liquids', Chapter 11 in Macosco, C. W. (ed.), *Rheology. Principles, Measurements and Applications*, VCH Publishers, New York, 1994, pp. 475–514.
34. Woodcock, L. V. *Mol. Simul.*, **2**, 253 (1989).
35. Angel, C. A., Clarke, J. H. R. and Woodcock, L. V., *Adv. Chem. Phys.*, **48**, 397 (1981).
36. Castle, J., Merrington, A. and Woodcock, L. V., *Prog. Colloid Polym. Sci.*, **98**, 111 (1995).
37. Mewis, J. and Macosco, C. W., 'Suspension rheology', Chapter 10 in Macosco, C. W.

(ed.), Rheology. Principles, Measurements and Applications, VCH Publishers, New York, 1994, pp. 425–474.

38. Antonietti, M., Pakula, T. and Bremser, W., *Macromolecules*, **28**, 4227 (1995).
39. Prosa, T. J., Bauer, B. J., Amis, E. J., Tomalia, D. A. and Scherrenberg, R. *J. Polym. Sci: Part B., Polym. Phys.*, **36**, 2913 (1997).
40. Uppuluri, S., Morrison, F. A. and Dvornic, P. R. *Macromolecules*, **33**, 2551 (2000).
41. Hawker, C. J., Farrington, P. J., Mackay, M. E., Wooley, K. L. and Fréchet, J. M. J. *J. Am. Chem. Soc.*, **117**, 4409 (1995).
42. Krieger, I. M., *Adv. Colloid Interface Sci.*, **3**, 111 (1972).
43. Eilers, H., *Kolloid-Z.*, **97**, 913 (1941); **102**, 154 (1943).
44. Mooney, M., *J. Colloid Sci.*, **6**, 162 (1951).
45. It should be noted that discrepancies amidst available data may result from different methods of determination. For example, selected hydrodynamic radii of PAMAM dendrimers of Table 14.1 were obtained by two different experimental methods: for generations 1 through 4 by dilute solution viscometry, while for generations 5 through 10 by size exclusion chromatography.
46. Glasstone, S., Laidler, K. J. and Eyring, H. *The Theory of Rate Processes*, McGraw-Hill, New York, 1941.
47. Vinogradov, G. V. and Malkin, A. Ya. *Rheology of Polymers*, Mir Publishers, Moscow, 1980, p. 121.
48. Tager, A. A. and Dreval, V. E. *J. Polym. Sci., Part C*, **23**, 181 (1968).
49. Farrington, P. J., Hawker, C. J., Fréchet, J. M. J. and Mackay, M. E. *Macromolecules*, **31**, 5043 (1998).
50. Sendijarevic, I. and McHugh, A. J. *Macromolecules*, **33**, 590 (2000).
51. Williams, M. L., Landel, R. F. and Ferry, J. D. *J. Am. Chem. Soc.*, **77**, 3701 (1955).
52. Larson, R. G. *Constitutive Equations for Polymeric Melts*, Butterworth, Stoneham, MA, 1988, pp. 46–53.
53. Ferry, J. D. *Viscoelastic Properties of Polymers*, 3rd edn, John Wiley & Sons, New York, 1980.
54. Stutz, H. *J. Polym. Sci., Part B: Polym. Phys.*, **33**, 333 (1995).
55. Wooley, K. L., Hawker, C. J., Pochan, J. M. and Fréchet, J. M. J. *Macromolecules*, **26**, 1514 (1993).
56. De Brabander-van der Berg, E. M. M., Nijenhuis, A., Mure, M., Keulen, J., Reintjens, R., Vandenbooren, F., Bosman, B., de Raat, R., Frijns, T., Wal, S. V. D., Castelijns, M., Put, J. and Meijer, E. W. *Macromol. Symp.*, **77**, 51 (1994).
57. Dvornic, P. R., de Leuze-Jallouli, A. M., Owen, M. J. and Perz, S. V. *Polym. Prepr.*, **39**(1), 473 (1998); **40**(1), 408 (1999).
58. Dvornic, P. R., de Leuze-Jallouli, A. M., Owen, M. J. and Perz, S. V. '*Radially layered poly(amidoamine-organosilicon) (PAMAMOS) copolymeric dendrimers*' in Clarson, S. J., Owen, M. J., Fitzgerald, J. J., Smith, S. D. (eds.), *Silicones and Silicone-Modified Materials*, ACS Symposium Series, Vol. 729, American Chemical Society, Washington, DC, 2000, pp. 241–269.
59. Aklonis, J. J., MacKnight, W. J. and Shen, M. *Introduction to Polymer Viscoelasticity*, Wiley-Interscience, New York, 1972, p. 67.
60. Fox, Jr., T. G., Flory, P. J. *J. Appl. Phys.*, **21**, 581 (1950).
61. Lescanec, R. L. and Muthukumar, M., *Macromolecules*, **23**, 2280 (1990).
62. Wege, V. U. and Grubbs, R. H. *Polym. Prepr.*, **36**(2), 239 (1995).

PART III

Properties and Applications of Dendritic Polymers

15

Dendritic and Hyperbranched Glycoconjugates as Biomedical Anti-Adhesion Agents

R. ROY
Department of Chemistry, University of Ottawa, Ottawa, ON, Canada

1   INTRODUCTION

Multiantennary cell surface carbohydrates are recognized as versatile motifs that direct adhesion to other cells, pathogens, or soluble proteins such as antibodies, hormones, toxins and enzymes [1–3]. These rather weak interactions, on a per saccharide basis, are transformed into very potent and effective attractive forces when multiple ligand copies are presented to similarly clustered receptors [4–6]. This phenomena, known as the *glycoside cluster effect* [7], has been initially observed with asialoglycoprotein receptors found on hepatocytes. This finding has stimulated the syntheses of several N-acetylgalactosamine/galactose clusters designed as vectors for liver targeting [7]. It should be pointed out, however, that glycobiologists have long observed the high inhibitory potencies of naturally occurring glycoproteins having dense carbohydrate decoys [8]. Unfortunately, such proteins are highly heterogeneous and immunogenic, thus they are capable of triggering unwanted anti-carbohydrate antibodies by virtue of the T-cell dependent nature of the protein scaffolds. On the contrary, polysaccharides, which are essentially natural carbohydrate polymers, are non- or only poorly immunogenic since they lack the peptide components. In this respect, they are considered T-independent antigens.

Since carbohydrates are ubiquitous in their interactions with the surroundings, it is very appealing to design carbohydrate-based drugs from a therapeutic

Dendrimers and Other Dendritic Polymers. Edited by Jean M. J. Fréchet and Donald A. Tomalia
© 2001 John Wiley & Sons Ltd

standpoint (Table 15.1). Deceptively, carbohydrate analogs have rarely been the candidate of choice since they generally show low K_Ds [9]. This behavior is due primarily to their largely hydrophilic nature and flexibility. Complementarily, carbohydrate protein receptors also show polar residues in their active sites. Therefore, both entities are highly hydrated and, consequently, large entropic loss occurs upon binding and release of water molecules to the bulk surroundings [10]. Except for a few successful glycohydrolase enzyme inhibitors, the syntheses of potent glycomimetics or conformationally restricted analogs has remained elusive [11–12].

Alternatively and fortunately, most mammalian carbohydrate ligands are formed by a rather limited array of mono- or oligosaccharides. It is estimated that the largest of these ligands encompass between 1 and 4 saccharide residues at most (see Table 15.1). It thus becomes apparent that a conceptually novel approach may be designed necessary to target a wide range of medicinally relevant applications involving saccharide binding (e.g. cancer, metastases, inflammation, fertilization, microbial infections, glycoprotein processing, etc.) [1–3].

Our group is focused upon applications dealing with inhibitions of: bacterial (*E. coli*, *Streptococcus suis*) and viral (flu) infections, selectins involved in inflammation processes, acute rejection following xenotransplantation, homing of

Table 15.1 Glycodendrimers showing the carbohydrate ligands and their applications

Sugar	Structure	Microbial/protein receptors[a]	Ref.
Sialic acid	Neu5Ac	Flu virus; *Limax flavus*	13–15
GM$_1$	Gal(β1–3)GalNAc(β1–4) [Neu5Ac(α2–3)]Lactose	Cholera toxin B; *E. coli* enterotoxin	16, 17
GM$_3$	Neu5Ac(α2–3)Lactose	Cancer vaccine	18
Sialyl-LewisX	Neu5Ac(α2–3)Gal(β1–4) [Fuc(α1–3)GlcNAcβ	Selectin	19
3′-Sulfo-LewisX	3′SO$_3$-Gal(β1–4)[Fuc(α1–3)] Glc	Selectin	20
T-antigen	Gal(β1–3)GalNAcα	Cancer vaccine	21
Tn antigen	α-GalNAc	Cancer vaccine Lectin (*Vicia villosa*)	22–25
Trimannoside	Man(α1–2)Man(α1–6)Manα	MBP^c	26
Galabiose	Gal(α1–4)Galβ	*Streptococcus suis*	27
Pk trisaccharide	Gal(α1–4)Gal(β1–4)Glc	Shiga toxin	28
Lactosamine	Gal(β1–4)GlcNAcβ	Lectin (*E. cristagalli*)	29
α-Mannoside	Manα	*E. coli*; lectin; collitis	26, 30, 31
Lactose	Gal(β1–4)Glcβ	Galectin-1	32–35
Melibiose	Gal(α1–6)Glc	Xenotransplantation	36
B-Disaccharide	Gal(α1–3)Galβ	Anti-B IgG	37

[a] As antiadhesin and inhibitors unless stated otherwise.
[b] MBP: Mannose binding protein.

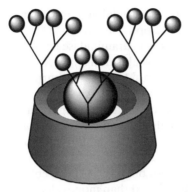

Figure 15.1 Schematic representation of hyperbranched glycocluster having inclusion complex capability, ex. calix[4]arene and β-cyclodextrin

metastatic cells, sperm egg fertilization, and several plant lectins used as models [4, 38]. The advantages of lectins (phytohemagglutinins) as models reside in the fact that several crystalline structures are known and their carbohydrate binding requirements are well established, at least on a per saccharide basis [39]. Moreover, their degree of association, i.e. whether they exist as dimers, tetramers, etc. is in a few cases, known and somewhat controllable. This chapter will highlight some recent findings in the design of multivalent glycoconjugates [4, 38, 40]. A particular emphasis will be dedicated toward the synthesis of dendritic or hyperbranched sialosides built on a biocompatible chitin polysaccharide isolated from the shell of crustaceans. We will also describe the syntheses of calix[4]arene and β-cyclodextrin neoglycoconjugates. These molecules are attractive starting materials since they are already highly branched with the added advantage that they can form inclusion complexes with drugs. In this way, these molecules may act as 'Trojan horses' to deliver medication to specific target cells (Figure 15.1).

2 ADHESION MECHANISMS INVOLVED IN INFLUENZA VIRUS AND RELATED MICROBIAL INFECTIONS

The initial step in the sequence of events leading to influenza virus infections in mammalian hosts is mediated by the multiple attachment of virus particles to host sialoside receptors in the nasopharynx [41]. These receptors consist largely of cell surface sialylated glycoproteins and gangliosides. The subsequent steps involve receptor–mediated endocytosis with ensuing release of the viral nucleoplasmid. The first event responsible for the receptor-virus interaction is therefore an attractive target for both antiviral and related microbial intervention.

Influenza virus particles are spheroidal and approximately 100 nm in diameter. The outer-membrane envelope contains $\sim$ 500 copies of hemagglutinin (HA) trimers and $\sim$ 100 copies of neuraminidase tetramers. The hemagglutinin constitutes the receptor sites for α-sialoside ligands. X-ray analyses show that the three sialic acid binding pockets reside $\sim$ 46 Å apart, each trimer being separated on the virion surface by about 65–110 Å [42].

Numerous researchers have attempted to map the arrangements of hemagglutinin combining sites through the syntheses of sialic acid analogs. Almost all of the sialic acid functionalities are essential for favorable binding interactions. The carboxylate group constitutes the critical site of attachment and any modifications to this function abolishes binding [43]. Except for few aromatic glycosides, there has been no better ligand than monovalent α-sialosides [44].

As mentioned above, individual carbohydrate protein interactions are generally of low affinity, and sialic acid-influenza virus interactions are no exception. Detailed ^{1}H-NMR experiments with methyl α-sialoside and bromealin-released hemagglutinin (BHA) have revealed an intrinsically weak dissociation constant of 2.8 mM [45]. In fact, it has been known for a long time that influenza viruses bind to these ligands with low affinity since naturally occurring sialomucins of high sialic acid contents were a thousandfold better inhibitors of human erythrocytes hemagglutination than monosaccharides or mucins of low sialic acid contents [8].

Evidence gained toward the requirements for multivalency in tight binding interactions has been accumulating. The synthesis of bidentate α-sialoside ligands has been reported [44, 46]. It was found that, while none of these clusters showed improved binding to isolated hemagglutinin (BHA), both demonstrated evidence of increased affinity with the intact virus particles as measured by inhibition of hemagglutination of human erythrocytes. However, several sialic acid containing polymers were shown to be powerful inhibitors [4, 12a, 14, 47].

As mentioned above, several pathogens or their released toxins use the above strategy as an initial infection mechanism. As illustrated in Figure 15.2, carbohydrate clusters, polymers, and hyperbranched polymers have the potential to behave as competitive inhibitors against the attachment of pathogens to host tissues. There have been several recent accounts of this fact and a few reviews cover the topic in depth [1, 2, 4]. The approach allows the amplification of local carbohydrate ligand concentrations to produce an enhanced 'velcro' effect, which augments the overall avidity of the binding interaction. It is likely that this effect has its origin in a favorable combination of variable thermodynamic and entropic terms in both the microscopic and macroscopic complexes. In the discussion that follows, several related strategies have been employed to generate suitable dendritic scaffolds for presentation of these carbohydrate motifs.

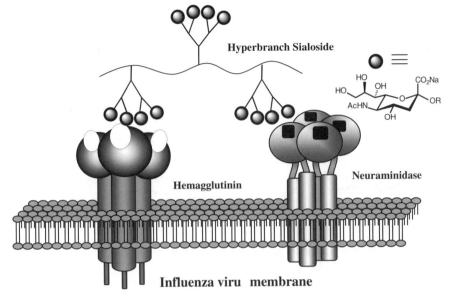

Figure 15.2 Hyperbranched sialosides and related glycopolymers can inhibit microbial attachment to host tissues by blocking their receptor sites

3 SUITABLY FUNCTIONALIZED CARBOHYDRATE PRECURSORS

Except for small clusters, our experience in multivalent neoglycoconjugate syntheses suggests that the ideal approach consists of using pre-made scaffolds onto which one can attach the required carbohydrate ligands of interest. This is true whether linear polymers, dendrimers or dendronized polymers were involved. This strategy has the advantage of giving access to alternative saccharide decoys that can be tested as negative control in subsequent biological assays. Moreover, a single scaffold can be fully or partially derivatized, thus allowing accurate evaluation of the sugar density effect on a given scaffold. In the present study, readily available sialosides were used in either fully protected or deprotected forms and thus, ready to be evaluated. In our hands, the preferred and most effective functionalities for conjugating these sugar moieties to scaffolding included: carboxylic acids, aldehydes, thiols (or thioacetates), bromo- or chloroacetamides or isothiocyanates (**1–6**, Scheme 1). Interestingly, amine-ending scaffolds gave highly reproducible results with aldehydes by reductive amination and with isothiocyanates by thiourea formation. Scheme 1 illustrates typical but not exclusive examples that are obviously applicable to other carbohydrate derivatives.

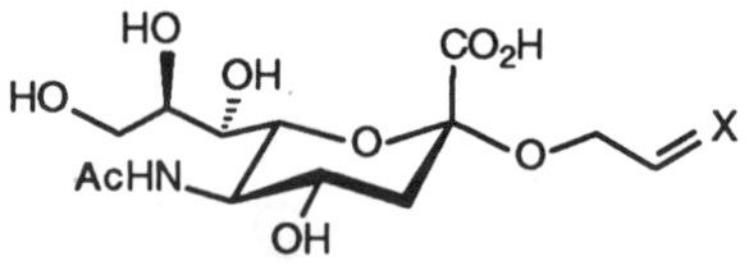

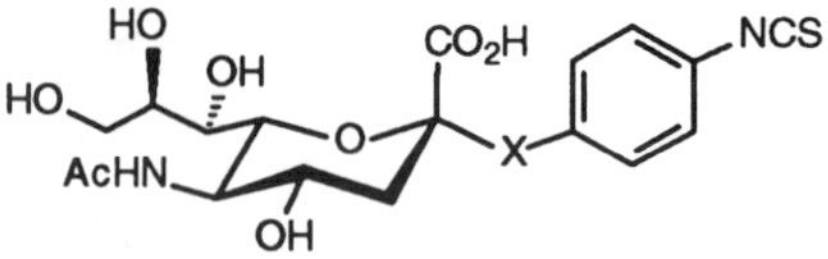

1 X = CH$_2$
2 X = O

3 X = O
4 X = S

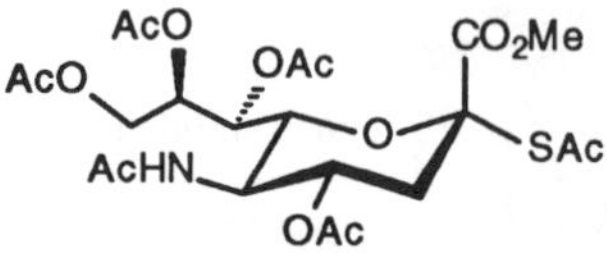

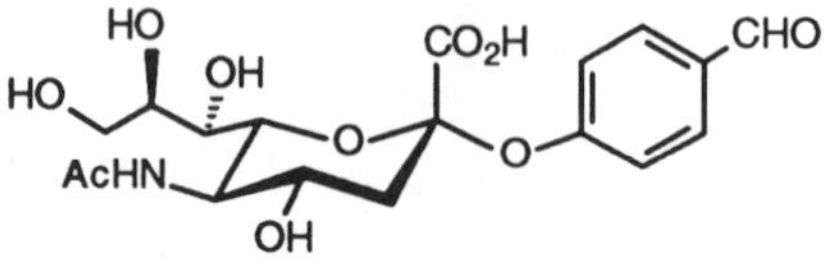

5   6

Scheme 1   Sialoside precursors having varied functionalities for direct attachment to hyperbranched or dendritic scaffolds

4   CALIX[4]ARENE AND β-CYCLODEXTRIN AS SIALOSIDE SCAFFOLDS

Calix[4]arenes and cyclodextrins (CDs) have found widespread applications as molecular scaffolds and templates in supramolecular chemistry (Scheme 2) [48]. In particular, CDs have been used as a core in the synthesis of medium-sized glycoconjugates with the aim of developing new vectors for site-specific delivery of therapeutics [49]. These structures combine both the inclusion capabilities of the hydrophobic cavity of CDs and the high biological receptor binding ability of multiple saccharide epitopes in the same molecule. Most CD-based glycoconjugates synthesized to date are monosubstituted derivatives at a single primary position of the CDs. Recently, the monoconjugation of β-mannosylated dendritic branches has been described [50]. Persubstituted CDs are scarce and only a few of them contain mono-or di-saccharidic monovalent branches, as well as divalent branches. Grafting of the saccharides to the CD core has been performed in the majority of cases by nucleophilic displacement or by means of amide or thiourea linkages obtained by reaction of isothiocyanates with amines [51]. In an effort to design multivalent glycoforms as a function of molecular weight, shape, valency and geometry, we prepared a variety of per-sialylated calix[4]arenes [13] and β-cyclodextrins [52].

Tert-butyl calix[4]arene possessing four antennary amine groups (**7**) were initially synthesized from commercially available phenolic *tert*-butyl calix[4]arene according to a published procedure [13]. After derivatization of the phenolic group by alkylation with ethyl bromoacetate, hydrolysis, acid

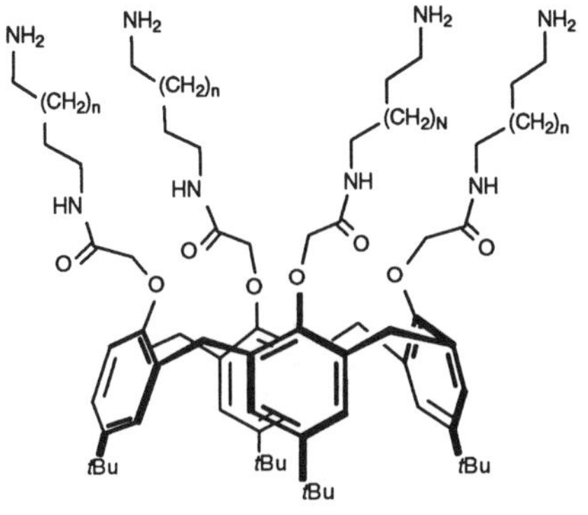

Tert-butyl calix[4]arene

7

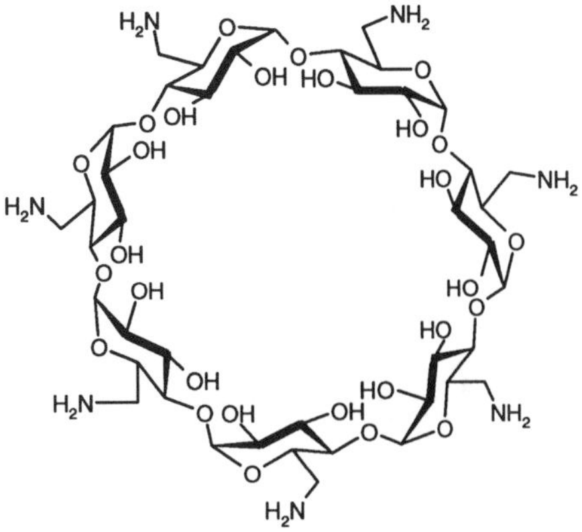

β-Cyclodextrin

8

Scheme 2 Examples of polyaminated calix[4]arene **7** and β-cyclodextrin scaffolds **8**

chloride formation, treatment with mono-Boc-protected hexamethylenediamine, and finally deprotection of the Boc-protecting groups, tetramine **7** (Scheme 2) was obtained. Using a novel strategy, each of the amine group from tetramine **7** was treated with sialic acid bromoacetamide derivative **9** to provide double *N*-alkylation of the primary amines [22] to generate an octameric derivative which upon full deprotection with sodium hydroxide afforded branched sialoside **10** (Scheme 3) [53].

In order to avoid the construction of congested architectures that would result from the direct attachment of sialic acid derivative such as **4** at the convergent CD's lower rim in amine **8**, a derivative having a long spacer arm was built. Based on our previous experience, β-CD containing an extended chloroacetamido function at the primary position appeared to be the template of choice upon which to perform such a substitution reaction [51]. The previously described *N*-cloroacetamide derivative, obtained by addition of chloroacetic anhydride to a suspension of per-6-amino-6-deoxy-β-CD (**8**, Scheme 2), was further O-acetylated prior to spacer arm extension to facilitate the isolation and characterization of the resulting intermediates. Successive treatment with thiolated 6-*N*-Boc-hexamethylenediamine derivative, trifluoroacetolysis and coupling with the protected form of isothiocyanate **4** afforded heptameric β-CD cluster **11** after protection group removal (Scheme 4) [52]. Preliminary biological assays of these novel sialoside conjugates with model receptors showed interesting protein cross-linking abilities.

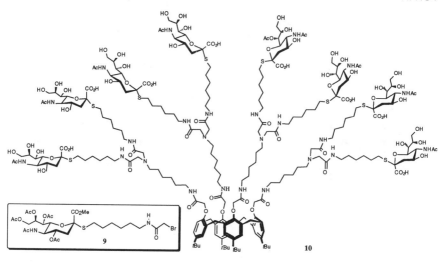

Scheme 3 Octameric *p-tert*-butyl calix[4]arene prepared by a double *N*-alkylation strategy of tetramine **7** with *N*-bromoacetamido sialoside **9**

5 GLYCODENDRIMERS BASED ON POLY(AMIDOAMINES) (PAMAM) DENDRIMERS

Prebuilt poly(amidoamine) dendrimers offer several advantages as scaffolding for biologically relevant glycodendrimer syntheses. First, the core structure has been shown to be nontoxic and nonimmunogenic [54], even when carbohydrate haptens are linked to them [23]. Secondly, they are readily available [55] and can be efficiently transformed by routine coupling methodologies. The exposed surface amine functionality can directly react with lactones [33], acids through amide coupling reagents [21, 23], aldehydes by reductive amination [16], and with isocyanates or isothiocyanates [4, 15b]. A wide range of carbohydrate ligands have been attached to PAMAM dendrimers, including sialic acid, sialyloligosaccharides, simple sugars and cancer markers such as the T and the Tn antigens (Table 15.1). Because of their spheroidal nature and the increased surface group congestion as a function of generation, higher generation glycodendrimers are difficult to fully substitute. On the other hand, we showed that G5 (128 surface sites) can still be completely functionalized [35]. In all cases observed so far, a plateau of inhibitory potency is reached as a function of dendrimer scaffolding generation above which the binding interactions begin to diminish [4]. This is simply due to the surface congestion induced inaccessibility of individual sugars toward their protein receptors. Moreover, we also observed that the structures (antibodies vs globular proteins) and valency (number of

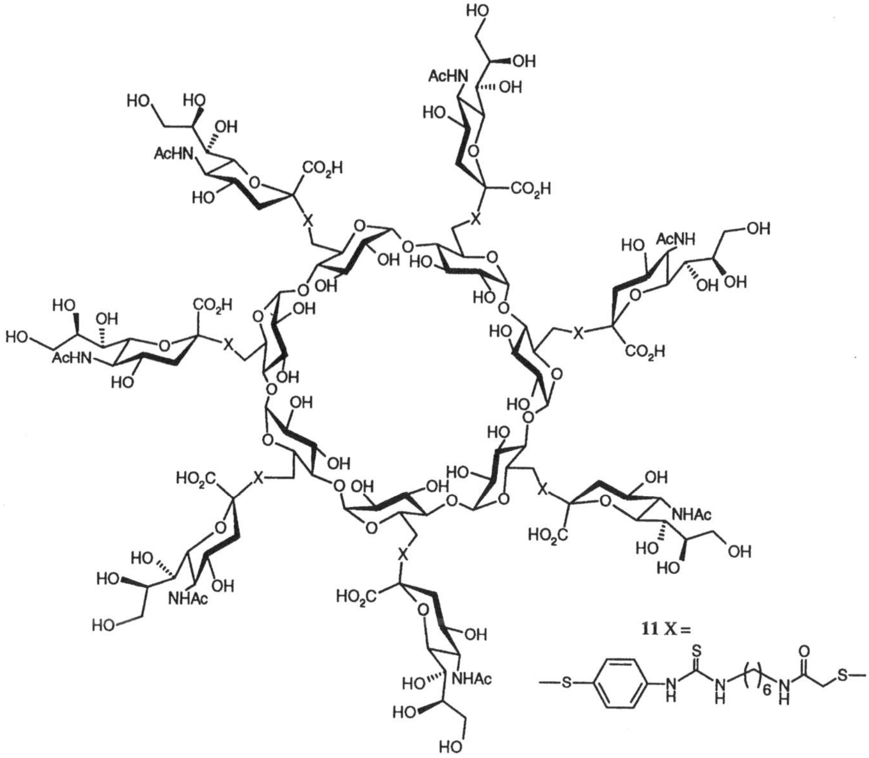

Scheme 4 Heptameric α-thioaryl sialosides at the convergent rim of β-cyclodextrin using thiourea linkages

binding sites per protein) were critical factors that determine maximize individual binding interactions [35].

Scheme 5 below illustrates the structure of a fully sialylated G3 PAMAM dendrimer (32-mer) [15b]. It was prepared by simply reacting peracetylated p-isothiocyanophenyl α-sialoside such as **4** with commercially available PAMAM. The extent of substitution could be easily monitored during the coupling reaction using a ninhydrin test. At this rather low level of molecular weight, high field proton NMR can still give an accurate level of incorporation. It was demonstrated that mono-dendron-type scaffolding could be equally well synthesized using solid-phase synthesis [15a, 15c, 56]. In fact, these sialo(dendrons) represented the first case of glyco(dendritics) made available in 1993. Moreover, an analogous version built on a 3,3'-iminobis(propylamine) core showed even better binding properties toward the slime lectin *Limax flavus* perhaps due to the exposed sialoside residues which were better interspaced [15d, e].

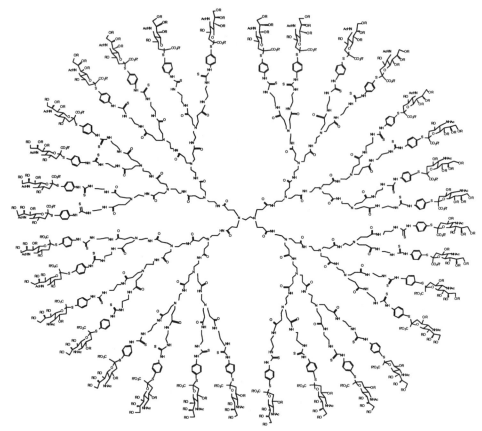

Scheme 5 Typical example of poly(amidoamine) PAMAM dendrimer (G3, 32-mer) fully substituted with α-sialosides. Several such glycodendrimers are readily available

6   POLY(ETHYLENEIMINE) SCAFFOLD TOWARD HYPERBRANCHED SIALOSIDES

When evaluated on a per saccharide basis, the modest inhibitory improvement generally observed with dense packed glycodendrimers suggests that individual carbohydrate ligands are experiencing steric crowding toward their homologous protein receptors. On the other hand, glycopolymers based on linear, random coil scaffolding were generally showing the desired strong cluster effects. These observations have led to the design of two novel approaches. In the first case, the direct consequences of ligand crowding was solved by synthesizing poly(amidoamine) dendrimer scaffolding possessing interspaced functionalities as schematized by structure **18** (Figure 15.3). The scaffolds for these glycodendrimers were readily constructed using the usual methyl acrylate/ethylenediamine strategy to PAMAM dendrimers, except that, at higher half generation levels, a few surface methyl ester groups were capped with tris(hydroxymethyl)amine [14]. This provided a reduction in surface amine functionality as one advanced to the next generation. The surface amine functionality were both better interspaced, as well as enhanced with water-solubilizing hydroxyl groups. Finally, the residual amines were treated with sialic acid p-isothiocyanatophenyl derivative **4** (Scheme 1) to provide heterobifunctional sialo-PAMAM dendrimers having varied levels of substitutions. For instance, G5 PAMAM series having OH/Neu5Ac ratios varying from 50: 50 to 67: 33 were produced [14].

In the second strategy, it was decided that both dendrimer and linear-polymer properties should be combined into a single architecture. To this end, comb-branched (**15**), dendrigraft (**16**), and hyperbranched (**17**) structures were prepared (Scheme 6) [14]. Using a linear- poly(ethyleneimine) (PEI) backbone (**14**), the three types of structures described above were prepared using further various divergent construction schemes. The first strategy involved the synthesis of dendrigraft scaffolding. This involved grafting poly(2-ethyl-2-oxazoline) (PEOX) onto a PEI backbone (**14**); followed by amide hydrolysis of the comb-branched precursor (**13**) (30%) and then regrafting to give PEI sequences as a part of the branches (**16**) [57]. These secondary amine sites were used to further build up PAMAM onto the PEI backbone (**17**) [58]. The PEIs were prepared by living cationic polymerization of 2-ethyl-2-oxazoline (**12**) with methyl tosylate, capping with morpholine, followed by amide hydrolysis of PEOX **13** (aq. H_2SO_4) to provide the PEIs (**14**) with various molecular weights (Scheme 6). By capping the accessible amine residues with sialoside derivatives **4**, various hyperbranched glycopolymer conjugates were obtained and demonstrated as successful inhibitors in the hemagglutination of erythrocytes by different influenza virus strains (Sendai, X-31 H3N2, H2N2 mouse, and H2N2 chicken). The most potent of the linear and spheroidal inhibitors (**18**) were 32-256-fold more effective than the monomeric sialoside against H2N2 virions. Linear-dendron copolymers were 1025–8200-fold more effective against H2N2, X-31 and sendai viruses. *The most potent were the comb-branched and dendrigraft inhibitors* which showed up to 50 000-fold increased activity against these viruses. Dose-dependent reductions of influenza infection in mammalian cells were demonstrated. These new biomaterials were not cytotoxic to mammalian cells at therapeutic levels. In order to further increase the therapeutic potential of such biomolecules, other polymeric backbones possessing amine functionality were investigated.

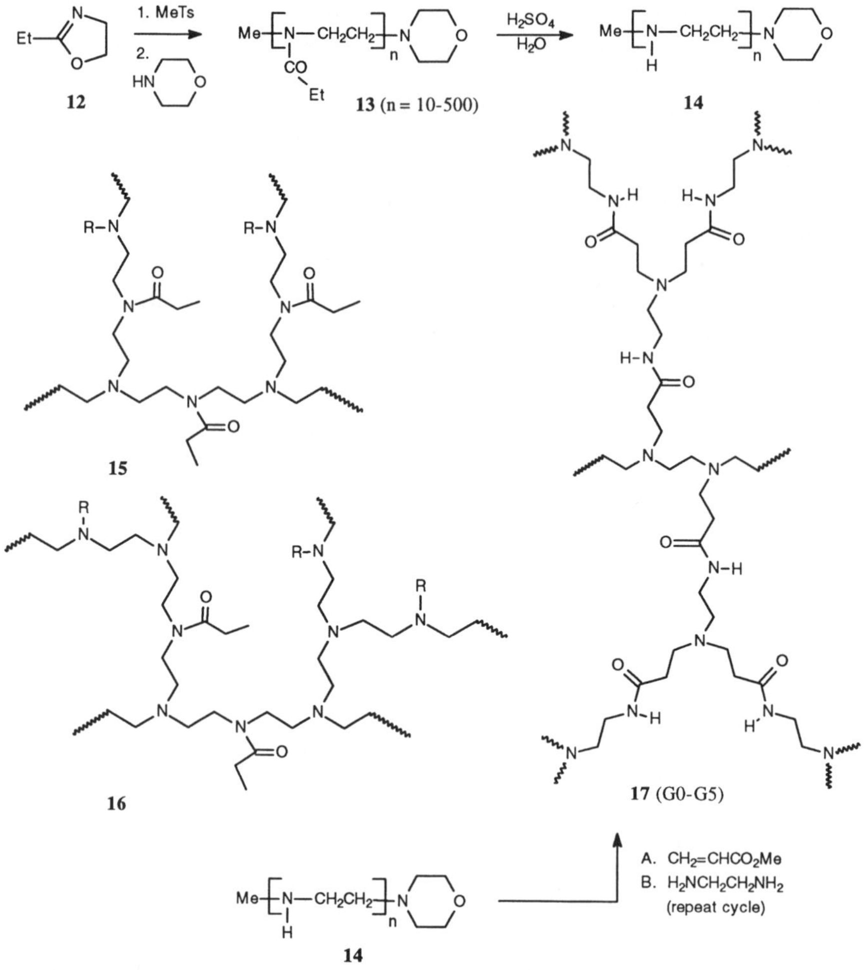

Scheme 6 Representative examples of hyperbranched base polymers used for sialic acid attachment and subsequent inhibition of flu virus infection. **14**: Poly(ethyleneimine) (PEI) backbone; **15**: comb-branch polymer; **16**: dendrigraft polymer; **17**: PAMAM scaffolded on PEI

7   CHITOSAN AS POLYSACCHARIDE SCAFFOLD TOWARD HYPERBRANCHED SIALOSIDES

Chitin and the deacetylated form, chitosan, are attractive linear amino polysaccharides found in the shell of crustaceans. These abundant biopolymers, composed mainly of poly(β-(1,4)-2-acetamido/2-amino-2-deoxy-D-glucopyranose)

repeating units constitute valuable biomass resources useful in the preparation of functional biomaterials. Since chitosan itself is nontoxic, biodegradable and shows widespread biological activities, it is an appealing bioactive polymer for further development [59]. Unfortunately, its poor solubility in both organic solvents and aqueous solutions has hampered its widespread development. Recently, this problem has been partly overcome by using counter anions of organic acids [60].

Dendronized polymers, on the other hand, are also attractive because of their rod-like nanostructures. Although several investigations have been published toward the synthesis of dendronized polymers [61], there are few reports on dendronized polysaccharide, especially related to chitosan backbone. The preparation of a chitosan–dendrimer hybrid for the purposes of generating novel hyperbranched chitosans is described herein. Figure 15.3 illustrates several existing strategies that may be used to obtain dendronized polymers. Recently, we prepared chitosan-sialodendrimer containing tetra(ethylene glycol) type spacers [62]. The synthesis of these hybrids (path A, Figure 15.3) was limited, however, due to low dendron attachment. This was particularly true at higher

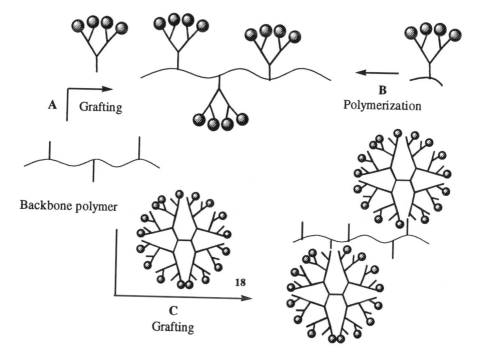

Figure 15.3 Various strategies leading to dendronized or hyperbranched glycopolymers. (A) grafting of dendrons to pre-formed polymers; (B) polymerization of dendron monomers; (C) grafting dendrimers on pre-formed polymers

generations, undoubtedly due to steric hindrance inherent to forming these sialodendrimers. The stepwise coupling of surface preformed dendrimers is expected to avoid such problems (path C, Figure 15.3). Moreover, this type of hybrid is a novel form of dendronized form of linear polymer wherein the preformed dendrimers are attached directly to the linear scaffolding. In this study, the successful preparation of surface bound chitosan–sialodendrimer hybrids using a doubly convergent approach (path C) is reported [63].

We initially prepared the first chitosan–sialoside hybrid **19** by treating 80% de-N-acetylated chitosan with p-formylphenyl α-sialoside **6** (Scheme 1) under reductive amination conditions (NaBH$_3$CN) (Scheme 7) [64]. The level of sialoside incorporation could be controlled by increasing the amount of **6** (Table 15.2). The reactivity of aldehyde **6** toward chitosan was found to be in the range 25–48% due to excessive reduction of the aldehyde under the acidic reaction conditions. Water-soluble materials were only achieved at high DS (DS $\geq$ 0.53). Sialo-hybrids with lower substitutions were further derivatized with succinic

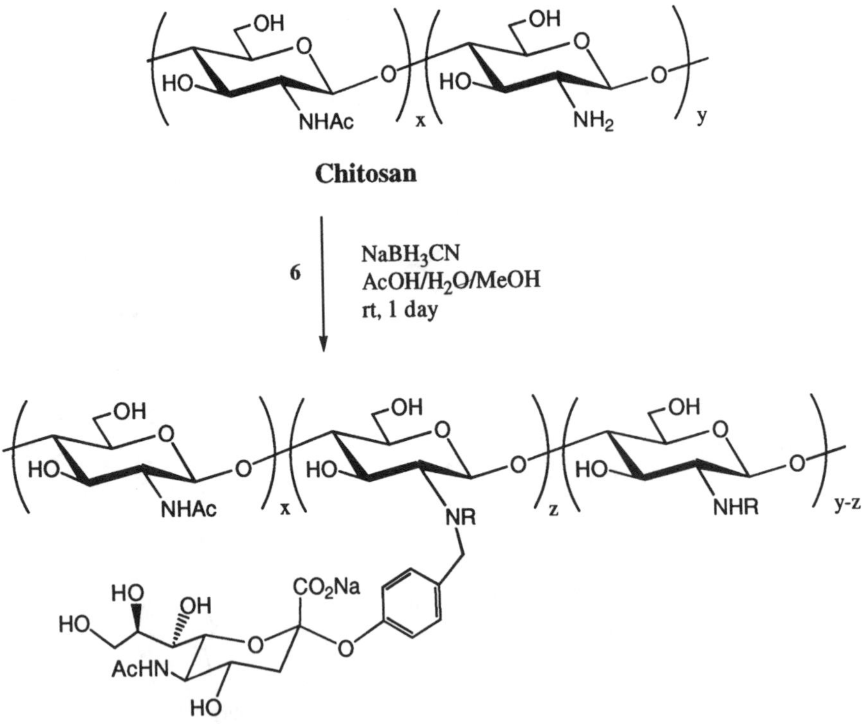

Chitosan

6 | NaBH$_3$CN
AcOH/H$_2$O/MeOH
rt, 1 day

19 R = H
20 R = H or COCH$_2$CH$_2$CO$_2$Na

Scheme 7 Chitosan-sialic acid hybrid as branched polysaccharide

Table 15.2 Preparation of chitosan-sialoside hybrids **19** with aldehyde **6** (Scheme 7)

Entry	6 (equiv.)	DS	Yield (%)	Water-solubility	MW (KDa)
1	0.2	0.06	100	No	27
2	0.4	0.10	77	No	29
3	0.6	0.29	76	No	39
4	0.9	0.44	74	No	48
5	1.2	0.53	84	Yes	53

anhydride to provide water solubility (**20**), a condition necessary for biological applications. All these derivatives **20** showed strong binding with the plant lectin wheat germ agglutinin (WGA), thus demonstrating the validity of the approach [64].

With the proof of concept in hand, the challenging synthesis of hyperbranched chitosan-sialodendrimer hybrids was undertaken. As discussed above (Figure 15.3), it is conceivable to synthesize these dendronized polymer hybrids using either convergent or semi-convergent approaches. In initial attempts, sialodendrons having an aldehyde functionality at the focal point were build up using the reiterative PAMAM strategy [63].

The desired long spacer necessary to allow multiple branch construction, was obtained with commercial tetra(ethylene glycol) which was readily modified into aminoacetal **21** in five steps (Scheme 8). Using **21** as an amine source, poly(amidoamine) (PAMAM) dendrons were prepared according to published protocol using reiterative cycles of methyl acrylate and ethylenediamine reactions. The terminal amines of each generation (G = 1, G = 2 and G = 3, **22a-c**) were obtained quantitatively from each of the half generation methyl ester precursors.

The sialic acid residues were initially attached onto each dendron by reductive N-alkylation with known p-formylphenyl α-sialoside **6a** using $NaCNBH_3$ in methanol. Since 3 equiv of aldehyde **6a** were used to ensure complete amine substitutions, partial N,N-disubstitution of **22a-c** inevitably occurred (on average, 10 mol sialosides were bound to 8 mol of amine in octameric dendron **23c**, with three sialosides in G1 **23a** and five in G2 **23b**. The deprotection of dendritic sialoacetal **23a–c** was carried out with 80% aqueous trifluoroacetic acid (CF_3CO_2H) at room temperature for 16 h to provide aldehydes **24a–c**. After purification, aldehydes **24a–c** were then used directly in the reductive amination of chitosan (Scheme 9). The deprotection of O-acetyl groups and methyl ester of the α-sialoside moieties was performed with 0.1 M NaOH at room temperature for 2 h to afford hyperbranched hybrids **25a–c**. The degree of substitutions (DS) were determined to be 0.08 (**25a**, 47%), 0.04 (**25b**, 25%), and 0.02 (**25c**, 25%), which suggests that only a small proportion of aldehyde **24a–c** reacted with chitosan. The low reactivity in the third generation could be ascribed to the steric

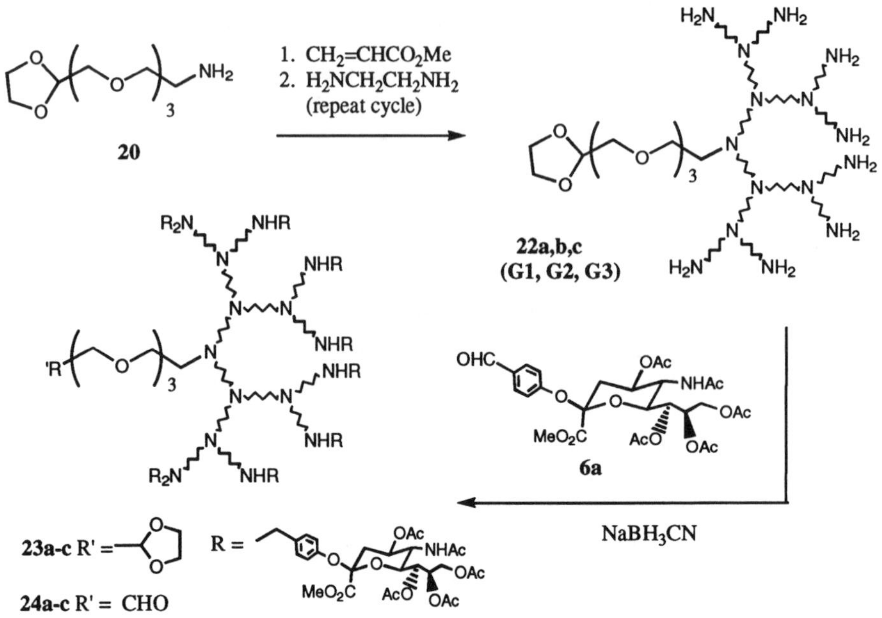

Scheme 8 Synthesis of PAMAM dendrons having aldehyde functionality at the focal point and sialylated dendrons using reductive amination

hindrance of the high molecular weight **24c** (MW = 7567). From the DS values (0.02–0.08) of the products and the DP [140] in the original chitosan, it was assumed that an average of 2.8–11.2 molecules of sialodendrimers were attached per molecule of the polysaccharide (approximately 28–33.6 sialoside residues/polysaccharide chain) [63].

3,4,5-Trihydroxybenzoic acid (gallic acid)-based sialodendrons were next synthesized using commercially available tri(ethylene glycol) as spacer arms and following our previous methodology [15f, 53]. The hydrophilic spacer was chosen to ensure advantageous water solubility of the resulting hybrid and to counteract the hydrophobic effect of the aromatic gallic acid used as the focal backbone. The terminal carboxyl group of the gallic acid core was modified into a focal acetal precursor of **26a** by simple coupling with commercially available aminoacetaldehyde diethyl acetal using carbodiimide (EDC) and hydroxybenzotriazole, followed by reiterative build up of the next generation of poly(amino) dendrons as described (Scheme 10). Attachment of peracetylated derivative of **3** using urea linkages and hydrolysis of acetal with aq. 80% CF_3CO_2H at room temperature (25°C) for 1 day provided fully sialylated aldehyde **26b**. Noteworthy is the fact that the O-sialoside linkage has resisted the hydrolysis step owing to the presence of the ester protecting groups that confer higher stability to the molecule.

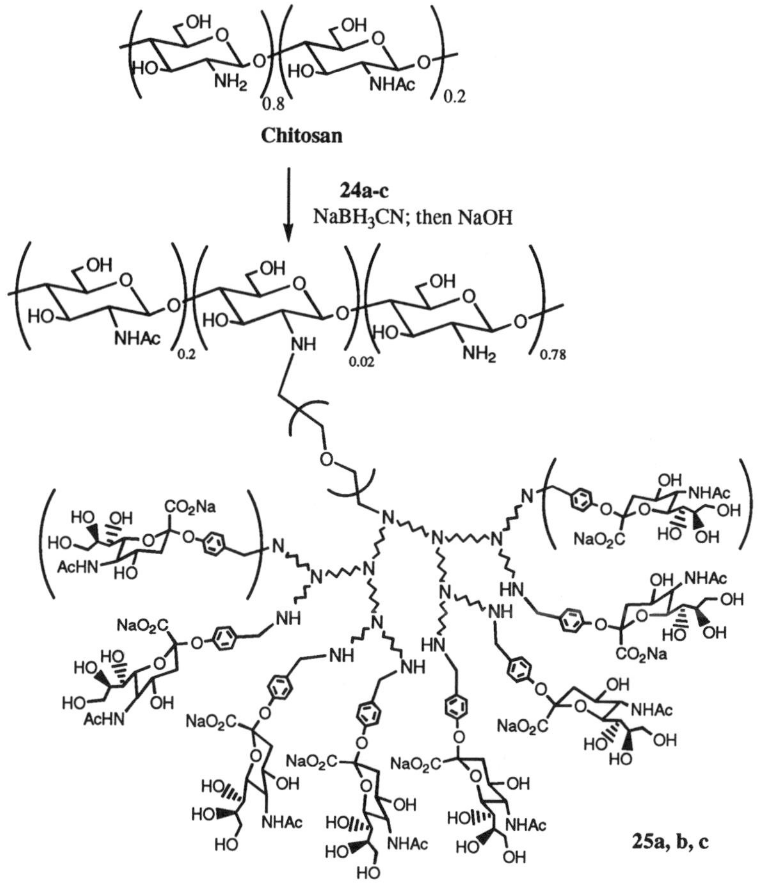

Scheme 9 Attachment of focal sialodendrons to chitosan backbone by reductive amination

Scheme 11 shows the structure of sialodendron aldehyde **26a** that was efficiently coupled to chitosan by reductive amination with $NaCNBH_3$ as shown above for **19** (Scheme 7) to give **27** after ester group hydrolysis. The degree of substitution (DS) of the trimer was shown to be 0.13 by NMR which indicates that 87% of aldehyde trimer equivalent to **26b** (not shown) conjugated to the primary amino groups of chitosan. The nonavalent dendronized chitosan hybrid **27** was also successfully prepared from aldehyde **26b** (0.15 equiv) in 80% yield. The DS of hybrid **27** was 0.06 and only 40% of aldehyde **26b** could be attached to chitosan. The lower reactivity of **26b** could be due to the steric hindrance of the higher molecular weight aldehyde (MW = 7866), compared with its trimeric analog (MW = 2361). Both trimeric and 9-mer hybrids **27** were only slightly soluble in neutral water and thus not as useful in biological evaluation. To

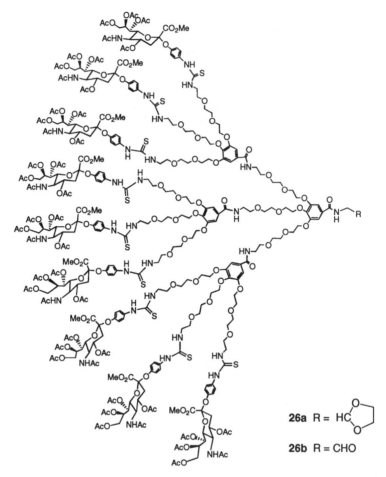

26a R = HC (with cyclic acetal/dioxolane)

26b R = CHO

Scheme 10 Sialylated dendron having a focal aldehyde group and gallic acid scaffold

further improve their water solubility, the remaining unmodified amino groups of the chitosan backbone were succinylated as before with excess succinic anhydride to provide **27** ($R = COCH_2CH_2CO_2Na$) [63].

Tomalia et al. reported that the surface amines of PAMAM dendrimer can successfully react with methyl esters of other PAMAM dendrimers to afford 'core–shell tecto-(dendrimer) molecules' [65]. Furthermore, they also reported the synthesis of rod-shaped cylindrical dendronized polymers from poly(ethyleneimine) cores without any crosslinking, albeit with the use of excess reagents [58]. These reports lead us to propose a new approach toward hybridized dendrimers and polymers (path C, Figure 15.3). As shown in Scheme 12,

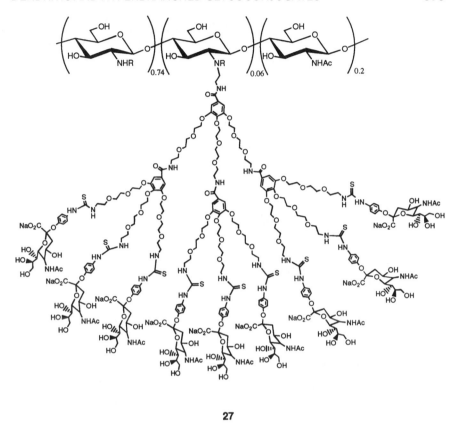

27

Scheme 11 Gallic acid sialylated dendron branched on chitosan backbone prepared by a convergent approach

high generation PAMAM dendrimers (G = 2–5) were indeed successfully attached to chitosan ester **28** to give chitosan–dendrimer hybrid **29** without gel formation. The absence of gel formation strongly suggests that no intramolecular crosslinking had occurred. The slow reaction under the heterogeneous conditions may have been responsible for the lack of intramolecular crosslinking. The results of these chitosan dendronizations are summarized in Table 15.3. As expected, the DS of the hybrids gradually decreased with increasing dendrimer generation. This indicates again that steric hindrance of the underivatized PAMAM dendrimers as a function of a generation affects the reactivity, but to a much lesser extent than those of previously synthesized preformed sialodendrimers. All hybrids were soluble in acidic water (0.2 M HCl). Especially, low generation (G = 1 and 2) hybrids were soluble even in neutral water.

Sialodendrimer-chitosan hybrid (**30a–e**) and their *N*-succinates (*R* =

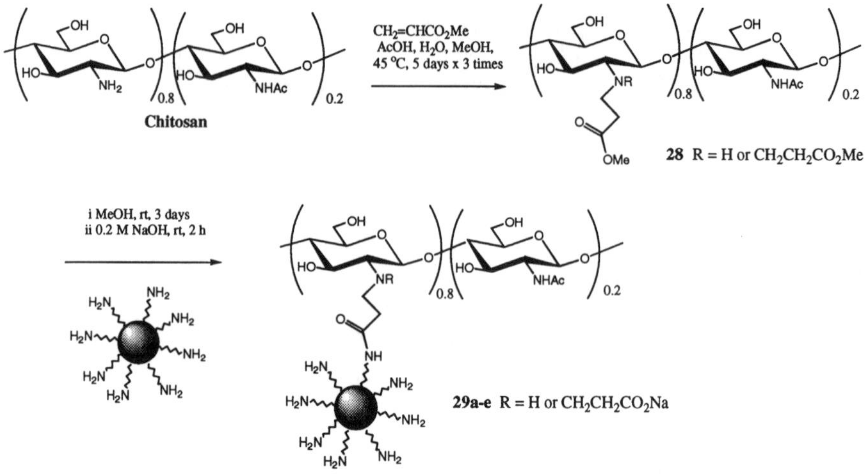

Scheme 12 PAMAM dendrimers attached to chitosan backbone modified with acrylate residues

Table 15.3 Reaction of N-carboxyethylchitosan ester **28** with PAMAM[a]

Dendrimer[b]						Solubility	
PAMAM	NH_2/CO_2Me	Product	% Yield[c]	DS		H_2O	0.2 M HCl
G = 1	4	**29a**	70	0.53		Yes	Yes
G = 2	8	**29b**	68	0.40		Yes	Yes
G = 3	16	**29c**	77	0.21		No	Yes
G = 4	32	**29d**	70	0.17		No	Yes
G = 5	64	**29e**	86	0.11		No	Yes

[a] Solvent: MeOH, rt, 3 days.
[b] Dendrimer (equiv/ CO_2Me) = 1.0.
[c] Yield was estimated on the basis of the weight of **28**.

$COCH_2CH_2CO_2Na$) were next prepared. Since several primary amino groups are available in these chitosan–dendrimer hybrids (**29a–e**), these amines would be useful for further modifications. We tested the attachment of peracetylated p-formylphenyl α-sialoside **6a** to the various hybrids by reductive amination (Scheme 13). The results are summarized in Table 15.4. Although an excess amount of sialoside **6a** (3.0 equiv/ NH_2) was used, 22–23% of it was attached to dendrimer G = 1–3 and 64–70% of primary amines were N-alkylated. The reactivity decreased with increasing generation, especially above G = 4. Several hybrids (**30c–e**) were insoluble in neutral water and thus were be useless for biological evaluation. The remaining amino groups of all hybrids (**30a–e**) were transformed by N-succinylation. The N-succinylated hybrids were obtained

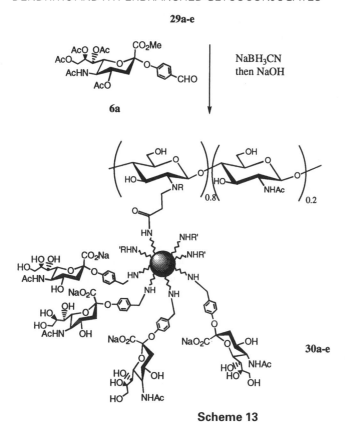

Scheme 13

Table 15.4 Preparation of sialodendrimer-chitosan hybrid **30a–d** (Scheme 12)[a]

PAMAM-Chitosan 29a–e	DS (NH_2)	Sialoside hybrid	DS (Neu5Ac)	Reactivity (%)[b]
G = 1	1.0	**30a**	0.70	23
G = 2	2.0	**30b**	1.40	23
G = 3	2.1	**30c**	1.35	22
G = 4	4.0	**30d**	1.27	11
G = 5	5.0	**30e**	1.21	8

[a] Aldehyde **6a**, 3.0 equiv/NH_2.
[b] % = (DS(Neu5Ac)/DS(NH_2)) × 3) × 100.

quantitatively and a dramatic increase in the water solubility was observed. Although the actual DS value of succinyl group in hybrids **30a–e** could not be directly estimated by ^1H NMR because of partial overlapping signals, most primary amines in the PAMAM–dendrimer moiety and part of the secondary amines in both dendrimer moiety (sialic acid branches) and chitosan backbone were *N*-succinylated [63]. These novel chitosan-sialo-PAMAM hybrids are presently being evaluated in various antimicrobial adhesin assays.

8 CONCLUSION

The design and successful syntheses of several clustered and hyperbranched and dendritic sialic acid neoglycoconjuagtes have been described. They are defining new routes as routes to novel biomacromolecules with the potential to inhibit microbial adhesion to host tissues. Many of the new structures described in this chapter differ from previous copolyacrylamides due to the following important attribute. Firstly, they are likely to be less toxic, and secondly represent completely new architectural presentations [14], [62–64]. Because most biomacromolecules described herein are initially targeted to respiratory, gastrointestinal and urinary tract infections, their large molecular weights will constitute a positive aspect, not a negative one, as usually required for small drug molecules. The strategies described herein are being applied to other important carbohydrate antigens listed in Table 15.1.

9 REFERENCES

1. Varki, A. *Glycobiology*, **3**, 97 (1993).
2. Karlsson, K. A. *Curr. Opin. Struct. Biol.*, **5**, 622 (1995).
3. Lis, H. and Sharon, N. *Chem. Rev.*, 637 (1998).
4. (a) Zanini, D. and Roy, R. in Chapleur, Y. (ed.), *Carbohydrate Mimics: Concepts and Methods*, Verlag Chemie, Weinheim, Germany, 1998, p. 385; (b) Roy, R. *Topics Curr. Chem.*, **187**, 241 (1997); (c) Roy, R. in Boons, G. J. (ed.), *Carbohydate Chemistry*, Chapman & Hall, London, 1998, p. 243; (d) Roy, R. in Khan, S. H. and O'Neil, R. (eds), *Modern Methods in Carbohydrate Synthesis*, Academic Publishers, Amsterdam, Harwood, 1996, p. 378.
5. Mammen, M., Choi, S. K. and Whitesides, G. M. *Angew. Chem. Int. Ed.*, **37**, 2754 (1998).
6. (a) Kiessling, L. L. and Pohl, N. L. *Chem. Biol.*, **3**, 71 (1996); (b) Gabius, H.-J. and Romero, A. *Carbohydr. Eur.*, **23**, 24 (1998); (c) Bovin, N. V. and Gabius, H.-J. *Chem. Soc. Rev.*, **24**, 413 (1995).
7. Lee, Y. C. and Lee, R. T. *Acc. Chem. Res.*, **28**, 321 (1995); (b) Lee, R. T. and Lee, R. T., in Lee, Y. C. and Lee, R. T. (eds), *Neoglycoconjugates: Preparation and Applications*, Academic Press, San Diego, 1994, p. 23.
8. (a) Matrosovich, M. N. *FEBS Lett.*, **252**, 1 (1989); (b) Pritchett, T. J. and Paulson, J. C. *J. Biol. Chem.*, **264**, 9850 (1989).
9. (a) Brewer, C. F. *Chemtracts-Biochem. Molec. Biol.*, **6**, 165 (1996); (b) Toone, E. J. *Curr. Opin. Struc. Biol.*, **4**, 719 (1994).
10. Lemieux, R. U. *Acc. Chem. Res.*, **29**, 373 (1996).
11. (a) McAuliffe, J. C. and Hindsgaul, O. *Chem. Industry*, 170 (1997); (b) Yarema, K. J.

and Bertozzi, C. R. *Curr. Opin. Chem. Biol.*, **2**, 49 (1998).
12. (a) Roy, R. in Witczak, Z. J. and Nieforth, K. A. (eds), *Carbohydrates in Drug Design*, Marcel Dekker, New York, 1997, p. 83; (b) Witczak, Z. B. *Curr. Med. Chem.*, **6**, 165 (1999); (c) von Itzstein, M. and Colman, P. *Curr. Opin. Struct. Biol.*, **6**, 703 (1996).
13. Meunier, S. J. and Roy, R. *Tetrahedron Lett.*, **37**, 5469 (1996).
14. Reuter, J. D., Myc, A., Hayes, M. M., Gan, Z., Roy, R., Qin, D., Yin, R., Piehler, L. T., Esfand, R., Tomalia, D. A. and Baker Jr, J. R. *Bioconjugate Chem.*, **10**, 271 (1999).
15. (a) Roy, R., Zanini, D., Meunier, S. J. and Romanowska, A. *ACS Symp. Ser.*, **560**, 104 (1994); (b) Zanini, D. and Roy, R. *J. Org. Chem.*, **63**, 3486 (1998); (c) Llinares, M. and Roy, R. *Chem. Commun.*, 2119 (1997); (d) Zanini, D. and Roy, R. *J. Am. Chem. Soc.*, **119**, 2088 (1997); (e) Zanini, D. and Roy, R. *J. Org. Chem.*, **61**, 7348 (1996); (f) Meunier, S. J., Wu, Q., Wang, S. N. and Roy, R. *Can. J. Chem.*, **75**, 1472 (1997).
16. (a) Thompson, J. P. and Schengrund, C. L. *Glycoconjugate J.*, **14**, 837 (1997); (b) Mitchell, J. P., Roberts, K. D., Langley, J., Koentgen, F. and Lambert, J. N. *Bioorg. Med. Chem. Lett.*, **9**, 2785 (1999).
17. Fan, E., Zhang, Z., Minke, W. E., Hou, Z., Verlinde, C. L. M. J. and Hol, W. G. J. *J. Am. Chem. Soc.*, **122**, 2663 (2000).
18. (a) Roy, R., Zanini, D., Romanowska, A., Meunier, S. J., Park, W. K. C., Gidney, M. A., Harrison, B., Bundle, D. R. and Williams, R. E. *XVIIth Int. Carbohydr. Symp.*, B2.31 (1994); (b) Gan, Z. and Roy, R. *XVIIIth Int. Carbohydr. Symp.*, BP 178 (1996); (c) Earle, M. A., Manku, S., Hultin, P. G., Li, H. and Palcic, M. M. *Carbohydr. Res.*, **301**, 1 (1997).
19. Palcic, M. M., Li, H., Zanini, D., Bhella, R. S. and Roy, R. *Carbohydr. Res.*, **305**, 433 (1998).
20. (a) Roy, R., Park, W. K. C., Zanini, D., Foxall, C. and Srivastava, O. P. *Carbohydr. Lett.*, **2**, 259 (1997); (b) Roy, R., Park, W. K. C., Srivastava, O. P. and Foxall, C. *Bioorg. Med. Chem. Lett.*, **6**, 1399 (1996).
21. Roy, R., Baek, M. G., Xia, Z. and Rittenhouse-Diakun, K. *Glycoconjugate J.*, **16**, S53 (1999).
22. Roy, R. and Kim, J. M. *Angew. Chem. Int. Ed. Engl.*, **38**, 369 (1999).
23. Toyokuni, T. and Singhal, A. K. *Chem. Soc. Rev.*, 231 (1995).
24. Roy, R. and Kim, J. M. *Polym. Mater. Sci. Eng.*, **77**, 195 (1997).
25. Bay, S., Lo-Man, R., Osinaga, E., Nakada, H., Leclerc, C. and Cantacuzène, D. *J. Peptide Res.*, **49**, 620 (1997).
26. (a) Langer, P., Ince, S. J. and Ley, S. V. *J. Chem. Soc., Perkin Trans. 1*, 3913 (1998); (b) Roy, R., Pagé, D., Figueroa Perez, S. and Verez Bencomo, V. *Glycoconjugate J.*, **15**, 251 (1998).
27. Hansen, H. C., Haataja, S., Finne, J. and Magnusson, G. *J. Am. Chem. Soc.*, **119**, 6974 (1997).
28. (a) Matsuoka, K., Terabatake, M., Esumi, Y., Terunuma, D. and Kuzuhara, H. *Tetrahedron Lett.*, **40**, 7839 (1999); (b) Kitov, P.V., Sadowska, J. M., Mulvey, G., Armstrong, G. D., Ling, H., Pannus, N. S., Read, R. J. and Bundle, D. R. *Nature*, **403**, 669 (2000).
29. Zanini, D. and Roy, R. *Bioconjugate Chem.*, **8**, 187 (1997).
30. (a) Pagé, D., Aravind, S. and Roy, R. *Chem. Commun.*, 1913 (1996); (b) Pagé, D. and Roy, R. *Glycoconjugate J.*, **14**, 345 (1997); (c) Pagé, D., Zanini, D. and Roy, R. *Bioorg. Med. Chem.*, **4**, 1949 (1996); (d) Ashton, P. R., Hounsell, E. F., Jayaraman, N., Nilsen, T. M., Spencer, N., Stoddart, J. F. and Young, M. *J. Org. Chem.*, **63**, 3429 (1998).
31. (a) Dubber, M. and Lindhorst, T. K. *Carbohydr. Res.*, **310**, 35 (1998); (b) Kieburg, C. and Lindhorst, T. K. *Tetrahedron Lett.*, **38**, 3885 (1997).
32. Lindhorst, T. K. and Kieburg, C. *Angew. Chem. Int. Ed. Engl.*, **35**, 1953 (1996).

33. Aoi, K., Itoh, K. and Okada, M. *Macromolecules*, **28**, 5391 (1995).
34. (a) Roy, R., Park, W. K. C., Wu, Q. and Wang, S. N. *Tetrahedron Lett.*, **36**, 4377 (1995); (b) Ashton, P. R., Boyd, S. E., Brown, C. L., Nopogodiev, S. A., Meijer, E. W., Peerlings, H. W. I. and Stoddart, J. F. *Chem. Eur. J.*, **3**, 974 (1997); (c) Ashton, P. R., Boyd, S. E., Brown, C. L., Jayaraman, N., Nopogodiev, S. A. and Stoddart, J. F. *Chem. Eur. J.*, **2**, 1115 (1996).
35. André, S., Cejas Ortega, P. J., Alamino Perez, M., Roy, R. and Gabius, H.-J. *Glycobiology*, **9**, 1253 (1999).
36. (a) Sashiwa, H., Thompson, J. M., Das, S. K., Shigemasa, Y., Tripathy, S. and Roy, R. *Biomacromolecules*, **1**, 303 (2000); (b) Roy, R., Thompson, J., Sashiwa, H., Das, S. K., Tripathy, S. and Gabius, H.-J. *Proceedings of Acfas Congress, Ottawa*, c 104, 1999.
37. Tsvetkov, D. E., Cheshev, P. E., Tuzikov, A. B., Pazynina, G. V., Bovin, N. V., Rieben, R. and Nifant'ev, N. E. *Mendeleev Commun.*, 47 (1999).
38. (a) Roy, R. *Trends Glycosci. Glycotechnol.*, **8**, 79 (1996); (b) Roy, R. *Polymer News*, **21**, 226 (1996); (c) Roy, R. *Curr. Opin. Struct. Biol.*, **6**, 692 (1996).
39. (a) Rini, J. M. *Annu. Rev. Biophys. Biomol. Struct.*, **24**, 551 (1995); (b) Loris, R., Hamelryck, T., Bouckaert, J. and Wyns, L. *Biochem. Biophys. Acta*, **1383**, 9 (1998).
40. (a) Magnusson, G., Chernyak, A. Y., Kihlberg, J. and Kononov, L. O. in Lee, Y. C. and Lee, R. T. (eds), *Neoglycoconjugates: Preparation and Applications*, Academic Press, San Diego, 1994, p. 53; (b) Bovin, N. V. *Glycoconjugate J.*, **15**, 431 (1998).
41. Wiley, D. C. and Skehel, J. J. *Annu. Rev. Biochem.*, **56**, 365 (1987).
42. Weiss, W., Cusack, S., Paulson, J. C., Skehel, J. J. and Wiley, D. C. *Nature*, **333**, 426 (1988).
43. (a) Pritchett, T. J., Brossmer, R., Rose, U. and Paulson, J. C. *Virology*, **160**, 502 (1987); (b) Kelm, S., Paulson, J. C., Rose, U., Brossmer, R., Schmid, W., Bandgar, B. P., Schreiner, E., Hartmann, M. and Zbiral, E. *Eur. J. Biochem.*, **205**, 147 (1992).
44. Glick, G. D., Toogood, P. L., Wiley, D. C., Skewel, J. J. and Knowles, J. R. *J. Biol. Chem.*, **266**, 23660 (1991).
45. Sauter, N. K., Bednarski, M. D., Wurzburg, B. A., Hanson, J. E., Whitesides, G. M., Skehel, J. J. and Wiley, D. C. *Biochemistry*, **28**, 8388 (1989).
46. Sabesan, S., Duus, J. Ø., Domaille, P., Kelm, S. and Paulson, J. C. *J. Am. Chem. Soc.*, **113**, 5865 (1991).
47. (a) Spaltenstein, A. and Whitesides, G. M. *J. Am. Chem. Soc.*, **113**, 686 (1991); (b) Matrosovich, M. N., Mochalova, L. V., Marinina, V. P., Byramova, N. E. and Bovin, N. V. *FEBS Lett.*, **272**, 209 (1990); (c) Roy, R., Andersson, F. O., Harms, G., Kelm, S. and Schauer, R. *Angew. Chem. Int. Ed. Engl.*, **31**, 1478 (1992); (d) Gamian, A., Chomik, M., Laferrière, C. A. and Roy, R. *Can. J. Microbiol.*, **37**, 233 (1991); (e) Kamitakahara, H., Suzuki, T., Nishigori, N., Suzuki, Y., Kanie, O. and Wong, C.-H. *Angew. Chem. Int. Ed.*, **37**, 1524 (1998).
48. (a) Nepogodiev, S. A. and Stoddart, J. F. in Boons, G. J. (ed.), *Carbohydrate Chemistry*, Blackie Academic & Professional, London, 1998, p. 322; (b) Conors, K. A. *Chem. Rev.*, **97**, 1325 (1997); (c) van Loon, J.-D., Verboom, W. and Reinhoudt, D. N. *Org. Prep. Proced. Int.*, **24**, 439 (1992).
49. (a) Lancelon-Pin, C. and Driguez, H. *Tetrahedron Lett.*, **33**, 3125 (1992); (b) Derobertis, L., Lancelon-Pin, C., Driguez, H., Attioui, F., Bonaly, R. and Marsura, A. *Bioorg. Med. Chem. Lett.*, **4**, 1127 (1994).
50. Baussanne, I., Benito, J. M., Ortíz-Mellet, C., Fernández, J. M. G., Law, H. and Defaye, J. *Chem. Commun.*, 1489 (2000).
51. N García-López, J. J., Hernández-Matéo, F., Isac-Gracía, J., Kim, J. M., Roy, R., Santoyo-González, F. and Vargas-Berenguel, *J. Org. Chem.*, **64**, 522 (1999).

52. Roy, R., Hernández-Matéo, F. and Santoyo-González, F. *J. Org. Chem.*, **65**, 8743 (2000).
53. Meunier, S., Ph. D. dissertation, University of Ottawa, Canada, 2000.
54. (a) Duncan, R. and Malik, N. *Proc. Int. Symp. Controlled Release Bioact. Mater.*, **23**, 105 (1996); (b) Roberts, J. C., Bhalgat, M. K. and Zera, R. T. *J. Biomed Mater. Res.*, **30**, 53 (1996).
55. Tomalia, A. and Durst, H. D. *Topics Curr. Chem.*, **165**, 193 (1993).
56. Roy, R., Zanini, D., Meunier, S. J. and Romanowska, A. *J. Chem. Soc. Chem. Commun.*, 1869 (1993).
57. Tomalia, D. A., Hedstrand, D. M. and Ferritto, M. S. *Macromolecules*, **24**, 1435 (1991).
58. Yin, R., Zhu, Y, Tomalia, D. A. and Ibuki, H. *J. Am. Chem. Soc.*, **120**, 2678 (1998).
59. (a) Tanigawa, T., Tanaka, Y., Sashiwa, H., Saimoto, H. and Shigemasa, Y. in Brine, C. J., Sandford, P. A. and Zikakis, J. P. (eds), *Advances in Chitin and Chitosan*, Elsevier, London, 1992, p. 206; (b) Minami, S., Okamoto, Y., Tanioka, S., Sashiwa, H., Saimoto, H., Matsuhashi, A. and Shigemasa, Y. in Yalpani, M. (ed.), *Carbohydrates and Carbohydrate Polymers*, ATL Press, Inc., IL, 1993, p. 141.
60. (a) Sashiwa, H., Shigemasa, Y. and Roy, R. *Chem. Lett.*, 862 (2000); (b) Sashiwa, H., Shigemasa, Y. and Roy, R. *Chem. Lett.*, 596 (2000).
61. (a) Schlüter, A. D. and Rabe, J. P. *Angew. Chem. Int. Ed.*, **39**, 864 (2000); (b) Zistler, A., Koch, S. and Schlüter, A. D. *J. Chem. Soc. Perkin Trans.* 1, 901 (1999); (c) Aoi, K., Itoh, K. and Okada, M. *Macromolecules*, **30**, 8072 (1997); (d) Aoi, K., Tsutsumiuchi, K., Yamamoto, A. and Okada, M. *Tetrahedron*, **53**, 15415 (1997); (e) Furuike, T., Nishi, N., Tokura, S. and Nishimura, S. I. *Chem. Lett.*, 823 (1995).
62. Sashiwa, H., Shigemasa, Y. and Roy, R. *Macromolecules*, **33**, 6913 (2000).
63. Sashiwa, H., Shigemasa, Y. and Roy, R. Unpublished data.
64. Sashiwa, H., Makimura, Y., Shigemasa, Y. and Roy, R. *Chem. Comm.*, 909 (2000).
65. (a) Uppuluri, S., Swanson, D. R., Brothers, H. M., Piehler, L. T., Li, J., Meier, D. J., Hagnauer, G. L. and Tomalia, D. A. *Polym. Mater. Sci. Eng.*, **80**, 55 (1999); (b) Li, J., Swanson, D. R., Qin, D., Brothers, H. M., Piehler, L. T., Tomalia, D. A. and Meier, D. J. *Langmuir*, **15**, 7347 (1999).

16

Some Unique Features of Dendrimers Based upon Self-Assembly and Host–Guest Properties

J.-W. WEENER, M.W. P. L. BAARS AND E. W. MEIJER
Laboratory of Macromolecular and Organic Chemistry, Eindhoven University of Technology, Eindhoven, The Netherlands

1 INTRODUCTION

Since the introduction of low molecular weight cascade molecules by Vögtle in 1978 [1] the rapid evolution of dendrimers has led to their application in an impressive, still growing, plethora of research areas [2–12]. This chapter describes the contribution of dendritic macromolecules in the rapidly developing fields of supramolecular host–guest chemistry and self-assembly by noncovalent interactions. In the first part, dealing with the self-assembly of dendrimers, the emphasis will be on the use of amphiphilic dendrimers as multifunctional building blocks in the construction of mono- and multilayers, liquid crystals and supramolecular architectures in solution. Functional materials that utilize specific dendritic properties will be discussed. The, very often unique, conformational behaviour of different dendrimer types within these supramolecular assemblies will be explained. In the second part, an overview of dendritic host–guest systems will be presented. These systems are arranged by the type of interactions from topological encapsulation to hydrophobic, hydrogen bonding, electrostatic and metal–ligand interactions. The critical question concerning the presence or absence of internal void space in dendrimers will be addressed and the ability of certain dendrimers to create a specific micro-environment will be exemplified.

Dendrimers and Other Dendritic Polymers. Edited by Jean M. J. Fréchet and Donald A. Tomalia
© 2001 John Wiley & Sons Ltd

2 SELF-ASSEMBLY OF DENDRIMERS

The design and synthesis of supramolecular architectures with parallel control over shape and dimensions is a challenging task in current organic chemistry [13, 14]. The information stored at a molecular level plays a key role in the process of self-assembly. Recent examples of nanoscopic supramolecular complexes from outside the dendrimer field include hydrogen-bonded rosettes [15, 16], polymers [17], sandwiches [18, 19] and other complexes [20–22], helicates [23], grids [24], mushrooms [25], capsules [26] and spheres [27].

As early as 1993, Tomalia presented the idea that dendrimers are ideal, well-defined building blocks for creating self-assembled nanostructures [28–31]. Meanwhile, dendrimers have indeed been shown to be an exciting class of multifunctional building blocks in the versatile construction of (a) mono- and multilayers, (b) Langmuir–Blodgett films, (c) liquid crystals and (d) supramolecular architectures in solution. In order to successfully assemble molecules in an organized array one generally needs well-defined, but flexible modules that control their shape but can assume different conformational states in the process of self-assembly [32]. Examples in this section will show that the successful application of dendrimers in the areas mentioned above rely not only on their uniformity, but also on their tuneable architectures. The focus will be on amphiphilic dendrimers at interfaces, in solution and in liquid-crystalline materials. The trend toward the construction of functional assemblies, that exploit unique dendritic properties, will be illustrated with recent examples.

2.1 DENDRIMERS ON SURFACES: CONFORMATIONAL BEHAVIOUR

The preparation of thin films and layers from dendrimers by self-assembly is a topic of great current interest since it allows the construction of functional interfaces that use specific dendritic properties such as: size, shape, porosity, end-group density and multifunctionality. It is also an area of research that benefits substantially from the flexible character intrinsic to most dendrimers.

Using Monte Carlo simulations, Mansfield *et al.* was the first to predict dendrimer deformations when deposited on a surface [33]. More importantly, these calculations predicted an enhanced flattening for higher generation dendrimers with increasing interaction strength as indicated in the phase diagram in Figure 16.1.

Predictions made by Mansfield *et al.* regarding the surface behaviour of dendrimers were confirmed by Tsukruk *et al.* who used poly(amidoamine) (PAMAM) dendrimers (generations G3.5–G10) to construct self-assembled monolayers (SAMs) [34, 35]. PAMAMs of two adjacent generations with surface amine and carboxylic acid end-groups, respectively, were used in an electrostatic

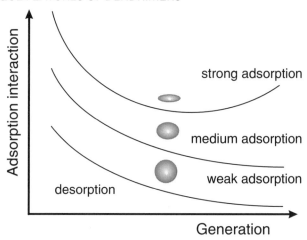

Figure 16.1 'Phase diagram' displaying the influence of the interaction-strength and generation number on the shape of a dendrimer at a surface [33]

layer-by-layer deposition. The resulting films were studied using scanning probe microscopy and X-ray reflectivity. The dendrimers were found to possess oblate shapes with axial ratios in a range from 1 : 3 to 1 : 6. Similar results were obtained upon deposition of anionic carboxylated PAMAMs on positively charged surfaces [36, 37]. Next to electrostatic interactions, various other types of secondary interactions have been used to adhere dendrimers to surfaces, ultimately leading to flattened dendrimer structures. Sheiko *et al.* observed strong deformations of hydroxyl terminated carbosilane dendrimers on mica due to multiple hydrogen bonding interactions [38]. These interactions were absent in case of substrates that were first coated with a semi-fluorinated polymer, resulting in a surface that was only partially wetted with dendrimers. Crooks *et al.* utilized amine–Au interactions in the construction of PAMAM monolayers on gold [39–43]. Three different methods were used leading to the successful confinement of dendrimers on Au, as depicted schematically in Figure 16.2.

The first approach involved direct attachment of PAMAM dendrimers to Au, without any intermediate (Figure 16.2A) [39, 40]. Close-packed, highly stable layers were obtained for generations 4–8 in which the individual dendrimers exhibit a disk-like shape. Upon addition of hexadecathiol ($C_{16}H_{33}SH$) the dendrimers were highly compressed from an oblate to prolate conformation. The change in conformation is ascribed to the stronger thiol–Au interactions compared to the dendritic amine–Au interactions. For surfaces of near-monolayer coverage, instead of single dendrimers, exposure to hexadecanethiol caused the dendrimers to gradually agglomerate, forming dendrimer 'pillars' up to 30 nm high [41]. In a second approach the PAMAM dendrimers were covalently attached to a self-assembled monolayer (SAM) of alkylthiols [42]. This surface

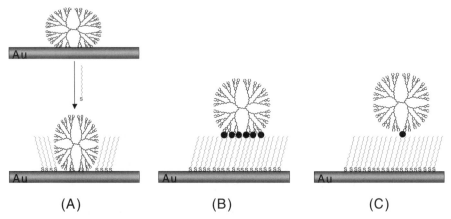

(A) (B) (C)

Figure 16.2   Three different approaches used by Crooks *et al.* to immobilize PAMAM dendrimers on Au surfaces: (A) Direct attachment, using amine–Au interactions [39–41]; (B) covalent attachment on a SAM prepared from alkylthiols using multiple interactions [42]; (C) covalent attachment on a SAM prepared from alkylthiols using only one or a few connections [43]

attachment protocol resulted again in a significant flattening of the dendrimers, as shown in Figure 16.2B. The last approach also involved the immobilization of PAMAM dendrimers on a SAM prepared from alkylthiols, but in this case the SAM was diluted with longer alkylthiol molecules terminated with reactive groups [43]. This resulted in linking of one or just a few dendrimer end-groups to the surface and hence distortion of the dendritic spherical shape was avoided (Figure 16.2C).

The conformational behaviour of several types of amphiphilic dendrimers at the air–water interface has also been investigated extensively [44–47]. Amphiphilic molecules have been prepared both from the polar PAMAMs and poly(propylene imine) dendrimers by functionalization with apolar alkyl chains [44, 45]. In the case of amphiphilic apolar poly(benzyl ether) Fréchet-type dendrons, the polar component of the amphiphile consisted of a single hydroxyl group [47, 48] or a hexa(ethylene glycol) tail [46] at the focal point. Fréchet dendrimers bearing eight alkyl chains at the periphery and alcohol or carboxylic acid functions at the focal point have also been reported [49]. For poly(propylene imine) dendrimers, PAMAM dendrimers and poly(benzyl ether) dendrimers a linear increase was found between the molecular area and the molecular weight [44, 46]. Moreover, in the case of alkyl chain modified poly(propylene imine) and PAMAM dendrimers the molecular areas obtained were equal to the sum of the molecular areas of the alkyl chains attached to the dendrimer for each generation. In addition to the generation number, other parameters such as core size and the length of the alkyl chains were varied, but these were found not to influence the dendritic molecular area [45]. These results support that the molecular area of a dendrimer at the air–water interface is dictated entirely by the total area of the substituents attached to the dendrimer as depicted schematically in Figure 16.3 [44].

For all three types of dendrimers described above, a flattened, disk-like conformation was observed for the higher generations. However, the molecular shape at the air–water interface is also intimately associated with the polarity, and hence the type of dendrimer used. In case of the poly(propylene imine) and PAMAM dendrimers the hydrophilic cores interact with the sub-phase and hence these dendrimers assume an oblate shape for all generations. The poly(benzyl ether) dendrimers, on the other hand, are hydrophobic and want to minimize contact with the water surface. This property results in a conformational shape change from ellipsoidal, for the lower generations, to oblate for the higher generations [46].

Studies on dendrimers at interfaces have shed new light on the conformational behaviour of these macromolecules. Conventionally, dendrimers were viewed as spherical macromolecules with a hollow-core, dense-shell character, according to the theoretical model of de Gennes [50]. Later, Mansfield, Muthukumar and others argued against this model based on their calculations which suggest that the end-groups were back-folded into the dendritic interior to a nonnegligible extent [51–54]. When we take recent literature into account, we realize that next to the dendrimer structure itself, the medium around the dendrimer manifests a profound influence on its conformational behaviour. Dendrimers may swell or contract dramatically in response to changes in solvent type [55, 56] or pH [57], a property which obviously affects the end-group localization as well. In this context, the behaviour of dendrimers at surfaces provides us with the most extreme examples of conformational changes in response to external stimuli.

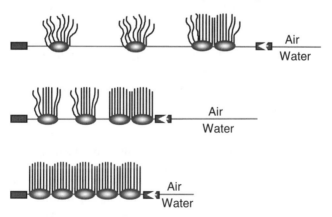

Figure 16.3 Schematic representation of the compression of alkyl modified dendrimers at the air–water interface, the dendrimers assume a flattened, disk-like conformation

Depending on the strength and type of interaction between the dendrimer and the surface, they can flatten out almost completely. Although the unique spherical shape of the dendrimer is very often lost in self-assembled films, other specific properties such as multifunctionality, porosity etc. are retained. This important feature has led to the birth of a new and exciting area of research: functional dendritic surfaces, which will be discussed in the next section.

2.2 FUNCTIONAL THIN FILMS USING DENDRIMERS

Recently, the focus of the dendrimer chemist has shifted towards the construction of functional interfaces. In this section a brief compilation of functional self-assembled dendritic surfaces will be given along with the specific dendritic property(ies) used in each application. The areas in which functional dendritic films have been applied so far include chemical- and biosensors, lithography, photoresponsive materials, magnetic resonance contrast agents and organic light-emitting diodes (LEDs).

Dendrimers are viewed as high potential materials for chemical sensing because of their morphological characteristics, synthetic flexibility, and highly amplified terminal functionality [58]. The self-assembled monolayers on Au developed by Crooks *et al.* are very effective chemical sensors for volatile organic compounds (VOCs). The PAMAM dendrimer shown in Figure 16.2B, for instance, displays ideal chemical sensor features towards butanol: rapid response, low signal-to-noise ratio and full reversibility [42]. The extent of penetration of the VOC in the dendrimer framework is governed by the size and surface density of the dendrimer, and hence, by the generation number. Gorman *et al.* used the same permeability strategy in the construction of monolayers on gold from organothiol dendrons of the first through third generation [59]. The ability to trap and hold small molecules within these layers was found to be generation dependent. Miller *et al.* modified poly(amido) dendrimers with oligothiophene cation radicals that could be used as 'molecular sponges' to detect VOCs [60]. The electrical conductivity of thin films prepared from these dendrimers showed rapid, dramatic (up to $800 \times$), and reversible increases in conductivity when exposed to certain VOCs.

Biosensors are fabricated by immobilizing enzymes or other functional proteins on the surface of an electrode [61]. A promising approach in this direction was developed by Anzai *et al.* who prepared multilayered thin films of avidin (Av) and biotin labeled PAMAM (G4) dendrimers (Figure 16.4) [62, 63]. Avidin is a glycoprotein (68 000 Da) containing four binding sites for biotin. The binding constant between avidin and biotin is very high ($K_a = c.\ 10^{15}\ M^{-1}$) and hence virtually irreversible [64, 65].

The layer formation was initiated by the deposition of avidin on a hydrophobic quartz slide, yielding a homogeneous monomolecular layer [66]. Subsequently, a multilayer structure was built up in a step-by-step fashion by immersing the quartz slide alternately in a PAMAM–biotin and avidin solution (Figure 16.4A). Similar experiments carried out with biotin labeled poly(ethylene imine) (PEI) and/or poly(allylamine) (PAA) resulted in the formation of less defined avidin multilayers (Figure 16.4B + C). The controlled multilayer formation obtained with dendrimers is ascribed to their unique spherical shape. It should be emphasized here that, from the viewpoint of practical applications of these layers, control over the loading of the films is considered to be a prerequisite. Although the avidin/PAMAM–biotin layers exhibit no intrinsic functionality, the approach is promising because of its versatility. Several avidin–enzyme conjugates are commercially available and can in principle be used in this immobilization concept.

Alonso *et al.* prepared ferrocenyl silicon dendrimers [67], which could be used as mediators in glucose biosensors, based on glucose oxidase [68, 69]. The ferrocenyl units are located at the end of long, flexible, silicon containing branches and serve to electrically connect the enzyme to the electrode. The flexibility of the dendrimer is proposed to play an important role in the interaction with the redox center of glucose oxidase.

Fréchet *et al.* showed that both ionically and covalently bound SAMs of poly(benzyl ether) dendrimers [70, 71] can serve as resists for scanning probe lithography [72, 73]. The tip of a scanning probe microscope (SPM) was used to write a pattern in a dendritic monolayer by oxidation of the underlying silicon substrate. The oxide pattern thus obtained could be removed selectively under aqueous hydrofluoric acid etching conditions. The dendrimer monolayers proved resistant to the HF (aq) etchant and in this way a positive tone image of the written pattern could be obtained with dimensions below 60 nm. The increased stability of the dendritic monolayers toward the etchant compared to traditional monolayers prepared from low molecular weight molecules is ascribed to the dense nature of the dendrimers used, resulting in a better protection of the

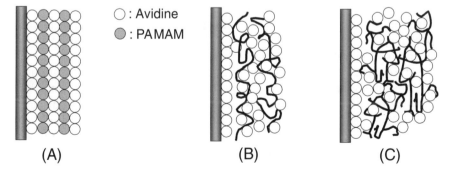

(A)　　　　　　　　(B)　　　　　　　　(C)

○ : Avidine
◉ : PAMAM

Figure 16.4 Proposed structures of multilayer films based on avidin–biotin interactions in case of: (A) PAMAM (G4) dendrimer, (B) PAA, (C) PEI [62, 63]

anchoring group. Future applications of these dendritic resists are foreseen in the area of high-density data storage systems.

Weener *et al.* prepared photo-responsive monolayers from azobenzene modified poly(propylene imine) dendrimers which also hold promise in the area of optical data storage [74]. A fifth generation poly(propylene imine) dendrimer was functionalized with equal amounts of palmitoyl and azobenzene containing alkyl chains, resulting in the formation of an amphiphilic copolymer with a random shell structure (Figure 16.5).

AFM and grazing incidence X-ray reflectivity measurements revealed an orientation of the dendrimers within the monolayers identical to the one previously established by Schenning *et al.* (Figure 16.3) [44]. The azobenzene groups displayed facile and reversible isomerization in Langmuir and Langmuir–Blodgett monolayers as shown in Figure 16.5. The dendrimer used can be viewed as a very well defined, headgroup polymerized, amphiphilic system. The use of the dendritic scaffold in this case resulted in the formation of stable monolayers and, additionally, prevented microphase separation of the azobenzene units within the monolayers, resulting in reversible switching behaviour.

Vögtle and Balzani *et al.* recently reported on poly(propylene imine) dendrimers up to generation four functionalized with azobenzene groups in the periphery with potential use as materials for holographic data storage [75]. Repetitive *cis-trans* isomerizations of the azobenzene groups in solvent cast films, resulted in a preferred orientation of the azobenzene moieties, perpendicular to the polarization direction of the incident light. AFM measurements performed on films after irradiation revealed the presence of relief gratings with heights up to 1500 nm.

Dendrimers are promising candidates in the area of organic LEDs [76–84]. Starburst molecules were shown to function very well as hole transport layers in organic LEDs and the increased lifetimes found have been attributed to the prevention of crystallization by the dendrimer [76–78]. Conjugated dendrimers were shown to be particularly interesting as materials for application in organic LEDs. The concept of a light-emitting core, conjugated branches and carefully chosen surface groups to increase processing properties has led to the successful preparation of blue-emitting LEDs [79–82]. Sooklal *et al.* reported on blue-emitting CdS/dendrimer nanocomposites [83]. It was shown that the molecular architecture of the PAMAM dendrimers used was necessary for the nucleation of the highly luminescent CdS clusters. Recently, Schenning *et al.* presented a universal approach to adjust the emission wavelength of LEDs using dendrimer/dye assemblies [84]. Poly(propylene imine) dendrimers up to generation five were functionalized with π-conjugated oligo(*p*-phenylene vinylene)s (OPVs) (Figure 16.6), yielding amphiphilic macromolecules capable of forming SAMs.

Moreover, these dendrimers proved to be good hosts for anionic dye molecules. Films prepared from these host–guest complexes showed efficient (> 90%) energy transfer from the OPV units to the encapsulated dye molecules

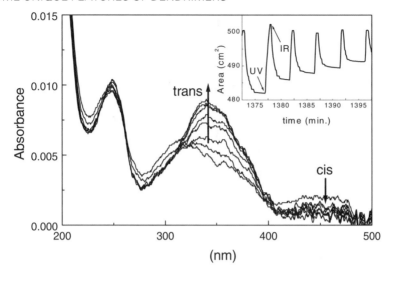

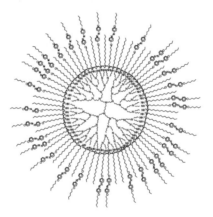

Figure 16.5 Reversible switching of amphiphilic azobenzene modified poly(propylene imine) dendrimer on quartz, and at the air–water interface (inset) [74]

(Figure 16.6). By choosing the appropriate dye, the emission wavelength could be tuned.

Dendrimers have also been found useful in the construction of thin films containing isolated fullerenes [85, 86]. Such materials could eventually find practical application in sensors and/or optoelectronic devices [87]. Generally, monolayers prepared from C_{60} are ill-defined due to the aggregation tendency of fullerenes. The covalent attachement of C_{60} to bulky dendritic frameworks

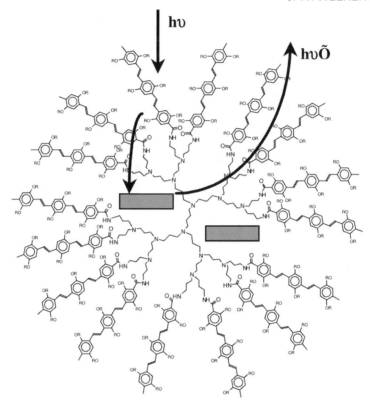

Figure 16.6 Schematic representation of the energy transfer process between OPV–poly(propylene imine) dendrimer and a dye guest molecule [84]

prevents such fullerene–fullerene aggregation. However, so far no reports have been made describing properties and/or applications of such films.

2.3 AMPHIPHILIC DENDRIMERS

In this section the emphasis will be on assemblies prepared from amphiphilic dendrimers. Amphiphilic dendrimers, carrying both hydrophobic and hydrophilic regions within one molecule, tend to self-assemble into a large variety of different aggregates depending on their structure. The dendritic amphiphiles investigated so far include unimolecular micelles, bolaamphiphiles, superamphiphiles and various other AB and ABA block copolymers.

Unimolecular micelles are defined as a class of dendritic macromolecules, wherein an interior hydrophobic core is surrounded by a hydrophilic surface layer. These structures closely resemble the shape of classical micelles (shape, size, number of end-groups) except that they are static, in contrast to the dynamic nature of micelles, with all end-groups attached to a central core. Newkome *et al.* prepared the first unimolecular micelle containing 36 carboxylic acid groups equidistant from a central neopentyl core [88]. The ability of guest molecules to penetrate the lipophilic interior of this cascade framework showed the unimolecular micelle concept [89]. In the same group three series of polyamide dendrimers were prepared with either acidic, neural or basic terminal groups starting from a pentaerythritol core-molecule [90]. Upon a change in the environmental pH, these dendrimers displayed swell/shrink behaviour leading to expansion/contraction of the internal void domains within these molecules [91, 92]. Fréchet *et al.* showed that dendritic polyethers based on 3,5-dihydroxybenzyl alcohol with 32 carboxylate surface groups also behaved as unimolecular micelles capable of solvating hydrophobic molecules [93]. More recently, the same group reported on the synthesis of poly(benzyl ether) dendrimers modified with apolar alkyl chains that could function as nanoscale catalytic systems [94]. Only a few reports have been made concerning the self-assembly of dendritic unimolecular micelles. Newkome *et al.* observed the association of dendritic polyols into aggregates containing *c.* 20 molecules [95]. A critical aggregation constant of 2 mM was found and dynamic light scattering (DLS) gave a Stokes radius for the aggregates of *c.* 95 nm. Recently, glucose substituted PAMAM dendrimers, with unimolecular micelle behaviour, were found to form superstructures in water judging from DLS and TEM measurements [96]. However, no comments were made concerning the exact nature of the aggregates in this case. The aggregation behaviour of palmitoyl and alkoxyazobenzene modified poly(propylene imine) dendrimers of different generations has been investigated extensively by Schenning *et al.* (Figure 16.7) [44].

Upon immersion in acidic water (pH = 1) these unimolecular inverted micelles form vesicles in which the dendrimer component has a highly distorted conformation with an axial ratio of 1:8 for the highest generations. The amphiphilic dendrimers within the aggregates in solution are thought to have a flattened shape similar to those proposed for dendrimers at the air–water interface (section 16.2.1).

Bolaamphiphiles are molecules carrying two polar end-groups separated by an hydrophobic spacer [97, 98]. Several different types of bolaamphiphiles have been reported (Figure 16.8).

Some of the bolaamphiphiles depicted in Figure 16.8 could form thermally reversible aqueous gels, depending not only on the length and rigidity of the spacer used, but also on the size of the polar dendritic groups. In case of an alkyl spacer for example (Figure 16.8a), long fibrous rods with a uniform diameter of 40 Å and variable lengths (*c.* > 2000 Å) were observed under the transmission electron microscope [99, 100]. Appearance of the fibres was explained with a model in which the hydrophobic spacers stack on top of each other in an orthogonal fashion while the polar dendritic head-groups shield them from the

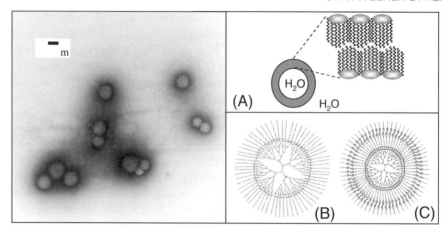

Figure 16.7 TEM picture (uranyl acetate staining) of vesicles reported by Schenning *et al.* [44]: (A) schematic representation of the bilayer, (B) palmitoyl- and (C) azobenzene-modified poly(propylene imine) dendrimers used in the construction of the aggregates

aqueous solution. Bolaamphiphiles containing a central triple bond (Figure 16.8b) were also shown to form aqueous gels, but in this case the individual strings (30–40 Å) self-organized into higher order helical rods with diameters up to 600 Å [101]. The difference in aggregation behaviour is proposed to arise from the rigid alkyne moiety which prefers nonorthogonal stacking (Figure 16.8). Bolaamphiphiles containing biphenyl or spirane central spacers failed to form aqueous gels [102, 103]. Incorporation of a tetrathiafulvalene (TTF) unit in the spacer (Figure 16.8c) led to the formation of gels in ethanol/water and DMF/ water mixtures in which the TTF units are stacked on top of each other [104].

Block copolymers are known to form a wide variety of different aggregates in solution [105–108]. Superamphiphiles are a special class of dendritic–linear diblock-copolymers that mimic, on a macromolecular level, traditional, low molecular weight, organic surfactants [109]. A variety of parameters can be varied (e.g. length and nature of the linear block, size and nature of the dendritic part) leading to a morphological wonderland. Chapman and co-workers reported on the formation of aggregates in aqueous solution constructed from linear poly(ethylene glycol) (PEG) and *tert*-butyl terminated poly-L-Lysine dendrimers called hydraamphiphiles [110]. Surface tension measurements indicated a critical micelle concentration (cmc) of 8×10^{-5} M. Above the cmc, the hydraamphiphiles could solubilize the orange-OT dye in water. Fréchet *et al.* functionalized hydrophilic PEG tails of different lengths with apolar poly(benzyl ether) dendrons of different generations and thus obtained superamphiphiles with various PEG/dendron mass ratios [111]. The PEG–dendron coupling was achieved using the Williamson ether synthesis. End-group analysis was carried

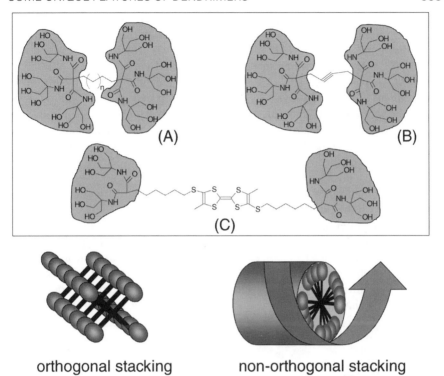

orthogonal stacking   non-orthogonal stacking

Figure 16.8 Three different kinds of bolaamphiphiles containing different apolar spacers and/or polar head groups. The cartoons represent the orthogonal and nonorthogonal arrangements proposed to explain the fiber structures observed in solution

out with the aid of MALDI-TOF and indicated that all of the PEG tails had indeed been functionalized with dendrimer [112]. Below the cmc, unimolecular micelles were observed in which the PEG chains formed a hydrophilic corona around the apolar poly(benzyl ether) dendrons, whereas above the cmc multimolecular micelles were formed [113]. In comparison to common surfactants, which have a polar head and an apolar tail, these superamphiphiles have a unique composition in that the polar and apolar parts are reversed. This property allowed for the hydrophilization of hydrophobic surfaces [114]. Recently, superamphiphiles constructed from PEG (MWts. 2000 and 5000) and PAMAM dendrimers (generation 1–4) have been reported [115, 116]. Their behaviour in solution, at the air–water interface and on solid substrates was studied, using intrinsic viscosity, Langmuir–Blodgett and AFM respectively. Similar to the findings of Fréchet and co-workers it was established that, in the case of PEG-5000-PAMAMs, the PEG-tail could wrap around the dendrimer, leading to the formation of unimolecular micelles. Van Hest *et al.* used the unique hydrophilic character of the poly(propylene imine) dendrimers in the preparation of amphiphilic block copolymers [117]. The dendrimer was grown from OH-functionalized polystyrene (PS) which in turn was prepared via an anionic polymerization ($M_n = 3 \times 10^3$ g.mol^{-1}, $M_w/M_n = 1.05$). Five different generations, from PS-*dendr*-NH$_2$ up to PS-*dendr*-(NH$_2$)$_{32}$ were prepared [118]. Dynamic light scattering, conductivity measurements and transmission electron microscopy (TEM) showed that in aqueous solution, PS-*dendr*-(NH$_2$)$_{32}$ formed spherical micelles (Figure 16.9a), PS-*dendr*-(NH$_2$)$_{16}$ formed micellar rods (Figure 16.9b), PS-*dendr*-(NH$_2$)$_8$ formed vesicular structures (Figure 16.9c) and PS-*dendr*-(NH$_2$)$_4$ formed inverted micellar structures [117]. The change in amphiphile geometry, influenced by the size of the dendrimer head-group, was in qualitative agreement with the theory of Israelachvili *et al.* [119].

Hence, this observation provided the bridge between the classical surfactants and amphiphilic block copolymers. The versatility of these amphiphiles is based not only on the ability to change the size of the head-group, but also on the possibility of chemically modifying of the dendritic head-group. Carboxylic acid terminated dendritic amphiphiles were prepared that exhibited a pH dependent aggregation behaviour due to the zwitter-ionic nature of their head-groups

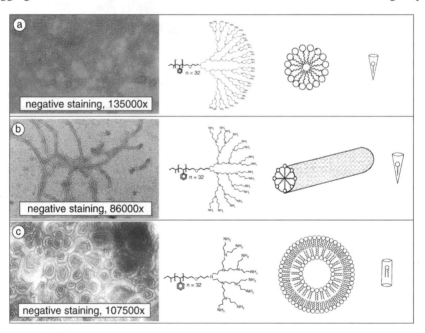

Figure 16.9 TEM photographs (negative staining) of different types of aggregates resulting from changes in the head-group size of superamphiphiles: (a) PS-dendr-(NH$_2$)$_{32}$; (b) PS-dendr-(NH$_2$)$_{16}$; (c) PS-dendr-(NH$_2$)$_8$ [117]

[120]. TEM measurements revealed clustering of aggregates in all cases except for PS-*dendr*-(COOH)$_8$ for which individual 'worm-like' micelles were observed.

More recently, Fréchet *et al.* reported on the solution behaviour of ABA and star-shaped block copolymers. The ABA block copolymers have a barbell-like shape, and consist of a linear, polar PEG block which is end-capped with apolar poly(benzyl ether) wedges at both sides [121]. Copolymers with different PEG chain lengths and of varying dendrimer generations were used. The star-shaped block copymers were prepared starting from star-shaped PEG with a pentaerythritol core to which four poly(benzyl ether) wedges were attached [122]. The influence of the solvent type on the aggregation behaviour of these macromolecules was investigated along with the effect of different polar/apolar ratios. Different micellar structures were observed as a function of the environment.

2.4 LIQUID CRYSTALLINE DENDRIMERS

Thermotropic liquid crystalline (LC) phases or mesophases are usually formed by rod-like (calamitic) or disk-like (discotic) molecules. Spheroidal dendrimers are therefore incapable of forming mesophases unless they are flexible, because this would allow them to deform and subsequently line up in a common orientation. However, poly(ethyleneimine) dendrimers were reported to exhibit lyotropic liquid crystalline properties as early as 1988 [123].

Percec was the first to report on flexible, non-spherical dendrimers with AB$_2$ mesogens in the branches, that exhibited nematic and smectic LC behaviour [124, 125]. Since then several reports have been made concerning the decoration of dendrimers with mesogenic units on the periphery. Frey *et al.* attached several mesogenic units to carbosilane dendrimers, such as cyanobiphenyl [126] and cholesteryl [127]. Besides these classical calamitic mesogens, carbosilane dendrimers carrying perfluoralkyl groups have also been also reported [128]. In all cases smectic mesophases were observed in which the dendrimers were deformed (stretched) in a cooperative way to adjust to the superstructure imposed by the mesogens [129]. However, reduction of the spacer length and/or increasing the generation number complicated the formation of well-developed smectic phases [126]. A comparable generation effect was observed by Latterman and co-workers who functionalized poly(propylene imine) dendrimers with mesogenic 3,4-bis-(decyloxy)benzoyl groups [130, 131]. Generations one to four exhibited a hexagonal columnar mesophase in which the dendrimers had a cylindrical conformation (Figure 16.10a), whereas generation five was not liquid crystalline. The lack of mesomorphism for the highest generation dendrimer was attributed to the incapability of this dendrimer to reorganize into a cylindrical shape.

Baars *et al.* functionalized poly(propylene imine) dendrimers (generation 1, 3, 5) with pentaoxycyanobiphenyl and decyloxycyanobiphenyl mesogens [132]. All dendrimers were found to exhibit smectic A mesophases. The S_A-layer spacings observed were independent of the dendrimer generation for both spacer lengths, indicating that the dendritic part was completely distorted, even for the higher generations (Figure 16.10b). These findings were later confirmed by Yonetake and Ueda *et al.*, who also used cyanobiphenyl substituted poly(propylene imine) dendrimers [133]. PAMAM dendrimers (generations 1–4) derivatized with mesogenic ester units (4-(4'-decyloxybenzyloxy)salicylaldehyde) also displayed a smectic A mesophase. The molecular picture to explain this behaviour is identical to the flattened dendrimer model previously described by Baars *et al.* [132]. For poly(propylene imine) dendrimers functionalized with palmitoyl alkyl chains ($-C_{15}H_{31}$) lamellar structures were present in the bulk according to XRD, TEM [134] and SANS [135] measurements. The inter-lamellar distance found was 5 nm and independent of dendrimer generation, a feature which again could be explained by assuming a flattened dendrimer shape. A similar observation has recently been described by Stün *et al.* in a SANS/SAXS study of perfluorinated carbosilane dendrimers. In this case the strong tendency of the perfluorinated end-groups to form layered structures led to strong deformations of the dendritic scaffolding [136].

The above-mentioned examples clearly illustrate that two opposing forces compete in the formation of liquid crystal phases from dendrimers terminated with mesogenic units. The functionalization of PAMAM, poly(propylene imine) and carbosilane dendrimers with mesogenic groups results in an initial spherical arrangement, while the mesogenic units tend to interact with each other to form a parallel arrangement at the same time. Therefore, the flexibility of dendrimers is a prerequisite for the successful preparation of LC materials from these macromolecules. Recent exciting developments in the field of LC dendrimers include control over polymer shape by rational design [32, 137–141] and applications of dendrimers in LC displays [142].

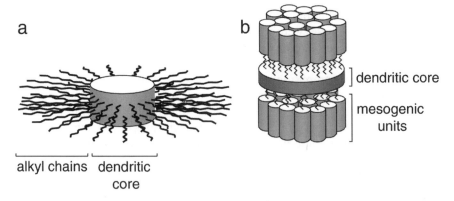

Figure 16.10 Conformational changes of poly(propylene imine) dendrimers in LC materials from spherical to (a) cylindrical, in hexagonal columnar mesophase [130], and (b) ellipsoid, in smectic A mesophase [132]

3   HOST–GUEST CHEMISTRY OF DENDRITIC MACROMOLECULES

Host–guest chemistry involves the binding of a substrate molecule (guest) in a receptor molecule (host). The design and construction of hosts that are capable of selectively binding guest molecules requires precise control over geometrical features and interacting complementarily. As early as 1982, Maciejewski presented a theoretical discussion of highly branched molecules as ideal molecular containers, showing the challenges in host–guest interactions of dendritic molecules [143]. In the mean time, the field of host–guest properties of dendritic molecules has developed significantly and now occupies a unique position within the area of supramolecular chemistry [144–146]. Many exciting applications have been suggested for dendrimers in this relatively new area, including their use as drug-delivery vehicles [147–150]. In this section a brief survey of dendritic host–guest systems will be presented arranged by type of interactions from topological encapsulation to electrostatic, hydrophobic, hydrogen bonding and metal–ligand interactions.

3.1   DO CAVITIES EXIST IN DENDRIMERS?

A very important question in the context of dendrimers and their utility as host molecules relates to the existence of cavities within these macromolecules. The presence of internal voids in dendrimers is closely related to their conformational behaviour and to the degree of back-folding of the terminal branches into the interior of the dendrimer. The issue of back folding was already briefly touched upon in section 16.2.1. Next to the purely theoretical calculations mentioned there, several calculations have been performed on specific dendrimer types.

Scherrenberg *et al.* modeled the conformational behaviour of poly(propylene imine) dendrimers in good and bad solvents [151]. In the latter case the authors found a homogeneous density distribution, but emphasize at the same time that it is precarious to generalize their findings to other dendrimers because the internal dendritic structure is determined by a variety of parameters such as segment length, core and branch multiplicity and rigidity of the building blocks. Next to these parameters several external factors, such as solvent type and pH, influence the conformational behavior of dendrimers significantly. Weener *et al.* showed a transient from a 'dense-core' to a 'hollow-core, dense-shell' situation for poly(propylene imine) dendrimers, upon lowering of the pH [57]. Welch and Muthukumar observed similar changes in the density profile of dendritic polyelectrolytes, with a constitution almost identical to that of the poly(propylene imine) dendrimers, which depended on the ionic strength of the solvent [55].

Murat and Grest performed a molecular dynamics study on PAMAMs in

low pH   high pH

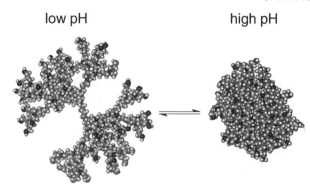

Figure 16.11 The pH-dependent conformational behaviour of poly(propylene imine) dendrimers. At low pH (left) the occurrence of a 'soft-core, dense-shell' dendrimer, whereas at high pH (right) severe back-folding occurs leading to a 'dense-core' structure [57]

solvents of various qualities and found significant back-folding [52]. Moreover, it was found that the amount of back-folding increased with decreasing solvent quality. The conformational behaviour of PAMAM dendrimers has also been investigated with size exclusion chromatography (SEC) in combination with intrinsic viscosity measurements [152, 153]. From these studies the authors conclude that the dendrimers have a hollow core and a densely packed outer layer, in agreement with the de Gennes model [50]. However, these inhomogeneous distributions are in contrast to the general findings for most other unmodified dendrimers, for which homogeneous density distributions were found.

Poly(benzyl ether) dendrimers synthesized by Fréchet *et al.* have been studied with many techniques in order to reveal their conformational properties. Size exclusion measurements performed by Mourey *et al.* [154], rotational-echo double resonance (REDOR) NMR studies by Wooley *et al.* [155] and spin lattice relaxation measurements by Gorman *et al.* [156] reveal that back-folding takes place and the end-groups can be found throughout the molecule. The observed trends are in qualitative agreement with the model of Lescanec and Muthukumar [54].

Several studies have been devoted to determine the localization of end-groups in modified poly(propylene imine) dendrimers. Goddard *et al.* [157] and Cavallo and Fraternali [158] investigated the properties of the dendritic box, a fifth-generation poly(propylene imine) dendrimer functionalized with (*t*-BOC)-protected L-phenylalanine residues (Figure 16.12a) [159].

Their results are indicative of a low-density region inside the dendrimer with a considerable end-group interaction in the shell. Similar results were obtained from CHARMm molecular mechanics calculations on the dendritic box (Figure 16.12b), which indicated a globular architecture in which the dendritic interior is almost completely shielded by the bulky amino-acid end-groups [160]. Chiroptical studies [161] and spin-lattice relaxation times (T_1), also pointed toward a solid phase behaviour of the shell around the dendrimer in solution. Presumably, intramolecular hydrogen bonding between several L-Phe residues in the shell is responsible for this rigid nature. Similar results were obtained in case of poly(propylene imine) dendrimers modified with second-order, nonlinear optical chromophores (4-dimethylaminophenylcarboxamide) [162]. The symmetry of these macromolecules was probed with hyper-Rayleigh scattering, and it was concluded that in case of generations 4 and 5 globular structures were present, in which the end-groups had a restricted mobility. On the other hand, SANS studies recently performed on poly(propylene imine) dendrimers, modified with deuterated acetyl groups, pointed to significant back-folding of the terminal groups [163].

Returning to the question formulated at the start of this section, concerning the presence of voids in dendrimers, the studies summarized above indicate that the molecular conformation of dendrimers is strongly influenced by several different factors. These factors not only involve the nature of the dendrimer under investigation and the type of end-groups, but also include the surrounding solvent and pH. Some studies indicate severe back-folding in case of unmodified dendrimers, whereas in the case of functionalized dendrimers, secondary interactions between end-groups can occur leading to phase separation or rigid shell formation through hydrogen bonding.

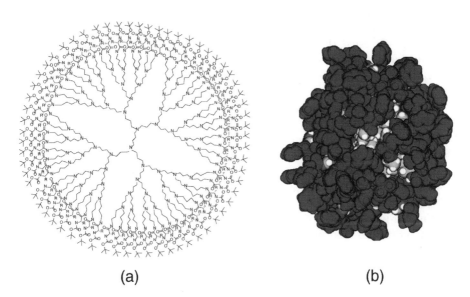

(a)   (b)

Figure 16.12 Chemical structure (a), and CHARMm molecular modeling picture (b), of the dendritic box [159]

3.2 TOPOLOGICAL ENCAPSULATION OF GUEST MOLECULES

The possibilities for encapsulating guest molecules in dendritic hosts were first proposed by Maciejewski in 1982 [143]. In 1990, Tomalia presented evidence for 'unimolecular encapsulation' of guest molecules in dendrimers and pointed out that it was one of the possible future research areas in dendrimer chemistry [164].

Jansen *et al.* demonstrated that it was possible to physically lock guest molecules in a dendritic host molecule, the so-called dendritic box [159]. The dendritic box was constructed from a fifth-generation poly(propylene imine) dendrimer with 64 amine end-groups (Figure 16.12a). Guest molecules were captured within the internal cavities of the box when an outer (*t*-BOC protected) phenylalanine shell was constructed in the presence of guest molecules. Several types of guest molecules were incorporated, yielding information that proved to be valuable in the understanding of the properties of this novel host.

The dendritic box loaded with Bengal Rose, an anionic xanthene dye, was studied with several spectroscopic techniques including UV [165], circular dichroism (CD) [161, 166] and fluorescence spectroscopy [159]. From UV measurements it was evident that the maximum number of guest molecules attainable was limited to four. Induced circular dichroism (ICD) was found for Bengal Rose in the chiral phenylalanine dendritic box. ICD is based on the transfer of chirality from the environment to an achiral guest. These results seemed to indicate the presence of chiral cavities in the dendritic box. Moreover, an exciton coupled CD spectrum was obtained for a dendritic box that was loaded, on the average, with 4 Bengal Rose molecules. The observed exciton coupling suggested that the guest molecules were in close proximity with a certain fixed orientation. The use of 3-carboxy-proxyl radicals as guest molecules in various concentrations led to the incorporation of 0.3–6 molecules per dendritic box as determined by electron spin resonance (ESR) spectroscopy [167]. The ESR spectra obtained from a 'radical-in-dendrimer' suggested that at least two different trapping sites existed, one that allowed relatively free motion and another where the radical was almost immobilized. The dye Eriochrome Black T was used to study possible diffusion of dye out of the box [159]. Even after prolonged heating, dialysis or sonication the dye remained inside. Therefore, it was concluded that the diffusion of the dye out of the box was unmeasurably slow.

Further investigation of this host–guest system showed not only that different guests could be incorporated in one host, but also that guest molecules of different size could be liberated in a shape selective fashion [168, 169].

3.3 RECOGNITION BASED ON HYDROPHOBIC INTERACTIONS

The fact that the dendritic shell can produce localized microenvironments has been used by Diederich *et al.* who developed water-soluble dendritic cyclophanes (dendrophanes) as models for globular proteins (Figure 16.13) [5, 170, 171]. These dendrimers contain well-defined cyclophane recognition sites as initiator cores for the complexation of small aromatic guests [172–174] and steroids [174–176]. Enlargement of the cyclophane core could be used as a tool to complex larger steroid molecules.

¹H-NMR binding titrations and fluorescence relaxation measurements in solution indicated fast host–guest kinetics and suggest a relatively open structure for the dendrimer at all generations. Fluorescent probes like 6-(*p*-toluidino)naphthalene-2-sulfonate (TNS) demonstrated that the micro-polarity around the binding cavity was significantly reduced with increasing dendritic size. The reduced micro-polarity in the cyclophane core and fast host–guest exchange kinetics make the water-soluble dendrophanes attractive targets as mimics of globular enzymes. More recently, Kenda and Diederich described the

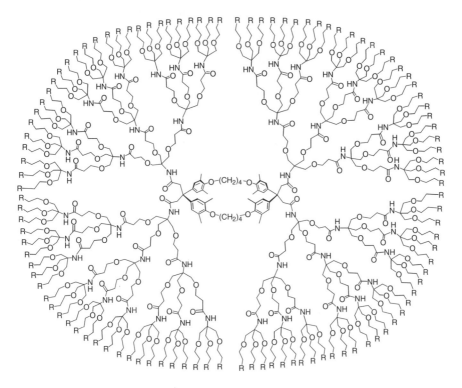

Figure 16.13 Example of a dendritic cyclophane used by Diederich *et al.* [172] as a receptor for hydrophobic guest molecules

threading of dendritic cyclophanes on molecular rods functionalized with steroid termini [177]. The threading of the dendrophanes onto testosterone termini is hydrophobically driven (apolar interactions, hydrophobic desolvation) and yields well-defined structures with molecular weights exceeding 14 kD.

Several research groups have been involved in the complexation of β-cyclodextrin (β-CD) to dendrimers. Cuadrado [178] and Kaifer [179] *et al.* described the synthesis of various generations of ferrocenyl-functionalized poly(propylene imine) dendrimers. Since ferrocene is an excellent substrate for the inclusion complexation by β-CD ($K_a = c.$ 1230 M^{-1}) [180], the host–guest properties of these dendrimers toward β-CD have been investigated. Although the solubility of the dendrimers in aqueous media decreased with generation, a significant solubility enhancement was observed in the presence of β-CD. This is rationalized by the formation of β-CD/ferrocene inclusion complexes on the surface of the dendritic structures. Interestingly, the availability of redox-active ferrocene end-groups made it possible to break up these supramolecular species electrochemically, since the binding affinity constant of the β-CD–ferrocene complex is strongly diminished after oxidation. The same concept of using dendrimers as three-dimensional, electrochemically switchable templates for the organization of β-CD has been shown for a series of poly(propylene imine) dendrimers functionalized with 4, 8, 16 and 32 peripheral cobaltocenium units [181]. These dendrimers represent a novel type of host–guest system in which the formation of multi-site β-CD/dendrimer complexes is driven by the reduction of the cobaltocene subunits of the dendrimers. Upon reduction, the charged end-groups are transformed into very hydrophobic species that efficiently complex in the β-CD cavity.

Newkome *et al.* described the synthesis of dendritic wedges of different generations attached to a β-cyclodextrin receptor [182]. Binding studies with phenolphthalein, adamantane-amine or a bis(adamantane) units showed that the binding cavities of the modified receptors retained their molecular recognition properties. Using a bi-functional adamantyl compound, a first step toward recognition-based assembly of dendrimers was established, comparable to Diederich's threading process.

Shinkai *et al.* described the synthesis of dendritic saccharide sensors based on a PAMAM dendrimer labeled with eight boronic acid residues [183]. The dendritic compound showed enhanced binding affinity for D-galactose and D-fructose. The fact that the dendritic boronic acid functions as a saccharide 'sponge' is ascribed primarily to the cooperative action of two boronic acids to form an intramolecular 2:1 complex. When one boronic acid binds a saccharide, its counterpart cannot participate in dimer formation and seeks a guest.

3.4 RECOGNITION BASED ON HYDROGEN BONDING INTERACTIONS

Hydrogen bonding interactions between dendrimers and guests have been achieved by incorporation of coordinating sites, that are complementary to the guest, at several different positions in the dendrimer including the focal point, interior and the periphery.

Fox *et al.* investigated hydrogen bonding interactions between PAMAM dendrimers (with up to eight end-groups) and several molecules of biological interest [184]. The guests employed (pyridine, quinoline, quinazoline, nicotine and trimethadione) were chosen, not only because of their biological activity, but also for their established reactivity as hydrogen bond acceptors. Only lower generations PAMAM dendrimers were used in this study, containing two active sites for potential complexation with hydrogen bonding partners: the external surface amino functions and the internal amido groups. It was found that complexation could in principal occur at both the external and the internal sites. The inability of some of the guest molecules to complex internally was surprisingly attributed to the tight local packing of the dendrimers.

Newkome *et al.* reported the construction of dendrimers in which four 2,6-diamidopyridine units were incorporated in the interior [185]. Several generations were synthesized containing up to 36 end-groups. The association behaviour with complementary guests such as glutarimide, barbituric acid and 3'-azido-3'-deoxythymidine (AZT) was investigated, yielding apparent association constants of $c.$ 70 M^{-1}. These constants are comparable with reported values for similar, non dendritic, host–guest complexes [186].

Zimmerman and Moore recently reported on dendrimers with hydrogen bonding units at the focal point [187]. Two classes of dendritic hosts were synthesized with naphthyridine units in the core capable of hydrogen bonding benzamidinium derivatives (Figure 16.14).

Since the strength of hydrogen bonding is dependent on the polarity of the solvent, the guest molecules serve as sensitive probes of the dendrimer's internal accessibility and polarity. From their measurements the authors conclude that the dendrimer exerts a negligible influence on the core nano-environment, even for the highest generations. Moreover, the insensitivity of the complexation strength to the dendrimer size and nature suggests that the hosts are highly porous. These data are in contrast to the observations by Hawker and Fréchet [188] who reported a change in the local polarizability at the core of similar dendrimers.

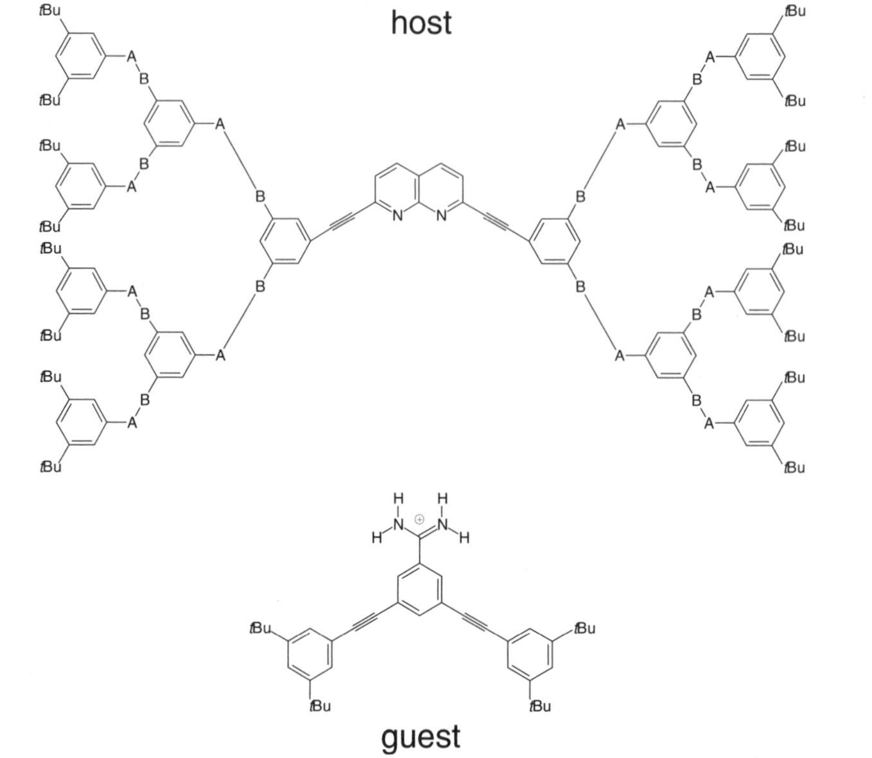

host

guest

Figure 16.14 To investigate hydrogen bonding in dendrimers Zimmerman and Moore [187] prepared dendritic wedges (A–B = CH$_2$O or C≡C) functionalized with anthypyridine units at the focal point that function as hosts for benzamidinium guest molecules

3.5 *RECOGNITION BASED ON ELECTROSTATIC INTERACTIONS*

The field in which the interaction between the dendritic host and the guest molecule(s) can be classified as electrostatic is very elaborate and therefore the focus in this section will be on the interaction of organic acids with dendrimers.

Tomalia *et al.* reported the complexation of guest molecules in dendritic structures by acid–base interactions and studied the change in guest ^{13}C spin-lattice relaxation times (T_1) [189]. PAMAM dendrimers with methyl ester termini were used as the dendritic host with aspirin and 2,4-dichloro-phenoxyacetic acid as guest molecules. The values for T_1 decreased, as the generation number increased from 0.5 to 3.5, but remained constant for the higher generations. The maximum concentration of the guests was roughly 3 : 1

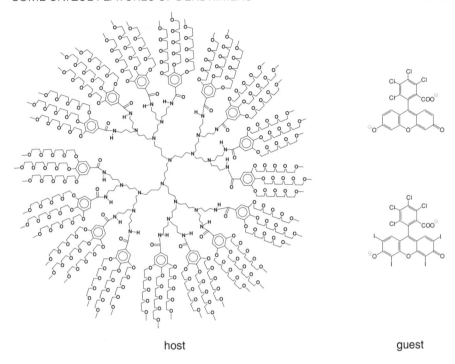

host   guest

Figure 16.15 Ethyleneglycol-functionalized poly(propylene imine) dendrimers as water-soluble hosts of TCF (top) and RB (bottom) [191]

based on a molar comparison of the carboxylic guest and the interior tertiary amines, confirming an acid–base interaction between host and guest. Twyman and Mitchell, who developed a convenient route to highly water-soluble dendrimers starting from PAMAM-dendrimers, obtained similar results [190]. These dendrimers were shown to be capable of binding and solubilizing small acidic, water insoluble, hydrophobic molecules, like benzoic acid, salicylic acid and 2,6-dibromonitrophenol.

Baars *et al.* recently investigated the host–guest properties of poly(propylene imine) dendrimers functionalized with tris-3,4,5-tri(tetraethyleneoxy)benzoyl units (Figure 16.15) [191]. These hosts are highly soluble in a broad range of solvents, from apolar solvents such as toluene to polar aqueous media.

The host–guest properties were studied in buffered aqueous media using two water-soluble anionic xanthene dyes (i.e. 4,5,6,7-Tetrachlorofluorescein (TCF) and Rose Bengal (RB)) as guest molecules. Interactions could again be rationalized by acid–base interactions between the acidic functionality of the guest and the tertiary amines of the dendritic host. SAXS measurements showed a preferential organization of the guests in the center regions of the dendritic hosts. The large size of this novel dendritic host (>6 nm) was exploited in the separation of non-encapsulated guest molecules from encapsulated ones, in a pH selective manner, using ultrafiltration.

Recently, Archut *et al.* reported the first attempts to develop a dynamic host–guest system of which the interactions could be tuned by an external stimulus, for instance light [192]. The interaction of a fourth generation azobenzene-functionalized poly(propylene imine) dendrimer with eosin Y was studied in DMF. The light absorbed by eosin was effective to promote the photoisomerization of azobenzene moieties from the E-form to Z-form. Fluorescence quenching experiments showed that eosin was hosted inside the dendrimer, as a consequence of acid–base interactions between host and guest, and suggested that the Z-form of the dendrimer was a better host than the corresponding E-form.

Polar dendrimers modified with apolar end-groups resemble unimolecular inverted micelles which can be used as molecular shuttles to transport guest molecules between two different phases. Modification of poly(propylene imine) dendrimers with palmitoyl ($-C_{15}H_{31}$) alkyl chains, for instance, resulted in the formation of a unimolecular micelle (Figure 16.7b) with a polar core and an apolar periphery as demonstrated by Stevelmans *et al.* [193]. The palmitoyl-dendrimers displayed single particle behaviour with a hydrodynamic radius of 2–3 nm in dichloromethane as was concluded from dynamic light scattering measurements. These compounds were able to encapsulate guest molecules like Rose Bengal in organic media. Recently, Baars *et al.* [194] extended research in this field and showed that these dendrimers actually represent a new family of tertiary amine extractants resembling the structures of low molecular weight tri-octylamine extractants. The dendritic extractants proved very effective and selective in the transfer of anionic solutes from an aqueous medium into an organic phase, typically dichloromethane or toluene. The interaction between the dendrimer (host) and solute (guest) is dominated by acid–base interactions. The dendrimer generation determines the number of tertiary amine sites, and as a consequence, the amount of solute molecules that can be extracted per dendrimer. In case of a fifth-generation dendrimer (carrying 62 tertiary amines) it was possible to extract up to 50 molecules of Rose Bengal, yielding an assembly with a molecular weight of 70 kD, *c.* 2.5 times the molecular weight of the dendrimer. Modification of poly(propylene imine) dendrimers with fluorinated chains made the extraction of water-soluble solutes into supercritical carbon dioxide possible [195, 196]. The mechanism of extraction is similar to that of the poly(propylene imine) dendrimers with apolar end-groups; however, this technology uses an environmentally friendly process design with environmentally compatible solvents and has the potential to replace hazardous organic solvents. Stephan and Vögtle *et al.* demonstrated that lipophilic urea-functionalized dendrimers are efficient carriers for oxyanions [197]. Extraction experiments were performed with diagnostically relevant anions like pertechnetate, perrhenate, ADP and ATP. The anions were bound to these unimolecular inverted dendritic micelles through electrostatic acid–base interactions and their pH dependency allowed controlled release of the guest molecules from the dendrimer host. The hydrophobic periphery of the dendrimer shielded the anions from hydrophilic attack. The urea groups located in the periphery of the dendrimer were believed to stabilize the complexed anions as well.

Crooks *et al.* reported the transfer of amine-functionalized poly(amidoamine) dendrimers into toluene containing dodecanoic acid [198]. The method is based on the formation of ion pairs between the fatty acids and the terminal amine-groups. These dendrimer-fatty acid complexes resemble unimolecular inverted micelles and could be used as phase transfer vehicles for the transport of Methyl Orange, an anionic dye molecule, into an organic medium.

3.6   RECOGNITION BASED ON METAL–LIGAND INTERACTIONS

The area of dendrimer–metal complexes is one of the most challenging areas in dendrimer research from an application point of view. Owing to their large size, promising applications are foreseen in the field of catalysis combined with ultrafiltration [199–201]. Recently, van Koten and co-workers reported on novel gas-sensor materials based on metallodendrimers [202]. The development of contrast agents for magnetic resonance imaging (MRI), to visualize the bloodstream in the body, is probably the most spectacular progress made so far in this field [203–209]. The focus in this section will be on dendrimer–metal interactions within the framework of host–guest chemistry. For dendrimers that use metal ions as building blocks, the reader is referred to specific reviews on this topic [12, 210, 211]. The complexation of metal ions in dendrimers has been investigated at various positions within the dendrimer framework including the core, interior and periphery.

In 1993, Fréchet *et al.* was the first to demonstrate that the dendritic architecture could alter certain physical properties at the core [188]. Dendrons of varying generation were functionalized with a solvatochromic probe at the focal point. UV/Vis absorption spectra of the central chromophore shifted to longer wavelengths with increasing generation, resulting from an increase in shielding from the environment by the dendrons. Since then, several attempts have been made to exploit the unique microenvironment created by the dendrimer in the field of host–guest chemistry.

Aida *et al.* reported the coordination of imidazoles with varying sizes to a dendritic zinc porphyrin (Figure 16.16) [212, 213]. With a 1:1 stoichiometry of host and guest, binding constants decreased significantly as the generation number of the porphyrin increased from 4 to 5, indicative of a decreased possibility for interpenetration of host and guest. Size-selective guest complexation was observed as the dendritic porphyrin binds preferentially a small vitamin K_3 molecule (2-methyl-1,4-naphthoquinone) in the presence of a larger por-

16:

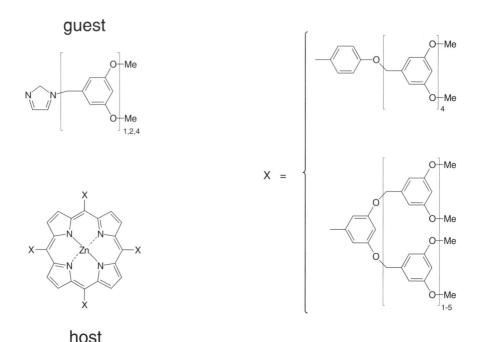

guest

host

Figure 16.16 Dendritic aryl ether zinc porphyrins and dendritic imidazoles used by Aida *et al.* [212, 213] to study the interpenetrating interactions between dendrimer molecules

phyrin guest. The dendritic substituents serve as a steric barrier preventing the larger molecule from binding close to the core of metalloporphyrin [213]. The same group recently reported on an iron porphyrin covalently encapsulated in a large aryl ether dendrimer cage representing the first monomolecular model of dioxygen-carrying haemoproteins [214, 215]. In the case of fifth-generation Fréchet dendrons, the dioxygen adduct was stable over a period of months, even in the presence of water. The long-lived dioxygen complex is believed to arise from steric and hydrophobic protection of the active site by the dendrons.

Meijer, Nolte and co-workers recently reported an example of peripheral metal–oxygen binding [216]. Poly(propylene imine) dendrimers (generation 1–4) were reacted with vinylpyridine yielding dendrimers with bis[2-(2-pyridyl)ethyl]-amine (PY2) ligands. UV/Vis titrations showed quantitative coordination with Cu(I) metal ions for all generations. Low-temperature (− 85 °C) UV/Vis measurements revealed that approximately 60–70% of the copper centers could bind dioxygen. These complexes can be viewed as synthetic analogs of hemocyanin, a copper-containing oxygen transport protein.

Bosman *et al.* reported on the use of amine-functionalized poly(propylene imine) dendrimers as polyvalent ligands for various transition metals, like Cu(II), Zn(II) or Ni(II) [217]. The bis(propylamine)amine pincer, present in the periphery of the parent poly(propylene imine) dendrimers, acts as a tridentate ligand for these metals as evidenced by UV/Vis, EPR and NMR measurements (Figure 16.17a). The strong dendrimer–metal interactions allowed for 'counting' of the dendrimer end-groups in a UV-Vis titration experiment. A first-generation amine-functionalized poly(propylene imine) dendrimer could also be used as a template for the assembly of two rigid Tröger's base dizinc(II) bis-porphyrin receptor molecules, yielding a self-assembled molecular capsule (Figure 16.17b) [218]. The same concept was also extrapolated to the higher generations poly(propylene imine) dendrimers, but in these cases steric interactions hampered complete loading of the dendrimers [219].

In contrast to Bosman *et al.*, who only found metal complexation in the periphery of poly(propylene imine) dendrimers, Tomalia and co-workers reported on the incorporation of copper ions into the interior of PAMAM dendrimers judging from EPR and UV/Vis studies [220, 221]. Metal binding in the dendrimer interior has also been observed for dendrimers carrying multiple ligands for metal complexation within their framework such as crown-ethers [222, 223] (Cs(I)-complexes), piperazine [224] (Pd(II)- and Cu(II)-complexes) or triazocyclononane [225] (Cu(II)- and Ni(II)-complexes). In most cases addition of the metal-salt to the dendrimer led to the formation of 1 : 1 complexes.

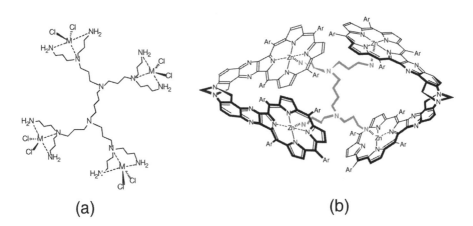

(a)   (b)

Figure 16.17 Amine-terminated poly(propylene imine) dendrimers act as tridentate ligands for the complexation of transition metals [217] (a), and can function as templates for the assembly of Tröger's base dizinc(II) bis-porphyrin molecules, (b) [218]

4 CONCLUSION AND PROSPECTS

The successful application of dendrimers as functional building blocks in the construction of mono- and multilayers, liquid crystals and supramolecular architectures in solution is mainly due to the flexible nature intrinsic to most dendrimers. The numerous examples involving assemblies of dendrimers described in this chapter clearly illustrate that the conventional viewpoint of dendrimers, as spherical macromolecules with a dense shell, often has to be adjusted when dendrimers are used as building blocks in the construction of higher order aggregates. The shape of most dendrimers changes drastically from spherical to oblate or prolate under the influence of secondary interactions when they are incorporated in monolayers, liquid crystalline materials or assemblies in solution. In this context it is important to realize that next to the dendrimer structure itself, the medium around the dendrimer influences its conformational behaviour significantly [44, 77, 84, 226]. External stimuli like hydrophobic effects, phase-separation, electrostatic interactions and hydrogen bonding interactions are driving forces that determine the localization of the end-groups and hence the dendrimer shape to a large extent. Unique dendritic properties such as multi-functionality, porosity and chemical tuneability are often retained within assemblies prepared from these macromolecules. This explains successful applications for dendrimers in the fields of chemical- and bio-sensors, lithography, photo-responsive materials, organic LEDs and LC materials. The conformational characteristics of dendrimers also play a predominant role in the field of host–guest chemistry, wherein the presence of internal voids is closely related to the degree of back-folding, which in turn is influenced by solvent type and pH. The fact that the dendritic shell can produce localized micro-environments has been shown in a number of elegant studies that are of significant interest in the area of bio-organic chemistry. The dendrophanes developed by Diederich and co-workers serve as models for globular proteins [174], whereas the dendritic porphyrins reported by Aida [214, 215] and Diederich [170] represent the first examples of artificial haemoproteins. The nanometer dimensions of dendrimers combined with their multifunctionality has led to exciting applications in the fields of catalysis [199–201], sensor materials [202] and more recently resulted in the development of a novel contrast agents for magnetic resonance imaging (MRI) [203–209]. The development of unimolecular micelles led to the birth of a novel class of amphiphiles with properties that are superior to those of conventional, low molecular weight surfactants. Their static nature, high local concentration of sites and cooperative effects render them effective phase-transfer agents [194] and make them promising candidates for future applications as drug-delivery vehicles [191]. The areas in which specific dendritic properties are exploited to obtain particular material characteristics will undoubtedly continue to grow and lead to many additional applications in the near future.

5 REFERENCES

1. Buhleier, E., Wehner, W. and Vögtle, F. *Synthesis*, 155 (1978).
2. Bosman, A. W., Janssen, H. M. and Meijer, E. W. *Chem. Rev.*, **99**, 1665 (1999).
3. Chow, H. F., Mong, T. K. K., Nongrum, M. F. and Wan, C. W. *Tetrahedron*, **54**, 8543 (1998).
4. Matthews, O. A., Shipway, A. N. and Stoddart, J. F. *Prog. Polym. Sci.*, **23**, 1 (1998).
5. Smith, D. K. and Diederich, F. *Chem. Eur. J.*, **4**, 1353 (1998).
6. Hawker, C. J. *Curr. Opin. Colloid Interface Sci.*, **4**, 117 (1999).
7. Archut, A. and Vögtle, F. *Chem. Soc. Rev.*, **27**, 233 (1998).
8. Fischer, M. and Vögtle, F. *Angew. Chem., Int. Ed. Engl.*, **38**, 884 (1999).
9. Zeng, F. W. and Zimmerman, S. C. *Chem. Rev.*, **97**, 1681 (1997).
10. Emrick, T. and Fréchet, J. M. J. *Curr. Opin. Colloid Interface Sci.*, **4**, 15 (1999).
11. Kim, Y. and Zimmerman, S. C. *Curr. Opin. Chem. Biol.*, **2**, 733 (1998).
12. Newkome, G. R., He, E. F. and Moorefield, C. N. *Chem. Rev.*, **99**, 1689 (1999).
13. Lehn, J. M., *Supramolecular Chemistry, Concepts and Perspectives*, VCH, Weinheim, 1995.
14. Whitesides, G. M., Mathias, J. P. and Seto, C. T. *Science*, **254**, 1312 (1991).
15. Whitesides, G. M., Simanek, E. E., Mathias, J. P., Seto, C. T., Chin, D. N., Mammen, M. and Gorden, D. M., *Acc. Chem. Res.*, **28**, 37 (1995).
16. Kolotuchin, S. V. and Zimmerman, S. C. *J. Am. Chem. Soc.* **120**, 9092 (1998).
17. Sijbesma, R. P., Beijer, F. H., Brunsveld, L., Folmer, B. J. B., Hirschberg, J., Lange, R. F. M., Lowe, J. K. L. and Meijer, E. W. *Science*, **278**, 1601 (1997).
18. Russell, V. A., Evans, C. C., Li, W. J. and Ward, M. D. *Science*, **276**, 575 (1997).
19. Zimmerman, S. C. *Science*, **276**, 543 (1997).
20. Prins, L. J., Huskens, J., Dejong, F., Timmerman, P. and Reinhoudt, D. N. *Nature*, **398**, 498 (1999).
21. Jolliffe, K. A., Timmerman, P. and Reinhoudt, D. N. *Angew. Chem. Int. Ed.*, **38**, 933 (1999).
22. Sijbesma, R. P. and Meijer, E. W. *Curr. Opin. Coll. Int. Sci.*, **4**, 24 (1999).
23. Koert, U., Harding, M. M. and Lehn, J. M. *Nature*, **346**, 339 (1990).
24. Hanan, G. S., Volkmer, D., Schubert, U. S., Lehn, J. M., Baum, G. and Fenske, D. *Angew. Chem. Int. Ed.*, **36**, 1842 (1997).
25. Stupp, S. I., Lebonheur, V., Walker, K., Li, L. S., Huggins, K. E., Keser, M. and Amstutz, A. *Science*, **276**, 384 (1997).
26. Conn, M. M. and Rebek, J. *Chem. Rev.*, 1647 (1997).
27. Macgillivray, L. R. and Atwood, J. L. *Nature*, **389**, 469 (1997).
28. Tomalia, D. A. and Durst, H. D. *Topics in Current Chemistry*, Springer Verlag, Berlin, 1993, p. 193.
29. Tomalia, D. A. *Aldrichimica Acta*, **26**, 91 (1993).
30. Tomalia, D. A. *Adv. Mater.*, **6**, 529 (1994).
31. Tomalia, D. A. *Macromol. Symp.*, **101**, 243 (1996).
32. Percec, V., Ahn, C. H., Ungar, G., Yeardly, D. J. P., Moller, M. and Sheiko, S. S. *Nature*, **391**, 161 (1998).
33. Mansfield, M. L. *Polymer*, **37**, 3835 (1996).
34. Tsukruk, V. V. *Adv. Mater.*, **10**, 253-257 (1998).
35. Tsukruk, V. V., Rinderspacher, F. and Bliznyuk, V. N. *Langmuir*, **13**, 2171 (1997).
36. Bliznyuk, V. N., Rinderspacher, F. and Tsukruk, V. V. *Polymer*, **39**, 5249 (1998).
37. Esumi, K. and Goino, M. *Langmuir*, **14**, 4466 (1998).
38. Sheiko, S. S., Muzafarov, A. M., Winkler, R. G., Getmanova, E. V., Eckert, G. and Reineker, P. *Langmuir*, **13**, 4172 (1997).

39. Tokuhisa, H., Zhao, M. Q., Baker, L. A., Phan, V. T., Dermody, D. L., Garcia, M. E., Peez, R. F., Crooks, R. M. and Mayer, T. M. *J. Am. Chem. Soc.*, **120**, 4492 (1998).
40. Zhao, M., Tokuhisa, H. and Crooks, R. M. *Angew. Chem., Int. Ed. Engl.*, **36**, 2596 (1997).
41. Hierlemann, A., Campbell, J. K., Baker, L. A., Crooks, R. M. and Ricco, A. J. *J. Am. Chem. Soc.*, **120**, 5323 (1998).
42. Wells, M. and Crooks, R. M. *J. Am. Chem. Soc.*, **118**, 3988 (1996).
43. Tokuhisa, H. and Crooks, R. M. *Langmuir*, **13**, 5608 (1997).
44. Schenning, A. P. H. J., Elissen-Román, C., Weener, J. W., Baars, M., Vandergaast, S. J. and Meijer, E. W. *J. Am. Chem. Soc.*, **120**, 8199 (1998).
45. Sayedsweet, Y., Hedstrand, D. M., Spinder, R. and Tomalia, D. A. *J. Mater. Chem.*, **7**, 1199 (1997).
46. Kampf, J. P., Frank, C. W., Malmstrom, E. E. and Hawker, C. J. *Langmuir*, **15**, 227 (1999).
47. Saville, P. M., Reynolds, P. A., White, J. W., Hawker, C. J., Fréchet, J. M. J., Wooley, K. L., Penfold, J. and Webster, J. R. P. *J. Phys. Chem.*, **99**, 8283 (1995).
48. Saville, P. M., White, J. W., Hawker, C. J., Wooley, K. L. and Fréchet, J. M. J. *J Phys. Chem.*, **97**, 293 (1993).
49. Bo, Z. S., Zhang, X., Yi, X. B., Yang, M. L., Shen, J. C., Rehn, Y. H. and Xi, S. Q. *Polym. Bull.*, **38**, 257 (1997).
50. De Gennes, P. G. and Hervet, H. *J. Phys. Lett. Fr.*, **44**, L351 (1983).
51. Mansfield, M. L. and Klushin, L. I. *Macromolecules*, **26**, 4262 (1993).
52. Murat, M. and Grest, G. S. *Macromolecules*, **29**, 1278 (1996).
53. Boris, D. and Rubinstein, M. *Macromolecules*, **29**, 7251 (1996).
54. Lescanec, R. L. and Muthukumar, M. *Macromolecules*, **23**, 2280 (1990).
55. Welch, P. and Muthukumar, M. *Macromolecules*, **31**, 5892 (1998).
56. Stechemesser, S. and Eimer, W. *Macromolecules*, **30**, 2204 (1997).
57. Weener, J. W., van Dongen, J. L. J. and Meijer, E. W. *J. Am. Chem. Soc.*, **121**, 10346 (1999).
58. Crooks, R. M. and Ricco, A. J. *Acc. Chem. Res.*, **31**, 219 (1998).
59. Gorman, C. B., Miller, R. L., Chen, K. Y., Bishop, A. R., Haasch, R. T. and Nuzzo, R. G. *Langmuir*, **14**, 3312 (1998).
60. Miller, L. L., Kunugi, Y., Canavesi, A., Rigaut, S., Moorefield, C. N. and Newkome, G. R. *Chem. Mater.*, **10**, 1751 (1998).
61. Anzai, J. I. and Osa, T., *Avidin-Biotin Supramolecular Complexation for Biosensor Application*, JAI Press Inc., Greenwich, Connecticut, 1997, p. 143.
62. Anzai, J. and Nishimura, M. *J. Chem. Soc. Perkin Trans.* **2**, 1887 (1997).
63. Anzai, J. I., Kobayashi, Y., Nakamura, N., Nishimura, M. and Hoshi, T. *Langmuir*, **15**, 221 (1999).
64. He, P., Takahashi, P., Hoshi, T., Anzai, J., Suzuki, Y. and Osa, T. *Mater. Sci. Eng.*, **C2**, 103 (1994).
65. Hoshi, T., Anzai, J. and Osa, T. *Anal. Chem.*, **34**, 770 (1995).
66. Ebersole, R. C., Miller, J. A., Moran, J. R. and Ward, M. D. *J. Am. Chem. Soc.*, **112**, 3239 (1990).
67. Alonso, B., Cuadrado, I., Moran, M. and Losada, J. *J. Chem. Soc. Chem. Commun.*, 2575 (1994).
68. Losada, J., Cuadrado, I., Moran, M., Casado, C. M., Alonso, B. and Barranco, M. *Anal. Chim. Acta*, **338**, 191 (1997).
69. Alonso, B., Moran, M., Casado, C. M., Lobete, F., Losada, J. and Cuadrado, I. *Chem. Mater.*, **7**, 1440 (1995).
70. Hawker, C. J. and Fréchet, J. M. J. *J. Am. Chem. Soc.*, **112**, 7638 (1990).

71. L'abbe, G., Forier, B. and Dehean, W. *Chem. Commun.*, 2143 (1996).
72. Tully, D. C., Trimble, A. R., Fréchet, J. M. J., Wilder, K. and Quate, C. F. *Chem. Mater.*, **11**, 2892 (1999).
73. Tully, D. C., Wilder, K., Fréchet, J. M. J., Trimble, A. R. and Quate, C. F. *Adv. Mater.*, **11**, 314 (1999).
74. Weener, J. W. and Meijer, E. W. *Adv. Mater.*, **12**, 741 (2000).
75. Archut, A., Vögtle, F., Decola, L., Azzellini, G. C., Balzani, V., Ramanujam, P. S. and Berg, R. H. *Chem. Eur. J.*, **4**, 699 (1998).
76. Kuwabura, Y., Ogawa, H., Inada, H., Noma, N. and Shirota, Y. *Adv. Mater.*, **6**, 677 (1994).
77. Bettenhausen, J., Greczmiel, M., Jandke, M. and Strohriegl, P. *Synth. Met.*, **91**, 223 (1997).
78. Bettenhausen, J. and Strohriegl, P. *Adv. Mater.*, **8**, 507 (1996).
79. Wang, P. W., Liu, Y. J., Devadoss, C., Bharathi, P. and Moore, J. S. *Adv. Mater.*, **8**, 237 (1996).
80. Halim, M., Pillow, J. N. G., Samuel, D. W. and Burn, P. L. *Adv. Mater.*, **11**, 371 (1999).
81. Halim, M., Pillow, J. N. G., Samuel, I. D. W. and Burn, P. L. *Synth. Met.*, **102**, 922 (1999).
82. Halim, M., Samuel, I. D. W., Pillow, J. N. G. and Burn, P. L. *Synth. Met.*, **102**, 1113 (1999).
83. Sooklal, K., Hanus, L. H., Ploehn, H. J. and Murphy, C. J. *Adv. Mater.*, **10**, 1083 (1998).
84. Schenning, A. P. H. J. P. H. J., Peeters, E. and Meijer, E. W. *J. Am. Chem. Soc.*, **122**, 4489 (2000).
85. Cardullo, F., Diederich, F., Echegoyen, L., Habicher, T., Jayaraman, N., Leblanc, R. M., Stoddart, J. F. and Wang, S. P. *Langmuir*, **14**, 1955 (1998).
86. Felder, D., Gallani, J. L., Guillon, B. H., Nicoud, J. F. and Nierengarten, J. F. *Angew. Chem., Int. Ed. Engl.*, **39**, 201 (2000).
87. Mirkin, C. A. and Caldwell, W. B. *Tetrahedron*, **52**, 5113 (1996).
88. Newkome, G. R., Moorefield, C. N., Baker, G. R., Johnson, A. L. and Behera, R. K. *Angew. Chem., Int. Ed. Engl.*, **30**, 1176 (1991).
89. Newkome, G. R., Moorefield, C. N., Baker, G. R., Saunders, M. J. and Grossman, S. H. *Angew. Chem., Int. Ed. Engl.*, **30**, 1178 (1991).
90. Young, J. K., Baker, G. R., Newkome, G. R., Morris, K. F. and Johnson, C. S. *Macromolecules*, **27**, 3464 (1994).
91. Newkome, G. R. and Moorefield, C. N. *Macromol. Symp.*, **77**, 63 (1994).
92. Newkome, G. R., Güther, R. and Gardullo, F. *Macromol. Symp.*, **98**, 467 (1995).
93. Hawker, C. J., Wooley, K. L. and Fréchet, J. M. J. *J Chem Soc Perkin Trans 1*, 1287 (1993).
94. Piotti, M. E., Rivera, F., Bond, R., Hawker, C. J. and Fréchet, J. M. J. *J. Am. Chem. Soc.*, **121**, 9471 (1999).
95. Newkome, G. R., Yao, Z., Baker, G. R., Gupta, V. K., Russo, P. S. and Saunders, M. J. *J. Am. Chem. Soc.*, **108**, 849 (1986).
96. Schmitzer, A., Perez, E., Ricolattes, I., Lattes, A. and Rosca, S. *Langmuir*, **15**, 4397 (1999).
97. Fuhrop, J. H. and Mathieu, J. *Angew. Chem., Int. Ed. Engl.*, **23**, 100 (1984).
98. Escamilla, G. H. and Newkome, G. R. *Angew. Chem., Int. Ed. Engl.*, **33**, 1937 (1994).
99. Newkome, G. R., Baker, G. R., Saunders, M. J., Russo, P. S., Gupta, V. K., Yao, Z.-q., Miller, J. E. and Bouillon, K. *J. Chem. Soc., Chem. Commun.*, 752 (1986).
100. Newkome, G. R., Baker, G. R., Arai, S., Saunders, M. J., Russo, P. S., Theriot, K. J.,

Moorefield, C. N., Rogers, L. E., Miller, J. E., Lieux, T. R., Murray, M. E. and Philips, B. *J. Am. Chem. Soc.*, **112**, 8458 (1990).

101. Newkome, G. R., Moorefield, C. N., Baker, G. R., Behera, R. K., Escamillia, G. H. and Saunders, M. J. *Angew. Chem., Int. Ed. Engl.*, **31**, 917 (1992).
102. Newkome, G. R., Lin, X. F., Yaxiong, C. and Escamilla, G. H. *J. Org. Chem*, **58**, 3123 (1993).
103. Newkome, G. R., Lin, X. F., Chen, Y. X. and Escamilla, G. H. *J. Org. Chem.*, **58**, 7626 (1993).
104. Jørgensen, M., Bechgaard, K., Bjørnholm, T., Sommer-Larsen, P., Hansen, L. G. and Schaumburg, K. *J. Org. Chem.*, **59**, 5877 (1994).
105. Eisenberg, A. and Zhang, L. *Science*, **268**, 1728 (1995).
106. Cameron, N. S., Corbierre, M. K. and Eisenberg, A. *Can. J. Chem.*, **77**, 1311 (1999).
107. Discher, B. M., Won, Y.-Y., Ege, D. S., Lee, J. C.-M., Bates, F. S., Disher, D. E. and Hammer, D. A. *Science*, **284**, 1143 (1999).
108. Won, Y.-Y., Davis, H. T. and Bates, F. S. *Science*, **283**, 960 (1999).
109. Cornelissen, J., Fischer, M., Sommerdijk, N. and Nolte, R. J. M. *Science*, **280**, 1427 (1998).
110. Chapman, T. M., Hillyer, G. L., Mahan, E. J. and Shaffer, K. A. *J. Am. Chem. Soc.*, **116**, 11195 (1994).
111. Gitsov, I., Wooley, K. L., Hawker, C. J., Ivanova, P. T. and Fréchet, J. M. J. *Macromolecules*, **26**, 5621 (1993).
112. Yu, D., Vladimirov, N. and Fréchet, J. M. J. *Macromolecules*, **32**, 5186 (1999).
113. Gitsov, I. and Fréchet, J. M. J. *Macromolecules*, **26**, 6536 (1993).
114. Fréchet, J. M. J., Gitsov, I., Monteil, T., Rochat, S., Sassi, J. F., Vergelati, C. and Yu, D. *Chem. Mater.*, **11**, 1267 (1999).
115. Iyer, J., Fleming, K. and Hammond, P. T. *Macromolecules*, **31**, 8757 (1998).
116. Iyer, J. and Hammond, P. T. *Langmuir*, **15**, 1299 (1999).
117. Van Hest, J. C. M., Delnoye, D. A. P., Baars, M., Van Genderen, M. H. P. and Meijer, E. W. *Science*, **268**, 1592 (1995).
118. Van Hest, J. C. M., Delnoye, D. A. P., Baars, M. W. P. L., Elissen-Román, C., van Genderen, M. H. P. and Meijer, E. W. *Chem. Eur. J.*, **2**, 1616 (1996).
119. Israelachvili, J. N., Mitchell, D. J. and Ninham, B. W. *J. Chem. Soc. Faraday Trans. 2*, **72**, 1525 (1976).
120. Van Hest, J. C. M., Baars, M., Elissen Román, C., van Genderen, M. H. P. and Meijer, E. W. *Macromolecules*, **28**, 6689 (1995).
121. Gitsov, I., Wooley, K. L. and Fréchet, J. M. J. *Angew. Chem., Int. Ed. Engl.*, **31**, 1200 (1992).
122. Gitsov, I. and Fréchet, J. M. J. *J. Am. Chem. Soc.*, **118**, 3785 (1996).
123. Friberg, S. E., Podzimek, M., Tomalia, D. A. and Hedstrand, D. M. *Mol. Cryst. Liq. Cryst.*, **164**, 157 (1988).
124. Li, J. F., Crandall, K. A., Chu, P. W., Percec, V., Petschek, R. G. and Rosenblatt, C. *Macromolecules*, **29**, 7813 (1996).
125. Percec, V., Chu, P. W., Ungar, G. and Zhou, J. P. *J. Am. Chem. Soc.*, **117**, 11441 (1995).
126. Lorenz, K., Holter, D., Stuhn, B., Mulhaupt, R. and Frey, H. *Adv. Mater.*, **8**, 414 (1996).
127. Frey, H., Lorenz, K. and Mülhaupt, R. *Macromol. Symp.*, **102**, 19 (1996).
128. Lorenz, K., Frey, H., Stuhn, B. and Mulhaupt, R. *Macromolecules*, **30**, 6860 (1997).
129. Frey, H., Lach, C. and Lorenz, K. *Adv. Mater.*, **10**, 279 (1998).
130. Cameron, J. H., Facher, A., Lattermann, G. and Diele, S. *Adv. Mater.*, **9**, 398 (1997).
131. Stebani, U. and Lattermann, G. *Adv. Mater.*, **7**, 578 (1995).

132. Baars, M., Söntjens, S. H. M., Fischer, H. M., Peerlings, H. W. I. and Meijer, E. W. *Chem. Eur. J.*, **4**, 2456 (1998).
133. Yonetake, K., Masuko, T., Morishita, T., Suzuki, K., Ueda, M. and Nagahata, R. *Macromolecules*, **32**, 6578 (1999).
134. Román, C., 'Amphiphilic dendrimers', Ph.D. thesis, Eindhoven University of Technology (1999).
135. Ramzi, A., Bauer, B. J., Scherrenberg, R., Froehling, P., Joosten, J. and Amis, E. J. *Macromolecules*, **32**, 4983 (1999).
136. Stark, B., Stuhn, B., Frey, H., Lach, C., Lorenz, K. and Frick, B. *Macromolecules*, **31**, 5415 (1998).
137. Percec, V., Cho, W.-D., Mosier, P. E., Ungar, G. and Yeardly, D. J. P. *J. Am. Chem. Soc.*, **120**, 11061 (1998).
138. Balagurusamy, V. S. K., Ungar, G., Percec, V. and Johansson, G. *J. Am. Chem. Soc.*, **119**, 1539 (1997).
139. Percec, V., Ahn, C.-H. and Barboiu, B. *J. Am. Chem. Soc.*, **119**, 12978 (1997).
140. Percec, V., Johansson, G., Ungar, G. and Zhou, J. *J. Am. Chem. Soc.*, **118**, 9855 (1996).
141. Yeardley, D. J. P., Ungar, G., Percec, V., Holerca, M. N. and Johansson, G. *J. Am. Chem. Soc.*, **122**, 1684 (2000).
142. Baars, M. W. P. L., van Boxtel, M. C. W., Bastiaansen, C. W. M., Broer, D. J., Söntjes, S. H. M. and Meijer, E. W. *Adv. Mater.* **12**, 715 (2000).
143. Maciejewski, M. *J. Macrom. Sci.-Chem.*, **17A**, 689 (1982).
144. Newton, S. P., Stoddart, J. F. and Hayes, W. *Supramolec. Sci.*, **3**, 221 (1996).
145. Vögtle, F., *Supramolecular Chemistry; an introduction*, John Wiley & Sons, Chichester, 1991.
146. Reinhoudt, D. N., *Supramolecular Materials and Technologies*, John Wiley & Sons, Chichester. 1999.
147. Langer, R. *Chem. Eng. Sci.*, 4109 (1995).
148. Duncan, R. *Abstr. Pap. Am. Chem. Soc.*, **217**, 141 (1999).
149. Duncan, R. and Kopecek, J. *Adv. Polym. Sci.*, **57**, 51 (1984).
150. Malik, N., Wiwattanapatapee, R., Klopsch, R., Lorenz, K., Frey, H., Weener, J. W., Meijer, E. W., Paulus, W. and Duncan, R. *J. Cont. Rel.*, **65**, 133 (2000).
151. Scherrenberg, R., Coussens, B., Vanvliet, P., Edouard, G., Brackman, J., De Brabander, E. and Mortensen, K. *Macromolecules*, **31**, 456 (1998).
152. Dubin, P. L., Edwards, S. L., Kaplan, J. I., Mehta, M. S., Tomalia, D. and Xia, J. L. *Anal. Chem.*, **64**, 2344 (1992).
153. Tomalia, D. A., Hall, V. B. and Hedstrand, D. M. *Macromolecules*, **20**, 1167 (1987).
154. Mourey, T. H., Turner, S. R., Rubinstein, M., Fréchet, J. M. J., Hawker, C. J. and Wooley, K. L. *Macromolecules*, **25**, 2401 (1992).
155. Wooley, K. L., Klug, C. A., Tasaki, K. and Schaefer, J. *J. Am. Chem. Soc.*, **119**, 53 (1997).
156. Gorman, C. B., Hager, M. W., Parkhurst, B. L. and Smith, J. C. *Macromolecules*, **31**, 815 (1998).
157. Miklis, P., Cagin, T. and Goddard, W. A. *J. Am. Chem. Soc.*, **119**, 7458 (1997).
158. Cavallo, L. and Fraternali, F. *Chem. Eur. J.*, **4**, 927 (1998).
159. Jansen, J. F. G. A., de Brabander van den Berg, E. M. M. and Meijer, E. W. *Science*, **266**, 1226 (1994).
160. Jansen, J. F. G. A., de Brabander-van den Berg, E. M. M. and Meijer, E. W. *New Macromolecular Architectures and Functions*, 99, Springer-Verlag, Berlin, Heidelberg, New York, 1995.
161. Jansen, J. F. G. A., Peerlings, H. W. I., de Brabander van den Berg, E. M. M. and

Meijer, E. W. *Angew. Chem., Int. Ed. Engl.,* **34**, 1206 (1995).

162. Put, R. J. H., Clays, K., Persoons, A., Biemans, H. A. M., Luijkx, C. P. M. and Meijer, E. W. *Chem. Phys. Lett.,* **260**, 136 (1996).

163. Bodnár, I., Silva, A. S., Deitcher, R. W., Weisman, N. E., Kim, Y. H. and Wagner, N. J. *J. Polym. Sci.,* **38**, 857 (2000).

164. Tomalia, D. A., Naylor, A. and Goddard III, W. A. *Angew. Chem., Int. Ed. Engl.,* **29**, 138 (1990).

165. Jansen, J. F. G. A., de Brabander van den Berg, E. M. M. and Meijer, E. W. *Abstr. Pap. Am. Chem. Soc.,* **210**, 64 (1995).

166. Jansen, J. F. G. A., de Brabander van den Berg, E. M. M. and Meijer, E. W. *Recl. Trav. Chim.-J. Roy. Neth. Chem.,* **114**, 225 (1995).

167. Jansen, J. F. G. A., Janssen, R. A. J., de Brabander van den Berg, E. M. M. and Meijer, E. W. *Adv. Mater.,* **7**, 561 (1995).

168. Jansen, J. F. G. A., Meijer, E. W. and de Brabander van den Berg, E. M. M. *J. Am. Chem. Soc.,* **117**, 4417 (1995).

169. Jansen, J. F. G. A. and Meijer, E. W. *Macromol. Symp.,* **102**, 27 (1996).

170. Dandliker, P. J., Diederich, F., Gisselbrecht, J. P., Louati, A. and Gross, M. *Angew. Chem., Int. Ed. Engl.,* **34**, 2725 (1996).

171. Dandliker, P. J., Diederich, F., Zingg, A., Gisselbrecht, J. P., Gross, M., Louati, A. and Sanford, E. *Helv. Chim. Acta,* **80**, 1773 (1997).

172. Sebastiano, M., Seiler, P. and Diederich, F. *Helv. Chim. Acta,* **78**, 1904 (1995).

173. Wallimann, P., Mattei, S., Seiler, P. and Diederich, F. *Helv. Chim. Acta,* **80**, 2368 (1997).

174. Mattei, S., Wallimann, P., Kenda, B., Amrein, W. and Diederich, F. *Helv. Chim. Acta,* **80**, 2391 (1997).

175. Wallimann, P., Seiler, P. and Diederich, F. *Helv. Chim. Acta,* **79**, 779 (1996).

176. Wallimann, P., Marti, T., Furer, A. and Diederich, F. *Chem. Rev.,* **97**, 1567 (1997).

177. Kenda, B. and Diederich, F. *Angew. Chem., Int. Ed. Engl.,* **37**, 3154 (1998).

178. Cuadrado, I., Moran, M., Casado, C. M., Alonso, B., Lobete, F., Garcia, B., Ibisate, M. and Losada, J. *Organometallics,* **15**, 5278 (1996).

179. Castro, R., Cuadrado, I., Alonso, B., Casado, C. M., Moran, M. and Kaifer, A. E. *J. Am. Chem. Soc.,* **119**, 5760 (1997).

180. Matsue, T., Evans, D. H., Osa, T. and Kobayashi, N. *J. Am. Chem. Soc.,* **107**, 3411 (1985).

181. Gonzalez, B., Casado, C. M., Alonso, B., Cuadrado, I., Moran, M., Wang, Y. and Kaifer, A. E. *Chem. Commun.,* 2569 (1998).

182. Newkome, G. R., Godinez, L. A. and Moorefield, C. N. *Chem. Commun.,* 1821 (1998).

183. James, T. D., Shinmori, H., Takeuchi, M. and Shinkai, S. *Chem. Commun.,* 705 (1996).

184. Santo, M. and Fox, M. A. *J. Phys. Org. Chem.,* **12**, 293 (1999).

185. Newkome, G. R., Woosley, B. D., He, E. F., Moorefield, C. N., Guther, R., Baker, G. R., Escamilla, G. H., Merrill, J. and Luftmann, H. *Chem. Commun.,* 2737 (1996).

186. Kotera, M., Lehn, J. M. and Vigneron, J. P. *J. Chem. Soc. Chem. Commun.,* 197 (1994).

187. Zimmerman, S. C., Wang, Y., Bharathi, P. and Moore, J. S. *J. Am. Chem. Soc.,* **120**, 2172 (1998).

188. Hawker, C. J., Wooley, K. L. and Fréchet, J. M. J. *J. Am. Chem. Soc.,* **115**, 4375 (1993).

189. Naylor, A. M., Goddard, W. A., Kiefer, G. E. and Tomalia, D. A. *J. Am. Chem. Soc.,* **111**, 2339 (1989).

190. Twyman, L. J., Beezer, A. E., Esfand, R., Hardy, M. J. and Mitchell, J. C. *Tetrahedron Lett.,* **40**, 1743 (1999).

191. Baars, M. W. P. L., Kleppinger, R., Koch, M., Yeu, S. L. and Meijer, E. W. *Angew. Chem., Int. Ed. Engl.*, **39**, 1285 (2000).
192. Archut, A., Azzellini, G. C., Balzani, V., Decola, L. and Vögtle, F. *J. Am. Chem. Soc.*, **120**, 12187 (1998).
193. Stevelmans, S., van Hest, J. C. M., Jansen, J. F. G. A., Van Boxtel, D., Van den Berg, E. and Meijer, E. W. *J. Am. Chem. Soc.*, **118**, 7398 (1996).
194. Baars, M., Froehling, P. E. and Meijer, E. W. *Chem. Commun.*, 1959 (1997).
195. Cooper, A. I., Londono, J. D., Wignall, G., McClain, J. B., Samulski, E. T., Lin, J. S., Dobrynin, A., Rubinstein, M., Burke, A. L. C., Fréchet, J. M. J. and Desimone, J. M. *Nature*, **389**, 368 (1997).
196. Goetheer, E. L. V., Baars, M. W. P. L., Vorstman, M. A. G., Meijer, E. W. and Keurentjes, J. T. F. in *Procedings on the 6th Meeting on Supercritical Fluids, Chemistry and Materials*, Nottingham, UK, 1999, p. 507.
197. Stephan, H., Spies, H., Johannsen, B., Klein, L. and Vögtle, F. *Chem. Commun.*, 1875 (1999).
198. Chechik, V., Zhao, M. Q. and Crooks, R. M. *J. Am. Chem. Soc.*, **121**, 4910 (1999).
199. Knapen, J. W. J., Van der Made, A. W., De Wilde, J. C., Van Leeuwen, P., Wijkens, P., Grove, D. M. and Van Koten, G. *Nature*, **372**, 659 (1994).
200. Gossage, R. A., Van de Kuil, L. A. and Van Koten, G. *Acc. Chem. Res.*, **31**, 423 (1998).
201. Reetz, M. T., Lohmer, G. and Schwickardi, R. *Angew. Chem., Int. Ed. Engl.*, **36**, 1526 (1997).
202. Albrecht, M. and Van Koten, G. *Adv. Mater.*, **11**, 171 (1999).
203. Wiener, E. C., Brechbiel, M. W., Brothers, H., Magin, R. L., Gansow, O. A., Tomalia, D. A. and Lauterbur, P. C. *Magn. Reson. Med.*, **31**, 1 (1994).
204. Wiener, E. C., Auteri, F. P., Chen, J. W., Brechbiel, M. W., Gansow, O. A., Schneider, D. S., Belford, R. L., Clarkson, R. B. and Lauterbur, P. C. *J. Am. Chem. Soc.*, **118**, 7774 (1996).
205. Wiener, E. C., Konda, S., Shadron, A., Brechbiel, M. and Gansow, O. *Invest. Radiol.*, **32**, 748 (1997).
206. Dong, Q., Hurst, D. R., Weinmann, H. J., Chenevert, T. L., Londy, F. J. and Prince, M. R. *Invest. Radiol.*, **33**, 699 (1998).
207. Tóth, E., Pubanz, D., Vauthey, S., Helm, L. and Mehrbach, A. E. *Chem. Eur. J.*, **2**, 1607 (1996).
208. Adam, G., Neuerburg, J., Spüntrup, E. and Mühler, A. *J. Magn. Reson. Imaging*, **4**, 462 (1994).
209. Tacke, J., Adam, G., Claßen, H. and Mühler, A. *J. Magn. Reson. Imaging*, **7**, 678 (1997).
210. Venturi, M., Serroni, S., Juris, A., Campagna, S. and Balzani, V. *Top. Curr. Chem.*, **197**, 193 (1998).
211. Gorman, C. *Adv. Mater.*, **10**, 295 (1998).
212. Tomoyose, Y., Jiang, D. L., Jin, R. H., Aida, T., Yamashita, T., Horie, K., Yashima, E. and Okamoto, Y. *Macromolecules*, **29**, 5236 (1996).
213. Jin, R. H., Aida, T. and Inoue, S. *J. Chem. Soc. Chem. Commun.*, 1260 (1993).
214. Jiang, D. L. and Aida, T. *Chem. Commun.*, 1523 (1996).
215. Jiang, D. L. and Aida, T. *J. Macromol. Sci. Pure Appl. Chem.*, **A34**, 2047 (1997).
216. Gebbink, R., Bosman, A. W., Feiters, M. C., Meijer, E. W. and Nolte, R. J. M. *Chem. Eur. J.*, **5**, 65 (1999).
217. Bosman, A. W., Schenning, A. P. H. J., Janssen, R. A. J. and Meijer, E. W. *Chem. Ber-Recl.*, **130**, 725 (1997).
218. Reek, J. N. H., Schenning, A. P. H. J., Bosman, A. W., Meijer, E. W. and Crossley, M. *J. Chem. Commun.*, 11 (1998).

219. Bosman, A. W. 'Dendrimers in Action', Ph.D. Thesis, Eindhoven University of Technology, 1999, p. 123.
220. Ottaviani, M. F., Bossmann, S., Turro, N. J. and Tomalia, D. A. *J. Am. Chem. Soc.*, **116**, 661 (1994).
221. Ottaviani, M. F., Montalti, F., Turro, N. J. and Tomalia, D. A. *J. Phys. Chem. B.*, **101**, 158 (1997).
222. Nagasaki, T., Ukon, M., Arimori, S. and Shinkai, S. *J. Chem. Soc. Chem. Commun.*, 608 (1992).
223. Nagasaki, T., Kimura, O., Masakatsu, U., Arimori, S., Hamachi, I. and Shinkai, S. *J. Chem. Soc. Perkin. Trans.*, 175 (1994).
224. Newkome, G. R., Gross, J., Moorefield, C. N. and Woosley, B. D. *Chem. Commun.*, 515 (1997).
225. Beer, P. D. and Gao, D. *Chem. Commun.*, 443 (2000).
226. Li, J., Piehler, L. T., Qin, D., Baker, Jr, J. R., Tomalia, D. A. and Meier, D. J. *Langmuir*, **16**, 5613 (2000).

17

Dendritic Polymers: Optical and Photochemical Properties

D-L. JIANG AND T. AIDA
Department of Chemistry and Biotechnology, Graduate School of Engineering,
The University of Tokyo, Hongo, Bunkyo-ku, Tokyo 113-8656, Japan

1 INTRODUCTION

To achieve photosynthesis with a totally artificial system is a supreme challenge in science and a dream of chemists. In nature, plants and photosynthetic bacteria depend on photosynthesis utilizing elaborate chromophore arrays to trap solar energy, followed by an efficient energy transfer to the reaction center. Although there have been many efforts to design molecular systems for light harvesting, they usually suffer from inadequate energy transfer efficiency. Synthetic macromolecules have attracted attention as potential photosynthetic antennae, since multiple chromophoric functionalities can be incorporated and organized. However, examples thus far reported are generally derived from linear-chain polymers which, unlike biological macromolecules, can adopt ill-defined morphologies, many of which lead to complicated photochemical events associated with intra- and interchain interactions. Moreover, broad molecular weight distributions and uncontrolled structures inherent in linear-chain synthetic polymers, make it difficult to develop meaningful correlation between their structures and photochemical function.

Recent developments in the design of dendritic molecules has provided both new methodology and molecular architecture which are structure controlled macromolecules, globular-shaped, dendritic-branched tree-like structures with nanoscscale dimensions [1]. Dendrimers generally consist of a focal core, many building blocks (monomer units) and a mathematically defined number of ex-

Dendrimers and Other Dendritic Polymers. Edited by Jean M. J. Fréchet and Donald A. Tomalia
© 2001 John Wiley & Sons Ltd

terior termini. The synthetic methods involve either divergent or convergent approaches which can provide nearly total control over critical molecular design parameters such as *molecular size, shape* and *disposition* of functionalities introduced at the interior core, building units, and/or the exterior surface. Recently, dendrimers have attracted great attention as artificial light-harvesting antennae [2], due to their morphological similarity to chromophore aggregates discovered in natural photosynthetic centers [3]. For example, Balzani *et al.* have reported a tridecanuclear ruthenium(II)-polypyridine supramolecular dendrimer as a light-harvesting complex [2a, b]. Fox *et al.* have investigated poly(benzyl ether) dendritic structures as intervening media for photoinduced electron transfer reaction [2c]. Fréchet *et al.* [2d] have reported site isolation and antenna effects related to luminescence properties of self-assembled lanthanide-core dendrimers. On the mechanism of energy transduction, Moore and Kopelman *et al.* have investigated the luminescence properties of a series of perylene-terminated highly conjugated dendrons consisting of aromatic and alkynyl units, and have highlighted the importance of an energy gradient for the efficient transfer of the excitation energy [2f, g]. Theoretical calculations on energy transduction events in dendritic architectures have also been investigated by Klafter [2g, h] and Mukamel *et al.* [2i].

In the present chapter, emphasis is placed on the authors' recent studies concerning the molecular design and optical/photochemical properties of Fréchet-type poly(benzyl ether) dendrimers.

2 LIGHT-HARVESTING ANTENNA FUNCTIONS OF DENDRIMERS

2.1 MORPHOLOGY DEPENDENCE OF EXCITED SINGLET ENERGY TRANSFER EVENTS

Poly(benzyl ether) dendritic molecules with a focal porphyrin functionality are good motives for investigating the intramolecular energy transfer events, since the dendrimer framework serves as a potential antenna, while the porphyrin core may function as an energy trap [4]. For a study on the excited singlet energy transfer characteristics [5], dendrimer porphyrins **1a–1f** bearing different numbers (1–4) of five-layered poly(benzyl ether) dendron subunits have been synthesized by an alkaline-mediated coupling of a four-layered poly(benzyl ether) dendron bromide with tetraphenylporphyrin derivatives having corresponding numbers of 3',5'-dihydroxyphenyl groups at the *meso*-positions. The products can be isolated by flash column chromatography and identified by ^{1}H NMR, MALDI-TOF-MS, and UV-Vis spectroscopies.

The dendron subunits, upon excitation at the characteristic 280 nm absorption band, emit a fluorescence centered at 310 nm, which is slightly overlapped

1a: $R^1 = L5$, $R^2 = R^3 = R^4 = $ tolyl
1b: $R^1 = R^2 = L5$, $R^3 = R^4 = $ tolyl
1c: $R^1 = R^3 = L5$, $R^2 = R^4 = $ tolyl
1d: $R^1 = R^2 = R^3 = L5$, $R^4 = $ tolyl
1e: $R^1 = R^2 = R^3 = R^4 = L5$
1f: $R^1 = R^2 = R^3 = R^4 = L4$

$L5 = $

with the Soret absorption band of the porphyrin core (421.5 (**1a**)–425.5 nm (**1e**)). Therefore, the dendron subunits appear to have the possibility to communicate with the porphyrin core by energy transfer. In fact, upon 280 nm excitation of tetra-substituted dendrimer porphyrin **1e**, the emission from the dendron subunits is significantly quenched, while a strong fluorescence from the porphyrin core (656, 718 nm) appears. This is clearly the result of an excited singlet energy transfer from the dendron subunits to the porphyrin core. By comparison of the excitation spectrum with the absorption spectrum, the energy transfer quantum efficiency (Φ_{ENT}) has been evaluated to be 80%. In sharp contrast, excitation of partially substituted dendrimer porphyrins **1a–1d** at 280 nm results in an emission predominantly from the dendron subunits with only a weak fluorescence from the porphyrin core. The Φ_{ENT} values for **1a**, **1b**, **1c** and **1d** are respectively 31, 19, 10 and 10%, which are obviously smaller than that of tetra-substituted **1e**. Also of interest is the difference in Φ_{ENT} between geometrical isomers of di-substituted **1b** and **1c**, where the Φ_{ENT} value of the *anti*-isomer (**1c**) is definitely lower than that of the *syn*-isomer (**1b**), and is almost comparable to that of mono-substituted dendrimer porphyrin **1a**. Such a morphology dependence of the energy transfer characteristics is indicative of a possible cooperation of the dendron subunits for the intramolecular singlet energy transfer. Fluorescence depolarization measurements have indicated that the four dendron subunits in

1e behave like a single, large chromophore, where the excitation energy migrates over the dendrimeric three-dimensional array of the aromatic building units around the energy trap. Consequently, the probability of energy transfer to the interior trap is highly enhanced. On the other hand, in partially substituted **1a–1d**, the dendrimeric arrays of the aromatic building units are more loosely packed and discontinuous, so that the excitation energy may migrate less efficiently and is more likely to be lost by radiation before transfer to the energy trap.

^{1}H NMR pulse relaxation time T_1 measurements have shown an egg-like structural resemblance to tetra-substituted **1e**: The T_1 value of the pyrrole-β signal of the porphyrin core is virtually intact to the number of the dendron subunits (1.50–1.58 sec), indicating that the interior environment is not constrained, even in the largest **1e**. In contrast, the T_1 of the exterior OMe signal, which stays in the range 0.96–0.86 s for **1a–1d**, displays a significant drop (0.25 sec) for **1e**. Thus, the dendron subunits in tetra-substituted **1e** are highly constrained in conformational change due to their dense packing, whereas those in partially substituted **1a–1d** are able to change their conformation. Relative to this trend, the cooperation of the dendron subunits (aromatic building units) for the energy migration process is expected to be reduced upon elevating the temperature to activate conformational motion. In this regard, temperature dependency of the energy transfer event in partially substituted **1d**, together with those of tetra-substituted **1e** and **1f**, has been investigated. When the temperature is raised from 20 to 80 °C in 1,2-dichloroethane, the Φ_{ENT} of **1d**, upon 280 nm excitation of the dendron subunits, drops off in a sigmoidal fashion from 31 to 12%, which is almost comparable to that of **1a** having a single dendron subunit. Tetra-substituted **1f** with one-generation smaller dendron subunits, displays a much clearer temperature dependency, where the Φ_{ENT} is decreased significantly from 79 to 35% upon elevating the temperature from 20 to 80 °C. In sharp contrast, the Φ_{ENT} of the largest **1e** does not drop but stays around 80% even at 80 °C. As mentioned above, the dendron subunits in **1e** are highly constrained in conformational change, whereas those in tri-substituted **1d** are able to change their conformation. Tetra-substituted **1f** is almost comparable to **1d** in terms of the conformational change activity, as judged from the ^{1}H NMR T_1 value of the exterior OMe signal (0.81 s). Since the fluorescence properties of neither the dendron subunits nor the porphyrin core are little affected by the temperature in the range 20–80 °C, the obvious reductions of Φ_{ENT}, observed for **1d** and **1f** at higher temperatures, have been attributed to the thermal-enhanced conformational motion of the molecule: The four dendron subunits (aromatic building units) at higher temperatures are likely to behave virtually independently or much less cooperatively for the energy migration process. In this sense, it is quite interesting that the four dendron subunits in the largest **1e** are still highly cooperative at 80 °C, indicating a high potential for the conformationally rigid, spherical poly(benzyl ether) dendrimeric framework to function as an artificial antenna for light harvesting.

The energy transduction profile observed for the large, spherical dendrimer porphyrin (**1e**) is related to the light-harvesting events in wheel-like supramolecular assemblies of chromophores found in a purple photosynthetic bacterium. In this case, the excitation energy migrates very rapidly and efficiently along the wheel, followed by transfer to the interior energy trap (special pair) to initiate the photosynthetic process [3].

2.2 MOLECULAR DESIGN OF BLUE-LUMINESCENT DENDRITIC RODS

Conjugated polymers have attracted special attention for their potential utility as molecular wire for the transfer of electron and excitation energy along the backbone and also as organic photo/electro-luminescent devices. However, because of their high conformational rigidity, conjugated polymers usually have a limited solubility and are difficult to process. On the other hand, from a photochemical point of view, such a strong aggregation tendency of the conjugated macromolecular chain also results in a collisional quenching of the excited state, and therefore spoils its photofunctional properties. From this point of view, soluble conjugated polymers with long hydrocarbon side chains have been developed, in which self-quenching of the photo-excited states cannot be suppressed efficiently.

A dendritic macromolecular rod **2c** is an interesting luminescent molecule [6], since the conjugated backbone is spatially isolated by large poly(benzyl ether) dendritic wedges possessing a light-harvesting function [7]; **2c** and its lower-generation homologues **2a** and **2b** have been synthesized by a Pd(0)/Cu(I)-catalyzed polycondensation of the corresponding dendritic diethynylbenzenes (**3a–3c**) with 1,4-diiodobenzene at 50°C in THF/iso-Pr$_2$NH [6]. Dendritic monomers **3a–3c** can be prepared in DMF at 60°C by an alkaline-mediated coupling of 2,5-diethynylhydroquinone with poly(benzyl ether) dendron bromides of different generation numbers. Higher-generation **2b** and **2c** are soluble in common organic solvents such as THF, CHCl$_3$, CH$_2$Cl$_2$, and benzene, whereas **2a** has a limited solubility; **2b** with a high molecular weight (MW = 280 000) can be obtained from **3b**, while the polycondensation with the largest **3c** takes place rather sluggishly to give lower-molecular-weight **2c** (MW = 40 000).

In THF at 20°C, dendritic monomers **3a–3c** show absorption bands at 335 and 278 nm, due to the focal 1,4-diethynylbenzene unit and the dendritic wedges, respectively. On the other hand, dendritic macromolecular rods **2a–2c** display a strong absorption band in the visible region (400–460 nm), characteristic of an extended electronic conjugation in the backbone. Upon excitation of the conjugated backbone at 425 nm in THF ($abs_{425\ nm} = 0.01$) at 20°C, **2a–2c** show a strong blue fluorescence at 454 nm, where the quantum yield (Φ_{FL}) has been evaluated to be virtually 100%. Of much interest is the fact that the Φ_{FL} value of **2c** stays at nearly 100%, even when the solution is concentrated until the absorbance at 425 nm ($abs_{425\ nm}$) is increased to 0.1. Such a high Φ_{FL} value over a rather wide concentration range has never been realized with other soluble poly(phenyleneethynylene) derivatives reported to date (35–40%). The luminescence activity of 2a–2c is substantially dependent on the size of the dendritic wedges: Although the Φ_{FL} value of one-generation smaller 2c is also very high under dilute conditions ($\sim$ 100% at $abs_{431\ nm}$ of 0.01), it shows a significant drop to 67% when the absorbance of the solution is increased to 0.1. Such a trend is more explicit in the case of the lowest-generation 2a, where the Φ_{FL} value is only 56% even under dilute conditions ($abs_{425\ nm}$ = 0.01), and further drops upon concentration of the solution. Therefore, the large dendrimer framework in 2c appears to encapsulate the conjugated backbone as an 'envelope' and prevent the photoexcited state from collisional quenching.

Even more interestingly, the dendrimer framework functions as a light-harvesting antenna, which can channel the excitation energy very efficiently to the focal luminescent unit. Upon excitation of the dendritic wedges at 278 nm (abs = 0.06) in THF at 20°C, 3c emits a fluorescence at 373 nm from the focal diethynylbenzene unit without any detectable fluorescence from the dendritic wedges at 310 nm. This observation indicates a highly efficient intramolecular singlet energy transfer (ENT) from the dendritic wedges to the focal chromophore unit (quantum efficiency $\Phi_{ENT} \sim 100\%$). Similarly, the dendritic macro-molecular rod **2c** displays highly efficient ENT characteristics: Upon excitation of the dendrimer framework at 278 nm ($abs = 0.06$), only a blue fluorescence at 454 nm from the conjugated backbone appears, whereas no luminescence from the dendrimer framework is observed. The fluorescence intensity is much higher than that upon direct excitation of the conjugated backbone at 425 nm. The excitation spectrum, monitored at 454 nm, is a perfectly superimposed image of the absorption spectrum, again indicating a 100% ENT quantum efficiency. Owing to the prominent light-harvesting function together with the efficient 'site isolation effect' of the large dendritic envelope, **2c** is much superior to lower-generation **2a** and **2b** in terms of the luminescence activity. For example, excitation of the dendritic wedges of **2c** at 278 nm in THF ($abs_{425\,nm} = 0.1$), the observed fluorescence is 11 times more intense than **2a** under identical conditions. Therefore, the highest-generation **2c** is an excellent candidate as an organic light emitter, which can efficiently collect photons of a rather wide wavelength range from ultraviolet to visible, and then convert them into the blue emission.

2.3   ISOMERIZATION OF AZODENDRIMERS BY LIGHT HARVESTING

Azobenzene is a well-known photochromic molecule, which isomerizes from the *trans* form to the *cis* form upon irradiation with ultraviolet light, while the *cis* form isomerizes to the *trans* form under irradiation with visible light or thermally (Scheme 1). A large poly(benzyl ether) azodendrimer **4b** with an azobenzene functionality at the focal point has been synthesized by an alkaline-mediated coupling of the corresponding poly(benzyl ether) dendron bromide with *trans*-3,3′,5,5′-tetrahydroxyazobenzene [8, 9]. ^{1}H NMR pulse relaxation time (T_1) measurement has shown that the T_1 value of the exterior MeO signal in **4b** is much shorter than that of nondendritic **1a**, while the T_1 values of the aromatic signals due to the interior azobenzene unit in **4b** are almost identical to those of **1a**. These contrasting trends indicate an egg-like structural resemblance to dendritic **4b**, where the interior environment has a freedom of conformational change, whereas the exterior surface region is rather constrained due to dense packing of the exterior dendritic units.

Similarly to the case of non-dendritic **4a**, the *trans* form of dendritic **4b**, upon irradiation of a CHCl$_3$ solution at 340 nm, isomerizes to the *cis* form, which undergoes a backward isomerization to the *trans* form upon irradiation at 450 nm or heating. On the other hand, when the dendritic wedges are irradiated at 280 nm, *cis*–**4b** undergoes an isomerization to the *trans* form, to reach a photostationary state. In contrast, nondendritic **4a** does not undergo such a *cis*-to-*trans* isomerization under similar conditions. Relative to these observations, a large but nonsymmetrically substituted azodendrimer such as **4c** in the *cis* form,

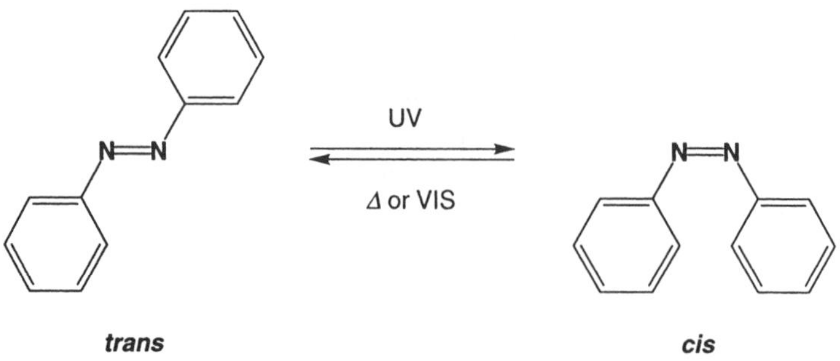

Scheme 1 Isomerization of azobenzene

upon 280 nm excitation of the dendritic wedge, hardly isomerizes to the *trans* form. Therefore, large **4b** with a spherical morphology is photochemically very unique. As described in the above two sections, poly(benzyl ether) dendrimers, upon excitation at 280 nm, emit a fluorescence centered at 310 nm. Since the azobenzene core in the *trans* form has an absorption band centered at 340 nm, the 280 nm excitation of the *trans*-azodendrimer **4b** at the dendritic wedges may possibly result in an energy transfer to the focal azobenzene unit to cause the *trans*-to-*cis* isomerization. However, the presence of the photostationary state in the isomerization of **4b**, upon 280 nm excitation at the dendritic wedges, suggests the occurrence of a *cis*-to-*trans* isomerization in competition with the expected *trans*-to-*cis* isomerization.

Relative to these interesting observations, dendritic **4b** has been shown to undergo the *cis*-to-*trans* isomerization upon irradiation with a 75 W Nichrome glowing light source. In contrast, nondendritic *cis*–**4a** hardly isomerizes to the *trans* form under similar conditions. On the basis of several experiments with an infrared monochromater, using an infrared light at 1600 cm^{-1}, which corresponds to the vibrational mode of aromatic rings, has been claimed to account for the isomerization of **4b**. From the temperature dependency of the thermal-induced isomerization rate, the activation free energy for the *cis*-to-*trans* isomerization of **4b** in CHCl$_3$ has been evaluated to be 0.84 eV. Since the energy of a single photon at 1600 cm^{-1} is only as low as 0.2 eV, the isomerization is likely to involve multiple low-energy photons. However, considering an extremely low photon density and an incoherency of the light source, simultaneous absorption of multiple photons by **4b** is unlikely [10], and one may consider some alternative mechanisms. In any case, this unusual isomerization event of large, spherical **4b** may possibly be related to the isomerization induced by the 280 nm UV excitation of the dendritic wedges. Further studies are necessary to clarify the mechanism of such a unique energy transduction event.

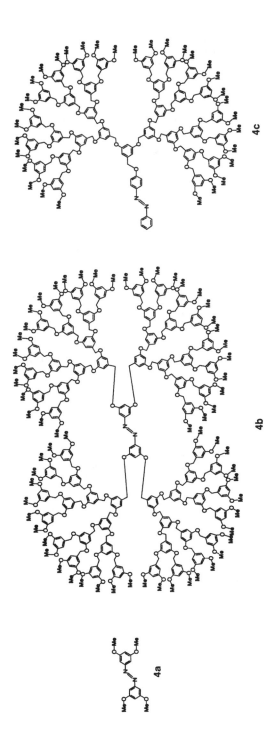

4c

4b

4a

3 PHOTOINDUCED ENERGY TRANSFER *THROUGH* DENDRIMER ARCHITECTURE

As described above, dendrimers are nanosized hyperbranched macromolecules with well-defined three-dimensional shapes, and are potential building blocks for the construction of organized functional materials with nanometric precision [1b]. A photofunctional dendritic supramolecular assembly, in which energy-donating and -accepting units are arranged in a highly controlled fashion (Scheme 2), has been prepared by electrostatic assembly of negatively and positively charged dendrimer porphyrins [11].

Poly(benzyl ether) dendrimer porphyrins with methyl ester functionalities on the exterior surface have been synthesized by an alkaline-mediated coupling of 5, 10, 15, 20-tetrakis(3′,5′-dihydroxyphenyl)porphyrin with methoxycarbonyl-terminated poly(benzyl ether) dendron bromides. The free-base dendrimer porphyrins, thus obtained, are metalated with $Zn(OAc)_2$, followed by alkaline hydrolysis of the ester groups, affording water-soluble dendrimer zinc porphyrin **6a** bearing 32 carboxylate ion functionalities on the exterior surface. Here **6a** can be transformed into a positively charged dendrimer zinc porphyrin **6b** upon amidation of the carboxylic acid functionalities, with N, N-dimethylethylene diamine followed by methylation of the dimethylamino groups. Similarly, alkaline hydrolysis of the exterior ester groups of the dendrimer porphyrin free-base affords **5a**, which can also be converted into **5b** in a manner similar to the above method for the preparation of **6b**. Likewise, lower-generation homologues **7a** and **7b** can be prepared.

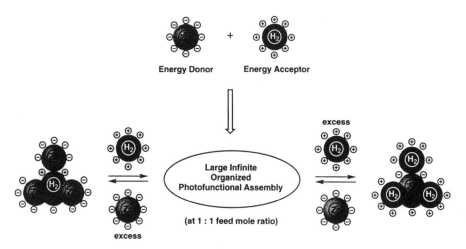

Scheme 2 Electrostatic assembly of positively charged dendrimers to form a well-defined photofunctional aggregate

Upon mixing of phosphate buffer solutions of dendrimer zinc porphyrin **6b** bearing a positively charged surface and dendrimer porphyrin free-base **5a** having a negatively charged surface, large fluorescent aggregates, sized 10–20 μm, appear and grow. Of interest, is the fact that aggregate formation is highly sensitive to the mole ratio of **6b** to **5a** (Scheme 2). Large aggregates are formed only when the mole ratio [**6b**]/[**5a**] is close to unity. However, when either **6b** or **5a** is added to this solution in excess with respect to the counterpart, the large aggregates spontaneously disappear. This is in sharp contrast to the assembly of linear-chain macromolecular electrolytes, in which the electrostatic interaction is usually irreversible due to chain entanglement.

The zinc and free-base porphyrin functionalities of compounds **6b** and **5a** in the aggregates are interactive photochemically by energy transfer from the former to the latter. In methanolic KOH (0.1 M), positively charged dendrimer zinc porphyrin **6b**, upon excitation of the Soret band at 439 nm, emits fluorescences at 613 and 664 nm from the zinc porphyrin core. On the other hand, in the presence of an equimolar amount of negatively charged dendrimer free-base porphyrin **5a**, the fluorescences from **6b** are quenched significantly, while **5a** emits fluorescences at 654 and 718 nm from its free-base porphyrin core. These results are clearly due to an intermolecular energy transfer from **6b** to **5a**. In sharp contrast, upon mixing of negatively charged **6a** and **5a** under the same

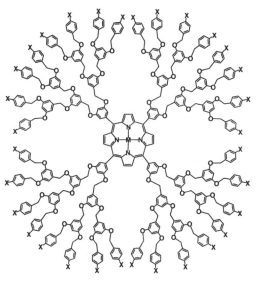

7a: M = Zn, X = CO_2^-
7b: M = Zn, X = $CONH(CH_2)_3N^+Me_3$

5a: M = 2H, X = CO_2^-
5b: M = 2H, X = $CONH(CH_2)_3N^+Me_3$

6a: M = Zn, X = CO_2^-
6b: M = Zn, X = $CONH(CH_2)_3N^+Me_3$

conditions, no aggregate formation takes place due to an electrostatic surface repulsion, and therefore no intermolecular energy transfer from **6a** to **5a** (fluorescence quenching of **6a**) occurs.

The energy transfer efficiency (ratio of fluorescence intensities at 718 and 613 nm) is enhanced as the molar ratio [**6b**]/[**5a**] is increased. However, when the ratio [**6b**]/[**5a**] exceeds 4, the energy transfer efficiency appears to reach a plateau at a level of about 72% and is no longer enhanced. The saturation tendency of the energy transfer efficiency suggests a possibility that **6b** can accommodate at most four molecules of **5a** on its exterior surface. This interaction profile has also been supported by the fluorescence polarization study on the **6b/5a** system. In general, when a chromophore with a limited rotation activity is excited by a polarized light, it emits a polarized fluorescence. Therefore, the degree of polarization should increase with the size of chromophoric molecule or aggregate. Excitation with a polarized light at 520 nm of **5a** in methanolic KOH results in a polarized fluorescence at 654 and 718 nm. On the other hand, addition of **6b** to the above solution containing **5a** results in an increase in the degree of polarization, where the titration curve shows a clear inflection point at a molar ratio [**6b**]/[**5a**] of 4/1. In sharp contrast, when electrostatically repulsive **6a** is mixed with **5a** under the same conditions, the degree of polarization increases only slightly upon increasing the concentration of **6a**. All the above observations indicate that the dendrimer porphyrins with positively and negatively charged surfaces self-assemble through electrostatic interaction.

Fluorescence lifetime measurements on the aggregate have shown that the rate constant of the intermolecular energy transfer from the zinc porphyrin unit to the free-base porphyrin unit has been evaluated to be 3.0×10^9 s^{-1}. This value is reasonable from a model in which dendritic donor **6b** and acceptor **5a** contact each other directly at their exterior surfaces (Scheme 2). Therefore, electrostatic assembly of positively and negatively charged dendrimers provides a promising supramolecular approach to construct photofunctional materials with nanometric precision.

4 PHOTOINDUCED ELECTRON TRANSFER *THROUGH* DENDRIMER ARCHITECTURE

Long-range electron transfer reactions have attracted attention relative to the initial stages of photosynthesis, and special interest has been focused on the design of noncovalently assembled donor–acceptor arrays. Dendrimers with spatially isolated porphyrin cores are potential motifs for investigating such long-range electron transfer reactions [12].

A water-soluble dendrimer zinc porphyrin **6a**, having 32 carboxylate anion

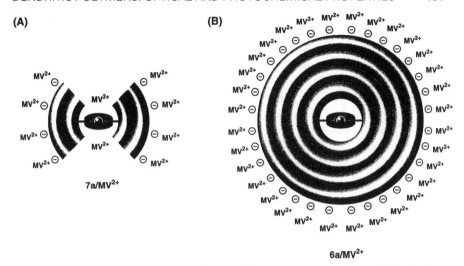

(A)   **(B)**

7a/MV^{2+}

6a/MV^{2+}

Scheme 3 Schematic representations of electrostatic assemblies of dendrimer porphyrins **6a, 7a** with MV^{2+}

functionalities, has been found to trap positively charged electron acceptor molecules such as methylviologen (MV^{2+}) on the exterior surface, to form a spatially separated donor–acceptor supramolecular assembly in aqueous media (Scheme 2B). When **6a** (2.9 μM) is titrated with MV^{2+}, the absorption spectrum, upon increase in [MV^{2+}] up to 2×10^4 μM, shows no substantial change associated with the interaction of the zinc porphyrin core with MV^{2+} at the ground state. In sharp contrast, titration of lower-generation **7a** (2.0 μM) with MV^{2+} results in a notable spectral change, even upon increase in [MV^{2+}] to 10^2 μM. Therefore, the interior zinc porphyrin functionality in lower-generation **7a** is directly interactive with MV^{2+} (Scheme 3A) at the ground state. This is due to the open architecture of the dendrimer framework, whereas in higher-generation **6a** this intimate donor–acceptor interaction is sterically prohibited by the large dendrimer framework.

Upon excitation of **6a** and **7a** (2.9 μM) at their Soret absorption bands in aqueous media, the zinc porphyrin cores emit fluorescences at 610 and 660 nm. On the other hand, these fluorescences are quenched by MV^{2+}, where the quenching profiles differ dramatically from one another depending on the generation number of the dendrimer framework. Stern–Volmer plots ($I_0/I \sim$ [MV^{2+}]) for the **7a**/MV^{2+} system shows a highly efficient fluorescence quenching even at a low [MV^{2+}]. In sharp contrast, the ratio I_0/I of higher-generation **6a** levels off when the concentration of MV^{2+} exceeds 10^3 μM, indicating that the negatively charged surface of **6a** is completely surrounded by positively charged MV^{2+} molecules (Scheme 3B). The fluorescence quenching of **6a** by MV^{2+} is obviously the result of a long-range, photoinduced electron transfer *through* the dendrimer framework: When the zinc porphyrin Soret of the **6a**/MV^{2+} system is continuously irradiated in the presence of triethanolamine as sacrificial donor, the solution turns blue, where the absorption due to the methylviologen radical ion appears at 605 nm. Time-resolved fluorescence spectroscopy shows that the lifetime of the **6a** fluorescence (1.5 ns) is considerably shortened to 0.3 ns in the presence of excess MV^{2+}. From the fluorescence lifetimes, the rate constant (k_{ET}) for the long-range electron transfer from the zinc porphyrin core to the trapped MV^{2+} molecule has been evaluated as 2.6×10^9 s^{-1}. The observed k_{ET} is sufficiently large, considering an estimated thickness of 20 Å for the dendritic shell of **6a**. Dendritic compound **6a** may be a potential candidate as a sensitizer for photochemical water splitting.

5 CONCLUSION

Poly(benzyl ether) dendrimers have several unique optical and photochemical properties. In particular, it is quite interesting that some photochemical events are considerably affected by the molecular size and morphology of the dendrimer molecules. These examples, together with those by other researchers [2], will provide a new strategy toward next-generation, nanoscopic photofunctional materials.

6 REFERENCES

1. (a) Fréchet, J. M. J. *Science*, **263**, 1710 (1994); (b) Tomalia, D. A. *Adv. Mater.*, **6**, 529 (1994); (c) Fischer, M. and Vögtle, F. *Angew, Chem. Int. Ed.*, **38**, 884 (1999).
2. (a) Campagna, S., Denti, G., Serroni, S., Ciano, M., Juris, A. and Balzani, V. *Inorg. Chem.*, **37**, 2982 (1992); (b) Balzani, V., Campagna, S., Denti, G., Juris, A., Serroni, S. and Venturi, M. *Acc. Chem. Res.*, **31**, 26 (1998); (c) Stewart, G. M. and Fox, M. A. *J. Am. Chem. Soc.*, **118**, 4354 (1996); (d) Kawa, M. and Fréchet, J. M. J. *Chem. Mater.*, **10**, 286 (1998); (e) Devadoss, C., Bharathi, P. and Moore J. S. *J. Am. Chem. Soc.*, **118**, 9635 (1996); (f) Kopelman, P., Shortreed, M., Shi, Z.-Y., Tan, W., Xu, Z., Moore, J. S., Bar-Haim, A. and Klafter, J. *Phys. Rev. Lett.*, **78**, 1239 (1997); (g) Bar-Haim, A., Klafter, J. and Kopelman, R. *J. Am. Chem. Soc.*, **119**, 6197 (1997); (h) Bar-Haim, A. and Klafter, J. *J. Phys. Chem. B*, **102**, 1662 (1998); (i) Tretiak, S., Chernyak, V. and Mukamel, S. *J. Phys. Chem. B*, **102**, 3310 (1998).
3. (a) Mcdermott, G., Prince, S. M., Freer, A. A., Hawthornthwaite-Lawless, A. M., Rapiz, M. Z., Codell, R. J. and Isaacs, N. W. *Nature*, **374**, 517 (1995); (b) Kuhlbrandt, W. *Nature*, **374**, 497 (1995).
4. (a) Jin, R.-H., Aida, T. and Inoue, S. *J. Chem. Soc., Chem. Commun.*, 1260 (1993); (b) Jiang, D.-L. and Aida, T. *Chem. Commun.*, 1235 (1996).
5. Jiang, D.-L. and Aida, T. *J. Am. Chem. Soc.*, **120**, 10895 (1998).
6. Sato, T., Jiang, D.-L. and Aida, T. *J. Am. Chem. Soc.*, **121**, 10658 (1999).
7. Examples of dendrimers with conjugated backbones: (a) Schenning, A. P. H. J., Martin, R. E., Ito, M., Diederich, F., Boundon, C., Gisselbrecht, J.-P. and Gross, M.

Chem. Commun., 1013 (1998); (b) Stocker, W., Karakaya, B., Schürmann, B. L., Rabe, J. P. and Schlüter, A. D. *J. Am. Chem. Soc.*, **120**, 7691 (1998); (c) Bao, Z., Amundson, K. R. and Lovinger, A. J. *Macromolecules*, **31**, 8647 (1998); (d) Malenfant, P. R. L., Groenendaal, L. and Fréchet, J. M. J. *J. Am. Chem. Soc.*, **120**, 10990 (1998).

8. Jiang, D.-L. and Aida, T. *Nature*, **388**, 454 (1997).

9. Azodendrimers with a different substitution pattern at the core unit have been reported in: Junge, D. M. and McGrath, D. Y. *Chem. Commun.*, 857 (1997).

10. Wakabayashi, Y., Tokeshi, M., Jiang, D.-L., Aida, T. and Kitamori, T. *J. Luminescence*, **83–84**, 313 (1999).

11. Tomioka, N., Takasu, D., Takahashi, D. and Aida, T. *Angew. Chem. Int. Ed.*, **37**, 1531 (1998).

12. Sadamoto, R., Tomioka, N. and Aida, T. *J. Am. Chem. Soc.*, **118**, 3978 (1996).

18

Bioapplications of PAMAM Dendrimers

J. D. EICHMAN, A. U. BIELINSKA,
J. F. KUKOWSKA-LATALLO, B. W. DONOVAN
AND J. R. BAKER, JR
University of Michigan, Center for Biologic Nanotechnology,
Department of Internal Medicine, Division of Allergy, Ann Arbor, MI, USA

1 INTRODUCTION

Interest in dendritic polymers (dendrimers) has grown steadily over the past decade due to use of these molecules in numerous industrial and biomedical applications. One particular class of dendrimers, Starburst® polyamidoamine (PAMAM) polymers, a new class of nanoscopic, spherical polymers that appears safe and nonimmunogenic for potential use in a variety of therapeutic applications for human diseases. This chapter will focus on investigations into PAMAM dendrimers for *in vitro* and *in vivo* nonviral gene delivery as these studies have progressed from initial discoveries to recent animal trials. In addition, we will review other applications of dendrimers where the polymers are surface modified. This allows the opportunity to target-deliver therapeutics or act as competitive inhibitors of viral or toxin attachment to cells.

2 DENDRIMER SYNTHESIS AND CHARACTERIZATION

As outlined in other chapters in this volume, Tomalia *et al.* first reported the successful well-characterized synthesis of dendrimers in the early 1980s [1]. These molecules range in size from 10 Å to 130 Å in diameter for generation 0 (G0) through generation 10 (G10). In the ideal situation, PAMAM dendrimers are monodispersed spherical conformation with a highly branched three-dimensional structure (Figure 18.1) that provides a scaffold for the attachment of

Dendrimers and Other Dendritic Polymers. Edited by Jean M. J. Fréchet and Donald A. Tomalia
© 2001 John Wiley & Sons Ltd

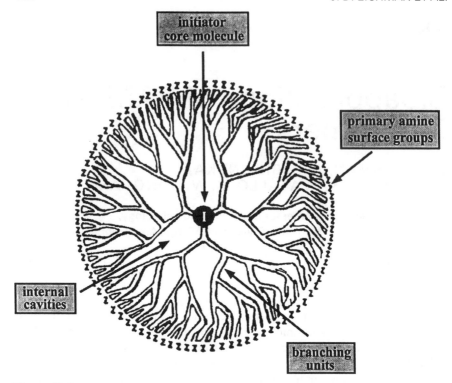

Figure 18.1

multiple biomolecules. With each new layer that is synthesized, the molecular weight of the dendrimer increases exponentially, the number of primary amine surface groups exactly double and the diameter increases by 10 Å (Table 18.1) [2]. As dendrimers grow in generation, the subsequent increase in exterior branching density begins to impart various structural effects to the polymer shape. Lower generation dendrimers (0 through 4) manifest a flexible, somewhat flat shape, while at the higher generations (5 through 10), the congested branching induces a persistent, robust spherical conformation [3]. Beginning at generation 4 (ethylenediamine core), the interior of the dendrimer develops internal void spaces that are accessible to molecules that may be encapsulated for drug delivery or other potential applications [4, 5]. Dendritic purity (isomolecularity) is typically around 98% due to small defects in branch formation during synthesis. These defects may be due to retro-Michael reactions or intramolecular macrocycle formation [6]. Dendrimer preparations are purified using ultrafiltration techniques. Subsequent structural characterization is performed by a number of analytical methods including: high performance liquid chromatography (HPLC), size exclusion chromatography (SEC), nuclear magnetic resonance

Table 18.1 Physical characteristics of PAMAM dendrimers (EDA core)

Dendrimer generation	Molecular weight	Primary amino surface groups	Diameter (Å)
0	517	4	15
1	1 430	8	22
2	3 256	16	29
3	6 909	32	36
4	14 215	64	45
5	28 826	128	54
6	58 048	256	67
7	116 493	512	81
8	233 383	1024	97
9	467 126	2048	114
10	934 787	4096	135

(NMR) techniques (^{1}H, ^{13}C, ^{15}N, ^{31}P), electron paramagnetic resonance (EPR), electrospray-ionization mass spectroscopy (ESI–MS), capillary electrophoresis, or gel electrophoresis [7–11]. Analytical methodology usually requires a combination of the above techniques in order to provide a detailed account of the exact identity and composition of the dendrimer sample. PAMAM dendrimers are currently commercially available with production occurring under good manufacturing procedures (GMP) to provide suitable samples for various biomedical and gene therapy applications [11b]. Atomic force microscope images of generation 9 (G9 EDA) dendrimers are shown in Figure 18.2. Li *et al.* have also recently obtained AFM images of PAMAM dendrimers ranging from G5 EDA to G10 EDA [12]. These studies clearly demonstrate the shape and consistency of these molecules.

3 DNA DELIVERY *IN VITRO* WITH UNMODIFIED DENDRIMERS

3.1 DENDRIMER/DNA INTERACTIONS: CHARACTERIZATION OF THE COMPLEX FORMATION

Formation of a complex between DNA and polycationic compounds appears to be the initial and quite possibly a critical parameter for nonviral gene delivery. Several synthetic vector systems, which are generally cationic in nature, including poly(lysines), cationic liposomes or various types of block copolymers and recently dendrimers, have been shown to self-assemble with plasmid DNA [13–15] [16]. Specific physicochemical properties manifested by these DNA complexes depend on the type of cationic agent used; however, interesting patterns for such interactions are beginning to evolve [17, 18]. Under certain conditions, the interaction of DNA with polyvalent cations results in

444 J. D. EICHMAN *ET AL.*

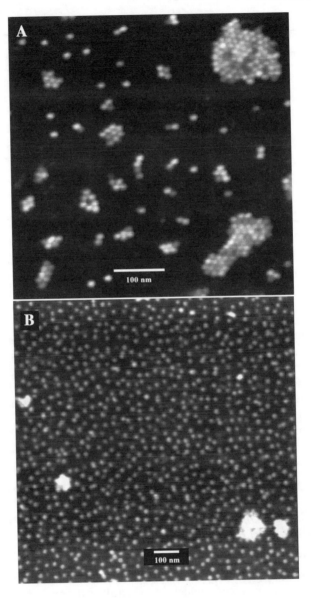

Figure 18.2

compaction of extended DNA structures to produce aggregation and precipitation from the solution [19–21]. Similar to other large polycationic compounds, PAMAM dendrimers form complexes with DNA through sequence-independent electrostatic interactions between negatively charged phosphate groups of the nucleic acid components and the cationic primary amino groups on the dendrimer surface (Figure 18.3). Charge neutralization of both components and alterations of the net charge of the complex lead to changes in both physicochemical and biologically relevant properties.

Complex formation analysis and characterization as soluble–insoluble or low–high density particles can be performed by various methods including UV light absorption, laser light scattering (LLS) and measurements that utilize radiolabeled DNA and/or dendrimers. The actual binding affinity constant (K_a) of DNA and dendrimers are difficult to determine in part because of the tendency to aggregation and precipitation [22]. However, the formation of high molecular weight and high-density complexes depends strongly on the DNA concentration (Figure 18.4). In salt-free water solutions, the precipitate formation increases as the DNA concentration rises from 10 ng/ml to 1 mg/ml. At DNA concentrations of 10 ng/ml, the complex formed with various dendrimer generations (i.e. G5, G7 and G9, with both NH_3 and EDA cores) form soluble, low density aggregates that remain suspended in water. Complex formation is facilitated by increasing the dendrimer concentration that effectively increases the dendrimer–DNA charge ratio [21].

Electrostatic charge effects (attraction or repulsion of charged molecules) appear to be modulated by the dendrimer generation (size of the polymer). Figure 18.5 illustrates that, although generally parallel, the precipitation curves shift as a result of the size (i.e. generation) of the dendrimer. Complexing of the

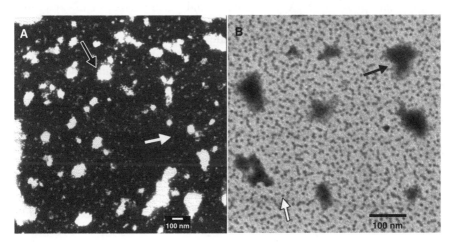

Figure 18.3

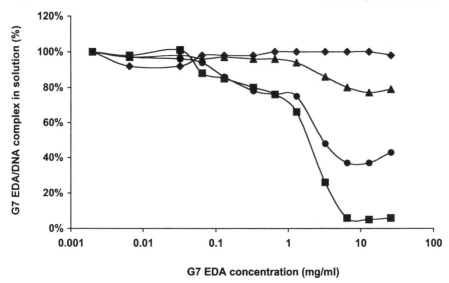

Figure 18.4

smaller G5 and G7 EDA dendrimers with DNA leads to highly aggregated forms of the complex at lower charge ratios than with larger, G9 EDA dendrimers [21]. This suggests that a purely electrostatic effect is modulated by the size and molecular weight of the dendrimer. PAMAM dendrimers readily form complexes with various forms of nucleic acids including single stranded oligonucleotides, circular plasmid DNA, linear RNA and various sizes of linear double stranded DNA. The larger the nucleic acid molecule, the lower the dendrimer concentration that is required to generate high-density complexes. With progressive increases in the dendrimer–DNA charge ratio (> 20), an increase in the quantity of low-density, soluble complexes is observed. Functional analysis revealed that the majority (> 90%) of transfection is carried by low-density, soluble, subpopulations of complexes, which may represent approximately 10–30% of the total complexed DNA (Figure 18.6).

The continuous distribution of the radioactively labeled dendrimer–DNA complexes in glycerol density gradients indicates the heterogeneous nature of the complex population. The size evaluation of dendrimer–DNA complexes (in water) using dynamic LLS, further reveals the polydispersed nature of particles with hydrodynamic diameters ranging from 30–100 nm to 20–200 μm depending on DNA concentration, size of dendrimer and charge ratio between each polymeric component. Complexes formed at very low DNA concentrations (e.g. 1–10 ng/ml) are usually smaller and more uniform than particles generated at high DNA concentrations. Complex formation at higher DNA concentrations results

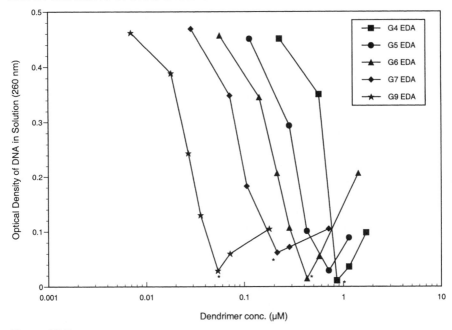

Figure 18.5

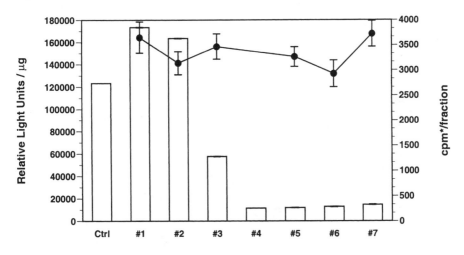

Fraction number: 0-5% step glycerol gradient

Figure 18.6

in the nonuniform distributions of larger, high-density aggregates and precipitates.

3.2 MECHANISM OF DENDRIMER-MEDIATED CELL ENTRY

Studies that focus on the cell entry mechanisms for several nonviral vectors, including liposomes, lipospermines, poly(ethylenimine) and PAMAM dendrimers have been previously reported [23–27]. Figure 18.7 shows a proposed dendrimer–DNA complex pathway into cells with subsequent processing [28, 29]. The cationic surface charge imparted to the complex through high dendrimer–DNA charge ratios (e.g. 5–20) is required for subsequent interaction with the anionic glycoproteins and phospholipids that reside on the cell membrane surface. This interaction initiates the interior movement of the dendrimer–DNA complex into the cell cytosol, either by passive transport through membrane or by endocytosis [30]. Complexes formed without an excess cationic surface charge do not mediate high gene transfection efficiency, which furnishes support for the importance of the initial electrostatic interaction between the complex and cell membrane. Studies following the incorporation of radiolabeled DNA and/or dendrimer components into cells established that the uptake in most cells was primarily via an active endocytosis mechanism [31]. Cells preincubated with inhibitors of endocytosis (i.e. cytochalasin B and deoxyglucose) or cellular metabolism (i.e. sodium azide) reduced the uptake that corresponded to lower transgene expression, regardless of cell type. After being entrapped within the endosome, complex release into the cytosol is essential

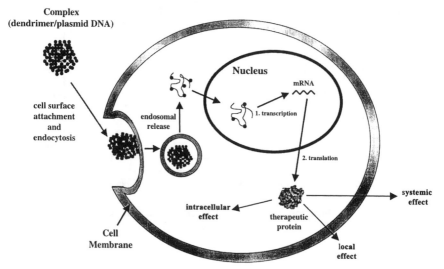

Figure 18.7

before acidic or enzymatic DNA degradation within the endosomal/lysosomal compartment takes place. The enhanced dendrimer-mediated transfection can be obtained in several cell types using chloroquine, an inhibitor of endosomal acidification.

It has been postulated that branched cationic polymers have a high buffer capacity owing to the basic amine groups [25, 32]. This characteristic enables dendrimers to act as a weak base and retard degradation caused by acidification within the endosome-lysosome. A reduction in pH might also lead to polymeric swelling within the endosome, thus disrupting the membrane barrier of the organelle and promoting DNA and/or complex release [23, 25]. After discharge from the endosome, DNA must penetrate the nuclear membrane for transcription and subsequent expression to occur. Translocation into the nucleus does occur within 30 min post-transfection, but the exact molecular and cellular mechanisms that mediate these events are unclear at this time, and are probably different for each cell type [33]. Recent studies tracing the movement of poly(ethylenimine) (PEI)–DNA complexes within cells indicated that complete separation of the polymer–DNA complex was not necessary for DNA entry into the nucleus [34]. Therefore, it might be possible that PAMAM dendrimers are also associated with DNA as it crosses into the nuclear compartment.

3.3 PLASMID DNA DELIVERY

The ability of PAMAM dendrimers to associate electrostatically with polyanionic DNA imparts a unique property to these polymers that enables them to be used as a synthetic vector for delivery of genes. Other cationic vectors such as poly(lysine), PEI, liposomes and block copolymers can also interact with plasmid DNA, resulting in successful gene transfer *in vitro* [13, 35–37]. Initially Haensler and Szoka performed studies to investigate the application of PAMAM den-drimers as nonviral vectors for *in vitro* gene transfer [32]. Their results demonstrated that complexes consisting of G5 PAMAM dendrimer and expression plasmid DNA had enhanced transfection efficiency over naked plasmid DNA in many cells, particularly cell lines derived from monkey and human neoplasms [32]. Further work from these investigators revealed that the transfection observed in their experiments was improved by thermally degraded dendrimers [23]. Thermally degraded dendrimers appear to be polydispersed in size and structure enhance transfection efficiency when compared with the intact G5 dendrimer. The enhanced transfection activity of the degraded polymer was attributed to increased flexibility of the structure. However, the exact structural changes that account for this have not been identified. Independent studies by Baker and co-workers documented the efficiency of intact dendrimers as synthetic vectors for the delivery of genetic material into cells

[31, 38]. They used different generations of intact dendrimers to transfect plasmid DNA in a variety of cells (Table 18.2) using luciferase, CAT (chloramphenicol acetyltransferase) and β-galactosidase reporter genes to quantify transfection efficiency.

In contrast to the results obtained with degraded PAMAM dendrimers, only intact dendrimers of G5 mediated significant transfection, and this required the addition of a dispersing agent such as (diethylamino)ethyl (DEAE) dextran [31]. In a number of cells analyzed, typified by the rat embryonal fibroblast cell line Rat 2, an exponential rise in transfection efficiency was observed using G5–G10 of either NH_3 or EDA core dendrimers [31]. High transfection levels of the luciferase reporter gene were obtained in non-adherent cells of lymphoid lineage (i.e. Jurkat and U937) and adherent cells (COS-1 and Rat 2). Expression levels of luciferase in Jurkat and U937 cells was one to two orders of magnitude greater than those obtained in experiments using the commercial lipid prepara-

Table 18.2   Cell lines transfected with PAMAM dendrimers

Cell line	Cell type
Rat 2	Rat embryonal fibroblast
Clone 9	Rat liver epithelium
rat Clone B	Rat mesothelium
YB2	Rat myeloma, nonsecreting
NRK52E	Rat kidney epithelial-like
NIH3T3	Mouse embryonal
10′	Mouse embryonal, 3T3-like, p53 deficient
EL4	Mouse lymphoma
D5	Mouse melanoma
Cos1	Monkey kidney fibroblast
Cos7	Monkey kidney fibroblast
CHO	Chinese hamster ovary
HMEC-1	Human microvascular endothelium
MSU1.2	Human fibroblast
NHFF3	Normal human foreskin fibroblast
QS	Human synoviocyte
HepG2	Human liver hepatoblastoma
Jurkat	Human T cell leukemia
JR	Human T cell leukemia
SW 480	Human colon adenocarcinoma
COLO 320 DM	Human colon adenocarcinoma
SW 837	Human rectum adenocarcinoma
YPE	Porcine vascular endothelial
BHK-21	Hamster kidney fibroblast-like
NHBE	Normal human bronchial epithelium
SAEC	Small airway epithelium
CCD-37Lu	Human normal lung fibroblast
A549	Human lung carcinoma epithelial-like

tions, Lipofectamine and DMRIE-C (1,2-dimyristyloxy-3-dimethylhydroxy ethyl ammonium bromde formulated with cholesterol) [33]. Typically, nonviral vectors do not mediate high levels of expression in primary cell lines. PAMAM dendrimers are highly efficient in transfecting a wide array of primary cells of various origins including human fibroblasts (HF1) and human lung epithelial cells [31]. Each cell line reacts differently to transfection with this agent, owing to subtle changes in physiological makeup, so the dendrimer generation that is optimal for each particular cell line must be determined experimentally.

The addition of other agents to the DNA–dendrimer complex can markedly alter transfection. For example, chloroquine or cationic DEAE dextran added to dendrimer–DNA complexes significantly increase transgene expression in a number of cell lines [31]. DEAE dextran alters the dendrimer–DNA complex by dispersing the complex aggregates [31]. However, it is cytotoxic and might prevent stable gene integration.

3.4   STABLE TRANSFORMED CELL LINES

One of the most important goals of gene therapy is the transfer of genetic material that is permanently integrated and expressed in cells. A large percentage of cells must be transfected for a few to retain stable gene integration. This is owing to the low efficiency of DNA integration that is observed with nonviral vector systems (as compared with retroviruses). To overcome these problems, *ex vivo* approaches have been used, in which a small quantity of tissue is removed from the patient and the cells within that tissue are placed into culture. After clonal expansion of the cells, transfection of the desired gene is performed *in vitro*. The genetically modified cells are then returned to the patient by transplantation or implantation to obtain a limited number of stable clones expressing the gene [39, 40]. Transfections with calcium phosphate–DNA precipitates, DEAE–dextran–DNA, or dendrimer–DNA complexes with expression plasmids carrying both β-galactosidase and neomycin resistance genes were performed on D5 mouse melanoma cells [41]. Dendrimer–DNA complexes produced approximately 90-fold more neomycin-resistant stable clones than other complexes, including calcium phosphate precipitation and DEAE dextran [41]. It was demonstrated that up to one-third of the neomycin resistant colonies produced β-galactosidase activity. This indicates that a nonselected gene was integrated into the cellular genomic DNA and expressed. It also suggested that if cells are transfected with dendrimer–DNA complexes *in vivo*, the transfected DNA will integrate into the host chromosome and be permanently expressed [41].

3.5 OLIGONUCLEOTIDE DELIVERY

Antisense oligonucleotides are short, gene-specific sequences of nucleic acids, typically 15–25 bases in length. These molecules are designed to interact with complementary sequences on a targeted mRNA and, in principle, prevent the message from being translated into a protein. Protein synthesis is inhibited by initiating enzymatic degradation of the mRNA by RNAse H or by interfering with ribosomal reading of the message to form the encoded protein [42]. There are approximately 10 different phosphorothioate oligonucleotides that are presently being tested in human clinical trials [42]. Additional experiments have also demonstrated that some antisense oligonucleotides can inhibit gene expression selectively [43]. The biological effect of an oligonucleotide will ultimately correspond to its intracellular concentration and the rate of degradation. One of the primary difficulties associated with the delivery of antisense nucleic acids is obtaining a sufficient intracellular concentration. The second problem is preventing its rapid degradation by cellular endo- and exonucleases. Methods such as enhancing the chemical stability through modifications of the phosphodiester bond attempt to increase the half-life of the oligonucleotide [44]. Other techniques aim to increase cellular uptake, including the use of high initial concentrations, use of cationic lipids as carriers and microinjection [44–46]. However, these methods have encountered various problems associated with their use *in vivo*. Therefore, more efficient delivery systems are needed to achieve successful applications of antisense technology.

Recent studies by Bielinska *et al.* have shown that PAMAM dendrimers can be complexed to antisense oligonucleotides or plasmid expression vectors coding antisense mRNA, to inhibit specific reporter luciferase gene expression by 30% to 60% (Figure 18.8) [38]. These results were obtained in several stable cell clones that demonstrated long-term luciferase expression of cDNA from mouse melanoma (D5) and rat embryonal fibroblast (Rat 2) cells. The wide variety of inhibition depends on DNA concentration, charge ratios of the dendrimer–DNA complex and generation of dendrimer used in the particular study. Specific inhibition of gene expression is obtained in picomolar concentrations with dendrimer delivery. When Bielinska *et al.* attempted to use Lipofectamine (Life Technologies, Gaithersburg, MD, USA) a commercial liposome preparation, consistent results were not obtained because of cytotoxicity. Dendrimers were not cytotoxic at the concentrations required for gene delivery [38]. Oligonucleotide stability is required for successful inhibition of gene expression through antisense application. Exposure of 'naked' oligonucleotide leads to its rapid degradation in serum and endosomes, or by cellular enzymes. An increase in phosphodiester oligonucleotide stability occurs when such oligonucleotides are complexed with dendrimers [38]. A later study was performed evaluating PAMAM dendrimers as an adjuvant to enhance the delivery of antisense phosphorothioate deoxyoligonucleotides (PODN) directed against chloramphenicol

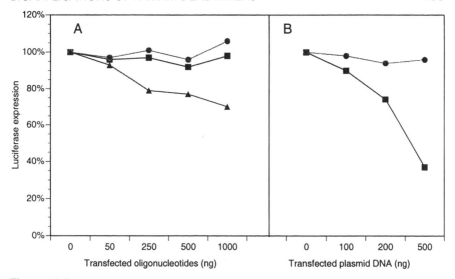

Figure 18.8

acetyltransferase (CAT) in Chinese hamster ovary (CHO) cells [47]. The level of antisense inhibition could be modulated as a function of DNA concentration, and by generation of dendrimer and dendrimer–DNA charge ratio. The use of G5 dendrimers resulted in a 35–40% reduction in CAT expression. G3 (EDA core) dendrimers also enhanced PODN uptake 50-fold in human astrocytoma cells (U251) and improved delivery to the nucleus [48]. Other transfection studies achieved targeted gene expression inhibition in nano- to micromolar oligonucleotide concentrations [49–51]. Thus the complexation of oligonucleotides with dendrimers might enable DNA delivery in a manner that could lead to *in vivo* uses for a variety of regulatory nucleic acid applications.

3.6   ENHANCEMENT OF IN VITRO GENE DELIVERY

The major limitation of nonviral-based gene delivery is low transfection efficiency when compared with viral vectors. Various techniques have been developed in an attempt to increase the quantity of therapeutic genes delivered intracellularly, including mechanical enhancements such as electroporation, the gene gun and membrane permeability enhancers [52–59]. It has been demonstrated that pulmonary surfactants increase *in vitro* adenoviral-mediated gene transfer, but have the reverse effect when used in conjunction with cationic liposomes [58, 60]. Recently, Exosurf Neonatal® (GlaxoWellcome, Research Triangle Park, NC, USA) a commercialized pulmonary surfactant preparation used for treatment of neonatal respiratory distress syndrome, was investigated for enhancement of dendrimer-mediated transgene expression *in vitro* [61]. Exosurf Neonatal® is a mixture of three components: dipalitoylphosphatidylcholine (DPPC), tyloxapol and cetyl alcohol.

Studies with Exosurf Neonatal® have demonstrated a significant enhancement in luciferase reporter gene expression in a variety of cells, including primary cells of different origins [61]. Primary human lung epithelial (NHBE) and porcine vascular endothelial (YPE) cells showed the highest enhancement with a 41.4- and 25.3-fold increase in luciferase expression, respectively, over the control [61]. Exosurf Neonatal® also significantly enhanced luciferase expression in a nonadherent lymphoid Jurkat cell line by 11.5-fold over control [61]. Jurkat cells are historically difficult to transfect by nonviral transfection techniques. It was determined that tyloxapol, a formaldehyde polymer of a polyoxyethylene condensate of *p*-(1,1,3,3-tetramethylbutyl)phenol, was the sole component inducing the den-drimer-mediated expression enhancement [61–63]. Tyloxapol is also classified as a nonionic surfactant. Substitution of Exosurf Neonatal® with tyloxapol in transfection experiments produced comparable transfection enhancement, while DPPC and cetyl alcohol alone had no effect [61]. Tyloxapol did not induce significant toxicity at concentrations required for enhanced transgene expression (0.25–0.50 mg/ml) [61].

Surfactants are commonly used as absorption enhancers and typically act by destabilizing the cell membrane [64–67]. Initial studies seeking the mechanism(s) of tyloxapol-mediated enhancement have been performed, but the mechanism has not been completely defined (J. D. Eichman, unpublished). Membrane permeability studies measuring intracellular enzyme release (i.e. lactate dehydrogenase; LDH) and fluorescent marker leakage (i.e. fluorescein diacetate) and uptake (i.e. propidium iodide) indicated in part an enhancement of permeability [61]. Preincubation of cells with tyloxapol with subsequent washing prior to transfections was shown to elevate transgene expression (A. U. Bielinska, unpublished). This indicates that tyloxapol might temporarily activate certain cell membrane properties, which promote DNA uptake even after surfactant removal [61].

4 DNA DELIVERY *IN VIVO*

Many human diseases, including cancer and inherited illness such as cystic fibrosis are caused by genetic disorders. The progress in biotechnology, such as being able to synthesize DNA constructs containing genes of interest, has effected dramatic changes in therapeutic agents. Recent advances in molecular biology leads to greater understanding of the molecular genetic basis for human diseases, as well as the promise for the development of novel treatment through gene therapy [68]. The feasibility of using gene therapy to correct human genetic disorders has led to a flurry of excitement in both the scientific community and biotechnology industry [68, 69]. However, a major obstacle for successful *in vivo* gene therapy has been the absence of suitable DNA delivery vectors [70]. The therapeutic application of genes requires non-toxic, cost effective methods for gene delivery that are safe, non-immunogenic and provide producible levels of the gene products [71]. Viral vectors such as adenovirus are highly efficient for gene transfer and have been used for various clinical trials such as cystic fibrosis [72, 73]. However, a number of problems with the adenoviral vector have limited its use. The major drawback is immunogenicity associated with the adenovirus that may have played an important role in the recent death of a patient enrolled in a gene therapy clinical trial. Owing to the potential limitations of viral vectors, non-viral *in vivo* strategies that employ plasmid DNA encoding a target gene are seen to be a promising alternative. Cationic lipids have been used in gene delivery to the lung. However, an inflammatory response to the lipid limits its use [74]. PAMAM dendrimers have recently been tested as a synthetic vector system for gene therapy to the lung [75].

Poly(amidoamine) (PAMAM) dendrimers, as well as other families of dendrimers are finding utility in a wide variety of uses, from particle size standards to a number of biological applications [5]. In order for PAMAM dendrimers to be useful for *in vivo* applications, studies establishing various parameters such as pharmacokinetics and biocompatibility are necessary. The criteria for choice of carrier molecules can be summarized as follows. The carrier should be (1) non-toxic, (2) non-immunogenic, (3) biocompatible, (4) have adequate functional groups for chemical fixation, (5) limited body accumulation and (6) maintain the original drug activity until reaching the site of action.

4.1 IN VIVO TOXICITY

Evaluation of dendrimer toxicity was recently performed utilizing specific pathogen-free, female CBA/J mice. The mice were injected intravasculary (i.v.) with G5, G7 or G9 EDA PAMAMs dissolved in physiological saline at doses of 10, 50, 100, 200, 300, 600 μg/mice. Control animals received saline alone. The animals were observed for behavioral abnormalities, activity level and eating/drinking behavior. Dendrimers were injected every four days up to four times per mouse. Changes in body weight over the 30-day observation period were recorded as a measure of overall health. Groups of mice were sacrificed on days 4, 8, 16, 20 and 30 after treatment. Results indicated that dendrimer toxicity is dependent on both the concentration and polymer generation. High doses of G5 EDA (200 μg/mouse) and G7 EDA (300 μg/mouse) dendrimers showed signs of toxicity. Tissue pathology and toxicity studies with G9 EDA dendrimer i.v. administration up to 600 μg/mouse has not indicated lung inflammation and pneumonitis. Dendrimers also did not induce an immune response when administered alone or in combination with adjuvant [76].

4.2 BIODISTRIBUTION

Biodistribution of G3 and G4 ^{125}I-labelled PAMAM dendrimers after administration by intravenous (i.v.) or intraperitoneal (i.p.) routes accumulated primarily in the liver (60–90%) with only 0.1–1.0% of the recovered dose detected in the blood after 60 min in Wistar rats [77]. This distribution is different from that obtained with studies using i.v. delivery of generation 3, 5 and 7 on Swiss-Webster mice. G3 dendrimers showed the highest accumulation in the kidney, with G5 and G7 dendrimers mainly localized in the pancreas [76]. The differences between the studies may be attributable to the choice of animal model and/or dendrimer generation used. When dendrimers interact with DNA to form complexes of varying charge ratios, complex distribution patterns would be expected to be entirely different from that obtained for naked dendrimer administration. This is due to changes in the size, density and surface charge of the complex compared to individual dendrimer molecules. To date, the *in vivo* distribution patterns of dendrimer/DNA complexes have not yet been established.

4.3 EXPERIMENTAL TRIALS

A number of experimental *in vivo* gene therapy trials on animals using PAMAM dendrimers are in their preliminary stages. Numerous *in vivo* experiments are currently being conducted in order to optimize the dendrimer generation, concentration, and complex charge ratio to obtain optimal transfection efficiency.

Complexes of dendrimer and DNA have been administered to the lungs of mice and rats by intranasal delivery. Preliminary results have indicated that administration by this route may not led to optimal gene expression when compared to 'naked' DNA delivery [78]. Polymerase chain reaction (PCR) confirmed long-term plasmid DNA survival when complexed with dendrimer. Amplification was performed directly with lung homogenate and/or nucleic acids isolated from lung tissue [33]. Semiquantitative PCR of plasmid DNA removed from lung tissue showed up to a 10^3-fold increase in its survival when complexed with dendrimer than when plasmid DNA was administered alone.

Recently, i.v. delivery of dendrimer/DNA complexes was evaluated on mice. When delivered by i.v. administration, the greatest amount of gene expression in Balb/C mice has been obtained in the bronchial and alveolar regions of the lung [75]. It is also possible that *in vivo* targeted gene delivery using antibody-conjugated dendrimers may eventually be achieved. Antibody–dendrimer conjugates have already been used in tumor imaging, diagnostics and radioimmunotherapy [79, 80] More recently, dendrimers were investigated as a method to increase plasmid DNA gene transfer in a murine cardiac transplantation model.

4.4 GENETIC APPROACHES TO THE THERAPY INFLAMMATORY AND FIBROTIC LUNG DISEASE

Pulmonary diseases are present in a heterogeneous group of lung disorders of known or unknown etiology. Gene therapy has been a useful tool against these diseases for a variety of reasons. First, the genetic basis for lung cancer and common inherited pulmonary disease such as cystic fibrosis has been identified. Secondly, there are no effective therapies for many lung diseases, especially for chronic inflammatory lung diseases such as idiopathic pulmonary fibrosis (IPF). IPF is a progressive disease resulting in respiratory failure. Recent progress in understanding the pathogenic mechanisms of lung fibrosis has provided insights into the inflammatory process underlying this disorder. IPF is commonly characterized by inflammation of the alveolar wall leading to disruption of the normal alveolar architecture and interstitial as well as intra alveolar fibrosis. The process involves cellular interactions via a complex cytokine network and heightened collagen gene expression with abnormal deposition in the lung. Recent studies have identified a myriad of cytokines with potential roles in pulmonary fibrosis (PF) including transforming growth factor β_1 (TGFβ_1), tumor necrosis factor α (TNF-α), platelet derived growth factor (PDGF), interleukin 5 (IL-5), and IL-8. Of the investigations so far, the TGFβ_1 family has the most potent stimulatory effect on extracellular matrix deposition. TGFβ_1 stimulates fibroblast procollagen gene expression and protein synthesis. The murine model of bleomycin induced lung fibrosis, increased TGFβ_1 gene expression and collagen elevation. This information provides a rational basis for developing anticytokine agents. Furthermore, recent studies have shown that *in vivo* administration of antisense oligonucleotides (ONDs) against TGFβ_1 reduced malignant mesthelioma tumor growth in mice. Antisense ODNs can block the translation of particular gene products within cells and represent a unique method of specifically inhibiting the effects of target proteins in the cells. The ability of the antisense technique to down-regulate the expression of specific genes is well documented both *in vitro* and *in vivo*. In animal models, gene transfer was achieved with a variety of vectors such as lipids, virus or naked plasmid DNA. In a recent study, Baker *et al.* investigated the ability of G9 (EDA core) dendrimers complexed with TGFβ_1 ODN to inhibit pulmonary fibrosis in a well-characterized model of bleomycin induced pulmonary fibrosis utilizing CBA/J mice. Intravenous G9 (EDA core) dendrimer/TGFβ_1 antisense ODN complex administration caused a significant reduction in lung fibrosis as indicated by TGFβ_1 mRNA suppression and inhibition of lung hydroxyproline contents, collagen synthesis, cytokine expression and pulmonary eosinophilia. Administration of either TGFβ_1 antisense ODNs alone or sense oligos complexed to G9 EDA dendrimers did not produce an equivalent inhibition in pulmonary fibrosis indicators. These results suggest the possibility for *in vivo* use of dendrimer vectors with antisense ODNs to combat pulmonary fibrosis as well as other diseases.

5 CONCLUSION

PAMAM dendrimers are a highly efficient non-viral vector for gene delivery into numerous cell lines *in vitro* and *in vivo*. Results obtained from *in vitro* studies do not always correlate to similar observations obtained in the *in vivo* experiments. The use of dendrimers for *in vivo* gene delivery is in its initial stages with studies primarily focusing on optimizing localized transgene expression. The ability of certain dendrimer generations to transfect cells without inducing biocompatibility issues or toxicity confers a significant advantage over other gene delivery vectors for use *in vivo*. Preliminary studies delivering plasmid DNA and antisense oligonucleotides in mice for gene therapy in cancer, pulmonary fibrosis and other diseases have resulted in a positive outlook for the future use of dendrimers as an *in vivo* synthetic gene delivery vector.

6 REFERENCES

1. Tomalia, D. A., Baker, H., Dewald, J., Hall, M., Kallos, G., Martin, S., Roeck, J., Ryder, J. and Smith, J. *Polymer J.*, **17**, 117–132 (1985).
2. Lothian-Tomalia, M. K., Hedstrand, D. M., Tomalia, D. A., Padia, A. B. and Hall, Jr. H. K. *Tetrahedron*, **53**(45), 15495–15513 (1997).
3. Pötschke, D., Ballauff, M., Lindner, P., Fischer, M. and Vögtle, F. *Macromolecules*, **32**, 4079–4087 (1999).
4. Tomalia, D. A., Hall, M. and Hedstrand, D. M. *J. Amer. Chem. Soc.*, **109**, 1601–1603 (1987).
5. Bieniarz, C. Dendrimers: applications to pharmaceutical and medicinal chemistry, in Swarbrick, J. and Boylan, J. C. (eds) *Encyclopedia of Pharmaceutical Technology*, **v. 18**, Marcel Dekker Inc, New York, 1999, pp. 55–89.
6. Tomalia, D. A., Uppuluri, S., Swanson, D. R., Brothers II, H. M., Piehler, L. T., Li, J., Meier, D. J., Hagnauer, G. L. and Balogh, L. *Mater. Res. Soc. Symp. Proc.*, **543**, 289–298 (1999).
7. Tomalia, D. A., Naylor, A. M. and Goddard III, W. A. *Angewandte Chem.*, **29**, 138–175 (1990).
8. Ottaviani, M. F., Montalti, F., Turro, N. J. and Tomalia, D. A. *J. Phys. Chem. B*, **101**, 158–166 (1997).
9. Pesak, D. J., Moore J. S. and Wheat, T. E. *Macromolecules*, **30**, 6467–6482 (1997).
10. Kallos, G. J., Tomalia, D. A., Hedstrand, D. M., Lewis, S. and Zhou, J. *Rapid Commun. Mass Spectrom.*, **5**, 383–386 (1991).
11. Weiner, E. C., Brechbiel, M. W., Brothers, H., Magin, R. L., Gansow, O. A., Tomalia, D. A. and Lauterbur, P. C. *Mag. Reson. Med.*, **31**, 1–8 (1994).
12. Li, J., Piehler, L. T., Qin, D., Baker, J. R. and Tomalia, D. A. *Langmuir*, **16**, 5613–5616 (2000).
13. Wolfert, M. A., Schacht, E. H., Toncheva, V., Ulbrich, K., Nazarova, O. and Seymour, L. W. *Hum. Gene Ther.*, **7**, 2123–2133 (1996).
14. Rädler, J. O., Koltover, I., Salditt, T. and Safinya, C. R. *Science*, **275**, 810–814 (1997).
15. Bielinska, A. U., Kukowska-Latallo, J. F. and Baker, J. R. Jr, *Biochim. Biophys. Acta*, **1353**, 180–190 (1997).

16. Ottaviani, M. F., Sacchi, B., Turro, N. J., Chen, W., Jockusch, S. and Tomalia, D. A. *Macromolecules*, **32**, 2275–2282 (1999).
17. Bloomfield, V. A., Ma, C. and Arscott, P. C. in K. S. Schmitz (ed.), *Macro-Ion Characterization from Dilute Solutions to Complex Fluids*, American Chemical Society, Washington DC; 1994, pp. 195–209.
18. Post, C. B. and Zimm, B. H. *Biopolymers*, **21**, 2123–2137 (1982).
19. Pelta, J., Livolant, F. and Sikorav, J. L. *J. of Biol. Chem.*, **271**, 5656–5662 (1996).
20. Dunlap, D. D., Maggi, A., Soria, M. R. and Monaco, L. *Nucl. Acids Res.*, **25**, 3095–3101 (1997).
21. Bielinska, A. U., Chen, C., Johnson, J. and Baker, J. R. Jr. *Bioconj. Chem.*, **10**, 843–850 (1999).
22. Rau, D. C. and Parsegian, V. A. *Biophys. J.*, **61**, 246–259 (1992).
23. Tang, M. X., Redemann, C. T. and Szoka, F. C. Jr. *Bioconj. Chem.*, **7**, 703–714 (1996).
24. Zabner, J., Fasbender, A. J., Moninger, T., Poellinger, K. A. and Welsh, M. J. *J. of Biol. Chem.*, **270**, 18997–19007 (1995).
25. Boussif, O., Lezoualc'h, F., Zanta, M. A., Mergny, M. D., Scherman, D., Demeneix, B. and Behr, J.-P. *Proc. Natl Acad. Sci. U.S.A.*, **92**, 7297–7301 (1995).
26. Remy, J.-S., Abdallah, B., Zanta, M. A., Boussif, O., Behr, J.-P. and Demeneix., B. *Adv. Drug Del. Rev.*, **30**, 85–95 (1998).
27. Pollard, H., Remy, J.-S., Loussouarn, G., Demolombe, S., Behr, J.-P. and Escande, D. *J. of Biol. Chem.*, **273**, 7507–7511 (1998).
28. Tomalia, D. A. and Brothers II, H. M. *Biological Molecules in Nanotechnology: The Convergence of Biotechnology, Polymer Chemistry and Material Science*. IBC Library Series Publication No. 1927, 1998, pp. 107–119.
29. Tomalia, D. A. in Walsh, B. (ed.), *Non-Viral Genetic Therapeutics Advances, Challanges and Applications for Self-Assembling Systems*, IBC USA Conferences, Inc., Westborough, MA, 1996, pp. 1.2–1.2.36.
30. Dvornic, P. R. and Tomalia, D. A. *Sci. Spectra*, **5**, 36–41 (1996).
31. Kukowska-Latallo, J. F., Bielinska, A. U., Johnson, J., Spindler, R., Tomalia, D. A. and Baker J. R. Jr. *Proc. Natl Acad. Sci. USA*, **93**, 4897–4902 (1996).
32. Haensler, J. and Szoka, F. C. Jr. *Bioconj. Chem.*, **4**, 372–379 (1993).
33. Kukowska-Latallo, J. F., Bielinska, A. U., Chen, C., Rymaszewski, M., Tomalia, D. A. and Baker J. R. Jr. 'Gene transfer using starburst dendrimers', in Kabanov, A. V., Felgner, P. L. and Seymour, L. W. (eds) *Self-Assembling Complexes for Gene Delivery*, John Wiley & Sons, New York, 1998, pp. 241–253.
34. Godbey, W. T., Wu, K. K. and Mikos, A. G. *Proc. Natl Acad. Sci. USA*, **96**, 5177–5181 (1999).
35. Wagner, E., Cotten, M., Foisner, R. and Birnstiel, M. L. *Proc. Natl Acad. Sci. USA*, **88**, 4255–4259 (1991).
36. Zhou, X. and Huang, L. *Biochim. Biophys. Acta*, **1189**, 195–203 (1994).
37. Astafieva, I., Maksimova, I., Lukanidin, E., Alakhov, V. and Kabanov, A. *FEBS Lett.*, **389**, 278–280 (1996).
38. Bielinska, A., Kukowska-Latallo, J., Johnson, J., Tomalia, D. A., Baker, J. R. Jr. *Nucl. Acids Res.*, **24**, 2176–2182 (1996).
39. Chowdhury, J. R., Grossman, M., Gupta, S., Chowdhury, N. R., Baker, J. R. Jr. and Wilson, J. M. *Science*, **254**, 1802–1805 (1991).
40. Anderson, W. F. *Scientific Amer.*, **273**, 124–128 (1995).
41. Baker, J. R. Jr. in Heller, M. J., Lehn, P. and Behr, P. (eds), *Conference Proceedings Series. Artificial Self-Assembling Systems for Gene Delivery*, American Chemical Society, Washington DC, 1996, pp. 129–145.
42. Phillips, M. I. and Gyurko, R. in Weiss, B. (ed.), *Antisense Oligodeoxynucleotides and*

Antisense RNA. Novel Pharmacological and Therapeutic Agents, CRC Press, New York, 1997, pp. 131–148.
43. Crooke, S. T. and Bennett, C. F. *Ann. Rev. Pharmacol. Toxicol.*, **36**, 107–129 (1996).
44. Bennett, C. F. in Stein, C. A. and Krieg, A. M. (eds), *Applied Antisense Oligonucleotide Technology*, Wiley-Liss, Inc., New York, 1998, pp. 129–145.
45. Lewis, J. G., Lin, K.-Y., Kothavale, A., Flanagan, W. M., Matteucci, M. D., DePrince, R. B., Mook, R. A., Hendren, R. W. and Wagner, R. W. *Proc. Natl Acad. Sci. USA*, **93**, 3176–3181 (1996).
46. Giles, R. V., Spiller, D. G. and Tidd, D. M. *Antisense Res. Develop.*, **5**, 23–31 (1995).
47. Hughes, J. A., Aronsohn, A. I., Avrutskaya, A. V. and Juliano, R. L. *Pharm. Res.*, **13**, 404–410 (1996).
48. Delong, R., Stephenson, K., Loftus, T., Fisher, M., Alahari, S., Nolting, A. and Juliano, R. L. *J. Pharm. Sci.*, **86**, 762–764 (1997).
49. Bennett, C. F., Chiang, M. Y., Chan, H., Shoemaker, J. E. and Mirabelli, C. K. *Mol. Pharmacol.*, **41**, 1023–1033 (1992).
50. Goodarzi, G., Gross, S. C., Tewari, A. and Watabe, K. *J. Gen. Virol.*, **71**, 3021–3025 (1990).
51. Colige, A., Sokolov, B. P., Nugent, P., Baserga, R. and Prockop, D. J. *Biochem.*, **32**, 7–11 (1993).
52. Yang, N. S., Burkholder, J., Roberts, B., Martinell, B. and McCabe, D. *Proc. Natl Acad. Sci. USA*, **87**, 9568–9572 (1990).
53. Matthews, K.E. and Keating, A. *Exp. Hematol.*, **22**, 702 (1994).
54. Takahashi, M., Furukawa, T., Nikkuni, K., Aoki, A., Nomoto, N., Koike, T., Moriyama, Y., Shinada, S. and Shibata, A. *Exp. Hematol.*, **19**, 343–346 (1991).
55. Wagner, E. *J. Controlled Rel.*, **53**, 155–158 (1998).
56. Freeman, D. J. and Niven, R. W. *Pharm. Res.*, **13**, 202–209 (1996).
57. Midoux, P., Kichler, A., Boutin, V., Maurizot, J.-C. and Monsigny, M. *Bioconj. Chem.*, **9**, 260–267 (1998).
58. Manuel, S. M., Guo, Y. and Matalon, S. *Am. J. Physiol.*, **273**, L741–L748 (1997).
59. Van Der Woude, I., Wagenaar, A., Meekel, A. A. P., Ter Beest, M. B. A., Ruiters, M. H. J., Engberts, J. B. F. N. and Hoekstra, D. *Proc. Natl Acad. Sci. USA*, **94**, 1160–1165 (1997).
60. Duncan, J. E., Whitsett, J. A. and Horowitz, A. D. *Hum. Gene Ther.*, **8**, 431–438 (1997).
61. Kukowska-Latallo, J. F., Chen, C., Eichman, J., Bielinska, A. U., Baker, J. R. Jr. *Biochem. Biophys. Res. Commun.*, **264**, 253–261 (1999).
62. Westesen, K., *Inter. J. Pharm.*, **102**, 91–100 (1994).
63. Westesen, K. and Koch, M. H. J. *Inter. J. Pharm.*, **103**, 225–236 (1994).
64. O'Hagan, D. T. and Illum, L. *Crit. Rev. Ther. Drug Carrier Syst.*, **7**, 35–97 (1990).
65. Muranishi, S., *Crit. Rev. Ther. Drug Carrier Syst.*, **7**, 1–33 (1990).
66. Lee, V. H., Yamamoto, A. and Kompella, U. B. *Crit. Rev. Ther. Drug Carrier Syst.*, **8**, 91–192 (1991).
67. Elbert, K. J., Schäfer, U. F., Schäfer, H. J., Kim, K. J., Lee, V. H. and Lehr, C. M. *Pharm. Res.*, **16**, 601–608 (1999).
68. Hug, P. and Sleight, R. G. *Biochim. Biophys. Acta*, **1097**, 1–17 (1991).
69. Ledley, F. D. *Hum. Gene Ther.*, **6**, 1129–1144 (1995).
70. Lee, R. J. and Huang, L. *Crit. Rev. Ther. Drug Carrier Syst.*, **14**, 173–206 (1997).
71. Huber, B. E. *Ann. N. Y. Acad. Sci.*, **716**, 1–5 (1994).
72. Miller, N. and Vile, R. *FASEB*, **9**, 190–199 (1995).
73. Uckert, W. and Walther, W. *Pharmacol. Ther.*, **63**, 323–347 (1994).
74. Logan, J. J., Bebok, Z., Walker, L. C., Peng, S., Felgner, P. L., Siegal, G. P., Frizzell, R. A., Dong, J., Howard, M., Matalon, *et al.*, *Gene Ther.*, **2**, 38–49 (1995).

75. Kukowska-Latallo, J. F., Raczka, E., Quintana, A., Chen, C., Rymaszewski, M. and Baker, J. R. Jr. *Hum. Gene Ther.*, **11**, 1385–1395 (2000).
76. Roberts, J. C., Bhalgat, M. K. and Zera, R. T. *J. Biomed. Mater. Res.*, **30**, 53–65 (1996).
77. Malik, N., Wiwattanapatapee, R., Klopsch, R., Lorenz, K., Frey, H., Weener, J. W., Meijer, E. W., Paulus, W. and Duncan, R. *J. Control. Rel.*, **65**, 133–148 (2000).
78. Raczka, E., Kukowska-Latallo, J. F., Rymaszewski, M., Chen, C. and Baker J. R. Jr, *Gene Ther.*, **5**, 1333–1339 (1998).
79. Singh, P., Moll III, F., Lin, S. H., Ferzli, C., Yu, K. S., Koski, R. K., Saul, R. G. and Cronin, P. *Clin. Chem.*, **40**, 1845–1849 (1994).
80. Wu, C. *Bioorg. Med. Chem. Lett.*, **4**, 449–454 (1993).
81. Lamb, R. A. and Krug, R. M. 'Orthomyxoviridae: the viruses and their replication', in Fields, B. N., Knipe, D. M. and Howley, P. M. (eds), *Fields and Virology*, Lippincott-Raven Publ., Philadelphia, 1996, pp. 1353–1395.
82. Schulze, I. T. *J. of Infectious Dis.*, **176** (Suppl. 1), S24–S28 (1997).
83. Weis, W., Brown, J. H., Cusack, S., Paulson, J. C., Skehel, J. J. and Wiley, D. C. *Nature*, **333**, 426–431 (1988).
84. Pritchett, T. J. and Paulson, J. C. *J. Biol. Chem.*, **256**, 9850–9858 (1989).
85. Reuter, J. D., Myc, A., Hayes, M. M., Gan, Z., Roy, R., Qin, D., Yin, R., Piehler, L. T., Esfand, R., Tomalia, D. A. and Baker, J. R. Jr. *Bioconj. Chem.*, **10**, 271–278 (1999).

19

Dendrimer-based Biological Reagents: Preparation and Applications in Diagnostics

P. SINGH
Dade Behring Inc., Newark, DE, USA

1 INTRODUCTION

Ever increasing demands on the healthcare industry offer challenging opportunities for producing products that are not only more cost effective and outperform existing products, but also provide ready access to a variety of new biological and analytical reagents. To achieve this goal it is necessary to explore emerging technologies that can provide a broad family of products from common sets of reagents and methods.

All commercial reagents, whether synthesized, isolated from natural sources or prepared by a combination thereof must meet rigid performance requirements for application to *in vitro* diagnostic test kits and for clearance by regulatory agencies. A number of criteria must be fulfilled for broad acceptance of these reagents, especially if the reagents or the components thereof represent new and novel synthetic structures or possess unique properties. These synthetic materials must have the following characteristics:

1. Ease of preparation in commercial quantities.
2. Defined lot-to-lot purity.
3. Defined structure with minimum inclusion properties to prevent preferential adsorption of specific components from plasma or serum.
4. Structural flexibility relative to functional groups and their placement.
5. Reasonable cost.

Starburst dendrimers™, commonly referred to as dendrimers [1, 2] can fulfill all

Dendrimers and Other Dendritic Polymers. Edited by Jean M. J. Fréchet and Donald A. Tomalia
© 2001 John Wiley & Sons Ltd

of these requirements as reagents or components thereof for any commercial application, including *in vitro* diagnostics. The dendrimers are synthetic, novel water-soluble polymers [1] with a well-defined architecture. These polymers have defined molecular weights, nanoscale molecular size (10–150 Å diameter) and a high density of surface functional groups.

2 ANALYTE-ANTIBODY INTERACTIONS

The selective binding of antigens with specific antibodies is routinely utilized in performing nonisotopic solid phase immunoassays. These techniques are used to measure circulating levels of a number of markers utilized in management of patients with specific clinical symptoms. For example, measurement of circulating levels of myoglobin, creatine kinase and Troponin I have been shown [3] to provide valuable diagnostic information for the evaluation of patients with chest pain and suspected heart trauma. This is particularly true in situations where other tests such as EKG do not provide a clear diagnosis.

Immunoassays fall into two broad categories; namely, 'homogeneous' and 'heterogeneous'. A homogeneous assay system involves reactions carried out in an aqueous monophasic solution. Both the reactants and the products formed are present in a homogeneous phase in this format. As expected, in this mode of immunoassay, equilibrium is often achieved quickly. The procedures for performing these assays are simpler and the reagents employed are generally less expensive than heterogeneous assays. Homogeneous assays do not require separation of free analyte from the analyte–antibody complex on a solid phase and are thus devoid of artifacts resulting from nonspecific binding of analytes or plasma proteins to solid surfaces. In addition, the reagents utilized for a homogeneous system are adaptable to a large number of automated systems.

The alternative heterogeneous system involves separation of the antibody-antigen complex on a solid phase. Heterogeneous assays are inherently higher in analytical sensitivity but can require longer incubation times, multiple wash steps and more complicated instrumentation, especially in automated systems. Many current immunoassay formats are limited either by high background signals as in homogeneous immunoassays or by rate-limiting incubation times as in heterogeneous immunoassays.

2.1 *SOLID PHASE IMMOBILIZATION OF IMMUNE COMPLEXES*

A number of automated systems, e.g. Stratus® CS (Dade Behring), IMx® (Abbott Laboratories) and COBAS® core II (Roche Diagnostics) are now commercially available for quantitation of a large variety of analytes of clinical significance. An essential feature, common to all these systems, is the solid-phase immobilization of an immunoreactive component, such as an analyte-specific antibody known as the primary or capture antibody. In the Stratus® technology [4] the capture antibody is immobilized onto a glass fiber matrix through passive adsorption of a double-antibody immune complex. The resulting solid phase is referred to as a 'tab'. In the double-antibody technique, carefully titrated amounts of the capture antibody are mixed with a secondary antibody, specific for the Fc region of the primary antibody. The secondary antibody employed depends on the nature of the primary antibody utilized for the specific assay, e.g. goat anti-mouse antibody or goat anti-rabbit antibody would be utilized if the primary antibody is a mouse monoclonal or a rabbit polyclonal antibody, respectively. This technique has proven to be versatile as demonstrated by the immense assay menu for a large variety of analytes that have been successfully commercialized using this technology.

Despite extensive efforts toward covalent immobilization on the solid phase, surface adsorption is still the most widely used method for immobilization. Most adsorptions are carried out by empirically adjusting conditions to avoid or minimize immunoreactivity loses. Other factors that may affect the success of immobilization include: (1) limited surface area availability, (2) nonuniform distributions of the immune complexes on the solid phase; (3) the nature of random absorption of the immunoreactive species on the solid surface.

From a manufacturing standpoint, preparation of the double-antibody immune complex can be very labor intensive. For optimal manufacturability and analytical performance of this system, it is important to have a secondary antibody with a moderate to high affinity so that a mixture of immune complexes of appropriate molecular weights is formed. The molecular size and shape of complexes formed depends on a number of parameters, such as temperature, buffer characteristics, ionic strength and the presence of other solution components such as detergents. These conditions must be carefully controlled or else species of very high molecular weight could be formed due to temperature or buffer interactions. Lot-to-lot variability in the primary and secondary antibody raw materials can also affect the solid phase performance if not properly controlled.

In order to develop a process that is more robust and free from some of the limitations of the double-antibody concept, various methods have been explored to immobilize the primary antibody onto the solid phase. Covalent immobilization of the immunoreactive species directly on a solid support would have the distinct advantage of overcoming the capacity limitations and the possible molecular changes that can take place before or after immobilization. However, this mode has not been widely utilized due to lack of successful and reproducible methods of covalent attachment. A further complication is the uncontrollable heterogeneity of the solid supports used currently for immobilization. Realizing these limitations, the approach used for Stratus® has been to choose a design based on strong electrostatic interactions between the solid phase and well-characterized chemically modified immunoreactive species.

2.2 USE OF DENDRIMERS AS PROTEIN (IgG) REPLACEMENT

Of the several methods tested, the most promising results have been obtained when the primary (capture) antibody is covalently coupled to a dendrimer. An optically clear solution of this dendrimer-coupled antibody can either be applied directly to the tab or premixed with the patient's sample, in a competitive manner, prior to immobilization on the tab. This chapter describes results on the preparation and applications of dendrimer-coupled protein complexes in immunodiagnostics. For this application both a single protein and multiple proteins have been coupled onto dendrimers and the advantageous performance of these complexes in an immunoassay system has been demonstrated. In addition, utilization of dendrimer-based reagents in analysis of oligonucleotides is provided as an example of broad capability of these reagents for diagnostic testing.

3 A COMMERCIAL EXAMPLE OF DENDRIMER-PROTEIN CONJUGATE-BASED REAGENT TECHNOLOGY

Commercial application of the dendrimer-based reagent technology has been demonstrated by the successful development of The Stratus® CS STAT fluorometric analyzer [5] marketed by Dade Behring Inc. This rapid automated point of care immunoassay system provides quantitative analysis of whole blood or preprocessed plasma samples via unit use assay test packs. Up to four test packs can be introduced for each sample. All reagents [5–9] required for specimen analyses are contained within the test packs.

The Stratus® CS system utilizes some of the technology that was originally developed at Dade Behring for the Stratus® family of immunochemistry analyzers. These analyzers utilize a patented technology referred to as radial partition immunoassay (RPIA). In its most common form, RPIA [4] is performed by using a reaction zone created by immobilizing the capture antibody, as a double-antibody immune complex, onto a glass fiber matrix. In the sandwich assay format for large molecular weight analytes, such as proteins, sample and enzyme–antibody conjugate are then applied to the reaction zone in a sequential fashion. The enzyme, calf-intestinal alkaline phosphatase, is coupled to the antibody that may be the same or different from the primary antibody, depending on the nature of the primary antibody. After controlled incubations, the unbound reactants are then removed via application of a substrate-wash fluid. The wash fluid partitions these unbound components into an area outside the reaction zone, where they cannot be detected. The amount of labeled material remaining in the reaction zone is then quantified by front surface fluorescence measurement of umbelliferrone that is produced by enzymatic hydrolysis of umbelliferryl phosphate. The fluorescent signal is related to the concentration of a sample by reference to a standard curve generated with calibrators.

The RPIA technology has been enhanced in the Stratus® CS system by utilization of a dendrimer–antibody complex in which the *analyte-specific capture antibody is covalently coupled onto a dendrimer.* The test packs in the Stratus® CS system include dendrimer–capture antibody complex reagent, the alkaline phosphatase labeled antibody conjugate reagent, the substrate-wash reagent and a piece of glass fiber filter paper as the solid phase. Preparation and unique properties associated with these dendrimer-coupled antibody complexes are described below.

4 DENDRIMER-COUPLED ANTIBODY COMPLEXES

4.1 PREPARATION

Dendrimers are synthetic 'structure controlled core–shell' macromolecules prepared by a repetitive two-step reaction series, involving a methyl acrylate–ethylenediamine reaction sequence. This produces precise mathematically defined shell arrays of β-alanine around an intitator core molecule such as ammonia or ethylenediamine. This two-step reaction is repeated to provide successive shells and higher generations of these polymers. One set of reactions represents each generation [1] thereby producing an amino-terminated full generation dendrimer. The carboxy-terminated half-generation dendrimers are produced by base hydrolysis of the intermediate methyl ester obtained after reaction with methyl acrylate.

The dendrimers have been derivatized with a number of heterobifunctional crosslinking agents. These derivatized dendrimers have been well characterized for their reactive group contents using a number of physical and chemical methods [10]. Free amino groups in the full generation dendrimers react almost instantly with a solution of sulfosuccinimidyl-(4-iodoacetyl)-amino-benzoate (sulfoSIAB) in water to introduce the electrophilic phenyliodoacetamido groups [11]. The activated dendrimers are then coupled to a molecule of interest, e.g. a protein such as an antibody or a small molecular weight hapten such as a drug. The general method of covalent coupling of the primary antibody with the fifth-generation ethylenediamine, EDA core dendrimer (E5), to produce a dendrimer–antibody complex (E5–Ab), involves dithioerythritol (DTE) reduction of the disulfide groups present in the hinge region of the antibody molecule. The free sulfhydryl groups that are thus generated are used for coupling with the phenyliodoacetamido groups that had been incorporated in the dendrimer molecule by reaction with sulfoSIAB. This series of reactions is shown in Scheme 1.

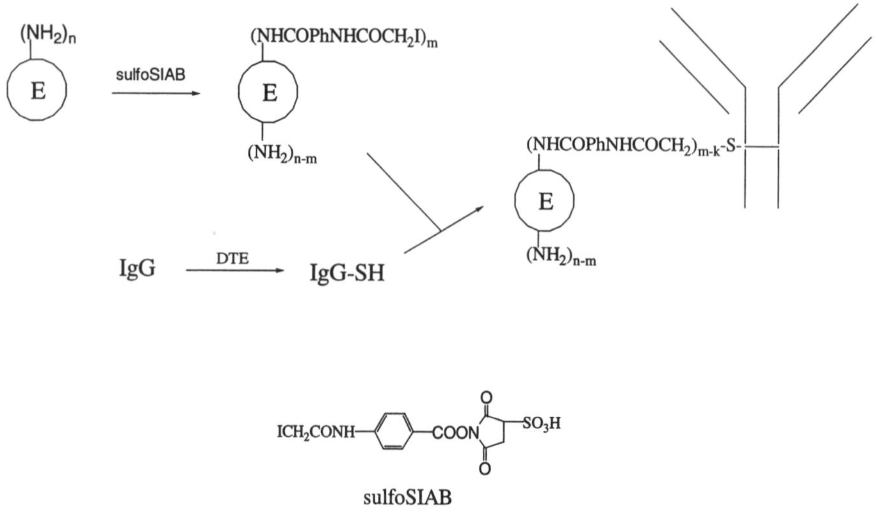

$(NH_2)_n$

E

sulfoSIAB

$(NHCOPhNHCOCH_2I)_m$

E

$(NH_2)_{n-m}$

IgG $\xrightarrow{\text{DTE}}$ IgG-SH

$(NHCOPhNHCOCH_2)_{m-k}$-S-

E

$(NH_2)_{n-m}$

ICH_2CONH—⬡—$COON$—SO_3H

sulfoSIAB

Scheme 1

4.2 *PERFORMANCE AS A REAGENT ON STRATUS® SYSTEMS*

4.2.1 Heterogeneous Assay Format

For immunochemical evaluation of dendrimer-coupled antibody complex, a buffer solution of the chromatographically pure E5–Ab stock was diluted and stored in a spotting buffer [12]. The spotting buffer is comprised of TRIS with added detergent and protein as stabilizers. All performance results described here were obtained using the storage/spotting buffer, although this system has not been optimized for each case of the modified antibody. These evaluations were carried out on Stratus®, Stratus® II or Stratus® Intellect automated fluorometric enzyme immunoassay systems, depending on the experimental design and instrumental capability.

Initial studies showed that to achieve response comparable to Stratus® II double-antibody system, at a reasonable protein concentration, the fifth-generation (N5, ammonia core) dendrimer (diameter size, 53 Å) was most preferable for coupling. A lower-generation dendrimer (G = 2, 23 Å) gave about 40% as much signal as the 53 Å dendrimer. Higher generation dendrimer (95 Å) gave about the same signal as that obtained from the 53 Å particles. This result suggested that a certain size (thus determining the number of reactive functional groups) and/or shape of the particles was essential for the optimum performance.

Comparative evaluation of anti-human thyroid stimulating hormone (hTSH) antibody, bound to the fifth-generation ammonia core (N5) or the fifth-generation ethylenediamine core (E5) dendrimer (1), did not show any differences in either the effective protein concentrations or the shape of the dose-dependent response curves (calibration curves) as determined from the recovery of standard controls. All the other experiments described here were thus carried out with the fifth-generation (i.e. dia. = 70 Å) particles of ethylenediamine core (E5) dendrimers. The later particles were selected for their ability to be produced reproducibly on a large scale.

The fifth-generation dendrimers (i.e. (N5) and (E5)) possess a globular shape, high solubility in water and molecular size roughly equivalent to the Fc region of an IgG class antibody [1]. In essence they appear to function as 'artificial proteins' in this particular application. The terminal amino groups in a dendrimer possess unusually high chemical reactivity [13] and high positive charge density at physiological pH. The E5–Ab complexes have unique solubility in aqueous buffers, high charge density, high affinity and specificity. These complexes provide for better presentation and functionality of the capture antibody on the glass fiber solid surface used in the assay. This in turn leads to more efficient capture of the specific analyte in question. In addition, these compounds have the important ability to remain in solution throughout the course of an analytical procedure allowing for flexibility in assay procedures. Application of reaction mixtures containing these antibody–dendrimer complexes onto negatively charged surfaces such as glass fibers showed very tenacious absorption presumably by means of electrostatic interaction through the dendrimer component of the antibody complexes.

The relative performance of a number of the E5–Ab complexes immobilized on the solid phase was studied by substitution of the corresponding classical double-antibody complexes with the dendrimer-based reagents. All other assay procedures and reagents, except the solid phase, were utilized from the Stratus® system without any changes. Analyses of creatine kinase MB-isoenzyme (CKMB) and hTSH were selected as test cases for initial technical feasibility demonstration. Data for CKMB and hTSH [12, 14] show that even without extensive optimizations, dose-dependent response curves [12] can be generated with the analyte-specific antibody present in the corresponding E5–Ab complexes immobilized on the solid phase. The shapes of these curves generated by the E5–Ab complexes immobilized on the solid phase were found to be very similar to those generated by the commercial reagents available on Stratus® system utilizing double-antibody immune complexes. The similarity in shapes of the calibration curves was also confirmed from the identical recovery of the standard serum-based control samples [12]. An additional advantage, observed by solid- phase immobilization of the E5–Ab complex, was a 40–50% reduction in the amount of the primary antibody requirement (Table 19.1) to perform these two assays on the Stratus® system.

Table 19.1 Primary antibody requirements to perform a test (μg/ test)

Assay	Double-antibody Immune Complex (4)	E5–Ab complex [12]	
		Solid phase format [12][a]	Solution phase format [12][b]
hTSH	3.8	2.3	0.76
CKMB	2.3	1.2	0.38
Myoglobin	3.75		0.17
T4	0.023		0.0095
Troponin I	10.0		0.05

[a] A buffered stock solution of the complex was prespotted on the tab. The dry tab with preimmobilized E5–Ab complex was then utilized to perform the assay.
[b] At the time of assay a buffered stock solution of the complex was spotted on the tab either by itself or as a mixture with the analyte depending on the specific assay format.

After the initial feasibility demonstration of this concept for the analysis of CKMB and hTSH utilizing a sandwich assay format, this concept was extended further to analyze samples with other possible assay formats. For small molecular weight analytes, such as drugs, E5–capture antibody complex was mixed with sample and a derivative or analog of the analyte modified with alkaline phosphatase (an enzyme conjugate). The sample analyte competes with the enzyme conjugate for binding sites on the E5–capture antibody. The mixture is transferred to the Stratus® tab where the E5–capture antibody becomes immobilized in the central portion of the solid phase, where the enzyme conjugate bound to immobilized E5 can be quantified. This liquid phase format can be run as a sequential assay as well, as is shown in Figure 19.1, for an assay to determine thyroxine (T4). In some instances the capture antibody may be an expensive reagent where procedural losses due to purification and coupling onto E5 are cost prohibitive. For these cases the anti-capture antibody (secondary antibody) has been coupled with the den-drimer. This dendrimer–secondary antibody may be prespotted onto the solid phase or it can be spotted onto the tab at the time of performing the assay. The capture antibody can then be used in the liquid phase in either a competitive or sequential mode. Upon transfer to the Stratus® tab the primary antibody is captured by the secondary antibody–dendrimer complex. Alternatively, the capture antibody can be precomplexed to the dendrimer–secondary antibody complex and be prespotted, spotted during the assay or used in the liquid phase. Also, a format has been developed in which the derivatized analyte instead of the antibody is linked to the dendrimer. Here, dendrimer–analyte is mixed with patient sample and the capture antibody, resulting in a competition between the dendrimer-analyte and patient sample analyte for the capture antibody binding sites. The composite mixture is applied to the tab where the primary antibody associated with the dendrimer-hapten will remain fixed to the center of the tab after application. Capture antibody is then quantified by the separate addition of a secondary antibody-alkaline phosphatase conjugate. This format is useful for

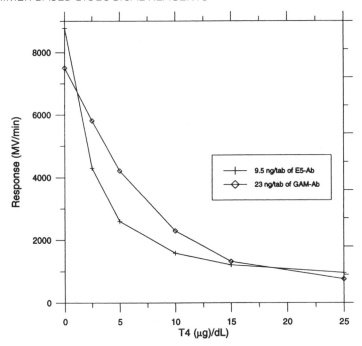

Figure 19.1 Dose-dependent response curve for thyroxine (T4) assay using E5–Ab and double-antibody immune complex. Solutions (20 μL) of the reference samples containing 0.0 to 25.0 μg/dL T4 were mixed with 180 μL of the releasing solution. Fractions (132 μL) of this mixture were combined with 38 μL of an E5–Ab solution. After standard incubation, 76 μL of this mixture was spotted on a blank Stratus® tab. Incubation times and reagent volumes used for the rest of the reagents, such as the conjugate and the substrate wash, were identical to those defined for an existing T4 assay on Stratus®

assays where the reaction buffer has a composition designed to extract a binding partner, but also has deleterious effects on the alkaline phosphatase conjugate.

Standard curves demonstrating required analytical sensitivity and range were obtained for ferritin, myoglobin and prostate specific antigen (PSA) using a sandwich assay format, for digoxin and cortisol with a sequential format, and testosterone and folate with a competitive format. In all these cases, only the capture antibody reagent, coupled onto E5, was changed while keeping all other reagents and assay parameters constant.

4.2.2 Enhanced Assay Formats

Reactions performed in a true homogeneous solution are expected to follow the law of mass action. In such cases, the binding reactions would reach equilibrium sooner by increasing the concentrations of the solution phase reactants. This was indeed found to be the case as is shown in Figures 19.2 and 19.3. Figure 19.2 shows a time-dependent effect on response observed from solutions containing various concentrations of an E5–Ab complex. For this study, an anti-CKMB antibody was coupled onto E5. This figure shows that after 10–60s of incubations, the response measured as indication of the analyte–antibody complex formation was dependent on the concentration of the E5–Ab complex. It is clear that a 30 s incubation between the analyte and E5–Ab (at 100–150 μg/mL) was required to capture the maximum amount of analyte.

A similar effect is observed when increased concentrations of the Ab–ALP conjugate were utilized to study the time-dependent effect on response, while keeping other reagents and their incubations constant. At any fixed time interval, the response observed (Figure 19.3) was found to be directly dependent on the protein concentration of the Ab–ALP conjugate.

This study suggests that by increasing the concentration of the reactants in

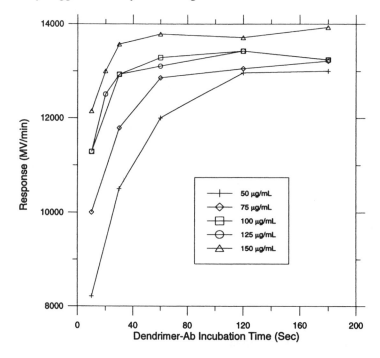

Figure 19.2 Kinetics of reaction of E5–Ab complex in a CKMB assay. A 132 μL solution of CKMB containing 128 ng/mL of the analyte was mixed with 38 μL of the E5–Ab complex containing various known concentrations of the antibody. After incubation for a fixed period of time the mixture was spotted on a Stratus® tab. Prior to addition of the substrate wash, a 60s incubation was carried out on the solid phase with the Ab–ALP conjugate containing 360 ng/mL of the protein.

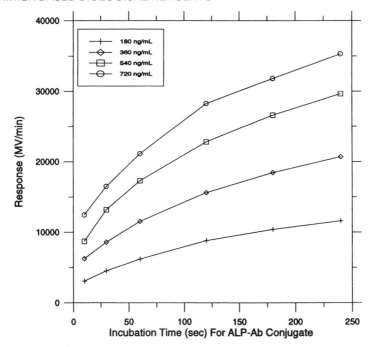

Figure 19.3 Kinetics of reaction of Ab–ALP conjugate. A 132 μL solution of CKMB containing 128 ng/mL of the analyte was incubated for 20 s with 38 μL of the E5–Ab complex containing 150 μg/mL protein. The mixture (76 μL) was spotted on a Stratus tab followed by addition of 45 μL of a solution of the Ab–ALP conjugate containing defined concentration of the protein

solution it should be possible to drive the binding reactions to completion sooner. This was indeed found to be the case [12]. It was thus possible to mix the sample with a solution of E5–Ab followed by completion of the assay on the glass fiber solid phase. Here rapid partition to remove the interfering components takes place followed by generation of a fluorescence signal dependent on the analyte concentration.

4.2.3 Performance Advantages of E5–Ab Conjugates

The flexibility to perform the analyte–antibody reaction in a homogeneous solution followed by separation of the unbound reactants on a solid phase offers new advantages. These advantages had previously been available only independently in the homogeneous and heterogeneous assay formats, respectively. Such a system would be expected to provide high speed, typical of a homogeneous system, and increased analytical sensitivity observed normally with a heterogeneous assay system. This was indeed found to be the case when dendrimer-based

P. SINGH

protein reagents were utilized to perform immunoassays. *Thus, it has been possible to complete an assay for CKMB and hTSH in half the time* [12] *it would normally take with the heterogeneous system.* During this work there was an indication that it would be possible to reduce further the time taken to complete an assay on an appropriately designed automated analytical instrument. A threefold to 27-fold increase in sensitivity has also been achieved (Table 19.2) when an E5–Ab reagent is utilized to replace the classical double antibody immune complex.

In nearly all cases studied, the amount of primary antibody required in the E5–Ab complex to perform an assay has been found to be substantially less than that required for the double antibody immune complex format. This was found to be the case (Table 19.1) when the E5–Ab complex was either directly immobilized on the solid phase, to imitate the double-antibody immune complex format, or utilized in a solution phase format [12].

4.3 STABILITY OF DENDRIMER–ANTIBODY CONJUGATES

4.3.1 Real Time Storage

For long- term storage, antibody solutions are often kept frozen at a temperature of $-20\,^{\circ}$C or lower. Successful application of dendrimer-based reagents in immunodiagnostics would require that a dendrimer-coupled antibody complex must maintain its immunoreactivity under similar storage conditions. To evaluate stability under these conditions, aliquots of the antibody-coupled dendrimer complexes in solution were stored at various temperatures and performance of these solutions were examined. Figure 19.4 shows the results with dendrimer-coupled anti-hTSH antibody after defined time intervals. The dendrimer-coupled antibody solutions, stored for one year at $4\,^{\circ}$C or at ambient temperature, showed less than 10% change in immunoreactivity. Also, the dose-depend-

Table 19.2 Sensitivity determination with analyte-specific primary antibody in the form of a dendrimer complex or a double-antibody immune complex

	Minimum detectable dose	
Assay	E5–Ab[a]	Double-Ab immune complex[b]
hTSH	0.01 μIU/mL	0.05 μIU/mL
CKMB	0.14 ng/mL	0.40 ng/mL
Myoglobin	0.05 ng/mL	0.37 ng/mL
Troponin I[c]	0.03 ng/mL	0.80 ng/mL

[a] Determined on Stratus® II by the solution phase format [12].
[b] Determined on Stratus® II as described by Giegel *et al.* [4].
[c] Determined on Stratus® CS by Heeschen *et al.* [25].

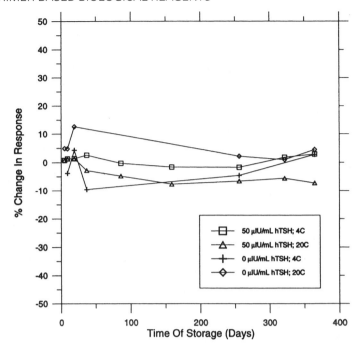

Figure 19.4 Stability of E5–Ab solution. A solution of the dendrimer-coupled anti-hTSH antibody containing 10 μg/mL of the protein in 100 mM Tris-2% BSA-detergents, pH 8.0 was stored at −20°C, 4°C and the ambient temperature. At appropriate time intervals, the aliquots stored at the different temperatures were tested on Stratus® using standard hTSH solutions containing 0.0 and 0.25 and 50 μIU/mL hTSH. Data shown are only for testing with standard solutions containing 0.0 and 50 μIU/mL hTSH. Percent change in response for the samples stored at 4°C or ambient temperature was calculated from response of the sample at −20°C for a specific storage time point

ent response curves generated (not shown here) by these samples stored at different temperatures for one year were indistinguishable from those of the freshly prepared solution of the dendrimer–antibody complexes.

4.3.2 Effects of Exposure to Hydrogen Peroxide, Bubbled Air, Oxygen and Nitrogen

It was of importance to determine the impact on performance of (E5)–dendrimer–Ab complexes should they be exposed to unintentional oxidizing conditions. To study such a challenge, the aqueous buffered solutions of naked (E5)–dendrimer at pH 2.5, 7.0 and 9.0 were exposed separately to a bubble stream of nitrogen, oxygen or air for 15 days at room temperature. These solutions were then analyzed for primary amine level with the fluorescamine assay [15], as well as for phenyliodoacetamido content [16] after reaction with sulfoSIAB. Performance of these treated and untreated samples was also evaluated after preparation of the corresponding dendrimer-coupled antibody conjugates. Quite remarkably, no essential differences were observed between the treated and the untreated dendrimer samples relative to their amino group and phenyl-iodoacetamido group contents (data not shown). Furthermore, the dose-dependent curves for the E5–Ab complex prepared with the treated E5–Ab conjugates showed no change in comparison to that of an untreated E5–Ab conjugate.

An aqueous solution of E5 was treated with 0.5% hydrogen peroxide in water and the resulting solution tested with the fluorescamine assay [15] for primary amino group content. This titration showed a 20–25% decrease in the amino group content of the dendrimer solution within 30 min to 4 h treatment. The dendrimer sample containing 25% less amino groups showed very little differences in capillary electrophoretic pattern [17], or phenyliodoacetamido content on treatment with sulfoSIAB (2.5 for the oxidized sample in comparison to 2.2 per mole of protein for the unoxidized dendrimer sample). Furthermore, when this oxidized E5 solution was used to prepare an E5–Ab complex utilizing anti-hTSH antibody, the response curve generated was identical to that of the corresponding normal E5–Ab conjugate.

These results show that, even with detectable decrease in the amino group content of E5 solutions, fully functional E5–Ab conjugates can be prepared with such solutions of dendrimers. This suggests that the number of amino groups in the E5–Ab conjugate is so large that even with a detectable decrease in amino groups there is very little impact on its binding to the negatively charged glass fiber matrix or target capture properties. This was confirmed by the fact that the dendrimer oxidation product does not cause any deterioration effect on the immunoreactivity of dendrimer-coupled antibody. Thus the method described here provides a very robust process for manufacturing E5–Ab conjugates even with samples of E5 that may not have been stored and handled under less than optimal conditions.

4.4 DEVELOPMENT OF THE NEW COMMERCIAL STRATUS® CS SYSTEM

As described above, the dendrimer-coupled antibody conjugates show uniquely enhanced properties compared to the classical double antibody systems. Important characteristics such as complete solubility in aqueous buffers, flexibility in immunoassay format, ability to improve assay sensitivity, consistent, reproducible manufacturing and favorable stability has driven the utilization of these dendrimer-based reagents in Stratus® CS, the latest member of the Stratus® family of immunochemistry analyzers. In this new analyzer system, the primary antibody in the form of an E5–Ab complex replaces the capture immune complex used in previous versions of this family of analyzers.

The Stratus® CS system is a fluorometric enzyme, immunoassay system designed for STAT testing applications; at present the test menu includes one pregnancy marker i.e. human chorionic gonadotropin (hCG) and three cardiac markers, i.e. CKMB, myoglobin and troponin I. All the reagents required for analysis are contained within test packs. The analytical process consists of adding dendrimer-coupled capture antibody, sample and the enzyme-labeled antibody conjugate in a sequential manner to a glass fiber solid support, followed by addition of the substrate-wash reagent. Unbound labeled antibody reagent is removed from the reaction zone by radial elution using the substrate-wash reagent. This permits the detection of fluorogenic product that is directly related to the concentration of the analyte present in the sample.

In the new Stratus® CS system results are available in less than 15 min after sample draw and the system has the capability to analyze four samples in less than 30 min. Ease of use, analytical sensitivity, accuracy, precision, and reproducibility makes this system suitable for use in chest pain centers, emergency departments, critical care units, observation wards and clinical laboratories.

5 DENDRIMER-MULTIFUNCTIONAL PROTEIN CONJUGATES

5.1 DENDRIMER-DOUBLE ANTIBODY CONJUGATES

5.1.1 Preparation and Performance as a Reagent on Stratus®

This concept of binding single proteins to poly(amidoamine) (PAMAM) dendrimers has now been extended to the binding of multiple proteins with either similar or dissimilar properties. As a model system for binding similar proteins, equal amounts of anti-CKMB antibody and anti-hTSH antibody were coupled simultaneously [18, 19] to the SIAB-activated E5-dendrimer by a procedure similar to that described earlier for the coupling of a single antibody. The single, multifunctional reagent prepared was used to analyze for multiple analytes; namely, CKMB and hTSH, on a Stratus® II analyzer utilizing the standard analyte- specific instrument parameters. The shapes of these dose-dependent standard curves were found to be remarkably similar [18] to those obtained with the commercial reagents available on the instrument. Similar shapes of the curves, even at the low analyte concentrations, suggests that the presence of a second antibody, specific to an unrelated analyte, has practically no impact on the immunoreactivity of the specific antibody in these multi-functional reagents.

5.2 DENDRIMER–ENZYME–ANTIBODY CONJUGATES

5.2.1 Preparation and Performance as a Reagent on Stratus®

To study the binding of proteins with dissimilar structures and reactivities, an enzyme alkaline phosphatase (ALP) and the Fab' fragment of anti-CKMB antibody were selected. Performance of the dendrimer complex containing simultaneously, coupled Fab' and ALP (ALP-E-Fab') could then be compared directly with the commercial enzyme conjugate [20] utilized to quantitate CKMB on the Stratus® system. The negatively charged glass fiber solid support of the Stratus® system shows a strong affinity for the positively charged primary terminal amino groups present in the full-generation dendrimers. For true evaluations of these multiprotein–dendrimer complexes it was necessary to minimize such an impact. For this reason the full-generation dendrimer–multifunctional protein complexes were prepared by a series of reactions. Dendrimers, activated with an excess of N-hydroxysuccinimidyl iodoacetate [10], were reacted successively with ALP-SH and Fab' to form the conjugate ALP–S–CH$_2$CONH–E–NHCOCH$_2$–SFab'.

The half-generation dendrimers, containing terminal carboxyl groups were activated to form the electrophilic N-hydroxysuccinimidyl esters. The active esters present on the activated dendrimer [10] were then reacted in sequence with ALP, 6-(bromoacetamido)-hexylamine and then an excess of Fab' to form the conjugate ALP–E–Fab' [10]. Quite remarkably, the performance characteristics of the ALP–E–Fab' conjugates prepared from the dendrimers of generations 1–4 amine terminated, as well as 1.5–4.5 (carboxyl terminated) were comparable [18] to those of the commercial conjugate. This was true with respect to both the effective protein concentration and standard dose–response curves. Any of these dendrimer-based bifunctional reagents could be substituted for the commercial enzyme conjugate without any change in the analyte recovery or compromising sensitivity of the assay. The bifunctional reagent prepared from the half-generation dendrimers, with terminal carboxyl groups, showed much lower non-specific binding [18] as would be expected from the net negative charges on the dendrimer and the glass fiber solid support used in the Stratus® system.

6 HYDROPHOBICITY OF DENDRIMER-COUPLED PROTEIN CONJUGATES

Further characterization of the dendrimer-coupled protein complexes was studied by hydrophobic interaction chromatography carried out by purification over an octyl-Sepharose column. The products obtained by reaction of SIAB with ALP and its complex with third-generation dendrimer, ALP-E3, were purified separately with this column. Hydrophobicity of the protein fractions eluted in the buffer systems were characterized by their phenyliodoacetamido group content (Table 19.3). This table shows that under identical activation reaction conditions, ALP produces a mixture of two groups of species with distinctly different number of phenyliodoacetamido groups. However, activation of the dendrimer-coupled enzyme complex (ALP–E3) produces a more homogeneous species. Also the phenyliodoacetamido content of the activated ALP–E3 was found to be higher than that obtained with the activated ALP. This would be expected to be due to the much higher reactivity of the dendrimer terminal groups. The high reactivity to the dendrimer terminal groups is not affected in the protein–dendrimer conjugate obtained by covalent coupling of the dendrimer with a protein such as ALP (molecular weight 140 kDa). It should be noted that this coupled protein is about 20-fold larger in molecular weight than the dendrimer scaffolding used for this study.

The dissimilar multiprotein complexes, ALP-Fab' and ALP-E3-Fab', were similarly purified over an octyl-Sepharose column. The protein fractions eluted from this column were characterized relative to the effective protein concentrations required to generate a specific response equivalent to that of ALP-Fab'. These results show (Table 19.4) that the more hydrophobic protein fraction

Table 19.3 Phenyliodoacetamido-content (per mole of protein) of proteins before and after passage over octyl-Sepharose

| | Phenyliodoacetamido-content in protein | | |
| | | Pools from column | |
	Before column separation	A^a	B^b
ALP-NH_2 + sulfoSIAB	5.6	5.8	8.8
ALP-E3 + sulfoSIAB	10.8	11.1	—

^{a} 10 mM triethanolamine-90 mM sodium chloride-1.0 mM magnesium chloride, pH 7.8.
^{b} 10 mM trietanolamine-90 mM sodium chloride-1.0 mM magnesium chloride-10% n-propanol, pH 7.8.

Table 19.4 Effective protein concentration to generate a response, for a sample containing 139 ng/mL CKMB, equivalent to that generated by the commercial (ALP-Fab') conjugate on Stratus®

| | Effective protein concentration (ng/mL) | | |
| | | Pools from column | |
	Before column separation	A^a	B^b
ALP-Fab'	465	553	606
ALP-E3-Fab'	465	560	386

$^{a,\,b}$ same as in Table 19.3.

present in ALP-E3-Fab' requires about 30% less protein in comparison to 10% more protein concentration required for the equivalent fraction in ALP-Fab'. The unfractionated protein mixture, as well as the protein fractions that had low binding affinity towards this column were effective at almost identical concentrations.

7 STOICHIOMETRY OF DENDRIMER–MULTI-PROTEIN CONJUGATES

It was of interest to determine the impact of the presence of a dendrimer linker on the relative stochiometry of the two proteins present in a conjugate that might be prepared by covalent coupling of these proteins. A conjugation reaction of the N-hydroxysuccinimidyl iodoacetate-activated ALP–E2 complex with the Fab' fragment of anti-CKMB antibody was used for this purpose [10]. The second-generation dendrimer, E2, was used as a linker molecule to give the ALP–E2 conjugate. ALP and F(ab')$_2$ were labeled with two differentiated fluorescent markers possessing nonoverlapping spectral characteristics. The markers used were fluorescein (Abs/Em 495/520 nm) and BODIPY$^®$ (Molecular Probes, Inc.; Abs/Em 580/590 nm), respectively. The presence of these labels did not show any impact on the enzyme activity of ALP or the immunoreactivity of F(ab')$_2$. The ratio of Fab'/ALP was calculated to be 6 and 4 in the conjugates ALP-Fab' and ALP-E2-Fab', respectively; this calculation was derived from the fluorescence content of these conjugates. It is thus possible to control the relative stoichiometry of the two proteins since the use of different dendrimer generations as linkers would be expected to lead to different relative protein ratios in the multifuctional protein–dendrimer conjugates. The molecular weights calculated for the two conjugates; ALP–Fab' and ALP–E2–Fab', using their size exclusion chromatography elution data, were found to be almost identical.

8 DENDRIMER DNA PROBE ASSAYS

More recently, the broad potential applicability of dendrimer-based reagents has been confirmed by demonstrating the feasibility of using dendrimer-DNA probes in an assay for Chlamydia. This assay system is based on the detection of the specific oligonucleotide as a marker for the bacterial infection. The assay mixture consists of a capture probe and a detector probe. The capture probe was prepared by coupling SIAB-activated E5 with the 5'-thiol derivative of the specific oligonucleotide. Thiolation at the 5' position of the oligonucleotide was carried out by a sequence of three reactions; namely, phosphorylation [21, 22], cystaminylation and a dithiothreitol reduction. The target sample (RNA) was

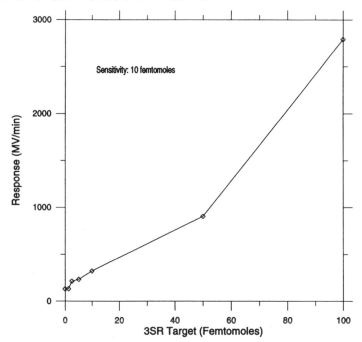

Figure 19.5 Dose-dependent response curve of dendrimer-based Chlamydia assay. Samples containing 0 to 100 mol of the target RNA were amplified by 3SR technology followed by detection on Stratus® (see text for details)

amplified off-line with self-sustained sequence technology [23, 24] followed by hybridization of the amplified sample with the capture probe and the detector probe. The detector probe was prepared by coupling of ALP at the 3′ position of the specific oligonucleotide. The oligonucleotides used for the capture and the detector probe were selected so as to hybridize at the 5′ and the 3′ positions of the amplified sample RNA, respectively. The hybridized mixture was spotted on the Stratus® tab followed by on-line detection on the instrument. The dose-dependent response curve generated is shown in Figure 19.5. This system was shown to have a detection limit of 10 femtomol for the specific target analyte.

9 CONCLUSION

Our results have shown that the use of STARBURST™ dendrimers for the covalent coupling of the molecules of biological interest can lead to a reproducibly performing product with very little impact on the biological activity of the immobilized protein. By careful optimization of the reaction parameters, it has been possible to consistently reproduce performance similar to the characteristics of the native protein. The dendrimer-based reagents have shown good lot-to-lot consistency. Clear solutions of the dendrimer- coupled protein complexes can be prepared in large quantities and stored both as a liquid (at $\leq 20\,°C$) or frozen solid for extended periods of time with little impact on the immunoreactivity of the coupled antibody. This feature allows robust manufacture of these dendrimer-based reagents and convenience in shipping as a product to customer.

The general method utilized to prepare E5–Ab solutions obviates the need for stocking large numbers of reagents which would be necessary if different activation methods were used for each antibody. A number of specific antibodies immobilized by this process have shown response similar to that of the same antibodies when adsorbed as immune complexes in the Stratus® system. In addition, the dendrimer-coupled antibodies have shown dramatic improvements in sensitivity, flexibility and precision for the enzyme immunoassay system. Feasibility demonstration of an assay for DNA probes is a prelude to what can possibly be achieved with these dendrimer-based reagents.

The methods developed by us for preparation of these dendrimer-based reagents are quite general and have the potential of application to a large variety of molecules that have utility in diagnostics, therapeutics, biocatalysis, enzymology and food processing industry.

The author is grateful for a large number of present and former co-workers without whose help and support this work could not be possible. Special thanks are due to Drs Spencer Lin and Susan Evans for their contributions and continuous support.

10 REFERENCES

1. Tomalia, D. A., Naylor, A. M. and Goddard, W. A. Starburst dendrimers: molecular-level control of size, shape, surface chemistry, topology and flexibility from atoms to macroscopic matter, *Angew. Chem., Int. Ed. Engl.*, **29**, 138–175 (1990).
2. Tomalia, D. A. and Durst, H. D. 'Geneologically directed synthesis: Starburst/ Cascade dendrimers and hyperbranched structures', *Topics Curr. Chem.*, **165**, 193–313 (1993).
3. Apple, F. S., Christenson, R. H., Valdes Jr., R., Andriak, A. J., Berg, A., Duh, S.H., Feng, Y. J., Jortani, S. A., Johnson, N. A., Koplen, B., Mascotti, K. and Wu, A. H. B. 'Simultaneous rapid measurment of whole blood myoglobin, creatine kinase MB and cardiac troponin I by the triage cardiac panel for detection of myocardial infarction'. *Clin. Chem.*, **45**, 199–205 (1999).
4. Giegel, J. L., Brotherton, M., Cronin, P., D'Aquino, M., Evans, S., Heller, Z. H., Knight W. S., Krishnan, K and Sheiman, M. 'Radial partition immunoassay'. *Clin. Chem.*, **28**, 1894–1888 (1982).
5. Cronin, P., Evers, T., Mozza, J., Bauer, R., Hickey, G. and Kamm, C. 'A flexible system for the quantitative analysis of cardiac markers with minimized total turnaround time: the Stratus® CS Analyzer', *Clin. Chem.*, **44**, 542 (Abstract) (1998).

6. McKeon, K., Hanavan, M., Murray, V. and Moll, F. 'A quantitative method for the determination of myoglobin for a whole blood immunochemistry analytical system'. *Clin. Chem.*, **43**, 102 (Abstract) (1997).
7. Hall, L., Janes, C. and Peddicord, J. 'A quantitative method for the determination of cardiac Troponin-I for a whole blood immunochemistry analyzer', *Clin. Chem.*, **43**, 100 (Abstract) (1997).
8. Stengelin, M., Hudson, K., Bedzyk, W. D., Singh, P., Desai, V., LeBlanc, M., Berg, M., Murray, V. and Bauer, R. 'Development and performance of a high-sensitivity C-Reactive Protein assay for the Stratus® CS STAT fluorometric analyzer', *Clin. Chem.*, **46**, 26 (Abstract) (2000).
9. Hall, L., Haehner, B., Swann, S., Feliciano, J. A., Feliciano, D., Desai, V., Singh, P. and Bauer, R. 'A quantitative method for the determination of total Beta-hCG from whole blood using the Stratus® CS STAT fluorometric analyzer', *Clin. Chem.*, **46**, 42 (Abstract) (2000).
10. Singh, P. 'Terminal groups in Starburst dendrimers: activation and reactions with proteins', *Bioconj. Chem.*, **9**, 54–63 (1998).
11. Weltman, J. K., Hohnson, S. A., Langevin, J. and Riester, E. F. 'N-Succinimidyl (4-iodoacetyl)aminobenzoate: a new heterobifunctional cross-linker', *Biotechniques*, **1**, 148–152 (1983).
12. Singh, P., Moll III, F., Ferzli, C., Yu, K. S., Koski, R. K., Saul, R. G. and Cronin, P. 'Starburst™ dendrimers: enhanced performance and flexibility for immunoassays', *Clin. Chem.*, **40**, 1845–1849 (1994).
13. Frechet, J. M. J. 'Functional polymers and dendrimers: reactivity, molecular architecture and interfacial energy', *Science*, **263**, 1710–1715 (1994).
14. Lin, S. H., Yu, K. S., Singh, P. and Diamond, S. E. 'Immobilization of specific binding assay reagents'. US Patent No. 5,861,319.
15. Lai, C. Y. 'Detection of peptides by fluorescence methods', in Hirs, C. H. W. and Timasheff, S. N. (eds), *Methods in Enzymology*, Vol. 47, Academic Press, New York, 1977, 236–243.
16. Ishikawa, E., Hashida, S., Kohno, T. and Tanaka, K. 'Methods for enzyme-labeling of antigens, antibodies and their fragments', in Ngo, T. T. (ed.), *Nonisotopic Immunoassay. Plenum Press, New York*, 1988, p. 37.
17. Moll III, F., Lin, S. H., Singh, P. and Cronin, P. 'Nanoscopic technology: characterization of Starburst dendrimer reagents for immunoassays', PITTCON '95. March 1995, New Orleans, La. Abstract 922.
18. Singh, P., Moll III, F., Lin, S. H. and Ferzli, C. 'Starburst™ dendrimers: a novel matrix for multifunctional reagents in immunoassays, *Clin. Chem.*, **42**, 1567–1568 (1996).
19. Singh, P., Moll III, F., Cronin, P., Lin, S. H., Ferzli, C., Koski, R. K. and Saul, R. 'Rapid detection of analytes with receptors immobilized on soluble submicron particles'. US Patent No. 5,898,005.
20. Goodnow, T., Leblanc, M., Koski, K., Welsh, M., Leung, K., Charie, L. and Evans, S. Radial partition immnuoassay: quantitating CKMB levels in human serum, *Clin. Chem.*, **33**, 988 (Abstract No. 528) (1987).
21. Cameron, V., Soltis, D. and Uhlenbeck, O. C. 'Polynucleotide kinase from a T4 mutant which lacks the 3′ phosphatase activity'. *Nucleic Acids Res.*, **5**, 825–833 (1978).
22. Chaconas, G. and van de Sande, J. H. '5′-^{32}P Labeling of RNA and DNA restriction fragments', in Grossman, L. and Moldave, K. (eds), *Methods in Enzymology*, Vol. 65, Academic Press, New York, 1980, pp. 75–85.
23. Guatelli, J. C., Whitfield, K. M., Kwoh, D. Y., Barringer, K. J., Richman, D. D. and Gingeras, T. R. 'Isothermal, in vitro amplification of nucleic acids by a multienzyme

reaction modeled after retroviral replication', *Proc. Natl Acad. Sci. USA*, **87**, 1874–1878 (1990).

24. Fahey, E., Kwoh, D. Y. and Gingeras, T. R. 'Self-sustained sequence replication (3SR): an isothermal transcription-based amplification system alternative to PCR', *PCR Methods Appl.*, **1**, 25–33 (1991).

25. Heeschen, C., Goldmann, B. U., Langenbrink, L., Matschuck, G. and Hamm, C. W. 'Evaluation of a rapid whole blood elisa for quantitation of troponin I in patients with acute chest pain', *Clin. Chem.*, **45**, 1789–1796 (1999).

20

Dendritic Polymer Applications: Catalysts

A. W. KLEIJ, A. FORD, J. T. B. H. JASTRZEBSKI
AND G. VAN KOTEN*
Utrecht University, Debye Institute, Utrecht, The Netherlands

1 INTRODUCTION

During the last decade there has been a remarkable increase in the number of reports describing the synthesis of dendritic polymers with a wide range of properties and/or applications such as host–guest interactions, redox activity, liquid crystalline behaviour, drug delivery systems, light-harvesting effects, molecular recognition, self-assembly and also catalytic activity. In recent years several groups have focused on the incorporation or complexation of (transition) metal fragments onto the dendrimer periphery and presently a broad spectrum of metallodendritic species are known. One of the most interesting applications of these species is found in the field of (homogeneous) catalysis. In general, these metallodendrimers are thought to be macromolecular materials that can combine the advantages of both homogeneous and heterogeneous catalysts. Because of their 'pseudo'-spherical nature and their resultant conformations the metal sites in these well-defined polymeric catalysts should be easily accessible for substrate molecules and reagents, and therefore exhibit characteristics usually encountered in homogeneous catalysis such as fast kinetics, specificity and solubility. Owing to their precise persistent nanoscale size they may be easily removed from product streams, e.g. by means of ultrafiltration techniques.

Metallodendrimers can be constructed via binding of groups with suitable donor atoms (e.g., polydentate ligands) on either the periphery or the core of the dendrimer and the subsequent complexation/coordination of these ligands to an appropriate metal salt. Ultimately, this binding can involve the formation of a direct σ bond linkage (i.e., a M–C bond). This chapter describes various

Dendrimers and Other Dendritic Polymers. Edited by Jean M. J. Fréchet and Donald A. Tomalia
© 2001 John Wiley & Sons Ltd

metallodendrimers reported to date, their specific catalytic applications as well as some examples of their use in membrane reactors under continuous operation conditions. In the last section future perspectives and developments will be briefly discussed.

2 METALLODENDRITIC CATALYSTS

2.1 MISCELLANEOUS DENDRITIC METAL CATALYSTS

The first catalytically active *metallo*dendrimer was made by the Van Koten group [1] using the first examples of dendritic carbosilane molecules prepared by Van Leeuwen, Van der Made and collaborators [2]. The synthesis of this metallodendrimer started from carbosilane (CS) molecules containing reactive silicon–chloride bonds on the periphery which were used for the binding of diamino aryl bromide ligand precursors. In this initial approach, the ligating site was placed away from the CS periphery by using a 1,4-butanediol linker to prevent possible interactions between the sites. The introduction of nickel was accomplished by oxidative addition of these peripherial aryl bromide groups to a zero-valent nickel source (e.g. $Ni(PPh_3)_4$). The resultant dendritic arylnickel species were tested as homogeneous catalysts in the atom transfer radical addition reaction (ATRA or Kharasch addition reaction) of CCl_4 to methyl methacrylate (= MMA), see Scheme 1. The catalytic data suggested that each nickel site in these metallodendrimers acts as an independent unit which is well accessible for incoming substrates with the activity per Ni-site being slightly lower in the case of the two dendritic species $G0–Ni_4$ and $G1–Ni_{12}$ (**1**, Scheme 1) as compared with the mononuclear nickel(II) catalyst. However, more importantly, in the case of these metallodendrimers all characteristics found for the monomeric nickel homogeneous catalyst are retained: i.e. it involved a clean, regioselective 1:1 addition without telomerization/polymerization of the alkene or the formation of other side products.

A simplified synthetic protocol developed by Kleij *et al.* [3] led to the preparation of different generations of carbosilane dendrimers which are directly functionalized at their periphery with organometal d^8 complexes. Metal introduction in the dendritic ligand precursors was accomplished through a sequence that involved direct lithiation followed by transmetallation of the lithiated dendrimer using an appropriate metal d^8 salt. This procedure was used to prepare a series of nickelated carbosilane dendrimers which were likewise tested as catalysts in the Kharasch addition reaction [4]. The latter work revealed an interesting dependency, a 'dendritic effect', of the catalytic activity on the generation number of the dendrimer catalyst. Whereas $G0–Ni_4$ has a comparable activity per metal atom as compared to a mononuclear model, the $G1–Ni_{12}$ (**2**, Scheme 2) and $G2–Ni_{36}$ dendrimers showed a tremendous decrease in (initial) catalytic activity. Of

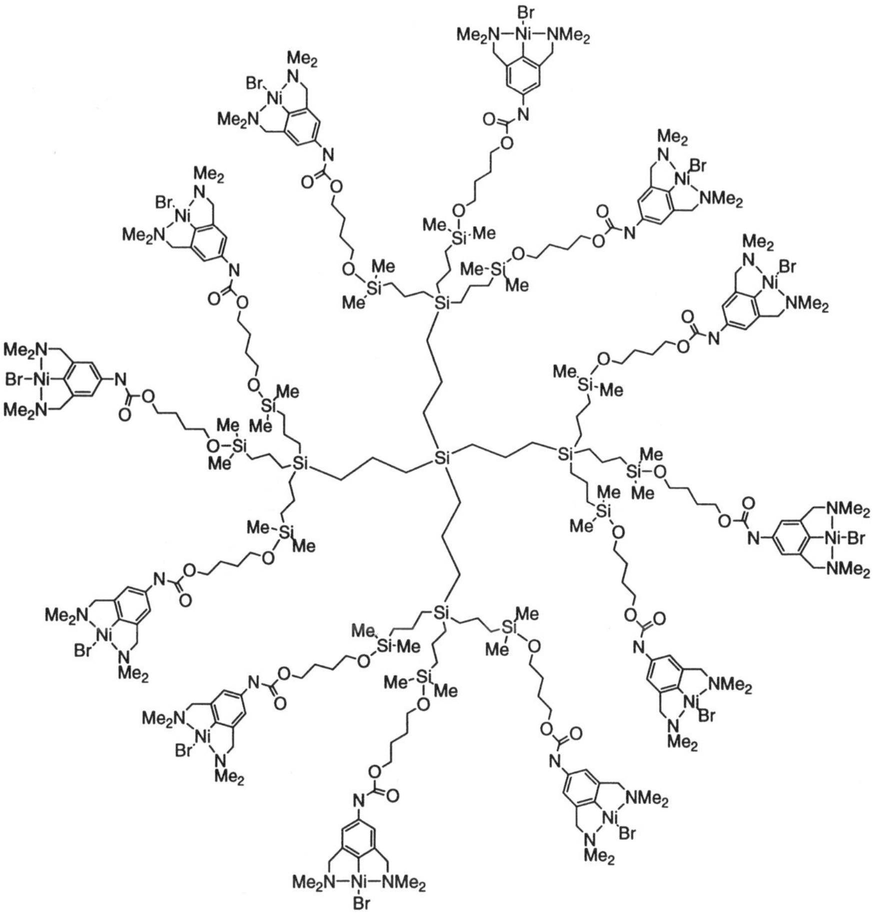

1

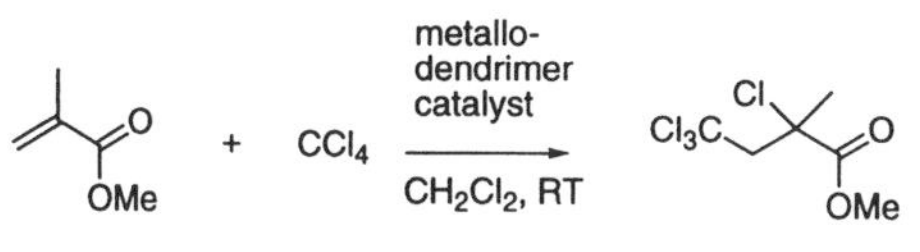

Scheme 1

particular interest in the catalytic runs with $G1-Ni_{12}$ and $G2-Ni_{36}$ is the total deactivation of the catalytic systems with maximum conversions of MMA being 18 and 1.5%, respectively. The authors showed that these results are most likely due to a proximity effect between the immobilized Ni^{II} sites [4] which is particularly apparent in this redox catalytic process that involves a Ni^{II}/Ni^{III} redox couple. Modification of the carbosilane support was carried out to test this hypothesis and different [G1] dendrimers (**3** and **4**, Scheme 2) with a less congested dendrimer periphery were prepared. Indeed, these species were successfully applied in catalysis and almost quantitative yield of the 1:1 adduct was achieved.

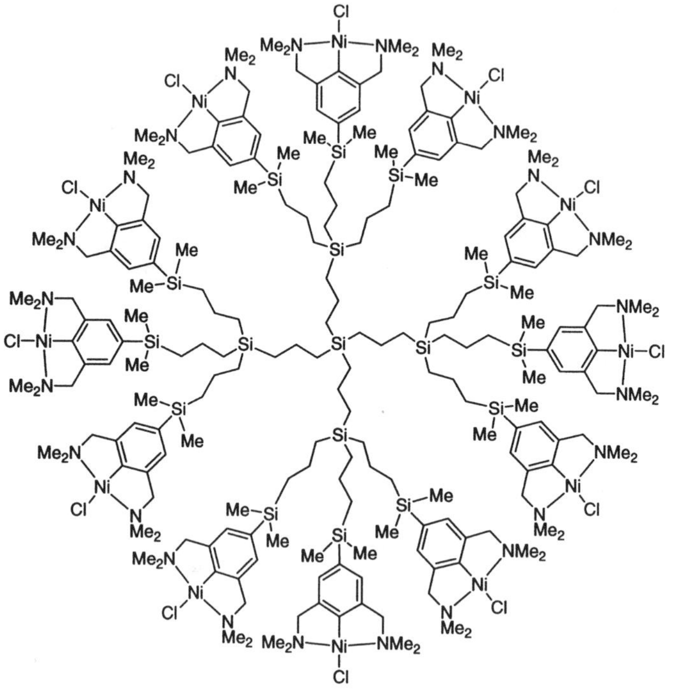

2

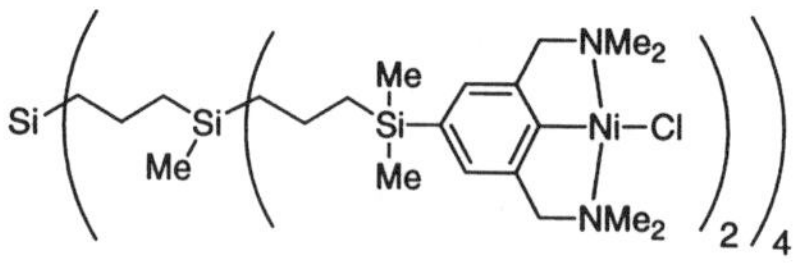

3

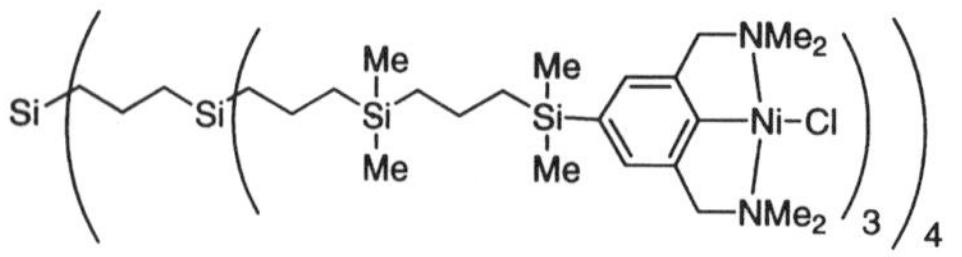

4

Scheme 2

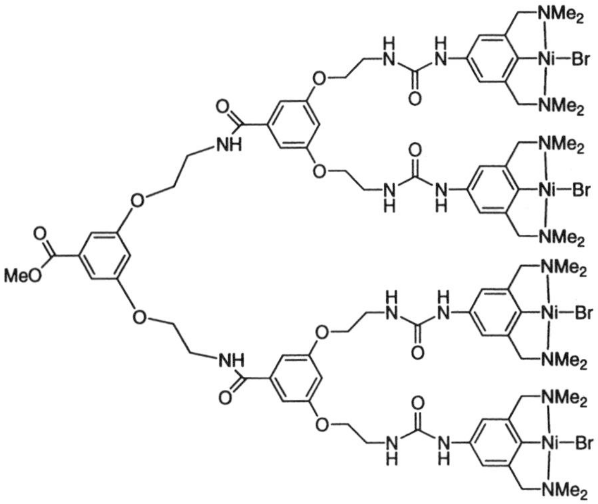

5

Scheme 3

Another approach directed toward the connection of organometallic complexes onto dendritic frameworks is the use of amino acid based materials (e.g. **5**, see Scheme 3). Gossage et al. [5] described the preparation of a series of highly polar dendritic supports with arylnickel(II) functionalities. The catalytic performance of these metallodendrons in the Kharasch addition of CCl_4 to MMA was not influenced by the presence of the polar functional groups in the dendrons. The catalytic activities of these metallodendrimers were in the same order of magnitude as reported earlier for the parent model compound $[NiBr(C_6H_3\{CH_2NMe_2\}_2\text{-}2,6)]$.

Moore and Suslick [6, 7] conducted shape selective catalysis with metalloporphyrins appended with sterically crowded polyester dendrons. They investigated the effect of the metallodendrimer size (see Scheme 4) on the substrate selectivity in two types of catalytic epoxidation reactions which were carried out with iodosylbenzene as the oxygen donor and different alkenes. The performance of the metalloporphyrin dendritic catalysts were compared with those of a parent manganese–porphyrin complex. A clear enhancement of selectivity of the metallodendrimers towards external (less hindered) double bonds vs internal ones was noted with increasing metallodendrimer size. Moreover, a higher affinity for more electron-rich olefins in the case of the dendrimer catalysts was observed. Furthermore, an increased stability of the metalloporphyrin core unit toward oxidation was noted with increasing size of the metallodendrimer species leading to higher total turnover numbers per metal atom.

Bis(oxazoline)copper(II) complexes elaborated with polyether dendrons (e.g.,

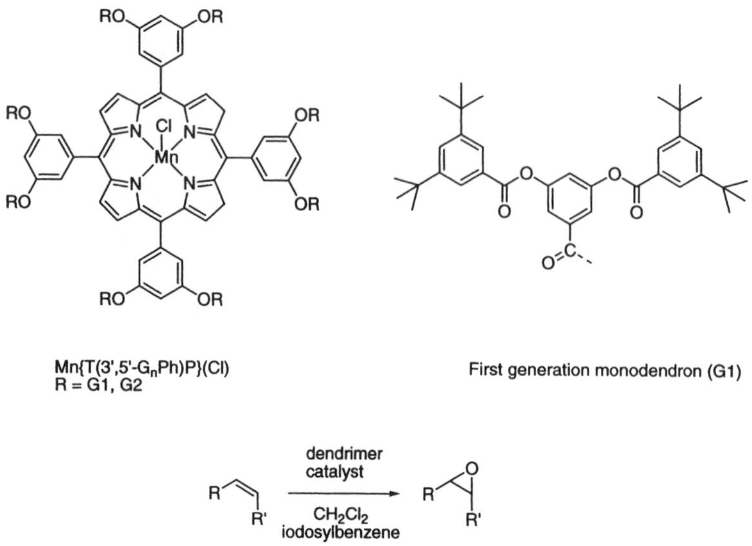

Mn{T(3',5'-G_nPh)P}(Cl)
R = G1, G2

First generation monodendron (G1)

Scheme 4

6, see Scheme 5) can be used as catalysts in Diels–Alder reactions between cyclopentadiene and different dienophiles [8, 9]. In order to investigate the mechanism of this reaction, the Diels-Alder reaction between cyclopentadiene and a crotonyl imide was studied (see Scheme 5). The initial reaction rate was found to be proportional with the catalyst concentration. From the catalytic data the binding constants of the dendrimer catalysts with the crotonyl imide as well as the reaction rates could be deduced. Two major conclusions were drawn from this work. Firstly, the binding constant between the dendrimer catalysts and the crotonyl imide decreased with increasing generation number of the catalyst. Secondly, the G0–G2 generation dendrimer catalysts all accelerated this cycloaddition reaction with similar rates, whereas a significant drop in catalytic activity was encountered for the G3 generation copper catalyst. These effects were rationalized by assuming that the increase of the steric congestion around the catalytic cavity when the dendron sections R increase in size, lead to back-folding of wedges because of steric repulsion resulting in steric blocking of the metal site. Similar to the work of Moore [6, 7], an increase in substrate selectivity for complexing of the relatively smaller crotonyl imide substrates by the copper(II) center in the third generation dendrimer was found.

The research group of Van Leeuwen has focused on catalysis at the core of a carbosilane dendrimer in an effort to be able to control stereoselectivity [10]. To this end, a ferrocenyl diphosphine backbone was functionalized with different generations of carbosilane dendrons producing a series of dendrimer phosphine ligands with an increasing steric demand (see **7** for an example, Scheme 6). *In situ*

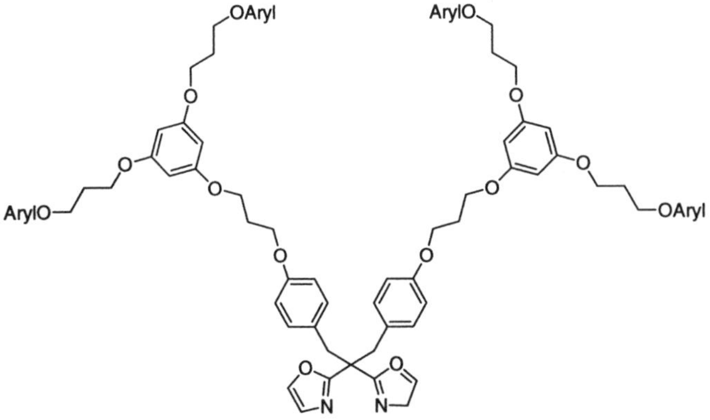

6

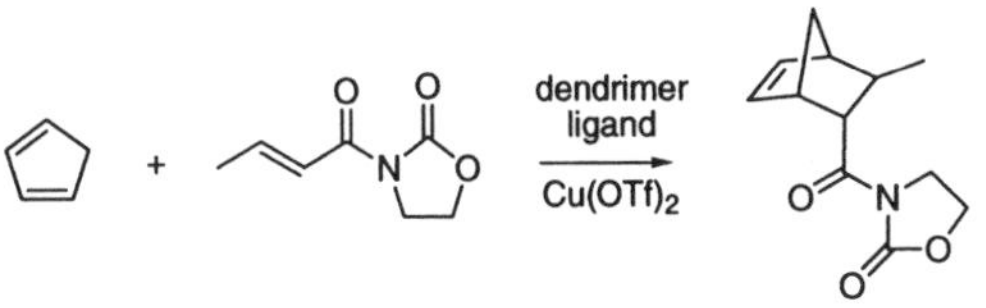

Scheme 5

complexation of these dendrimer phosphines with a suitable Pd^{II}-precursor followed by an incubation period gave the $PdCl_2$-complexed dendrimers. The Pd-dendrimers were used as catalysts in the allylic alkylation of 3-phenylallyl acetate with diethyl 2-sodio-2-methylmalonate (see equation, Scheme 6). Two major observations were made. Firstly, as may be expected, the catalytic activity dropped significantly with increasing generation number of the dendrons in the phosphine ligand. Secondly, the selectivity for the *trans* product slightly decreased from 90% in the case of a parent model compound (i.e. diphenylphosphino ferrocene, dppf) to 79% for the largest dendrimer catalyst system employed (i.e. with G3 dendrons attached to the central diphenylphosphine ferrocenyl ligand).

3 DENDRIMER CATALYSTS DERIVED FROM REACTIVE METAL ENCAPSULATION

Recently, two new approaches for the preparation of metallodendritic catalysts were described. The first one involves the use of PAMAM dendrimers [11] as both template and stabilizer of metal ions. The cavities of these dendrimers can serve as host type 'nano-reactors' for metal ion guests. This strategy is referred to as 'reactive encapsulation'. The concept was first elegantly demonstrated by

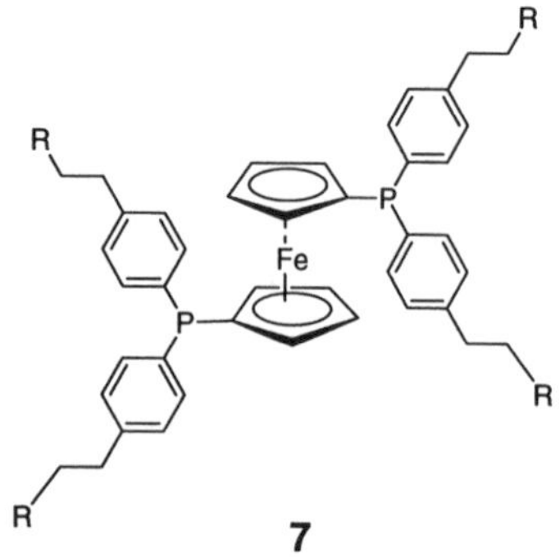

7

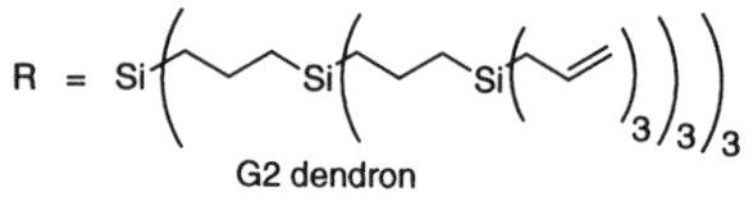

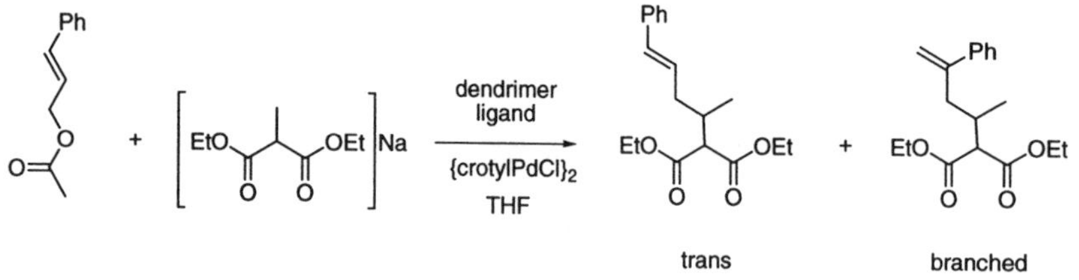

G2 dendron

trans   branched

Scheme 6

Tomalia and Balogh [12]. Crooks *et al.* [13] loaded these so-called dendrimer 'nanotemplates' with suitable platinum(II) salts (Scheme 7) and by means of *in situ* chemical reduction of this composite, dendrimer-encapsulated zerovalent metal clusters were formed. In this way, the dendrimer stabilizes the zerovalent metal clusters by preventing on the one hand agglomeration of these clusters to larger metal species while on the other hand leaving enough space for incoming substrate molecules to react at the metal cluster surface. It must be noted that the sequence of synthetic steps that the authors developed gives access to stable and soluble $[Pt^0]_n$-dendrimer species containing spherical metallic platinum clusters with sizes of approximately 1.6 nm. The size of these metal clusters most probably depends on the nature of the dendrimer cavities. The catalytic properties of these platinum modified dendrimers in the electrochemical reduction of O_2 were studied by cyclic voltammetry and revealed that in these cases a much larger current is observed in combination with a significant shift of the peak potential to higher values. Preliminary studies have indicated that this general approach can be employed to prepare bimetallic metal clusters as well.

The same authors have extended this work also to a preliminary study of the

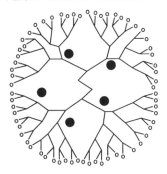

● = Stabilized platinum nanoparticles

Scheme 7

synthesis of dendritic homogeneous catalysts based on hydroxyl-terminated PAMAM dendrimers with encapsulated Pd^0 and Pt^0 nanoparticles (see also Scheme 7) [14]. X-ray photoelectron microscopy (XPS) provided supporting information about the composition and oxidation state of the encapsulated d^8 metal particles. These nanoparticle catalysts were tested in the hydrogenation of both N-isopropyl acrylamide and allyl alcohol in aqueous solution. The kinetic data obtained for the Pd/Pt^0 dendrimer catalysts compare favorable with those measured for water-soluble, polymer-bound Rh^I catalysts. It was found that the fourth generation dendrimer, $G4–OH(Pd_{40})$, is a very highly active catalyst. Its platinum analogue, (i.e. $G4\text{-}OH(Pt_{40})$), was found to have only a moderate hydrogenation activity. In addition, it was demonstrated that with the size of the dendrimer the hydrogenation activity can be controlled. Whereas the fourth-generation dendrimer catalyst is still very active for the hydrogenation of N-isopropyl acrylamide, the sixth and eighth generation dendrimers exhibited a reduced catalytic hydrogenation activity: only 10 and 5%, respectively, of that of $G4–OH(Pd_{40})$. It was concluded that the size of the dendrimers can act as a selective nanoscopic filter for substrate molecules.

Another method of producing dendrimer catalysts was illustrated by the group of Alper [15], who made use of PAMAM dendrimers which were grafted onto silica surfaces. These new heterogeneous, polyaminoamido diphosphonated dendritic materials (when complexed to Rh^I and connected to a silica gel support) are good hydroformylation catalysts (e.g. **8**, see Scheme 8). The Rh^I content was determined by ICP analysis, while the dendritic substructure was studied by solid state ^{31}P NMR spectroscopy. A wide range of aryl olefins and vinyl esters could selectively be converted into branched chain aldehydes. The activity of these catalysts as measured by turnovers per hour increased at higher temperatures but decreased when the reaction time was prolonged at these temperatures.

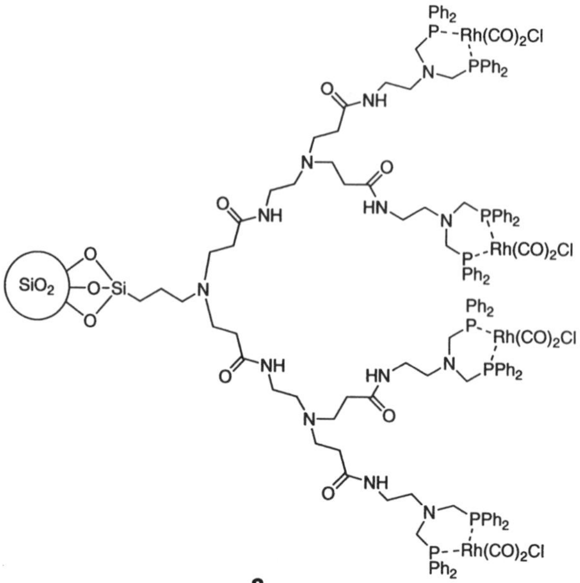

8

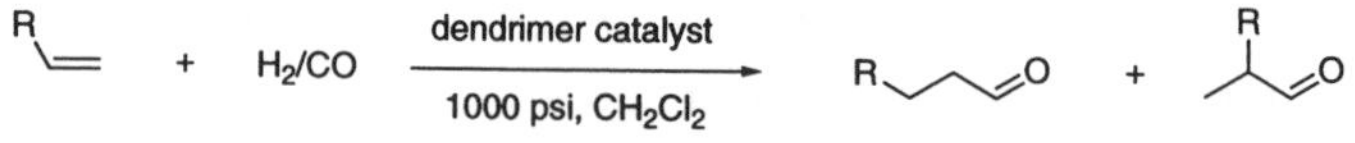

Scheme 8

4 CATALYSIS WITH PHOSPHINE-BASED DENDRIMERS

One of the first results on the use of phosphine dendrimers in catalysis was reported by Dubois and co-workers [16]. They prepared dendritic architectures containing phosphorus branching points which can also serve as binding sites for metal salts. These terdentate phosphine-based dendrimers were used to incorporate cationic Pd centers in the presence of PPh_3. Such cationic metallodendritic compounds were successfully applied as catalysts for the electrochemical reduction of CO_2 to CO (e.g. **9**, Scheme 9) with reaction rates and selectivities comparable to those found for analogous monomeric palladium–phosphine model complexes suggesting that this catalysis did not involve cooperative effects of the different metal sites.

Another elegant example of phosphine-containing dendrimers is represented by dendrimer **10** in Scheme 10 [17]. These structures are build up via nitrogen branching points while the final nitrogen branching point is used to make a

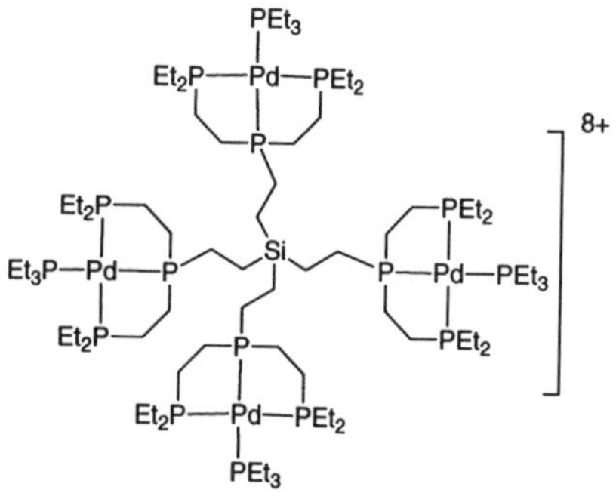

9

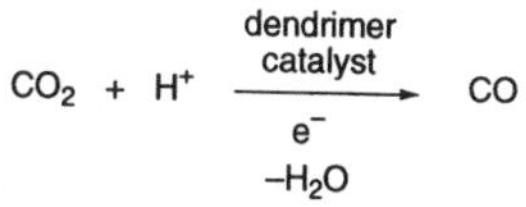

$$CO_2 + H^+ \xrightarrow[\substack{e^- \\ -H_2O}]{\substack{\text{dendrimer} \\ \text{catalyst}}} CO$$

Scheme 9

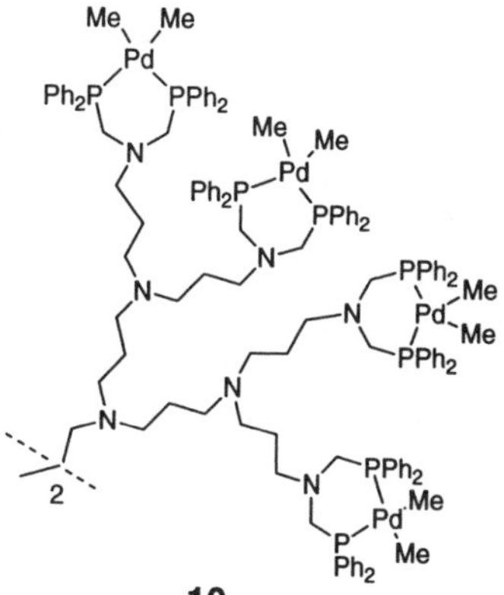

10

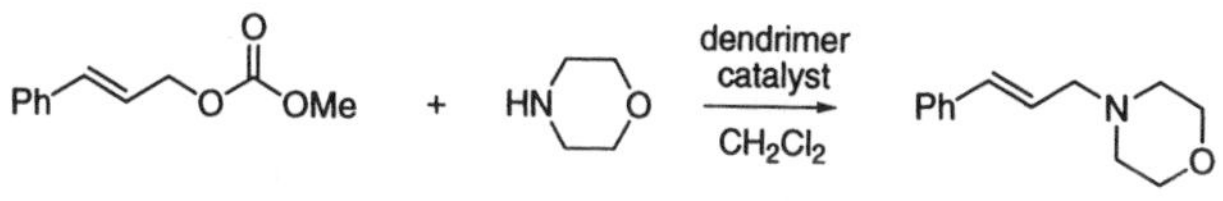

Scheme 10

potentially terdentate *P,N,P*-ligand. The latter ligands were reacted with an appropriate 'PdMe$_2$' precursor reagent to afford neutral PdMe$_2$-containing dendrimers. These metallodendrimers were used as precursors and showed good activity as precatalysts in the allylic substitution reaction of (3-phenyl-2-propenyl)acetate methyl ester with morpholine (see equation in Scheme 10). An interesting effect of the binding of these Pd complexes to a dendrimer periphery is the increased kinetic stability of the palladium sites during catalysis. Under the applied reaction conditions, unlike their nondendritic analogues, these palladodendritic catalysts *do not* degrade to metallic palladium. This is an interesting example which clearly demonstrates a positive effect of the use of a dendritic support on catalyst stability in homogeneous catalysis. Moreover, recent work [18] has demonstrated that the application of metallodendrimer catalysts in an ultrafiltration membrane reactor is feasible which provides a means to separate the homogeneous catalyst from the product after the reaction or even continuously (see last section of this chapter).

Recently Togni *et al.* [19] focussed on the preparation of asymmetric dendrimer catalysts derived from ferrocenyl diphosphine ligands anchored to dendritic backbones constructed from benzene-1,3,5-tricarboxylic acid trichloride and adamantane-1,3,5,7-tetracarboxylic acid tetrachloride (e.g. **11**, Scheme 11). *In situ* catalyst preparation by treatment of the dendritic ligands with [Rh(COD)$_2$]BF$_4$ afforded the cationic Rh-dendrimer, which was then used as a homogeneous catalyst in the hydrogenation reaction of, for example, dimethyl itaconate in MeOH. In all cases the measured enantioselectivity (98.0–98.7%) was nearly the same as observed for the ferrocenyl diphosphine (Josiphos) model compound (see Scheme 11).

Finally, several examples of metal catalysts derived from dendritic ligands containing potential chiral ligating sites (see approach (3), next section) were reported. Kakkar *et al.* [20] succeeded in the synthesis of various generations of metallophosphino dendrimers by acid–base hydrolysis of aminosilanes to produce molecules with terminal OH groups. Rhodium centers could be efficiently introduced throughout each generation of the dendrimer by means of Rh–N coordination. These metallodendrimers are active as hydrogenation catalysts. Decene was hydrogenated with activities similar to that of the monomeric rhodium-adduct (e.g. **12**, Scheme 12). Furthermore, the authors demonstrated that recycling of the catalytic species (by a simple work-up procedure) in a batch-reactor approach did not appreciably lower the catalytic activity of the rhodium–phosphine dendrimer catalyst.

Beside these catalytically active metallophosphine dendrimers (see above), preliminary studies on the chemical properties of phoshorus-based dendrimers complexed to metals such as platinum, palladium and rhodium have been described by Majoral, Caminade and Chaudret [21]. They showed that these macromolecules (see Scheme 13) could be useful for the (*in situ*) generation of metallodendrimer catalysts.

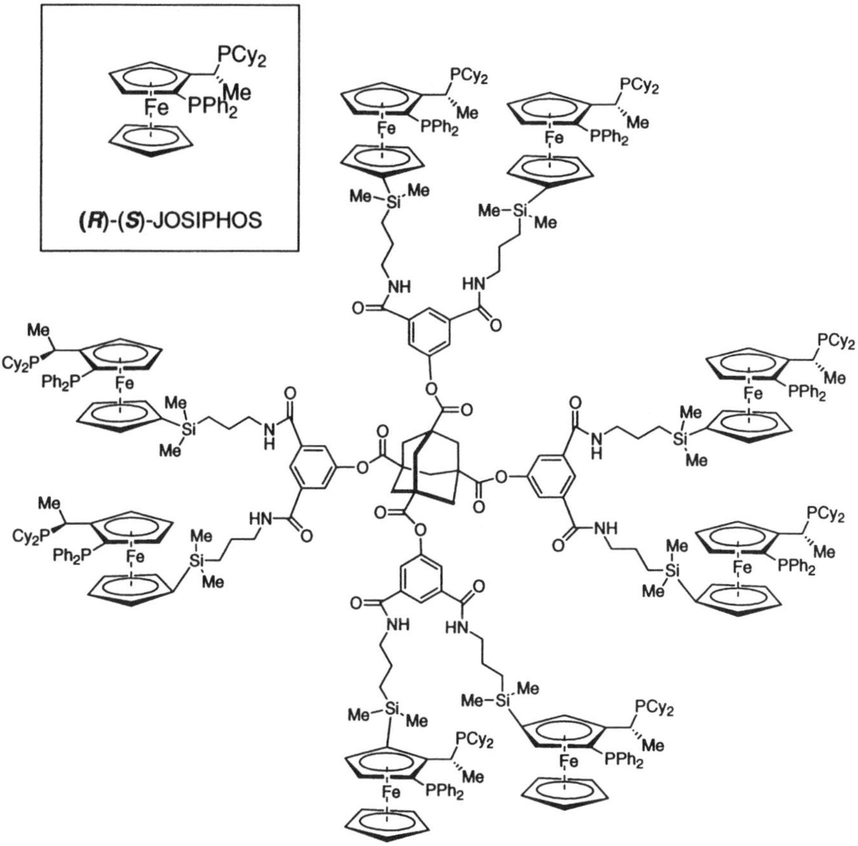

11

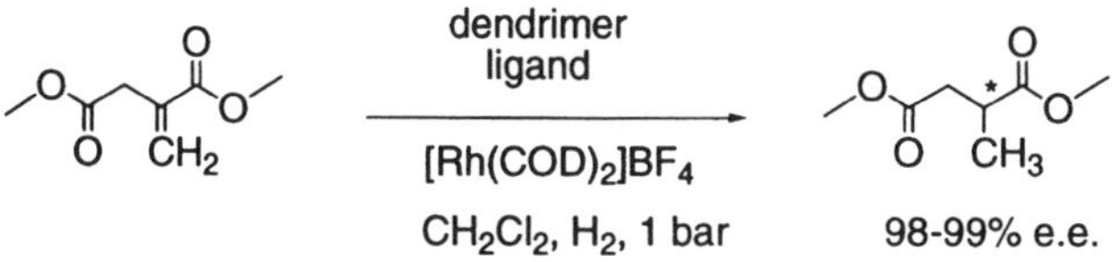

Scheme 11

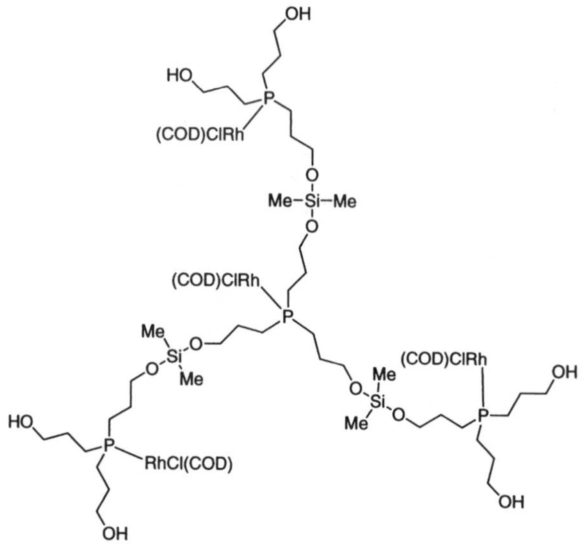
OH
HO
P
(COD)ClRh
O
Me–Si–Me
O
(COD)ClRh
P
Me
O
Si
O
Me
Me
(COD)ClRh
OH
O
Si
O
Me
P
HO
P
RhCl(COD)
OH
OH

12

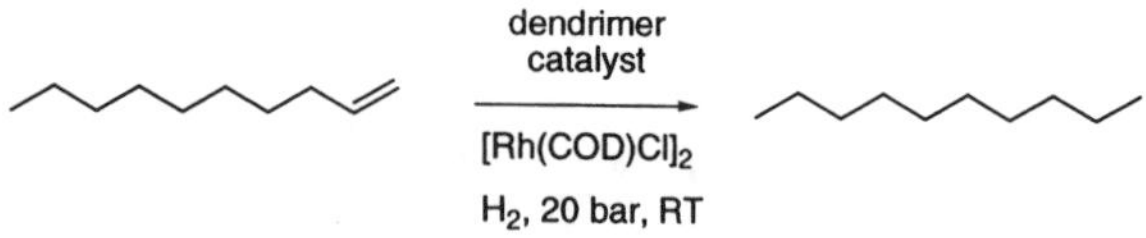
dendrimer
catalyst
[Rh(COD)Cl]2
H2, 20 bar, RT

Scheme 12

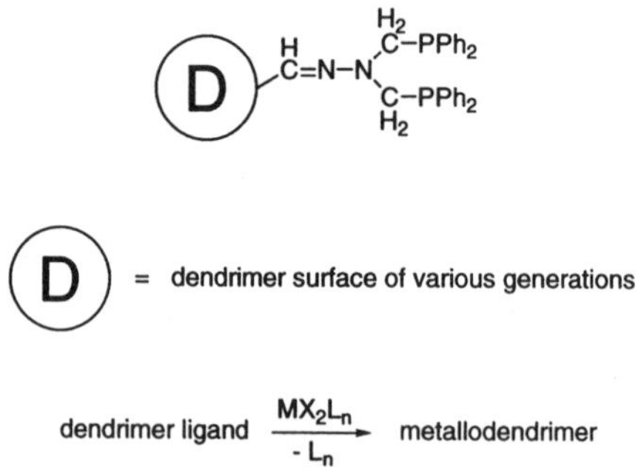
D
H
C=N–N
H2
C–PPh2
C–PPh2
H2
D
= dendrimer surface of various generations
dendrimer ligand
MX2Ln
- Ln
metallodendrimer

Scheme 13

5 CATALYSIS WITH (METALLO)DENDRIMERS CONTAINING CHIRAL LIGANDS

A number of groups have reported the preparation and *in situ* application of several types of dendrimers with chiral auxiliaries at their periphery in asymmetric catalysis. These chiral dendrimer ligands can be subdivided into three different classes based on the specific position of the chiral auxiliary in the dendrimer structure. The chiral positions may be located at, (1) the periphery, (2) the dendritic core (in the case of a dendron), or (3) throughout the structure. An example of the first class was reported by Meijer *et al.* [22] who prepared different generations of poly(propylene imine) dendrimers which were substituted at the periphery of the dendrimer with chiral aminoalcohols. These surface functionalities act as chiral ligand sites from which chiral alkylzinc aminoalcoholate catalysts can be generated *in situ* at the dendrimer periphery. These dendrimer systems were tested as catalyst precursors in the catalytic 1,2-addition of diethylzinc to benzaldehyde (see *e.g.* **13**, Scheme 14).

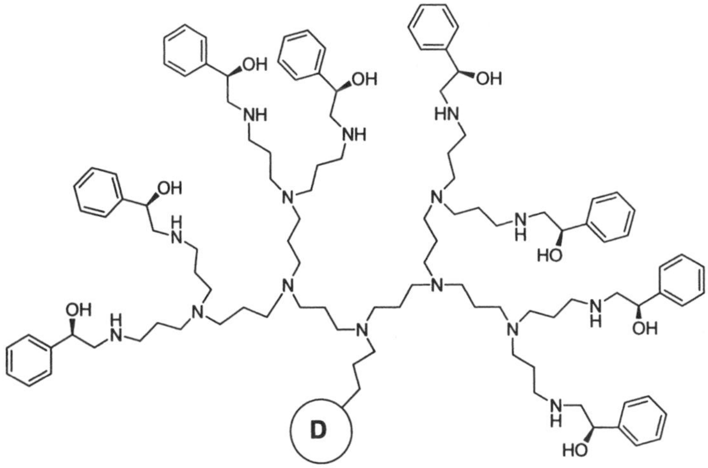

13

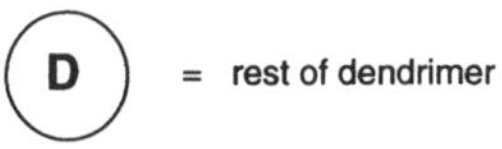

= rest of dendrimer

Part of 3rd generation poly(propylene imine) dendrimer

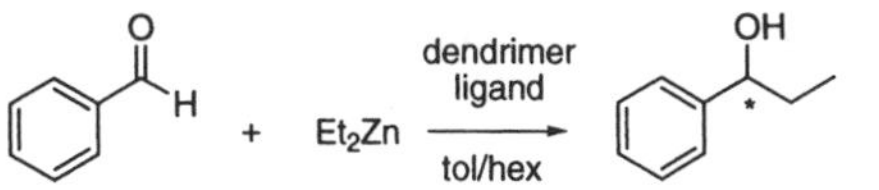

G_n	e.e (%)
1	5.5
2	4.5
3	2.5
4	0.8
5	0.5

Scheme 14

The fifth generation dendrimer showed virtually no enantioselectivity in this particular reaction and no measurable optical rotation for these chiral dendrimers were estimated. The decrease in product yield as well as selectivity was explained in terms of multiple interactions between the terminal groupings on the periphery resulting from increasing steric congestion. Similar results were reported by Soai *et al.* who used PAMAM dendrimers as a scaffolding for chiral ephedrine peripheral groups [23]. The latter dendrimer ligands exhibited only a moderate effect on the product e.e.s in the chiral addition of diethylzinc to *N*-diphenylphosphinylimines (see *e.g.* **14**, Scheme 15) whereas a high chiral induction is found for the 'model' species. Furthermore, it must be noted that the

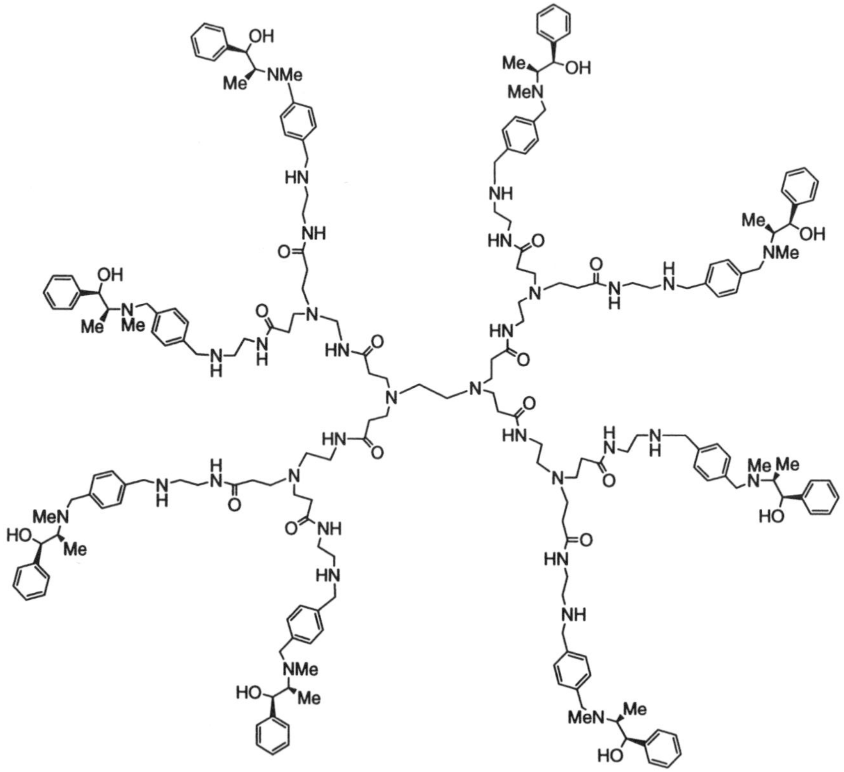

14

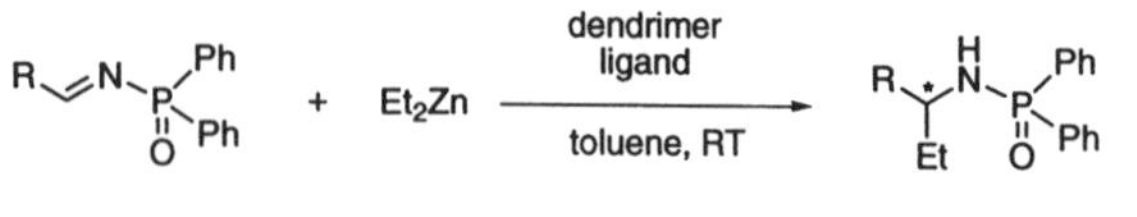

Gn	Yield(%)	e.e (%)
model	46	92
0	18	40
1	8	30

Scheme 15

chiral dendrimer ligands were used in relatively large catalytic amounts ranging from 17 to 50 mol%.

Another approach was followed by Bolm et al. [24], who prepared dendron ligands consisting of a chiral pyridyl alcohol connected to the core of Fréchet type dendrons [25]. The chiral dendron ligands were used for the in situ generation of ethylzinc dendron ligand complexes which catalyze the addition of diethylzinc to benzaldehyde (see e.g., **15**, Scheme 16). The size of the dendron appeared to have practically no influence on the enantioselectivity of this reaction.

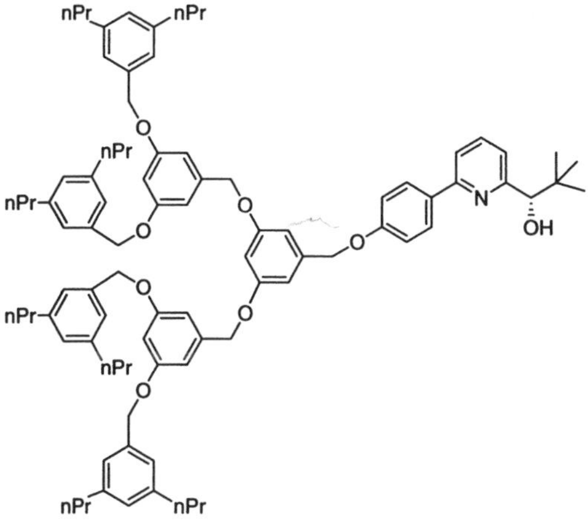

15

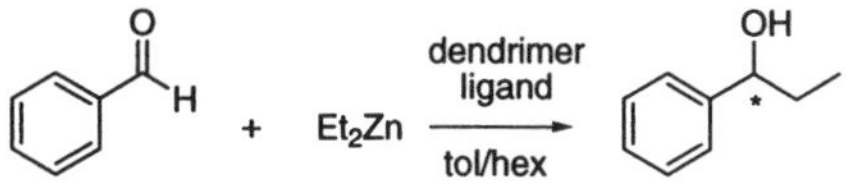

G_n	Yield(%)	e.e (%)
0	49	86
1	86	85
2	84	86

Scheme 16

Brunner et al. [26] synthesized and applied so-called 'dendrizymes' in enantioselective catalysis. These catalysts are based on dendrimers which have a functionalized periphery that carries chiral subunits, (e.g. dendrons functionalized with chiral menthol or borneol ligands). The core phosphine donor atoms can be complexed to (transition) metal salts. The resultant dendron-enlarged 1,2-diphosphino-ethane (e.g. **16**, see Scheme 17) Rh[I] complexes were used as catalysts in the hydrogenation of acetamidocinnamic acid to yield N-acetyl-phenylalanine (Scheme 17) [26]. A small retardation of the hydrogenation of the substrate was encountered, pointing to an effect of the meta-positioned dendron substituents. No significantly enantiomerically enriched products were isolated. However, a somewhat improved enantioselectivity (up to 10–11% e.e.) was observed in the cyclopropanation of styrene with ethyl diazoacetate using dendritic pyridine derivatives in combination with Cu(I)OTf (see e.g. **17**, Scheme 18) [27].

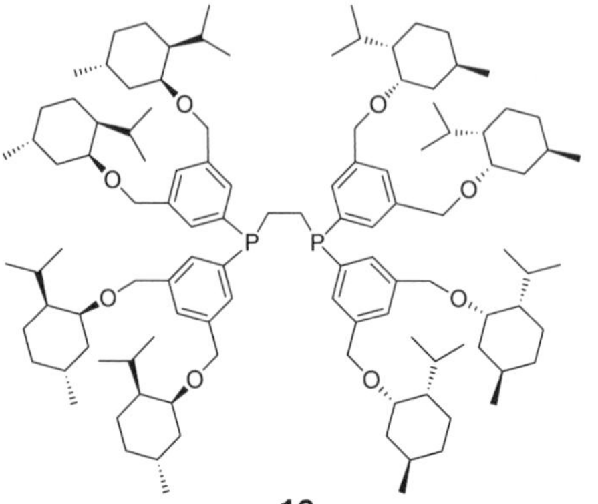

16

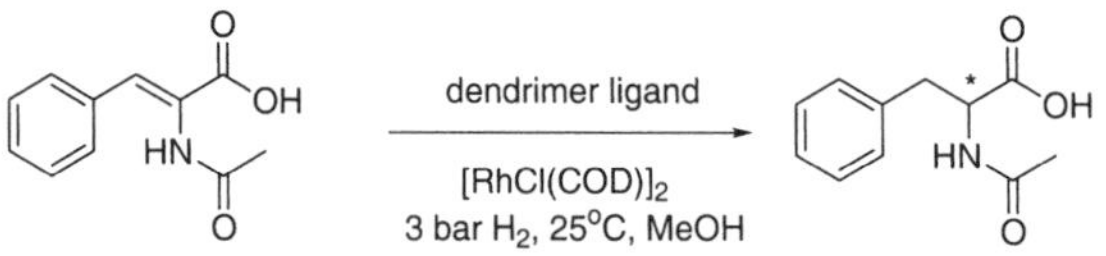

Scheme 17

Good enantioselectivities were obtained by Seebach *et al.* who applied α,α,α',α'-tetra-aryl-1,3-dioxolane-4,5-dimethanols (TADDOLS) as a dendritic ligand for titanium-based dendrimer catalysis [28]. By using two different types of chiral dendrimers (combining the periphery approach as well as the 'focal point' approach plus the additonal attachment of chiral dendrons), the dendrimeric Ti-TADDOLates were good catalysts for the 1,2-addition of diethylzinc to benzaldehyde yielding the secondary alcohols with e.e.s up to 98% (e.g. **18**, Scheme 19), almost identical to the results obtained with model TADDOL complexes of this catalysis type.

A *stochiometric* approach was applied by Van Koten and co-workers [29], who used chiral carbosilane dendrimers as soluble supports in the *in situ* ester enolate-imine condensation in the synthesis of β-lactams (e.g. **19**, Scheme 20). The formation of the β-lactam products proceeded with high *trans* selectivity, and with the same level of stereoinduction as was earlier established in reactions without the dendritic supports, (i.e. the use of the enantiopure dendritic support did not affect the enantioselectivity of the C–C bond formation). After the reaction, the dendrimer species could be separated from the product by precipitation or GPC techniques and reused again.

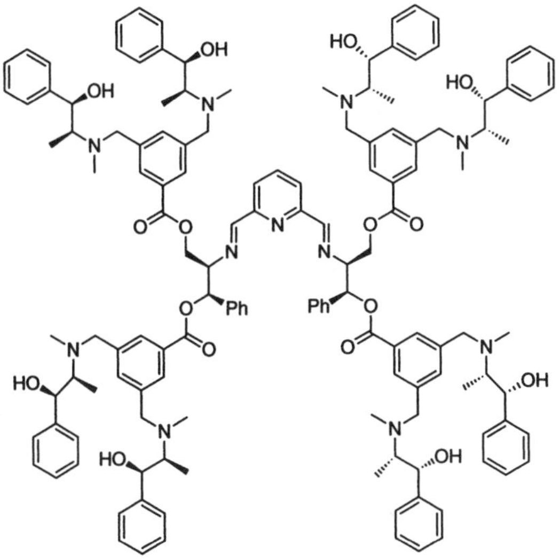

17

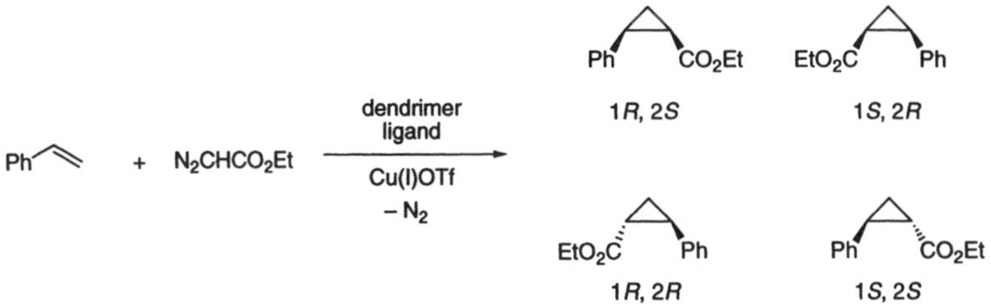

Scheme 18

6   NON-METAL CONTAINING DENDRIMERS

One of the earliest reports on the use of dendrimers in catalysis is the unimolecular decarboxylation of 6-nitro-benzisoxazole-3-carboxylate in the presence of a dendrimer comprising ether dendrons which are functionalized at their periphery with tetra-alkylammonium cations (e.g. **20**, Scheme 21) [30]. In aqueous media, the quaternary ammonium groupings promote the reactivity of organic anions which presumably bind in high concentration to the polycationic periphery of the dendrimer. The latter species enhances the rate of the bimolecular hydrolysis of p-nitrophenyl diphenyl phosphate catalyzed by o-iodosobenzoate ion.

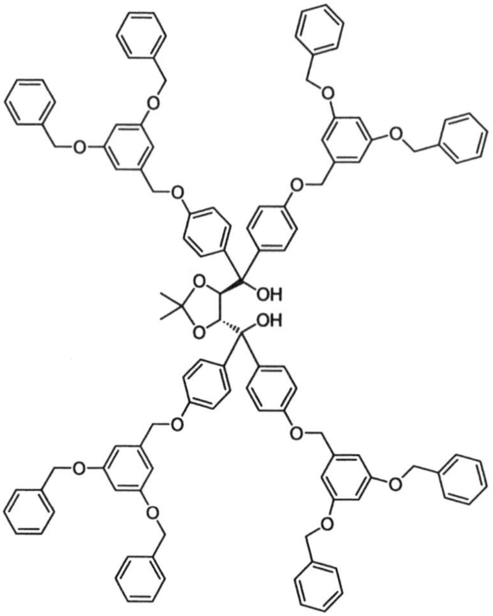

18

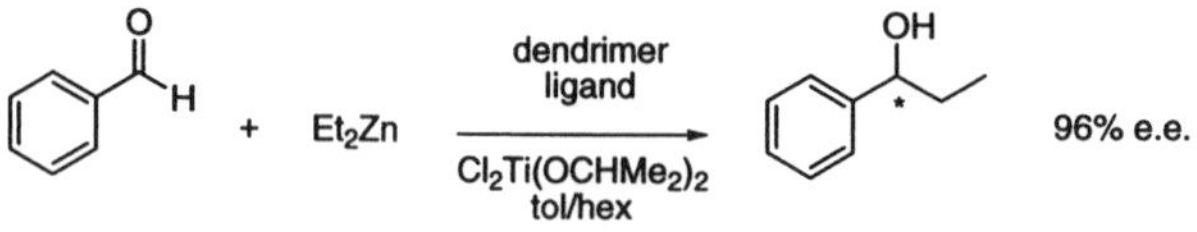

Scheme 19

Another, more recent example of the use of a non-metal-containing dendrimer is the nitroaldol (Henry) reaction between aromatic aldehydes and nitroalkanes catalyzed by a single triethylene amine core unit functionalized with branched Fréchet-type polyether segments (e.g. **21**, Scheme 22) [31]. Molecular studies indicated that the cavities around the core nitrogen center should be able to facilitate the base-catalyzed aldol reaction, although the higher generation dendrimers showed an appreciably lower activity, suggesting that the reactive site is relatively more 'shielded' by the attached polyether dendrons in the higher generation species.

High asymmetric induction by amphiphilic dendrimers was reported by Rico-Lattes and co-workers [32]. These water-soluble but THF-insoluble dendrimers (e.g. **22**, Scheme 23) consist of useful, readily available chiral auxiliaries and can be used in the homogeneous (when H_2O is the solvent) or heterogeneous (in the case of THF as the reaction medium) catalyzed reduction of prochiral aromatic

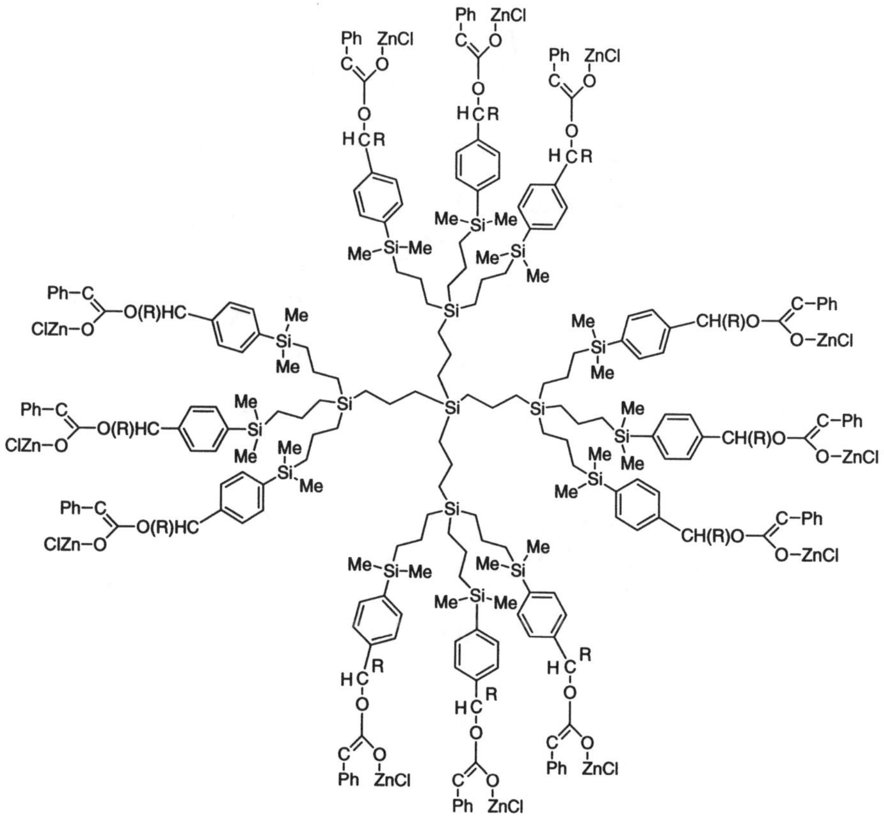

19

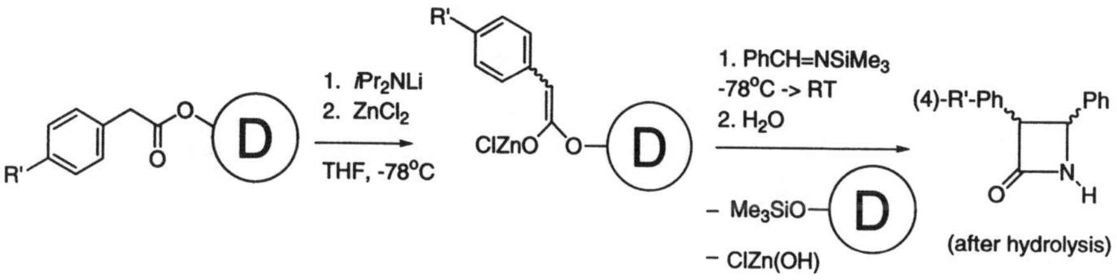
1. iPr2NLi
2. ZnCl2
THF, -78°C
ClZnO
D
1. PhCH=NSiMe3
-78°C -> RT
2. H2O
– Me3SiO
– ClZn(OH)
(4)-R'-Ph
Ph
(after hydrolysis)

Scheme 20

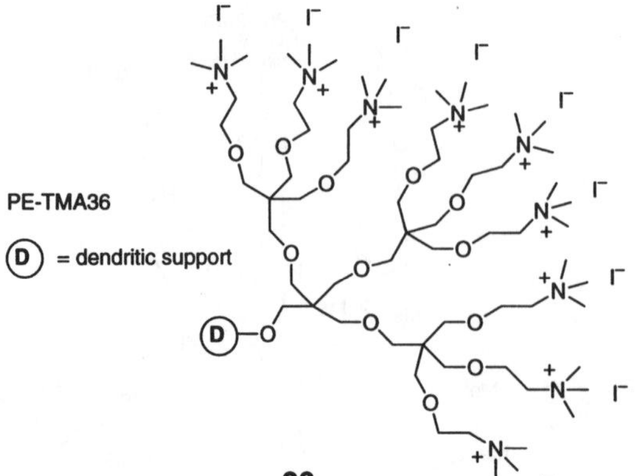
PE-TMA36
D = dendritic support

20

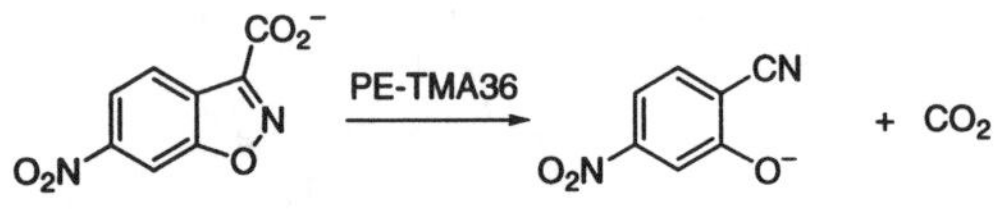
PE-TMA36
+ CO2

Scheme 21

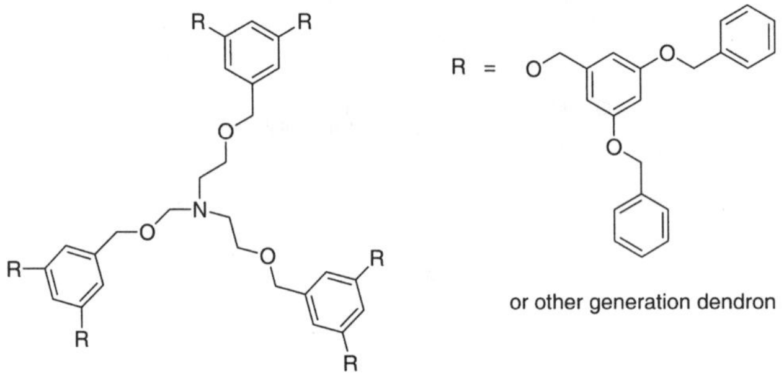
R =
or other generation dendron

21

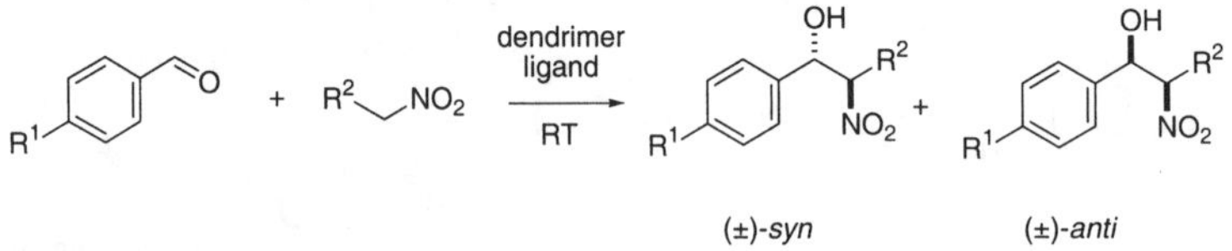
dendrimer
ligand
RT
(±)-*syn*
(±)-*anti*

Scheme 22

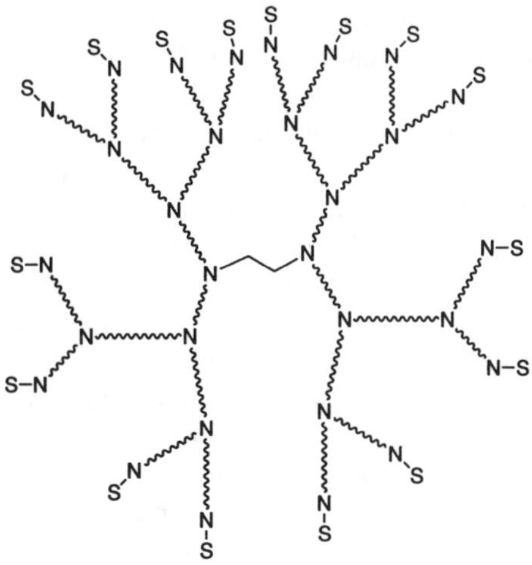

22

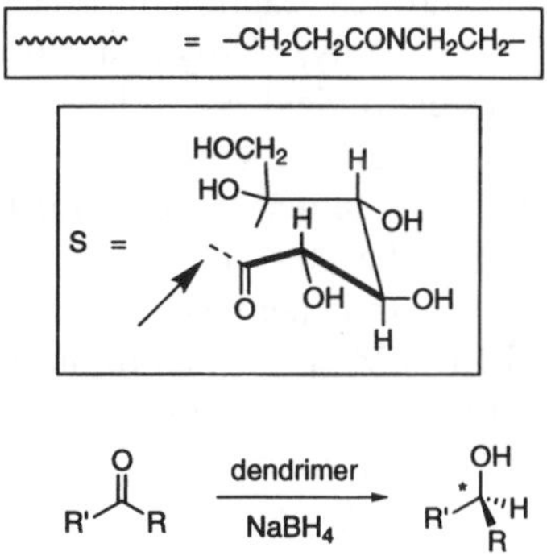

$\sim\sim\sim$ = $-CH_2CH_2CONCH_2CH_2-$

S =

Scheme 23

ketones by sodium borohydride to yield their corresponding alcohol derivatives in high optical purity. Interestingly, in the homogeneous system these organic reactions take place at the 'pseudo-micellar chiral interface' in an aqueous medium.

7 METALLODENDRITIC CATALYSTS AND MEMBRANE CATALYSIS: CATALYST RECOVERY

The globular shape of dendritic macromolecules with a persistent nanosize and radius should allow easy separation or retention by ultra- or nanofiltration membranes. This concept of separating the catalysts from the product/substrate stream [33] is of keen interest because of the ever expanding need for cleaner and more efficient industrial and chemical processes. Recently, a few groups have reported separately their findings in this promising area.

One of the first pioneering studies involving membrane catalysis was reported by Kragl and coworkers [34]. They used, rather than well-defined dendrimer catalysts, polymer enlarged oxazaborolidines for the (homogeneously catalyzed) enantioselective reduction of ketones by borane (see Scheme 24) in a continuous operating membrane reactor. Two types of polymeric catalysts were employed (i.e. **23a** and **23b**, see Scheme 24) and the results show that not only very high conversions are obtained in these reduction reactions, but also high e.e. of the chiral alcohol products (84–99% e.e.) can be achieved. This is in contrast with enantioselective reductions of ketones in batch type reactors, in which the ketone is added in one portion. By using such a procedure, the uncatalyzed reduction of the substrate can be of significance at low conversion and this leads to lowering of the e.e. of the final product. A continuous high conversion of the ketone in a membrane reactor with a low stationary adduct (i.e. BH_3) concentration therefore efficiently suppresses the uncatalyzed reaction in this case.

Another example is the palladium catalyzed allylic substitution of 3-phenyl-2-propenyl-carbonic acid methyl ester to yield *N*-(3-phenyl-2-propenyl)morpholine reported by Reetz, Kragl and co-workers. This reaction was performed in the presence of phosphino-terminated amine dendrimers [17, 18] loaded with Pd^{II} cations as shown in Scheme 10. For this particular dendrimer with a molecular weight of 10 212 g/mol, a retention of 0.999 per residence time [35] was estimated in a membrane reactor with a SELRO MPF-50 membrane. It must be noted that a very high retention is a prerequisite for a continuous operating system, since a small leaching of the dendrimer leads to an exponential decrease in the amount

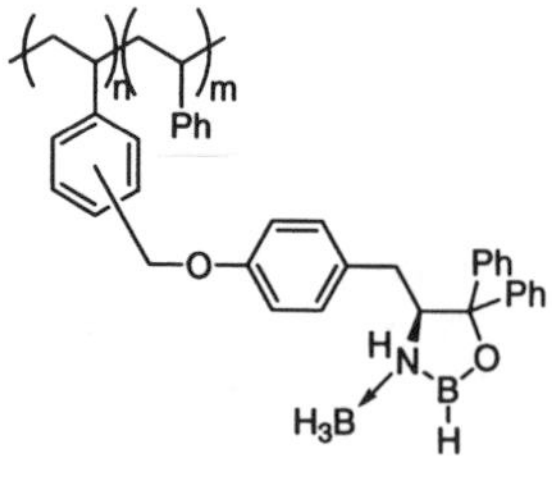

23a 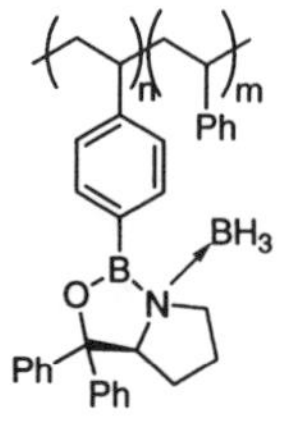   **23b**

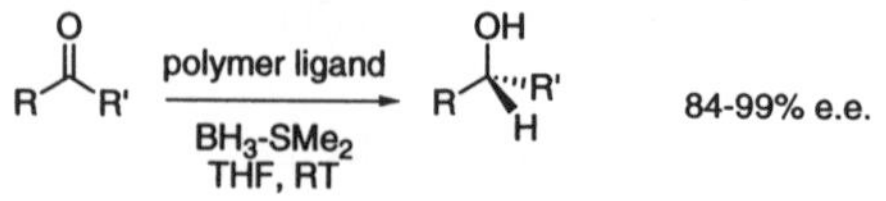

Scheme 24

of catalyst. Using this type of catalyst during a period equivalent to a 100 residence times, the conversion of substrate into product dropped from quantitatively to *ca.* 80% which is equivalent to a leaching of palladium of about 0.07 to 0.14% per residence time. During this period some deactivation of the Pd catalyst was also noted which most probably arises from reaction with the solvent (*i.e.* CH_2Cl_2) to give catalytically inactive ligated $PdCl_2$ fragments.

An elegant demonstration of the use of membrane technology for the effective recovery of metallodendritic catalysts and selective product formation was detailed by Van Koten, Vogt and coworkers [36]. In this work, carbosilane dendrimers were functionalized at the periphery with various ω-diphenylphosphino carboxylic acid ester end groups (i.e. **24**, Scheme 25), which can act as hemi-labile bidentate ligands to metal d^8 fragments. The metal-containing systems were prepared *in situ* by addition of $[(\eta^3\text{-}C_4H_7)Pd(cod)]BF_4$ and were subsequently tested in the Pd^{II}-catalyzed hydrovinylation of styrene. One of the major problems in this reaction could be solved by this approach. At higher conversions, isomerization of the product (i.e. 3-phenyl-1-butene) to internal olefins (both *E* and *Z* isomers) occurs (see Scheme 25). As a consequence, this reaction has to be run at low conversion with continuous removal of the 3-phenyl-1-butene or carried out at high styrene concentrations. A strategy was developed to selectively produce the desired 3-phenyl-1-butene at *low* conversions under membrane reactor conditions. Under these specific conditions using the G0–Pd_4 catalyst, a highly selective conversion of styrene is achieved with no significant isomerization or generation of other side products, although the yield per time unit of the desired product was very low. A modest retention in a MPF-60 NF membrane system (85%) of this small dendritic species was encountered, which is far from being ideal for continuous operations. Surprisingly, using the G0–Pd_4 catalyst, 3-phenyl-1-butene was produced during a period of 80 h with concomitant formation of palladium black inside the reactor. Unfortunately, in the case of the G1–Pd_{12} analog similar results were obtained and this appears to be related to the stability of this type of dendrimer catalysts under the conditions employed. The decomposition of the Pd-catalyst is connected to the intrinsic properties of palladium catalysis and has also been observed in experiments carried out by Reetz *et al.*

The research group of Van Leeuwen reported the use of carbosilane dendrimers appended with peripherial diphenylphosphino end groups (i.e. **25**, Scheme 26) [37]. After *in situ* complexation with allylpalladium chloride, the resultant metallodendrimer **25** was used as catalyst in the allylic alkylation of sodium diethyl malonate with allyl trifluoroacetate in a continuous flow reactor. Unlike in the batch reaction, in which a very high activity of the dendrimer catalyst and quantitative conversion of the substrate was observed, a rapid decrease in space time yield of the product was noted inside the membrane reactor. The authors concluded that this can most probably be ascribed to catalyst decomposition. The product flow (i.e. outside the membrane reactor) was also investigated and it was shown that no active catalyst had gone through the membrane.

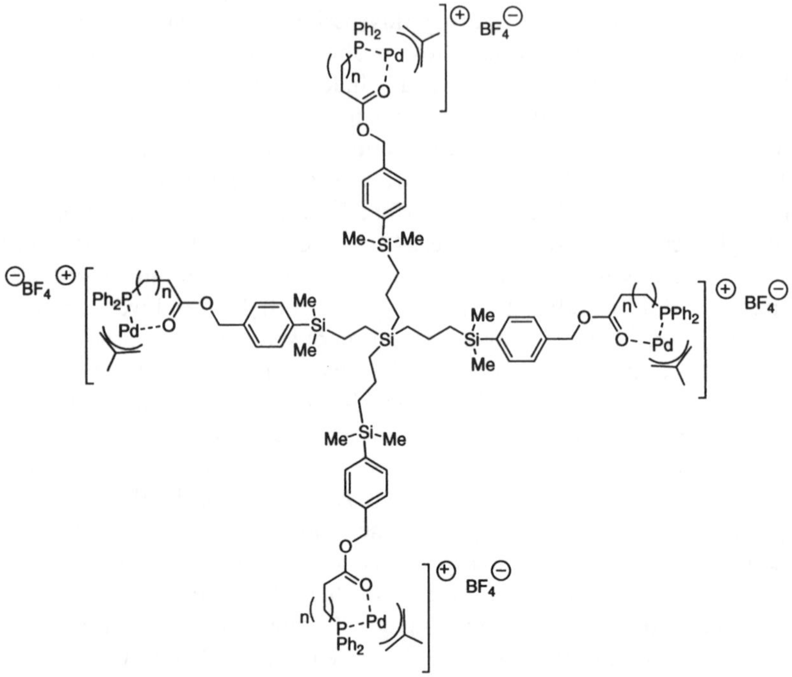

24

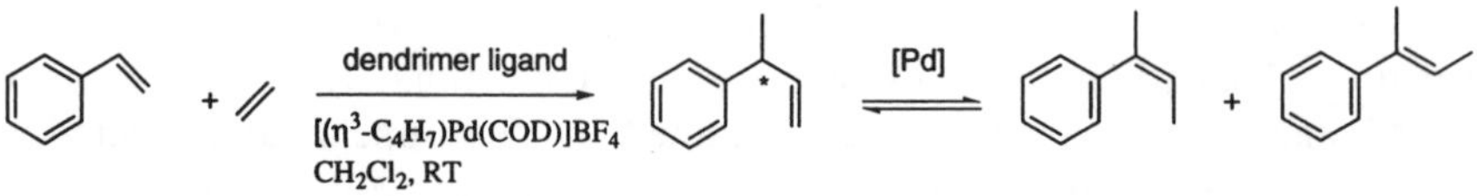

1 ***E*-2** ***Z*-2**

	Conv. (%)	Yield of **1** (%)	select. for **1-2** (%)	select. for **1** (%)
model compound	99.9	4.4	93.4	0.2
[G0]-Pd_4	8.1	7.6	96.3	98.3

Scheme 25

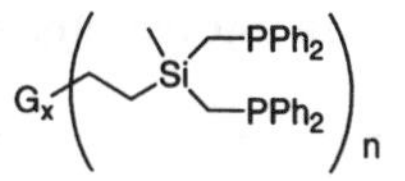

25

For **25**: x = 1 and n = 12

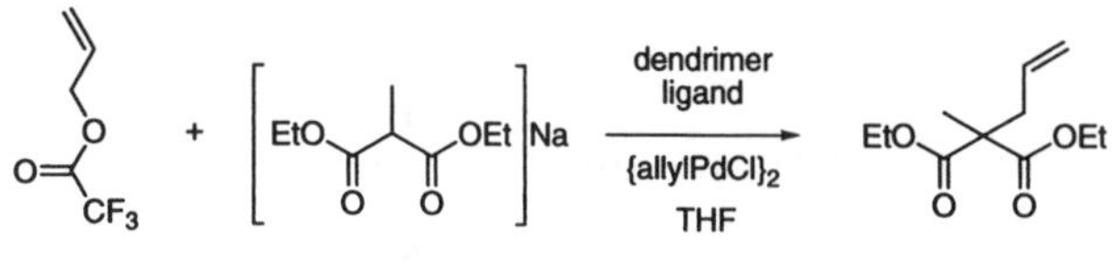

Scheme 26

8 SUMMARY

This brief survey clearly shows that the field of 'dendrimer catalysis' represents both great promise, as well as certain chemical challenges. Obviously, this field of chemistry still needs appropriate time to become optimized before working systems are achieved. In Table 20.1, some relevant characteristics of the dendrimer species in homogeneous dendrimer catalysis (i.e. preparation, amount of catalyst, stability) are presented. From Table 20.1, several conclusions may be drawn. Presently, most of the synthesized and applied dendrimer catalysts in fact are *metallo*dendrimeric species, while only a few examples of nonmetal-containing dendrimer catalysts are known [30–32]. Furthermore, from Table 20.1 it can be deduced that the range of metals employed for the construction of metallodendrimer catalysts is still limited. Furthermore, the majority of these metallodendrimer catalysts are prepared *in situ* and represent an intrinsically less defined category of metallodendrimer catalysts. An interesting difference between the metallodendrimer catalysts is the nature of the dendrimer ligand-to-metal binding: up to now, the greater part of the metallodendrimers have the metal bonded via coordination of the dendrimer donor atoms in a monodentate or bidentate fashion. Obviously, this metal–dendrimer ligand bonding mode is subject to dissociation/association processes and therefore affects the long-term stability of the metallodendrimeric species. More importantly, this also influences the catalyst stability toward leaching of the metal site under severe reaction conditions, for example, in membrane reactions applications. For this reason, the use of metallodendrimers involving organometallic groups containing a M–C bonded metal catalyst site seems to be the most robust system (cf. the carbosilane pincer-nickel(II) catalysts, see Schemes 1–3).

Table 20.1 Comparison of some known characteristics of the (metallo) dendrimercatalysts 1–25

Compound(s)	Reference	Complexed metal	Catalyst (mol %)	In situ prepared	Discrete catalyst	Metal 'chelaton'
1	[1]	Ni	0.3	No	Yes	N,C,N
2	[4]	Ni	0.3	No	Yes	N,C,N
3	[4]	Ni	0.3	No	Yes	N,C,N
4	[4]	Ni	0.3	No	Yes	N,C,N
5	[5]	Ni	1.8	No	Yes	N,C,N
6	[8,9]	Cu	0.4–0.7	Yes	No	N,N
7	[10]	Pd	0.05	Yes	No	P,P
8	[15]	Rh	0.03–0.56	No	No	P,P
9	[16]	Pd	—	No	Yes	P,P',P
10	[17,18]	Pd	4	No	Yes	P,P
11	[19]	Rh	1	Yes	No	P,P
12	[20]	Rh	0.5	No	Yes	P
13	[22]	Zn	2	Yes	No	O
14	[23]	Zn	17–50	Yes	No	O
15	[24]	Zn	5	Yes	No	O
16	[26]	Rh	—	Yes	No	P,P
17	[27]	Cu	—	Yes	No	N,N',N
18	[28]	Ti	—	Yes	No	O,O
19	[29]	Zn	100	Yes	No	O
20	[30]	—	—	No	Yes	—
21	[31]	—	15	No	Yes	—
22	[32]	—	3.1	No	Yes	—
23a+b	[34]	B	14–19	Yes	No	N,O
24	[36]	Pd	0.1–0.2	Yes	No	P,O
25	[37]	Pd	0.05	Yes	No	P,P

An interesting novel approach to the synthesis of (metallo)dendrimer catalysts could be the use of random hyperbranched polymers [38]. Obviously, these hyperbranched polymers have comparable but less defined structures, but to arrive at dendrimers with similar sizes, a larger number of preparative steps are required, which may be an economic disadvantage. Furthermore, materials involving heterogeneous supports with well-defined metallodendritic subunits [15] can be a promising future direction giving rise to new types of supramolecular catalysts that can easily be recovered from production streams.

The future will undoubtedly show considerable activity in the field of (metallo)dendrimers, not only because of the beauty of the structures and the synthetic challenge their preparation involves, but also because of their usefulness in fundamental applications and applied science. Finally, while we are learning to design, synthesize and apply these multimetallic catalytic objects (either soluble or insoluble), other challenges may include preparation of multifunctional catalytic prototypes by including, e.g. a substrate recognition function next to the catalytically active site.

9 REFERENCES

1. Knapen, J. W. J., van der Made, A.W., de Wilde, J. C., van Leeuwen, P. W. M. N., Wijkens, P., Grove, D. M. and van Koten, G. *Nature*, 372, 659 (1994).
2. (a) Van der Made, A.W., van Leeuwen, P. W. M. N., de Wilde, J. C. and Brandes, A. C. *Adv. Mater.*, 5, 466 (1993); (b) van der Made, A. W. and van Leeuwen, P. W. M. N., *J. Chem. Soc. Chem. Commun.*, 1400 (1992).
3. Kleij, A. W., Kleijn, H., Jastrzebski, J. T. B. H., Smeets, W. J. J., Spek, A. L. and van Koten, G. *Organometallics*, 18, 268 (1999).
4. Kleij, A. W., Gossage, R. A., Jastrzebski, J. T. B. H., Lutz, M., Spek, A. L. and van Koten, G. *Angew. Chem. Int. Ed. Engl.*, 39, 176 (2000).
5. Gossage, R. A., Jastrzebski, J. T. B. H., van Ameijde, J., Mulders, S. J. E., Brouwer, A. J., Liskamp, R. M. J. and van Koten, G. *Tetrahedron Lett.*, 40, 1413 (1999).
6. Bhyrappa, P., Young, J. K., Moore, J. S. and Suslick, K. S. *J. Mol. Cat. A: Chemical*, 113, 109 (1996).
7. Bhyrappa, P., Young, J. K., Moore, J. S. and Suslick, K. S. *J. Am. Chem. Soc.*, 118, 5708 (1996).
8. Chow, H. F. and Mak, C. C. *J. Org. Chem.*, 62, 5116 (1997).
9. Mak, C. C. and Chow, H. F., *Macromolecules*, 30, 1228 (1997).
10. Oosterom, G. E., van Haaren, R. J., Reek, J. N. H., Kamer, P. C. J. and van Leeuwen, P. W. N. M., *Chem. Commun.*, 1119 (1999).
11. Zeng, F. and Zimmerman, S. C. *Chem. Rev.*, 97, 1681 (1997).
12. (a) Balogh, L. and Tomalia, D. A., *J. Am. Chem. Soc.*, 120, 7355 (1998); (b) Balogh, L., Tomalia, D. A. and Hagnauer, G. L. *Chemical Innovation*, 30(3), 19 (2000). (c) Gröhn, F., Bauer, B. J., Akpalu, Y. A., Jackson, C. L. and Amis, E. J. *Macromolecules*, 33(16), 6042 (2000).
13. Zhao, M. and Crooks, R. M. *Adv. Mater.*, 11, 217 (1999).
14. Zhao, M. and Crooks, R. M. *Angew. Chem.*, 111, 375 (1999); *Angew. Chem. Int. Ed. Engl.*, 38, 364 (1999).
15. Bourque, S. C., Maltais, F., Xiao, W. J., Tardiff, O., Alper, H., Arya, P. and Manzer, L. E., *J. Am. Chem. Soc.*, 121, 3035 (1999).
16. Miedaner, A., Curtis, C. J., Barkley, R. M. and Dubois, D.L. *Inorg. Chem.*, 33, 5482 (1994).
17. Reetz, M. T., Lohmer, G. and Schwickardi, R. *Angew. Chem. Int. Ed. Eng.*, 36, 1526 (1997).
18. Brinkmann, N., Giebel, D., Lohmer, G., Reetz, M. T. and Kragl, U. *J. Cat.*, 183, 163 (1999).
19. Köllner, C., Pugin, B. and Togni, A. *J. Am. Chem. Soc.*, 120, 10274 (1998).
20. Petrucci-Samija, Guillemette, V., Dasgupta, M. and Kakkar, A. K. *J. Am. Chem. Soc.*, 121, 1968 (1999).
21. Bardaji, M., Monika, K., Caminade, A. M., Majoral, J. P. and Chaudret, B. *Organometallics*, 16, 403 (1997).
22. Sanders-Hovens, M. S. T. H., Jansen J. F. G. A., Vekemans, J. A. J. M. and Meijer, E. W. *Polym. Mater. Sci. Eng.*, 73, 338 (1995).
23. Suzuki, T., Hirokawa, Y., Ohtake, K., Shibata, T. and Soai, K. *Tetrahedron: Assymetry*, 8, 4033 (1997).
24. Bolm, C., Derrien, N. and Seger, A. *Synlett.*, 387 (1996).
25. For relevant literature see for instance: (a) Hawker, C. J. and Fréchet, J. M. J. *J. Am. Chem. Soc.*, 112, 7638 (1990); (b) Wooley, K. L., Hawker, C. J. and Fréchet, J. M. J. *J. Am. Chem. Soc.*, 113, 4253 (1990); (c) Hawker, C. J., Wooley, K. L. and Fréchet, J. M. J. *J. Chem. Soc. Chem. Commun.*, 925 (1994); (d) Gitsov, I. and Fréchet, J. M. J. *Macromolecules*, 27, 7309 (1994).

26. Brunner, H. *J. Organomet. Chem.*, **500**, 39 (1995).
27. Brunner, H. and Altmann, S. *Chem. Ber.*, **127**, 2285 (1994).
28. Seebach, D., Marti, R. E. and Hintermann, T. *Helv. Chim. Acta*, **79**, 1710 (1996).
29. Hovestad, N. J., Jastrzebski, J. T. B. H. and van Koten, G. *Polym. Mater. Sci. Eng.*, **80**, 53 (1999).
30. Lee, J. J., Ford, W. T., Moore, J. A. and Li, Y. *Macromolecules*, **27**, 4632 (1994).
31. Morao, I. and Cossío, F. P. *Tetrahedron Lett.*, **38**, 6461 (1997).
32. Schmitzer, A., Perez, E., Rico-Lattes, I. and Lattes, A. *Tetrahedron Lett.*, **40**, 2947 (1999).
33. (a) Kragl, U., Dreisbach, C. and Wandrey, C. 'Membrane reactors in homogeneous catalysis', Cornils, B. and Herrmann, W. A. (eds), *Applied Homogeneous Catalysis with Organometallic Compounds*, VCH, Weinheim, 1996, p. 833; (b) Kragl, U. and Dreisbach, C. *Angew. Chem. Int. Ed. Engl.*, **35**, 642 (1996).
34. Giffels, G., Beliczey, J., Felder, M. and Kragl, U. *Tetrahedron: Asymm.*, **9**, 691 (1998).
35. A residence time is defined as the time needed to fully replace the reaction volume of the continuous operating, membrane reactor system.
36. Hovestad, N. J., Eggeling, E. B., Heidbüchel, H. J., Jastrzebski, J. T. B. H., Kragl, U., Keim, W., Vogt, D. and van Koten, G. *Angew. Chem. Int. Ed. Engl.*, **38**, 1655 (1999).
37. De Groot, D., Eggeling, E. B., de Wilde, J. C., Kooijman, H., Haaren, R. J., van der Made, A. W., Spek, A. L., Vogt, D., Reek, J. N. H., Kamer, P. C. J. and van Leeuwen, P. W. N. M. *Chem. Commun.*, 1623 (1999).
38. For some recent examples of hyperbranched molecules see: (a) Chang, H. T. and Fréchet, J. M. *J. Am. Chem. Soc.*, **121**, 2313 (1999); (b) Jikei, M., Chon, S. H., Kakimoto, M., Kawauchi, S., Imase, T. and Watanebe, J. *Macromolecules*, **32**, 2061 (1999); (c) Sunder, A., Hanselmann, R., Frey, H. and Mülhaupt, R. *Macromolecules*, **32**, 4240 (1999).

Note added in proof:

After submission of this manuscript, the field of dendritic catalysis has further evolved. Some representative examples are given below.

(a) Piotti, M. E., Rivera, F. Jr., Bond, R., Hawker, C. J., Fréchet, J. M. J. *J. Am. Chem. Soc.*, **121**, 9471 (1999).
(b) Schlenk, C., Kleij, A. W., Frey, H., van Koten, G. *Angew. Chem. Int. Ed. Engl.*, **39**, 3445 (2000). Kleij, A. W., Gossage, R. A., Klein Gebbink, R. J. M., Brinkmann, N., Reijerse, E. J., Kragl, U., Lutz, M., Spek, A. L., van Koten G. *J. Am. Chem. Soc.*, **122**, 12112 (2000).
(c) Wijkens, P., Jatsrzebski, J. T. B. H., van der Schaaf, P. A., Jolly, R., Hafner, A., van Koten, G. *Org. Lett.* **2**, 1621, (2000).
(d) Fan, Q.-H., Chen, Y.-M., Chen, X.-M., Jiang, D.-Z., Xi, F., Chan, A. S. C. *Chem. Commun*, 789 (2000).
(e) Reetz, M. T., Giebel, D. *Angew. Chem. Int. Ed. Engl.*, **39**, 2498 (2000).
(f) Bourque, S. C., Alper, H., Manzer, L., Arya, P. *J. Am. Chem. Soc.*, **122**, 956 (2000). Garber, S. B., Kingsbury, J. S., Gray, B. L., Hoveyda, A. H. *J. Am. Chem. Soc.*, **122**, 8168 (2000).
(g) Maraval, V., Laurent, R., Caminade, A.-M., Majoral, J.-P. *Organometallics*, **19**, 4025 (2000).
(h) Sato, I., Shibata, T., Ohtake, K., Kodaka, R., Hirokawa, Y., Shirai, N., Soai, K. *Tetr. Lett.*, **41**, 3123 (2000).
(i) Zeng, H., Newkome, G. R., Hill, C. L. *Angew. Chem. Int. Ed. Engl.*, **39**, 1842 (2000).
(j) Kimura, M., Kato, M., Muto, T., Hanabusa, K., Shirai, H. *Macromolecules*, **33**, 1117 (2000).
(k) Pan, Y., Ford, W. T. *Macromolecules*, **33**, 3731 (2000).
(l) Breinbauer, R., Jacobsen, E. N. *Angew. Chem.*, **112**, 3750 (2000).
(m) Hecht, S., Fréchet, J. M. J. *J. Am. Chem. Soc.*, **123**, 6959 (2001).
(n) Crooks, R. M., Zhao, M., Sun, L., Chechik, V., Yeung, L. K. *Acc. Chem. Res.*, **34**, 181 (2001).
(o) Francavilla, C., Drake, M. D., Bright, F. V., Detty, M. R. *J. Am. Chem. Soc.*, **123**, 57 (2001).

21

Optical Effects Manifested by PAMAM Dendrimer Metal Nano-Composites

T. GOODSON III
Department of Chemistry, Wayne State University, Detroit, MI, USA

1 INTRODUCTION

The focus of this chapter is to overview some of the most recent findings concerning the optical properties in PAMAM dendrimer metal nanocomposites. These novel materials are of immense interest to those concerned with probing the fundamental optical properties of metal nanoparticles. The present research has attracted a large degree of synthetic and fabrication interests as well as linear and nonlinear optical research efforts. In order to cover the higher priority research investigations in the metal-PAMAM dendrimers nanocomposites field, this chapter has been divided into those areas related to the fabrication as well as the photophysical properties of these new and important materials. The introductory section discusses some of the fundamental issues involved in fabricating these novel materials as well as their chemical characterization. The second section discusses the linear and nonlinear optical properties of these materials, whereas the third section gives details of the ultra-fast emission properties of gold- and silver-dendrimer nanocomposites.

The connection between the optical properties and technological applications of macromolecules has been a subject of intense research and development for many years [1–5]. There is presently an emphasis on using materials and molecular architectures based on a very small size scale of less than 100s of nanometers [6–10]. Revolutionary ideas and concepts have emerged which may lead to the creation of superior miniature size materials for a variety of applications. Some of these concepts are directed at synthetic schemes to recreate

Dendrimers and Other Dendritic Polymers. Edited by Jean M. J. Fréchet and Donald A. Tomalia
© 2001 John Wiley & Sons Ltd

the basic building blocks of nature in a relatively inexpensive and flexible manner [11, 12]. Other ideas are motivated by using nanometer scale materials to fabricate devices to produce miniaturized effects found in conventional electronic and mechanical functions [13–15]. Presently, nanoparticles are attracting immense research attention due to their potential applications in areas such as catalysis, optics and electronics to just name a few [16–20]. If the progress of technology is to continue at an accelerated pace, new materials with new functions must be both fabricated and investigated in a precise manner. This will require the enhanced understanding of the fundamental physics and chemistry behind small particles, while at the same time observing and quantifying their differences from larger conventional devices.

The initial issues determining the direction of nanoparticle research were guided by the necessity of using common synthetic schemes to prepare novel nanoparticle materials, and by the requirement of fabricating materials in a self-assembly fashion [21, 22]. Much attention was given initially toward the fabrication of ordered assemblies of nanocrystallites commonly called 'quantum dots' [23]. Ultimately, the proper procedures were discovered for controlling the shape and size distribution of semiconductor nano-crystals such as CdSe, CdS and Ag_2S quantum dots [24]. In terms of optical investigations these are highly desirable materials since their band-gap transitions are shifted to the visible part of the spectrum, and not in the infrared as is the case for their bulk semiconductor counterparts. With these novel structures, information concerning the optical and electronic properties of semiconductor quantum dots was enhanced to a great extent. New methodologies and concepts have emerged that can now be translated to other (nonsemiconducting) chemical nanoparticles [25, 26]. While there was a large interest in the creation and characterization of semiconductor quantum dots, less has been done with other nanoparticles systems such as transition metal particles or with other particle geometries. Meanwhile, in another area, efforts in the field of clusters and colloids have grown dramatically over the last two decades [27]. Organic synthesis (utilizing particular ligand systems) have made it possible to produce both clusters and small colloid particles. From the perspective of optical and electronic effects much of the interest in this field has evolved from the huge variation in the number of surface atoms found on larger particles vs. nanometer scale particles. It is well known that this difference in percentage of surface atoms found in each of these particles sizes dramatically alters both the physical and chemical properties of the respective particles [28, 29].

A major obstacle in making precise structures with metal colloids has been the control of aggregation and particle size distribution. The use of micelles has allotted some success in this regard with the formation of different metal colloid geometries [30]. It is known that the nanoparticles must be stabilized by organic molecules attached to their surface [31] and in general must be embedded in a solid matrix [32]. This is done to prevent agglomeration and precipitation as well as the formation of aggregates. The long chains of the organic molecules prevent the particles from coming close to each other. Much is already known about how electrostatic or steric stabilization processes can aid in the formation of homogenous nanoscale metal particles. By use of electrical double layers or through polymeric or surfactant molecules the air oxidation and isolation can be achieved in a relatively controlled manner with high yield [33]. Recently, a new methodology for stabilizing metal nanoparticles has attracted great attention. This methodology utilizes the macromolecular architecture of organic dendrimers and will be discussed in this chapter.

Dendrimers are now an important macromolecular architecture in chemistry, physics and materials science [34]. Macromolecular architectures such as conjugated organic dendrimers and polymers attract a large degree of interest in the physical sciences for a variety of reasons. Contrary to linear polymers, which are composed of linear building blocks, dendrimers are synthesized by repeating branched units in a hierarchical self-similar fashion [34–37]. Dendrimer are core–shell-type macromolecules that are generally characterized by their shells or number of generations. Also, the dendrimer structure has the potential of developing branched (fractal) architectures, which could enhance the understanding of the relationship of dimensionality (geometric restriction) with excitation localization length [37–39]. Dendrimers have received increasing attention due to their interesting electronic properties and their potential optical applications [37–44]. The nature of optical excitations in dendrimers is still a matter of keen debate. It has been suggested that after excitation the electron (and hole) is localized on linear segments of the dendrimer [37]. Optical excitations can be modeled as a set of weakly (Coulombic) interacting chromophores described by a Frenkel exciton Hamiltonian [39]. However, the character of these interactions and the extent of delocalization of the wave-function of excitons in dendrimers is not clear. Also the character of energy migration between segments (hopping or coherent) is not certain [37–39, 45]. To clarify the character of excited state interactions between different segments of dendrimer, time-resolved ultra-fast measurements are of great necessity [38, 46, 47].

From intense synthetic efforts it is now known that nano-structured materials can be fabricated by a variety of methods, one of which involves the use of synthetic dendrimers as host container or scaffolding molecules [48]. It has been discovered that dendrimers have the ability to form organic–inorganic nanocomposites [49]. The initial studies showed that the reduction of copper or gold ions in the presence of generation 4 (G4) polyamidoamine (PAMAM) dendrimer leads to stable colloidal solutions, rather than macroscopic formation of metal precipitates [50]. The formation of these nanometer scale copper particles was confirmed by UV-Vis spectroscopy [50]. Dendrimer nanostructured materials have been suggested as prime candidates for drug delivery systems, quantum-confined structures, and nano-level storage units [51, 52]. It was also discovered that there was a great deal of flexibility in the fabrication (synthesis) of the dendrimer nanocomposites. For example, it has been demonstrated that different zero-valent transition metals may be encapsulated inside PAMAM dendrimers to produce a variety of metal–dendrimer topologies [53, 54].

Dendrimers have also been used for the formation of thin film metal colloids. A novel approach for Au and Ag colloid monolayer formation on different silicon oxide surfaces such as glass, silicon, or ITO using PAMAM dendrimers has been described [55]. The relatively facile method yields monolayers with easily controlled spacing within the monolayer without aggregation of metal particles. The colloid monolayers are prepared in two steps: (1) modification of the substrates with PAMAM dendrimers, and (2) noble metal colloid deposition onto the dendrimer layer [55]. Different Au and Ag colloids, ranging from 15 to 80 nm in particle diameter, have been deposited onto the dendrimer-modified surfaces. The structure and properties of the resulting particle arrays and dendrimer nanocomposite solutions have been studied by atomic force microscopy (AFM), scanning electron microscopy (SEM), X-ray photoelectron spectroscopy (XPS), UV-Vis spectroscopy, and surface-enhanced Raman scattering(SERS) [55, 56]. XPS data show that the dendrimers spontaneously adsorb to various silicon oxide surfaces. SEM and AFM data show that the colloids spontaneously form continuous films on the dendrimer-modified surfaces [55]. The noble metal particles are well isolated and confined to a single layer, and aggregation does not occur on the surface. The interparticle spacing (74–82 nm) and surface coverage can be controlled over a wide range by colloid size, colloid concentration and immersion time. UV-Vis spectroscopic data show that the microstructure directly controls the optical properties of the layer. This result was one of the first justifications of the wide range of applications with dendrimer-metal nanostructured materials.

The motivation for using dendrimers as scaffolding molecules has evolved into a number of different directions. From the synthetic (fabrication) point of view, the motivation stems from the desire to understand the general features of polymer nanotemplating with model systems. This is quite reasonable since the well-known PAMAM dendrimers are monodisperse and well-characterized macromolecules with sizes ranging from 1 to 15 nm for generation 2–10 [57, 58]. This discrete, systematic size (continuum spacers) has characteristic dimensions of low molecular weight molecules, polymers and colloids. Access to such a systematic succession of structural dimensions may help bring about the fundamental understanding of organic–inorganic hybrid materials, including their nanoscale characterization, formation mechanisms and optical properties. It has already been shown, in detail, that the dendrimer-stabilized nanoclusters can result in effective high surface area catalysts in solution that allow substrates to penetrate the dendrimer interior and access the cluster surface [59]. It may also be true that the dendrimer architecture may play an important role in understanding the mechanisms of the linear and nonlinear optical properties of transition metal nanoparticles.

Information concerning the emission properties of gold nanoparticles is of great importance to future applications in nanotechnology. Mooradian [60] first observed the fluorescence from bulk metal (Au) with an extremely low efficiency of 10^{-10}. While it has been recently demonstrated that metal nano-particles exhibit an increase (compared to bulk) of the fluorescence quantum efficiency [61] when the shape is somewhat elongated, the quantum efficiency of metal nanoparticles is still very low so that the actual fluorescence signal is difficult to measure. Less is known about the time-resolved fluorescence from transition metals. The characteristic size and separation of core clusters in dendrimer nanoparticle systems may play a large role in the ability to observe the emission resulting from the optical excitation process. Although several selected systems have produced various dynamical models for ultra-fast excitations in metal nano-particles [62, 63], there is a widely accepted general model that accounts for the majority of the events in the excitation and decay processes. This involves the optical excitation of the electrons by interband and intraband transitions. These processes are usually followed by a loss of coherence, which is largely due to electron–electron and electron–surface scattering processes that result in a quasi-equilibrated electron system. This relaxation to the quasi-equilibrated electron system normally has duration of approximately 100 fs. The hot electron system can also lose its energy through electron–phonon coupling [62, 64]. There have been several reports of the time-resolved transient absorption effects in metal nanoparticles. Utilizing time-resolved transient absorption, several reports have clarified a large degree of the complexity of ultrafast optical excitations in nanoparticles [62–65]. However, the signals obtained in transient absorption measurements are a superposition of several nonlinear optical processes. For example, the interplay of bleaching and excited state absorption leads to complicated features in the pump-probe experiments. Certainly, there is more to be learned from the emission properties of metal nano-particles.

The remaining sections outline recent findings on the optical properties of metal-dendrimer nano-composites. The sections have been organized into areas of interest related to the fabrication as well as photophysical properties of these new and important materials. The first section discusses some of the important issues concerning the fabrication of these novel materials and their chemical characterization. The second section discusses the linear and nonlinear optical properties of these materials. The third section provides details related to the ultrafast emission properties of gold- and silver-dendrimer nanocomposites.

2 FABRICATION OF METAL–DENDRIMER NANOCOMPOSITES

The fabrication of metal-dendrimer nanocomposites has now been demonstrated by a number of research groups using a variety of metal constituents. The synthesis of Cu, Ag, Au, Pt, Pd and CdS dendrimer nanocomposites have all been reported in the literature [66–70]. In all cases to date, the primary host dendrimer has been variations of the PAMAM dendrimer (schematically illustrated in Figure 21.1a). Pristine PAMAM dendrimers are now available commercially at a vari-ety of generation levels [71]. The host PAMAM dendrimers can be tailored with appropriate functional groups (either on surface or internally) for further chemical alteration. Host dendrimers of the EDA core type have tertiary amine functional groups in the interior, while COOH groups can be introduced on the surface [72, 73]. Details for the fabrication method vary depending on the architecture type desired for the respective metal. However, the principal procedure requires that the host dendrimers provide appropriate legation or charged sites (Figure 21.1B). These dendrimers are then combined with the appropriate metal salts. The dendrimer–metal complexes are then reduced to zero valent dendrimer nanocomposites by addition of a reducing agent such as $NaBH_4$. The rate, dendrimer size (generation), weight ratio and concentration are all important parameters in the fabrication of the dendrimer nanocomposites. The reported syntheses of these structures have certainly verified the complexity of preparing reproducible composite systems, even though the methodology appears relatively straightforward.

The type of metal used is very important in the fabrication of the particles. Crooks *et al.* [74] demonstrated that metal nanoclusters ranging in size from 1 to 2 nm could be prepared within dendrimer templates by a two-step synthesis process. It was found that metal ions such as Cu^{2+} and Pd^{2+} and Pt^{2+} initially partition into the interior of certain PAMAM (–OH terminated) dendrimers [74, 75]. The resulting structure contains a zero valent metal nanocluster (see Figure 21.1c). Crooks *et al.* [74, 75] have shown that the procedure involves the partitioning of a particular number of metal ions into the dendrimer's interior. The driving force of the reaction is suggested to be due to associative properties of amines or other terminal groups. Crooks and co-workers also reported that dendrimer-encapsulated metal nanoclusters can undergo multiple *in situ* displacement reactions [66]. Pd and Pt dendrimer encapsulated nanoparticles prepared by direct reduction, as well as by primary or secondary displacement reactions, are catalytically active for electrochemical reduction of O_2 [66]. These structures were characterized by TEM, XPS, FT–IR as well as UV-Vis spectroscopy [66, 76].

With precise functionalization, it is possible to prepare different types of dendrimer nanocomposite topologies. Balogh and Tomalia have utilized the functionalization of PAMAM dendrimers to create both external and internal dendrimer nanocomposite topologies (for example the external topology is shown in Figure 21.1D) [70, 77]. Here, it was suggested that the dendrimer host are monodisperse nano-reactors, possessing architecture and ligand sites that allow the pre-organization of metal ions within their interiors. It was reported that the use of PAMAM dendrimers as templated containers allows *in situ* generated reaction products to be dispersed as amorphous or slightly ordered domains within the dendrimer interior [77]. Several interesting applications have been suggested due to the flexibility of the synthesis in this manner. These include encapsulating semiconductors and noble metals, environmental cleanup, as well as magnetic and electronic applications. In this particular particle synthesis, 1 ml solution of G4 PAMAM dendrimer was added to 1 ml of 10 mM Cu(II) acetate solution [77]. The Cu^{2+} ion complexing capacity of a particular PAMAM dendrimer was followed by spectrophotometric titrations. According to the analysis of the spectrophotometric data, a surface plasmon resonance at 590 nm suggested that the metal shape was spherical and the particle size was smaller than the host dendrimer template [78].

Perhaps the most detailed characterization of dendrimer nanocomposites was given recently by Amis and co-workers in a study where dendrimers were used as templates for gold nanoclusters [58]. Several generations (G2–G10) of PAMAM dendrimers were used as nanotemplates for the formation of inorganic–organic hybrid colloids in aqueous solution [58]. The reduction of gold with sodium borohydride was used in the synthesis since it was previously shown that the reaction speed could be controlled. While it was not proven that $AuCl^{4-}$ ions exhibit any specific coordination to amine groups, it has been shown with polyelectrolyte microgels that the $AuCl^{4-}$ has sufficient electrostatic attraction to precursor ions to form inorganic–organic hybrid structures. Results of TEM, SAXS and SANS were used to characterize the different structures formed with the various dendrimer–metal combinations. This work was very inclusive and several different types of topologies were found [58]. Gold/dendrimer ratios, as well as dendrimer mass fraction, and reaction rate are all important in determining the desired structures. It was also found that under certain experimental conditions PAMAM dendrimers in the range of G2 to G4 may behave like molecular mass colloid stabilizers. It was suggested that in these systems, several dendrimers surround the surface of the metal particles formed. On the other hand, the higher generation systems were found to produce different results. For G6 to G9, the PAMAM dendrimers act as effective 'polymeric' templates [58]. Metal particles appear to be completely encapsulated inside individual dendrimers. The metal particles appear to aggregated in to well defined domains within the dendrimer host (see Figure 21.1E). This work demonstrated that the size of the metal can be precisely controlled by the number of precursor ions added per dendrimer molecule. This may be performed without changing the morphology of the hybrid particle [58].

While more systematic studies are still necessary to completely characterize all the possible structures obtained from metal–dendrimer nanocomposites, the reported work clearly demonstrates that the PAMAM dendrimer may be viewed as a nanoscale template. From these studies metal particles have been shown to be encapsulated inside the higher generation PAMAM dendrimer host, whereas there is evidence that the metal particles can be stabilized on the surface of other PAMAM dendrimers. However, the detailed nature of the interaction of the PAMAM dendrimers with the metal colloids has many unanswered questions. Amis's data showing the metal particles inside the PAMAM dendrimer are aggregated slightly off center is an interesting result that requires further characterization. Such work might define new features concerning electrostatic interactions within dendrimer-nanocomposites [58]. The understanding of these characteristics may also prove very critical in explaining and predicting the electronic properties of the metal–dendrimer nanocomposites as well. As described in the next sections, the various dendrimer nanocomposite topologies appear to exhibit different optical behavior. The consequence of placing approximately the same number of metal (gold or silver) atoms either on the outside or the interior has proven to be very important in the case of the nonlinear optical effects.

3   LINEAR AND NONLINEAR OPTICAL PROPERTIES IN METAL–DENDRIMER NANOCOMPOSITES

While there has been great success in the fabrication of many new types of dendrimer nanocomposites (DNC) understanding the optical properties of these novel materials is still in its infancy. Linear absorption studies have demonstrated the connection of the spectral position of the maximum and full width half maximum (FWHM) of the surface plasmon resonance peak with the size of the metal nanoparticles. It is known that reducing the size of the nanoparticle has a pronounced effect on the energy level spacing as the system becomes more confined. As it is often taught in undergraduate physical chemistry, for a particle confined to a specified container (a 'box'), the energy separation between adjacent levels increases with decreasing dimensions. In a metal, the conduction band is half-filled and the density of energy levels is so high that a noticeable separation in energy levels within the conduction band is only observed when the nanoparticle is made of just a few atoms [79]. The surface plasmon resonance is the coherent excitation of all the 'free' electrons within the conduction band. This ultimately leads to an in-phase oscillation [80]. For the larger particles, of several tens of nanometers in which their size is still small compared with the wavelength of light, excitation of the surface resonance can take place with visible light. Thus it can surely be seen that the surface plays a crucial role for the observation of the surface plasmon resonance as it alters the boundary conditions for the polarizibility of the metal and therefore shifts the resonance to

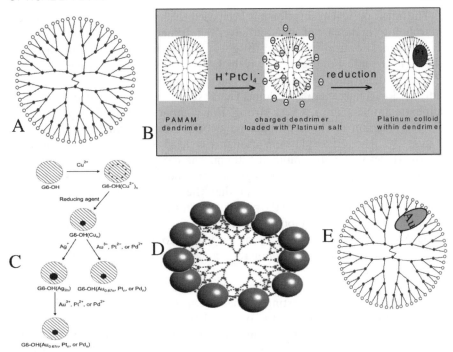

Figure 21.1 A-Structure of the host PAPAM dendrimer. B-Fabrication procedure for metal-dendrimer nanocomposites. C- Pd and Pt dendrimer encapsulated nanoparticles prepared by displacement reactions [66]. D, E -External and internal dendrimer nanocomposite topologies, respectively [70, 77].

optical frequencies accessible with normal spectroscopic methods.

For dendrimer metal nanocomposites the connection between the size of the particles and character of the surface plasmon resonance can be well characterized [81]. The electric field of an incoming light wave induces a polarization of the 'free' conduction electrons with respect to the much heavier ionic core of a spherical nanoparticle [81]. The net charge difference occurs at the nanoparticle boundaries (the surface) which in turn acts as a restoring force. In this manner a dipolar oscillation of the electrons is created within a particular time period. It was shown by Mie that a series of multipole oscillations may be responsible for the extinction cross-section of metal nanoparticles [80]. For nanoparticles much smaller than the wavelength of light only the dipole oscillation contributes significantly to the extinction cross-section [82]. For larger nanoparticles the dipole approximation may no longer be valid, and the plasmon resonance will depend explicitly on the particle size. The larger the particles become, the more important the higher order polar contributions become. The plasmon bandwidth increases with increasing particle size.

The linear optical absorption has been used to characterize the metal particle morphology in the case of gold- and silver-dendrimer nanocomposites. In Figure 21.2 the linear optical spectra for the gold (Au) dendrimer nanocomposite. Also shown in the insert of Figure 21.2 is the linear absoroption spectra for the pure PAMAM dendrimer (generation 5). The absorption of the pure PAMAM dendrimer has a maximum at 230 nm. The absorption of the highly concentrated aliphatic amine groups is very strong. For comparison, the absorption of the surface plasmon of the dendrimer nanocomposite is normalized to unity in Figure 21.2. The surface plasmon resonance is clearly seen for the gold dendrimer nanoparticles with a maximum at 530 nm. The FWHM of the surface plasmon resonance absorption peak for this particular dendrimer nanocomposite sample was ~ 0.36 eV [83]. It was found that the width of the surface plasmon absorption peak changes with the particular dendrimer nanocomposite architecture. Silver-dendrimer nanocomposites have also been prepared and show a surface plasmon resonance at wavelengths shifted to the blue of the gold systems. The FWHM of the surface plasmon resonance also ranged over a relatively large span of wavelengths.

As it can also be seen from Figure 21.2, there is another linear absorption resonance at 270 nm. Interestingly, when the spectra of the pure dendrimer and the dendrimer nanocomposite are compared, the resonance at 270 nm is not seen for the pure dendrimer. At this point, it is not completely certain as to the origin of this resonance, but it is possibly associated with the dendrimer–metal electrostatic association, or the metal band transition itself. The surface plasmon resonance becomes broader with larger size particles in the dendrimer metal hybrid architecture. For the absorption spectra shown in Figure 21.2, the TEM results suggested that the size of the organic–inorganic hybrid structures ranged from 5 nm to 20 nm. The surface plasmon resonance has a FWHM of 0.36 eV which is consistent with particles of this size.

For other metal–dendrimer nanoparticles, such as those made of Pt^{2+}, strong absorption in the ultraviolet has also been observed. Crooks. *et al.* [84] have investigated changes in the linear absorption spectra of $PtCl_4^{2-}$ and its combination with G4–OH PAMAM dendrimer. When the two are mixed for sufficient time a new band at 250 nm is formed. This band was interpreted as being proportional to the number of Pt^{2+} ion in the dendrimer, with the number of atoms ranging from 0 to 60. In this case, the titration-like behavior also suggests that it is possible to control the dendrimer–metal ratio [84]. The linear absorption spectra were also used to show that Pt particles arising from Pt^{2+} ions bound to the terminal primary amine ligands agglomerate and only the intra-dendrimer bound Pt^{2+} ions yield stable soluble clusters. Crook's absorption results support the suggestion of the very high degree of stability in these dendrimer nanoparticles in that no agglomeration was detected for up to 150 days and the material could be redissolved after repeated dryings.

The linear absorption of dendrimer nanocomposite thin films has also shown interesting features. Tripathy *et al.* [85] have utilized the processes of electrostatic multilayer deposition to form thin films of gold–dendrimer nanocomposites mixed with layers of poly(sodium 4-styrenesulfonate) (PSS). UV-Vis spectra of PSS/gold-dendrimer nanocomposites showed that in the UV region of the spectrum the amplitude of the absorption increased with increasing number of bi-layers. For these thin films the absorption of the gold could be monitored by UV-Vis spectroscopy as well. The number of cycles of adsorption was critical in observing the surface plasmon resonance [85]. The UV data suggested that the gold–dendrimer nanocomposite could be adsorbed onto the PSS film after a certain number cycles had been completed. A uniform array of the gold–dendrimer nanocomposite on the PSS layer was further confirmed by AFM studies.

New materials for nonlinear optics are important for a variety of reasons. With very efficient second and third harmonic generation exhibited in novel materials new avenues are now possible for application in optical switching and optical computing architectures [86]. Materials fabricated in a waveguide structure have given rise to very efficient electro-optical effects for faster optoelectronic design [87]. Recently, there has also been intense interest in materials that possess large third- and fifth-order nonlinear responses in new materials. These effects include optical phase conjugation, two- and three-photon absorption, optical limiting, as well as certain nonlinear phenomena necessary to observe quantum optical effects [88, 89]. Optical limiting effects are observed when the material's integrated absorption increases with increasing input intensity. Novel polymeric materials have reached the point of magnitude and thermal stability comparable to inorganic single crystal systems, for the case of second order nonlinear optical effects. However, dendrimers and dendrimer nanocomposites are a new class of materials for nonlinear optics and have shown interesting nonlinear refraction and nonlinear absorption properties. The nonlinear optical processes in a material can be best characterized by a the macroscopic polarization given by [90].

$$P_i = \chi_{ij}^1 E_j + \chi_{ijk}^2 E_j E_k + \chi_{ijkl}^3 E_j E_k E_l + \tag{1}$$

The first term on the right of equation (1) is the linear electrical susceptibility. The second terms is the second-order nonlinear susceptibility which gives information responsible for such effects as second harmonic generation and the linear electro-optic effect. The third term is the third-order nonlinear susceptibility, which is responsible for third harmonic generation and two-photon absorption. The third order nonlinear processes which give rise to large optical limiting effects are important for a variety of applications including eye protection from intense laser beams, sensor protection and possible biomedical applications. Organic chromophores and polymers have enjoyed a large degree of success in providing materials with large nonlinear susceptibilities, which are comparable (and in some cases larger than) to their inorganic crystalline counterparts. It is now known that organic dendrimers can also exhibit strong nonlinear optical properties as well [91]. Due to the possibility of combining different molecular components that are important for NLO properties within a single structure, and the fact that the organic dendrimers are themselves amorphous and form nice thin films, these new materials offer a wide range of new applications for nonlinear optics. The nonlinear absorption and refraction in dendrimer nano-composites have also been investigated [83]. It appears that combining the synthetic architecture of amorphous dendrimers with metal nanocomposites may be a new avenue for preparing superior, stable and robust nonlinear optical materials.

As described in the previous section, the optical properties of nano-scale metal domains are strongly influenced by the surface plasmon (SP) resonance [80]. This is relevant in the case of optical limiting where the material's absorption is dependent upon the intensity of the input beam. It has been reported that the external (E) type {Ag(0)}-dendrimer topology, structure known as {Ag(0)}$_E$:({(Ag(0))$_{96.97}$-PAMAM_E5.5COOH}) has a strong surface-plasmon resonance at 414 nm. The maxima and the FWHM of the surface plasmon gives information concerning the size distribution of the metal particle domains, and has exhibited interesting nonlinear optical properties [83]. For the case of

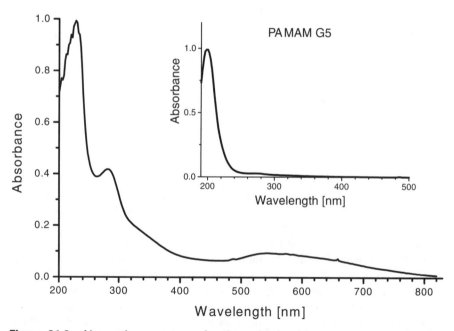

Figure 21.2 Absorption spectrum for the gold-dendrimer nanocomposite. Inset: absorption spectrum for PAMAM dendrimer (generation 5).

the $\{Ag(o)\}_E$ the size of the particles range from 5 to 25 nm. The metal-dendrimer nanocomposites in this report were prepared in water (aqueous) solutions [92].

The linear absorption spectra of $\{Ag(0)\}_E$ and the pure dendrimer host were described above. The $\{Ag(0)\}_E$ structure exhibits an SP resonance absorption peak in the visible region at around 414 nm. The pure dendrimer absorption consists mainly of a strong peak at around 230 nm. The nonlinear optical properties of these novel compounds were first investigated with 6.5 ns laser pulses from a frequency doubled Nd:YAG laser at 532 nm. The repetition rate of the laser pulses was varied between 2 Hz and 10 Hz. The optical setup consisted of an f/8 lens focusing the Gaussian beam (initial radius of 2.2 mm) to a spot radius of $\sim 40~\mu$m, and a power-meter positioned close to the output facet of the 1 mm-thick sample cell [83]. The control of the input energy was obtained by a combination of quarter-wave plates and polarizers. The result of the nonlinear transmission measurement with a pulse repetition rate of 10 Hz for a solution in a relatively low concentration (i.e. 2.0×10^{-4} mole/kg) is shown in Figure 21.3. When the input fluence varies from 0.7 to 10.0 J/cm^2 (equivalent to the increase of the peak irradiance from 0.2 to 1.3 GW/cm^2) the transmission decreases by 62%. The threshold fluence for optical limiting is around 2.0 J/cm^2. The optical limiting performance of $\{Ag(0)\}_E$ compares well to the results obtained with novel organic structures such as: EHO–OPPE [93] (where a transmission drop of 65% was obtained between 0 and 0.6 GW/cm^2); the AF-380 dye [94] (exhibiting a threshold fluence of ~ 2 J/cm^2 and a transmission loss of 60% for an input fluence increase up to 13 J/cm^2). On the other hand single-walled carbon nano- tube suspensions [95] (showed a threshold fluence of ~ 2 J/cm^2 and transmission decrease of 70% for fluence increasing up to 6 J/cm^2). Thus the Ag den- drimer nanocomposite system exhibits strong optical limiting effects at 532 nm.

Measurements with synthetic organic dendrimers, where organic chromophores are attached to the surface groups, have shown NLO properties that are enhanced in comparison to the guest–host systems [96, 97]. This implies that dendritic architecture may provide a novel environment for enhanced intramolecular interactions for NLO effects. This may be crucial for nonlinear optical effects that may involve orientational dipolar order [98, 99]. The orientational order of the dendritic architecture may be stabilized by interactions of the surface groups and possibly by locking (crosslinking) the groups for thermal stability. In terms of the DNC's, the nonlinear optical effect is significantly larger than the pristine organic PAMAM dendrimer host. In fact, for the host system no detectable optical limiting effect was observed at 532 nm. This strongly implies that the NLO effects are strongly dependent upon the stabilized nanoparticles inside the dendrimer host.

While the mechanisms of the NLO properties in dendrimer nanocomposties have been experimentally approached, they are still not entirely understood.

However, certain important characteristics may give a plausible explanation for the mechanisms. For example, the dependence of the solution concentration on the nonlinear transmission behavior has been investigated. Measurements of the optical limiting effect as a function of increasing concentration or ratio of the metal has been reported [83]. As expected, the transmission decrease was less effective with the reduction of the solution concentration. The report also showed the repeated measurement of several cycles of increasing and decreasing the peak irradiance between 0 and ~ 1.3 GW/cm^2 and the results are the same as seen in Figure 21.3. The information about the optical limiting processes occurring in the $\{Ag(0)\}_E$ structure at 532 nm can also be obtained by utilizing the open z-scan technique [97] in the same optical setup configuration, with 10 Hz pulse repetition rate. From the open z-scan results it can be concluded that drastic optical extinction, by a factor of 115, occurs for a concentrated solution $(5.9 \times 10^{-4}$ mole/kg) of $\{Ag(0)\}_E$ at the laser beam focus for a fluence of 3.3 J/cm^2 (relatively moderate input irradiance of 0.42 GW/cm^2) [83].

It is interesting to note that the dendrimer metal nanocomposite results not only vary from conventional organic conjugated dendrimers but they also vary from thin films of metal particles. Smith *et al.* [100] used the z-scan technique at a similar wavelength (532 nm) to measure the nonlinear absorption coefficient of continuous, approximately 50 Å-thick gold films, deposited on surface modified quartz substrates. As it was found with the dendrimer metal nanocomposites, the highly absorbing metal films required analysis of both the real and imaginary parts of the susceptibility, utilizing both open and closed z-scan measurements. Interestingly, when peak intensities of the order of 0.1 GW/cm^2 were used a transmission decrease of $\sim 8\%$ was detected [100]. For intensities > 0.160 GW/cm^2 ablative damage occurred which was observed by a sharp peak in the z-scan transmission which increased with each scan. This instability to intense laser pulses of the metal thin film material was also noted by an asymmetry in the baseline of the z-scan. While the magnitude of the effect may have been small in the metal thin film, the measured nonlinear response could be described relatively well by mean field theories and a Fermi smearing mechanism [100]. The metal dendrimer nanocomposites appear to have much better stability to intense laser pulses and larger nonlinear optical responses as well. The dendrimer nanocomposite has a smooth nonlinear transmission and open z-scan curves. The silver dendrimer nanocomposite also seems to be quite stable up to 1.00 GW/cm^2. This high stability was also seen in the ultrafast time-resolved measurements as well. Indeed, with the large nonlinear optical effects observed in the metal dendrimer nanocomposites, coupled with their high stability to intense laser pulses, new applications in optoelectronics and biomedical procedures may be possible [83].

Understanding the mechanisms of the optical limiting effect in metal dendrimer nanocomposites may also require understanding the timescale of the effect. In general, for optical excitation close to the linear absorption band, such as at 532 nm for $\{Ag(0)\}_E$, the cross-sections for reverse saturable absorption (RSA) and nonlinear scattering should be much higher than the cross-section for two-photon absorption [83]. Time-resolved photoluminescence measurements on $\{Ag(0)\}_E$ have shown that the excited state lifetimes in metal dendrimer nanocomposites are very short, i.e. of the order of picoseconds (see next section) [101]. RSA processes were seen to develop on a timescale similar to the excited state lifetime [102, 103]. For example, in the case of phthalocyanine complexes, this is usually in the nanosecond range [104]. It was thus suggested that the contribution of RSA is small for the $\{Ag(0)\}_E$ since the lifetime of the excited state is extremely short. Another explanation of the relatively slow optical limiting effect observed in dendrimer nanocomposites is to consider processes that originate due to absorption-induced nonlinear scattering [105].

Indeed, the timescale of the optical limiting effect in dendrimer nanocomposites is somewhat different than that found in other materials and this may be crucial to the understanding of the mechanism. Recent reports have investigated

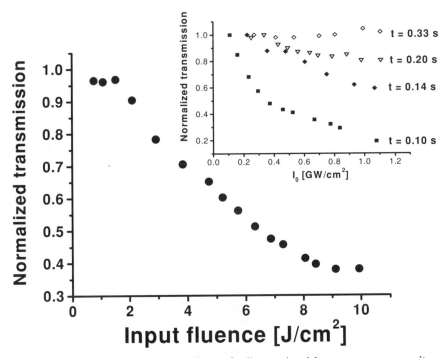

Figure 21.3 Nonlinear transmission of silver -dendrimer nanocomposites $\{Ag(0)\}_E$ at 532 nm. Pulse repetition rate- 10 Hz. Inset: nonlinear transmission results for $\{Ag(0)\}_E$ in a concentration of 2.95×10^{-4} mole/kg at 532 nm, with the variation of the pulse repetition rate; t indicates the specific repetition periods.

the relative timescale of the optical limiting effect. This has been investigated by probing the dependence of the nonlinear transmission on the pulse repetition rate (see inset of Figure 21.3) [83]. The transition from the actual optical limiting behavior, for repetition rates higher than 5 Hz, to the transparency regime seen at a repetition rate of 3 Hz indicates that the mechanism governing the optical limiting is indeed slow. The strong absorption at 532 nm can lead to local heating of the aqueous solution at the focal spot which is then followed by the reversible creation of scattering centers, most probably microbubbles, approaching boiling temperatures [95]. Increased beam scattering at large angles could be observed with the naked eye at the power-meter aperture. Such thermally assisted scattering processes are indeed expected to occur on a millisecond timescale and to be very sensitive to solvent boiling point [95, 102]. The nano-templating topology of the DNC determines the nonlinear optical properties. Both {Ag(o)}I and the {Ag(0)}E have been chemically characterized in the literature and the properties of these two topologies have been shown to vary greatly. As is expected, these properties are also dependent upon the probe excitation wavelength.

The nonlinear transmission results of the metal nanocomposites at 532 nm are totally different from those at 1064 nm. As illustrated above, the silver external type metal nanocomposites showed an impressive optical limiting effect as seen in Figure 21.3. The transmission decreased 48% for the {Ag(0)} nanocomposites when the input energy fluence varies from 1[J/cm^2] to 11[J/cm^2] (which equals to the input irradiance from 0.3 to 1.5 [GW/cm^2]) [106]. It can seen (in Figure 21.4) that there is a remarkably different behavior for the internal type samples {Ag(0)}I when probed at 532 nm. An increase in the transmission with the input energy fluence was indicative of a strong saturable absorption mechanism at the resonance wavelength of 532 nm. We have analyzed the dependence of the absorption on the incident irradiance I according to the model of homogeneous broadening [106]:

$$\alpha L = \alpha_0 L / [1 + I / I_{sat}] \qquad (2)$$

where α is the irradiance-dependent absorption coefficient, α_0 the initial absorption coefficient without at zero irradiance, L the sample thickness and I_{sat} the saturation irradiance. For both samples the estimated saturation irradiances was 1.3 GW/cm^2. The saturable absorption effect at 532 nm can be directly related to the linear absorption resonance for both samples in that spectral region. The reason that the internal type topology did not show the same optical limiting behavior as the external structures may be due to the difference in metal domain sizes. The internal type may form as little as a single nanoparticle per dendrimer inside the dendrimer architecture and consequently has a smaller surface area [107]. On the other hand the metal particle domains in the external topology are somewhat larger [92]. The thermal conductivity may be decreased due to the smaller surface area of the agglomerated metal particles in internal type.

The linear and nonlinear optical properties of the dendrimer nanocomposites have exhibited interesting properties that are strongly dependent on the fabrication of the structures. The external Ag structure showed a very impressive optical limiting effect at 532 nm. The optical limiting properties of the internal dendrimer nanocomposites were very different from those observed for the external structures. The concentration dependence of the nonlinear optical effects supports the suggestion that increased concentration of the dendrimer metal nanocomposites results with increased magnitude of the transmission decrease. Measurements of the timescale of the optical limiting effects showed that the mechanism may involve a relatively slow process including strong nonlinear absorption of the metal dendrimer nanoparticles. The effects at the infrared wavelength (1064 nm) are also very interesting. Here the internal and external dendrimer nanocomposites seem to be governed by a similar mechanism. This novel nonlinear optical behavior, and control of properties by changing the metal dendrimer nanocomposite topology offers interesting possibilities for applications in the future. The strong optical limiting effects may be used for highly localized heating and possibly thermal imaging effects in biological or cellular hosts. The localized site specific heating due to the excitation by the laser pulses which was observed in some cases, may be useful for certain drug/chromophore release functions inside biological host systems as well.

4 ULTRAFAST EXCITED STATE DYNAMICS AND PHOTO-LUMINESCENCE PROPERTIES OF DENDRIMER METAL NANOCOMPOSITES

While the linear absorption and nonlinear optical properties of certain dendrimer nanocomposites have evolved substantially and show strong potential for future applications, the physical processes governing the emission properties in these systems is a subject of recent high interest. It is still not completely understood how emission in metal nanocomposites originates and how this relates to their (CW) optical spectra. As stated above, the emission properties in bulk metals are very weak. However, there are some processes associated with a small particle size (such as local field enhancement [108], surface effects [29], quantum confinement [109]) which could lead in general to the enhancement of the fluorescence efficiency as compared to bulk metal and make the fluorescence signal well detectable [110, 111].

Reports on several selected systems have produced various dynamical models for ultrafast excitations in nanoparticles [62, 107, 112, 113]. Many have used a widely accepted model to account for both the excitation and decay processes in metals. This involves the optical excitation of the electrons by interband and intraband transitions. These transitions are followed by a loss of coherence, which is largely due to electron–electron and electron–surface scattering processes that result in a quasi-equilibrated electron system that normally has a duration of approximately 100 fs [63, 114]. The hot electron system can also lose its energy through electron–phonon coupling. There have been several reports of the time-resolved transient absorption effects in metal nanoparticles [63, 114–116]. It has been found that the time constant of the transient changes of absorption and reflectivity was strongly dependent on the pump-pulse energy [62]. It has been verified [62] that the time-dependent induced transmission could be enhanced near the surface plasmon resonance of the nanoparticles. A retardation of the electron cooling process to the lattice temperature was observed at the plasmon resonance [62]. Reports of time-resolved transient absorption have also demonstrated the importance of electron–surface scattering processes to the ultra-fast dynamics. Other authors have found that the origin of the time response of the transient decay can be ascribed to the cooling process of nonequilibrium electrons through the electron–phonon coupling processes. Von Plessen *et al.* [113] have investigated the dynamics of metal nanoparticles in relation to the large size particles that may be formed. Investigations of the nonequilibrium relaxation processes in optically excited large gold and silver clusters. Time-resolved pump-probe experiments and model calculations show that optical excitation of the clusters by femtosecond laser pulses results in a heating of the electron system, which is followed by electron cooling via phonon emission. The electron heating leads to an enhanced damping of the surface-

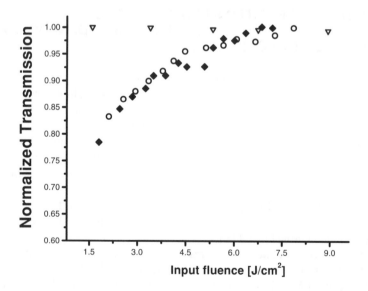

Figure 21.4 Nonlinear transmission for the internal type [70, 77] samples {Ag(0)}ᵢ when probed at 532 nm.

plasmon resonance in the clusters. This enhanced damping is caused by an enhancement of the Landau damping and electron scattering rates at high electron temperatures. It was also found that the rate of electron cooling in the clusters changes with electron temperature. This was attributed to the consequence of the temperature-dependent specific heat of the conduction electrons.

It appears that the electron–phonon coupling process is indeed critical to understanding the excitation in metal nanoparticles. Hartland *et al.* [116] have utilized femtosecond pump-probe methods to investigate the electron–phonon coupling process in metal particles. In their report, ultrafast laser spectroscopy was used to characterize the low frequency acoustic breathing modes of Au particles, with diameters between 8 and 120 nm. It was shown that these modes are impulsively excited by the rapid heating of the particle lattice that occurs after laser excitation. This excitation mechanism is a two-step process; the pump laser deposits energy into the electron distribution, and this energy is subsequently transferred to the lattice via electron–phonon coupling. The measured frequencies of the acoustic modes are inversely proportional to the particle radius. Analysis of the data showed that a inhomogeneous decay dominates the damping, even for our high quality samples (8–10% dispersion in the size distribution). The size dependence of the electron–phonon coupling constant was also examined for these particles. The results show that, to within the inherent error due to the signal-to-noise ratio in the measurements, the electron–phonon coupling constant does not vary with size for particles with diameters between 4 and 120 nm. Furthermore, the value obtained is the same as that measured for bulk gold [117].

Although the reports utilizing time-resolved transient absorption have clarified a large degree of the complexity of ultrafast optical excitations in nanoparticles, the interpretation of these results could be ambiguous in some cases as the signals obtained in transient absorption measurements are a superposition of several nonlinear optical processes. A better understanding of the excitation and emission pathways would lead to a clearer picture of the dynamics in metal nanoparticles. Until recently there have been no time-resolved photoluminescence (PL) studies with metal nanoparticles in any other architecture except dendrimer nanocomposites reported in the literature. It has been shown that time-resolved PL properties of metal nanoparticles may be investigated utilizing the properties of the dendrimer nanocomposite topology.

Time-resolved PL measurements in organic dendrimers have already given some important information on the nature of the excitations in these topologies [38, 46, 47]. Many have suggested that organic dendrimers could be used as artificial light harvesting systems [37, 45, 118]. These earlier systems were normally conjugated systems in which the surface of the dendrimers contains organic functional groups that absorb light energy that is higher than the component at the core of the dendrimer structure [37]. A mechanism for excitation energy transport in dendrimers is not completely clear at the present time, and there have been few ultra-fast experimental investigations to resolve these important parameters. Utilizing dendrimer nanocomposites may help to give more detailed information about excitation in both metals and metal nanoparticles. PL measurements may allow correlation of nanoparticle size and shape with the optical decay.

Mooradian's initial results on fluorescence from bulk metal (Au) illustrated the emission process occurred with a quantum efficiency of 10^{-10} [60]. El-Sayed and co-workers have recently demonstrated a large increase (compared to bulk) in the fluorescence quantum efficiency for elongated gold nanoparticles [110]. The characteristic size and separation of core clusters in dendrimer nanoparticle systems may play a large role in the ability to observe the emission resulting from the optical excitation process. Recently, there have been reports of the ultrafast emission decay in metal dendrimer nanocomposites [101]. The metal dendrimer nanocomposites under investigation were $\{Au(0)\}_I$ formally called $\{(Au(0)_{10.01}$-PAMAM_E5.NH$_2\}$, and $\{Ag(0)\}_I$, formally called $\{Ag(0)_{96.9}$-PAMAM_E4.TRIS$\}$, in which the dendrimer template was polyamidoamine (PAMAM) [92]. These two structures had already been extensively characterized in the literature [70, 77, 92]. High-resolution transmission electron microscopy (HrTEM) revealed that the guest-metal domains were located predominately at the core of the dendrimer host. The average size of the metal domains, also estimated by HrTEM, was found to vary between 5 and 20 nm [77].

Because of the very inefficient emission in metal nanoparticles continuous wave (CW) emission measurements give very little information about the excitations and dynamics in these systems. However, with ultra-fast time-resolved measurements it was possible to observe the very short-lived excited states in these dendrimer nanocomposites. Recently, femtosecond up-conversion spectroscopy was employed to temporally resolve the polarized fluorescence of the metal dendrimer nanocomposites [101]. The optical arrangement for a typical fluorescence up-conversion experiments has been reported in the literature [96]. For the measurements the laser source was a Ti:sapphire laser with an average pulse width of 100 fs tuned at 790–860 nm and a repetition rate of 82 MHz (Tsunami, Spectra Physics). The sample was excited with light pulses delivered by a frequency-doubled output of the laser at 395–430 nm. Fluorescence emitted from the sample was focused with an achromatic lens into a nonlinear crystal made of betabarium borate (BBO). A temporal profile of the fluorescence was monitored by sum-frequency generation with reference pulse from the laser at 790–860 nm that was first passed through a variable delay line. Sum frequency light was dispersed by a monochromator and detected by a single photon counting system. The time resolution was determined by the pulse width of the laser and group velocity dispersion within optical elements of the system. The FWHM of the cross-correlation function at 790/395 nm was estimated to be 190 fs. It is important to note that we could observe the fluorescence dynamics on a timescale covering almost four decades (200 fs–1 ns) in one measurement. All experiments in these studies were performed at room temperature. The rotating sample cell and holder (1 mm – thickness in case of solutions, ~ 1 μm thick in case of thin film) were used to avoid thermal and photochemical accumulative effects. The energy of the excitation pulse did not exceed 0.5 nJ/pulse. We found that there was no excitation intensity dependence of the decay dynamics. The measured fluorescence decay curves were fit by the result of the convolution of the instrument response function with an exponential decay model in order to minimize the sum of weighted residuals (χ^2) [96]. The quality of the fit was monitored by the values of the reduced χ^2, inspection of the residuals, and monitoring of the auto-correlation function of the residuals.

Shown in Figure 21.5 is the time-resolved emission from $\{Au(0)\}_I$ for excitation at 395 nm [101]. It is seen that the dynamics of the emission signal has at least two decay components. One of these decay components is comparable in duration with the width of the instrument response function (IRF) (also shown in Figure 21.5 by the dashed line) while the second one is relatively long, on a timescale of several picoseconds. The emission dynamics of the nanocomposites was directly compared with the time-resolved emission from the host (PAMAM) dendrimer. The host dendrimer revealed a weak and slowly decaying fluorescence signal similar to the long decay component of the nanocomposite emission [106]. This measurement, together with the polarization measurements suggest that the relatively long-lived fluorescence decay component may be a function of the host dendrimer, whereas the fast decaying component is associated with the metal emission.

In previous reports of CW fluorescence from bulk metals [60, 119] as well as from metal nanoparticles embedded in micelles [110] the fluorescence originates from the recombination of electrons in the s–p band with holes in the d band. In Figure 21.6 we show a schematic diagram summarizing the sequence of excitation and relaxation mechanisms involving transitions between the d and s–p bands in Au nanoparticles. The surface plasmon (SP) resonance peak in the absorption spectrum of the Au dendrimer nanocomposite is shown (solid line) in the inset, together with the spectral distribution of the peak amplitude of the time-resolved luminescence signal (dotted line and diamonds) from the $\{Au(0)\}_I$ nanocomposite. It is important to note that there is no spectral shift between the SP absorption and the luminescence peak amplitude. This suggests that the sharp peak in the luminescence decay is associated with the metal nanoparticle.

It is interesting to note that the fast emission dynamics of $\{Ag(0)\}_I$ (shown in Figure 21.7) differs from that of $\{Au(0)\}_I$ [101]. The decay curve for $\{Au(0)\}_I$ could be reasonably fit by a two-exponential decay function with time constants of 74 fs, 5.5 ps and relative amplitudes 0.95, 0.05, respectively (best fit curve shown in Figures 21.5 and 21.7). The Ag nanoparticles initial (71 fs with 0.91 amplitude) and final (5.3 ps with 0.01 amplitude) decay components were similar to those of gold; however, an additional component of 650 fs (with 0.08 amplitude) was also detected [101]. It has been shown that both electron-electron (e–e) and electron–surface (e–s) scattering processes occur on a timescale less than 100 fs in metal nanoparticles. There are various channels for nonradiative decay following the quenching of fluorescence. As illustrated in Figure 21.7, these processes involve electron–phonon coupling as well as the coupling of the metal nanoparticles to the thermal reservoir of the dendrimer host [62, 65]. An alternative mechanism for the observed ultrafast emission could be the interaction of the emission dipole moment of the metal SP resonance with the emission dipole moment of the dendrimer host (surface-enhanced fluorescence). This interaction could lead to quantum yield enhancement as well as to a much shorter fluorescence decay time [108, 120]. However, the almost depolarized character (see below) of the initial emission makes this assumption much less probable in comparison to the first mechanism proposed above, at least in the case of spherical metal particles.

Fluorescence anisotropy is generally used to provide information about the dipolar orientational dynamics occurring after excitation of a system. This technique has successfully been used to probe ultrafast dynamics of energy transfer in organic conjugated dendrimers. The detected emission intensities I_{par} and I_{per} for parallel and perpendicularly polarized excitation respectively, were used to construct an observable emission anisotropy $R(t)$ in accordance with the equation [121]:

$$R(t) = \frac{I_{par}(t) - GI_{par}(t)}{I_{par}(t) - 2GI_{par}(t)} \tag{3}$$

The factor G accounts for the difference in sensitivities for the detection of emission in the perpendicular and parallel polarized configurations.

The anisotropy decay result for $\{Au(0)\}_1$ is shown in Figure 21.8 [101]. The temporal profiles of the emission in the parallel plane (dashed line) and the plane perpendicular (dotted line) to the polarization of the excitation pulse are shown. It is clearly seen that the polarization state of the emission during the first 300 fs after excitation is quite different from that for longer times. The 'dip' and 'rise' in the anisotropy curve usually suggest that there is more than one species in the system whose contributions to the effective $R(t)$ have different timescales [121]. Indeed, depolarized emission has been observed for metals [60] while highly polarized emission is typical for organics [121]. From these observations we can attribute the polarized long decay to the emission of the dendrimer template and the fast (and nearly depolarized) component to the emission of the metal nanoparticle [101]. This result is in agreement with the measurements of the pure dendrimer fluorescence decay described above. The results showed that the fast emission increases linearly with excitation intensity. This observation as well as the nearly depolarized fast component in the emission dynamics suggests an incoherent character of the metal PL. All these results strongly support the conclusion that the observed fast PL is indeed the fluorescence of the metal nanoparticle.

Anisotropy in metal particles has also been investigated by other spectroscopic techniques. Pileni and co-workers [122] compared the optical properties of spherical particles organized in a two-dimensional structure with disordered and coalesced particles. Both particle preparations were deposited on cleaved graphite substrates. When particles are arranged in a hexagonal array, the optical measurements under p-polarization show a new high-energy resonance which is interpreted as a collective effect, resulting from optical anisotropy due to the mutual interactions between particles [122]. For disordered and coalesced particles, a low-energy resonance appears instead of the high-energy resonance observed for spherical and organized particles. This is interpreted as optical shape anisotropy due to the asymmetrical shape of coalesced particles.

It is indeed clear that the shape of the nanoparticles plays a strong role in the absorption and emission properties, even on a femtosecond timescale. It may also be true that the femtosecond pulses may be involved in the final shape of the particles as well. El-Sayed and co-workers [123] have found that gold nanorods change their shape after excitation with intense pulsed laser irradiation. The final irradiation products strongly depend on the energy of the laser pulse as well as on its width. A series of measurements in which the excitation power was varied over the range of the output power of an amplified femtosecond laser system producing pulses of 100 fs duration and a nanosecond optical parametric oscillator (OPO) laser system having a pulse width of 7 ns were performed [123]. The shape transformations of the gold nanorods are followed by two techniques: (1) visible absorption spectroscopy by monitoring the changes in the plasmon

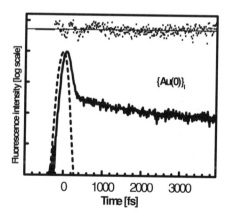

Figure 21.5 Fluorescence dynamics of {Au(0)}$_I$, for excitation at 395 nm, and emission at 570 nm. The corresponding numerical fits to the data are indicated by the thin solid lines. The residuals of the fit are shown at the top of graph.

absorption bands characteristic for gold nanoparticles; (2) transmission electron microscopy (TEM) in order to analyze the final shape and size distribution. While at high laser fluences (of ~ 1 J/cm^2) the gold nanoparticles fragment, a melting of the nanorods into spherical nanoparticles (nanodots) was observed when the laser energy is lowered. Upon decreasing the energy of the excitation pulse, only partial melting of the nanorods takes place [123]. Shorter but wider nanorods were observed in the final distribution as well as a higher abundance of particles having odd shapes. The threshold for complete melting of the nanorods with femtosecond laser pulses is about 0.01 J/cm^2. Comparing the results obtained using the two different types of excitation sources (femtosecond vs nanosecond laser), it was found that the energy threshold for a complete melting of the nanorods into nanodots is about two orders of magnitude higher when using nanosecond laser pulses than with femtosecond laser pulses. This is explained in terms of the successful competitive cooling process of the nanorods when the nanosecond laser pulses are used. For nanosecond pulse excitation, the absorption of the nanorods decreases during the laser pulse because of the

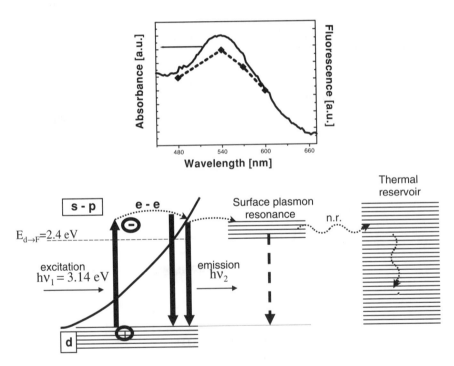

Figure 21.6 Schematic illustration of the dynamics of the photoluminescence from the $\{Au(0)\}_I$ system. The inset shows the comparison of the $\{Au(0)\}_I$ surface plasmon absorption peak with the spectral distribution of the $\{Au(0)\}_I$ emission peak for excitation at 3.14 eV (395 nm).

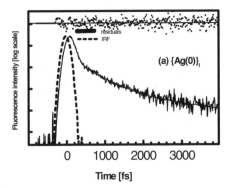

Figure 21.7 Fluorescence dynamics of {Ag(0)}$_I$, for excitation at 395 nm, and emission at 480 nm. The corresponding numerical fits to the data are indicated by the thin solid lines.

bleaching of the longitudinal plasmon band. In addition, the cooling of the lattice occurring on the 100 ps timescale can effectively compete with the rate of absorption in the case of the nanosecond pulse excitation, but not for the femtosecond pulse excitation. When the excitation source is a femtosecond laser pulse, the involved processes (absorption of the photons by the electrons (100 fs), heat transfer between the hot electrons and the lattice (< 10 ps), melting (30 ps), and heat loss to the surrounding solvent (> 100 ps)) are clearly separated in time.

However, for the dendrimer nanocomposite metallic systems this change in shape was not observed. Again, due to the high stability to intense laser pulses, the anisotropy value of the gold dendrimer nanocomposite, which can be viewed as a measure of the symmetry of the particle, did not change after several repeated cycles of measurements. It is possible that the initial optical pumping of the electron–phonon modes of the metal particles is partially absorbed by the encapsulating PAMAM dendrimer.

The emission of the metal particles may thus originate from a band-to-band transition in the metal particle, which occurs at about 516 nm for gold [60, 119]. As stated above, the nature of the interaction of the dendrimer (PAMAM) host is still uncertain, there could be very strong electrostatic interactions that may play a part in the enhancement of the metal particles quantum efficiency for emission. However, one would expect that this enhancement would result in slightly distorted emission spectra, different from what was observed for the gold dendrimer nanocomposite. Further work is necessary to completely characterize the manner in which the dendrimer encapsulation enhances the emission of the metal nanoparticles. With further synthetic work in preparation of different size nanoparticles (in other words elongated and nonspherical shape particles, including nanorods) it may be possible to develop the accurate description of a

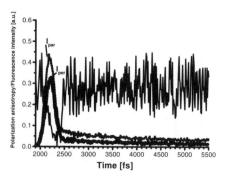

Figure 21.8 The dynamics of the polarization anisotropy for the {Au(0)}ᵢ emission. The temporal profiles of the emission in the parallel plane (dashed line) and the plane perpendicular (dotted line) to the polarization of the excitation pulse are also shown.

general mechanism for different nanoparticles by utilizing time-resolved emission analysis.

5 SUMMARY

The study of metal PAMAM–dendrimer nanocomposites is evolving rapidly and is expected to offer many new architectures and optoelectronic effects. There is still a critical need for more systematic studies to understand the effects that parameters such as generation level, surface functionality and interior composition manifest on nanocomposite topologies. While it was well demonstrated that PAMAM dendrimers can be used as nanoscale templates or reactors, the nature of the interaction of metal particles with dendrimers is not certain. It has been well established that zero valent metals may be encapsulated inside PAMAM dendrimer hosts as well as be stabilized on the surface of the dendrimers. Both the linear and nonlinear optical properties are strongly dependent on the topology of the nanocomposite. Optical limiting properties of the internal topologies were very different from those observed for the external structures. The novel nonlinear optical behavior exhibited by the dendrimer nanocomposites, and control of the properties by changing the architecture, will undoubtedly be very useful for a variety of applications.

Ultrafast emission measurements are possible with the dendrimer metal nanocomposites. The gold and silver internal dendrimer nanocomposites showed a fast emission decay of approximately 0.5 ps, which was followed by a slower decay process. The fast decay emission is attributed to decay processes of the gold (or silver) metal nanoparticles. Ultrafast emission anisotropy measurements suggest that the initial fast decay component is relatively depolarized. The fact that the wavelength dependence of the ultrafast emission is very close to the spectrum of the surface plasmon of the metal particle and that the initial fast anisotropy is relatively depolarized strongly suggests that this emission is predominately from the metal domains. The nature of the zero valent metal dendrimer host interaction is still uncertain, however these electrostatic interactions that may play a role in the enhancement of the metal particles quantum efficiency for emission. Further work is necessary to completely understand the manner in which the dendrimer encapsulation enhances the emission and other optical properties of these unique metal nanoparticles.

6 REFERENCES

1. Koide, Y., Wang, Q. W., Cui, J., Benson, D. D. and Marks, T. J. *J. of Amer. Chem. Soc.*, **122**, 11266 (2000).
2. Ho, P. K. H., Kim, J. S., Burroughes, J. H., Becker, H., Li, S. F. Y., Brown, T. M., Cacialli, F. and Friend, R. H. *Nature*, **404**, 481 (2000).
3. Dodabalapur, A., Torsie, L. and Katze, H. E. *Science*, **268**, 270 (1995).
4. Collier, C. P., Wong, E. W., Belohradsky, M., Raymo, F. M., Stoddart, J. F., Kuekes, P. J., Williams, R. S. and Heath, J. R. *Science*, **285**, 391 (1999).
5. KukowskaLatallo, J. F., Bielinska, A. U., Johnson, J., Spindler, R., Tomalia, D. A. and Baker, J. R. *Proc. of the Nat. Acad. of Sci. of the USA*, **93**, 4897 (1996).
6. Kroto, H. W., Heath, J. R., Obrien, S. C., Curl, R. F. and Smalley, R. E. *Nature*, **318**, 162 (1985).
7. Tomalia, D. A., Baker, H., Dewald, J., Hall, M., Kallos, G., Martin, S., Roeck, J., Ryder, J. and Smith, P. *Polymer Journal*, **17**, 117 (1985).
8. Brus, L. E. *J. of Chem. Phys.*, **80**, 4403 (1984).
9. Tsui, D. C., Stormer, H. L. and Gossard, A. C. *Phys. Rev. Letts.*, **48**, 1559 (1982).
10. Quake, S. R. and Scherer, A. *Science*, **290**, 1536 (2000).
11. Gust, D. and Moore, T. A. *Science*, **244**, 35 (1989).
12. Preece, J. A. and Stoddart, J. F. *Nanobiology*, 3, **149** (1994).
13. Tour, J. M., Kozaki, M. and Seminario, J. M. *J. of the Amer. Chem. Soc.*, **120**, 8486 (1998).
14. Bennett, C. H. *Physics Today*, **48**, 24 (1995).
15. Bennett, C. H. and Wiesner, S. J. *Phys. Rev. Letts.*, **69**, 2881 (1992).
16. Gates, B. C. *Chem. Rev.*, **95**, 511 (1995).
17. Wong, E. W., Sheehan, P. E. and Lieber, C. M. *Science*, **277**, 1971 (1997).
18. Borrelli, N. F., Hall, D. W., Holland, H. J. and Smith, D. W. *J. of Appl. Phys.*, **61**, 5399 (1987).
19. Tour, J. M. *Chem. Rev.*, **96**, 537 (1996).
20. Domb, A. J. and Nudelman, R. *J. of Polymer Sci. Part A*, **33**, 717 (1995).
21. Nikoobakht, B., Wang, Z. L. and El-Sayed, M. A. *J. of Phys. Chem. B*, **104**, 8635 (2000).
22. Mirkin, C. A., Letsinger, R. L., Mucic, R. C. and Storhoff, J. J. *Nature*, **382**, 607 (1996).
23. Murray, C. B., Norris, D. J. and Bawendi, M. G. *J. of the Amer. Chem. Soc.*, **115**, 8706 (1993).

24. Vossmeyer, T., Katsikas, L., Giersig, M., Popovic, I. G., Diesner, K., Chemseddine, A. and Weller, H. *J. of Phys. Chem.*, **98**, 7665 (1994).

25. Tolbert, S. H., Herhold, A. B., Johnson, C. S. and Alivisatos, A. P. *Phys. Rev. Letts.*, **73**, 3266 (1994).

26. Alivisatos, A. P., Harris, A. L., Levinos, N. J., Steigerwald, M. L. and Brus, L. E. *J. of Chem. Phys.*, **89**, 4001 (1998).

27. Schmid, G. *Chem. Rev.*, **92**, 1709 (1992).

28. Gotz, T., Buck, M., Dressler, C., Eisert, F. and Trager, F. *Applied Physics A*, **60**, 607 (1995).

29. Kreibig, U., Bour, G., Hilger, A. and Gartz, M. *Phys. Stat. Sol.*, **175**, 351 (1999).

30. Yu Y.-Y., Chang, S.-S. and Cris Wang, C. R. *J. Phys. Chem. B*, **101**, 6661 (1997).

31. Carrot, G., Scholz, S. M., Plummer, C. J. G., Hilborn, J. G. and Hedrick, J. L. *Chem. of Mater.*, **11**, 3571 (1999).

32. Klabunde, K. J., Youngers, G., Zuckerman, E. J., Tan, B. J., Antrim, S. and Sherwood, P. M. *European J. of Solid State and Inorganic Chem.*, **29**, 227 (1992).

33. Svergun, D. I., Shtykova, E. V., Kozin, M. B., Volkov, V. V., Dembo, A. T., Shtykova, E. V., Bronstein, L. M., Platonova, O. A., Yakunin, A. N. and Valetsky, P. M. *J. of Phys. Chem. B*, **104**, 5242 (2000).

34. Tomalia, D. A., Naylor, A. M. and Goddard III., W. A. *Angew. Chem. Int. Ed. Engl.* **29**, 138 (1990).

35. Fréchet, J. M. J. *Science*, **263**, 1710 (1994).

36. Naylor, A. M., Goddard, W. A., Keifer, G. and Tomalia, D. A. *J. Am. Chem. Soc.*, **111**, 2339 (1989).

37. Kopelman, R., Shortreed, M., Zhong-You Shi, Tang, W., Zhifu Xu., Moore, J., Bar-Haim, A. and Klafter, J. *Phys.Rev.Lett.*, **78**, 1239 (1997).

38. Hofkens, J., Larettini, L., DeBelder, G., Gensch, T., Maus, M., Vosch, T., Karni, Y., Schweitzer, G., DeSchryver, F. C., Herrmann, A. and Mullen, K. *Chem. Phys. Lett.*, **304**, 1 (1999).

39. Poliakov, E. Y., Chernyak, V., Tretiak, S. and Mukamel, S. *J. Chem. Phys.*, **110**, 8161 (1999).

40. Devadoss, C., Brahathi, P. and Moore, J. S. *J. Am. Chem. Soc.*, **118**, 9635 (1996).

41. Baars, M. W. P. L., van Boxtel, M. C. W., Bastiaansen, C. W. M., Broer, D. J., Sontjens, S. H. M. and Meijer, E. W. *Advanced Materials*, **12**, 715 (2000).

42. Stevelmans, S., vanHest, J. C. M., Jansen, J. F. G. A., vanBoxtel, D. A. F. J., vandenBerg, E. M. M. D. and Meijer, E. W. *J. of the Amer. Chem. Soc.*, **118**, 7398 (1996).

43. Jansen, J. F. G. A., Meijer, E. W. and Debrabandervandenberg, E. M. M., *J. of the Amer. Chem. Soc.*, **117**, 4417 (1995).

44. Hofkens, J., Maus, M., Gensch, T., Vosch, T., Cotlet, M., Kohn, F., Herrmann, A., Mullen, K. and De Schryver, F. *J. of the Amer. Chem. Soc.*, **122**, 9278 (2000).

45. Jiang, D. L. and Aida, T., *J. of the Amer. Chem. Soc.*, **120**, 10895 (1998).

46. Varnavski, O., Menkir, G., Goodson, T. and Burn, P. L. *Appl. Phys. Letts..*, **77**, 1120 (2000).

47. Varnavski, O., Leanov, A., Liu, L., Takacs, J. and Goodson III, T. *Phys. Rev. B.*, **15**, 114 (2000).

48. Stark, B., Lach, C., Frey, H. and Stuhn, B. *Macro. Symp.*, **146**, 33 (1999).

49. Stathatos, E., Lianos, P., Stangar, U. L., Orel, B. and Judeinstein, P. *Langmuir*, **16**, 8672 (2000).

50. Otta Viani, M. F., Bossmann, S., Turro, N. J. and Tomalia, D. A. *J. of the Amer. Chem. Soc.*, **116**, 661 (1994).

51. Wakabayashi, Y., Tokeshi, M., Jiang, DL., Aida, T. and Kitamori, T., *J. of Lumines-*

cence, **83**, 313 (1999).

52. Archut, A., Vogtle, F., De Cola, L., Azzellini, G. C., Balzani, V., Ramanujam, P. S. and Berg, R. H. *Chem. A- Euro. Journ.*, **4**, 699 (1998).

53. Tan, N. C. B., Balogh, L., Trevino, S. F., Tomalia, D. A. and Lin, J. S. *Polymer*, **40**, 2537 (1999).

54. Zhao, M. Q., Sun, L. and Crooks, R. M. *J. of the Amer. Chem.*, **120**, 4877, 1998.

55. Bar, G., Rubin, S., Cutts, R. W., Taylor, T. N. and Zawodzinski, T. A. *Langmuir*, **12**, 1172 (1996).

56. Nisato, G., Ivkov, R. and Amis, E. J. *Macromolecules*, **33**, 4172 (2000).

57. Topp, A., Bauer, B. J., Tomalia, D. A. and Amis, E. J. *Macromolecules*, **32**, 7232 (1999).

58. Grohn, F., Bauer, B. J., Akpalu, Y. A., Jackson, C. L. and Amis, E. J. *Macromolecules*, **33**, 6042 (2000).

59. Breinbauer, R. and Jacobsen, E. N. *Angewandte Chemie-International Ed.*, **39**, 3604 (2000).

60. Mooradian, A. *Phys. Rev. Lett.*, **22**, 185 (1969).

61. Mohamed, M. B., Volkov, V., Link, S. and El-Sayed, M. A. *Chem. Phys. Lett.*, **317**, 517 (2000).

62. Bigot, J. Y., Halté, V., Merle, J. C. and Daunois, A. *Chem. Phys.*, **251**, 181 (2000).

63. Hodak, J., Martini, I. and Hartland, G. V. *Chem. Phys. Lett.*, **284**, 135 (1998).

64. Bonn, M., Denzler, D. N., Funk, S., Wolf, M., Wellershoff, S. S. and Hohlfeld, J. *Phys. Rev. B.*, **61**, 1101 (2000).

65. Hailwail, E. J. and Hochstrasser, R. M. *J. Chem. Phys.*, **82**, 4762 (1985).

66. Zhao, M. Q. and Crooks, R. M. *Chem. Mat.*, **11**, 3379 (1999).

67. Hovestad, N. J., Hoare, J. L., Jastrzebski, J. T. B. H., Canty, A. J., Smeets, W. J. J., Spek, A. L. and van Koten, G. *Organometallics*, **18**, 2970 (1999).

68. Tokuhisa, H., Zhao, M. Q., Baker, L. A., Phan, V. T., Dermody, D. L., Garcia, M. E., Peez, R. F., Crooks, R. M. and Mayer, T. M. *J. of the Amer. Chem. Soc.*, **120**, 4492 (1998).

69. Esumi, K., Suzuki, A., Yamahira, A. and Torigoe, K. *Langmuir*, **16**, 2604 (2000).

70. Balogh, L. and Tomalia, D. A. *J. of the Amer. Chem. Soc.*, **120**, 7355 (1998).

71. The Polyamidoamine (PAMAM) dendrimer generation 4 can be purchased from Aldrich Co.

72. Imae, T., Ito, M., Aoi, K., Tsutsumiuchi, K., Noda, H. and Okada, M. *Colloids and Surfaces A*, **175**, 225 (2000).

73. Larsen, G., Lotero, E. and Marquez, M. *J. of Phys. Chem. B*, **104**, 4840 (2000).

74. Zhao, M. Q., Tokuhisa, H. and Crooks, R. M. *Angewandte Chemie-International Ed.*, **36**, 2596 (1997).

75. Zhao, M. Q. and Crooks, R. M. *Angewandte Chemie-International Ed.*, **38**, 364 (1999).

76. Garcia, M. E., Baker, L. A. and Crooks, R. M. *Anal. Chem.*, **71**, 256 (1999).

77. Balogh, L., Tomalia, D. A., Hagnauer, G. L. *Chem. Innov.*, **30**, 19 (2000).

78. Diallo, M. S., Balogh, L., Shafagati, A., Johnson, J. H., Goddard, W. A. and Tomalia, D. A. *Environmental Sci. and Tech.*, **33**, 820 (1999).

79. Palpant, B., Prevel, B., Lerme, J., Cottancin, E., Pellarin, M., Treilleux, M., Perez, A., Vialle, J. L. and Broyer, M. *Phys. Rev. B*, **57**, 1963 (1998).

80. Mie, G. *Ann. Phys.*, **25**, 377 (1908).

81. Mulvaney, P. *Langmuir*, **12**, 788 (1996).

82. Inaoka, T. *Surface Sci.*, **273**, 191 (1992).

83. Ispasoiu, R. G., Balogh, L., Varnavski, O. P., Tomalia, D. A. and Goodson, T. *J. Amer. Chem. Soc.*, **122**, 11005 (2000).

84. Zhao, M. Q. and Crooks, R. M. *Adv. Materials*, **11**, 217 (1999).
85. He, J. A., Valluzzi, R., Yang, K., Dolukhanyan, T., Sung, C. M., Kumar J., Tripathy, S. K., Samuelson, L., Balogh, L. and Tomalia, D. A. *Chem. of Materials*, **11**, 3268 (1999).
86. Volodin, B. L., Kippelen, B., Meerholz, K., Javidi, B. and Peyghambarian, N. *Nature*, **383**, 58 (1996).
87. Shi, Y. Q., Zhang, C., Zhang, H., Bechtel, J. H., Dalton, L. R., Robinson, B. H. and Steier, W. H. *Science*, **288**, 119 (2000).
88. Hughes, S., Wherrett, B. S. and Spruce, G. *J. of the Opt. Soc. of Amer. B*, **14**, 526 (1997).
89. Ispasiou, R. G. and Goodson, T. *Optics Communications*, **178**, 371 (2000).
90. Shen, Y. R. *The Principles of Nonlinear Optics*, John Wiley & Sons, 1976.
91. Varnavski, O., Leanov, A., Liu, L., Takacs, J. and Goodson, T. *J. of Phys. Chem. B*, **104**, 179 (2000).
92. Balogh, L., Valluzzi, R., Hagnauer, G. L., Laverdure, K. S., Gido, S. P. and Tomalia, D. A. *J. of Nanoparticle Res.*, **1**, 353 (1999).
93. He, G. S., Weder, C., Smith, P. and Prasad, P. *IEEE J. Quantum Electron.* **34**, 2279 (1998).
94. Joshi, M. P., Swiatkiewicz, J., Xu, F., Prasad, P., Reinhardt, B. A. and Kannan, R. *Opt. Lett.*, **23**, 1742 (1998).
95. Mishra, S. R., Rawat, H. S., Mehendale, S. C., Rustagi, K. C., Sood, A. K., Bandyopadhyay, R., Govindaraj, A. and Rao, C. N. R. *Chem. Phys. Lett.*, **317**, 510 (2000).
96. Varnavski, O. and Goodson, T. *Chem. Phys. Letts.*, **320**, 688 (2000).
97. Ispasiou, R. G., Narawal, M., Varnavski, O., Yan, J., Pass, H., Fugaro, J. and Goodson, T. *Macromolecules*, **33**, 4013 (2000).
98. Goodson, T. and Wang, C. H., *Macromolecules*, **26**, 1837 (1993).
99. Dureiko, R. D., Schuele, D. E. and Singer, K. D. *J. of the Opt. Soc. of Amer. B*, **15**, 338 (1998).
100. Smith, D. D., Yoon, Y., Boyd, R. W., Campbell, J. K., Baker, L. A., Crooks, R. M. and George, M. *J. of Appl. Phys.*, **86**, 6200 (1999).
101. Varnavski, O., Ispasoiu, R. G., Balogh, L., Tomalia, D. A. and Goodson III, T. *J. Chem. Phys.*, **114**, 1962, (2001).
102. Tutt, L. W. and Boggess, T. F. *Progr. Quant. Electr.*, **17**, 299 (1999).
103. Harilal, S. S., Bindhu, C. V., Nampoori, V. P. N. and Vallabhan, C. P. G. *J. of Appl. Phys.*, **86**, 1388 (1999).
104. Perry, J. W. 'Organic and metal-containing reverse saturable absorbers for optical limiting', in Nalwa, H. S. and Miyata, S. (eds), *Nonlinear Optics of Organic Molecules and Polymers*, 1997, CRC Press, Boca Raton, FL.
105. François, L., Mostafavi, M., Belloni, J., Delouis, J. F., Delaire, J. and Feneyrou, P. *J. of Phys. Chem. B*, **104**, 6133 (2000).
106. (a) Goodson, T, Proceedings of SPIE annual meeting San Diego, CA, July (2001), (in press); (b) Larson, C. L. and Tucker, S. A., *Appl. Spectrosc.*, **55**, 679, (2001).
107. Link, S. and El-Sayed, M. A. *International Rev. of Phys. Chem.*, **19**, 409 (2000).
108. Moskovits, M., *Rev. Mod. Phys.*, **57**, 783 (1985).
109. Alvarez, M. M., Khoury, J. T., Gregory Shaaf, T., Shafigullin, M. N., Vezmar, I. and Whetten, R. L., *J. Phys. Chem B*, **101**, 3706 (1997).
110. Mohamed, M. B., Volkov, V., Link, S. and El-Sayed, M. A. *Chem. Phys. Letts.*, **317**, 517 (2000).
111. Wilcoxon, J. P., Martin, J. E., Parsapaour, F., Wiedenman, B. and Kelley, D. F. *J. Chem. Phys.*, **108**, 9137 (1998).

112. Lamprecht, B., Krenn, J. R., Leitner, A. and Aussenegg, F. R. *Phys. Rev. Letts.*, **83**, 4421 (1999).
113. von Plessen, G., Perner, M. and Feldmann, J. *Appl. Phys. B.*, **71**, 381 (2000).
114. Bigot, J. Y., Merle, J. C., Cregut, O. and Daunois, A. *Phys. Rev. Lett.*, **75**, 4702 (1995).
115. Inouye, H., Tanaka, K., Tanahashi, I. and Hirao, K. *Phys. Rev. B*, **57**, 11334 (1998).
116. Link, S. and El-Sayed, M. *J.Phys. Chem. B*, **103**, 8410 (1999).
117. Hodak, J. H., Henglein, A. and Hartland, G. V. *J Chem. Phys.*, **111**, 8613 (1999).
118. Raychaudhuri, S., Shapir, Y., Chernyak, V. and Mukamel, S. *Phys. Rev. Lett.*, **85**, 282 (2000).
119. Boyd, G. T., Yu, Z. H. and Shen, Y. R. *Phys. Rev. B.*, **33**, 7923 (1986).
120. Tomas, K. G. and Kamat, P. V. *J. Am. Chem. Soc.*, **122**, 2655 (2000).
121. Lakowicz, J. R. *Principles of Fluorescence Spectroscopy*, Ch. 11, Kluwer, New York, 1999.
122. Taleb, A., Russier, V., Courty, A. and Pileni, MP. *Appl. Surf. Sci.*, **162**, 655 (2000).
123. Link, S., Burda, C., Nikoobakht, B., El-Sayed, M. A. *J. Phys. Chem. B*, **104**, 6152 (2000).

22

Dendrimers in Nanobiological Devices

S. C. LEE
Monsanto Company, 700 Chesterfield Village Parkway, North AA4C; St Louis, MO, USA

1 BIOLOGY FOR NANOTECHNOLOGY

Nanotechnology aims to construct materials and devices that owe their valuable properties to some nanoscale aspect of their structure. Relatively precise nanostructures have been made using chemical and physical methods, and the field has been largely the province of physicists and engineers from its inception [12]. Biology focuses on surveying the diversity of living things, deriving mechanistic explanations for their properties, activities and interactions. With the advent of biotechnology, it has involved the 'engineering' of organisms and biological macromolecules for human use. While biology and nanotechnology might appear to be disparate fields, their mutual utility has been increasingly appreciated [1, 8, 9, 16, 26–29, 39–41, 44, 45, 64]. Biological macromolecules provide a precedent in that they reside in the nanometer size range, and biological systems provide innumerable examples of nanoscale macromolecular assemblies that perform useful work [72] Additionally, organismal biology illustrates the power that integration of multiple nanodevices and systems can provide. Biological structures also provide models for individual synthetic nanodevices ('bioinspired' structures). Since biological processes occur in the nanometer range, biology also provides an important *raison d'être* for nanotechnology. Interactions occurring between nanoscale components underlie most disease processes, and so medicine provides a steady stream of therapeutic applications for nanotechnology [5, 10, 11, 63]. Certain methodologies used in biotechnology, protein chemistry and biochemistry (chromatography, gel electrophoresis, ultracentrifugation, mass spectroscopy, amino acid composition analysis,

Dendrimers and Other Dendritic Polymers. Edited by Jean M. J. Fréchet and Donald A. Tomalia
© 2001 John Wiley & Sons Ltd

immunoblotting) are directly applicable to synthetic (nanoscale) macromolecules. Additionally, specific biological affinity reagents (peptides, nucleic acids, antibodies) that recognize synthetic macromolecules can be made and used for separation and detection of synthetic macromolecules as well as assembly of complexes of synthetic nanomaterials.

Most importantly, biotechnology provides a ready source of prefabricated functional components (biological macromolecules) for use in nanodevices. Proteins can be made in gram to kilogram quantities, and proteins can perform a truly dazzling diversity of chemical activities. *Devices containing both biological and synthetic components organized at the nanoscale to do useful work may be described as nanobiological devices* [26–29] *or nanobiotechnological devices* [1, 16, 64]. Biotechnology offers a unique source of functional nanocomponents that cannot be obtained using synthetic technology. Nanodevices built entirely from biological macromolecules may be best suited to some tasks and working environments, see refs 30, 36, 37 for example. Additionally, individual functional biologic nanostructures and devices might be integrated into higher order (perhaps even macroscopic) structures, much as organisms are integrated aggregations of nanoscale subsystems.

Nanosynthetics serve primarily as structural components, providing the higher-order structure to the biological components that allows their individual activities to sum to a work process. Conversely, biomolecules can be scaffolds around which to organize synthetic molecules – 'biogenic structures' [3, 41, 45, 54, 70]. These two approaches are complementary, and there is no reason that individual components cannot contribute to both function and morphology. This paper (arbitrarily) focuses on devices in which synthetic materials are used primarily as inert 'scaffolds' for functional biological components.

2 BUILDING NANOBIOLOGICAL DEVICES

2.1 DESIGN

Currently we do not understand the organizational logic of multicomponent biological systems. For that reason, nanobiological device design is in its infancy and remains a challenge. However, there is growing interest in understanding how multiple biological components are functionally integrated to produce useful work [13, 19, 56, 66, 68]. Understanding these issues will not only provide new insights on the treatment of disease or biological dysfunction, but will eventually allow incorporation of analogous organizational features into nanobiological devices. Still, nanobiotechnologists today are left to their own devices as far as design is concerned. The initial step in design is definition of a desired task that is targeted to a nanodevice. The task is then divided into subroutines that constitute steps in the work process, followed by identification of biological macromolecules that carry out those subroutines.

2.2 ENGINEERING COMPONENTS

A range of data should be considered to select biological components of nanobiological devices (genetic, biological, biochemical, physical, therapeutic, regulatory, historical and patent), but no 'searchable' database integrating this information is presently available. The information must be mined laboriously from the literature for each molecule individually. Hence, it is often difficult to identify 'device-ready' components from nature. Desired activities can be engineered into existing molecules or identified *de novo* from libraries (display libraries [6]). While there has been dramatic progress in protein engineering in the last 10 years [2, 23, 29, 43, 57], there still remain many unanswered questions concerning folding properties and integration with abiotic components. However, enough is known to be encouraged to exploit existing protein species when possible [73].

Similar exercises must be performed to identify synthetic nanocomponents. Considerations include suitability for the proposed environment, synthesis and handling properties, polydispersity, structural/chemical properties, as well as amenability to assembly into higher order structures. For medical devices, tolerability and safety of the structural materials is also an issue. Current materials technology offers powerful, but limited capacity to engineer an 'off the shelf' approach to nanostructures.

2.3 ASSEMBLY

Parallel assembly processes to make small numbers of 'hand made' nanostructures by direct intervention of an operator have been known for years, however these technologies do not appear amenable to mass production manufacturing in the near future [9, 26–29, 40]. Ideal assembly processes will undoubtedly be tailored to allow rapid bulk device construction in aqueous solutions. Construction driven by component self-assembly properties is attractive, but is not sufficient to build all nanostructures [3, 15, 34, 54, 55, 69]. For that reason site-specific conjugation of chemically dissimilar components in solution is desirable. The problem has two challenges: control of the position of conjugation in the biological and control of position of conjugation in the nanosynthetic. Positional control of conjugation in the biological is essential for bioactivity [44]. Promiscuous conjugation chemistry (such as carbodiimide methods) inactivates proteins, whereas orthogonal coupling methods [35] often allow retention of biological activity (see [31–33, 46]). Chemoselective chemistries that operate efficiently in aqueous solution allow one to manufacture specific biologic nanostructures in bulk [26–29, 35, 44, 53].

Proteins generally have either a pocket (clefts) which bind substrates (in enzymes) or a face that interacts with other macromolecules or surfaces (in receptors); if these sites are occluded, bioactivity suffers. In nanobiological devices, orthogonal coupling of the protein to the device may not fully address these concerns and recourse to a linker or global reorganization of the biomolecule (protein circular permutation, 20–22) may be needed. Positional control of conjugation within a synthetic nanomaterial is more difficult, though controlled nanoscale arrays are potentially useful in many applications. Unfortunately, repetitive nanomaterial surface features can be difficult to distinguish chemically from one another for conjugation. No general solution to this problem is available at this time, though methodology specific for particular nanomaterials have been developed. Ruoff and colleagues [49] proposed that mechanical stress can make the carbon–carbon bonds of nanotubes differentially labile, allowing site-specific 'mechanosynthesis'. Introducing mechanical strain at regular intervals along nanotubes might allow functional molecules to be arranged in linear arrays. Dendritic polymers can be synthesized to contain a single reactive group [18, 74], and conjugation at that position gives a specific, homogeneous product. The 'packing' properties of dendrimers might be exploited to give so-called 'regiospecific control' of conjugation [61]. Lithographic approaches have been extended to patterning surfaces coated with proteins or dendrimers to produce surfaces that can be site-specifically derivitized [47, 48, 49, 60] and biological molecules can also be regularly deployed using soft lithography [68]. Minimum feature size using these methods is currently limited to hundreds of nanometers, however, this may be reduced substantially using dendrimers [75]. Even if nanomaterial surface residues cannot be differentiated chemically, adducts might be placed site-specifically using assembly jigs [41]. Specific affinity reagents could also be used to protect regions of synthetic surfaces while the unprotected regions are activated or passivated. Antibodies to nanostructures (i.e. antidendimer antibodies [33]) are exquisitely specific and could be used to 'footprint' dendrimers for subsequent site-specific orthogonal conjugation.

2.4 ANALYSIS

After device construction, structural and functional analysis are critical. One might argue that only the second issue matters, but structural data often give insights into why devices perform suboptimally, and provide important clues about how to improve device function. We routinely use protein analytics (matrix-assisted laser desorption-ionization mass spectroscopy, amino acid composition analysis, gel electrophoresis, Western blotting, circular dichroism, various high-performance liquid chromatography methods, etc.) and biological analytical techniques (receptor binding, cell proliferation, enzyme activity assays, enzyme-linked immunosorbent assays, flow cytometry, etc.) to characterize protein–nanoparticle conjugates. Conjugate troubleshooting is driven by the entirety of the analytical package. Robust analytical capabilities are particularly important early on, when knowledge of material properties and design logic is limited. An understanding of how biological nanosystems integrate productively is rudimentary at best. For that reason, the capacity to make multiple design variants in parallel in small quantities (i.e. as few as one copy of each variant) is important, and analytical assay systems must be highly sensitive, and able to handle multiple test objects simultaneously.

3 CHARACTERISTICS OF NANOBIOLOGICAL DEVICES

Nanobiological devices exist today [25, 30, 36, 37]. See [16, 27, 28, 64] for reviews], and some of them perform complex, multistep work processes. For example, the bacmid molecular cloning system [30, 36, 37] produces recombinant baculoviral expression vectors by a multistep process that is the result of the summed activities of molecular components taken from dozens of distinct biological sources. The sophistication of the device is possible since the logic of molecular genetics is well defined, and the engineering required to make and assemble components to carry out the work process is straightforward, though laborious. High operational sophistication is less common in functional protein devices (though not unprecedented, [10, 11]). This is essentially a design and construction issue. Undoubtedly, the problem will be mitigated as our understanding of integration of biological systems grows [13, 19, 55, 67, 69].

Biological molecules are labile compared to most abiotic synthetics. Proteins and other molecules isolated from extremophiles (microorganisms that live at extremes of temperature, osmolarity, pH, radiation, organic solvents, etc.) are more robust than their homologues from mesophiles and so may be well suited for devices operating in similar environments. Additionally, naturally occurring nanobiological devices (living things) will treat synthetic components as resources and attempt to utilize them, however, enzymatic degradation of biological components will be a key challenge. Labile components of nanobiological devices might be encapsulated in a membrane. Alternatively, genetic and chemical methods to increase protein stability are available [2, 17, 23, 29, 35, 43, 58]. Synthetic proteins containing amino acids other than those used by living things can be highly resistant to enzymatic degradation (i.e. protease insensitivity exhibited by D-proteins [7]). Still, biological macromolecules are less stable than many synthetic materials. Devices will be as robust as their most labile components. As device complexity increases, so will the difficulty of engineering all biological components for stability increase. Nanobiological devices will have to be designed to fit their intended use or environment and may fail at high temperature, in chaotropes or solvents, at high radiation levels, or in the presence living things. Unless components can be stabilized, tasks involving extreme conditions may not be amenable to a nanobiological solution. Still, there are conditions where biological macromolecules are more or less stable, corresponding to the human world. This regime is of interest to humans, and is where nanobiological devices may offer the greatest benefits.

4 DENDRIMERS IN NANOBIOLOGICAL DEVICES

Dendrimers are well suited for use in many nanobiological devices. They are highly monodisperse, and while expensive on a per gram basis compared to other synthetic macromolecules, are *less expensive per unit mass than commercially produced proteins*. PAMAMs are water soluble and can be modified for solubility in organic solvents. PAMAMs are also well tolerated in animals [50] and are non-immunogenic [5, 32, 50, 62]. However, some PAMAM dendrimer–protein bioconjugates are not [32–33]. Dendrimer and dendrigraft synthesis allows a degree of control over *size, surface topography/functionality* and *particle shape*. Dendritic architecture accommodates chemically diverse types of core and subunit compositions, allowing the control of surface chemistry/composition, and giving further control of particle size, topography and shape. PAMAM dendrimers support multiple orthogonal chemistries for protein conjugation [31–33]. Limited regio-specific positional control over conjugation sites within dendritic polymers is available by limiting biomolecule conjugation to a single position in a dendron [18] or by regiospecific control of conjugation that takes advantage of 'packing' properties of dendrimers [61]. Antibody interactions with PAMAM dendrimers have been demonstrated and might be used in a 'footprinting' approach to pattern dendrimer surfaces [33].

Many simple nanobiological devices have been made with dendrimers [14]. Antibodies have been conjugated to dendrimers for use in immunoassays and the conjugates enhance the sensitivity, precision, accuracy and speed of the assays [58, 59]. Multiple antibody species can be conjugated simultaneoulsly to produce multifunctional reagents [59]. The conjugates retain the specific binding properties of the antibodies, but can be conveniently isolated by virtue of the high charge density of the dendrimers. (see Chapter 19)

Gadolinium–dendrimer conjugates have been used as blood pool contrast agents *in vivo* for nuclear magnetic resonance imaging (MRI) of tumors [65]. The efficacy of the conjugates in such applications is dependent on their biodistribution properties, and these properties vary as a function of dendrimer molecular weight and chemical composition [50]. Dendrimer architecture and synthesis allow these properties to be tuned, and so tailored contrast agents are possible. Targeting proteins have been used to further control the biodistribution of dendrimer-metal chelates [66].

PAMAM dendrimers are taken up efficiently by eukaryotic and prokaryotic cells, and are used as carriers for nucleic acids [5, 24]. These reagents are candidates for gene therapy, but are not particularly selective with respect to the cell types they transfect, implying the need for a targeting function (antibodies or other protein targeting agents) to deliver DNA to desired sites. To date, the reported propensity of PAMAM dendrimers to accumulate in specific organs as a function of their generation [50] has not been exploited for gene therapy.

Dendrimers are attractive vectors for the delivery of chemotherapeutics to tumors. Much like the polymer therapeutic PK1 [10, 11], dendrimer–cytotoxin/dendrimer–metal conjugates expolit the enhanced permeability and retention properties of tumor tissue to deliver potentially systemically toxic materials preferentially to the desired site of therapeutic action [49, 67]. Antibodies to tumor specific antigens can be conjugated to dendrimers as a means to enhance the tumor targeting specificities of dendrimer antitumor agents [49]. Control of biodistribution and activity of cytotoxins by polymer therapeutic strategies radically reduces dose-limiting systemic toxicity, in part because the polymer-linked antitumors are significantly less toxic than are the free toxins, making more intensive therapy possible [10, 11].

Glycodendrimers allow the presentation of high valency sugar residues on their surfaces [53]. They are powerful reagents to study the biology of sugar interactions with their counter-receptors, lectins. Monovalent sugar–lectin interactions are weak, and technically difficult to study. Glycodendrimers are also powerful competitive inhibitors for sugar-lectin binding, and may be used therapeutically to inhibit processes that depend on that binding. Glycodendrimers are potential anti-infectives, competing with cell surfaces for pathogen binding and potential anti-metastatics by inhibiting attachment of cancer cells to sites distal to tumors [53]. Utility will depend on the pharmacology of glycodendrimers (biodistribution, circulating half-life, immunogenicity, etc.).

PAMAM dendrimer–cytokine conjugates have been made [31–33], and orthogonal conjugation methods were used to limit the position of conjugation to the N-terminal amine of the protein. Multiple copies of hIL-3 were linked to each dendrimer, and the conjugates were active in an hIL-3-dependent cell proliferation assay [31, 32]. However, the potency of the conjugate was less than that of the free protein, despite the higher valency of the conjugate. Also, dendrimer-specific antibody responses occurred in animals receiving the conjugates [32]. Antibodies capable of specifically recognizing unmodified PAMAM dendrimers, as well as PAMAM dendrimers modified to have oxiamine and sulfhydryl surfaces have now been generated [33]. A covalent linkage of the protein to dendrimer is necessary to trigger the responses: presumably the proteins 'haptenize' the dendrimers [32].

These antibodies are useful in immunological detection and quantitation of dendrimers, and we expect they will be useful to pattern dendrimers to support site-specific conjugation [33]. However, the data demonstrate that covalent linkage of dendrimers to proteins may render them immunogenic. In applications where proteins are used to target therapeutic dendrimer conjugates, anti-dendrimer antibody responses may interfere with biodistribution, clearance and therapeutic efficacy. This should be considered for dendrimer nanobiological devices intended for long-term *in vivo* use.

5 SUMMARY AND PROSPECTS

Biological macromolecules are excellent candidates as functional components of nanodevices. Dendritic polymers are particularly attractive as nanodevice components (they exhibit high monodispersity, tunable particle size, shape and surface chemistry, self-assembly properties, etc.) and PAMAM dendrimers are well tolerated in animals. Many dendrimer-containing nanobiological devices have been made, demonstrating that dendrimers are suitable frameworks to be decorated with functional proteins and nucleic acids. Dendrimers are most often structural components, though they contribute directly to the work process in some cases (by binding pathogens or conveying materials across membranes, for instance). Most of these devices are relatively simple, though there are several examples of devices incorporating multiple functionalities into a single nanoassembly. These sophisticated devices perform multistep work processes (for example, targeting to a site, transiting the cell membrane and making intracellular delivery of effector molecules). As knowledge of the integration of biological nanosystems into functional macroscale organisms matures, that learning will be applied to more complex nanobiological devices. The self-assembly properties of dendrimers make them particularly attractive for use in sophisticated, complex nanosystems. See Chapters 6 and 27. The propensity of covalently linked proteins to haptenize dendrimers to generate antidendrimer antibody responses will have to be considered for devices that will be administered to immunocompetent hosts. This caveat aside, as the large scale manufacturing of biologic nanodevices emerges over the next 10–20 years, dendritic polymers will undoubtedly be key components.

6 REFERENCES

1. Anonymous. Announcement for the Nara Institute of Science and Technology International Symposium: Nanotechnology and biotechnology for future devices. Nara Institute of Science and Technology, Nara, Japan, 1998.
2. Arnold, F. H. *Curr. Opin. Biotechnol.* **4**, 450 (1993).

3. Askay, I. Nanostructured ceramics through self-assembly, in Seigel, R. W., Hu, E. and Roco, M. C. (eds) *R & D Status and Trends in Nanoparticles, Nanostructured Materials and Nanodevices in the United States*, International Technology Research Institute, Baltimore, MD, USA, 1998.
4. Baker, J., Jr Therapeutic nanodevices, in Lee, S. C. and Savage, L. (eds) *Biological Molecules in Nanotechnology: the Convergence of Biotechnology, Polymer Chemistry and Materials Science*, IBC Press, Southborough, MA, USA, 1998, pp. 173–183.
5. Barth, R. F., Adams, D., Soloway, A. H., Alam, F. and Darby, M. V. *Bioconjugate Chem.*, **5**, 58 (1994).
6. Clackson T. and Wells J. A. *TIBTech.*, **12**, 173 (1994).
7. deLisle, M. R., Milton, S., Schnozler, M. and Kent, S. B. H. Synthesis of proteins by chemical ligation of unprotected peptide segments: mirror image enzyme molecules, D and L-HIV protease analogues, in Angeletti, R. (ed.) *Techniques in Protein chemistry IV*, Academic Press, New York, NY, USA, 1993, pp. 257–267.
8. Drexler, K. E. *Engines of Creation: the Coming Era of Nanotechnology*. Anchor Books, New York, NY, USA, 1986.
9. Drexler, K. E. *TIBTech.*, **17**, 5 (1999).
10. Duncan, R. *Chemistry & Industry*, **7**, 262 (1997).
11. Duncan, R., Seymour, L. W., O'Hare, K. B., Flanagan, P. A., Wedge, S., Hume, I. C., Ulbrich, K., Stolholm, J., Subr, V., Spreafico, F., Forou, M. and Suarato, A. *J. Controlled Release*, **19**, 331 (1994).
12. Feynman, R. P. There's plenty of room at the bottom: an invitation to enter a new field of physics. Http//nano.xerox.com/nanotech/feynman.html, 1959.
13. Goldenfeld, N. and Kadanoff, L. P. *Science*, **284**, 87 (1999).
14. Grohn, F., Bauer, B. J. and Amis, E. J. Characterization, modelling and applications of dendritic polymers, in *Abstracts to the NIST Workshop on Properties and Applications of Dendritic Polymers*. National Institute of Standards and Technology, Gaithersburg, MD, USA, 1998.
15. Heller, M. J. Utilization of synthetic DNA for molecular electronic and photonic-based device applications, in Lee, S. C. and Savage, L. (eds), *Biological Molecules in Nanotechnology: the Convergence of Biotechnology, Polymer Chemistry and Materials Science*, IBC Press, Southborough, MA, USA, 1998, pp. 59–66.
16. Jelinski, L. Overview: biological, carbon and theory issues, in Seigel, R. W., Hu, E. and Roco, M. C. (eds), *R & D Status and Trends in Nanoparticles, Nanostructured Materials and Nanodevices in the United States*, International Technology Research Institute, Baltimore, MD, USA, 1997, pp. 161–169.
17. Kent, S. B. H. Building proteins through chemistry: total synthesis of protein molecules by chemical ligation of unprotected peptide segments, in Lee, S. C. and Savage, L. (eds), *Biological Molecules in Nanotechnology: the Convergence of Biotechnology, Polymer Chemistry and Materials Science*, IBC Press, Southborough, MA, USA, 1998, pp. 75–92.
18. Klimash, J. W., Brothers, H. M., Swanson, D. R., Yin, R., Spindler, R., Tomalia, D. A. Hsu, Y. and Cheng, R. C. Disulfide-containing dendritic particles. US Patent 6,020,457 (2000).
19. Koch, C. and Laurent, G. *Science*, **284**, 96 (1999).
20. Kreitman R. J., Puri R. K., McPhie P. and Pastan I. *Cytokine*, **7**, 311 (1995).
21. Kreitman R. J., Puri R. K. and Pastan I. *Cancer Res.*, **55**, 3357 (1995).
22. Kreitman, R. J., Puri, R. K. and Pastan, I. *Proc. Nat'l. Acad. Sci., USA*, **91**, 6889 (1994).
23. Kuchner, O. and Arnold, F. H. *TIBTech.*, **15**, 523 (1997).
24. Kukoska-Latallo, J. F., Bielinska, A. U. Johnson, J. Tomalia, D. and Baker, J. R., Jr. PNAS **93**, 4897 (1996).

25. Lakshmi, B. B. and Martin, C. R. *Nature*, **388**, 753 (1997).
26. Lee, S. C. *TIBTech.*, **16**, 239 (1998).
27. Lee, S. C. How molecular biologists can wind up organizing nanotechnology meetings, in Lee, S. C. and Savage, L. (eds), *Biological Molecules in Nanotechnology: the Convergence of Biotechnology, Polymer Chemistry and Materials Science*, IBC Press, Southborough, MA, USA, 1998, pp. i–v.
28. Lee, S. C. The nanobiological strategy for construction of nanodevices. in Lee, S. C. and Savage, L. (eds), *Biological Molecules in Nanotechnology: the Convergence of Biotechnology, Polymer Chemistry and Materials Science*, IBC Press, Southborough, MA, USA, 1998, pp. 3–14.
29. Lee, S. C. Molecular genetic techniques for engineering the protein components of nanobiological devices, in Lee, S. C. and Savage, L. (eds), *Biological Molecules in Nanotechnology: the Convergence of Biotechnology, Polymer Chemistry and Materials Science*, IBC Press, Southborough, MA, USA, 1998, pp. 67–74.
30. Lee, S. C., Leusch, M. S., Luckow, V. A. and Olins, P. O. Method of producing recombinant eukaryotic viruses in bacteria. US Patent No. 5, 348,886, (1994).
31. Lee, S. C., Parthasarathy, R. and Botwin, K. *Polymer Preprints*, **40**, 449 (1999).
32. Lee, S. C., Parthasarathy, R., Botwin, K., Kunneman, D., Rowold, E., Zobel, J., Beck, T., Miller, T. and Voliva, C. A. Biochemical and immunological properties of cytokines conjugated to dendritic polymers. Manuscript in preparation.
33. Lee, S. C., Parthasarathy, R., Botwin, K., Rowold, E., Miller, T. and Voliva, C. A. Characterization of antisera to dendritic polymers and their use in immunological detection of PAMAM dendrimers. (See Chapter 23).
34. Leonard, D., Krishamurthy, M., Reaves, C. M., Denbaars, S. P., and Petroff, P. M. *Applied Physics Letters*, **63**, 3205 (1997).
35. Lemeux, G. A. and Bertozzi, C. R. *TIBTech.*, **16**, 506 (1998).
36. Leusch, M. S., Lee, S. C. and Olins, P. O. *Gene*, **160**, 191 (1995).
37. Luckow, V. A., Lee, S. C., Barry, G. F. and Olins, P. O. *J. Virol.*, **67**, 4566 (1993).
38. Lindsay, S. M. The scanning probe microscope in biology, in Bonell, D. (ed.), *Scanning Tunnelling Microscopy and Related Techniques*, John Wiley & Sons, London, in press.
39. Merkle, R. C. Whither nanotechnology? in Seigel, R. W., Hu, E. and Roco, M. C. (eds) *R & D Status and Trends in Nanoparticles, Nanostructured Materials and Nanodevices in the United States*, International Technology Research Institute, Baltimore, MD, USA, 1997, pp. 156–160.
40. Merkle, R. C. *TIBtech.*, **17**, 271 (1999).
41. Morse, D. E. *TIBTech.*, **17**, 230 (1999).
42. Noji, H., Yasida, R., Yoshida, M. and Kinosita, K., Jr. *Nature*, **386**, 299 (1997).
43. O'Fagain, C. *Biochim. Biophys. Acta.*, **528**, 1252 (1995).
44. Offord, R. and Rose, K. Multicomponent synthetic constructs: controlling the specificity of activation of biological molecules for coupling to polymeric supports and analysis of coupled products, in Lee, S. C. and Savage, L. (eds), *Biological Molecules in Nanotechnology: the Convergence of Biotechnology, Polymer Chemistry and Materials Science*, IBC Press, Southborough, MA, USA, 1998, pp. 93–106.
45. Parkinson, J. and Gordon, R. *TIBTech.*, **17**, 190 (1999).
46. Parthasarathy, R. and Lee, S. C. Site-specific ligation of proteins to synthetic particles. US Patent application, pending (1999).
47. Pum, D. and Sleytr, U. Using crystalline bacterial surface layers as patterning elements in molecular nanotechnology, in Lee, S. C. and Savage, L. (eds), *Biological Molecules in Nanotechnology: the Convergence of Biotechnology, Polymer Chemistry and Materials Science*, IBC Press, Southborough, MA, USA, 1998, pp. 139–144.
48. Pum, D. and Sleytr, U. E. *TIBTech.*, **17**, 8 (1999).

49. Roberts, J., Adams, Y. E., Tomalia, D., Mercer-Smith, J. A. and Lavallee, D. K. *Bioconjugate Chemistry*, **2**, 305 (1990).
50. Roberts, J., Bhalgat, M. and Zera, R. T. (1996) Preliminary evaluation of poly-amidoamine (PAMAM) starburst dendrimers. *J. Biomedical. Materials Res.*, **30**, 53 (1996).
51. Rohrs, H. W. and Ruoff, R. S. Use of carbon nanotubes in hybrid nanometer scale devices, in Lee, S. C. and Savage, L. (eds), *Biological Molecules in Nanotechnology: the Convergence of Biotechnology, Polymer Chemistry and Materials Science*, IBC Press, Southborough, MA, USA, 1998, pp. 33–38.
52. Roy, R. *Polymer News* **21**, 226 (1996).
53. Rose, K., Zeng, W., Brown, L. E. and Jackson, D. C. *Mol. Immunol.*, **32**, 1031 (1995).
54. Sarikaya, I. and Askay, I. A. *Design and Processing Materials by Biomimicking.* American Institute of Physics, New York, NY, USA, 1994.
55. Seeman, N. C., Chen J., Zhang Z., Lu, B., Qui, H., Fu, T.-J., Wang, Y., Li, X., Qi, J., Lu, F., Wenzler. L. A., Du, S., Mueller, J. E., Wang, H., Mao, C., Sun, W., Shen Z., Wong, M. H. and Sha, R. A bottom up approach to nanotechnology using DNA. in Lee, S. C. and Savage, L. (eds) *Biological Molecules in Nanotechnology: the Convergence of Biotechnology, Polymer Chemistry and Materials Science*, IBC Press, Southborough, MA, USA, 1998, pp. 45–58.
56. Service, R. *Science*, **284**, 80 (1999).
57. Shaw, A. and Bott, R. *Curr. Opin. Struct. Biol.*, **6**, 546 (1996).
58. Singh, P., Moll, F., Lin S. H., Ferzli, C., Yu, K. S., Koski, R. K., Saul, R. G. and Cronin, P. *Clin Chem.*, **40**, 1845 (1996).
59. Singh, P., Moll, F., Lin S. H., and Ferzli, C. *Clin Chem.*, **42**, 1845 (1996).
60. Sleyter, U. B. and Sara, M. *TIBTech.*, **15**, 20 (1997).
61. Tomalia, D. A. and Brothers, H. M., II. Regiospecific conjugation to dendritic polymers to produce nanodevices, in Lee, S. C. & Savage, L. (eds) *Biological Molecules in Nanotechnology: the Convergence of Biotechnology, Polymer Chemistry and Materials Science*, IBC Press, Southborough, MA, USA, 1998, pp. 107–122.
62. Toyokuni, T. and Singhal, A. K. *Chem. Soc. Rev.*, **22**, 231 (1995).
63. Uhrich, K. *TRIP*, **5**, 388 (1997).
64. US–EC Task Force on biotechnology research. Thinking small on a global scale: the international outlook for nanobiotechnology. http//www.bio.Cornell.edu. nanobiotech/nbt.htm, 1998.
65. Wiener E. C., Brechbeil, M. W., Brothers, H. M. Magin, R. L., Gansow, O. A., Tomalia, D. A., and Lauterber, P. C. *Magn. Reson. Med.*, **31**, 1 (1994).
66. Weiner, E., Konda, S., Brechbeil, M. and Gansow, O. *Invest. Radiol.*, **32**, 748 (1997).
67. Weng, G., Bhalla, U. S. and Iyengar, R. *Science*, **284**, 92 (1999).
68. Whitesides, G. M. Soft lithography. in Seigel, R. W., Hu, E. and Roco, M. C. (eds), *R & D Status and Trends in Nanoparticles, Nanostructured Materials and Nanodevices in the United States*, International Technology Research Institute, Baltimore, MD, USA, 1998.
69. Whitesides, G. M. and Ismagilov, R. F. *Science*, **284**, 89 (1999).
70. Wu, C. Brechbiel, M. W. Kozak, R. and Gansow, O. *Bioorg. and Med. Chem Letter*, **3**, 449 (1994).
71. Zhang, Y. and Seeman, N. C. *J. Am. Chem. Soc.*, **114**, 2656 (1992).
72. Tomalia, D. A. *Scientific American*, **272**(5), 62 (1995).
73. Lvov, Y. and Möhwald, H. (eds.) in *Protein Architecture*, Marcel Dekker, New York, 2000.
74. Fréchet, J. M. J. *Science*, **263**, 1710 (1994).
75. Tully, D. C., Wilder, K., Fréchet, J. M. J., Trimble, A. R. and Quate, C. F. *Adv. Mater.*, **11**, 314 (1999).

23

Antibodies to PAMAM Dendrimers: Reagents for Immune Detection, Patterning and Assembly of Dendrimers

S. C. LEE*, R. PARTHASARATHY, T. D. DUFFIN,
K. BOTWIN, J. ZOBEL, T. BECK, R. JANSSON,
G. LANGE D. KUNNEMAN, E. ROWOLD AND
C. F. VOLIVA
Monsanto Company, 700 Chesterfield Village Parkway, North AA4C; St Louis,
MO, USA

1 INTRODUCTION

Dendrimers are nanoscale polymers with fractal architecture [4, 13], and are among the most monodisperse synthetic nanomaterials known. They are attractive for applications in which structural uniformity is desirable, such as in pharmaceuticals and nanotechnology. Antibodies are vertebrate immune proteins that bind foreign macromolecules (in various contexts referred to as immunogens, antigens or epitopes) in a highly specific fashion, and are immensely useful in biotechnology for antigen detection, quantitation and separation [3, 5]. Anti-dendrimer antibodies would be expected to be interesting reagents, but attempts to generate these antibodies using unmodified PAMAM dendrimers as immunogens have been unsuccessful [1, 2, 12, 14].

Recently three murine antisera to PAMAM dendrimers have been generated using dendrimer-protein conjugates as immunogens [9, 10]. The recognition properties of the three sera are exquisitely specific, recognizing unmodified

Dendrimers and Other Dendritic Polymers. Edited by Jean M. J. Fréchet and Donald A. Tomalia
© 2001 John Wiley & Sons Ltd

PAMAM dendrimers or modified PAMAM dendrimers (with oxiamine or sulfhydryl surface functionalities) exclusive of one another. These antisera recognize den-drimers in ELISA (enzyme-linked immunosorbent assays), dot blots and Western blots. The immunogenicity of dendrimer–protein conjugates has implications for therapeutic use of dendrimers as vaccines and we anticipate that antidendrimer antibodies will have applications in patterning and assembling nanostructures containing dendrimers.

2   GENERATING ANTI-DENDRIMER ANTIBODIES AND THEIR SPECIFICITY

Dendritic polymers are intrinsically poor immunogens, though dendrimer–protein conjugates can be quite immunogenic (Figure 23.1) [9]. Covalently linked proteins seem to haptenize dendrimer epitopes [9, 10], which are presumably not effectively presented to T-helper cells on their own. We have generated antibodies to PAMAM dendrimers using conjugates to multiple carrier proteins [9, 10]

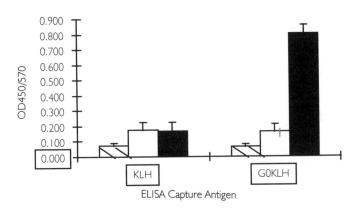

Figure 23.1   An antibody response to PAMAM dendrimer-BSA conjugates in mice [9, 10], as measured by capture ELISA assay. Capture antigens (KLH or G_0–KLH conjugate) were deployed on the wells of microtiter plates. Serum from unimmunized mice (cross-hatched bars), or mice immunized with either BSA (white bars) or dendrimer-BSA conjugates (black bars) were added to the wells and allowed to bind the capture antigens. Murine antibodies in the test sera that bound the capture antigen were detected using a commercially available secondary antibody directed to murine IgG antibodies and conjugated to horse radish peroxidase (HRP). Bound HRP activity is read spectrophotometrically using commercially available chromogenic substrates. Optical density (OD) 450nM/OD 570nM ratio is directly proportional to specifically bound enzyme activity (on the Y axis) and is directly proportional to the titer of antibodies in the tested sera that recognize the capture antigens

and believe that many protein PAMAMs may be immunogenic, making anti-dendrimer antibody responses a consideration for therapeutic protein–dendrimer bioconjugates. We have generated antibodies to both G_0 and G_5 PAMAMs, though we have not attempted to produce antibodies to higher generation PAMAM dendrimers.

Our anti-dendrimer sera recognize specific chemical functionalities on the dendrimer surface [10] . Sera raised to unmodified PAMAM dendrimers recognize the primary amines of the dendrimer surface, but do not recognize PAMAM dendrimers whose surfaces amines are replaced with either oxiamine or sulfhydryl groups. Sera raised to PAMAMs whose surfaces amines are derivatized to either oxiamine or sulfhydryl groups, recognize their cognate immunogens, but not dendrimers with other surface functionalities. However, surface chemical functionalities alone do not determine antibody binding. Sera generated to PAMAM den-drimers do not recognize POPAM dendrimers, despite the presence of multiple primary amines on the surfaces of both polymers.

Antibodies raised to one generation of PAMAM dendrimers antibodies recognize other generations of PAMAM dendrimers with the same surface chemistry. for example, antibodies raised to a G_0 PAMAM dendrimer–protein conjugate additionally recognize other generations of PAMAM dendrimers [10]. This may reflect the fractal nature of dendrimers, in that G_0 dendrimers have epitopes in common with higher generation dendrimers. *In toto*, the observations suggest that while surface residues are necessary to antibody-dendrimer binding, additional factors like three-dimensional arrangement of those residues may also play a role in specific recognition.

3 IMMUNE DETECTION OF DENDRIMERS

Immune detection is a key utility of antibodies in biotechnology [3, 5]. Antidendrimer sera efficiently detect dendrimers in multiple assay formats, including enzyme-linked immunosorbent assays (ELISA), and in Western and dot blots [3, 5]. ELISA assays are commonly used to quantitate proteins, and a quantitative ELISA could be developed for dendrimers using our sera, though doing so would require development of dendrimer standards of known concentration that could be used for calibration.

Our antisera readily recognize dendrimers immobilized on nitrocellulose or other solid surfaces. The primary utility of these types of assays, particularly in Western blot formats, is to generate qualitative structural/chemical information when making dendrimer bioconjugates. For instance, the sera can be used to follow the efficiency of surface modifications. As amines are more completely replaced with other chemical groups, the amine-specific antibody binding declines and the binding of antibodies that recognize the new surface increases.

Western blots of dendrimer–protein conjugates can also be used to characterize the valency of protein–dendrimer conjugates.

Antibodies are powerful tools in pharmaceutical development, and as dendrimers are incorporated into various drugs and medical devices, antibodies such as described should be useful to study dendrimer biodistribution and pharmacokinetic properties. We have not attempted to detect PAMAM dendrimers in biological fluids (blood, urine, sputum, etc.), but expect that it would work. Additionally, the antibodies might be used in immunohistochemistry to study the tissue and cellular distribution of dendrimer therapeutics.

4 ANTIBODIES AS ASSEMBLY REAGENTS

The immune detection methods we have described for dendrimers involve powerful, though fairly commonplace, applications of antibody reagents. Antibodies might additionally satisfy current technological needs in nanotechnology, specifically for controlled assembly of nanoscale components. This could include nanopatterning to give positional control for conjugation of macromolecules to chemically homogeneous synthetic surfaces [6–8]. For these applications, monoclonal antibodies [3, 5] are preferable to the polyclonal sera currently in hand, and we are presently generating antidendrimer monoclonals.

Naturally occurring antibodies are monospecific, bivalent antigen-binding molecules [3, 5], (Figure 23.2). Their capacity to form relatively disordered antigen/antibody lattices is well known [3, 5], and will be applicable to forming lattices of dendrimers, but more regular supramolecular structures can also be made (Figure 23.2). For instance, under appropriate stoichiometric conditions, antibodies will drive assembly of two copies of their cognate dendrimer antigens (Figure 23.2). Engineered bispecific antibodies can also be used to assemble distinct heterogeneous dendrimers (for instance, PAMAM dendrimers with chemically distinct surfaces). Individual dendrimers might be decorated with different functional biological molecules, and thus antibodies could be used to assemble multicomponent nanobiological devices. This application may in some cases obviate the need for individually patterned nanoparticles (see below), but if these approaches are used with appropriately patterned and derivatized dendrimers, specific supramolecular structures of high complexity could be assembled.

5 ANTIBODIES AS NANOPATTERNING REAGENTS

Positional control of conjugation is essential at two levels in the assembly of nanobiological devices [6–8]. Recently developed chemoselective conjugation

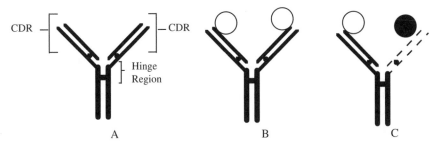

CDR — Hinge Region — CDR

A   B   C

Figure 23.2 Antibody structure and the use of antidendrimer antibodies to construct supramolecular assemblies of dendrimers. (A) Antibodies [3,5] are 150 000 molecular weight complexes composed of four separate polypeptide chains cross-linked by disulfide linkages. Each antibody has two complementary determining regions (CDRs), which in native antibodies recognize identical epitopes. (B) Two identical PAMAM dendrimers (indicated by the open circles) assembled as a homodimer by binding one to each identical CDR of the antibody. (C) Two distinct PAMAM dendrimers (for example, one with an amine surface, open circle, and one modified to have a sulfhydryl surface, black circle) assembled as a heterodimer by binding one to each CDR of the antibody. In this case, the antibody is bispecific [3], having been engineered to have two distinct CDRs, each recognizing a different epitope (in this example, recognizing amine-terminated PAMAMS, indicated by solid lines, and recognizing sulfhydryl terminated PAMAMs, indicated by the dashed lines)

chemistries [11] largely satisfy the need for positional control within biological macromolecules. However, due to the relative chemical homogeneity of many nanosynthetic surfaces, positional control of conjugation is generally more difficult to achieve. For instance, one amine on the surface of a PAMAM dendrimer is difficult to distinguish from all the others. Antibodies offer a way to pattern dendrimers to support chemoselective conjugation limited to specific regions of the dendrimer. As with assembly applications, high affinity monoclonal antibodies are preferable to the polyvalent antisera currently available for dendrimer patterning.

Antibodies reproducibly bind their cognate epitopes: consequently, they occlude specific sites on proteins and other nanostructures reproducibly [3, 5]. The occluded region is sometimes called the 'footprint' of the antibody on its antigen. The footprinted region of the antigen is protected from further chemical or enzymatic modification. Since antibodies are bivalent but monospecific, in a 1:1 complex of antigen to antibody, two areas of the antigen can be protected, each corresponding to the binding of one of the antibody complementarity determining regions (CDR). The shape and size of the protected region is a function of the shape and size of the CDRs, as well as the relative position of the protected regions. This is a function of the CDRs in three-dimensional space, as determined by the structure of the antibody. Thus, antibodies are potentially useful for patterning nanostructures to obtain limited positional control, and application of antibodies to dendrimer patterning is depicted schematically in Figure 23.3.

Monoclonal antibodies to sulfhydryl-terminated PAMAM dendrimers [9, 10] could be bound to sulfhydryl-terminated PAMAM dendrimers under stoichiometric conditions favoring complexes of one antibody to one dendrimer. After binding, the exposed sulfhydryl surface may be derivatized to eliminate SH groups. Antibody and dendrimer complexes are then dissociated and the patterned dendrimers are isolated. Only areas masked by the antibody retain sulfhydryl groups, and can therefore be differentially derivatized (to a gold adduct, or to maleimide-linked macromolecules, for instance).

Consistent patterning requires consistent masking during derivatization, requiring that individual CDRs stay in contact with the nanoantigen surface throughout derivatization. Therefore, high off-rate antibodies are undesirable in these patterning applications. Additionally, flexibility of the hinge region of antibodies [3, 5] may introduce variability in the relative positions of the protected regions of the dendrimer surface. It may be necessary to modify the hinge region of the antibody so that it is sufficiently rigid to support reproducible protection and patterning. Additionally, the suitability of dendrimers as patterning substrates may vary with the size (generation) of the dendritic polymer. Low-generation dendrimers may be so flexible as to thwart reproducible masking and therefore, reproducible patterning.

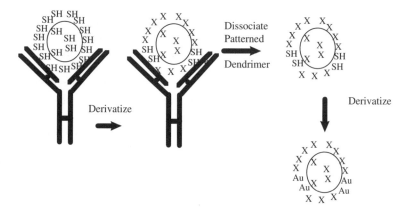

Figure 23.3 Proposed scheme for antibody-dependent patterning of a PAMAM dendrimer. A monoclonal IgG antibody (Y shape) that recognizes SH-terminated PAMAM dendrimers is bound to a SH-terminated PAMAM dendrimer (circle). Unprotected SH groups are eliminated (for instance, using a mild oxidative reagent, and resulting in a distinct surface group in unprotected areas, indicated by X), following which the antibody is dissociated from the dendrimer. The protected areas of the dendrimer exhibit the starting surface SH groups, which can be differently derivatized (with gold, indicated by Au, in this example)

6 CONCLUSION AND PROSPECTS

Though unmodified PAMAM dendrimers are intrinsically poorly immunogenic, certain protein conjugates of dendrimers can trigger antibody responses. These antibodies have numerous uses for immune detection of dendrimers and characterization of dendrimer bioconjugates. Antibodies to dendrimers might also be used to assemble both lattices, as well as more precise supramolecular assemblies of dendrimers. Additionally, the antibodies might be used to pattern dendrimers to constrain the relative positions available for decorating a dendrimer surface. Ultimately, this approach might be used to build up complex arrays of adducts on dendrimer surfaces, and therefore be of interest to nanotechnologists. Though this article focuses on anti-dendrimer antibodies, our experiences suggests that not only are dendrimers immunogenic in some contexts, so are other nanomaterials (carbon nanotubes, silicon materials, etc.) likely to be [15]. This has implications for use of these materials *in vivo*. Clearly, the powerful immune detection, separation, patterning and assembly applications of antidendrimer antibodies delineated here suggest the use of antibodies complimentary to other nanostructures. Antibodies are thus an important new tool for nanotechnologists, and will almost certainly find applications beyond dendrimer manipulations.

7 REFERENCES

1. Baker, J., Jr. Therapeutic nanodevices, in Lee, S. C. & Savage, L. (eds) *Biological Molecules in Nanotechnology: the Convergence of Biotechnology, Polymer Chemistry and Materials Science*, IBC Press, Southborough, 1998, pp. 173–183.
2. Barth, R. F., Adams, D., Soloway, A. H., Alam, F. and Darby, M. V. *Bioconjugate Chem.*, **5**, 58 (1994).
3. Breitling, F. and Dubel, S. *Recombinant Antibodies*, John Wiley & Sons, New York, 1999.
4. Dvornic, P. R. and Tomalia, D. A. *Current Opinion Coll. Interface Science*, **1**, 221 (1996).
5. Harlow, E. and Lane, D. *Antibodies: A Laboratory Manual.* Cold Spring Harbor Laboratory, Cold Spring Harbor, 1988.
6. Lee, S. C. *TIBTech.*, **16**, 239 (1998).
7. Lee, S. C. The nanobiological strategy for construction of nanodevices, in Lee, S. C. & Savage, L. (eds), *Biological Molecules in Nanotechnology: the Convergence of Biotechnology, Polymer Chemistry and Materials Science*, IBC Press, Southborough, 1998.
8. Lee, S. C. 2000. Dendrimers in nanobiological devices, in Tomalia, D. and Frechet, J. M. J. (eds). Dendritic Polymers-A New Macromolecular Architecture Based on the Dendritic State, John Wiley & Sons, in press.
9. Lee, S. C., Parthasarathy, R., Botwin, K., Kunneman, D., Rowold, E., Zobel, J., Beck, T., Duffin, T. and Voliva, C. F. Biochemical and immunological properties of cytokines conjugated to dendritic polymers. Manuscript in preparation.
10. S. C. Lee, Parthasarathy, R., Duffin, T., Botwin, K., Beck, T., Lange, G., Zobel, J.,

Kunneman, D., Rowold, E. and Voliva, C. F. Recognition properties of antibodies to PAMAM dendrimers and their use in immune detection of dendrimers. Manuscript in preparation.

11. Lemeux, G. A. and Bertozzi, C. R. *TIBTech.*, **16**, 506 (1998).
12. Roberts, J., Bhalgat, M. and Zera, R. T. *J. Biomedical. Materials Res.*, **30**, 53 (1996).
13. Tomalia, D. A., Baker, H., Dewald, J., Hall, M., Kallos, G., Roeck, J., Ryder, J. and Smith., P. B. *Polymer J.*, **17**, 117 (1985).
14. Toyokuni, T. and Singhal, A. K. *Chem. Soc. Rev.*, **22**, 231 (1995).
15. Chen, B.-X., Wilson, S. R., Das, M., Coughlin, D. J. and Erlanger, B. F. *Proceedings of the National Academy of Sciences of the United States of America*, Vol. 95, No. 18. (Sep. 1, 1998), pp. 10809–10813.

PART IV

Laboratory Preparation of Dendrimers and Conclusion

24

Preparation of 'Fréchet-type' Polyether Dendrons and Aliphatic Polyester Dendrimers by Convergent Growth: An Experimental Primer

J. M. J. FRÉCHET, H. IHRE AND M. DAVEY
Department of Chemistry, University of California, Berkeley CA, USA

1 INTRODUCTION

The convergent growth approach to dendrimers [1] first introduced [2] in 1989 at the IUPAC meeting on macromolecules in Seoul, Korea, has provided a useful alternative to the divergent methods exemplified by the work of Tomalia *et al.* on PAMAM dendrimers [3] and Meijer *et al.* on poly(propylene imine) dendrimers [4]. Today several hundred papers have exploited the convergent approach to dendrimers to prepare a variety of synthetic functional macromolecules of unparalleled structural precision.

Perhaps the three most significant features [5] of the convergent method are:

1. Its simplicity and great precision, since growth steps generally involve only the coupling of two dendrons to a monomer unit (rather than the ever-increasing number of coupling steps required in the divergent procedure), thereby reducing the excess of reagent needed to obtain high yields and facilitating purification at each step of growth [1].
2. Its functional versatility enabling the preparation of dendrons with differentiated [6, 7], and usually orthogonal, functionalities located respectively at the

Dendrimers and Other Dendritic Polymers. Edited by Jean M. J. Fréchet and Donald A. Tomalia
© 2001 John Wiley & Sons Ltd

focal point and at the chain ends of the dendrons. Combining different dendrons into a dendrimer leads to novel dendritic 'copolymers' with an unsymmetrical arrangement [8, 9] of chain-end functionalities or alternating building blocks that are not accessible by the divergent route.

3. Its design versatility, as 'generic' dendrons may be prepared to be used later as building blocks in conjunction with other reactive molecules, or coupled to a multifunctional core to afford functional dendrimers, dendritic-linear hybrids, dendronized polymers, etc. This may be a particularly significant advantage if the coupled reactive or core molecule is itself sensitive to the reaction conditions used in the multiple steps of the iterative synthesis of a dendrimer.

The procedures described below, based partly on our original accounts [1, 2], have been scaled up and updated with additional details and notes to facilitate duplication, while original references to the characterization [1] of various dendrons have been included. Although many types of convergent dendrimers have been prepared over the past decade, we only focus on two families: the polyether dendrimers [1] derived from the 3,5-dihydroxybenzyl alcohol moiety, and the aliphatic polyester dendrimers [10] derived from the 2,2-bis-hydroxy-methylpropionic acid repeat unit. Over the past dozen years, the polyether dendrons, often designated as 'Fréchet-type' dendrons, have been used extensively in the preparation of a variety of structures including for example organic light harvesting antennas [11], self-assembling dendrimers [12, 13], molecular 'machines' [14] dendritic-linear hybrid polymers [15, 16], nanoscale catalysts [17], and a variety of encapsulated [18, 19] functional macromolecules.

2 'FRÉCHET-TYPE' POLYETHER DENDRONS BASED ON 3,5-DIHYDROXYBENZYL ALCOHOL

A variety of small convergent polyether dendrons are available commercially from Tokyo Kasei Co., Ltd, see http://www.tokyokasei.co.jp. Unfortunately, at the time of this writing, sales were limited to Japan. Figure 24.1 outlines the preparation of these dendrons to generation four.

2.1 PREPARATION OF METHYL 3,5-BIS(BENZYLOXY)BENZOATE [1]

Methyl 3,5-dihydroxybenzoate (25.39 g, 0.151 mol, Aldrich) and potassium carbonate (52.69 g, 0.381 mol) were placed into a 1 L round bottom flask with 200 mL of acetone and a magnetic stirrer. Benzyl bromide (55.04 g, 0.322 mol) was added and washed into the flask with an additional 150 mL of acetone, followed by the addition of a catalytic amount of 18-crown-6 (0.57 g, 2.2 mmol).

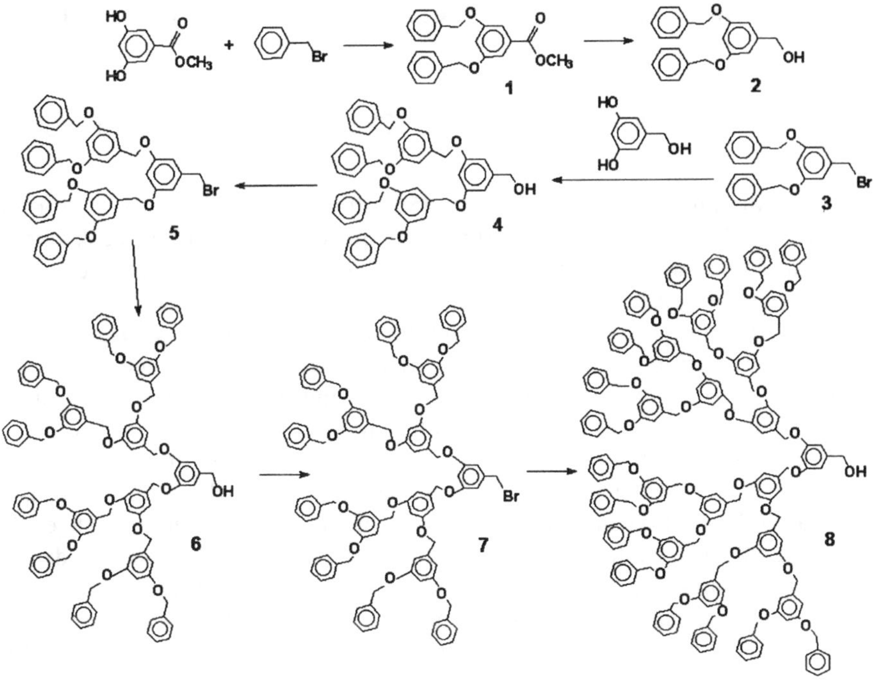

Figure 24.1

A reflux condenser was added and the solution heated to reflux with stirring in air. After 17 h, the reaction was checked by TLC (70% dichloromethane / hexanes). A small amount of the starting diol was still present as evidenced by a spot at the origin of the TLC plate. Additional potassium carbonate (7.0 g, 0.050 mol) and 18-crown-6 (0.57 g, 2.2 mmol) was added and the solution was kept at reflux with stirring for an additional 24 h. Further monitoring by TLC showed the reaction to be complete as evidenced by disappearance of the starting diol ($R_f = 0.0$) and appearance of the product ($R_f = 0.42$). The reaction mixture was cooled to room temperature and filtered. The filtered salts were further washed twice with dichloromethane and the organic phase was combined with the acetone solution. The solvent was then removed under reduced pressure using a rotary evaporator to afford an oil that turned into a solid upon standing. The solid was crystallized from methanol and the mother liquor was concentrated to yield a second crop of crystals of compound **1**. The combined yield of the white crystals of **1** was 48.93 g (93%); mp 69–70°C. Spectral characterization: IR 3120–2980, 2980–2800, 1715, 1600 cm^{-1}; ^{1}H NMR (CDCl$_3$) δ 3.90 (s, 3 H, OCH$_3$), 5.06 (s, 4 H, CH$_2$Ph), 6.80 (t, 1 H, J = 2 Hz, ArH), 7.22–7.50 (m, 12 H, ArH); mass spectrum (EI), m/z 348 (M)$^+$, 317 (M–OCH$_3$)$^+$, 181, and 91

$(CH_2Ph)^+$. Anal. Calcd for $C_{22}H_{20}O_4$: C, 75.85; H, 5.79. Found: C, 75.65; H, 5.75.

2.2 PREPARATION OF 3,5-DI(BENZYLOXY)BENZYL ALCOHOL [2]

Lithium aluminum hydride (5.51 g, 0.145 mol) was suspended into 100 mL of freshly distilled THF in a dry 500 mL three-neck round-bottom flask under argon and fitted with a magnetic stirrer, an addition funnel and a reflux condenser. Ester **1** (47.69 g, 0.137 mol) was dissolved in 150 mL of freshly distilled THF and added dropwise to the lithium aluminum hydride solution causing a very exothermic reaction to occur. Following complete addition, the reaction mixture was heated to reflux with stirring for 2 h. Monitoring by TLC (70% dichloromethane / hexanes) showed the reaction to be complete. The THF solution was cooled to room temperature and placed into a 4 L beaker fitted with a large magnetic stirrer. Water was added dropwise and very slowly to the vigorously stirred THF solution until the gray color of the lithium aluminum hydride disappeared and a white solid was formed. More THF may be added if needed to facilitate stirring. After filtration and washing of the solids with THF, the solvent was removed under reduced pressure to afford the crude product as a solid. The crude product was crystallized from 95% methanol / water to afford 37.93 g (86%) of white crystals of **2** with mp = 78–80 °C. Spectral characterization: IR 3610, 2920, 1600, 1160 cm^{-1}; ^{1}H NMR (CDCl$_3$) δ 1.64 (t, 1 H, $J = 9$ Hz, CH$_2$O*H*), 4.61 (d, 2 H, $J = 9$ Hz, C*H*$_2$OH), 5.02 (s, 4 H, C*H*$_2$Ph), 6.54 (t, 1 H, $J = 2$ Hz, Ar*H*), 6.60 (d, 2 H, J = 2 Hz, Ar*H*), 7. 33–7.42 (m, 10 H, Ph*H*); mass spectrum (EI), m/z (%) 347 (M$^+$, 33), 181 (26), 91 (100).

2.3 PREPARATION OF 3,5-DI(BENZYLOXY)BENZYL BROMIDE [3]

Prior to starting the reaction, 1.5 equivalents (relative to the amount of alcohol **2** to be used) each of carbon tetrabromide and triphenylphosphine were weighed and set aside. An additional 0.5 equivalents of each was also weighed and set aside. Alcohol **2** (37.93 g, 0.118 mol) dissolved in 50 mL of THF was placed in a 250 mL round bottom flask containing a magnetic stirrer and the solution was stirred while carbon tetrabromide (1.5 equiv.) (58.91 g, 0.178 mol) was added. Once dissolution was complete, triphenylphosphine (1.5 equiv.) (42.95 g, 0.178 mol) was added quickly but portionwise (i.e. – a large, full spatula at a time) as the reaction is very exothermic. The solution turned to a yellowish color, which usually occurs when the reaction is near completion. Monitoring by TLC (1% methanol / dichloromethane) confirmed that the reaction was indeed complete.

Note: *In some instances CBr$_4$ and CHBr$_3$ may obscure the product spot during*

TLC analysis. After running TLC, wait several minutes to allow them to evaporate. Recheck TLC to observe the product or to determine if $CBr_4/CHBr_3$ is present. The reaction was immediately quenched by the addition of water ($\sim$ 50 mL). In the event that the reaction had not gone to completion, the additional 0.5 equiv. of carbon tetrabromide set aside earlier would have been added and the solution stirred until it had dissolved. The 0.5 equiv. of triphenylphosphine set aside earlier would then have been added immediately afterward. Stir the reaction and watch for a sudden change in color to a very deep, dark yellow color. Once this color change is seen, the reaction should immediately be quenched with water to limit side reactions. This also applies to the addition of the initial dose of PPh_3; if the solution turns a deep, dark yellow, it should be immediately quenched with water.

After quenching with water, the THF was removed under reduced pressure (rotary evaporator) and the aqueous phase was extracted three times with 75 mL of dichloromethane. The combined organic extracts were dried over magnesium sulfate, filtered and dried directly onto silica gel. This impregnated silica gel was dry loaded onto a prepacked silica gel column and elution was first carried out with neat hexanes ($\sim$ 2L) to remove excess carbon tetrabromide, followed by a gradient elution using 10%, 25% and 50% dichloromethane/ hexanes. The appropriate fractions were collected and the solvent removed under reduced pressure. The resulting white solid was recrystallized from hexanes. The mother liquor was again dried onto silica, chromatographed, and recrystallized as described above affording 40.84 g (90%) of the desired product; mp 95–96°C. Characterization [1]: IR 1596, 1343, 1298, 1167, 1157, 1049, 692 cm^{-1}; ^{1}H NMR (CDCl$_3$) δ 4.39 (s, 2 H, CH_2Br), 5.01 (s, 4 H, ArCH_2O), 6.55 (d, 1 H, $J = 2$ Hz, ArH), 6.64 (d, 2 H, $J = 2$ Hz, ArH), 7.30–7.50 (m, 10 H, PhH); ^{13}C NMR (CDCl$_3$) δ 33.6(CH$_2$Br), 70.0 (CH$_2$O), 102.1, 108.0, 127.5, 128.0, 128.6, 136.5, 139.7, 160.0.

2.4 PREPARATION OF [G-2]-OH [4]

The procedure used was exactly as described for **1** using **3** (37.05 g, 96.7 mmol), 3,5-dihydroxybenzyl alcohol (6.59 g, 47.0 mmol), acetone (375 mL), potassium carbonate (19.52 g, 141 mmol) and a catalytic amount of 18-crown-6 (1.67 g, 6.3 mmol). After 24 h, additional potassium carbonate (5.15 g, 37.3 mmol) and 18-crown-6 (0.62 g, 2.3 mmol) were added and refluxing was continued for an additional 12 h at which point the reaction was judged to be complete (TLC monitoring, 50% dichloromethane/hexanes). After the standard workup as above, the resulting solids were dissolved in a minimum amount of dichloromethane (200 mL) and added to 1 L of methanol in a 2 L container. The solution was heated to a gentle boil while being stirred to distill out most of the dichloromethane until slight cloudiness appeared. Upon cooling to room temperature, an oil formed at the bottom of the container. Upon standing, the oil became solid and additional solids precipitated from the solution. The mixture was then placed in a refrigerator overnight. The solids were filtered out and dried in a vacuum oven (low heat) overnight. The remaining solution was concentrated to yield an oily residue. Both TLC and NMR analysis of this material showed that it contained none of the desired product and it was therefore discarded. The final material **4**, removed from the vacuum oven, was then recrystallized from toluene / hexanes to afford 34.12 g (97%) of a white solid: mp 110–111 °C. Characterization [1]: IR 1600, 1430, 1365, 1160, 1070 cm^{-1}; ^{1}H NMR (CDCl$_3$) δ 4.62 (d, 2 H, $J = 6$ Hz, CH_2OH), 4.97 (s, 4 H, ArCH_2O), 6.52 (t, 1 H, $J = 2$ Hz, ArH), 6.57 (t, 2 H, $J = 2$ Hz, ArH), 7.29–7.42 (m, 20 H, PhH); ^{13}C NMR (CDCl$_3$) δ 65.29 (CH$_2$OH), 69.93, 70.10 (CH$_2$O), 101.32, 101.54, 105.74, 106.33 (Ar C), 127.54, 127.99, 128.57 (Ph CH), 126.75, 127.26, 143.40, 160.05, 160.15 (Ar and Ph C); mass spectrum (FAB), m/z 744. Anal. Calcd for C$_{49}$H$_{44}$O$_7$: C, 79.01; H, 5.95. Found: C, 78.86; H, 6.25.

2.5 PREPARATION OF [G-2]-BR [5]

The procedure followed was exactly as described for **3** using **4** (33.38 g, 44.8 mmol), a minimum amount of THF (60 mL), carbon tetrabromide (24.31 g, 67.2 mmol) and triphenylphosphine (17.63 g, 44.8 mmol) in air. The reaction was monitored by TLC (neat dichloromethane) and quenched with water upon completion or if deep, dark yellow color was observed. The solution was worked up as for **3** and chromatographed using a gradient elution of 10%, 25%, 50%, 75% dichloromethane / hexanes. The appropriate fractions were collected and the solvent removed under reduced pressure (rotary evaporator). The resulting white solid was recrystallized by placing the white solids in $\sim$ 1.4 L of hexane, heating to a light reflux, followed by the gradual addition of toluene ($\sim$ 150 mL) until the solids dissolved in the solution. Allowing the solution to cool to room temperature lead to the formation of a white solid. The mother liquor was again dried onto silica, chromatographed and recrystallized as described above to give a white solid material. Total yield: 32.58 g (90%); mp 129–130.5 °C. Characterization [1]: IR 1595, 1435, 1370, 1160, 1075 cm^{-1}; ^{1}H NMR (CDCl$_3$) δ 4.41 (s, 2 H, CH_2Br), 4.97 (s, 4 H, ArCH_2O), 5.04 (s, 8 H, PhCH_2O), 6.53 (t, 1 H, $J = 2$ Hz, ArH), 6.59 (t, 2 H, $J = 2$ Hz, ArH), 6.63 (d, 2 H, $J = 2$ Hz, ArH), 6.68 (d, 4 H, $J = 2$ Hz, ArH), 7.30–7.44 (m, 20 H, PhH); ^{13}C NMR (CDCl$_3$) δ 33.57 (CH$_2$Br), 69.98, 70.08 (CH$_2$O), 101.60, 102.16, 106.35, 108.16 (Ar C), 127.53, 127.99, 128.56 (Ph CH), 136.71, 139.00, 139.73, 159.92, 160.14 (Ar and Ph C); mass spectrum (FAB), m/z 806/808 (c. 1:1). Anal. Calcd for C$_{49}$H$_{43}$BrO$_6$: C, 72.86; H, 5.36. Found: C, 73.00; H, 5.36.

2.6 PREPARATION OF [G-3]-OH [6]

The procedure followed was as described for **4** using **5** (38.10 g, 47.2 mmol),

3,5-dihydroxybenzyl alcohol (3.21 g, 22.9 mmol), acetone (300 mL), potassium carbonate (10.50 g, 76.0 mmol) and a catalytic amount of 18-crown-6 (0.68 g, 2.6 mmol). After 24 h, additional potassium carbonate (3.00 g, 21.7 mmol) and 18-crown-6 (0.50 g, 1.9 mmol) were added and refluxing was continued for an additional 12 hours at which point the reaction was deemed to be complete (TLC monitoring, 50% dichloromethane/hexanes). After the standard workup, the resulting solids were dissolved in a minimum amount of dichloromethane (50 mL) and precipitated into 1.5 L of diethyl ether with stirring. After the precipitate had settled to the bottom, the ether was decanted and the precipitate was allowed to dry at the bottom of the container. The white solid was then dissolved in the minimum amount of dichloromethane, transferred to a 500 mL round bottom flask and flash dried under high vacuum. This material was suitable for use without further purification; however, it could be rigorously purified by the procedure outlined below for the material remaining in the mother liquor. The ether was removed under reduced pressure (rotary evaporator) and the resulting solids dissolved in dichloromethane and dried onto silica gel. This material was then dry-loaded onto a silica gel column and first eluted with a large volume of hexanes followed by a gradient column of 10%, 25%, 50%, 75%, 100% dichloromethane/hexanes, followed by 1% and 3% diethyl ether / dichloromethane. Appropriate fractions were collected and the solvent removed under reduced pressure. The desired material was isolated as a white solid: 34.78 g (95%). Characterization[1]: IR 1595, 1470, 1370, 1165, 1070 cm^{-1}; ^{1}H NMR (CDCl$_3$) δ 4.57 (d, 2 H, $J = 6$ Hz, CH_2OH), 4.94 (s, 12 H, ArCH_2O), 5.00 (s, 16 H, PhCH_2O), 6.36 (m, 3 H, ArH), 6.48 (d, 4 H, $J = 2$ Hz, ArH), 6.50 (d, 8 H, $J = 2$ Hz, ArH), 7.27–7.41 (m, 40 H, PhH); ^{13}C NMR (CDCl$_3$) δ 65.09 (CH$_2$OH), 69.83, 69.88, 70.01 (CH$_2$O), 101.15, 101.52, 105.64, 106.31 (Ar C), 127.50, 127.93, 128.51 (Ph CH), 136.68, 139.17, 139.26, 143.50, 159.97, 160.07 (Ar and Ph C); mass spectrum (FAB), m/z 1592, 1593. Anal. Calcd for C$_{105}$H$_{92}$O$_{15}$: C, 79.12; H, 5.82. Found: C, 79.02; H, 5.96.

2.7 PREPARATION OF [G-3]-BR [7]

The procedure utilized was as described for **5** using **6** (32.86 g, 20.6 mmol), a minimum amount of THF (50 mL), carbon tetrabromide (11.18 g, 30.9 mmol) and triphenylphosphine (8.11 g, 30.9 mmol) in air. Standard work up was performed as described above to afford 30.48 g (89%) of the desired product. Spectral characterization [1]: IR 1600, 1460, 1370, 1170, 1065 cm^{-1}; ^{1}H NMR (CDCl$_3$) δ 4.36 (s, 2 H, CH_2Br), 4.93 (s, 4 H, ArCH_2O), 4.95 (s, 8 H, ArCH_2O), 5.00 (s, 16 H, PhCH_2O), 6.52 (m, 3 H, ArH), 6.55 (t, 4 H, $J = 2$ Hz, ArH), 6.60 (d, 2 H, $J = 2$ Hz, ArH), 6.63 (d, 4 H, $J = 2$ Hz, ArH), 6.65 (d, 8 H, $J = 2$ Hz, ArH), 7.27–7.41 (m, 40 H, PhH); ^{13}C NMR (CDCl$_3$) δ 33.57 (CH$_2$Br), 69.96, 69.99, 70.07 (CH$_2$O), 101.56, 101.62, 102.15, 106.34, 106.41, 108.19 (Ar C), 127.53,

127.97, 128.55 (Ph CH), 136.73, 139.01, 139.16, 139.77, 159.93, 160.03, 160.13 (Ar and Ph CH); mass spectrum (FAB), m/z 1655, 1656, 1657, 1658. Anal. Calcd for $C_{105}H_{91}BrO_{14}$: C, 76.12; H, 5.54. Found: C, 76.18; H, 5.72.

2.8 PREPARATION OF [G-4]-OH [8]

The procedure utilized was as described for **6** using **7** (35.39 g, 21.4 mmol), 3,5-dihydroxybenzyl alcohol (1.46 g, 10.4 mmol), acetone (250 mL), potassium carbonate (4.32 g, 31.3 mmol) and a catalytic amount of 18-crown-6 (0.28 g, 1.1 mmol). After 24 h, additional potassium carbonate (2.21 g, 16.0 mmol) and 18-crown-6 (0.29 g, 1.1 mmol) were added and refluxing was continued for an additional 12 h at which point the reaction was deemed to be complete (TLC monitoring, 50% dichloromethane/hexanes). The standard workup and precipitation processes were followed as described. Rigorous purification using column chromatography can be performed as described for **6** if and as necessary. Following purification, the desired product was obtained as a white solid: 31.55 g (92%). Characterization [1]: IR 1595, 1470, 1360, 1170, 1065 cm^{-1}; ^{1}H NMR (CDCl$_3$) δ 4.54 (d, 2 H, $J = 6$ Hz, CH_2OH), 4.93 (s, 28 H, ArCH_2O), 5.00 (s, 32 H, PhCH_2O), 6.53–6.57 and 6.63–6.67 (m, 42 H, ArH), 7.29–7.42 (m, 80 H, PhH); ^{13}C NMR (CDCl$_3$) δ 64.99 (CH_2OH), 69.85, 69.95 (CH_2O), 101.09, 101.51, 105.66, 106.30 (Ar C), 127.46, 127.88, 128.47 (Ph CH), 136.69, 139.14, 139.18, 139.31, 143.51, 159.92, 159.96, 160.05 (Ar and Ph C). Anal. Calcd for $C_{217}H_{188}O_{31}$: C, 79.17; H, 5.76. Found: C, 79.22; H, 6.00.

2.9 PREPARATION OF HIGHER GENERATION DENDRONS

The procedures utilized to prepare higher generation dendrons, up to [G-6]–OH and [G-6]–Br are similar to those used for lower dendrons. Characterization details and yields can be found in the original literature [1]. It should be noted that dendrons of generation higher than six were not prepared using this procedure due to severe lowering of yields as a result of steric effects.

3 PREPARATION OF POLYETHER DENDRIMERS BY ASSEMBLY OF FRÉCHET-TYPE DENDRONS AROUND A CORE

A general procedure is given for the coupling of [G-n]–Br dendrons around a polyfunctional core molecule [1]. Once again, coupling involves a very simple and high-yielding Williamson ether synthesis (Figure 24.2). This procedure has been carried out with dendrons as large as [G-6]–Br to afford a dendrimer with a

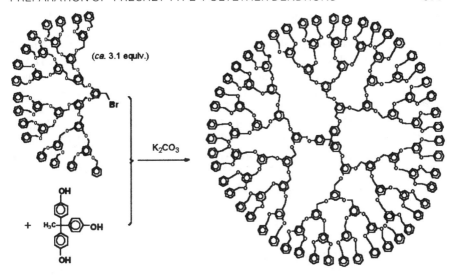

(*ca.* 3.1 equiv.)

Br

K₂CO₃

$+$ H₃C ... OH

OH

Figure 24.2

molecular weight of 40 689 Daltons. A thorough characterization of these dendrons by light scattering and viscometry has been carried out [20] providing for the first time experimental evidence for the unusual viscosity behavior of dendrimers, in agreement with the theoretical predictions of Tomalia *et al.* [21].

A mixture of the selected dendritic benzyl bromide (for example **9**, 31 mmoles), 1,1,1-tris(4'-hydroxyphenyl)ethane (10 mmoles), dried potassium carbonate (40 mmoles), and 18-crown-6 (3 mmoles) in dry acetone was heated at reflux and stirred vigorously under nitrogen for 48 h; a large excess of 1,1,1-tris(4'-hydroxyphenyl)ethane **10**, was then added, and stirring and heating were continued for 24 h. The mixture was allowed to cool and evaporated to dryness under reduced pressure. The residue was partitioned between water and CH_2Cl_2, the aqueous layer was extracted with CH_2Cl_2 (3X), and the combined organic layers were then dried and evaporated to dryness. The addition of an excess of 1,1,1-tris(4'-hydroxyphenyl)ethane at the end of the reaction is useful to capture any remaining dendron and enable its easy removal from the less polar dendrimer (e.g. by chromatographic separation). The dendrimer is readily separated from 1,1,1-tris(4'-hydroxyphenyl)ethane or its partly dendronized derivative due to the presence of phenolic moieties in the undesired material. If desired, the excess of 1,1,1-tris(4'-hydroxyphenyl)ethane can be easily recycled.

3.1 PREPARATION OF [G-4]₃-[C] (11)

This was prepared from [G-4]–Br, **9** (itself prepared from **8** as described for **7** above) and purified by flash chromatography, eluting with 1 : 3 hexane / CH_2Cl_2 to give **11** as a colorless glass: yield 84%; IR 1600, 1470, 1360, 1170, 1065 cm^{-1}; ^{1}H NMR (CDCl$_3$) δ 2.04 (s, 3 H, CH_3), 4.92 (s, 90 H, ArCH_2O), 4.99 (s, 96 H, PhCH_2O), 6.50–6.55, 6.62–6.66 (m, 135 H, ArH), 6.82 (d, 6 H, J = 9 Hz, core Ar′H), 6.97 (d, 6 H, J = Hz, core Ar′H), 7.27–7.42 (m, 240 H, PhH); 13C NMR (CDCl$_3$) 30.50 (CH$_3$), 50.60 (C–CH), 69.89, 69.99 (Ar and PhCH$_2$O), 101.56, 106.33, 106.45, ((Ar C), 113.97 (Core Ar′ C), 127.50, 127.91, 128.50 (Ph CH), 129.60 (Core Ar′C), 136.76, 139.21, 139.54, 142.01, 156.73, 160.00, 160.09, (Ar, Ar′, and Ph C). Anal. Calculated for $C_{671}H_{576}O_{93}$: C, 79.57; H, 5.75. Found: C, 79.29; H, 6.05.

4 ALIPHATIC POLYESTER DENDRONS AND DENDRIMERS BASED ON 2,2-BIS-HYDROXYMETHYLPROPIONIC ACID

The family of aliphatic polyester dendrimers based on 2,2-bis-hydroxymethyl-propionic acid, originally developed by Ihre *et al.* [10], and later modified for faster linear growth [22], is particularly interesting as it complements the Fréchet-type polyether dendrimers described earlier in this chapter. While the polyethers are more chemically rugged and perhaps easier to transform into functional nanodevices [17, 18], the aliphatic polyesters are especially attractive for their ability to be degraded by a simple hydrolytic process and for their high intrinsic biocompatibility, lack of toxicity [23], and ease of solubilization in water. This is especially important as recent studies by Duncan *et al.* [24] have shown that the PAMAM dendrimers, frequently advocated for their potential use in biomedical applications, may have significant intrinsic toxicity when in their cationic form, a finding that is not too unexpected in view of the known toxicity of poly(lysine) and many other cationic polymers.

Considering the large number of steps involved in the synthesis and purification of higher generation dendrimers, it is desirable to reduce the number of linear synthetic steps to simplify their preparation and to improve the overall yield of the final dendrimer. This may be achieved using a double stage convergent growth approach [25] where the focal point of a monodendron are coupled to the reactive chain ends of a complementary monodendron or dendrimer. Such an approach that combines some of the features of both convergent and divergent growth is illustrated below (Figure 24.3) with the preparation of a fourth generation monodendron [22]. While the method contributes to a reduction in the total number of linear steps, its modularity also enhances the versatility of the synthesis. It should be noted, however, that the same structures can also be prepared employing a strict convergent approach [10]. Once obtained, the

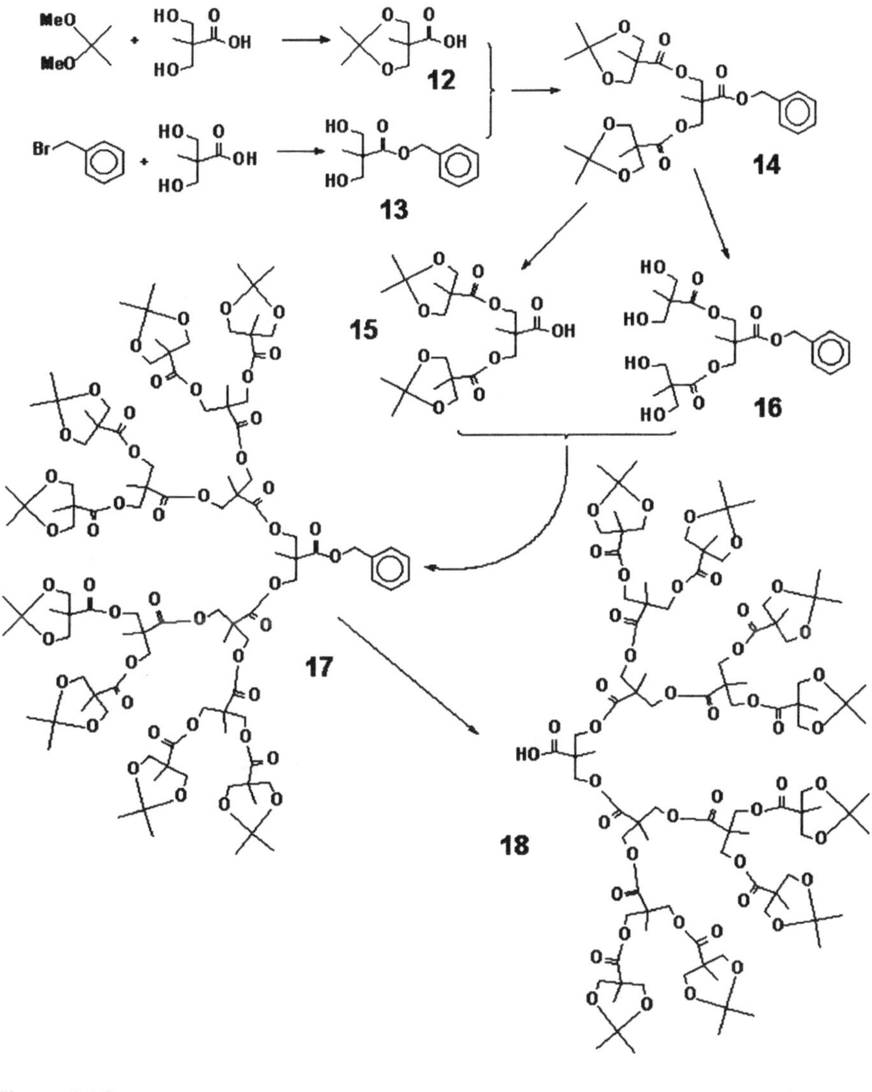

Figure 24.3

fourth generation dendron containing 16 ketal-protected hydroxyl groups may readily be coupled to a polyfunctional core. In the case of a trifunctional core, subsequent deprotection affords a tridendron dendrimer containing 48 hydroxyl groups (Figure 24.4). Reaction of the peripheral hydroxyl groups with different acid chlorides results in fully substituted 'surface-modified' dendrimers that possess a great diversity of solution and solid-state properties.

We have recently developed a novel, highly efficient, divergent synthesis [25]

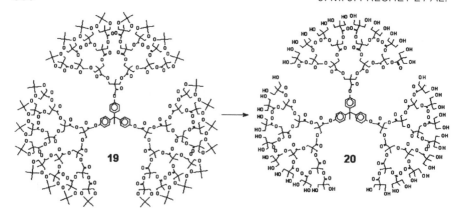

Figure 24.4

for the preparation of these and similar aliphatic polyester dendrimers with end-reactive moieties; this new approach is particularly attractive for the preparation of novel dendrimer–drug conjugates and a variety of hybrid or dendronized structures that may be used for targeted delivery [23].

4.1 PREPARATION OF ISOPROPYLIDENE-2,2-BIS(HYDROXYMETHYL)PROPIONIC ACID [9]

2,2-Bis(hydroxymethyl)propionic acid (bis-MPA) (10.00 g, 74.55 mmol), 2,2-dimethoxypropane (13.8 mL, 111.83 mmol) and p-toluenesulfonic acid monohydrate (0.71 g, 3.73 mmol) were dissolved in 50 mL acetone (Figure 24.3) and the mixture was stirred for 2 h at room temperature. After neutralizing the catalyst by adding 1 mL of a NH_3/EtOH (50: 50) solution, the solvent was evaporated at room temperature and the residue was then dissolved in ethyl acetate (250 mL) and subsequently extracted twice with 50 mL water. The organic phase was then dried with anhydrous magnesium sulfate ($MgSO_4$) and evaporated to afford ketal **9** as a white crystalline material: 12.0 g (92%). Characterization [22]: ^1H NMR ($CDCl_3$) δ 1.20 (s, 3H, $-CH_3$), 1.39 (s, 3H, $-CH_3$), 1.42 (s, 3H, $-CH_3$), 3.65 (d, 2H, $-CH_2O$), 4.18 (d, 2H, $-CH_2O$); ^{13}C NMR ($CDCl_3$) δ 18.48, 22.17, 25.03, 41.78, 65.84, 98.37, 180.40. Anal. Calcd for $C_8H_{14}O_4$: C, 55.16; H, 8.10. Found: C, 54.97; H, 8.17.

4.2 PREPARATION OF BENZYL 2,2-BIS(HYDROXYMETHYL)PROPIONATE [10]

Bis-MPA (9.00 g, 67.11 mmol) and KOH (4.30 g, 76.79 mmol) were dissolved in 50 mL dimethylformamide (DMF). The potassium salt was allowed to form at 100°C for 1 h, and benzyl bromide (13.80 g, 80.71 mmol) was then added. After 15 h of stirring at 100°C the DMF was evaporated under reduced pressure and the residue was dissolved in dichloromethane (200 ml) and extracted with two 50 mL portions of water. The organic phase was dried with $MgSO_4$ and evaporated to afford the crude product that was recrystallized from toluene. Pure **10** was obtained a white crystalline material: 10.0 g (67%). Characterization [22]: ^{1}H NMR (CDCl$_3$) δ 1.09 (s, 3H, $-CH_3$), 3.73 (d, 2H, $-CH_2OH$), 3.92 (d, 2H, $-CH_2OH$), 5.19 (s, 2H, $-CH_2Ar$), 7.36 (m, 5H, ArH); ^{13}C NMR (CDCl$_3$) δ 17.16, 49.35, 66.64, 67.49, 127.84, 128.28, 128.63, 175.69. Anal. Calcd for $C_{12}H_{16}O_4$: C, 64.27; H, 7.19. Found: C, 64.00; H, 7.17.

4.3 PREPARATION OF ACETONIDE PROTECTED [G2]-CO$_2$CH$_2$C$_6$H$_5$ [11] AND GENERAL ESTERIFICATION PROCEDURE

All esterifications were performed in dichloromethane by N,N'-dicyclohexylcarbodiimide (DCC) coupling using the catalyst 4-(dimethylamino)pyridinium p-toluenesulfonate (DPTS) developed by Moore and Stupp [26]. Other catalysts investigated resulted in low yields and the formation of significant amounts of N-acylurea. Isopropylidene-2,2-bis(hydroxymethyl)propionic acid (**9**, 3.26 g, 18.70 mmol), benzyl-2,2-bis(hydroxymethyl)propionate (**10**, 2.00 g, 8.92 mmol) and DPTS (1.05 g, 3.57 mmol) were mixed in 30 mL dichloromethane. DCC (4.60 g, 22.30 mmol) was added, and the mixture was stirred at room temperature overnight. Once the reaction was complete the urea side-product was filtered off using a glass filter and washed with a small amount of dichloromethane (10 mL). The crude product was purified by column chromatography on silica gel eluting with hexane gradually increasing to 40:60 ethyl acetate/hexane to afford acetonide protected dendron **11** as a colorless viscous oil: 4.00 g (84%). Characterization [22]: ^{1}H NMR (CDCl$_3$) δ 1.08 (s, 6H, $-CH_3$), 1.29 (s, 3H, $-CH_3$), 1.33 (s, 6H, $-CH_3$), 1.39 (s, 6H, $-CH_3$), 3.56 (d, 4H, $-CH_2O$), 4.09 (d, 4H, $-CH_2O$), 4.33 (s, 4H, $-CH_2C$), 5.14 (s, 2H, $-CH_2Ar$), 7.29–7.34 (m, 5H, ArH); ^{13}C NMR (CDCl$_3$) δ 17.74, 18.49, 22.28, 25.00, 42.04, 46.85, 65.37, 65.95, 65.97, 66.99, 98.12, 128.25, 128.45, 128.65, 135.51, 172.43, 173.56. Anal. Calcd for $C_{28}H_{40}O_{10}$: C, 62.67; H, 7.51. Found: C, 62.48; H, 7.55.

4.4 PREPARATION OF ACETONIDE PROTECTED [G2]–COOH [12] AND GENERAL PROCEDURE FOR THE REMOVAL OF THE BENZYL ESTER PROTECTING GROUP

Unmasking the carboxyl functional focal point was achieved by hydrogenolysis at atmospheric pressure using 10 wt% of Pd/C (10% Pd) as the catalyst. The benzyl ester derivatives were dissolved in ethyl acetate, the flask was evacuated, then and filled with hydrogen using a rubber balloon to maintain a positive pressure of hydrogen. The mixture was then stirred rapidly (1000 rpm) for 3 h until deprotection was complete. Therefore Pd/C (10%) 0.36 g was added to a solution of [G2]–$CO_2CH_2C_6H_5$ (3, 3.60 g, 6.71 mmol) in 30 mL of ethyl acetate. After 3 h of rapid stirring, the catalyst was filtered off using a glass filter and carefully washed with ethyl acetate (10 mL). The filtrate was evaporated to give **12** a white crystalline material: 2.90 g, (97%). Characterization [22]: ^{1}H NMR (CDCl$_3$) δ 1.11 (s, 6H, –CH$_3$), 1.27 (s, 3H, –CH$_3$), 1.31 (s, 6H, –CH$_3$), 1.37 (s, 6H, –CH$_3$), 3.58 (d, 4H, –CH$_2$O), 4.12 (d, 4H, –CH$_2$O), 4.29 (s, 4H, –CH$_2$C); ^{13}C NMR (CDCl$_3$) δ 17.67, 18.47, 22.25, 24.92, 42.04, 46.51, 65.09, 65.89, 65.92, 98.26, 173.51, 177.27. Anal. Calcd for $C_{21}H_{34}O_{10}$: C, 56.49; H, 7.68. Found: C, 56.60; H, 7.70.

4.5 PREPARATION OF $(OH)_4$- [G2]-$CO_2CH_2C_6H_5$ [13] AND GENERAL PROCEDURE FOR THE REMOVAL OF THE ACETONIDE PROTECTING GROUPS

The acetonide groups can be easily removed under mild conditions by stirring the acetonide derivatives in methanol in presence of an acidic Dowex® 50 WX2 resin. The 2,2-dimethoxypropane (b.p. 80 °C) formed during the trans-acetalization of the acetonide groups with methanol, is easily evaporated with the rest of the solution. The deprotection reaction is best monitored by ^{1}H NMR spectroscopy (DMSO-d_6) and the product is isolated quantitatively after removal of the ion exchange resin by filtration. Therefore, acetonide protected [G2]–$CO_2CH_2C_6H_5$ (**11**, 4.00 g, 7.45 mmol) was dissolved in 50 ml methanol. After addition of 2 g of Dowex H$^+$ resin® the reaction mixture was stirred for 3 h at room temperature. Upon completion of the reaction (monitoring by ^{1}H NMR spectroscopy) the Dowex, H$^+$resin® was removed using a glass filter and was then carefully washed with 50 mL of methanol. The methanol was evaporated to give **13** a white crystalline material: 3.35 g (98%). Characterization [22]: ^{1}H NMR (CDCl$_3$) δ 0.97 (s, 6H, –CH$_3$), 1.32 (s, 3H, –CH$_3$), 3.63–3.81 (m, 8H, –CH$_2$OH), 4.29 (d, 2H, –CH$_2$), 4.45 (d, 2H, –CH$_2$C), 5.18 (s, 2H, –CH$_2$-Ar), 7.35 (m, 5H, ArH); ^{13}C NMR (CDCl$_3$) δ 17.67, 18.47, 42.04, 46.51, 65.09, 65.89, 65.92, 173.51, 177.27. Anal. Calcd for $C_{22}H_{32}O_{10}$: C, 57.89; H, 7.07. Found: C, 57.82; H, 7.00.

4.6 PREPARATION OF ACETONIDE PROTECTED [G4]-CO₂CH₂C₆H₅ [14]

Acetonide protected [G2]–COOH (**12**, 1.174 g, 2.63 mmol), $(OH)_4$–[G2]–$CO_2CH_2C_6H_5$ (**13**, 0.200 g, 0.44 mmol), DPTS (0.516 g, 1.75 mmol) and DCC (0.588 g, 2.85mmol) in 10 ml dry CH_2Cl_2 were allowed to react for 48 h according to the general esterification procedure. The crude product was purified by column chromatography on silica gel eluting with hexane gradually increasing to 80:20 ethyl acetate/hexane to afford **14** as a colorless viscous oil: 0.870 g, (91%). Characterization [22]: ^1H NMR ($CDCl_3$) δ 1.09 (s, 24H, $-CH_3$), 1.14 (s, 6H, $-CH_3$), 1.22 (s, 12H, $-CH_3$), 1.26 (s, 3H, $-CH_3$), 1.30 (s, 24H, $-CH_3$), 1.37 (s, 24H, $-CH_3$), 3.57 (d, 16H, $-CH_2O$), 4.10 (d, 16H, $-CH_2O$), 4.15–4.20 (m, 16H, $-CH_2C$), 4.23–4.26 (m, 12H, $-CH_2C$), 5.12 (s, 2H, $-CH_2Ar$), 7.26–7.32 (m, 5H, ArH); ^{13}C NMR ($CDCl_3$) δ 17.41, 17.50, 17.68, 18.50, 22.02, 25.25, 33.97, 42.03, 46.65, 46.74, 46.81, 64.82, 65.49, 65.92, 66.40, 67.19, 98.09, 128.11, 128.39, 128.53, 128.71, 135.43, 171.40, 171.81, 173.48. Anal. Calcd for $C_{106}H_{160}O_{46}$: C, 58.66; H, 7.43. Found: C, 58.29; H, 7.43.

4.7 PREPARATION OF KETAL PROTECTED [G4]-COOH [15]

The ketal protected dendron [G4]–$CO_2CH_2C_6H_5$ (**14**, 4.14 g (1.80 mmol) was dissolved in 60 mL ethyl acetate and 0.41 g Pd/C (10%) was added following the general procedure for the removal of the benzyl ester group to give dendron **15** as a colorless viscous oil: 3.87 g (97%). Characterization [22]: ^1H NMR ($CDCl_3$) δ 1.10 (s, 24H, $-CH_3$), 1.24 (s, 18H, $-CH_3$), 1.28 (s, 3H, $-CH_3$), 1.31 (s, 24H, $-CH_3$), 1.38 (s, 24H, $-CH_3$), 3.59 (d, 16H, $-CH_2O$), 4.12 (d, 16H, $-CH_2O$), 4.20–4.31 (m, 28H, $-CH_2C$); ^{13}C NMR ($CDCl_3$) δ 17.47, 17.56, 17.68, 18.49, 21.86, 25.39, 42.10, 46.06, 46.70, 46.88, 64.97, 65.71, 65.94, 65.97, 66.79, 76.79, 77.10, 77.42, 98.20, 171.49, 171.81, 173.06, 173.62. Anal. Calcd for $C_{99}H_{154}O_{46}$: C, 57.16; H, 7.46. Found: C, 56.35; H, 7.48.

4.8 PREPARATION OF ACETONIDE TERMINATED [G4] TRIDENDRON DENDRIMER [16]

Dendron [G4]–COOH (**15**, 7.22 g, 3.19 mmol), 1,1,1-tris(4-hydroxyphenyl)ethane (0.272 g, 0.89 mmol), DPTS (0.783 g, 2.66 mmol), and DCC [26] (0.732 g, 3.55 mmol) were placed in 10 ml dry dichloromethane and allowed to react for 24 h according to the general esterification procedure. The crude product was purified by column chromatography on silica gel eluting with hexane, gradually increasing to 100% ethyl acetate to afford dendrimer **16** as a colorless viscous oil: 5.30 g, (85%). Characterization [22]: ^1H NMR ($CDCl_3$) δ 1.10 (s, 72H, $-CH_3$), 1.24 (s, 36H, $-CH_3$), 1.26 (s, 18H, $-CH_3$), 1.30 (s, 72H, $-CH_3$),

1.37 (s, 72H, –CH$_3$), 1.40 (s, 9H, –CH$_3$), 2.14 (s, 3H, –CH$_3$), 3.58 (d, 48H, –CH$_2$C), 4.11 (d, 48H, –CH$_2$C), 4.21-4.28 (m, 72H, –CH$_2$C), 4.31–4.39 (m, 12H, –CH$_2$C), 6.96 (d, 6H, ArH), 7.09 (d, 6H, ArH); ^{13}C NMR (CDCl$_3$) δ 17.62, 17.73, 18.50, 22.02, 25.99, 42.04, 46.74, 46.83, 47.03, 64.79, 65.38, 65.92, 65.97, 98.09, 120.78, 129.85, 146.44, 148.58, 170.69, 171.48, 171.86, 173.48. Anal. Calcd for C$_{317}$H$_{474}$O$_{138}$: C, 58.64; H, 7.36. Found: C, 58.55; H, 7.29.

4.9 PREPARATION OF HYDROXYL TERMINATED [G4] TRIDENDRON DENDRIMER [17]

The acetonide terminated [G4] tridendron dendrimer (**16**, 4.50 g, 0.64 mmol) was dissolved in a minimum amount of distilled THF and the solution was then diluted with 100 ml methanol. Using to the general procedure for removal of the acetonide protective group, dendrimer **17** with 48 terminal hydroxyl groups was obtained (Figure 24.4) as a white glass after 48 h of reaction: 3.60 g, (92%). Characterization [2]: ^{1}H NMR (DMSO-d_6) δ 1.00 (s, 72H, –CH$_3$), 1.15 (s, 36H, –CH$_3$), 1.21 (s, 18H, –CH$_3$), 1.36 (s, 9H, –CH$_3$), 2.15 (s, 3H, –CH$_3$), 3.43 (q, 96H, –CH$_2$OH), 4.08-4.15 (m, 48H, –CH$_2$C), 4.22-4.21 (m, 24H, –CH$_2$C), 4.57 (br.s, 48H, -OH), 4.38 (m, 12H, –CH$_2$C), 7.06 (d, 6H, ArH), 7.13 (d, 6H, ArH); ^{13}C NMR (DMSO-d_6) δ 16.61, 16.85, 17.06, 46.17, 46.46, 50.15, 63.59, 63.77, 64.26, 65.15, 120.89, 129.35, 146.14, 148.18, 170.67, 171. 37, 171.77, 173.97. Anal. Calcd for C$_{245}$H$_{378}$O$_{138}$: C, 53.20; H, 6.89. Found: C, 52.96; H, 7.01.

4.10 PREPARATION OF BENZOATE TERMINATED [G4] TRIDENDRON DENDRIMER AND GENERAL PROCEDURE FOR 'SURFACE' MODIFICATION OF THE HYDROXYL TERMINATED DENDRIMERS

A solution of benzoyl chloride (1.040 g, 7.41 mmol) in dry dichloromethane (3 mL), was added dropwise to a solution of the hydroxyl terminated [G4] tridendron dendrimer (**17**, 0.750 g, 0.124 mmol) containing DMAP (15 mg, 0.124 mmol) and triethylamine (0.875 g, 8.65 mmol) in 10 ml dry dichloromethane at 0 °C under argon atmosphere. After stirring at 0 °C for 1 h the temperature was raised to 25 °C and the mixture was allowed to stand overnight. The dichloromethane was evaporated and the yellow crude product was purified by column chromatography on silica gel eluting with hexane, gradually increasing to 20: 80 hexane/ethyl acetate to give the desired tridendron polyester dendrimer with 48 terminal benzoate groups as a colorless glass: 1120 g, (82%). Characterization [22]: ^{1}H NMR (CDCl$_3$) δ 1.11 (s, 63H, –CH$_3$), 1.29 (s, 48H, –CH$_3$), 1.94 (s, 3H, –CH$_3$), 4.01–4.33 (m, 84H, –CH$_2$C), 4.44–4.51 (k, 96H, –CH$_2$C), 6.89 (d, 6H, ArH), 6.98 (d, 6H, ArH), 7.30 (t, 96H, ArH), 7.44 (t, 48H, ArH), 7.93 (d, 96H, ArH);

^{13}C NMR (CDCl$_3$) δ 17.48, 17.91, 46.53, 46.63, 46.80, 46.95, 65.29, 65.84, 120.80, 128.50, 129.63, 133.26, 146.36, 148.57, 165.82, 170.67, 171.37, 171.97. Anal. Calcd for C$_{581}$H$_{570}$O$_{186}$: C, 66.28; H, 5.46. Found: C, 66.11; H, 5.35.   Financial support of this work by the National Science Foundation (DMR-9816166) and DOE (LBNL-Polymer Program) is acknowledged with thanks.

5 REFERENCES

1. Hawker, C. J. and Fréchet, J. M. J. *J. Am. Chem. Soc.*, **112**, 7638 (1990); Hawker, C. J. and Fréchet, J. M. J. *J. Chem Soc. Chem. Commun.*, 1010 (1990).
2. Fréchet, J. M. J., Jiang, Y., Hawker, C. J. and Philippides, A. E. *Proc. IUPAC Int. Symp., Macromol.*, Seoul, 1989, pp. 19–20; see also Fréchet, J. M. J.; Hawker, C. J. and Philippides, A. E. US Patent 5,041,516 Dendritic molecules and method of production.
3. Tomalia, D. A., Dewald, J. R., Hall, M. J., Martin, S. J. and Smith, P. B. *Preprints of the 1st SPSJ Int'l Polymer Conf., Soc. of Polym. Sci.*, Japan, Kyoto, 65 (1984); Tomalia, D. A., Baker, H., Dewald, J., Hall, M., Kallos, G., Martin, S., Roeck, J., Ryder, J. and Smith, P. *Polym. J., Tokyo*, **17**, 117 (1985).
4. De Brabander-van den Berg, E. M. M. and Meijer, E. W. *Angew. Chem.*, **105**, 1370 (1993).
5. Fréchet, J. M. J. *Science*, **263**, 1710 (1994).
6. Hawker, C. J. and Fréchet, J. M. J. *Macromolecules*, **23**, 4726 (1990).
7. Wooley, K. L., Hawker, C. J. and Fréchet, J. M. J. *J. Chem. Soc. Perkin I*, 1059 (1991).
8. Hawker, C. J. and Fréchet, J. M. J. *J. Am. Chem. Soc.*, **114**, 8405 (1992).
9. Hawker, C. J., Wooley, K. L. and Fréchet, J. M. J., *J. Chem. Soc. Perkin I*, 1287 (1993).
10. Ihre, H., Hult, A. and Soderlind, E. *J. Am. Chem. Soc.*, **118**, 6388 (1996).
11. Adronov, A. and Fréchet, J. M. J. *Chem. Commun.*, 1701 (2000).
12. Zimmerman, S. C., Zeng, F. W., Reichert, D. E. C. and Kolotuchin, S. V. *Science*, **271**, 1095 (1996).
13. Kawa, M. and Fréchet, J. M. J. *Chem. Mater.*, **10**, 286 (1998).
14. Amabilino, D. B., Ashton, P. R., Balzani, V., Brown, C. L., Credi, A., Fréchet, J. M. J., Leon, J. W., Raymo, F. M., Spencer, N., Stoddart, J. F. and Venturi, M. *J. Am. Chem. Soc.*, **118**, 12012 (1996).
15. Gitsov, I., Wooley, K. L. and Fréchet, J. M. J. *Angew, Chem. Int. Ed. Engl.*, **31**, 1200 (1992).
16. Leduc, M. R., Hawker, C. J., Dao, J. and Fréchet, J. M. J. *J. Am. Chem. Soc.*, **118**, 11111 (1996).
17. Piotti, M. E., Rivera, F., Bond, R., Hawker, C. J. and Fréchet, J. M. J. *J. Am. Chem. Soc.*, **121**, 9471 (1999).
18. Hecht, S. and Fréchet, J. M. J. *Angew. Chem. Int. Ed.*, **40**, 74 (2001).
19. Gorman, C. B. and Smith, J. C. *Acc. Chem. Res.*, **34**, 60 (2001).
20. Mourey, T. H., Turner, S. R., Rubinstein, M., Fréchet, J. M. J, Hawker, C. J. and Wooley, K. L. *Macromolecules*, **25**, 2401–2406 (1992).
21. Tomalia, D. A., Naylor, A. M. and Goddard III, W. A. *Angew. Chem. Int. Ed. Engl.*, **29**, 138 (1990).
22. Ihre, H., Hult, A., Fréchet, J. M. J. and Gitsov, I. *Macromolecules*, **31**, 4061 (1998).
23. Padilla de Jesus, O. L., Ihre, H., Fréchet, J. M. J., Gagnon, L. and Szoka, F., unpublished data.

24. Malik, N., Wiwattanapatapee, Klopsch, R., Lorentz, K., Frey, H., Weener, J. W., Meijer, E. W., Paulus, W. and Duncan, R. *J. Controlled Release*, **65**, 133 (2000).
25. Ihre, H., Padilla de Jesus, O. L. and Fréchet, J. M. J. *J. Am. Chem. Soc.*, **123**, 5908 (2001), US patent pending.
26. Moore, J. S. and Stupp, S. I. *Macromolecules*, **23**, 65 (1990).

25

Laboratory Synthesis of Poly(amidoamine) (PAMAM) Dendrimers

R. ESFAND AND D. A. TOMALIA
University of Michigan, Center for Biologic Nanotechnology, Ann Arbor, MI, USA

1 INTRODUCTION

The divergent growth strategy, now widely used for dendrimer synthesis, was discovered independently in parallel events that occurred in the Vögtle (University of Bonn) and the Tomalia laboratories (the Dow Chemical Company) in 1978–79. A brief communication from the Vögtle group [1] described the divergent synthesis of several low molecular weight (i.e. < 1.5 kD) dendritic structures derived from iterative reactions using acrylonitrile monomer. As described by the authors [2] and others [3], this critical chemistry was plagued with low yields, product isolation and analytical difficulties. More recently, this process was modified to furnish a commercial route to poly(propyleneimine) (PPI) dendrimers (see Chapter 26).

An analogous divergent methodology based on acrylate monomers was discovered in 1979 and developed in the Dow Chemical Laboratories during the period of 1979–84. It was first reported at the 1st International (JSPS) Conference, Japan Society of Polymer Science in Kyoto (1984) and published in 1985 [4]. This approach provided high yields of poly(amidoamine) (PAMAM) dendrimers with molecular weights ranging from several hundred to over 1 million Daltons (i.e. Generations 1–12) and is presently the preferred commercial route to Starburst® PAMAM dendrimers. Several significant advantages offered by the divergent method include:

1. Allows direct dendritic growth of dendrons from a wide variety of atomic,

Dendrimers and Other Dendritic Polymers. Edited by Jean M. J. Fréchet and Donald A. Tomalia
© 2001 John Wiley & Sons Ltd

molecular, polymeric as well as physical objects as cores. Does not require a second core anchoring step which is sterically limited via the convergent method.

2. Adaptable to large volume scale-up (e.g. *dendri*-PAMAM and *dendri*-PPI are produced in multi-kilogram quantities).
3. Low-cost, readily available commodity monomers (i.e. acrylates, acrylonitrile, alkyleneamines) may be used for synthesis.
4. May be used to prepared high generation (i.e. G = 0–12) dendrimers that precede and exceed the 'de Gennes dense packed' state.

The procedures described are based on improved modifications from original publications [4–8]. They focus on the divergent 'excess reagent' syntheses of *dendri*-poly(amidoamines) using various alkylenediamine cores. Examples of both divergent *in situ* branch cell methods, as well as divergent 'preformed' branch cell methods are presented.

Many of these dendrimeric nanostructures have shown commercial promise as gene transfection, drug delivery agents, immunodiagnostics reagents, nano-catalysts, magnetic resonance imaging contrast agents, nanoreactors and nano-calibrators. Dendrimers are expected to play a significant role in the systematic development of nanoscale chemistry architecture and properties both in biological, as well as abiotic areas of interest.

2 GENERAL COMMENTS

Laboratory procedures are presented for two divergent methodologies, namely: (1) the *in situ* branch cell (BC) method, and (2) the *preformed* (BC) coupling method (Scheme 1). The divergent, *in situ* (BC) method is a two-step iterative process for constructing poly(amidoamine) (PAMAM) dendrimers possessing either terminal ester or amine groups. A shorthand designation for these structures is as follows: [core]; (Generation)-*dendri*-PAMAM- ($-CO_2Me/ -NH_2)_n$. This method involves (a) alkylation with methylacrylate, and (b) amidation with ethylenediamine (Scheme l). The alkylation step produces ester terminated (sub-shells) that are referred to as 'half generations' and are designated (Gn. 5). The second step involves amidation of the ester terminated (Gn. 5) intermediates with large excesses of ethylenediamine to produce amine terminated, full generations, referred to as (Gn). These iteration sequences together with reaction conditions are catalogued in Figure 25.1. The structures obtained from the first two reaction sequences are 'starbranched' PAMAMs. They are designated: [EDA](G:-.5) *star*-PAMAM–$(CO_2Me)_4$ and [EDA](G:0)*star*-PAMAM–$(NH_2)_4$, respectively. The next iteration produces dendritic structures that are designated; [EDA](G:.5)*dendri*-PAMAM $-(CO_2Me)_8$ and [EDA](G:l)*dendri*-PAMAM $-(NH_2)_8$, respectively.

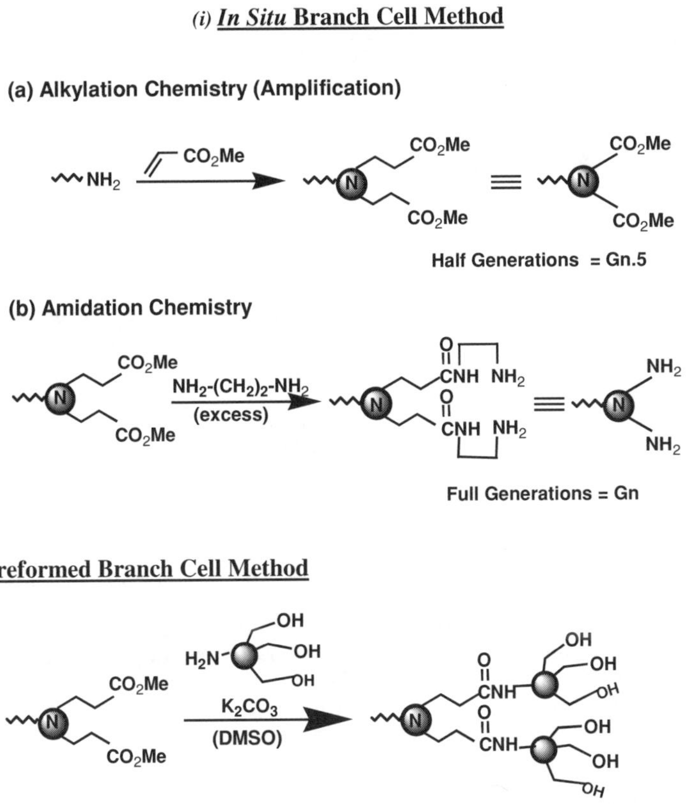

(i) *In Situ* Branch Cell Method

(a) Alkylation Chemistry (Amplification)

Half Generations = Gn.5

(b) Amidation Chemistry

Full Generations = Gn

(ii) Preformed Branch Cell Method

Scheme 1

The number of surface groups (Z), branch cells (BC) and molecular weights for a dendrimer series can be calculated with the math expressions shown below. These parameters, as well as hydrodynamic dimensions for the series [EDA](G:0-10)*dendri*-PAMAM–$(NH_2)_n$ are presented in Figure 25.1. The experimental procedures are general for a wide range of alkylenediamine initiator cores (e.g., $NH_2-(CH_2)-_nNH_2$). Characterization data for *dendri*-PAMAMs derived from these cores are included, where: $n = 2, 3, 4, 5, 6$.

The *preformed* (BC) method is a one-step process used in this case to introduce high multiplicity of terminal hydroxy groups. The method involves direct coupling of branch cell reagents (i.e. tris(hydroxymethyl)aminomethane (Tris-)) by amidation of ester terminated PAMAM dendrimers. Advancement to the next generation of branch cells occurs in one step.

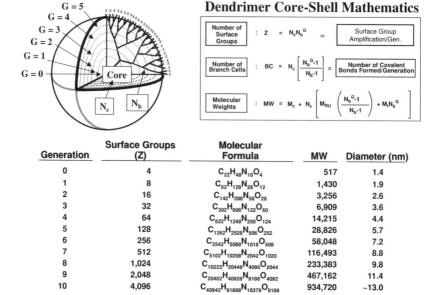

Generation	Surface Groups (Z)	Molecular Formula	MW	Diameter (nm)
0	4	$C_{22}H_{48}N_{10}O_4$	517	1.4
1	8	$C_{62}H_{128}N_{26}O_{12}$	1,430	1.9
2	16	$C_{142}H_{288}N_{58}O_{28}$	3,256	2.6
3	32	$C_{302}H_{608}N_{122}O_{60}$	6,909	3.6
4	64	$C_{622}H_{1248}N_{250}O_{124}$	14,215	4.4
5	128	$C_{1262}H_{2528}N_{506}O_{252}$	28,826	5.7
6	256	$C_{2542}H_{5088}N_{1018}O_{508}$	58,048	7.2
7	512	$C_{5102}H_{10208}N_{2042}O_{1020}$	116,493	8.8
8	1,024	$C_{10222}H_{20448}N_{4090}O_{2044}$	233,383	9.8
9	2,048	$C_{20462}H_{40928}N_{8186}O_{4092}$	467,162	11.4
10	4,096	$C_{40942}H_{81888}N_{16378}O_{8188}$	934,720	~13.0

Figure 25.1 (a) Dendrimer propagation mathematics; where: N_c, N_b = core, branch cell multiplicities; respectively, and G = generation; (b) mathematically defined values for surface groups (Z), molecular formulae and molecular weights (MW) as a function of generation for the (ethylenediamine core), poly(amidoamine) (PAMAM) dendrimer family.

3   EXPERIMENTAL METHODS

All reagents were purchased from Aldrich, Fluka, Sigma or Lancaster Synthesis, and were used after purification, unless otherwise stated. Solvents for reactions, work-up and purification procedures were used as supplied, unless otherwise stated. Solvent purification procedures were generally adopted from the book by Perrin and Armarego. Solutions were dried using anhydrous sodium sulfate or magnesium sulfate, unless otherwise stated. Combustion analyses were measured on a Perkin Elmer 240B or 240C elemental analyzer. Infrared spectra were recorded on a Perkin-Elmer 983 InfraRed spectrophotometer. Mass spectra were recorded by a VG Autospec mass spectrometer with ionization in a matrix of 3-nitrobenzyl alcohol. NMR spectra were recorded in deuterochloroform, unless otherwise stated, with a JEOL GX270 nuclear magnetic resonance spectrometer (270.5 MHz). GPC was carried out on a Plgel mixed E column, with tetrahydrofuran (THF) as eluent, connected to a Knaur refractive index detector and a PyeUnicam PU 410 computing integrator. The column was calibrated using narrow dispersity linear polystyrene standards.

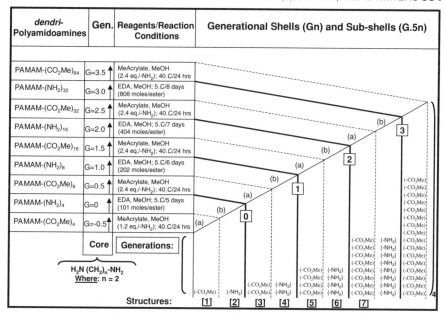

dendri- Polyamidoamines	Gen.	Reagents/Reaction Conditions	Generational Shells (Gn) and Sub-shells (G.5n)
PAMAM-(CO₂Me)₆₄	G=3.5	MeAcrylate, MeOH (2.4 eq./-NH₂); 40.C/24 hrs	
PAMAM-(NH₂)₃₂	G=3.0	EDA, MeOH; 5.C/8 days (808 moles/ester)	
PAMAM-(CO₂Me)₃₂	G=2.5	MeAcrylate, MeOH (2.4 eq./-NH₂); 40.C/24 hrs	
PAMAM-(NH₂)₁₆	G=2.0	EDA, MeOH; 5.C/7 days (404 moles/ester)	
PAMAM-(CO₂Me)₁₆	G=1.5	MeAcrylate, MeOH (2.4 eq./-NH₂); 40.C/24 hrs	
PAMAM-(NH₂)₈	G=1.0	EDA, MeOH; 5.C/6 days (202 moles/ester)	
PAMAM-(CO₂Me)₈	G=0.5	MeAcrylate, MeOH (2.4 eq./-NH₂); 40.C/24 hrs	
PAMAM-(NH₂)₄	G=0	EDA, MeOH; 5.C/5 days (101 moles/ester)	
PAMAM-(CO₂Me)₄	G=-0.5	MeAcrylate, MeOH (1.2 eq./-NH₂); 40.C/24 hrs	

Figure 25.2 Core–shell reaction sequence steps: (a) alkylation and (b) amidation steps for preparation of G.5n and Gn. [NH₂–(CH₂)$_{2–6}$–NH₂]; (G = 0–2.5)-*dendri*-poly(amidoamines)

3.1 DIVERGENT SYNTHESIS OF PAMAM DENDRIMERS VIA EXCESS REAGENT METHOD PREPARATION OF ESTER TERMINATED PAMAM STAR-BRANCHED PRECURSOR; [NH₂-(CH₂)$_{2-6}$-NH₂]; (G = -0.5); STAR-PAMAM(CO₂ME)₄; [1]

A solution of freshly distilled 1,2-diaminoethane (5 g, 5.5 ml, 0.083 mol) in methanol (20 ml) was added dropwise to a stirred solution of methylacrylate (35 g, 37 ml, 0.407 mol) in methanol (20 ml), under nitrogen, over a period of 2 h. The final mixture was stirred for 30 min at 0 °C and then allowed to warm to room temperature and stirred for a further 24 h. The solvent was removed under reduced pressure at 40 °C using a rotary evaporator and the resulting colourless oil dried under vacuum (10^{-1} mm Hg, 50 °C) overnight to give the final product (32 g, 95%).

(^{a}CH₂^{a}CH₂)[N(^{b}CH₂^{c}CH₂^{d}CO₂^{e}CH₃)₂]₂

Spectral characterization: IR v_{max}/cm^{-1}: 1740 (C=O); ^{1}H NMR (CDCl₃) δ_H: 3.61 (12H, s, e), 2.63 (8H, t, b), 2.41 (4H, s, a), 2.29 (8H, t, c); ^{13}C NMR (CDCl₃) δ_c: 174.12 (d), 52.15 (e), 50.27 (b), 45.23 (a), 31.56 (c); Mass spectrum (FAB): m/z 405 (M + H), 427 (M + Na); Elemental analysis: C₁₈H₃₂N₂O₈ Calcd (%) C, 53.5; H,

8.0; N, 6.9. Found (%) C, 53.2; H, 8.4; N, 7.1. Gel chromatography: $M_w = 816.2$, $M_n = 800.2$, PD = 1.08.

Dendrimer [core]	%Yield	MW	FAB-Ms (M + H)	GPC M_w/M_n
$H_2N-C_2H_4-NH_2$	99	404	405	1.08
$H_2N-C_3H_6-NH_2$	95	418	419	1.01
$H_2N-C_4H_8-NH_2$	96	432	433	1.03
$H_2N-C_5H_{10}-NH_2$	91	445	447	1.01
$H_2N-C_6H_{12}-NH_2$	98	460	471	1.01

Analytical data for *star*-PAMAM$(CO_2Me)_4$.

3.2 PREPARATION OF AMINE TERMINATED PAMAM STAR-BRANCHED PRECURSOR; $[NH_2-(CH_2)_{2-6}-NH_2]$; (G = 0); star-PAMAM$(NH_2)_4$; [2]

A solution of (G = − 0.5); *star*-PAMAM]$(CO_2Me)_4$; [1] precursor (10 g, 0.025 mol) in methanol (20 ml) was carefully added to a vigorously stirred solution of 1,2-diaminoethane (75 g, 85 ml, 1.248 mol) in methanol (100 ml) at 0 °C. The rate of addition was such that the temperature did not rise above 40 °C. After complete addition the mixture was stirred for 96 h at room temperature at which time no ester groups was detectable by NMR spectroscopy. The solvent was removed under reduced pressure maintaining the temperature no higher than 40 °C. The excess 1,2-diaminoethane was removed using an azeotropic mixture of toluene and methanol (9: 1). The remaining toluene was removed by azeotropic distillation using methanol. Finally, removal of the remaining methanol under vacuum (10^{-1} mm Hg, 40 °C, 48 h) gave the tetra-amine terminated G = 0 precursor as a colorless oil (12.5 g, 98%).

$(^aCH_2{}^aCH_2)[N(^bCH_2{}^cCH_2{}^dCO^eNH^fCH_2{}^gCH_2{}^hNH_2)_2]_2$

Spectral characterization: IR v_{max}/cm^{-1}: 1640 (C=O), 3200 (NH_2), 3400 (NH); ^1H NMR $(CDCl_3)$ δ_H: 8.05 (4H, bt, e), 3.31 (8H, bq, f), 2.71 (8H, t, b), 2.57 (8H, t,.c), 2.53 (4H, s, a), 2.45 (8H, t, h), 2.25 (8H, p, g); ^{13}C NMR $(CDCl_3)$ δ_c: 175.15 (d), 51.31 (b), 48.25 (a), 45.32 (f), 35.18 (c), 31.25 (g); Mass spectrum (FAB): m/z 518 (M + H), 540 (M + Na); Elemental analysis: $C_{22}H_{48}N_{10}O_4$ Calcd (%) C, 51.1; H, 9.3; N, 27.1. Found (%) C, 52.5; H, 9.7; N, 28.2.

Dendrimer [core]	% Yield	MW	FAB-Ms (M + H)
$H_2N-C_2H_4-NH_2$	98	516	517
$H_2N-C_3H_6-NH_2$	95	530	531
$H_2N-C_4H_8-NH_2$	94	544	545
$H_2N-C_5H_{10}-NH_2$	92	558	559
$H_2N-C_6H_{12}-NH_2$	92	572	573

Analytical data for *star*-PAMAM(NH$_2$)$_4$.

Note: It is essential to carefully examine the analytical data for the presence of cyclic compounds as a result of side reactions involved. The cyclization occurs as dendrimer amidation versus bridging amidation are kinetically similar. To prevent the intradendrimeric cyclization a large excess (50-fold) of 1,2-ethylenediamine is required. The excesses are removed to an undetectable level by azeotropic techniques.

Ideal and Non-ideal Amidation of the *dendri*-PAMAM(CO$_2$Me)$_4$

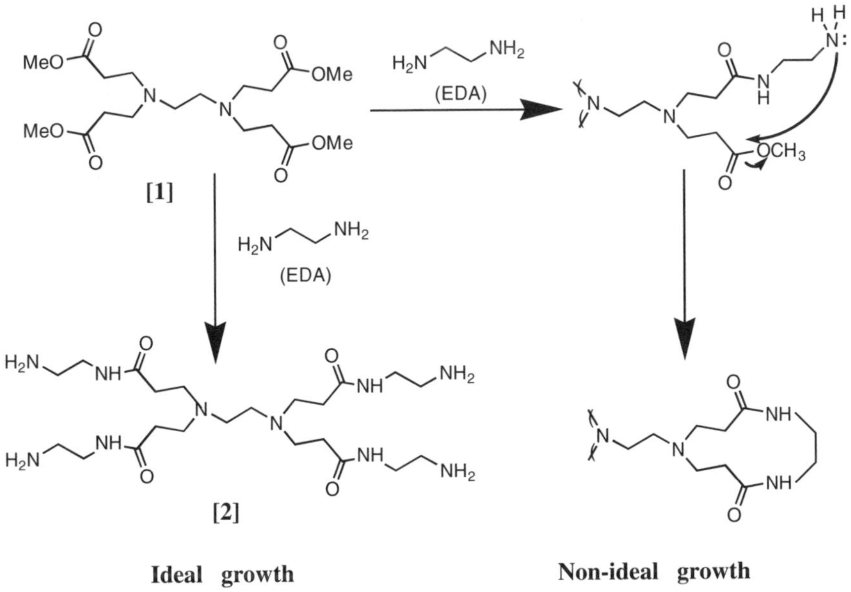

Scheme 2

Synthesis of PAMAM Dendrimers

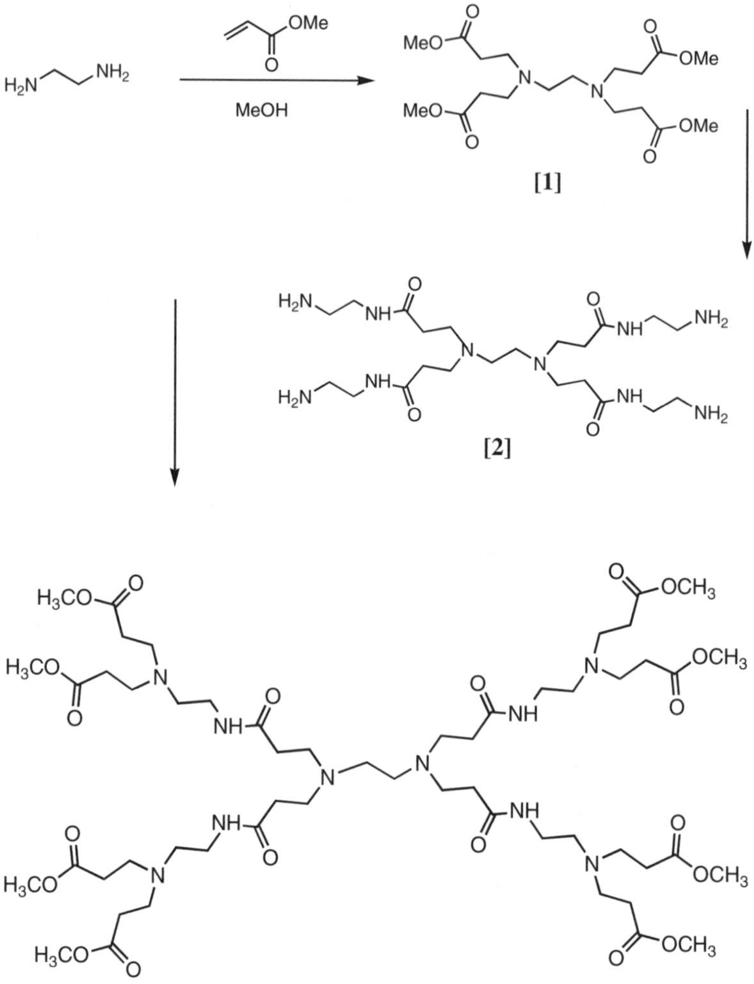

[1]

[2]

[3]

Scheme 2 (*cont.*)

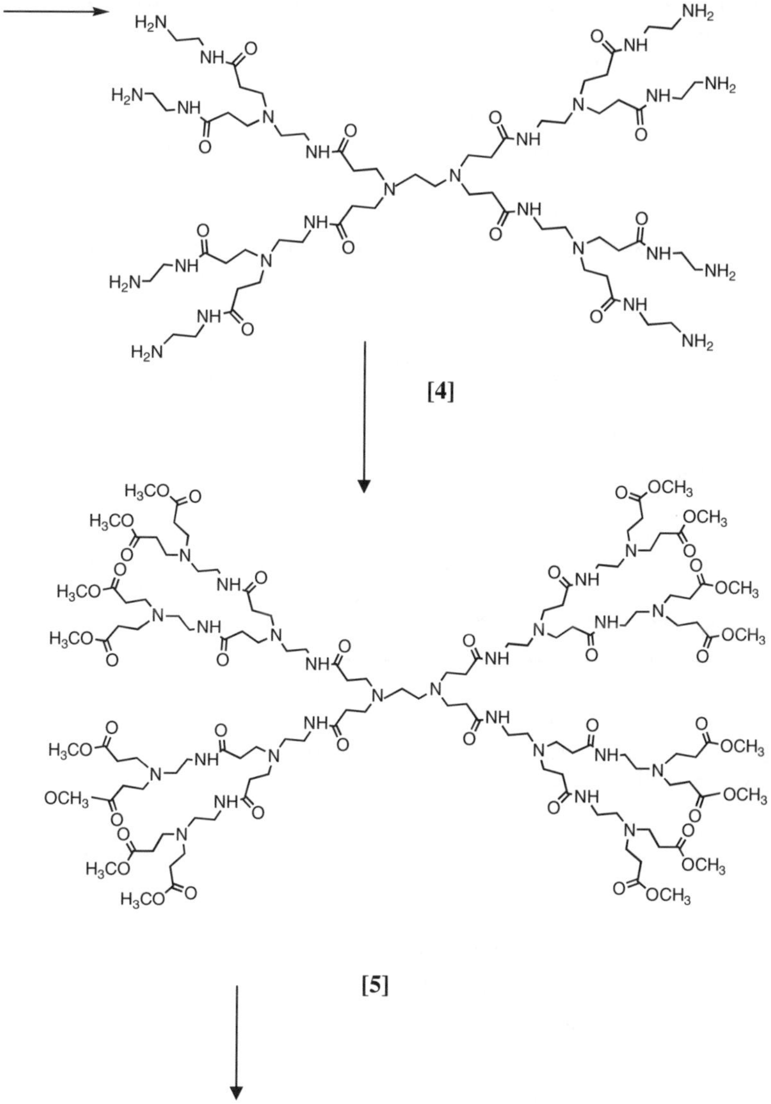

[4]

[5]

Scheme 2 (*cont.*)

3.3 PREPARATION OF ESTER TERMINATED PAMAM DENDRIMER; [NH₂-(CH₂)₂₋₆-NH₂]; (G = 0.5); dendri-PAMAM(CO₂Me)₈; [3]

A solution of (G = 0); dendri-PAMAM(NH₂)₄₊; [2] precursor (8 g, 0.015 mol) in methanol (20 ml) was added to a stirred solution of methylacrylate (12.9 g, 13.5 ml, 0.15 mol) in methanol (20 ml), under nitrogen, over a period of 1 h. The final mixture was stirred at 0°C for 1 h and then allowed to warm to room temperature and was stirred for a further 24 h. The solvent was removed under reduced pressure at 40°C and the resultant colorless oil vacuum dried (10⁻¹ mm Hg, 50°C) overnight to give the final product (17.5 g, 92%).

(^{a}CH₂^{a}CH₂)[N(^{b}CH₂^{c}CH₂^{d}CO^{e}NH^{f}CH₂^{g}CH₂N(^{h}CH₂^{i}CH₂^{j}CO₂^{k}CH₃)₂)₂]₂

Spectral characterization: IR v_{max}/cm⁻¹: 1750 (C=O), 3400 (NH); ¹H NMR (CDCl₃) δ_H: 7.21 (4H, t, e̲), 3.65 (24H, s, k̲), 3.25 (8H, q, f̲), 2.65–2.95 (20H, m,.a̲, b̲, g̲), 2.55 (16H, t, h̲), 2.45 (16H, t, i̲), 2.31 (8H, t, c̲);); ¹³C NMR (CDCl₃) δ_c: 173.45 (d̲), 172.23 (j̲), 53.22 (k̲), 53.21 (f̲), 52.22 (b̲), 51.24 (g̲); Mass spectrum (FAB): m/z 1207 (M + H), 1229 (M + Na); Elemental analysis: C₅₄H₉₆N₁₀O₂₀ Calcd (%) C, 53.8; H, 8.1; N, 11.6. Found (%) C, 53.5; H, 8.2; N, 11.5. Gel chromatography: M_w = 2465.1, M_n = 2471.3, PD = 1.0.

Dendrimer [core]	%Yield	MW	FAB-Ms (M + H)	SEC M_w/M_n
H₂N–C₂H₄–NH₂	97	1205	1206	1.00
H₂N–C₃H₆–NH₂	92	1219	1220	1.01
H₂N–C₄H₈–NH₂	93	1233	1234	1.05
H₂N–C₅H₁₀–NH₂	96	1247	1248	1.05
H₂N–C₆H₁₂–NH₂	92	1261	1262	1.05

Analytical data for dendri-PAMAM(CO₂Me)₈.

3.4 PREPARATION OF AMINE TERMINATED PAMAM DENDRIMER; [NH₂-(CH₂)₂₋₆–NH₂]; (G = 1.0); dendri-PAMAM (NH₂)₈; [4]

To a vigorously stirred solution of 1,2-diaminoethane (60 g, 65 ml, 0.994 mol) in methanol (100 ml), at 0°C under nitrogen, was added a solution of (G = 0.5); dendri-PAMAM(CO₂Me)₈; [3] (5 g, 0.004 mol) in methanol (20 ml). The addition was controlled such that the temperature did not rise above 40°C. The mixture was stirred at room temperature for 96 h, after which time no ester groups were detectable by NMR spectroscopy. The methanol was removed by vacuum distillation at < 40°C, and excess 1,2-diaminoethane was removed by azeotropic distillation using a mixture of toluene and methanol (9:1). The remaining toluene was removed by azeotropic distillation with methanol and finally the methanol removed under vacuum (10^{-1} mm Hg, 50°C, 48 h) to give the amine terminated G = 1.0 dendrimer as a pale yellow oil (5.7 g, 96%).

(^{a}CH$_2$^{a}CH$_2$)[N(^{b}CH$_2$^{c}CH$_2$^{d}CO^{e}NH^{f}CH$_2$^{g}CH$_2$N
(^{h}CH$_2$^{i}CH$_2$^{j}CO^{k}NH^{l}CH$_2$^{m}CH$_2$^{n}NH$_2$)$_2$)$_2$]$_2$

Spectral characterization: IR v_{max}/cm^{-1}: 1640 (C=O), 3420 (NH) ; ^{1}H NMR (DMSO) δ_H: 8.03 (12H, bt, e, k), 3.31–3.15 (24H, bq, f, l), 2.65 (8H, t, g), 2.61–2.51 (52H, m,.a, b, c, h, i), 2.45 (16H, t, n), 2.25 (16H, bp, m); ^{13}C NMR (DMSO) δ_c: 178.21 (d), 174.48 (j), 52.13 (f), 49.15 (l), 46.21 (g), 45.11 (a), 43.18 (b), 42.11 (c), 38.21 (h), 35.51 (i); Mass spectrum (FAB): m/z 1430 (M + H), 1452 (M + Na); Elemental analysis: C$_{62}$H$_{128}$N$_{26}$O$_{12}$ Calcd (%) C, 52.1; H, 8.9; N, 25.5. Found (%) C, 49.3; H, 10.2; N, 24.2.

Dendrimer [core]	%Yield	MW	FAB-Ms (M + H)
H$_2$N–C$_2$H$_4$–NH$_2$	96	1429	1430
H$_2$N–C$_3$H$_6$–NH$_2$	92	1443	1444
H$_2$N–C$_4$H$_8$–NH$_2$	92	1457	1458
H$_2$N–C$_5$H$_{10}$–NH$_2$	93	1471	1472
H$_2$N–C$_6$H$_{12}$–NH$_2$	93	1485	1486

Analytical data for *dendri*-PAMAM(NH$_2$)$_8$.

3.5 PREPARATION OF ESTER TERMINATED PAMAM DENDRIMER;
[NH$_2$-(CH$_2$)$_{2-6}$-NH$_2$]; (G = 1.5); dendri-*PAMAM(CO$_2$ME)$_{16}$;* [5]

A solution of (G = 1); *dendri*-PAMAM(NH$_2$)$_4$; [4] (6.5 g, 0.005 mol) in methanol (20 ml) was added to a stirred solution of methylacrylate (6.7 g, 7 ml, 0.077 mol) in methanol (20 ml), under nitrogen, over a period of 2 h. The final mixture was stirred at 0°C for 1 h and then allowed to warm to room temperature and stirred for a further 24 h. The solvent was removed under reduced pressure at 40°C and the resultant pale yellow oil vacuum dried (10^{-1} mm Hg, 50°C) overnight to give the final product (11.5 g, 91%).

(^{a}CH$_2$^{a}CH$_2$)[N(^{b}CH$_2$^{c}CH$_2$^{d}CO^{e}NH^{f}CH$_2$^{g}CH$_2$N(^{h}CH$_2$^{i}CH$_2$^{j}CO^{k}NH^{l}CH$_2$^{m}C
H$_2$N(^{n}CH$_2$^{o}CH$_2$^{p}CO$_2$^{q}CH$_3$)$_2$)$_2$)$_2$]$_2$

Spectral characterization: IR v_{max}/cm^{-1}: 1750 (C=O), 3600 (NH); ^{1}H NMR (CDCl$_3$) δ_H: 7.23 (12H, bt, e, k), 3.58 (48H, s, q), 3.31 (24H, bm, f, l), 2.95 (8H, bt,.g), 2.75–2.85 (52H, m, a, b, c, h, i), 2.63 (32H, bt, o), 2.45 (32H, bt, n), 2.25 (16H, bt, m); ^{13}C NMR (CDCl$_3$) δ_c: 173.15 (d), 172.55 (j), 171.95 (p); Mass spectrum (FAB): m/z

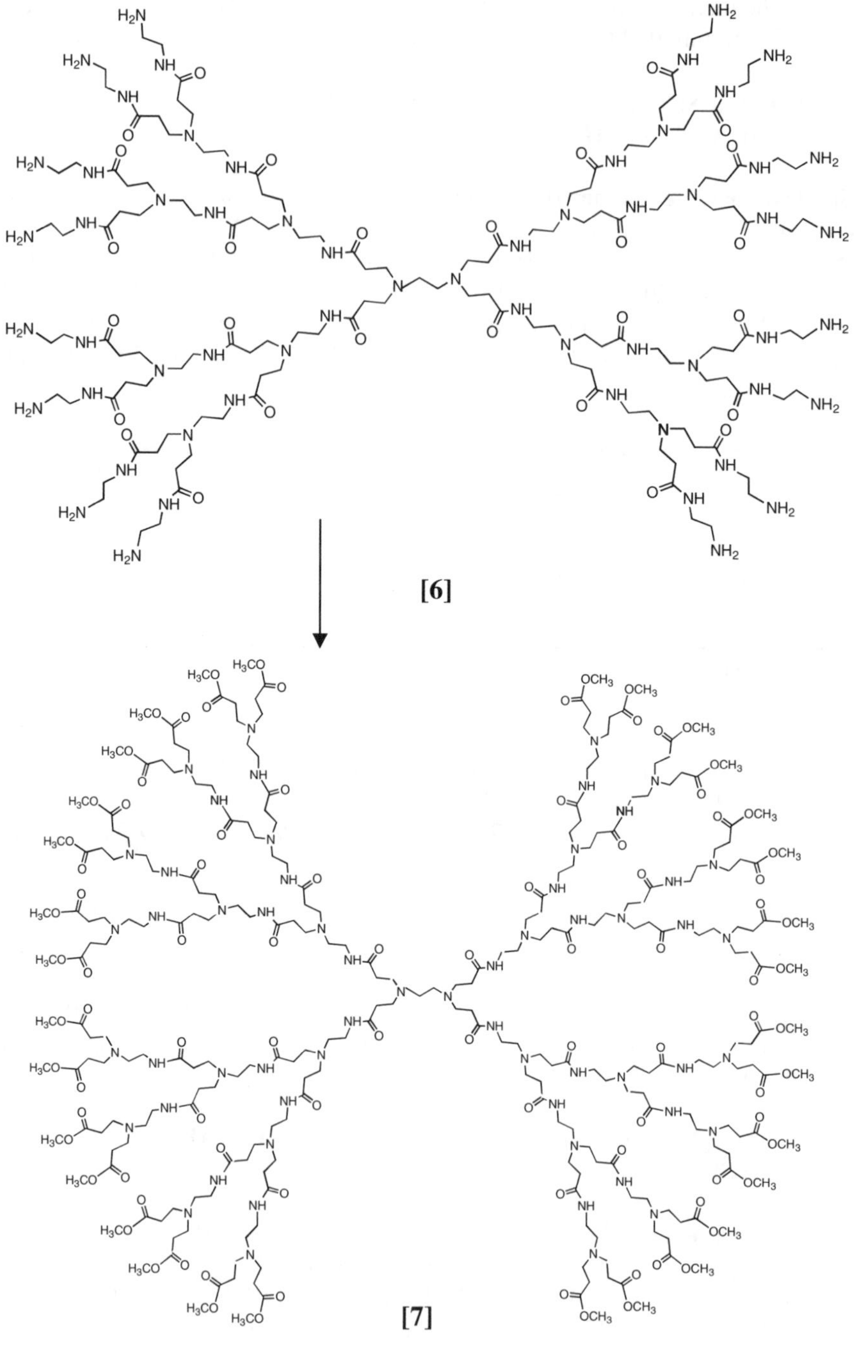

[6]

[7]

Scheme 3

2809 (M + H), 2830 (M + Na); Elemental analysis: $C_{126}H_{226}N_{26}O_{44}$ Calcd (%) C, 53.9; H, 8.1; N, 12.9. Found (%) C, 53.7; H, 8.8; N, 13.2.

Dendrimer [core]	%Yield	MW	FAB-Ms (M + H)
$H_2N-C_2H_4-NH_2$	97	2808	2809
$H_2N-C_3H_6-NH_2$	92	2822	2823
$H_2N-C_4H_8-NH_2$	93	2836	2837
$H_2N-C_5H_{10}-NH_2$	96	2850	2851
$H_2N-C_6H_{12}-NH_2$	92	2863	2864

Analytical data for *dendri*-PAMAM(NH$_2$Me)$_{16}$.

3.6 PREPARATION OF AMINE TERMINATED PAMAM DENDRIMER; [NH$_2$–(CH$_2$)$_{2-6}$–NH$_2$]; dendri-PAMAM (NH$_2$)$_{16}$; [6]

To a vigorously stirred solution of 1,2-diaminoethane (107 g, 118 ml, 1.781 mol) in methanol (150 ml), at 0 °C under nitrogen, was added a solution of (G = 1.5); *dendri*-PAMAM(CO$_2$Me)$_{16}$; [5] (10 g, 0.004 mol) in methanol (30 ml). The addition was controlled such that the temperature did not rise above 40 °C. The mixture was stirred at room temperature for 96 h, after which time no ester groups were detectable by NMR spectroscopy. The methanol was removed by vacuum distillation at < 40 °C, and the excess 1,2-diaminoethane was removed by azeotropic distillation using a mixture of toluene and methanol (9:1). The remaining toluene was removed by azeotropic distillation with methanol and finally the methanol removed under vacuum (10^{-1} mm Hg, 50 °C, 48 h) to give the amine terminated G = 2.0 dendrimer as a pale yellow oil (10.9 g, 94%).

(^{a}CH$_2$^{a}CH$_2$)[N(^{b}CH$_2$^{c}CH$_2$^{d}CO^{e}NH^{f}CH$_2$^{g}CH$_2$N(^{h}CH$_2$^{i}CH$_2$^{j}CO^{k}NH^{l}CH$_2$^{m}C H$_2$N(^{n}CH$_2$^{o}CH$_2$^{p}CO$_2$^{q}CH$_3$)$_2$)$_2$)$_2$]$_2$

Spectral characterization: IR v_{max}/cm^{-1}: 3350 (NH$_2$); ^{1}H NMR (CDCl$_3$) δ_H: 8.01 (28H, bm, e, k, q), 3.19–3.45 (56H, bm, f, l, r), 2.61–2.81 (116H, bm, a, b, c, h, i, n, o), 2.98–2.93 (24H, bm, g, m), 2.45 (32H, bt, t), 2.23 (32H, bp, s); ^{13}C NMR (DMSO) δ_c: 171.65 (d), 171.45 (j), 171.34 (p); Mass spectrum (FAB): m/z 3255 (M–H); Elemental analysis: $C_{142}H_{288}N_{58}O_{28}$ Calcd (%) C, 52.4; H, 8.9; N, 24.9. Found (%) C, 47.2; H, 9.7; N, 21.5.

3.7 PREPARATION OF ESTER TERMINATED PAMAM DENDRIMER; [NH$_2$–(CH$_2$)$_{2-6}$–NH$_2$]; (G = 2.5); dendri-PAMAM(CO$_2$Me)$_{32}$; [7]

A solution of (G = 2); *dendri*-PAMAM(NH$_2$)$_1$^{6}; [6] (10 g, 0.003 mol) in methanol (20 ml) was added to a stirred solution of methylacrylate (10.6 g, 11 ml, 0.123 mol) in methanol (20 ml), under nitrogen, over a period of 2 h. The final mixture was stirred at $0\,^\circ$C for 1 h and then allowed to warm to room temperature and stirred for a further 24 h. The solvent was removed under reduced pressure at $40\,^\circ$C and the resultant pale yellow oil vacuum dried (10^{-1} mm Hg, $50\,^\circ$C) overnight to give the final product (15.6 g, 84%).

$(^aCH_2{}^aCH_2)[N(^bCH_2{}^cCH_2{}^dCO^eNH^fCH_2{}^gCH_2N(^hCH_2{}^iCH_2{}^jCO^kNH^lCH_2{}^mCH_2N(^nCH_2{}^oCH_2{}^pCO^qNH^rCH_2{}^sCH_2N(^uCH_2{}^vCO_2{}^wCH_3)_2)_2)_2)_2]_2$

Spectral characterization: IR v_{max}/cm^{-1}: 1700 (C=O); ^{1}H NMR (CDCl$_3$) δ_H: 7.26 (28H, bm, e, k, q), 3.62 (96H, s, w), 3.29–3.41 (56H, bm, f, l, r), 2.77–2.56 (56H, bm, g, m, s), 2.25–2.45 (116H, bm, a, b, c, h, i, n, o); ^{13}C NMR (CDCl$_3$) δ_c: 177.34 (d), 173.07 (j), 172.32 (p), 171.51 (v); Mass spectrum (FAB): m/z 5978 (M-29); Elemental analysis: C$_{270}$H$_{480}$N$_{58}$O$_{92}$ Calcd (%) C, 53.9; H, 8.1; N, 13.5. Found (%) C, 51.3; H, 8.4; N, 13.6.

3.8 PREPARATION OF HYDROXY TERMINATED PAMAM DENDRIMER; [NH$_2$-(CH$_2$)$_{2-6}$-NH$_2$]; (G = 1.0); dendri-PAMAM(OH)$_{12}$; [8]

An oven-dried flask, with a septum capped nitrogen inlet, was charged with a solution of (G = − 0.5); dendri-PAMAM(CO$_2$Me)$_4$; [1] precursor (2.5 g, 6.2 mmol) in anhydrous dimethyl sulfoxide (10 ml). A mixture of Tris (tri-hydroxymethyl aminoethane) (3.76 g, 0.031 mol) and anhydrous potassium carbonate (2.7 g, 0.031 mol) was added to the suspension. The solution was heated in a paraffin oil bath maintained at $40\,^\circ$C for 96 h. After filtration the excess solvent was removed by vacuum pump (10^{-1} mm Hg, $50\,^\circ$C). The resulting thick, white oil was dissolved in a minimum quantity of distilled water and precipitated with propanone. The suspension was cooled in the freezer, and the solid filtered and dried in a vacuum oven (10^{-1} mm Hg, $40\,^\circ$C) to give the hygroscopic terminated dendrimer as a white powder (4.5 g, 95%).

$(^aCH_2{}^aCH_2)[N(^bCH_2{}^cCH_2{}^dCO^eNHC(^fCH_2{}^gOH)_3)_2]_2$
Spectral characterization: IR v_{max}/cm^{-1}: 1635 (C=O); ^{1}H NMR (DMSO) δ_H: 7.56 (4H, bt, e), 3.47 (24H, s, f), 2.65 (8H, t, b), 2.52 (4H, bt, a), 2.29 (8H, t, c); ^{13}C NMR (CDCl$_3$) δ_c: 172.86 (d), 50.25 (f), 50.21 (b), 35.62 (a), 30.21 (c); Mass spectrum (FAB): m/z 762 (M + H).

3.9 PREPARATION OF HYDROXY TERMINATED PAMAM DENDRIMER; [NH$_2$-(CH$_2$)$_{2-6}$-NH$_2$]; (G = 2.0); dendri-PAMAM(OH)$_{24}$; [9]

An oven-dried flask, with a septum-capped nitrogen inlet, was charged with a solution of (G = 0.5); dendri-PAMAM(CO$_2$Me)$_8$; [3] (5 g, 4.2 mmol) in anhyd

Synthesis of Polyhydroxy Terminated PAMAM Dendrimers

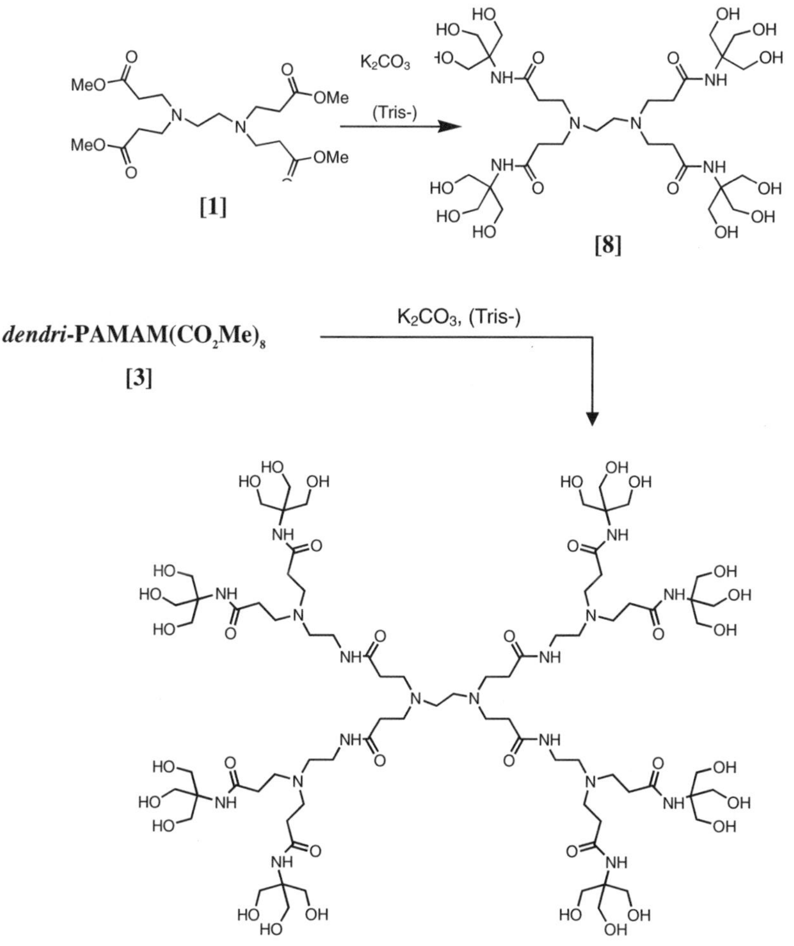

[1] → (K₂CO₃, (Tris-)) → **[8]**

dendri-PAMAM(CO₂Me)₈   —— K₂CO₃, (Tris-) ——→

[3]

[9]

Scheme 4

rous dimethyl sulfoxide (20 ml). A mixture of Tris (tri-hydroxymethyl amino-ethane) (5.1 g, 0.042 mol) and anhydrous potassium carbonate (5.7 g, 0.042 mol) was added to the suspension. The solution was heated in a paraffin oil bath maintained at 40°C for 96 h. After filtration the excess solvent was removed by vacuum pump (10^{-1} mm Hg, 50°C). The resulting thick, white oil was dissolved in a minimum quantity of distilled water and precipitated with propanone. The suspension was cooled in the freezer, and the solid filtered and dried in a vacuum oven (10^{-1} mm Hg, 40 °C) to give the hygroscopic terminated dendrimer as a white powder (6.2 g, 82%).

Note: Same procedure can be used when varying the central core unit from 2C to 6C.

(^{a}CH$_2$^{a}CH$_2$)[N(^{b}CH$_2$^{c}CH$_2$^{d}CO^{e}NH^{f}CH$_2$^{g}CH$_2$N
(^{h}CH$_2$^{i}CH$_2$^{j}CO^{k}NHC(^{l}CH$_2$^{m}OH)$_3$)$_2$)$_2$]$_2$

Spectral characterization: IR v_{max}/cm^{-1}: 1660 (C=O), 3500 (NH); ^{1}H NMR (DMSO) δ_H: 7.85 (12H, bt, e, k), 4.98 (24H, bs, m), 3.55 (48H, s, l), 3.18 (8H, bm, f), 2.73 (20H, bm, a, b, g), 2.43 (16H, bt, h), 2.32–2.15 (24H, m, i, c); ^{13}C NMR (DMSO) δ_c: 174.25 (d), 62.15 (f); Mass spectrum (FAB): m/z 1919 (M + H), 1941 (M + Na); Elemental analysis: C$_{78}$H$_{152}$N$_{18}$O$_{35}$ Calcd (%) C, 48.8; H, 7.9; N, 13.1. Found (%) C, 49.9; H, 8.2; N, 14.1.

Dendrimer [core]	%Yield	MW	FAB-Ms
H$_2$N–C$_2$H$_4$–NH$_2$	99	1918	1919 (M + H)
H$_2$N–C$_3$H$_6$–NH$_2$	95	1932	1931 (M − H)
H$_2$N–C$_4$H$_8$–NH$_2$	96	1946	1947 (M + H)
H$_2$N–C$_5$H$_{10}$–NH$_2$	91	1960	1998 (M + 38)
H$_2$N–C$_6$H$_{12}$–NH$_2$	98	1974	1974 (M)

Analytical data for *dendri*-PAMAM(OH)$_{24}$.

3.10 PREPARATION OF HYDROXY TERMINATED PAMAM DENDRIMER; [NH$_2$-(CH$_2$)$_{2-6}$-NH$_2$]; (G = 3.0); dendri-PAMAM(OH)$_{48}$; [10]

An oven-dried flask, with a septum-capped nitrogen inlet, was charged with a solution of (G = 1.5); *dendri*-PAMAM(CO$_2$Me)$_{16}$; [5] (3.1 g, 1.1 mmol) in anhydrous dimethyl sulfoxide (10 ml). A mixture of Tris (tri-hydroxymethyl aminoethane) (2.3 g, 0.019 mol) and anhydrous potassium carbonate (2.6 g, 0.019 mol) was added to the suspension. The solution was heated in a paraffin oil bath maintained at 40 °C for 96 h. After filtration the excess solvent was removed by vacuum pump (10^{-1} mm Hg, 50 °C). The resulting thick, white oil was dissolved in a minimum quantity of distilled water and precipitated with propanone. The suspension was cooled in the freezer, and the solid filtered and dried in a vacuum oven (10^{-1} mm Hg, 40 °C) to give the hygroscopic terminated dendrimer as a white powder (3.9 g, 83%).

(^{a}CH$_2$^{a}CH$_2$)[N(^{b}CH$_2$^{c}CH$_2$^{d}CO^{e}NH^{f}CH$_2$^{g}CH$_2$N(^{h}CH$_2$^{i}CH$_2$^{j}CO^{k}NH^{l}CH$_2$^{m}C
H$_2$N(^{n}CH$_2$^{o}CH$_2$^{p}CO^{q}NHC(^{r}CH$_2$^{s}OH)$_3$)$_2$)$_2$)$_2$]$_2$

Spectral characterization: IR v_{max}/cm^{-1}: 1600 (C=O); ^{1}H NMR (DMSO) δ_H:

8.15 (28H, bm, e, k, q), 4.84 (48H, bs, s), 3.55 (96H, s, r), 3.35 (24H, bm, f, l), 2.75 (56H, bm, g, h, m, n), 2.53 (32H, bt, o), 2.45 (24H, bm, c, i), 2.25 (12H, bm, a, b); ^{13}C NMR (DMSO) δ_c: 173.12 (d), 172.88 (j), 171.18 (p); Mass spectrum (FAB): m/z 4248 (M + 16); Elemental analysis: $C_{174}H_{336}N_{42}O_{76}$ Calcd (%) C, 48.9; H, 7.9; N, 13.9. Found (%) C, 51.2; H, 7.2; N, 13.2.

Dendrimer [core]	%Yield	MW	FAB-Ms
$H_2N–C_2H_2–NH_2$	83	4232	4248 (M + 16)
$H_2N–C_3H_6–NH_2$	96	4246	4247 (M − H)
$H_2N–C_4H_8–NH_2$	91	4260	4259 (M − H)
$H_2N–C_5H_{10}–NH_2$	82	4274	4269 (M − 5)
$H_2N–C_6H_{12}–NH_2$	85	4288	4287 (M − H)

Analytical data for *dendri*-PAMAM(OH)$_{48}$.

dendri-**PAMAM(CO$_2$Me)$_{16}$** $\xrightarrow{\text{K}_2\text{CO}_3, \text{ (Tris)}}$

[5]

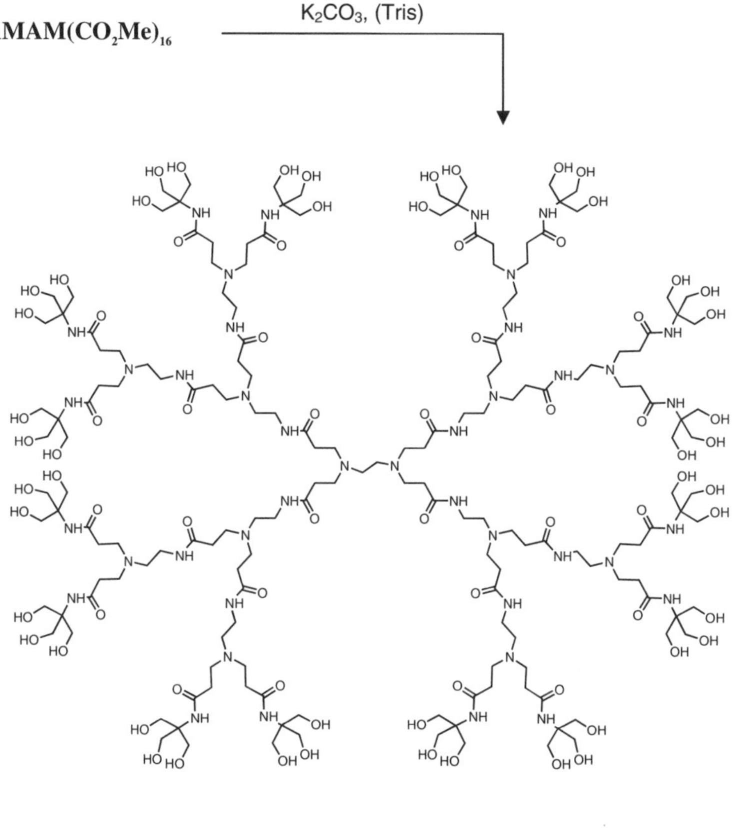

[10]

Scheme 5

4 REFERENCES

1. Buhleier, E., Wehner, W. and Vögtle, F. *Synthesis*, 155–58 (1978).
2. Moors, R. and Vögtle, F. *Chem. Ber.*, **126**, 2133–2135 (1993).
3. Wörner, C. and Mulhaupt, R. *Angew. Chem. Int. Ed. Engl.*, **32**(9), 1306–1308 (1993).
4. Tomalia, D. A., Baker, H., Dewald, J., Hall, M., Kallos, G., Martin, S., Roeck, J., Ryder, J. and Smith, P. *Polym. J.* (*Tokyo*), **17**, 117–32 (1985).
5. Tomalia, D. A., Baker, H., Dewald, J., Hall, M., Kallos, G., Martin, S., Roeck, J., Ryder, J. and Smith, P. *Macromolecules*, 2466–2468 (1986).
6. Smith, P. B., Martin, S. J., Hall, M. J. and Tomalia, D. A. A characterization of the structure and synthetic reactions of polyamidoamine 'STARBURST®' polymers in Mitchell, J. (ed.), *Appl. Polym. Analysis. Characterization*, Hanser Publishers, Munich, Vienna, New York, pp. 357–385 (1987).
7. Tomalia, D. A., Naylor, A. M. and Goddard III, W. A. *Angew. Chem.*, **102**(2), 119–57 (1990); *Angew. Chem. Int. Ed. Engl.*, **29**(2), 138–175 (1990).
8. Padias, A. B., Hall, H. K. Jr, Tomalia, D. A. and McConnell, J. R. *J. Org. Chem.*, **52**, 5305–5312 (1987).

26

Synthesis and Characterization of Poly(Propylene imine) Dendrimers

M. H. P. VAN GENDEREN[1], M. H. A. P. MAK[2],
E. M. M. DE BRABANDER-VAN DEN BERG[2] AND
E. W. MEIJER[1]
[1]Laboratory of Macromolecular and Organic Chemistry, Eindhoven University of Technology, Eindhoven, The Netherlands. [2]DSM Research, Geleen, The Netherlands

1 INTRODUCTION

The oligo(propylene imine) cascade structures made by Vögtle et al. in 1978 are perhaps the first, low molecular weight (i.e. < 1.5 kD) dendritic small molecules reported [1]. Their synthetic approach is based on a repetitive reaction sequence of the double Michael addition of an amine to acrylonitrile, followed by reduction of the nitriles to primary amines. The yields in each step are reasonable; however, the isolation procedures are tedious and no structures beyond what we now call the second-generation dendrimer were synthesized. The oligoamines prepared are highly effective ligands for all kinds of metal ions, thus hampering the purification after the $Co(II)/NaBH_4$ reduction step. Until recently, this elegant reaction sequence was regarded as useless for the synthesis of well-defined high-generation dendrimers, despite the simplicity of the structure as well as the chemistry involved.

Fifteen years later, in 1993, two research groups, Wörner/Mülhaupt (Freiberg Univ.) and de Brabander-van den Berg/Meijer (DSM), disclosed their modifications of the Vögtle approach [2, 3]. Major improvements in the synthesis of

Dendrimers and Other Dendritic Polymers. Edited by Jean M. J. Fréchet and Donald A. Tomalia
© 2001 John Wiley & Sons Ltd

poly(propylene imine) dendrimers were made by utilizing principles involved in the efficient industrial manufacturing of bulk chemicals for nylons. With these principles, a reiterative two-step process was developed which involved (a) a Michael addition of acrylonitrile, and (b) the heterogeneous hydrogenation of nitriles to primary amines. This resulted in the first synthesis of dendrimers on a large pilot-plant scale of approximately 10 kg per batch, up to the fifth generation. This process has provided commercial quantities of structurally well-defined poly(propylene imine) dendrimers of all generations up to 64 amine or nitrile endgroups. These materials are commercially available from DSM, The Netherlands, as Astramol™. In this chapter, we report on the details of the large-scale synthesis of the poly(propylene imine) dendrimers and their physical properties.

2   LARGE-SCALE SYNTHESIS

The prerequisites for an acceptable process to dendrimers involve the following requirements:

1. High yields and selectivities are required, especially when the divergent approach is used. Average selectivities of $> 99\%$ per conversion are needed.
2. Only simple, readily accessible and cheap raw materials, solvents and reagents can be used.
3. Simple procedures to isolate or purify the different generations are required.

In 1993 we have published a method which met most of these requirements and for the first time allowed for the large-scale synthesis of dendrimers [2]. Since that time, each of the steps have been optimized in the reaction scheme. In this chapter, we present 'state-of-the-art' procedures for the large-scale production of the poly(propylene imine) dendrimers.

A synthetic scheme for the so-called poly(propylene imine) dendrimers is shown in Figure 26.1 and consists of a repetition of a double Michael addition of acrylonitrile to primary amines, followed by the heterogeneously catalyzed hydrogenation of the terminal nitrile groups. This sequence results in a doubling of the number of end groups at each generation. 1,4-Diaminobutane (DAB) – the adduct of HCN to acrylonitrile – has been used as the core molecule in the studies presented in this chapter, but a variety of molecules with primary or secondary amino groups can be used as well. In most cases poly(propylene imine) dendrimers up to the fifth generation are prepared on a routine basis. The following nomenclature for the dendrimers is used: DAB-*dendr*-$(NH_2)_n$ and DAB-*dendr*-$(CN)_n$ for the amine- and nitrile-functionalized dendrimers, respectively. 'DAB' stands for the core molecule 1,4-diaminobutane, 'dendr' refers to the dendritic structure and $(NH_2)_n$ and $(CN)_n$ represent the type and number of

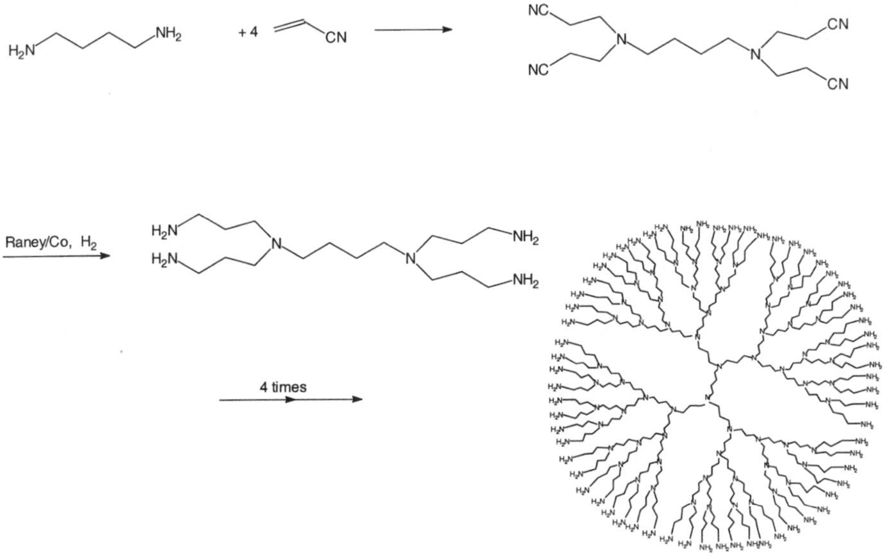

Figure 26.1 Synthetic scheme for poly(propylene imine) dendrimers using 1,4-diaminobutane as core

terminal groups.

All Michael reactions are performed in water at a concentration of up to 50 wt% of amine, to which 2.5–4 mol of acrylonitrile per mol of primary amine are added. In the double cyanoethylation, the first acrylonitrile molecule reacts at room temperature, while the reaction temperature has to be raised to 80 °C to accomplish the double Michael adduct. The reaction time required to achieve complete conversion increases with generation from 1 h for DAB-*dendr*-(CN)$_4$ to 3 h for DAB-*dendr*-(CN)$_{64}$. After completion of the reaction, the excess acrylonitrile is removed by distillation, making use of the water–acrylonitrile azeotrope, leaving a clear two-phase system. Decantation of the upper water layer affords the pure nitrile-terminated dendrimers. If necessary, the layer of nitrile-terminated dendrimers can be washed with water, to remove water-soluble side products as e.g. HOCH$_2$CH$_2$CN (the Michael addition product of H$_2$O to acrylonitrile) or incompletely cyanoethylated products. Analysis of the products showed that in some cases the double Michael adduct of acrylonitrile to water is present. This compound remains in the dendrimer layer during the phase separation procedure, but it can be efficiently removed via extraction. Further studies showed that the formation of this undesired product strongly depends on the pH at the beginning of the reaction. Slight neutralization of the reaction mixture with, for example, acetic acid reduces its formation in such a way that the extraction procedures are no longer necessary (values of below 0.1 wt% are present in the dendrimer, as was concluded from HPLC analysis).

Hydrogenation of the cyanoethylated structures, with Raney Co as catalyst and H_2 pressures of more than 30 bar, were initially performed in water as well. However, the selectivity, the productivity and the yields are all enhanced dramatically by changing the solvent from water to toluene and adding ammonia to the reaction mixture. After reduction, the reaction mixtures are filtered and the primary amine-terminated dendrimers are isolated by evaporating the solvent. The reaction time required for complete hydrogenation increases with generation, but even DAB-*dendr*-$(NH_2)_{64}$ can be obtained via this procedure within 3 h. The main drawbacks of the initial procedure were the high amounts of catalyst required and the low concentration of dendrimer that could be used. Furthermore, the catalyst had to be activated with hydroxide, causing additional side-product formation during the subsequent Michael additions.

The use of ammonia during hydrogenation of the nitriles to primary amines has been described in the literature for a long time. The main function of adding ammonia in the hydrogenation step is to prevent secondary amine formation, which occurs via the addition of a primary amine to an aldimine, followed by hydrogenation [4]. Secondary amine formation is possible both intra- and intermolecularly, leading to eight-membered rings and dendrimer dimerization, respectively. Detailed analysis shows that there is no evidence for intermolecular reactions. The intramolecular reaction is diminished significantly by adding NH_3. Generally, the formation of eight-membered rings is smaller than 1% (see later). The most important reason for using ammonia, however, is that it increases selectivity by suppressing the retro-Michael reaction of the nitrile-terminated dendrimers during the hydrogenation. Retro-Michael reactions can occur by elimination of acrylonitrile to yield defect dendrimers missing one or more branches. However, this side reaction is found to be less than 1% (see later). Under the hydrogenation conditions applied, acrylonitrile will either react with the newly formed amino groups resulting in the formation of a next generation dendrimer, or it can be reduced to propylamine, which can act as a dendrimer core itself. Retro-Michael reaction, therefore, leads to defective dendrimers, either having a lower degree of branching and/or being a mixture of generations. We have strong evidence that retro-Michael reaction at the nitrile-terminated dendritic sites is strongly enhanced in the presence of the amine-terminated dendritic products. Hence, at low reaction rates, nitrile and amine functions are in contact for long periods. By adding ammonia to the reaction mixture, the rate of the reaction is strongly increased and with this advantage the retro-Michael reaction is suppressed. In the presence of ammonia, the amount of catalyst required can be lowered by an order of magnitude. Furthermore, the catalyst needs no pretreatment with sodium hydroxide and the dendrimer concentration can be increased eight times, thus making this procedure very suitable for large-scale synthesis.

Using this improved procedure, even hydrogenation of DAB-*dendr*-$(CN)_{64}$ is complete within a couple of hours. This is surprising, if you realize that for a fifth generation dendrimer, 64 functional groups have to be converted by reaction with 128 molecules of H_2 in a heterogeneous system.

The optimization of the process is achieved by making use of advanced characterization techniques. The results are given in the next section.

3 CHARACTERIZATION

Dendrimers are regarded as macromolecules with a structural precision comparable to proteins or organic compounds. Accurate analysis and quantitative identification of side products are required to optimize and adjust the reaction conditions for the synthesis of DAB-*dendr*-$(NH_2)_n$ and DAB-*dendr*-$(CN)_n$. Therefore, it is a prerequisite to characterize the products obtained unambiguously. To achieve complete molecular characterization of the poly(propylene imine) dendrimers and the possible side-products, NMR- and IR-spectroscopy, HPLC, GPC and electrospray mass spectrometry are used.

NMR spectroscopy appears to be a very suitable technique to detect and assign failures in the outermost layer of the dendrimer structure at each generation. All three nuclei present in the dendrimers have been used including ^1H-, ^{13}C- and ^{15}N-NMR spectroscopy [5–7].

IR spectroscopy is used on a routine basis to follow the hydrogenation of DAB-*dendr*-$(CN)_n$. It is a useful technique for the detection of functional groups, e.g. nitrile absorptions at 2247 cm^{-1} or the characteristic double peaks due to NH_2 groups at 3356 and 2280 cm^{-1}. The IR spectra of the various generations with identical end groups are very similar.

Electrospray MS was performed on both series of nitrile- and amine-terminated poly(propylene imine) dendrimers [8]. The actual and reconstructed spectra of DAB-*dendr*-$(NH_2)_{64}$ are given in Figure 26.2. The measured spectrum shows a repetition of different clusters of peaks, corresponding to dendrimers with 4–12 charges per dendrimer (m/z with $z = 4$–12), while no counterion interactions are observed. Deconvolution yields a spectrum in which the largest peak at $M_r = 7168$ corresponds to the perfect DAB-*dendr*-$(NH_2)_{64}$ and a series of peaks with lower abundance is found with regular intervals of $\Delta M = 57.1$ (missing propylamine units) from the perfect dendrimer. Furthermore, the peak at $M_r = 7151$ (missing ammonia via the intramolecular ring closure) is assigned to the dendrimer with 62 primary amine groups and one cyclic secondary amine functionality, while there is also a series of peaks with intervals of 57.1 from $M_r = 7151$. After performing this analysis for all dendrimers within the synthetic scheme, it was possible to simulate the electrospray MS spectra in detail. These simulations show that the dendrimers possess statistical defects in their struc-

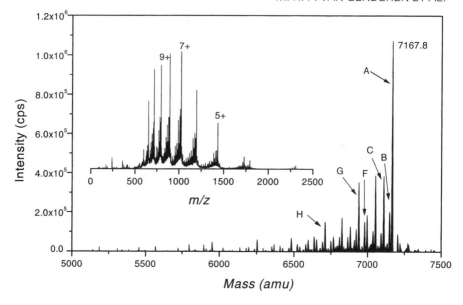

Figure 26.2 Deconvoluted ES–MS spectrum of DAB-*dendr*-(NH₂)₆₄. Inset: actual data

tures as the result of the many consecutive reactions that are performed to prepare these highly branched structures. The poly(propylene imine) dendrimers with 64 amine end groups, the result of 248 of these reactions, possess a dendritic purity of around 20% and have a polydispersity of around 1.002. These results show that the divergent synthesis, once optimized to the extreme, compares to the best results obtained in the Merrifield synthesis of polypeptides, etc. while the polydispersities are unprecedented in polymer synthesis.

4 PHYSICAL PROPERTIES

Apart from DAB-*dendr*-(CN)₄, which is a white crystalline solid, all generations are colorless to slightly yellow oils. The amine-terminated dendrimers are transparent, whereas the nitrile-terminated products are somewhat turbid. The solubility of the dendrimers is determined primarily by the nature of the end-group: DAB-*dendr*-(NH₂)ₙ is soluble in H_2O, methanol and toluene, whereas DAB-*dendr*-(CN)ₙ is soluble in a variety of common organic solvents.

The viscosity of the poly(propylene imine) dendrimers is investigated both in THF (intrinsic viscosity, Figure 26.3) and in the neat form. For the intrinsic

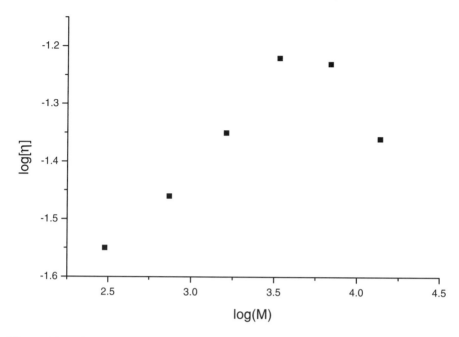

Figure 26.3 Intrinsic viscosity of the various nitrile-terminated poly(propylene imine) dendrimers in THF as a function of molar mass

viscosity, a maximum in $\log[\eta]$ is observed, which is characteristic for dendrimers [9]. The melt viscosities do not exhibit a maximum, but are strongly dependent on the end groups: they typically run from 0.03 to 6.7 Pa × s for the amine-terminated generations, and from 2.6 to 50 Pa × s for the nitrile-terminated species.

Thermal analysis of DAB-*dendr*-(CN)$_n$ and DAB-*dendr*-(NH$_2$)$_n$ shows a number of interesting aspects. The glass transition temperature (T_g) has been derived from differential scanning calorimetry and the results are presented in Figure 26.4. In both the amine- and nitrile-terminated dendrimer series, the T_g-values observed are low and from generation 2 onwards independent of the generation. In all cases the nitrile-terminated dendrimers show higher T_g-values than the amine-terminated structures, which can be attributed to the dipole–dipole interactions between the nitrile groups. There exists some confusion as to the meaning of a T_g for this kind of molecules. Since in all cases only one T_g is observed, we are tempted to assume that this value is linked to the mobility of segments within a single molecule. In any case, the low T_g-values indicate that the dendrimers possess a large degree of conformational freedom.

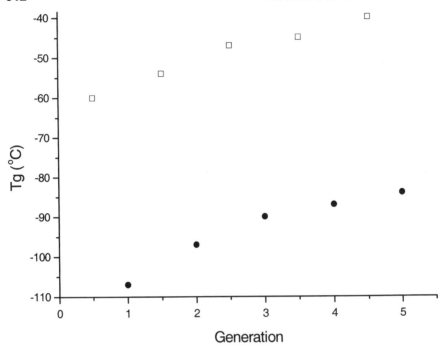

Figure 26.4 Plot of the glass transition temperatures of the poly(propylene imine) dendrimers with nitrile (□) and amine (•) end-groups

The thermal stability, determined by TGA, of the amine-terminated dendrimers is unexpectedly high and increases with increasing molar mass. The TGA$_{max}$ (temperature at which the weight loss is maximal) for DAB-*dendr*-(NH$_2$)$_4$, DAB-*dendr*-(NH$_2$)$_8$, DAB-*dendr*-(NH$_2$)$_{16}$, DAB-*dendr*-(NH$_2$)$_{32}$ are 330, 378, 424 and 470°C, respectively (measured at a heating rate of 20°C/min under nitrogen). For DAB-*dendr*-(NH$_2$)$_8$ less than 1.0% weight loss is observed at 310°C. The nitrile-terminated dendrimers are less stable, but in this case the stability also increases slightly with higher generations. Using TGA–MS, a thermally induced retro-Michael reaction could be identified to be the major degradation mechanism for the nitrile-terminated dendrimers.

All dendrimers consist of inner tertiary amines, located at the branching points of the various dendritic shells (layers). The amine-terminated dendrimers, furthermore, have basic primary amine end-groups. Basicity is therefore one of the most dramatic properties of the poly(propylene imine) dendrimers, and has been studied via titration experiments and calculations. Titration experiments of the dendrimers have been performed in water using 1 M hydrochloric acid. Only two equivalence points are observed for DAB-*dendr*-(NH$_2$)$_4$ in a ratio of 2:1. From these titrations, pK_a values of 10.0 (primary amine groups) and 6.7 (tertiary amine groups) have been derived. These results agree with the well-documented solvent dependence of the relative basicity of primary and tertiary amine groups. For DAB-*dendr*-$(NH_2)_8$ and DAB-*dendr*-$(NH_2)_{16}$ there is no big difference in the observed pK_a values compared to DAB-*dendr*-$(NH_2)_4$. Titration experiments in water showed that it is possible to protonate all nitrogen atoms up to DAB-*dendr*-$(NH_2)_{64}$. Also, ^{15}N-NMR titrations of the amine-terminated dendrimers in water, in combination with Ising model calculations on the protonation of branched polyamines, show that the protonation sequence of the various shells (layers) of nitrogens [7] is as follows: first the primary amines are protonated, followed by the most central tertiary amines, and finally the other tertiary amines in such a way that Coulomb interactions are minimized.

Titration experiments on the nitrile-terminated dendrimers in water show for DAB-*dendr*-$(CN)_4$ pK_a values of 3.2 and 4.8. The corresponding calculated pK_a values are 3.1 and 4.1 respectively (using the pKalc program, version 2.0, Compudrug chemistry). For DAB-*dendr*-$(CN)_8$ only the two inner nitrogen atoms can be protonated in acetonitrile, due to the low basicity of the four other ones. This is confirmed with calculated pK_a values of the four outer tertiary nitrogen atoms in DAB-*dendr*-$(CN)_8$, ranging from 2.0 to 3.2. The presence of the electropositive nitrile-functions and the protonated inner tertiary amines can account for this phenomenon.

5   DENDRIMER SHAPE

Following the first synthetic attempts toward dendrimers, theoretical investigations [10] predicted molecular dimensions, limits of growth, and intramolecular configurational details of dendrimers, which later were indeed confirmed experimentally. For example, the maximum of the intrinsic viscosity as a function of the generation number [9, 11] has been predicted and confirmed. In the same spirit, simulation techniques were employed by Lescanec *et al.* [12]. The internal structure predicted by the simulations differs strongly from the proposed analytical model of de Gennes *et al.* [10]. The Lescanec model predicts a (segment) density maximum near the center of the dendrimer, whereas de Gennes argues for a density minimum in the center. Recent results based on Monte Carlo simulations [13], calculations of the intrinsic viscosity [14], as well as ^{2}H and ^{13}C NMR studies [15] tended to favor the model of Lescanec *et al.* Recent work by Amis and co-workers [16] using SANS measurements on poly(amidoamine) dendrimers favor the de Gennes model. This work appears to clearly indicate that there is very little backfolding in poly(amidoamine) dendrimers and they do indeed behave much as de Gennes originally described [10].

There has been little experimental data on the generation dependence of molecular size to discriminate between the two models until recently. Some experimental studies on various other dendrimers have been reported using, for example, size exclusion chromatography (SEC), viscometry [11, 17], small-angle neutron scattering (SANS) [18] and diffusion-ordered spectroscopy [19], but the overall quantity of data is limited. Scherrenberg *et al.* have studied 1% (v/v) solutions of the amine-terminated poly(propylene imine) dendrimers in deuterated water at the SANS facility of the Risø National Laboratory, Roskilde, Denmark [20a]. The obtained radii of gyration for the nitrile- as well as for the primary amine-terminated dendrimers are given in Table 26.1. An excellent linear relationship is found between the radius of gyration (R_g) and the generation number. Such a linear relationship has already been predicted by Monte Carlo simulations for other dendrimer systems [13].

Molecular modeling techniques are a powerful tool to obtain a very detailed insight in the three-dimensional structure of dendrimer molecules at the atomic level. They have been applied to calculate sizes of the poly(propylene imine) dendrimers and radial density profiles in order to estimate the free volume inside the dendrimers, as well as to make predictions about 'de Gennes dense-packed' generations. The molecular modeling work by Coussens and co-workers [20] was focused on the generations 1–5 of the DAB-*dendr*-$(CN)_n$ and DAB-*dendr*-$(NH_2)_n$ ($n = 4, 8, 16, 32, 64$).

The sizes of the dendrimers have been determined by calculating the molecular volumes, as defined by the van der Waals radii of the atoms, and by calculating the radii of gyration for several configurations of the dendrimers, as obtained from a molecular dynamics simulation at room temperature. The solvent influence on the calculated radii was estimated by scaling the nonbonded interactions between the atoms. Molecular volumes and average radii for ensembles of 500 conformations of the DAB-*dendr*-$(NH_2)_n$ dendrimers have been collected in Table 26.2.

The calculated radii with all interactions included are somewhat smaller than the radii measured with SANS, whereas the radii obtained with only the van der Waals repulsions taken into account are somewhat larger. As could be anticipated, the sizes of the dendrimers are dependent on the pH of the solution. Since both the primary and the tertiary amine groups may be protonated, repulsions begin when the pH of the solution is decreased.

Table 26.1 Radii of gyration of DAB-*dendr*-$(CN)_n$ and DAB-*dendr*-$(NH_2)_n$, as determined from neutron scattering [20a]

DAB-*dendr*-$(NH_2)_n$ (n)	Radius/nm	DAB-*dendr*-$(CN)_n$ (n)	Radius/nm
4	0.44	—	—
8	0.69	8	0.60
16	0.93	16	0.80
32	1.16	32	1.01
64	1.39	64	1.22

Table 26.2 Molecular volumes and radii of gyration of the DAB-*dendr*-$(NH_2)_n$ dendrimers as obtained by molecular modeling in the gas phase [20a]

Generation dendrimer	$V^{a,b}/nm^3$	$<R_g>^a/nm$	$<R_g>^c/nm$
1	1.06	0.49	0.50
2	1.95	0.60	0.76
3	3.65	0.74	1.01
4	9.01	1.00	1.29
5	17.60	1.25	1.59

[a] All interactions (Coulombic and van der Waals) taken into account.
[b] The hydrodynamic radius ($= (5/3)^{1/2} R_g$) was used for calculating this volume.
[c] Only repulsive van der Waals interactions taken into account.

6 CONCLUSION

As has been shown, large-scale synthesis of the poly(propylene imine) dendrimers is possible, and leads to well-characterized products. Physical properties of poly(propylene imine) dendrimers are determined by the nature of the end-group, as well as by generation number. The effect of the end-group is clearly noticed from, for example, melt viscosities, DSC and TGA measurements, in which the amine-terminated structures show a higher thermal stability and a lower T_g than the nitrile-terminated dendrimers. Naturally, a different pH-dependent behavior between the two terminal group types is observed. Generation dependence is noticed with a maximum in the plot of intrinsic viscosity vs molar mass. This is a unique feature noted for all dendrimers. The cavities observed with molecular modeling studies are characteristic for high generation dendrimers. The radius of gyration increases linearly with generation. This is in harmony with the predictions based on Monte Carlo simulations of dendrimers.

7 REFERENCES

1. Buhleier, E., Wehner, W. and Vögtle, F. *Synthesis*, 155 (1978).
2. De Brabander-van den Berg, E. M. M. and Meijer, E. W. *Angew. Chem.*, **105**, 1370 (1993).
3. Wörner, C. and Mühlhaupt, R. *Angew. Chem.*, **105**, 1367 (1993).
4. (a) Freifelder, M. *Catalytic Hydrogenation in Organic Synthesis, Procedures and Commentary*, Wiley, New York, 1978, p. 43; (b) Rylander, P. N. *Best Synthetic Methods, Hydrogenation Methods*, Academic Press, London, 1985, p. 94.
5. Van Genderen, M. H. P., Baars, M. W. P. L., van Hest, J. C. M., de Brabander-van den Berg, E. M. M. and Meijer, E. W. *Recl. Trav. Chim. Pays-Bas*, **113**, 573 (1994).
6. Van Genderen, M. H. P., Baars, M. W. P. L., Elissen-Román, C., de Brabander-van den Berg, E. M. M. and Meijer, E. W. *Polym. Mater. Sci. Eng.*, **73**, 336 (1995), *Abstr. Pap. Am. Chem. Soc.*, **20**, 179 (1995).
7. Koper, G. J. M., van Genderen, M. H. P., Elissen-Román, C., Baars, M. W. P. L.,

Meijer, E. W. and Borkovec, M. *J. Am. Chem. Soc.*, **119**, 6512 (1997).

8. Hummelen, J. C., van Dongen, J. L. J. and Meijer, E.W. *Chem. Eur. J.*, **3**, 1489 (1997).
9. Fréchet, J. M. J., Hawker, C. J., Gitsov, I. and Leon, J. W. *J. Macromol. Sci. – Pure Appl. Chem.*, **A33**, 1399 (1996).
10. De Gennes, P. G. and Hervet, H. J. *Phys. Lett.* (*Paris*), **44**, 351 (1983).
11. Mourey, T. H., Turner, S. R., Rubinstein, M., Fréchet, J. M. J., Hawker, C. J. and Wooley, K. L. *Macromolecules*, **25**, 2401 (1992).
12. Lescanec, R. L. and Muthukumar, M. *Macromolecules*, **23**, 2280 (1990).
13. Mansfield, M. L. and Klushin, L. I. *Macromolecules*, **26**, 4262 (1993).
14. Mansfield, M. L. and Klushin, L. I. *J. Phys. Chem.*, **96**, 3994 (1992).
15. (a) Meltzer, A. D., Tirrell, D. A., Jones, A. A., Inglefield, P. T., Hedstrand, D. M. and Tomalia, D. A. *Macromolecules*, **25**, 4541 (1992); (b) Meltzer, A. D., Tirrell, D. A., Jones, A. A. and Inglefield, P. T. *Macromolecules*, **25**, 4549 (1992).
16. Topp, A., Bauer, B. J., Klimash, J. W., Spindler, R., Tomalia, D. A. and Amis, E. J. *Macromolecules*, **32**, 7226 (1999).
17. Tomalia, D. A., Naylor, A. M. and Goddard III, W. A. *Angew. Chem.*, **102**, 119 (1990).
18. (a) Bauer, B. J., Briber, R. M., Hammouda, B. and Tomalia, D. A. *Polym. Mater. Sci. Eng.*, **67**, 428 (1992); (b) Briber, R. M., Bauer, B. J., Hammouda, B. and Tomalia, D. A. *Polym. Mater. Sci. Eng.*, **67**, 430 (1992).
19. Young, J. K., Baker, G. R., Newkome, G. R., Morris, K. F. and Johnson, C. S. Jr. *Macromolecules*, **27**, 3464 (1994).
20. (a) Scherrenberg, R. L., Coussens, B., Mortensen, K. and de Brabander-van den Berg, E. M. M. *Macromolecules*, **31**, 456 (1998); (b) de Brabander, E. M. M., Brackman, J., Muré-Mak, M., de Man, H., Hogeweg, M., Keulen, J., Scherrenberg, R., Coussens, B., Mengerink, Y. and van der Wal, Sj. *Macromol. Symp.*, **102**, 9 (1996).

27

Laboratory Synthesis and Characterization of Megamers: Core-Shell Tecto(dendrimers)

D. A. TOMALIA AND D. R. SWANSON
Dendritic Nanotechnologies Limited, Central Michigan University,
Mt. Pleasant, MI, USA

1 INTRODUCTION

Dendrimer synthesis strategies now provide virtual control of macromolecular nanostructures as a function of size [1, 2], shape [3] and surface/interior functionality [4]. These strategies involve the covalent assembly of hierarchical components such as: (A) reactive monomers [5], (B) branch cells [6, 7] or (C) dendrons [8] around atomic or molecular cores according to divergent/convergent dendritic branching principles [7, 9]. Systematic filling of space around a core with shells (layers) of branch cells (i.e. generations) produces discrete core–shell dendrimer structures (D). Dendrimers are quantized bundles of mass possessing amplified surface functionality that are mathematically predictable [7]. Predicted molecular weights and surface stoichiometry have been confirmed experimentally by mass spectroscopy [7, 10, 11], gel electrophoresis [12] (see Chapter 10) and other analytical methods [1, 2]. It is now recognized that empirical structures; such as: (B) → (C) → (D) may be used to define these hierarchical constructions (Figure 27.1). Such synthetic strategies have produced traditional dendrimers with dimensions that extend well into the lower nanoscale region (i.e. 1–20 nm) [13]. The precise structure control and unique new properties exhibited by these dendrimeric architectures has yielded many interesting advanced material properties [14–16]. Nanoscale dendrimeric containers [14, 17, 18] and scaffolding [4, 17] have been used to template zero valent metal

Dendrimers and Other Dendritic Polymers. Edited by Jean M. J. Fréchet and Donald A. Tomalia
© 2001 John Wiley & Sons Ltd

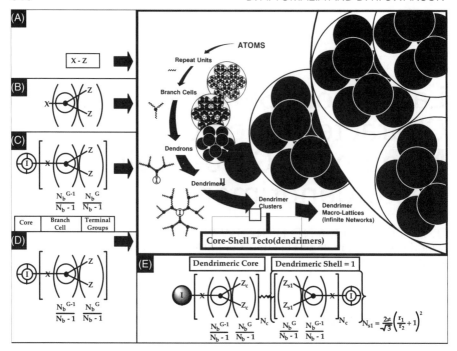

Figure 27.1 Hierarchy of empirical construction components: (A) monomers, (B) branch cells, (C) dendrons and (D) dendrimers leading to (E) core–shell tecto(dendrimers)

nanodomains [19, 20], nanoscale magnets [21–23], electron conducting matrices [24, 25], as well as provide a variety of novel opto-electronic properties [26, 27].

However, the use of such traditional strategies for the synthesis of precise nanostructures (i.e. dendrons (C) and dendrimers, (D)) larger than ∼ 15–20 nm has several serious disadvantages. Firstly, it is hampered by the large number of reiterative synthetic steps required to attain these higher dimensions (e.g. poly(amidoamine) (PAMAM) dendrimer (Generation 9); dia. ≅ 10 nm; requires 18 reaction steps). Secondly, these constructions are limited by the 'de Gennes dense packing' phenomenon which precludes ideal dendritic construction beyond certain limiting generations [7, 28]. For these reasons, our attention has turned to the use of dendrimers as reactive modules for the rapid construction of controlled nanoarchitectures possessing higher complexity and dimensions beyond the dendrimer. We refer to these generic poly(dendrimers) as *megamers*. Both randomly assembled megamers [29], as well as structure-controlled megamers [30–33] have been demonstrated. Recently new mathematically defined megamers (dendrimer clusters) or *core–shell tecto(dendrimers)* have been reported [15, 30–33]. The principles involved for these structure controlled megamer syntheses, mimic those used for the traditional core–shell construction of dendrimers. First a dendrimer core reagent is selected as a target substrate (i.e. usually a spheroid). Next a limited amount of the reactive core reagent is combined with an excess of a dendrimer shell reagent. The objective is to completely saturate the target spheroid surface with covalently bonded dendrimer shell reagent. Since the diameter of the dendrimer core and shell reagents, respectively, are very well defined, it is possible to mathematically predict the number of dendrimer shell molecules required to saturate a targeted dendrimer core [34].

These relationships were analyzed mathematically as a function of core (r_1) and shell (r_2) radii ratios [34]. At low r_1/r_2 values (i.e. 0.1–1.2) very important symmetry properties emerge as shown in Figure 27.2. It can be seen that when the core reagent (r_1) is small and the shell reagent (r_2) is larger, only a very limited number of shell-type dendrimers can be attached to the core dendrimer based on available space. However, when the ratio of core radius to shell radius is $\geqslant 1.2$ the space becomes available to attached many more shell reagents up to a discrete saturation level. The saturation number (N_{max}) is well defined and can be predicted from a general expression that is described by the Mansfield–Tomalia–Rakesh equation (Figure 27.2).

2 GENERAL COMMENTS

Laboratory procedures are presented for two divergent approaches to covalent structure controlled dendrimer clusters or more specifically – *core–shell tecto(dendrimers)*. The first method, namely: (1) the 'self assembly/covalent bond formation method' produces structure controlled 'saturated shell' products (see Scheme 1). The second route, referred to as (2) 'direct covalent bond formation method', yields 'partial filled shell' structures, as illustrated in Scheme 2. In each case, relatively monodispersed products are obtained. The first method yields precise 'shell saturated' structures [31, 32]; whereas the second method gives semi-controlled partially shell filled products [30, 33].

Structural notations describing these core–shell structures utilized an extension of nomenclature invoked to describe the traditional PAMAM dendrimer reactants (see Chapter 23). For example, a *dendrimer-core reagent*; such as: (EDA core), generation = 5, amine-terminated poly(amidoamine) (PAMAM) dendrimer (i.e. [EDA]-*dendri*-PAMAM-(NH₂)₁₂₈ reacted with a *PAMAM dendrimer shell reagent*, generation = 3, carbomethoxy terminated (i.e. [EDA]-*dendri*-PAMAM–(CO$_2$Me)$_{32}$ would be expected after amidation of the terminal groups to produce a *core–shell tecto(dendrimer)* designated as follows: [[EDA]-*dendri*-PAMAM–(NH$_2$)₁₂₈]-*amide*-{[EDA]-*dendri*-PAMAM–(CO$_2$Me)$_{32}$}$_n$, where *n*

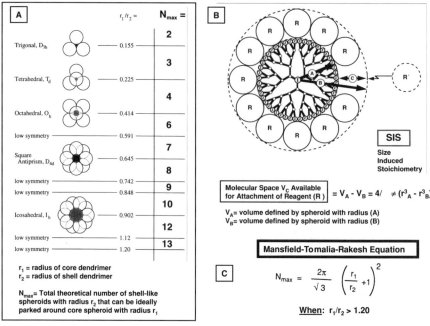

Figure 27.2 (a) Symmetry properties for core–shell structures where $r_1/r_2 <$ 1.20; (b) sterically induced stoichiometry (SIS) based on respective radii (r_1) and (r_2) core and shell dendrimers respectively; (c) Mansfield–Tomalia–Rakesh equation for calculation of maximum shell filling when $r_1/r_2 < 1.20$

denotes the number of $G = 3$; PAMAM shell reagents that are covalently bonded to a $G = 5$; PAMAM core reagent. Alternatively, a shorthand notation may be used for this structure which is designated as follows: [[EDA]; $G = 5$; –(NH$_2$)]-*amide*- {(EDA); $G = 3$; –(CO$_2$Me)}$_n$.

3   SELF-ASSEMBLY/COVALENT BOND FORMATION METHOD

The general chemistry used in this approach involves the combination of a limited amount of an amine-terminated dendrimer core reagent with an excess of carboxylic acid terminated dendrimer shell reagent [31]. These two charge differentiated species are allowed to self-assemble into the electrostatically driven supramolecular core–shell tecto(dendrimer) architecture. After equilibration, covalent bond formation at these charge neutralized dendrimer contact sites is induced with carbodiimide reagents (Scheme 1).

Various amine-terminated poly(amidoamine) (PAMAM) dendrimers were used as core reagents (i.e. $G = 5$, 6 and 7; –NH$_2$ terminated), whereas various

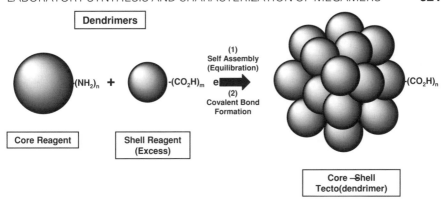

Scheme 1 Reaction scheme for *saturated shell model* where all surface dendrimers are carboxylic acid terminated

carboxylic acid terminated PAMAM dendrimers (i.e. G = 3 and 5), were used as the shell reagent. These shell reagents were synthesized by reaction of appropriate amine terminated PAMAM dendrimer using the ring opening of succinic anhydride to produce carboxylic acid terminated dendrimers (see section 3.1.4.1).

All reactions leading to core–shell tecto(dendrimer) were performed in the presence of LiCl at room temperature as dilute solutions (~ 0.5 wt%) in water. Equilibration times between 16 and 20 h were required to complete the charge neutralization and self-assembly of excess shell reagent around the limited den-drimer core reagent. Following this self-assembly and equilibration, a linking reagent; *1-(3-dimethylamino propyl)3-ethylcarbodiimide hydrochloride*, was added to covalently bond the assembly of dendrimer shell reagents to a single dendrimer core reagent at the amine-carboxylic acid interaction sites. These sites are presumed to reside primarily at the exterior of the dendrimer core reagent [35].

Remarkably monodispersed products were obtained by performing the core–shell self-assembly reactions in the presence of LiCl. In the absence of LiCl, these reactions yielded bimodal or trimodal product mass distributions (i.e. as observed by SEC). Core–shell products formed in the absence of LiCl were multimodal and are presumed to be due to clustering of the amine-terminated core reagent into various domain sizes. Such clustering of amine terminated PAMAM dendrimers has been noted in earlier work [2]. Attempts to subsequently charge neutralize these polydispersed domains with anionic dendrimer shell reagent produced a broad product distribution. Reversing the terminal functional groups on the core and shell reagents, respectively (i.e. using carboxylic acid-terminated PAMAM dendrimers as the core and excess amine-terminated PAMAM dendrimers as the shell reagent) under identical reaction conditions, did not yield the desired product. The reason for this is not evident from our studies so far.

Table 27.1

Core-shell compound	1	2	3	4
Formula: $[G'X'A]/\{G'Y'C\}_n a$	$[G5A]/\{G3C\}_n$	$[G6A]/\{G3C\}_n$	$[G7A]/\{G3C\}_n$	$[G7A]/\{G5C\}_n$
MW (calculated)	137 350	197 700	301 270	537 300
Number of shell tectons, Observed c/Theoretical d:	10/12	13/15	15/19	9/12
Ratio:	0.83	0.87	0.79	0.75
MALDI–MS (MW):	114 000	172 000	238 000	403 000
PAGE (MW):	120 000	168 000	250 000	670 000
AFM: Diameter (nm)	—	23.5 ± 6.7	28.2 ± 4.5	28.0 ± 4.2
Height (nm)	—	2.15 ± 0.5	2.64 ± 0.8	3.84 ± 1.0
Estimated MW^e	—	195 000	369 000	469 000

^{a} In the notation $[G'X'A]/\{G'Y'C\}\underline{n}$ above, $\underline{n}$ = number of dendrimer shell molecules $\{G'Y'C\}$ surrounding the dendrimer core molecule $[G'X'A'$, where: $[G'X'A]$ represents amine-terminated, EDA core, generation 'X' PAMAM dendrimer core reagent, and $[G'X'C]$ represents carboxylic acid-terminated, EDA core, generation 'X' PAMAM dendrimer core reagent, and $[G'Y'C]$ represents carboxylic acid-terminated, EDA core, generation 'Y', PAMAM dendrimer shell reagent. ^{b} MW of theoretical number of shell dendrimer tectons plus core tecton, as [34] determined from MALDI-MS data for individual core and shell tecton molecular weights. Values used for n in these calculations were obtained by using the Mansfield–Tomalia–Rakesh equation described. ^{c} Based on the experimental MALDI-MS molecular weights of the tecto(dendrimer)s as isolated. ^{d} The theoretical number (n) of dendrimer shell molecules that would be expected to surround a specific dendrimer core molecule according to the Mansfield– Tomalia–Rakesh equation [34]. ^{e} See ref. [30].

The progress of a typical core–shell tecto(dendrimer) reaction was monitored by both size exclusion chromatography (SEC) as well as by gradient poly(acrylamide) gel electrophoresis (PAGE) [12]. Formation of tecto(dendrimer), $\underline{3}$ (Table 27.1), appeared to be nearly complete within 1 h after addition of the carbodiimide reagent. An SEC analysis taken 2 h after addition of the carbodiimide reagent indicated a slight narrowing of the product distribution band without any significant changes in the eluogram. Excess shell reagent $\{G = 3, -CO_2H\}$, was removed by dialysis from the respective reaction runs to give cream colored, solid products in isolated yields that ranged from 88 to 91 % for core–shell tecto(dendrimers) ($\underline{1}$–$\underline{4}$) (Table 27.1).

As might be expected, this method provides for more efficient parking of the dendrimer shell reagents around the dendrimer core, thus yielding very high saturation levels (i.e. 75–83 %) as shown in Table 27.1. Our present experimentation indicates that this method will allow the assembly of additional shells in a very systematic fashion to produce larger nanostructures that may transcend the entire nanoscale region (1–100 nm).

3.1 EXPERIMENTAL

3.1.1 Materials

All amine-terminated poly(amidoamine) (PAMAM) dendrimers [i.e. [EDA]-*dendri*-PAMAM-$(NH_2)_n$] utilized as core reagents in this work were generously provided by Dendritic Sciences, Inc., Mt. Pleasant, Michigan. They were synthesized according to methods described elsewhere [36–38]. The carboxylic acid terminated dendrimer shell reagents were synthesized by ring opening succinic anhydride with appropriate amine terminated dendrimer reagents as described later. Dendrimers were thoroughly characterized by ^{13}C-NMR [39], MALDI-TOF mass spectroscopy [11, 40], SEC [4], poly(acrylamide) gel electrophoresis (PAGE) [12] and atomic force microscopy (AFM) [30].

3.1.2 Analytical Methods

3.1.2.1 MALDI-TOF Mass Spectrometer Conditions for PAMAM Dendrimers and Core–Shell Tecto (Dendrimers)

A stock solution of the dendrimer reagent or core–shell tecto(dendrimer) is prepared by dissolving 3–5 mg of material in methanol. A 1 μL aliquot of the stock solution is mixed with 9 μL of matrix solution. The matrix solution consists of 10 mg/mL 2,5-dihydroxybenzoic acid in 20:80 acetronitrile/water containing 0.1% trifluoracetic acid. The dendrimer/matrix mixture is spotted on the target plate (1 μL/spot) and the solvent allowed to evaporate under ambient conditions. Analysis is performed with a ThermoBioanalysis Vision 2000 mass spectrometer operating in the reflector mode. The value of the laser power setting is generally set between the value where matrix ions first appear and 110% of that value.

3.1.3 Size Exclusion Chromatography

The core–shell tecto(dendrimer) samples (**1–4**) were dissolved in water and passed through a 0.2 μm syringe-filter (Whatman Anotop 25). They were then injected into an SEC consisting of a Waters 510 HPLC pump, a TSK pre-column followed by three analytical columns (namely, a G4000PW, G3000PW and G2000PW) and equipped with a Wyatt Refracting Index dual detector at 30 °C (run time 40 min). The mobile phase was 0.5 M acetic acid/0.5 M sodium nitrate in water. The molecular weights (MW) and MW distributions were obtained by comparing directly to standard PAMAM dendrimer samples (i.e. G = 0–10). Carboxylic acid terminated (PAMAM) dendrimers [i.e. [EDA]-*dendri*-PAMAM-$(CO_2H)_m$)], used as dendrimer shell reagents in this work, were synthesized as described below.

3.1.4 Synthesis

3.1.4.1 Preparation of [EDA]-dendri-PAMAM –(CO$_2$H)$_m$ Shell Reagents

Amine-terminated, G3; (PAMAM) dendrimer, (0.316 g; 45.7 moles) was dissolved in anhydrous methyl sulfoxide (5 ml) in a 100 ml round-bottom flask flushed with dry nitrogen. After dendrimer had completely dissolved, succinic anhydride (Aldrich) (0.363 g; 3.6 mmol) was added to the reaction mixture with vigorous stirring, and the mixture was allowed to react for 24 h at room temperature. The product solution was diluted with deionized water, transferred to 3500 MWCO dialysis tubing (Spectrum) and dialyzed against deionized water (18 MΩ) for 3 d. The retentate solution was clarified by filtration through Whatman No. 1 filter paper, concentrated with a rotary evaporator, and lyophilized to yield a colorless powder (0.435 g, 94%). The product was analyzed by ^{13}C–NMR, FT–IR, SEC and MALDI–MS. The analytical data were consistent with the expected carboxylic acid-terminated product.

3.1.4.2 Preparation of [[EDA]-dendri-PAMAM–(NH$_2$)$_{128}$]-amide-{[EDA]-dendri-PAMAM–COH$_{32}$}$_{10}$ tecto(dendrimer); [G = 5, –NH$_2$]-amide-{G = 3; –CO$_2$H}$_{10}$

To a 100 mL round-bottomed flask was added 0.170 g (~ 18 μmol) of an [EDA core], (G = 3), carboxylic acid-terminated, (PAMAM) dendrimer shell reagent; (G = 3;CO$_2$H) and 5.5 g of water. This mixture was stirred until homogeneous. To this mixture was added dropwise an aqueous solution of [EDA core], (G = 5) amine-terminated, (PAMAM) dendrimer core reagent (G = 5; NH$_2$) (0.017 g, ~ 0.6 μmol of the amine-terminated dendrimer in 3 g of water containing 0.057 g of LiCl) over 10 min. The mixture was allowed to equilibrate for 20 h at room temperature. Next, the linking reagent; 1-(3-dimethylaminopropyl)3-ethylcarbodiimide hydrochloride (0.17 g) was added to the reaction mixture. After 6 h, the reaction was terminated and dialyzed, using a 10 000 Dalton molecular weight cut-off membrane (Spectrum). The sample was dialyzed against 2 L of 18 MΩ deionized (DI) water for 2 d, replacing the DI water five times during this period. After dialyzing, the sample was filtered through Whatman No. 1 filter paper, concentrated with a rotary evaporator, and lyophilized to give 80 mg (90% yield) of PAMAM [G = 5;-NH$_2$]-amide-{G = 3;-CO$_2$H}$_{10}$ core-shell tecto(dendrimer) (1) as a cream colored, solid product.

4 DIRECT COVALENT BOND FORMATION METHOD

The second method referred to as the 'direct covalent bond formation method', produces semi-controlled, partial shell filled structures. It involves the reaction of a limited amount of nucleophilic dendrimer core reagent with an excess of electrophilic dendrimer shell reagent as illustrated in Scheme 2. This route involves the random parking of reactive shell reagent on a target core substrate surface. As a consequence, partially filled shell products are obtained which possess relatively narrow, but not precise molecular weight distributions. These distributions are determined by the core-shell parking efficiency prior to covalent bond formation.

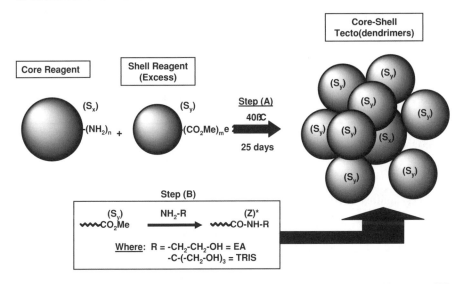

Scheme 2 Reaction scheme for *partial shell filled model*, step (A); step (B) describes surface capping reactions

Various poly(amidoamine) (PAMAM) dendrimer core reagents (i.e. either amine or ester functionalized) were allowed to react with an excess of appropriate PAMAM dendrimer shell reagents. The reactions were performed at 40°C in methanol and monitored by FTIR, ^{13}C-NMR, SEC and gel electrophoresis. Conversions in step (a) were followed by observing the formation of shorter retention time products consistent with higher molecular weight structures, using SEC. Additional evidence was gained by observing loss of the migratory band associated with the dendrimer core reagent present in the initial reaction mixture, accompanied by the formation of a higher molecular weight product, which displayed a much shorter migratory band position on the electrophoretic gel. In fact, the molecular weights of the resulting core–shell tecto(dendrimer) could be estimated by comparing the migratory time of the core–shell product (PAGE results, Table 27.2) with the migration distances of the PAMAM dendrimer reagents (e.g. G = 2–10) used for their construction [12].

It was important to perform capping reactions on the surface of the resulting ester terminated core–shell products in order to pacify the highly reactive

Table 27.2

X;[(Y)(Z*)]ₙ	G4[(G3); (EA)]ₙ	G5(G3); (TRIS)]ₙ	G6(G4); (TRIS)ₙ	G7(G5); (TRIS)ₙ
	(**1**)	(**2**)	(**3**)	(**4**)
Theoretical Shell saturation levels(n)		15	15	15
Observed Shell saturation levels (n*)	4	8–10	6–8	6
% Theoretical Shell saturation levels	44%	53–66%	40–53%	40%
MALDI-TOF-MS (MW):	56 496	120 026	227 606	288 970
PAGE (MW):	58 000	116 000	233 000	467 000
AFM: Observed dimensions	25 × 0.38 nm	33 × 0.53 nm	38 × 0.63 nm	43 × 1.1 nm
CALC.	(D,H)	(D,H)	(D,H)	(D,H)
(MW):	56 000	136 000	214 000	479 000

Where: X = dendrimer core reagent; generation Y = dendrimer shell reagent and Z = surface group functionality.

surfaces against further reaction. Preferred capping reagents were either 2-amino-ethanol or *tris*-hydroxymethyl aminomethane. The capping reaction, step (b), was monitored by following the disappearance of an ester band at 1734 cm^{-1}, using FTIR. Isolation and characterization of these products proved that they were indeed relatively mono(dispersed) spheroids as illustrated by AFM. It was very important to perform the AFM analysis at very high dilution to avoid undesirable core–shell molecular clustering [30].

Typical mass spectroscopy, AFM and gel electrophoresis (PAGE) results are shown in Table 27.2 for structures (*1*)–(*4*) in the core–shell series; [X] {(Y)(Z)}ₙ, where: [X] = core, (Y) = shell, (Z) = capping groups and n = number of shell-type dendrimers bonded to the core (Table 27.2).

A distinct core–shell dimensional enhancement was observed as a function of the sum of the core-shell generation values used in the construction of the series (e.g. core–shell: G4/G3, < G5/G3, < G6/G4 and < G7/G5).

Molecular weights for the final products were determined by MALDI-TOF-MS or (polyacrylamide) gel electrophoresis (PAGE). They were corroborated by calculated values from AFM dimension data and were found to be in relatively good agreement within this series (Table 27.2). Calculations based on these experimentally determined molecular weights allowed the estimation of shell filling levels for respective core–shell structures within this series. A comparison with mathematically predicted shell saturated values reported earlier [34], indicates these core–shell structures are only partially filled (i.e. 40–66% of fully saturated shell values, see Table 27.2).

4.1 EXPERIMENTAL

4.1.1 Materials and General Methods

All PAMAM dendrimers (both NH_3 and EDA core G = 0–12) were provided by Dendritic Sciences, Inc., Mt. Pleasant, Michigan. Efforts were made to utilize several different generational lot samples for the mass spectral data.

Mass spectral data were obtained either on electrospray infusion (EDI) Model TSQ 700 mass spectrometer (Finnigan MAT) or a Matrix Assisted Laser Desorption Time-of-Flight (MALDI-TOF) (Finnigan, Model 2000).

4.1.1.1 Preparation of Core-Shell Tecto(dendrimers); (PAMAM)Core (X); (PAMAM) Shell (Y); Surface Functionality (Z); X = (G6); Y = (G4); Z* = tri(hydroxylmethyl) aminomethane [33]

Step (a)   To a 500 mL round bottom flask containing a stir bar was added shell reagent (Y) G = 3.5 methyl ester PAMAM dendrimer, EDA core, (32 g, 2.6×10^{-3} mol, 164 mmol ester, 25 equivalents per core dendrimer (X) and 32 g of methanol. This mixture was stirred until homogeneous. To this mixture was added lithium chloride (7 g, 166 mmol, 1 equivalent per ester) and stirred until homogenous. To this mixture was added (drop-wise) PAMAM dendrimer, EDA core, G = 6 (6 g, 1.0×10^{-4} mol) in 20 g of methanol in ~ 10 min. This mixture was warmed to 25 °C and placed in a constant temperature bath at 40 °C for 25 days.

Step (b)   After this time, this mixture was cooled to 25 °C as tris(hydroxylmethyl) aminomethane (42 g, 347 mmol), followed by anhydrous K_2CO_3 (22 g, 159 mmol), was added. This resulting mixture was vigorously stirred for 18 h at 25 °C. This mixture was diafiltered in deionized water using an Amicon stainless steel tangential flow ultrafiltration unit containing a 30 K regenerated cellulose membrane to give 6 L of permeate and ultrafiltered in 800 mL retentate to give 10 L permeate (~ 12 recirculations). The retentate was filtered through Whatman No. 1 filter paper, evacuated free of volatiles on a rotary evaporator, followed by high vacuum at 25 °C to give 20 g of core–shell tecto(dendrimer) product.

5 REFERENCES

1. Li, J., Piehler, L. T., Qin, D., Baker Jr., J. R. and Tomalia, D. A. *Langmuir*, **16**, 5613–5616 (2000).
2. Jackson, J. L., Chanzy, H. D., Booy, F. P., Drake, B. J., Tomalia, D. A., Bauer, B. J. and Amis, E. J. *Macromolecules*, **31**, 6259–6265 (1998).
3. Yin, R. and Tomalia, D. A. *J. Am. Chem. Soc.*, **120**, 2678–2679 (1998).
4. Tomalia, D. A., Naylor, A. M. and Goddard III, W. A. *Angew. Chem. Int. Ed. Engl.*, **29**, 138–175 (1990).
5. Tomalia, D. A. *Sci. Am.*, **272**, 62–66 (1995).

6. Newkome, G. R., Moorfield, C. N. and Vögtle, F. *Dendritic Molecules*, VCH, Weinheim, 1996.
7. Lothian-Tomalia, M. K., Hedstrand, D. M. and Tomalia, D. A. *Tetrahedron*, **53**, 15495–15513 (1997).
8. Zeng, F. and Zimmerman, S. C. *Chem. Rev.*, **97**, 1681 (1997).
9. Matthews, O. A., Shipway, A. N. and Stoddard, J. F. *Prog. Polym. Sci.*, **23**, 1 (1998).
10. Tomalia, D. A. *Adv. Mater.*, **6**, 529–539 (1994).
11. Kallos, G. J., Tomalia, D. A., Hedstrand, D. M., Lewis, S. and Zhou, J. *Rapid Commun. Mass Spectrom.*, **5**, 383–386 (1991).
12. Brothers II, H. M., Piehler, L. T. and Tomalia, D. A. *J. Chromatogr. A*, **814**, 233–246 (1998).
13. Caminade, A.-M. and Majoral, J.-P. *Chem. Rev.*, **99**, 845 (1999).
14. Jansen, J. F. G. A., de Brabander-van den Berg, E. M. M. and Meijer, E. W. *Science*, **266**, 1226–1229 (1994).
15. Freemantle, M. *Chem. Eng. News*, **77**(44), 27–35 (1999).
16. Hecht, S. and Frechet, J. M. J. *Angew. Chem. Int. Ed.*, **40**, 74–91 (2001).
17. Balogh, L., Tomalia, D. A. and Hagnauer, G. L. *Chemical Innovation*, **30**, 19–26 (2000).
18. Tomalia, D. A. and Esfand, R. *Chem. Ind.*, **11**, 416–420 (1997).
19. Balogh, L. and Tomalia, D. A. *J. Am. Chem. Soc.*, **120**, 7355 (1998).
20. Crooks, R. M., Lemon III, B., Sun, L., Yeung, L. K. and Zhao, M. *Dendrimer-Encapsulated Metals and Semiconductors: Synthesis, Characterization, and Applications*, Springer-Verlag, Berlin-Heidelberg, 2001.
21. Shull, R. D., Balogh, L., Swanson, D. R. and Tomalia, D. A. *Magnetic Dendrimers and Other Nanocomposites*; MACR-069 Publisher, ACS, Washington.
22. Rajca, A. and Utampanya, S. *J. Am. Chem. Soc.*, **115**, 10688 (1993).
23. Rajca, A., Wongsriratanakul, J., Rajca, S. and Cerny, R. *Angew. Chem. Int. Ed.*, **37**, 1229 (1998).
24. Tabakovic, I., Miller, L. L., Guan, R. G., Tully, D. C. and Tomalia, D. A. *Chem. Mater.*, **9**, 736–745 (1997).
25. Miller, L. L., Duan, R. G., Tully, D. C. and Tomalia, D. A. *J. Am. Chem. Soc.*, **119**, 1005–1010 (1997).
26. Kawa, M. and Fréchet, J. M. *J. Chem. Mater.*, **10**, 286 (1998).
27. Sato, T., Jiang, D.-L. and Aida, T. *J. Am. Chem. Soc.*, **121**, 10658–10659 (1999).
28. de Gennes, P. G. and Hervet, H. J. *J. Physique-Lett.* (*Paris*) **44**, 351 (1983).
29. Tomalia, D. A., Hedstrand, D. M. and Wilson, L. R. *Dendritic Polymers*, 2nd edn, John Wiley & Sons, Vol. Index Volume, 1990, pp. 46–92.
30. Li, J., Swanson, D. R., Qin, D., Brothers II, H. M., Piehler, L. T., Tomalia, D. A. and Meier, D. J. *Langmuir*, **15**, 7347–7350 (1999).
31. Uppuluri, S., Piehler, L. T., Li, J., Swanson, D. R., Hagnauer, G. L. and Tomalia, D. A. *Adv. Mater.*, **12**(11), 796–800 (2000).
32. Tomalia, D. A., Uppuluri, S., Swanson, D. R. and Li, J. *Pure Appl. Chem.*, **72** (12), 2343–2358 (2000).
33. Tomalia, D. A., Esfand, R., Piehler, L. T., Swanson, D. R. and Uppuluri, S. *High Performance Polymers*, **13**, S1–S10 (2001).
34. Mansfield, M. L., Rakesh, L. and Tomalia, D. A. *J. Chem. Phys.*, **105**, 3245–3249 (1996).
35. Topp, A., Bauer, B. J., Klimash, J. W., Spindler, R. and Tomalia, D. A. and Amis, E. J. *Macromolecules*, **32**, 7226–7231 (1999).
36. Tomalia, D. A., Baker, H., Dewald, J., Hall, M., Kallos, G., Martin, S., Roeck, J., Ryder, J. and Smith, P. *Polym. J.* (*Tokyo*), **17**, 117–132 (1985).

37. Tomalia, D. A., Baker, H., Dewald, J., Hall, M., Kallos, G., Martin, S., Roeck, J., Ryder, J. and Smith, P. *Macromolecules*, **19**, 2466–2468 (1986).
38. Meltzer, A. D., Tirrell, D. A., Jones, A. A., Inglefield, P. T., Hedstrand, D. M. and Tomalia, D. A. *Macromolecules*, **25**, 4541–4548 (1992).
39. Smith, P. B., Martin, S. J., Hall, M. J. and Tomalia, D. A. in Mitchell J. Jr. (ed.), *A Characterization of the Structure and Synthetic Reactions of Polyamidoamine Starburst Polymers*; Hanser Publishers: Munich, 1987, pp. 357–385.
40. Hummelen, J. C., van Dongen, J. L. J. and Meijer, E. W. *Chem. Eur. J.*, **3**, 1489–1493 (1997).

28

Conclusion/Outlook – Toward Higher Macromolecular Complexity in the Twenty-first Century

D. A. TOMALIA[1] AND J. M. J. FRÉCHET[2]
[1]Dendritic Nanotechnologies Limited, Central Michigan University,
Mt. Pleasant, MI, USA
[2]University of California, Berkeley CA, USA

During the twentieth century at least five major technology ages have emerged and matured. In approximate chronological order, they have been referred to as the *chemical, nuclear, plastics, materials* and *biotechnology* ages. In all cases, significant new developments and commerce evolved that dramatically enhanced the gross national products of many countries and improved the human condition internationally. An obvious pattern in the emergence of these technologies has been the quest for new properties that might be derived from the ability to understand, synthesize and control higher-level molecular complexity. Each successive technology age builds on the successful developments of the previous age. Presently, the emergence of the *nanotechnology age* is proposed.

Within the polymer science field, a similar pattern exists. It is apparent that there is a keen interest in controlling polymerization processes and resulting macromolecular structures as polymer science progresses toward more complex objectives. Three times this past century, chemists have developed major polymer architectures that have launched significant new industries and commerce [1, 2]. The first two architectural classes (i.e. Class I (linear) and Class II (crosslinked) topologies) literally defined the origins of tradition polymer science as well as major polymer property differences (i.e. *thermoplastics* vs *thermosets*). The third architectural class (i.e. Class III, branched) is presently the focal point of dramatic growth related to new poly(olefins) topologies derived from 'single-

Dendrimers and Other Dendritic Polymers. Edited by Jean M. J. Fréchet and Donald A. Tomalia
© 2001 John Wiley & Sons Ltd

site', metallocene type catalysts [3]. Historically it has been widely recognized that macromolecular topologies significantly influence the determination of polymer properties and related new developments. This fact is well documented as one traces contributions associated with the elite group of scientists who have been associated with the explosive evolution of the field of polymer science.

Presently an international focus is emerging on 'nanotechnology'. It has been described as the 'ultimate scientific frontier' that will both define and lead the world into the next industrial revolution [4]. While this description is surely exaggerated as today's ultimate challenges become tomorrow's routine accomplishments, a very significant challenge facing this movement toward nanotechnology will be the development of structure-controlled methodologies that will enable the cost-effective production and controlled assembly of nanostructures in a very routine manner. Nature solved these problems and shattered this nanoscale synthesis barrier with its evolutive biological strategy for macromolecular structure control billions of years ago. This sets the stage for the sort of dimensional scaling that today determines essentially all significant molecular level factors dealing with life. These same parameters which include: *nanoscale sizes, nanosurfaces/interfaces, nanocontainment, nanoscale-transduction/amplification* and *information storage* have important implications, not only in biology [5], but also in critical abiotic areas such as catalysis, computer miniaturization, nanotribology, sensors and new materials. 'Bottom-up' synthetic strategies that produce size-monodispersed, well-defined organic and inorganic nanostructures with dimensions ranging between 1 and 100 nm will be of utmost importance. It will be essential that these strategies allow the systematic construction of nanoscale structures and devices with precise atom-by-atom control as a function of: *size, shape and surface chemistry* [6–11].

Dendritic polymers are expected to play a key role as an 'enabling technology' in this challenge during the next century, just as the first three traditional architectural classes of synthetic polymers have so successfully fulfilled critical material and functional needs for society during the past half-century. Their controlled shape, size, and differentiated functionality, their ability to provide both isotropic and anisotropic assemblies, their compatibility with many other nanoscale building blocks such as DNA, nanocrystals and nanotubes, their potential for ordered self-assembly, their capacity to form surfaces and interfaces [12] and to combine both organic and inorganic components, and their ability to either encapsulate or be engineered into unimolecular functional devices [11], make dendrimers uniquely versatile amongst existing nanoscale building blocks and materials. While it is doubtful that, with very few notable exceptions, dendritic polymers will ever emerge as major commodity materials, it is likely that they will both contribute to and inspire numerous new commercial ventures in the broad area of nanotechnology. It is from this perspective that it is appropriate to be optimistic about the future of the *dendritic state* as this new major polymer class enters the twenty-first century.

REFERENCES

1. Morawetz, H. *Polymers. The Origin and Growth of a Science*, John Wiley and Sons, New York, 1985.
2. Dvornic, P. R. and Tomalia, D. A. *Science Spectra*, **5**, 36 (1996).
3. Scheirs, J.and Kaminsky, W. (eds), *Metallocene-Based Polyolefins*, Vols 1 and 2, John Wiley & Sons, Brisbane, 2000.
4. Downey, M. L., Moore, D. T., Bachula, G. R., Etter, D. M., Carey, E. F. and Perine, L. A. National Science and Technology Council, Washington, DC, February, 2000.
5. Goodsell, D. S. *American Scientist*, **88**, 230 (2000).
6. Tomalia, D. A., Naylor, A. M. and Goddard III, W. A. *Angew. Chem. Int. Ed. Engl.*, **29**, 138 (1990).
7. Fréchet, J. M. J. *Science*, **263**, 1710 (1994).
8. Tomalia, D. A. and Esfand, R. *Chem. Ind.*, **11**, 416 (1997).
9. Tully, D. C., Wilder, K., Fréchet, J. M. J., Trimble, A. R. and Quate, C. F. *Adv. Mater.*, **11**, 314 (1999).
10. Tomalia, D. A. *Adv. Mater.*, **6**, 529 (1994).
11. Hecht, S. and Fréchet, J. M. J. *Angew. Chem. Int. Ed.*, **40**, 74 (2001).
12. Tully, D. C. and Fréchet, J. M. J. *Chem. Commun.*, **1229** (2001).

Index

T